T0189496

Springer Monographs in Mathematics

This series publishes advanced monographs giving well-written presentations of the "state-of-the-art" in fields of mathematical research that have acquired the maturity needed for such a treatment. They are sufficiently self-contained to be accessible to more than just the intimate specialists of the subject, and sufficiently comprehensive to remain valuable references for many years. Besides the current state of knowledge in its field, an SMM volume should ideally describe its relevance to and interaction with neighbouring fields of mathematics, and give pointers to future directions of research.

More information about this series at http://www.springer.com/series/3733

Tullio Ceccherini-Silberstein ·
Michele D'Adderio

Topics in Groups and Geometry

Growth, Amenability, and Random Walks

Foreword by Efim Zelmanov

Tullio Ceccherini-Silberstein
Dipartimento di Ingegneria
Università degli Studi del Sannio
Benevento, Italy

Michele D'Adderio
Département de Mathématique
Université Libre de Bruxelles
Bruxelles, Belgium

ISSN 1439-7382 ISSN 2196-9922 (electronic)
Springer Monographs in Mathematics
ISBN 978-3-030-88111-5 ISBN 978-3-030-88109-2 (eBook)
https://doi.org/10.1007/978-3-030-88109-2

Mathematics Subject Classification: 20F65, 05C25, 05C81, 34E18, 43A07

This Springer imprint is published by the registered company Springer Nature Switzerland AG
The registered company address is: Gewerbestrasse 11, 6330 Cham, Switzerland

To Olivia, Saverio, and nonna Lena
To Eleonora and Anna

Foreword

It is a privilege to introduce this book on the celebrated theorem of Misha Gromov and everything that is related to it.

In 1981, Gromov stunned the mathematical world with his beautiful proof of Milnor's Conjecture on groups of polynomially bounded growth. The proof combined ideas from many areas of mathematics and was at least as important as the result itself. It ushered in a new era in the history of infinite groups. The subject became a crossroad of different areas of mathematics and different mathematical cultures. It became a fertile ground where mathematical areas competed and showed their prowess.

The book captures this "multicultural" spirit, systematically and generously outlining the related areas including measures on groups, hyperbolic geometry, random walks, relevant harmonic analysis etc., in addition to being a wonderful introduction to the subject of infinite groups itself.

In the exposition, the authors have always selected ideas and examples over technical detail.

The book, with well selected numerous exercises, could be the basis for a graduate course.

The authors clearly enjoyed writing the book. I hope that the reader will enjoy reading it as well.

San Diego, May 2021 Efim I. Zelmanov

Preface

The main purpose of these lecture notes is to present Gromov's theorem on groups of polynomial growth, which is one of the milestones in the theory of infinite groups. The idea of looking at a finitely generated group with a geometer's eye by investigating the properties of the family consisting of all its word metrics is due to Misha Gromov and constitutes the central, ingenious idea for the proof of this beautiful theorem. One often refers to 1981, the year of its publication, as the date of birth of the flourishing branch of mathematics which is commonly known as geometric group theory.

Motivated by their studies in differential geometry, Efremovich and Schwarz in the 1950s and, independently, Milnor and Wolf in the 1960s, introduced the concept of growth of finitely generated groups (typically, fundamental groups of manifolds) viewed as metric spaces. With a finitely generated group G, together with a symmetric generating subset $X \subseteq G$, one associates the Cayley graph $\mathrm{Cay}(G,X)$ with vertex set $V := G$ and edge set $E := \{(g,gx) : g \in G, x \in X\}$; the corresponding graph distance $d = d_X$ makes G into a metric space for which the set $B_r := \{g = x_1 x_2 \cdots x_n : x_i \in X, 0 \leq n \leq r\} \subset G$ is the ball of radius $r \in [0,+\infty)$ centered at the identity element $1_G \in G$.

This led to the definition of groups of (sub-)polynomial growth, of sub-exponential growth, and of exponential growth (of the cardinality of B_r, as $r \to \infty$) and to the study of the algebraic properties of the groups in the corresponding classes. For instance, the free abelian group $\mathbb{Z}^d$ has polynomial growth of degree d and, more generally, a finitely generated nilpotent group has polynomial growth whose corresponding polynomial degree is expressed (Guivar'ch–Bass formula) in terms of the free-ranks of the consecutive quotients from its lower central series. On the other hand, the nonabelian free group F_2 (and therefore any finitely generated group G having nonabelian free subgroups) has exponential growth. Moreover (Milnor–Wolf theorem), finitely generated solvable groups of sub-exponential growth are virtually nilpotent (i.e. have a finite-index nilpotent subgroup) and therefore have polynomial growth.

Gromov's theorem (1981) then characterizes the groups of polynomial growth exactly as the virtually nilpotent groups.

The importance of this theorem, besides its statement, lies in its beautiful proof and the fascinating techniques developed to achieve it.

In his original proof, Gromov developed a notion of "limit of metric spaces", with respect to what is now called the "Gromov-Hausdorff" distance, which turned out to play a key role in the proof. Given a finitely generated group G generated by a symmetric subset $X \subseteq G$, one then considers the sequence of metric spaces $((G, d/n))_{n \in \mathbb{N}}$, where $d = d_X$ is the associated distance function: this corresponds, roughly speaking, to viewing G from farther and farther away. For instance, if $G = \mathbb{Z}$ (or more generally, $G = \mathbb{Z}^d$, for some $d \in \mathbb{N}$) then the associated limit is $(\mathbb{R}, \delta)$ (resp. $(\mathbb{R}^d, \delta)$), where δ is the ℓ^1-distance. Now, this limit metric space inherits a natural G-action (the limit of the Cayley action of G on itself) and, provided one chooses an appropriate subsequence, if needed, it has "nice" topological and geometric properties (it is connected, arc-wise connected, locally connected and locally arc-wise connected, and – if G has polynomial growth – it is separable, locally compact, and finite dimensional) so that, by virtue of an important result of Montgomery and Zippin (a solution to Hilbert's 5th problem), G is essentially a Lie group (and therefore a linear group). A deep and far reaching theorem of Tits (1970) states that a finitely generated linear group is either virtually solvable or contains a nonabelian free subgroup (and therefore has exponential growth). At this point, a combination of the Tits alternative and the Milnor–Wolf theorem then yields that if G has polynomial growth it is necessarily virtually nilpotent, concluding the proof.

In 1984 van den Dries and Wilkie, using methods from nonstandard Analysis, namely the use of ultrafilters and their associated ultraproducts and ultralimits, revisited the construction of Gromov's limits, which in this context are now called "asymptotic cones". This approach greatly simplified the proof by the use of the elegant formalism of ultrafilters. This is the one we present in these lecture notes.

More recently, new proofs of Gromov's theorem, of a more "analytical flavour" have appeared. We mention, for instance, the proofs of Kleiner (in terms of suitable spaces of harmonic functions), Shalom–Tao (a finitary version), Breuillard–Green–Tao, Hrushovsky, and Ozawa (in terms of the analysis of reduced cohomology and Shalom's property H_{FD}). The advantage of these new proofs consists in the fact that they rely neither on the Montgomery–Zippin theorem nor on Tits' alternative. However, the beauty of the geometric construction due to Gromov (and simplified by van den Dries and Wilkie) is completely lost. As our main target is to enlighten the "geometry" of (finitely generated) groups, we preferred to present the Gromov–vdD–W approach for our presentation.

It seems to us quite appropriate to quote here Zelmanov's statement: "It would have been a tragedy if Kleiner's proof had appeared before Gromov's proof: geometric group theory would have never been born!".

Growth of groups has developed as a central topic in geometric and analytic group theory: it is related to (and in fact motivated the flourishing of the theory of) automata groups (e.g. the famous group of intermediate growth constructed by Grigorchuk (1980): its growth function is super-polynomial but sub-exponential), branch groups, and fractal groups, as well as to numerous analytic and combinatorial questions, power series, asymptotic expansions and zeta-functions. In addition to this central topic, we present a self-contained and reasonably exhaustive treatment of more "algebraic group theory" such as free groups (including Klein's ping-pong lemma, which plays a crucial role in the search for free subgroups, for instance in the Tits alternative), nilpotent groups (with Malcev's beautiful theory culminating with their linearity), residually finite groups, solvable groups (with Malcev virtual triangularizability theorem), polycyclic groups (including Malcev's theorems and the Auslander-Swan theorem), and the Burnside problem and the Golod–Shafarevich construction; of more "geometric theory" such as some basic hyperbolic plane geometry as an introduction to a particular (yet significant) case of the Tits alternative, a quick introduction to topological groups (locally compact groups, their Haar measure, locally compact Abelian groups and the Pontryagin duality, Lie groups and Hilbert's fifth problem), a self-contained exposition of dimension theory (we use the approach of Hurewicz and Wallman in their classical monograph based on the inductive dimension) including Hausdorff dimension, and a comprehensive treatment of the necessary background of non-standard analysis (filters and ultrafilters on $\mathbb{N}$, ultralimits, ultrapowers, ultraproducts, asymptotic cones) with a particular emphasis on the important example consisting of hyperbolic metric spaces; of more "analytic group theory" such as amenability (a notion introduced by von Neumann in his studies of the Hausdorff–Banach–Tarski paradox) with its several characterizations – including the Tarski alternative, the Følner criterion, Kesten's theorem, the Grigorchuk co-growth criterion (this is expressed in terms of "relative growth" of the normal subgroup $N \leq F$ of a finitely generated free group such that $G \cong F/N$) – Følner functions and isoperimetric profiles; and "probabilistic group theory". The last chapter, indeed, is devoted to "random walks on groups" and to a presentation of the Polya–Varopoulos theorem. A celebrated result of Polya asserts that the symmetric random walk on $\mathbb{Z}^d$ is recurrent (i.e., with probability one, a random walker visits every $x \in \mathbb{Z}^d$, infinitely many times) for $d = 1,2$ and transient (i.e., for any finite subset $\Omega \subset \mathbb{Z}^d$, there exists a time $t(\Omega) \in \mathbb{N}$, such that, with probability one, the position $x(t)$ of the random walker at time $t \geq t(\Omega)$ satisfies that $x(t) \in \mathbb{Z}^d \setminus \Omega$). A celebrated theorem of Varopoulos generalizes Polya's result and characterizes the infinite finitely-generated groups for which the simple random walks are recurrent: these are exactly the groups admitting a finite index subgroup isomorphic to either $\mathbb{Z}$ or $\mathbb{Z}^2$. It is a remarkable fact that Gromov's theorem plays a crucial role in this characterization.

***

In mid February 2003 and in February–March 2004, TCS visited Efim Zelmanov at UCSD and attended a few lectures from his graduate course on "Groups and Combinatorics" which included the proof of Gromov's theorem along Gromov's original lines. In the academic year 2006–7, as a first-year graduate student at UCSD, MDA attended the whole course by Zelmanov: the proof of Gromov's theorem, this time, followed the Wilkie and van den Dries non-standard analysis approach. The present book was started under the strong encouragement by Zelmanov: it is based on the notes taken at his course, duly detailed, and expanded with several worked-out examples, exercises, and related material. The chapters on Non-Standard Analysis, Dimension Theory, on the Tits alternative, on topological groups and Hilbert's Fifth

Problem, on Amenability, and on the Polya–Varopoulos theorem were not covered in the original course.

***

Our main targets are mature undergraduate and graduate students in mathematics (familiar with the first rudiments in basic group theory and topology) interested in geometric, analytic, and probabilistic aspects of infinite groups. The material covered in this book constitutes, in our opinion, a possible approach to these topics. In particular, given the strong interdisciplinary aspect of the treatment, we believe that the reader will be stimulated to deepen – or even to approach for the first time – related areas such as Harmonic and Functional Analysis (spectral theory, L^2-spaces and their related invariants, isoperimetric questions, amenability and invariant measures), Probability Theory (Random Walks, Markov Chains, and Discrete Potential Theory, percolation), Analysis and Logic (ultrafilters, ultraproducts, and asymptotic cones), Geometric Analysis (dimension theory), the theory of Algebras (Lie rings and group rings, Gelfand–Kirillov dimension, PI-algebras), and Combinatorics (graph theory, matching theory, etc). This way, the book may be used as a textbook for a three-quarters (equivalently, two-semesters) course on "Infinite groups: algebra, geometry, and probability". Eventually, this book may also be useful to the mature researcher as a source for several important results and examples.

***

We heartily thank Efim Zelmanov for introducing us to the beautiful mathematics around Gromov's theorem, for allowing and in fact strongly encouraging us to write down, expand, and publish the notes from his course, and for honoring our modest work with his foreword.

We express our deepest gratitude also to many other friends and colleagues, in primis to Pierre de la Harpe and Slava Grigorchuk, as well as to Laurent Bartholdi, Florin Boca, Corentin Bodart, Alexander I. Bufetov, Matteo Cavaleri, Michel Coornaert, Daniele D'Angeli, Alfredo Donno, Larissa Horn, Donatella Iacono, Alex Lubotzky, Avinoam Mann, Mauro Mariani, Tatiana Nagnibeda, Pierre Pansu, Mark Sapir, Fabio Scarabotti, Paul Schupp, Zoran Sunic, Filippo Tolli, Alain Valette, and Wolfgang Woess for several remarks, suggestions, and encouragement.

Last but not least, we wish to thank our editor, Elena Griniari from Springer Verlag, for her precious advice and constant enthusiastic encouragement at all stages of the production of this book.

Rome, September 2021 TCS and MDA

Contents

Notation

Throughout this book, the following notational conventions are used:

- $\mathbb{N}$ is the set of nonnegative integers, so that $0 \in \mathbb{N}$;
- a *countable* set is a set which admits a bijection onto a subset of $\mathbb{N}$, so that finite sets are countable;
- given a set X, we denote by $\mathscr{P}(X)$ (resp. $\mathscr{P}^*(X)$) the set of all subsets (resp. nonempty subsets) of X;
- the notation $A \subset B$ means that each element in the set A is also in the set B, so that A and B may coincide;
- given two sets A and B, we denote by A^B the set consisting of all maps $f\colon B \to A$;
- $|A|$ denotes the *cardinality* of a (possibly infinite) set A;
- in a group G, we denote by $\langle X \rangle \subset G$ the *subgroup generated* by the subset $X \subset G$;
- given a group G and two subsets $A, B \subset G$, we set $AB := \{ab : a \in A \text{ and } b \in B\}$. If $A = \{a\}$ (resp. $B = \{b\}$) is a singleton set, then we simply write aB (resp. Ab) instead of $\{a\}B$ (resp. $A\{b\}$);
- given a group G and a subgroup H, we write $H \trianglelefteq G$ when H is normal in G;
- given a group G and a subgroup $H \leq G$ we denote by $G/H := \{gH : g \in G\}$ the set of all *left* cosets and by $H\backslash G := \{Hg : g \in G\}$ the set of all *right* cosets of H in G.
- all *rings* are assumed to be associative (but not necessarily commutative) with a unity element;
- a *field* is a nonzero commutative ring in which each nonzero element is invertible.

Part I
Algebraic Theory

Chapter 1
Free Groups

Free groups are the free objects in the category of groups: a group F is free if there exists a subset $X \subseteq F$ (called a basis) such that every map $f\colon X \to G$, where G is any group, extends in a unique way to a group homomorphism $\varphi\colon F \to G$. Free groups do exist! Indeed, given any set X there exists a free group F_X, unique up to isomorphism, which admits X as a basis (Theorem 1.5). As a consequence, every group is isomorphic to a quotient group of a free group. We prove the celebrated Nielsen–Schreier theorem (Theorem 1.15) which states that subgroups of free groups are free. In Section 1.7 we prove Klein's Ping-Pong lemma (Theorem 1.17), a useful criterion to establish that certain transformations generate free groups, and present some examples of applications. Finally, in Section 1.8 we study free abelian groups: these are the abelianizations of free groups. We show that subgroups of free abelian groups are free abelian (Theorem 1.29) and deduce the structure theorem for finitely generated abelian groups (Corollary 1.30).

1.1 Words

Let A be a set, and for $n \in \mathbb{N}$ denote by $A^{\times n}$ the Cartesian product of A with itself n times. An element $w = (a_1, a_2, \dots, a_n)$ of $A^{\times n}$ is called a *word* on A. We then set

$$A^* := \bigcup_{n \in \mathbb{N}} A^{\times n}.$$

The unique element of $A^{\times 0}$, denoted by ε, is called the *empty word*.

The *concatenation* of two words on A, $w = (a_1, a_2, \dots, a_m) \in A^{\times m}$ and $w' = (b_1, b_2, \dots, b_n) \in A^{\times n}$, is the word $w_1 w_2 \in A^{\times (m+n)}$ defined by

$$ww' := (a_1, a_2, \dots, a_m, b_1, b_2, \dots, b_n).$$

For all $w, w', w'' \in A^*$, we have $(ww')w'' = w(w'w'')$, and by convention $\varepsilon w = w\varepsilon = w$. It follows that A^* is a monoid for the concatenation product, with the empty word ε as identity element. It is called the *free monoid* over A.

T. Ceccherini-Silberstein and M. D'Adderio, *Topics in Groups and Geometry*,
Springer Monographs in Mathematics, https://doi.org/10.1007/978-3-030-88109-2_1

A^* satisfies the following universal property: if M is a monoid, then any map $f\colon A \to M$ uniquely extends to a monoid homomorphism $\varphi\colon A^* \to M$.

Note that each word $w = (a_1, a_2, \ldots, a_n) \in A^{\times n}$ may be uniquely written as a product of elements of $A = A^{\times 1}$, namely $w = a_1 a_2 \cdots a_n$. In the sequel, we shall adopt this notation. The non-negative integer $\ell(w) := n$ is then called the *length* of the word $w = a_1 a_2 \cdots a_n$.

Given two words $u, w \in A^*$, one says that u is a *subword* (resp. *prefix*) of w if there exist $u', u'' \in A^*$ such that $w = u'uu''$ (resp. $w = uu''$).

1.2 Definition of Free Groups

Definition 1.1. A group F is called a *free group* if there exists a subset $X \subseteq F$ satisfying the following universal property: for every group G and any map $f\colon X \to G$, there exists a unique group homomorphism $\varphi\colon F \to G$ extending f, that is, such that $\varphi(x) = f(x)$ for every $x \in X$. One then says that F is *based* at X or that X is a *(free) basis* for F.

The proofs of the following remarks are left as **exercises**.

Remark 1.2. Let X (resp. X_1, resp. X_2) be a set and let F (resp. F_1, resp. F_2) be a free group based at X (resp. X_1, resp. X_2).

(a) By the universal property, the unique homomorphism $\varphi\colon F \to F$ extending the inclusion map $i\colon X \to F$ is the identity id_F of F.

(b) Suppose that there exists a bijective map $f\colon X_1 \to X_2$. Then the free groups F_1 and F_2 are isomorphic.

(c) F is generated by X.

As a consequence, one also says that F is *the free group generated* by X, or that X *freely generates* F. We shall also write $F = F_X$. Moreover, we say that F is *the* free group generated by X, since, by (b), such a group, if it exists, is unique up to isomorphism.

(d) Let $\psi\colon F \to G$ be an isomorphism from F onto a group G. Then G is a free group as well, based at $\psi(X)$.

(e) Suppose that there exist subsets $X, Y \subset F$ such that F is free based on both X and Y, then there exists a bijection $f\colon X \to Y$. The cardinality $|X|$ of the (= any) base $X \subset F$ is called the *rank* of the free group F.

(f) Let $Y \subseteq X$ and let K denote the subgroup of F generated by Y. Then K is a free group (based at Y).

1.3 Reduced Forms

Let X be a set. Let $\gamma\colon X \to X'$ be a bijective map from X onto a disjoint copy X' of X (thus $X \cap X' = \varnothing$) and set $A := X \cup X'$. For each $a \in A$, define the element $\widetilde{a} \in A$ by setting

$$\widetilde{a} := \begin{cases} \gamma(x) & \text{if } a = x \in X, \\ \gamma^{-1}(x') & \text{if } a = x' \in X'. \end{cases}$$

Observe that the map $a \mapsto \widetilde{a}$ is an involution of A: $\widetilde{\widetilde{a}} = a$ for all $a \in A$.

Consider now the set A^* consisting of all words on A (see Section 1.1). Recall that A^* is a monoid for the product given by the concatenation of words, whose identity element is the empty word ε.

We say that a word $w \in A^*$ may be obtained from a word $w' \in A^*$ by an *elementary reduction* if there exist an element $a \in A$ and words $u, v \in A^*$ such that $w = uv$ and $w' = ua\widetilde{a}v$. Given words $w, w' \in A^*$, we write $w \approx w'$ if either w may be obtained from w' by an elementary reduction or, vice versa, w' may be obtained from w by an elementary reduction. Finally, we define a relation $\equiv$ on A^* (this is called the *transitive closure of the relation* $\approx$) by writing $w \equiv w'$ for $w, w' \in A^*$ if there exist an integer $n \geq 1$ and a sequence of words $w_1, w_2, \dots, w_n \in A^*$ with $w_1 = w$, $w_n = w'$ and $w_i \approx w_{i+1}$ for $i = 1, 2, \dots, n-1$. It is an **exercise** to check that $\equiv$ is an equivalence relation on A^*.

Definition 1.3. A word $w \in A^*$ is said to be *reduced* if it contains no subword of the form $a\widetilde{a}$ with $a \in A$, that is, if there is no word $w' \in A^*$ which can be obtained from w by applying an elementary reduction.

Note that the empty word ε is reduced and that every subword of a reduced word is itself reduced.

Theorem 1.4. *Every equivalence class for $\equiv$ contains a unique reduced word.*

Proof. Since any word $w \in A^*$ can be transformed into a reduced word by applying a suitable finite sequence of elementary reductions, we have that every equivalence class for $\equiv$ contains at least one reduced word.

Let us prove uniqueness. Denote by $R \subseteq A^*$ the set of all reduced words on A. For $a \in A$ and $r \in R$, define the word $\alpha_a(r) \in A^*$ by

$$\alpha_a(r) := \begin{cases} w & \text{if } r = \widetilde{a}w \text{ for some } w \in A^*, \\ ar & \text{otherwise.} \end{cases}$$

Note that in fact $\alpha_a(r) \in R$. This defines a map $\alpha_a \colon R \to R$ for each $a \in A$. We claim that

$$\alpha_a \circ \alpha_{\widetilde{a}} = \alpha_{\widetilde{a}} \circ \alpha_a = \mathrm{id}_R \tag{1.1}$$

for all $a \in A$. Indeed, let $a \in A$ and $r \in R$. If $r = aw$ for some $w \in A^*$, then $\alpha_{\widetilde{a}}(r) = w$ and hence $\alpha_a(\alpha_{\widetilde{a}}(r)) = \alpha_a(w) = aw = r$ (observe that w cannot start with $\widetilde{a}$ since r is reduced). Otherwise, we have $\alpha_{\widetilde{a}}(r) = \widetilde{a}r$ and therefore $\alpha_a(\alpha_{\widetilde{a}}(r)) = \alpha_a(\widetilde{a}r) = r$. It follows that $\alpha_a \circ \alpha_{\widetilde{a}} = \mathrm{id}_R$. Since $\widetilde{\cdot}$ is an involution on A, by replacing a by $\widetilde{a}$, we get $\alpha_{\widetilde{a}} \circ \alpha_a = \mathrm{id}_R$, completing the proof of the claim.

From (1.1), we deduce that $\alpha_a \colon R \to R$ is a bijection, equivalently, is a permutation of R, for all $a \in A$.

By setting

$$\alpha_w := \alpha_{a_1} \circ \alpha_{a_2} \circ \cdots \circ \alpha_{a_n}$$

for every word $w = a_1 a_2 \cdots a_n \in A^*$, we get a monoid homomorphism $w \mapsto \alpha_w$ from A^* into $\mathrm{Sym}(R)$, where $\mathrm{Sym}(R)$ denotes the *symmetric group* of R, consisting of all permutations of R equipped with the composition operation.

Note that

$$\alpha_r(\varepsilon) = r \quad \text{for all } r \in R. \tag{1.2}$$

On the other hand, it immediately follows from (1.1) that $\alpha_w = \alpha_{w'}$ whenever $w, w' \in A^*$ satisfy $w \approx w'$. Thus, if $r_1, r_2 \in R$ are in the same equivalence class, we have $\alpha_{r_1} = \alpha_{r_2}$ and therefore, from (1.2) we deduce that $r_1 = r_2$. □

1.4 Existence of Free Groups

We have already observed in Remark 1.2.(c) that given a set X, if a free group F based at X exists, then it is unique up to isomorphisms. We now show that such a free group always exists.

Theorem 1.5. *Let X be a set. Then there exists a free group F based at X.*

Proof. With the notation of Section 1.3, we denote by F the set of all reduced words on $A = X \cup X'$. Given an element $u \in A^*$, by virtue of Theorem 1.4 there exists a unique reduced word denoted $[u]_F$ such that $[u]_F \equiv u$. We observe that $[r]_F = r$ for every $r \in F$. Let now $u_1, u_2, v_1, v_2 \in A^*$ and suppose that $u_1 \equiv v_1$ and $u_2 \equiv v_2$. We then have $u_1 u_2 \equiv u_1 v_2 \equiv v_1 v_2$, so that $[u_1 u_2]_F = [v_1 v_2]_F$. In particular,

$$[u_1 u_2]_F = [[u_1]_F [u_2]_F]_F. \tag{1.3}$$

This shows that the product of r_1 and r_2 in F given by

$$r_1 \cdot r_2 := [r_1 r_2]_F \tag{1.4}$$

is well defined.

Note that this product differs from the concatenation of words: for instance, given $x \in X$, if $r_1 = x$ and $r_2 = x^{-1}$, we have $r_1 \cdot r_2 = [xx^{-1}]_F = [\varepsilon]_F = \varepsilon$, while $r_1 r_2 = xx^{-1}$, which is distinct from ε in A^*.

Let us show that the multiplication defined in (1.4) gives a group structure on F. For all $r_1, r_2, r_3 \in F$, using (1.3), we have

$$\begin{aligned}
(r_1 \cdot r_2) \cdot r_3 &= [r_1 r_2]_F \cdot r_3 \\
&= [[r_1 r_2]_F [r_3]_F]_F \\
&= [(r_1 r_2) r_3]_F \\
&= [r_1 (r_2 r_3)]_F \\
&= [[r_1]_F [r_2 r_3]_F]_F \\
&= r_1 \cdot [r_2 r_3]_F \\
&= r_1 \cdot (r_2 \cdot r_3).
\end{aligned}$$

This shows that the multiplication is associative. Moreover, for all $r \in F$, we have

$$\varepsilon \cdot r = [\varepsilon r]_F = [r]_F = r \quad \text{and} \quad r \cdot \varepsilon = [r\varepsilon]_F = [r]_F = r.$$

This shows that $1_F := \varepsilon$ is an identity element. In order to prove that F is a group, we are only left to show that every element in F admits an inverse. For $w = a_1 a_2 \cdots a_n \in A^*$, define the word $\widetilde{w} \in A^*$ by setting

$$\widetilde{w} := \widetilde{a_n} \widetilde{a_{n-1}} \cdots \widetilde{a_1}.$$

Observe that $w \in A^*$ is reduced if and only if $\widetilde{w}$ is reduced. Moreover, if $r = a_1 a_2 \cdots a_n \in F$ we have

$$\begin{aligned} r \cdot \widetilde{r} &= [a_1 a_2 \cdots a_n \widetilde{a_n} \cdots \widetilde{a_2} \widetilde{a_1}]_F \\ &= [a_1 \cdots a_{n-1} \widetilde{a_{n-1}} \cdots \widetilde{a_1}]_F \\ &\ \vdots \\ &= [a_1 a_2 \widetilde{a_2} \widetilde{a_1}]_F \\ &= [a_1 \widetilde{a_1}]_F \\ &= [\varepsilon]_F \\ &= 1_F. \end{aligned}$$

Thus, we have $r \cdot \widetilde{r} = 1_F$. Similarly, $\widetilde{r} \cdot r = 1_F$. In other words, $\widetilde{r}$ is an inverse of r. This proves that F is a group.

Let us show that F is free, based at X. First observe that every $a \in A$ is a reduced word. Thus, $X \subseteq F$. Also observe that since $x' = (\widetilde{x'})^{-1} \in X^{-1} := \{x^{-1} : x \in X\}$ for every $x' \in X'$, we have that X generates F as a group. Let now $f \colon X \to G$ be a map from X into a group G and let us prove that there exists a unique homomorphism $\varphi \colon F \to G$ extending f.

Uniqueness follows from the fact that X generates F.

In order to construct φ, we first extend f to a map $\overline{f} \colon A \to G$ by setting, for all $a \in A$,

$$\overline{f}(a) := \begin{cases} f(a) & \text{if } a \in X \\ f(\widetilde{a})^{-1} & \text{if } a \in X. \end{cases}$$

Note that $\overline{f}(\widetilde{a}) = \overline{f}(a)^{-1}$ for all $a \in A$. Consider the map $\varphi \colon A^* \to G$ defined by setting

$$\varphi(w) := \overline{f}(a_1) \overline{f}(a_2) \cdots \overline{f}(a_n)$$

for all $w = a_1 a_2 \cdots a_n \in A^*$. We have

$$\varphi(ww') = \varphi(w)\varphi(w') \tag{1.5}$$

for all $w, w' \in A^*$. Moreover, if $w \approx w'$, say $w = uv$ and $w' = ua\widetilde{a}v$ for some $a \in A$ and $u, v \in A^*$, then

$$\varphi(w') = \varphi(u)\overline{f}(a)\overline{f}(\widetilde{a})\varphi(v) = \varphi(u)\overline{f}(a)\overline{f}(a)^{-1}\varphi(v) = \varphi(u)\varphi(v) = \varphi(w).$$

By an inductive argument, we deduce that $\varphi(w) = \varphi(w')$ for all $w, w' \in A^*$ such that $w \equiv w'$. In particular, $\varphi(w) = \varphi([w]_F)$ for all $w \in A^*$. As a consequence of this, we get, for all $r_1, r_2 \in F$,

$$\varphi(r_1 \cdot r_2) = \varphi([r_1 r_2]_F) = \varphi(r_1 r_2) = \varphi(r_1)\varphi(r_2).$$

Therefore, φ is a group homomorphism. On the other hand, for all $x \in X$, we have

$$\varphi(x) = \overline{f}(x) = f(x),$$

which shows that φ extends f.

All this proves that F is a free group based at X. □

Corollary 1.6. *Let F be a free group and let X be a free basis of F. Then every element $r \in F$ can be uniquely written in the form*

$$r = x_1^{h_1} x_2^{h_2} \cdots x_n^{h_n} \tag{1.6}$$

with $n \geq 0$, $h_1, h_2, \ldots, h_n \in \mathbb{Z} \setminus \{0\}$, and $x_1, x_2, \ldots, x_n \in X$ satisfying $x_i \neq x_{i+1}$ for $1 \leq i \leq n-1$. □

Definition 1.7. The expression (1.6) is called the *reduced form* of the element r in the free group F relative to the free basis X.

Corollary 1.8. *Let G be a group. Let $U \subseteq G$ be a subset such that*

$$u_1^{k_1} u_2^{k_2} \cdots u_n^{k_n} \neq 1_G, \tag{1.7}$$

for all $n \geq 1$, $k_1, k_2, \ldots, k_n \in \mathbb{Z} \setminus \{0\}$, and $u_1, u_2, \ldots, u_n \in U$ satisfying $u_i \neq u_{i+1}$ for $1 \leq i \leq n-1$. Then the subgroup of G generated by U is free, with basis U.

Proof. Let F be the free group based at U (cf. Theorem 1.5). Denote by H the subgroup of G generated by U and by $i \colon U \to H$ the inclusion map. Then there exists a unique homomorphism $\varphi \colon F \to H$ that extends i. Note that φ is surjective since $\varphi(F) \supseteq \varphi(U) = i(U) = U$ and U generates H.

Let us show that φ is also injective. Consider an element $r \in F$ written in reduced form, say $r = u_1^{h_1} u_2^{h_2} \cdots u_n^{h_n}$ with $n \geq 0$, $h_j \in \mathbb{Z} \setminus \{0\}$ and $u_j \in U$ for $1 \leq j \leq n$ such that $u_j \neq u_{j+1}$ for $1 \leq j \leq n-1$. Then, since $\varphi(u_j) = i(u_j) = u_j$ for $1 \leq j \leq n$, we have

$$\varphi(r) = \varphi(u_1)^{h_1} \varphi(u_2)^{h_2} \cdots \varphi(u_n)^{h_n} = u_1^{h_1} u_2^{h_2} \cdots u_n^{h_n} = r.$$

It follows from (1.7) that $\varphi(r) = 1_H$ if and only if $n = 0$, that is, if and only if $r = 1_F$. This shows that φ is also injective, and hence it is an isomorphism.

It follows from Remark 1.2.(d) that H is the free group based at U. □

1.5 Subgroups, Quotients, and Extensions of Finitely Generated Groups

Let G be a group. Recall that a subset $X \subseteq G$ *generates* G, and one writes $G = \langle X \rangle$, provided that every element $g \in G$ can be expressed as a product of elements in $X \cup X^{-1}$, where $X^{-1} = \{x^{-1} : x \in X\}$, that is, there exist $n \geq 0$, $x_1, x_2, \ldots, x_n \in X$ and $\varepsilon_1, \varepsilon_2, \ldots, \varepsilon_n \in \{1, -1\}$ such that

$$g = x_1^{\varepsilon_1} x_2^{\varepsilon_2} \cdots x_n^{\varepsilon_n}.$$

A generating subset $X \subset G$ is *symmetric* if $X = X^{-1}$, i.e., $x^{-1} \in X$ for all $x \in X$. If this is the case, then we can simply express an element $g \in G$ as $g = x_1 x_2 \cdots x_n$ for suitable $x_1, x_2, \ldots, x_n \in X$.

Let now $H \leq G$ be a subgroup of G. Consider the set $H \backslash G := \{Hg : g \in G\}$ of all right cosets of H in G. For each coset Hg we fix a *representative* $\overline{g} \in Hg$. As a representative of the coset H we choose the identity element 1_G. Note that

$$\overline{g} g^{-1},\ g(\overline{g})^{-1} \in H \text{ for every } g \in G. \tag{1.8}$$

Let $S := \{\overline{g} : g \in G\} \subseteq G$ denote the corresponding set of all such representatives. This set is called a *(right) transversal* of H in G.

Observe that for all $h \in H$ and $g \in G$ we have $\overline{hg} = \overline{g}$. Also, since $\overline{g_1} g_2 \in H g_1 g_2$,

$$\overline{\overline{g_1} g_2} = \overline{g_1 g_2} \text{ for all } g_1, g_2 \in G. \tag{1.9}$$

Proposition 1.9. *Let G be a group and let $H \leq G$ be a subgroup. Let $X \subseteq G$ be a generating subset of G and let $S \subseteq G$ be a right transversal of H in G. Then the set*

$$Y := \{sx(\overline{sx})^{-1} : s \in S,\ x \in X\} \tag{1.10}$$

is contained in H and generates it.

Proof. The inclusion $Y \subseteq H$ immediately follows from (1.8).

Consider now $h \in H$. Since X generates G, we can find $m \geq 0$ and $a_1, a_2, \ldots, a_m \in X \cup X^{-1}$ such that

$$h = a_1 a_2 \cdots a_m.$$

Keeping in mind (1.9), we have

$$\begin{aligned}
&\left(1_G a_1 \left(\overline{1_G a_1}\right)^{-1}\right) \cdot \left(\overline{a_1} a_2 \left(\overline{\overline{a_1} a_2}\right)^{-1}\right) \cdot \left(\overline{a_1 a_2} a_3 \left(\overline{\overline{a_1 a_2} a_3}\right)^{-1}\right) \cdot \cdots \\
&\qquad \cdots \cdot \left(\left(\overline{a_1 a_2 \cdots a_{m-1}} a_m\right) \cdot \left(\overline{\overline{a_1 a_2 \cdots a_{m-1}} a_m}\right)^{-1}\right) \\
&\qquad = a_1 a_2 \cdots a_m \left(\overline{a_1 a_2 \cdots a_m}\right)^{-1} \\
&\qquad = h,
\end{aligned}$$

where the last equality follows from the fact that $\overline{a_1 a_2 \cdots a_m} = 1_G$, since $a_1 a_2 \cdots a_m = h \in H$. This shows that we can write h as a product of elements from the set $\{sa(\overline{sa})^{-1} : s \in S,\ a \in X \cup X^{-1}\}$. In order to show that Y generates H, we will prove that in such a product we can replace any factor of the form $sx^{-1}(\overline{sx^{-1}})^{-1}$, with $s \in S$ and $x \in X$, by the inverse of an element of Y.

Let then $s \in S$, $x \in X$ and set $t := \overline{sx^{-1}} \in S$. By using (1.8), we get

$$\overline{tx} = \overline{\overline{sx^{-1}}x} = \overline{sx^{-1}x} = \overline{s} = s$$

so that $tx(\overline{tx})^{-1} = txs^{-1}$. It follows that

$$tx(\overline{tx})^{-1} \cdot sx^{-1}(\overline{sx^{-1}})^{-1} = txs^{-1} \cdot sx^{-1}(\overline{sx^{-1}})^{-1} = 1_G,$$

and therefore

$$sx^{-1}(\overline{sx^{-1}})^{-1} = (tx(\overline{tx})^{-1})^{-1} \in Y^{-1}. \tag{1.11}$$

This shows that $Y = \{sx(\overline{sx})^{-1} : s \in S,\ x \in X\}$ generates H. □

Definition 1.10. Let G be a group and let $H \leq G$ be a subgroup. Let $X \subseteq G$ be a generating subset of G and let $S \subseteq G$ be a right transversal of H. The elements in the set Y (cf. (1.10)) are called the *Schreier generators* for H (relative to S and X).

If G is a finitely generated group, we denote by $\mathrm{rk}(G)$ the *rank* of G, that is, the minimal cardinality $|X|$ of a finite generating subset $X \subseteq G$.

Corollary 1.11. *Let G be a group and let $H \leq G$ be a subgroup of finite index. Then G is finitely generated if and only if H is finitely generated. In particular,*

$$\mathrm{rk}(G) \leq [G:H] + \mathrm{rk}(H) - 1 \tag{1.12}$$

and

$$\mathrm{rk}(H) \leq [G:H] \cdot \mathrm{rk}(G). \tag{1.13}$$

Proof. Let us fix a right transversal $S \subseteq G$ of H in G with $1_G \in S$, and observe that $|S| = [G:H]$.

Suppose that H is finitely generated and let $Y \subseteq H$ be a generating subset for H of minimal cardinality, that is, $|Y| = \mathrm{rk}(H)$. Then the set $X := Y \cup (S \setminus \{1_G\})$ clearly generates G. This shows (1.12).

Conversely, if G is finitely generated and $X \subseteq G$ is a finite generating subset for G, then the set Y in (1.10) generates H. Hence $\mathrm{rk}(H) \leq |Y| \leq |S| \cdot |X|$, and (1.13) follows as well. □

We leave the proof of the following proposition as an **exercise**.

Proposition 1.12. *Let G be a group and let N be a normal subgroup of G.*

(1) *If G is finitely generated, then the quotient group G/N is finitely generated, and* $\mathrm{rk}(G/N) \leq \mathrm{rk}(G)$.

(2) *If N and the quotient group G/N are both finitely generated, then G is finitely generated, and* $\mathrm{rk}(G) \leq \mathrm{rk}(N) + \mathrm{rk}(G/N)$.

1.6 Subgroups of Free Groups

Definition 1.13. Let F be a free group. Let $X \subseteq F$ be a free basis and let $H \leq F$ be a subgroup. A right transversal S of H in F is called a *Schreier system* of representatives of $H\backslash F$ with respect to X provided that it is closed under the operation of taking prefixes, that is, for every element (expressed as a reduced word) $s = a_1 a_2 \cdots a_m \in S$, where $a_i \in X \cup X^{-1}$, $i = 1,2,\ldots,m$, one has $a_1 a_2 \cdots a_{m-1} \in S$.

Lemma 1.14. *Let F be a free group. Let $X \subseteq F$ be a free basis and let $H \leq F$ be a subgroup. Then there exists a Schreier system of representatives of $H\backslash F$ with respect to X.*

Proof. Let $X = \{x_i : i \in I\}$, and assume first that $I = \{1,2,\ldots,n\}$ for some $n \in \mathbb{N}$, or $I = \mathbb{N}$. We introduce a total order on $X \cup X^{-1}$ by setting $1_F < x_1 < x_1^{-1} < x_2 < x_2^{-1} < \cdots$. We then extend this order to the whole of F (viewed as the set of all reduced words on X) as follows. Given two elements $g_1, g_2 \in F$, we say that $g_1 < g_2$ if either $\ell(g_1) < \ell(g_2)$, or we have decompositions (of reduced words!) $g_1 = u_1 x^{\varepsilon_1} u$ and $g_2 = u_2 y^{\varepsilon_2} u$ with $x, y \in X$, $\varepsilon_1, \varepsilon_2 \in \{1,-1\}$, $u_1, u_2, u \in F$ such that $\ell(u_1) = \ell(u_2)$ and $x^{\varepsilon_1} < y^{\varepsilon_2}$. For instance, if $X = \{x_1, x_2\}$, the elements of length 2 are ordered as

$$\begin{aligned} x_1^2 &< x_2x_1 < x_2^{-1}x_1 < x_1^{-2} < x_2x_1^{-1} < x_2^{-1}x_1^{-1} \\ &< x_1x_2 < x_1^{-1}x_2 < x_2^2 < x_1x_2^{-1} < x_1^{-1}x_2^{-1} < x_2^{-2}. \end{aligned}$$

Observe that for every $g \in F$ the set $\{g' \in F : g' \leq g\}$ is finite so that every subset of F admits a minimum.

Hence we construct a set S by taking for every coset its minimum with respect to this order. It is an **exercise** to show that S is a Schreier system of representatives of $H\backslash F$ with respect to X.

We leave it as an **exercise** to adapt the proof to the uncountable case (hint: use the fact that every set I admits a well-order, i.e. an order in which every subset $J \subseteq I$ has a minimum (this fact is equivalent to the axiom of choice)). □

Theorem 1.15 (Nielsen–Schreier theorem on subgroups of free groups). *Let F be a free group and let $H \leq F$ be a subgroup. Then H is free. Moreover, if $X \subseteq F$ is a free basis of F and S is any Schreier system of representatives of $H\backslash F$ with respect to X, then the set $Y := \{sx(\overline{sx})^{-1} : s \in S, x \in X\} \setminus \{1_F\}$ is a basis of H.*

Proof. In order to show that Y is a free basis of H, we consider a nonempty reduced product of elements in $Y \cup Y^{-1}$. By Corollary 1.8, it is enough to prove that this product is different from 1_F.

By virtue of equality (1.11), we can assume that this reduced product is of the form

$$\ldots \cdot sx^{\varepsilon}(\overline{sx^{\varepsilon}})^{-1} \cdot ty^{\delta}(\overline{ty^{\delta}})^{-1} \cdot \ldots \tag{1.14}$$

where $s, t \in S$, $x, y \in X$, and $\varepsilon, \delta = \pm 1$.

Claim 1. *Suppose that $sa(\overline{sa})^{-1} \neq 1_F$, where $s \in S$ and $a \in X \cup X^{-1}$. If s and $\overline{sa}$ are written in reduced form, then the product $s \cdot a \cdot (\overline{sa})^{-1}$ is reduced.*

We argue by contradiction. First, suppose that a cancellation in the product $s \cdot a$ occurs. In this case, in reduced form, $s = ta^{-1}$, where $t \in F$. Then $t \in S$ because S is a Schreier system. But then $\overline{sa} = \overline{ta^{-1}a} = \overline{t} = t$, which gives $sa(\overline{sa})^{-1} = ta^{-1}at^{-1} = 1_F$, contradicting our assumptions.

Suppose instead that a cancellation occurs in the product $a \cdot (\overline{sa})^{-1}$. Then, in reduced form, $(\overline{sa})^{-1} = a^{-1}t^{-1}$, for a suitable $t \in F$, so that $\overline{sa} = ta$. As before, $t \in S$. Moreover, from $\overline{sa} = ta$ we deduce that $Hsa = Hta$, hence $s \in Ht$. This implies that $t = s$. This gives $sa(\overline{sa})^{-1} = saa^{-1}s^{-1} = 1_F$, yielding again a contradiction. The claim follows.

Claim 2. *Any cancellation occurring in* (1.14) *stops before reaching x^ε or y^δ.*

There are three cases:

Case 1: the cancellation reaches x^ε first, i.e., $t = \overline{\overline{sx^\varepsilon} \cdot x^{-\varepsilon}} \cdot w$. But now, since S is a Schreier system, $\overline{sx^\varepsilon}x^{-\varepsilon} \in S$, hence $\overline{sx^\varepsilon}x^{-\varepsilon} = \overline{\overline{sx^\varepsilon}x^{-\varepsilon}} = \overline{sx^\varepsilon x^{-\varepsilon}} = \overline{s} = s$. We deduce that $sx^\varepsilon(\overline{sx^\varepsilon})^{-1} = 1_F$, a contradiction.

Case 2: the cancellation reaches y^δ first: the proof is the same as in the previous case, so it is left as an **exercise**.

Case 3: the cancellation reaches x^ε and y^δ simultaneously, that is, $\overline{sx^\varepsilon} = t$ and $x^\varepsilon = y^{-\delta}$. As a consequence, $\overline{ty^\delta} = \overline{\overline{sx^\varepsilon}x^{-\varepsilon}} = \overline{sx^\varepsilon x^{-\varepsilon}} = \overline{s} = s$, which gives

$$sx^\varepsilon(\overline{sx^\varepsilon})^{-1} = \overline{ty^\delta}y^{-\delta}t^{-1} = \left(ty^\delta(\overline{ty^\delta})^{-1}\right)^{-1},$$

contradicting the fact that the product in (1.14) was reduced. This proves the claim.

From Claim 1 and Claim 2 we deduce that our product cannot be equal to 1_F, completing the proof of the theorem. □

In the remainder of this section we determine the rank of a finite index subgroup of a finitely generated free group.

Theorem 1.16. *Let F be a free group of rank m and let $H \leq F$ be a subgroup of finite index $[F : H] = n$. Then $\mathrm{rk}(H) = 1 + (m-1)n$.*

Proof. Let $X \subseteq F$ be a free basis of F, so that $|X| = m$, and let $S \subseteq F$ be a Schreier system of representatives of $H \backslash F$ with respect to X, so that $|S| = n$. We set $Z := \{(s,x) \in S \times X : sx(\overline{sx})^{-1} = 1_F\}$. Also, for $x \in X$ and $a \in X \cup X^{-1}$, we denote by S_a the set of elements of S whose reduced form ends with a and we set $Z_x := \{s \in S : sx(\overline{sx})^{-1} = 1_F\}$. Note that for $s \in S$, we have $s \in Z_x$ if and only if $(s,x) \in Z$, so that

$$|Z| = \sum_{x \in X} |Z_x|. \tag{1.15}$$

Let $s \in S$, $x \in X$ and set $t := sx$. We have $s \in Z_x$ if and only if $sx = \overline{sx}$, i.e., if and only if $t \in S$. So either s is a prefix of t, and hence the word sx is reduced and $t = sx \in S_x$, equivalently, $s \in (S_x)x^{-1}$, or t is a prefix of s, equivalently $s \in S_{x^{-1}}$.

It follows that $|Z_x| = |S_{x^{-1}}| + |(S_x)x^{-1}| = |S_{x^{-1}}| + |S_x|$. From (1.15) we deduce

$$|Z| = \sum_{x \in X} (|S_{x^{-1}}| + |S_x|) = |S \setminus \{1_F\}| = n - 1.$$

From Theorem 1.15, by using the notation for Y therein, we deduce that

$$\operatorname{rk}(H) = |Y| = |X| \cdot |S| - |Z| = mn - (n-1) = 1 + (m-1)n.$$

The proof is complete. □

1.7 The Ping-Pong Lemma

The following theorem is often used to prove that a group is free.

Theorem 1.17 (Klein's Ping-Pong lemma). *Let G be a group. Let X be a generating subset of G having at least two distinct elements. Suppose that G acts on a set E and that there is a family $(A_x)_{x \in X}$ of nonempty subsets of E such that*

$$A_y \not\subseteq A_x \quad \text{for all distinct elements } x, y \in X \tag{1.16}$$

and

$$x^k A_y \subseteq A_x \text{ for all } x, y \in X \text{ s.t. } x \neq y, \text{ and } k \in \mathbb{Z} \setminus \{0\}. \tag{1.17}$$

Then G is a free group with basis X.

Proof. Consider an element $g \in G$ written as a nontrivial reduced word on the generating subset X, that is, in the form

$$g = x_1^{k_1} x_2^{k_2} \dots x_n^{k_n},$$

where $n \geq 1$, $x_i \in X$, $k_i \in \mathbb{Z} \setminus \{0\}$ for $1 \leq i \leq n$, and $x_i \neq x_{i+1}$ for $1 \leq i \leq n-1$. By Corollary 1.8, it is enough to show that $g \neq 1_G$.

Suppose first that either X contains at least three distinct elements or X has exactly two elements and $x_1 = x_n$. In this case, we can find an element $y \in X$ such that $y \neq x_1$ and $y \neq x_n$. By successive applications of (1.17), we get

$$\begin{aligned}
gA_y &= x_1^{k_1} x_2^{k_2} \dots x_{n-2}^{k_{n-2}} x_{n-1}^{k_{n-1}} x_n^{k_n} A_y \\
&\subseteq x_1^{k_1} x_2^{k_2} \dots x_{n-2}^{k_{n-2}} x_{n-1}^{k_{n-1}} A_{x_n} \\
&\subseteq x_1^{k_1} x_2^{k_2} \dots x_{n-2}^{k_{n-2}} A_{x_{n-1}} \\
&\dots \\
&\subseteq x_1^{k_1} x_2^{k_2} A_{x_3} \\
&\subseteq x_1^{k_1} A_{x_2} \\
&\subseteq A_{x_1}.
\end{aligned}$$

Since $A_y \not\subseteq A_{x_1}$ and $A_y \neq \varnothing$ by our hypotheses, we deduce that $g \neq 1_G$.

It remains to treat the case when X has exactly two elements and $x_1 \neq x_n$. Then

$$x_1^{k_1} g x_1^{-k_1} = x_1^{2k_1} x_2^{k_2} x_3^{k_3} \dots x_n^{k_n} x_1^{-k_1}$$

is a reduced form of $x_1^{k_1} g x_1^{-k_1}$. We have $x_1^{k_1} g x_1^{-k_1} \neq 1_G$ by the first case, and therefore $g \neq 1_G$. □

Corollary 1.18. *Let F be a free group of rank 2 and let $2 \leq n \leq \infty$. Then F contains a free subgroup of rank n.*

Proof. Let $\{a,b\}$ be a free basis for F. Let $I = \{0,1,\ldots,n-1\}$ (resp. $I = \mathbb{N}$) and let G be the (free) subgroup of F generated by the subset

$$X := \{x_i := a^i b a^{-i} : i \in I\}.$$

Let G act on F by left multiplication. For each $i \in I$, denote by $A_{x_i} \subseteq F$ the set of elements of F whose reduced form starts by $a^i b^h$ for some $h \in \mathbb{Z} \setminus \{0\}$. For $i, j \in I$ distinct we have $A_{x_i} \cap A_{x_j} = \varnothing$ and, moreover, for every $k \in \mathbb{Z} \setminus \{0\}$, all elements of $x_i^k A_{x_j}$ have a reduced form starting by $a^i b^k a^{j-i} b^h$, for some $h \in \mathbb{Z} \setminus \{0\}$, so that $x_i^k A_{x_j} \subseteq A_{x_i}$. Thus, the family $(A_x)_{x \in X}$ of subsets of F satisfies the hypotheses of Theorem 1.17. This shows that G is free based at X (and therefore has rank $|I|$). □

Example 1.19. Let $\mathrm{SL}(2,\mathbb{Z}) := \{A \in \mathrm{M}_2(\mathbb{Z}) : \det A = 1\}$, where $\mathrm{M}_n(R)$ is the ring of $n \times n$ matrices with coefficients in a ring R, and $\det A$ is the determinant of a matrix A. Let

$$x_1 = \begin{pmatrix} 1 & 2 \\ 0 & 1 \end{pmatrix}, \quad x_2 = \begin{pmatrix} 1 & 0 \\ 2 & 1 \end{pmatrix}, \quad E = \mathbb{R}^2 = \left\{ \begin{pmatrix} \alpha \\ \beta \end{pmatrix} : \alpha, \beta \in \mathbb{R} \right\}$$

and

$$A_{x_1} = \left\{ \begin{pmatrix} \alpha \\ \beta \end{pmatrix} : |\beta| > |\alpha| \right\}, \quad A_{x_2} = \left\{ \begin{pmatrix} \alpha \\ \beta \end{pmatrix} : |\beta| < |\alpha| \right\}.$$

We have

$$x_1^k \begin{pmatrix} \alpha \\ \beta \end{pmatrix} = \begin{pmatrix} \alpha + 2k\beta \\ \beta \end{pmatrix} \quad \text{and} \quad x_2^k \begin{pmatrix} \alpha \\ \beta \end{pmatrix} = \begin{pmatrix} \alpha \\ 2k\alpha + \beta \end{pmatrix}$$

for all $k \in \mathbb{Z}$, thus showing that $x_1^k A_{x_2} \subseteq A_{x_1}$ and $x_2^k A_{x_1} \subseteq A_{x_2}$ for all $k \in \mathbb{Z} \setminus \{0\}$. It follows from Theorem 1.17 that x_1 and x_2 generate a subgroup of $\mathrm{SL}(2,\mathbb{Z})$ which is free of rank 2.

1.8 Free Abelian Groups

Let F be an abelian group.

Definition 1.20. A subset $X \subseteq F$ is said to be a *basis* of F provided that X generates F and, for distinct $x_1, x_2, \ldots, x_k \in F$ and $n_1, n_2, \ldots, n_k \in \mathbb{Z}$, one has

$$n_1 x_1 + n_2 x_2 + \cdots + n_k x_k = 0 \Rightarrow n_1 = n_2 = \cdots = n_k = 0.$$

Proposition 1.21. *Let F be an abelian group. The following conditions are equivalent:*

(a) *F admits a nonempty basis;*
(b) *F is the direct sum of infinite cyclic subgroups;*
(c) *there exists an index set X such that $F \cong \bigoplus_{x\in X} \mathbb{Z}$;*
(d) *there exists a nonempty subset $X \subseteq F$ such that given any abelian group G and any map $f\colon X \to G$ there exists a unique group homomorphism $\varphi\colon F \to G$ which extends f, i.e. such that $\varphi(x) = f(x)$ for all $x \in X$.*

Proof. We will prove, in order, the implications: (a) $\Rightarrow$ (b) $\Rightarrow$ (c) $\Rightarrow$ (a), and (a) $\Rightarrow$ (d) $\Rightarrow$ (c).

(a) $\Rightarrow$ (b). Let $X \subseteq F$ be a basis of F. Then for every $x \in X$ and $n \in \mathbb{Z}$ we have $nx = 0$ if and only if $n = 0$. It follows that the subgroup $F_x = \langle x \rangle$ generated by $x \in X$ is infinite cyclic. As $F = \langle X \rangle$ we also have $F = \langle \bigcup_{x\in X} F_x \rangle$. Suppose that $z \in X$ is such that $F_z \cap \langle \bigcup_{\substack{x\in X\\ x\neq z}} F_x \rangle \neq \{0\}$. Then we can find $n, n_1, n_2, \ldots, n_k \in \mathbb{Z}$, $n \neq 0$, and distinct elements $x_1, x_2, \ldots, x_k \in X \setminus \{z\}$ such that $nz = n_1x_1 + n_2x_2 + \cdots n_kx_k$. But this contradicts the fact that X is a basis. It follows that $F_z \cap \langle \bigcup_{\substack{x\in X\\ x\neq z}} F_x \rangle = \{0\}$ and therefore $F = \bigoplus_{x\in X} F_x$.

(b) $\Rightarrow$ (c). This follows immediately from the fact that $F_x \cong \mathbb{Z}$ for all $x \in X$.

(c) $\Rightarrow$ (a). Suppose that $F \cong \bigoplus_{x\in X} \mathbb{Z}$ and let $\theta\colon F \to \bigoplus_{x\in X} \mathbb{Z}$ be a group isomorphism. For each $x \in X$ denote by $f_x \in \bigoplus_{x\in X} \mathbb{Z}$ the element defined by setting $f_x(y) = 1$ if $x = y$ and $f_x(y) = 0$, otherwise. It is immediate that the set $\{f_x : x \in X\}$ is a basis of $\bigoplus_{x\in X} \mathbb{Z}$. As a consequence, the set $\{\theta^{-1}(f_x) : x \in X\} \subseteq F$ is a basis of F.

(a) $\Rightarrow$ (d). Let $X \subseteq F$ be a basis of F and suppose that we are given a map $f\colon X \to G$, where G is an abelian group. Since X is a basis of F, for every $u \in F$ there exist unique $k \in \mathbb{N}$, $x_1, x_2, \ldots, x_k \in X$ (distinct), and $n_1, n_2 \ldots, n_k \in \mathbb{Z}$ such that $u = n_1x_1 + n_2x_2 + \cdots + n_kx_k$. We then define $\overline{f}\colon F \to G$ by setting $\overline{f}(u) = n_1 f(x_1) + n_2 f(x_2) + \cdots + n_k f(x_k)$. Note that $\overline{f}$ is well defined, that it extends f and, since G is abelian, it is a group homomorphism. Uniqueness follows from the fact that X generates F.

(d) $\Rightarrow$ (c). Suppose (d) holds and let $G = \bigoplus_{x\in X} \mathbb{Z}$. Let $f_x \in G$ be defined by $f_x(y) = 1$ if $x = y$, and $f_x(y) = 0$ otherwise, and consider the map $f\colon X \to G$ defined by $f(x) = f_x$ for all $x \in X$. Then f extends uniquely to a group homomorphism $\overline{f}\colon F \to G$. Note that $\overline{f}$ is surjective since $\overline{f}(F) \supset \overline{f}(X) = \{f_x : x \in X\}$ and the latter generates G. On the other hand the map $h\colon \{f_x : x \in X\} \to F$ defined by $h(f_x) = x$ for all $x \in X$ also extends to a unique homomorphism $\overline{h}\colon G \to F$. Since $h \circ f = \mathrm{Id}_X$ we deduce that $\overline{h} \circ \overline{f} = \mathrm{Id}_F$ so that $\overline{f}$ is an isomorphism. It follows that $F \cong \bigoplus_{x\in X} \mathbb{Z}$. □

Definition 1.22. An abelian group F satisfying one of the equivalent conditions (a) – (d) in Proposition 1.21 is said to be a *free abelian group*.

In the sequel, if X is a basis of a free abelian group F we shall denote by $\mathbb{Z}x$ the infinite cyclic subgroup $\langle x \rangle \subseteq F$, for $x \in X$, and we shall therefore write $F = \bigoplus_{x\in X} \mathbb{Z}x$.

Proposition 1.23. *Let F be a free abelian group. Then any two bases $X, Y \subseteq F$ have the same cardinality.*

Proof. Consider the subgroup $2F := \{2a : a \in F\}$ of F. If X is a basis of F, then $F = \bigoplus_{x \in X} \mathbb{Z}x$ and $2F = \bigoplus_{x \in X} 2\mathbb{Z}x$. Therefore $F/2F \cong \bigoplus_{x \in X} \mathbb{Z}/2\mathbb{Z}$. Hence $F/2F$ is finite if and only if X is finite.

From this we deduce the following alternative: either all the bases of F are finite, or they are all infinite.

In the first case, the cardinality of any basis is clearly $\log_2[F : 2F]$.

Otherwise, the statement is then achieved by observing that if $X \subseteq F$ is an infinite basis then $|X| = |F|$. Indeed $|X| \leq |F|$, trivially. Conversely, denoting by $\mathscr{P}_f(X)$ the set of all finite subsets of X, which is ordered by inclusion, we have $F = \bigcup_{Y \in \mathscr{P}_f(X)} \langle Y \rangle$, and $|\langle Y \rangle| = |\bigoplus_{y \in Y} \mathbb{Z}y| = |\mathbb{Z}|^{|Y|} = \aleph_0$ for every $Y \in \mathscr{P}_f(X)$. As $|\mathscr{P}_f(X)| = |X|$ we deduce that $|F| \leq \aleph_0 |X| = |X|$. Thus $|X| = |F|$ by the Cantor–Bernstein theorem (see [155, Chapter 22] and/or [59, Corollary H.3.5]). □

The cardinality of a basis of a free abelian group F is called the *(free abelian) rank* of F, denoted $\mathrm{rk}(F)$. When F is finitely generated, $\mathrm{rk}(F)$ coincides with the rank of F.

We leave the proof of the following proposition (whose proof is analogous to that of Proposition 1.23) as an **exercise**.

Proposition 1.24. *Two free abelian groups are isomorphic if and only if they have the same (free abelian) rank.*

We now study the relation between free groups and free abelian groups.

Definition 1.25. Let G be a group. The *commutator* of two elements $x, y \in G$ is the element $[x,y] := x^{-1}y^{-1}xy \in G$. The *commutator subgroup* of G is the subgroup generated by all the commutators $[x,y]$ with $x, y \in G$, and it is denoted by G'.

For all $x, y, g \in G$, we have $g^{-1}[x,y]g = [g^{-1}xg, g^{-1}yg]$. We deduce that G' is a normal subgroup of G.

Let X be a set. We denote by $\mathrm{Ab}(X) := \bigoplus_{x \in X} \mathbb{Z}$ the free abelian group on the index set X.

Proposition 1.26. *Let F be a free group and let $X \subseteq F$ be a free basis. Then $F/F' \cong \mathrm{Ab}(X)$.*

Proof. Let $\pi \colon F \to F/F'$ denote the canonical quotient homomorphism, and observe that, by Corollary 1.6, it induces a bijection between $X \subseteq F$ and $\pi(X) \subseteq F/F'$.

Given any map $f \colon \pi(X) \to G$, where G is an abelian group, consider the composition $f \circ \pi \colon X \to G$. Since F is free based at X, there exists a unique homomorphism $\varphi \colon F \to G$ extending $f \circ \pi$.

As all the commutators of F are in the kernel $\ker(\varphi)$ of φ, we have $F' \subseteq \ker(\varphi)$. This yields a unique homomorphism $\overline{\varphi} \colon F/F' \to G$ such that $\overline{\varphi} \circ \pi = \varphi$. Clearly $\overline{\varphi}$ extends f since

$$\overline{\varphi}(\pi(x)) = (\overline{\varphi} \circ \pi)(x) = \varphi(x) = (f \circ \pi)(x) = f(\pi(x))$$

for any $x \in X$.

It follows that F/F' satisfies the universal property of free abelian groups based at $\pi(X)$ (see Proposition 1.21.(d)) and therefore $F/F' \cong \mathrm{Ab}(\pi(X))$. Since we already observed that $|X| = |\pi(X)|$, Proposition 1.24 implies $\mathrm{Ab}(\pi(X)) \cong \mathrm{Ab}(X)$, completing the proof. $\square$

We can now deduce a fundamental result about free groups.

Corollary 1.27. *Let F_1 (resp. F_2) be a free group and let $X_1 \subseteq F_1$ (resp. $X_2 \subseteq F_2$) be a free basis. Then F_1 is isomorphic to F_2 if and only if $|X_1| = |X_2|$. In particular, two finitely generated free groups are isomorphic if and only if they have the same rank.*

Proof. Suppose that F_1 and F_2 are isomorphic. Then any isomorphism between them induces an isomorphism between F_1/F_1' and F_2/F_2'. Proposition 1.26 gives $F_i/F_i' \cong \mathrm{Ab}(X_i)$, $i = 1,2$. But then by Proposition 1.24 we deduce that X_1 and X_2 have the same cardinality.

The converse follows from Remark 1.2.(b). $\square$

We now present a characterization of finitely generated abelian groups.

Lemma 1.28. *Let $X = \{x_1, x_2, \ldots, x_k\}$ be a basis of a free abelian group F, and let $z \in \mathbb{Z}$. Then for all $1 \leq i \neq j \leq k$ the set $X' = \{x_1, x_2, \ldots, x_{j-1}, x_j + zx_i, x_{j+1}, \ldots, x_k\}$ is a basis of F.*

Proof. Since $x_j = (x_j + zx_i) - zx_i$, we have that X' generates F. Moreover, if $n_1, n_2, \ldots, n_k \in \mathbb{Z}$ and $n_1x_1 + n_2x_2 + \cdots + n_{j-1}x_{j-1} + n_j(x_j + zx_i) + n_{j+1}x_{j+1} + \cdots + n_kx_k = 0$, then (say $i < j$) we also have $n_1x_1 + n_2x_2 + \cdots + n_{i-1}x_{i-1} + (n_i + zn_j)x_i + n_{i+1}x_{i+1} + \cdots + n_{j-1}x_{j-1} + n_jx_j + n_{j+1}x_{j+1} + \cdots + n_kx_k = 0$ so that, since X is a basis, $n_1 = n_2 = \cdots = n_{i-1} = (n_i + zn_j) = n_{i+1} = \cdots = n_j = \cdots = n_k = 0$ (and therefore also $n_i = 0$). It follows that X' is a basis of F. $\square$

Theorem 1.29 (Dedekind). *Let F be a free abelian group of finite rank k and let $G \leq F$ be a nontrivial subgroup. Then G is free abelian of rank $r \leq k$. In fact, there exist a basis $X = \{x_1, x_2, \ldots, x_k\}$ of F and positive integers $d_1, d_2, \ldots, d_r$, with d_i dividing d_{i+1} for $i = 1, 2, \ldots, r-1$, such that the subset $Y = \{d_1x_1, d_2x_2, \ldots, d_rx_r\}$ is a basis of G.*

Proof. We prove the statement by induction on the rank k of F. If $k = 1$ we have $F = \langle x_1 \rangle \cong \mathbb{Z}$ so that there exists a positive integer d_1 such that $G = \langle d_1x_1 \rangle \cong \mathbb{Z}$. So in this case $r = 1$, $X = \{x_1\}$ and $Y = \{d_1x_1\}$.

Suppose that we have proved the statement for all free abelian groups of rank $\leq k-1$.

For every basis $Y = \{y_1, y_2, \ldots, y_k\} \subseteq F$ we denote by Z_Y the set of all $n \in \mathbb{Z}$ such that there exists $1 \leq i \leq k$ and coefficients $n_1, n_2, \ldots, n_{i-1}, n_{i+1}, \ldots, n_k \in \mathbb{Z}$ such that

$$n_1y_1 + n_2y_2 + \cdots + n_{i-1}y_{i-1} + ny_i + n_{i+1}y_{i+1} + \cdots + n_ky_k \in G.$$

Let d_1 denote the least positive integer belonging to some Z_Y's, where Y runs over all bases of F. Then we can find a basis $Y = \{y_1, y_2, \ldots, y_k\}$ and an element $g \in G$ such that $g = d_1y_1 + n_2y_2 + \cdots + n_ky_k$. Applying Euclidean division we have $n_i = d_1q_i + r_i$ where $q_i \in \mathbb{Z}$ and $0 \leq r_i < d_1$ for all $i = 2, \ldots, k$. It follows that

$$g = d_1(y_1 + q_2y_2 + \cdots + q_ky_k) + (r_2y_2 + r_3y_3 + \cdots + r_ky_k).$$

Let $x_1 := y_1 + q_2y_2 + \cdots + q_ky_k$. Then, by Lemma 1.28, the set $Y' = \{x_1, y_2, \ldots, y_k\}$ is also a basis of F. Since $g = d_1x_1 + r_2y_2 + r_3y_3 + \cdots + r_ky_k \in G$, we deduce that $r_2, r_3, \ldots, r_k \in Z_{Y'}$ and therefore, by minimality of d_1, we necessarily have $r_2 = r_3 = \cdots = r_k = 0$. In particular, $d_1x_1 = g \in G$. Consider now the group H generated by $y_2, y_3, \ldots, y_k$. It is free abelian of rank $k-1$ and $F = \langle x_1 \rangle \oplus H$.

We claim that $G = \langle d_1x_1 \rangle \oplus (G \cap H)$. First observe that $\langle d_1x_1 \rangle \cap (G \cap H) = \{0\}$ since Y' is a basis of F. Suppose now that $g = m_1x_1 + m_2y_2 + \cdots + m_ky_k \in G$, where $m_1, m_2, \ldots, m_k \in \mathbb{Z}$. By applying again the Euclidean algorithm we can write $m_1 = d_1q_1 + r_1$ where $q_1 \in \mathbb{Z}$ and $0 \leq r_1 < d_1$. Thus $g - q_1d_1x_1 = r_1x_1 + m_2y_2 + \cdots + m_ky_k \in G$ so that, necessarily, $r_1 = 0$ by minimality of d_1. It follows that $m_2y_2 + \cdots + m_ky_k \in G \cap H$ so that $g = (d_1x_1) + (m_2y_2 + \cdots + m_ky_k) \in \langle d_1x_1 \rangle \oplus (G \cap H)$. The claim follows.

Now, either $G \cap H = \{0\}$, in which case $G = \langle d_1x_1 \rangle$ and the statement follows, or $G \cap H \neq \{0\}$. But then, by the inductive hypothesis (recall that H has rank $k-1$), we can find a basis $X' = \{x_2, x_3, \ldots, x_k\}$ of H and positive integers r, with $2 \leq r \leq k$, and $d_2, d_3, \ldots, d_r$ such that d_i divides d_{i+1} for $i = 2, 3, \ldots, r-1$, and $G \cap H$ is free abelian with basis $\{d_2x_2, d_3x_3, \ldots, d_rx_r\}$. We thus have $F = \langle x_1 \rangle \oplus H = \langle x_1, x_2, \ldots, x_k \rangle$ and $G = \langle d_1x_1 \rangle \oplus G \cap H = \langle d_1x_1, d_2x_2, \ldots, d_rx_r \rangle$. We are only left to show that $d_1 \mid d_2$. We have $d_2 = qd_1 + r$ where $q \in \mathbb{Z}$ and $0 \leq r < d_1$. By Lemma 1.28 we have that $\overline{X} = \{x_1 + qx_2, x_2, x_3, \ldots, x_k\}$ is a basis of F and $d_1(x_1 + qx_2) + rx_2 = d_1x_1 + d_2x_2 \in G$ so that $r \in Z_{\overline{X}}$. By minimality of d_1 we necessarily have $r = 0$, therefore d_1 divides d_2. □

From this theorem we deduce the following structure theorem for finitely generated abelian groups. We leave the details of the proof as an **exercise** (see also [67, Section 1.3]).

Corollary 1.30 (Invariant factor decomposition). *Let G be a finitely generated abelian group. Then there exist $k \in \mathbb{N}$ and a finite abelian group T such that $G \cong \mathbb{Z}^k \oplus T$. In addition, there exist $r \in \mathbb{N}$ and positive integers $d_1, d_2, \ldots, d_r$, with d_i dividing d_{i+1} for $i = 1, 2, \ldots, r-1$ such that, in fact, G is isomorphic to $\mathbb{Z}^k \oplus (\mathbb{Z}/d_1\mathbb{Z}) \oplus (\mathbb{Z}/d_2\mathbb{Z}) \oplus \cdots \oplus (\mathbb{Z}/d_r\mathbb{Z})$.*

1.9 Notes

Free groups arose at the end of the 19th century in the study of hyperbolic geometry, as examples of *Fuchsian groups* (these are discrete groups acting by isometries on the hyperbolic plane: they were first studied by Henri Poincaré in [280] who named them after Lazarus Fuchs [119]). Walther von Dyck in 1882 in his paper [99] remarked that free groups are the groups with the simplest possible presentation: a group is free if and only if it admits a presentation with no (defining) relations. The algebraic study of free groups was initiated by the topologist Jakob Nielsen in 1924, who, in a series of fundamental papers [253, 254, 255], gave them their name and established many of their basic properties. Kurt Reidemeister included a comprehensive treatment of free groups in his pioneering book [289] on combinatorial topology.

The Nielsen–Schreier theorem (Theorem 1.15) is a non-abelian analogue of an older result of Richard Dedekind, namely that "every subgroup of a free abelian group is free abelian" (Theorem 1.29). First it was proved by Max Dehn in an unpublished work where he exploited a graph-theoretic method for describing free groups. Nielsen [254] proved it for finitely-generated groups and Otto Schreier proved it in its full generality in his habilitation thesis published in [303].

A topological proof, based on the fact that "free groups are fundamental groups of bouquets of circles" was given in 1930 by Reinhold Baer and Friedrich Levi [11].

Another proof of the Nielsen–Schreier theorem can be immediately deduced from the following result: "A group is free if and only if it can act freely by automorphisms on a tree". This statement, which can be found in Reidemeister's book as well as in Serre's [310, 311], constitutes the initial fundamental result of the *Bass–Serre theory*. This theory, developed in the 1970s by Jean-Pierre Serre (originally motivated by his studies of certain algebraic groups whose Bruhat–Tits buildings are trees) and Hyman Bass [20], resulted in a fundamental tool of geometric group theory and geometric topology, particularly in the study of 3-manifolds. We finally mention that John Stallings [321] put forward a topological approach to the study of the algebraic structure of subgroups of free groups based on the methods of covering space theory that also used a simple graph-theoretic framework. The paper introduced the notion of a *Stallings subgroup graph* for describing subgroups of free groups, and also introduced a folding technique (used for approximating and algorithmically obtaining the subgroup graphs) and the notion of a *Stallings folding*. Stallings subgroup graphs and Stallings foldings have been used as key tools in the approach of the *Hanna Neumann conjecture*. This conjecture – posed by Hanna Neumann in 1957 [250] and motivated by a theorem of Howson [178] who proved in 1954 that the intersection of any two (nontrivial) finitely generated subgroups H and K of a free group is always finitely generated – states that

$$\mathrm{rk}(H \cap K) - 1 \leq (\mathrm{rk}(H) - 1)(\mathrm{rk}(K) - 1).$$

The Hanna Neumann conjecture (and in fact a strengthened version) was proved in 2011 by Joel Friedman [117] and, independently, by Igor Mineyev [240].

The proof of Theorem 1.5 is based on the so-called *van der Waerden trick* from [345] which indeed applies to the more general setting of free products of groups: a free group (based at X) is a free product of copies (indexed by X) of $\mathbb{Z}$.

The criterion in Theorem 1.17 was largely used by Felix Klein [200] in his studies of Schottky groups, though its present formulation is more recent.

Around 1945, Alfred Tarski asked whether free groups on two or more generators have the same first-order theory (also called *elementary theory*), and whether this theory is decidable. The first question was answered by Zlil Sela [309] in 2006 by showing that any two nonabelian free groups have the same elementary theory. Independently and in the same year, Olga Kharlampovich and Alexey Myasnikov [199] answered both questions, showing that this theory is indeed decidable.

A similar, yet unsolved question asks whether or not the von Neumann group algebras of any two non-abelian finitely generated free groups are isomorphic. This led to the foundation and development of *free probability theory* initiated by Dan Voiculescu around 1986 (see [344]).

The structure theorem for finitely generated abelian groups (Corollary 1.30) that we present is in terms of the *invariant factor decomposition*; there is also another version in terms of the *primary factor decomposition*, where the finite part is expressed as a direct sum of primary (= of order a power of a prime) cyclic groups. The two versions are equivalent because of the Chinese remainder theorem, which here states that $\mathbb{Z}/m\mathbb{Z} \simeq \mathbb{Z}/j\mathbb{Z} \oplus \mathbb{Z}/k\mathbb{Z}$ if and only if j and k are coprime and $m = jk$. The structure theorem for *finite* abelian groups was proved by Leopold Kronecker [205] in 1870 who generalized an earlier result of Carl Friedrich Gauss from *Disquisitiones Arithmeticae* (1801). The theorem was stated and proved in the language of the theory of groups by Ferdinand Georg Frobenius and Ludwig Stickelberger [118] in 1878. The structure theorem for *finitely generated* abelian groups was proved by Henri Poincaré [281] in 1900, using matrix methods (which generalize to principal ideal domains) for his computations of the homology (Betti numbers and torsion coefficients) of a complex. Another proof, based on and generalizing Kronecker's finite case proof, was found by Emmy Noether [257] in 1926.

1.10 Exercises

Exercise 1.1. Let F_1 and F_2 be free groups based on the sets X_1 and X_2, respectively. Suppose that there exists a bijective map $f\colon X_1 \to X_2$. Show that the free groups F_1 and F_2 are isomorphic.

Exercise 1.2. Let F be a free group based at X. Show that F is generated by X.

Exercise 1.3. Show that the group $\mathbb{Z}$ is free (with basis $X = \{1\}$).

Exercise 1.4. Show that a free group is torsion-free. Deduce that no finite (nontrivial) group can be free.

Exercise 1.5. Let F be a free group and let $\psi\colon F \to \widetilde{F}$ be an isomorphism from F onto a group $\widetilde{F}$. Show that $\widetilde{F}$ is a free group as well, based at $\widetilde{X} := \psi(X)$.

Exercise 1.6. Let F be a free group with basis $X \subseteq F$. Let $Y \subseteq X$ and let K denote the subgroup of F generated by Y. Show that K is a free group with basis Y.

Exercise 1.7. A group P is termed *projective* if the following holds: if G and H are two groups, $\pi\colon G \to H$ is a surjective group homomorphism, and $\varphi\colon F \to H$ is a group homomorphism, then there exists a group homomorphism $\psi\colon P \to G$ such that $\varphi = \pi \circ \psi$. Show that a group is projective if and only if it is free (cf. Exercise 1.20).

Exercise 1.8. Show that the system S defined in the proof of Lemma 1.14 is a Schreier system of representatives for $H\backslash F$.

Exercise 1.9. Show that the statement of the Nielsen–Schreier theorem does not hold for free monoids.

Exercise 1.10. Find an example of a group G with a (finite) generating subset $X \subset G$, a group H, and a map $f: X \to H$ which does not extend to a group homomorphism $G \to H$.

Exercise 1.11. Let $n \in \mathbb{N}$. Find a free group F together with a surjective group homomorphism $F \to \mathrm{Sym}(n)$, the symmetric group of degree n.

Exercise 1.12. Prove that for every $m, n \in \mathbb{N}$ with $m \leq n$, any free group F of rank n has, as homomorphic image, a group G such that: (i) G is free of rank m; (ii) G is a subgroup of F.

Exercise 1.13. Let F be the free group of rank 2, freely generated, say by the elements $a, b \in F$. Let $G = \mathbb{Z}^2$ (written additively). The map $f: \{a, b\} \to G$ defined by $f(a) = (1, 0)$ and $f(b) = (0, 1)$ extends to a unique group homomorphism $\varphi: F \to G$. Show that (i) φ is surjective; (ii) $\ker(\varphi) = [F, F]$, the commutator subgroup of F.

Exercise 1.14. Prove that a free group of rank n has a subgroup of index m for all $n, m \in \mathbb{N}$, $n \geq 1$.

Exercise 1.15. Let F be the free group of rank 3, say freely generated by a, b, c. Let $N \leq F$ denote the normal subgroup generated by c (i.e., the intersection of all normal subgroups containing c). Show that (i) if $g \in N$, then there exist $m \in \mathbb{N}$, $\varepsilon_i \in \{1, -1\}$, $f_i \in F$, $i = 1, 2, \ldots, m$, such that $g = g_1 c^{\varepsilon_1} g_1^{-1} g_2 c^{\varepsilon_2} g_2^{-1} \cdots g_m c^{\varepsilon_m} g_m^{-1}$; and (ii) F/N is free of rank 2 (indeed, freely generated by aN and bN).

Exercise 1.16. Using the Nielsen–Schreier theorem prove that if F is a free group and $u, v \in F$ commute then they are powers of a common element in F. Deduce that the relation $u \sim v$ if $uv = vu$ is an equivalence relation on $F \setminus \{1\}$.

Exercise 1.17. Show that if F is free non-abelian, then its center $Z(F) := \{z \in F : zg = gz \text{ for all } g \in F\}$ is trivial.

Exercise 1.18. Show that if F is a free group, then for every nontrivial element $g \in F$, g cannot be conjugate to g^{-1}.

Exercise 1.19. Show that for $k \geq 2$ the matrices $A = \begin{pmatrix} 1 & k \\ 0 & 1 \end{pmatrix}$ and $B = \begin{pmatrix} 1 & 0 \\ k & 1 \end{pmatrix}$ in $\mathrm{SL}(2, \mathbb{Z})$ generate a free group. Show that, however, for $k = 1$, this is no longer the case.

Exercise 1.20. An abelian group P is termed *projective* (as an abelian group) if the following holds: if G is an abelian group, $H \leq G$ is a subgroup, and $\varphi: P \to G/H$ is a group homomorphism, then there exists a homomorphism $\psi: P \to G$ such that $\varphi = \pi \circ \psi$, where $\pi: G \to G/H$ is the canonical quotient homomorphism. Show that an abelian group is projective if and only if it is free abelian (cf. Exercise 1.7).

Exercise 1.21. Show that a free abelian group is a free group if and only if it is either trivial (in this case the basis is $X = \varnothing$) or isomorphic to $\mathbb{Z}$ (cf. Exercise 1.3).

Exercise 1.22. Let G be a finite abelian group. Using the structure theorem for finitely generated abelian groups (cf. Corollary 1.30), show that if $H \leq G$ is a subgroup, then there exists a subgroup $K \leq G$ which is isomorphic to the quotient G/H.

Exercise 1.23. Show that the multiplicative group of the positive rational numbers $\mathbb{Q}^+ := \{q \in \mathbb{Q} : q > 0\}$ is free abelian.

Exercise 1.24. Let G be an abelian group generated by n elements. Let $H \leq G$ be a subgroup. Show that H is generated by m elements with $m \leq n$.

Chapter 2
Nilpotent Groups

A group G is called nilpotent if it has a finite central series, that is, a sequence of normal subgroups $\{1_G\} = G_0 \leq G_1 \leq \cdots \leq G_n = G$ such that G_{i+1}/G_i is contained in the center of G/G_i for all $i = 0, 1, \ldots, n-1$. Equivalently, G is nilpotent if its lower central series $G = \gamma_1(G) \geq \gamma_2(G) \geq \cdots$, where $\gamma_{i+1}(G) := [G, \gamma_i(G)]$ is the commutator subgroup of G and $\gamma_i(G)$, $i = 1, 2, \ldots$, eventually reaches the trivial subgroup $\{1_G\}$ of G. The least integer $c \geq 1$ such that $\gamma_{c+1}(G) = \{1_G\}$ is called the nilpotency class of G.

The group $\mathrm{UT}(n, R)$ of upper unitriangular $n \times n$ matrices with coefficients in a commutative ring R is nilpotent; in contrast, the group $\mathrm{B}(n, R)$ of upper triangular $n \times n$ matrices with coefficients in a commutative ring R with $2 \in R$ invertible is not nilpotent. In Section 2.4 we compute their lower central series (as well as their upper central series).

The first fundamental result that we present is due to Malcev (Theorem 2.20) and gives a linear representation of finitely generated torsion-free nilpotent groups: every such group is embeddable in $\mathrm{UT}(m, \mathbb{Z})$ for some $m \geq 1$. The proof includes, as a side result of independent interest, the fact (also due to Malcev) that the coefficients of the product of two group elements in a finitely generated torsion-free nilpotent group, when expressed in terms of a canonical generating system (called a Malcev basis), have a polynomial expression in terms of the coefficients of the two group elements (Lemma 2.31). In the last section, we study finitely generated nilpotent groups with torsion and show that every finitely generated nilpotent group is virtually torsion-free (Lemma 2.42). From Theorem 2.20, we then deduce that every finitely generated nilpotent group is linear, in fact embeddable into $\mathrm{GL}(m, \mathbb{Z})$ for some $m \geq 1$ (Corollary 2.44).

2.1 Commutator Identities

Let G be a group. Let a and b be two elements of G. The *commutator* of a and b is the element $[a, b] \in G$ defined by

T. Ceccherini-Silberstein and M. D'Adderio, *Topics in Groups and Geometry*,
Springer Monographs in Mathematics, https://doi.org/10.1007/978-3-030-88109-2_2

$$[a,b] := a^{-1}b^{-1}ab.$$

We also set

$$a^b := b^{-1}ab.$$

More generally, given $n \geq 3$ and elements $a_1, a_2, \ldots, a_n \in G$, we recursively define

$$[a_1, a_2, \ldots, a_n] := [[a_1, a_2, \ldots, a_{n-1}], a_n]. \tag{2.1}$$

An element as in (2.1) is called a *simple commutator of weight n*.

The proof of the following lemma is left as an **exercise**.

Lemma 2.1. *Let G be a group. For all $a, b, c \in G$ the following* basic identities *hold:*

(1) $[a,b]^{-1} = [b,a]$;
(2) $[ab,c] = [a,c]^b[b,c] = [a,c][a,c,b][b,c]$;
(3) $[a,bc] = [a,c][a,b]^c = [a,c][b,a,c]^{-1}[a,b]$;
(4) $[a,b^{-1}] = ([a,b]^{b^{-1}})^{-1}$;
(5) $[a^{-1},b] = [b,a]^{a^{-1}}$;
(6) $[a,b^{-1},c]^b[b,c^{-1},a]^c[c,a^{-1},b]^a = 1$ (Hall–Witt identity).

Recall that a subgroup H of G is said to be *normal* in G, and we write $H \trianglelefteq G$, if $h^g \in H$ for all $h \in H$ and $g \in G$. Also, H is said to be *characteristic* in G if $\varphi(h) \in H$ for all $h \in H$ and all automorphisms φ of G.

Let H and K be subgroups of G. We denote by $[H,K]$ the subgroup of G generated by all commutators $[h,k]$, where $h \in H$ and $k \in K$. Notice that by Lemma 2.1.(1) we always have $[H,K] = [K,H]$.

It is an **exercise** to see that $[H,K] \subseteq K$ (resp. $[H,K] \subseteq H$) if K (resp. H) is normal in G, and that $[H,K]$ is normal (resp. characteristic) in G if H and K are both normal (resp. characteristic) in G.

Lemma 2.2 (The three subgroup lemma). *Let G be a group, let $N \trianglelefteq G$ be a normal subgroup, and let $H, K, L \leq G$ be three subgroups. If $[[H,K],L]$ and $[[K,L],H]$ are contained in N, then $[[L,H],K]$ is also contained in N.*

Proof. First observe that by Lemma 2.1.(2), the subgroup $[[H,K],L]$ (resp. $[[K,L],H]$, resp. $[[L,H],K]$) is generated by the commutators $[h,k,\ell]$ (resp. $[k,\ell,h]$, resp. $[\ell,h,k]$), where $h \in H$, $k \in K$ and $\ell \in L$. The statement then follows from the Hall–Witt identity (Lemma 2.1.(6)). □

2.2 The Lower Central Series

Let G be a group. A *descending* (resp. *ascending*) *series* in G is a sequence $(H_k)_{k \geq 1}$ of subgroups of G such that

$$G = H_1 \geq H_2 \geq \cdots \geq H_k \geq H_{k+1} \geq \cdots$$

(resp, $\{1_G\} = H_1 \leq H_2 \leq \cdots \leq H_k \leq H_{k+1} \leq \cdots$). The subgroups $H_1, H_2, \ldots$ are called the *terms* of the series.

A descending (resp. ascending) series $(H_k)_{k\geq 1}$ in G is called *finite* provided that there exists a $k \in \mathbb{N}$ such that $H_k = \{1_G\}$ (resp. $H_k = G$). The minimal integer k for which this happens is called the *length* of the series.

A descending (resp. ascending) series $(H_k)_{k\geq 1}$ in G is called *normal* if H_k is normal in G for all k.

Definition 2.3. Let G be a group. The *lower central series* of G is the sequence $(\gamma_k(G))_{k\geq 1}$ of subgroups of G recursively defined by

$$\gamma_1(G) := G \quad \text{and} \quad \gamma_{k+1}(G) := [\gamma_k(G), G] \quad \text{for all } k \geq 1.$$

It is an **exercise** to show that $\gamma_k(G)$ is characteristic (and therefore normal) in G and that $\gamma_{k+1}(G) \subseteq \gamma_k(G)$ for all k. It follows that the lower central series is a descending normal series.

Definition 2.4. A group G is said to be *nilpotent* if its lower central series is finite, that is, if there is an integer $k \geq 1$ such that $\gamma_k(G) = \{1_G\}$. If this is the case, the smallest integer $c \geq 1$ such that $\gamma_{c+1}(G) = \{1_G\}$ is then called the *nilpotency class* of G.

Example 2.5. (a) Every abelian group is nilpotent, of nilpotency class $c = 1$.

(b) Let R be a unital commutative ring with identity $1 := 1_R$, and let $n \geq 2$ be an integer. An $n \times n$ matrix M with coefficients in R is said to be *upper unitriangular* provided that its entries are equal to 1 on the diagonal and vanish below the diagonal. More explicitly, a matrix $M = \|x_{ij}\|_{1\leq i,j\leq n}$, with $x_{ij} \in R$ for all $1 \leq i, j \leq n$, is upper unitriangular if for all $i, j = 1, 2, \ldots, n$ we have $x_{ii} = 1$ and $x_{ij} = 0$ provided $i > j$. In other words, an upper unitriangular matrix is of the form:

$$M = \begin{pmatrix} 1 & x_{12} & x_{13} & \cdots & x_{1(n-1)} & x_{1n} \\ 0 & 1 & x_{23} & \cdots & x_{2(n-1)} & x_{2n} \\ \vdots & \vdots & \vdots & \vdots & \vdots & \vdots \\ 0 & 0 & 0 & \cdots & 1 & x_{(n-1)n} \\ 0 & 0 & 0 & \cdots & 0 & 1 \end{pmatrix}.$$

It is an **exercise** to see that the set $\mathrm{UT}(n,R)$ consisting of all $n \times n$ upper unitriangular matrices with coefficients in R forms a group with the usual multiplication of matrices. It is a slightly harder **exercise** to see that for $k \geq 1$, the k-th term $\gamma_k(\mathrm{UT}(n,R))$ of the lower central series of $\mathrm{UT}(n,R)$ consists of all upper unitriangular matrices $M = \|x_{ij}\|_{1\leq i,j\leq n}$ satisfying $x_{ij} = 0$ provided $1 \leq j - i \leq k$. In particular, $\gamma_n(\mathrm{UT}(n,R)) = \{I_n\}$, where I_n denotes the $n \times n$ identity matrix, but $\gamma_{n-1}(\mathrm{UT}(n,R)) \neq \{I_n\}$, so that $\mathrm{UT}(n,R)$ is nilpotent of nilpotency class $n-1$.

Lemma 2.6. *Let G be a group. Then*

$$[\gamma_i(G), \gamma_j(G)] \subseteq \gamma_{i+j}(G) \tag{2.2}$$

for all $i, j \geq 1$.

Proof. We prove the statement by induction on i. For $i = 1$, in (2.2) we have that $[\gamma_1(G), \gamma_j(G)] = [G, \gamma_j(G)]$ equals $\gamma_{j+1}(G)$ for all $j \geq 1$, by definition. Suppose the statement is true for $i \geq 1$ and let us show that for all j

$$[\gamma_{i+1}(G), \gamma_j(G)] \subseteq \gamma_{i+j+1}(G). \tag{2.3}$$

We first claim that given $a, b \in \gamma_{i+1}(G)$ and $c \in \gamma_j(G)$ we have

$$[ab, c] = [a, c][b, c] \text{ modulo } \gamma_{i+j+1}(G) \tag{2.4}$$

and

$$[a^{-1}, c] = [a, c]^{-1} \text{ modulo } \gamma_{i+j+1}(G). \tag{2.5}$$

Indeed, we have $a \in \gamma_{i+1}(G) \subseteq \gamma_i(G)$ so that, by the inductive hypothesis, $[a, c] \in [\gamma_i(G), \gamma_j(G)] \subseteq \gamma_{i+j}(G)$. As a consequence, $[a, c, b] = [[a, c], b] \in [\gamma_{i+j}(G), \gamma_j(G)] \subseteq [\gamma_{i+j}(G), G] = \gamma_{i+j+1}(G)$. Also, by Lemma 2.1.(2), we have

$$[ab, c] = [a, c][a, c, b][b, c] = [a, c][b, c]([b, c]^{-1}[a, c, b][b, c]).$$

Hence, by normality of $\gamma_{i+j+1}(G)$, also $[b, c]^{-1}[a, c, b][b, c] = [a, c, b]^{[b,c]}$ is in the set $\gamma_{i+j+1}(G)$, so this proves (2.4).

By taking $b = a^{-1}$ in (2.4) one deduces (2.5). This proves the claim.

As a consequence, in order to prove (2.3) we may limit ourselves to check it on the generators of $\gamma_{i+1}(G)$, i.e. to show that

$$[a, x, b] \in \gamma_{i+j+1}(G) \text{ for all } a \in \gamma_i(G), x \in G, \text{ and } b \in \gamma_j(G). \tag{2.6}$$

Now, the Hall–Witt identity (Lemma 2.1.(6)) gives

$$[a, x, b] = \left(([b, a^{-1}, x^{-1}]^a)^{-1} ([x^{-1}, b^{-1}, a]^b)^{-1} \right)^x.$$

On the other hand, observing that, by the inductive hypothesis, $[b, a^{-1}, x^{-1}] \in [\gamma_{i+j}(G), G] = \gamma_{i+j+1}(G)$ and $[x^{-1}, b^{-1}, a] \in [\gamma_{j+1}(G), \gamma_i(G)] \subseteq \gamma_{i+j+1}(G)$, recalling that $\gamma_{i+j+1}(G)$ is a normal subgroup, we deduce (2.6). □

A descending series $(G_k)_{k \geq 1}$ in G

$$G = G_1 \geq G_2 \geq \cdots \geq G_k \geq G_{k+1} \geq \cdots$$

is said to be *central* if $[G_i, G_j] \subseteq G_{i+j}$ for all $i, j \geq 1$. Observe that this condition implies that each G_i is normal in G, since $[G_i, G] = [G_i, G_1] \subseteq G_{i+1} \subseteq G_i$.

Thus we can reformulate Lemma 2.6 by saying that the lower central series of a group G is a central series. The following lemma explains the etymology for the term “lower”.

Lemma 2.7. *Let G be a group. Suppose that $(G_k)_{k \geq 1}$ is a descending central series of G. Then $G_i \supseteq \gamma_i(G)$ for all $i \geq 1$.*

Proof. We proceed by induction on i. For $i = 1$ this is obvious since $G_1 = G = \gamma_1(G)$. Suppose the statement holds for $i \geq 1$. From the defining property of a central series and the inductive hypothesis, we deduce

$$G_{i+1} \supseteq [G_i, G_1] \supseteq [\gamma_i(G), G] = \gamma_{i+1}(G).$$

This completes the proof of the lemma. □

These considerations immediately imply the following characterization of nilpotency: we leave the details as an **exercise**.

Proposition 2.8. *A group is nilpotent if and only if it admits a finite central series.*

We end this section by recording two important hereditary properties of the class of nilpotent groups.

Proposition 2.9. *Let G be a nilpotent group of nilpotency class c. Then any subgroup and any quotient of G is nilpotent of nilpotency class $\leq c$.*

Proof. Let $H \subset G$ be a subgroup. It immediately follows by induction that $\gamma_i(H) \subseteq \gamma_i(G)$ for all $i = 1, 2, \ldots, c$. In particular, $\gamma_{c+1}(H) \subseteq \gamma_{c+1}(G) = \{1_G\}$, thus showing that H is nilpotent with nilpotency class $\leq c$.

Let $\varphi \colon G \to \overline{G}$ be a surjective homomorphism onto a group $\overline{G}$. Let $H, K \subseteq G$ be two subgroups. For all $h \in H$ and $k \in K$ we have $\varphi([h,k]) = [\varphi(h), \varphi(k)]$ so that $\varphi([H,K]) = [\varphi(H), \varphi(K)]$. It follows that

$$\gamma_{i+1}(\overline{G}) = \gamma_{i+1}(\varphi(G)) = [\gamma_i(\varphi(G)), \varphi(G)] = \varphi([\gamma_i(G), G]) = \varphi(\gamma_{i+1}(G))$$

for all $i = 1, 2, \ldots, c$. In particular, $\gamma_{c+1}(\overline{G}) = \varphi(\gamma_{c+1}(G)) = \{1_G\}$, thus showing that $\overline{G}$ is nilpotent with nilpotency class $\leq c$. □

2.3 The Upper Central Series

We now introduce another important series in a group G.

Definition 2.10. Let G be a group. The *upper central series* of G is the ascending series $(Z_k(G))_{k \geq 0}$ in G defined by setting

$$Z_0(G) := \{1_G\} \quad \text{and} \quad Z_{i+1}(G) := \pi_i^{-1}(Z(G/Z_i(G))) \quad \text{for } i \geq 0,$$

where $\pi_i \colon G \to G/Z_i(G)$ denotes the quotient homomorphism, and, for any group H, we denote by $Z(H) := \{z \in H : zh = hz \text{ for all } h \in H\}$ the *center* of H.

Thus, for all $i \in \mathbb{N}$ we have

$$Z(G/Z_i(G)) = Z_{i+1}(G)/Z_i(G).$$

In particular, $Z_1(G) = Z(G)$ is the center of G.

The proof of the following lemma is left as an **exercise**.

Lemma 2.11. *Let G be a group. Suppose that the upper central series of G is finite, say $\{1_G\} = Z_0(G) \leq Z_1(G) \leq \cdots \leq Z_k(G) = G$. Then the descending series $(Z_{k-i+1}(G))_{i=1}^{k+1}$ is central.*

The following lemma explains the etymology for the term "upper".

Lemma 2.12. *Let G be a group. Suppose that G admits a finite central series $G = G_1 \geq G_2 \geq \cdots \geq G_r = \{1_G\}$. Then $Z_i(G) \supseteq G_{r-i}$ for all $i \geq 0$. In particular, $Z_{r-1}(G) = G$.*

Proof. We prove the statement by induction on i. For $i = 0$ this is obvious: $Z_0(G) = \{1_G\} = G_r$. Suppose the statement holds for some $i \geq 0$ and let us show that $Z_{i+1}(G) \supseteq G_{r-i-1}$. By definition, we have to show that $\pi_i(G_{r-i-1}) \subseteq Z(G/Z_i(G))$, equivalently, that $[G_{r-i-1}, G] \subseteq Z_i(G)$. But this follows from $[G_{r-i-1}, G] \subseteq G_{r-i}$, since the series is central, and from $G_{r-i} \subseteq Z_i(G)$, by induction. □

Remark 2.13. From what we have just proved, if the lower central series has length c, then $Z_{c-1}(G) \supseteq \gamma_{c-(c-1)}(G) = \gamma_1(G) = G$. Hence the upper central series, denoted $(Z_{k-i+1}(G))_{i=1}^{k+1}$, reaches the whole group G at least as "fast" as the lower central series reaches the trivial group $\{1_G\}$, i.e. $k \leq c-1$, that is $k+1 \leq c$.

On the other hand, by Lemma 2.11, the upper central series in this case is central. Therefore, if the length of the upper central series is $k+1 = c-r$, with $r \geq 1$, then already $Z_{c-r-1}(G) = G$. But by Lemma 2.7 we have $\gamma_i(G) \subseteq Z_{c-r-i}(G)$ for all i's. In particular, $\gamma_{c-r}(G) \subseteq Z_{c-r-(c-r)}(G) = Z_0(G) = \{1_G\}$, contradicting the definition of c. This shows that the lower central series cannot be "faster" than the upper central series.

Hence these two series must have the same length. In particular, if the upper central series of G has length c, then G is nilpotent of nilpotency class c.

From the above discussion, we deduce the following characterization of nilpotency.

Proposition 2.14. *A group is nilpotent if and only if its upper central series is finite.* □

2.4 Two Examples

In this section we study two important examples of groups, and we compute their lower central series and upper central series.

Example 2.15. Let R be a commutative ring with identity $1 = 1_R$, and let $\mathrm{UT}(n,R)$ be the group of $n \times n$ upper unitriangular matrices (see Example 2.5.(b)).

Let $(\gamma_i)_{i\geq 1}$ be the lower central series of $\mathrm{UT}(n,R)$. Let $G_1 := \mathrm{UT}(n,R)$, and, for $k = 2, \ldots, n$, let G_k be the group of upper unitriangular matrices with all (i,j)-th coefficients equal to zero when $1 \leq j-i \leq k-1$. So G_n is the trivial subgroup of $\mathrm{UT}(n,R)$; also by convention $G_h := G_n$ for $h \geq n+1$.

We want to show that $\gamma_i = G_i$ for all i.

The inclusion $\gamma_i \subseteq G_i$ is clear. To show the other inclusion, we need some notation: let $I^{(n)}$ be the identity matrix of order $n \times n$ and let $N_1^{(n)}$ be the $n \times n$ matrix with 1 in the positions $(i, i+1)$ and 0 elsewhere. For example

$$N_1^{(4)} = \begin{pmatrix} 0&1&0&0 \\ 0&0&1&0 \\ 0&0&0&1 \\ 0&0&0&0 \end{pmatrix}.$$

Also, we denote by $E_{i,j}^{(n)}$ the $n \times n$ matrix with 1 in the position (i, j), and 0 elsewhere.

We leave it as an **exercise** to check that for $m \geq 3$

$$[I^{(m)} + N_1^{(m)},\ I^{(m)} + E_{2,m}^{(m)}] = I^{(m)} + E_{1,m}^{(m)}. \tag{2.7}$$

Using this observation, looking at the block diagonal matrices with one block of the form $I^{(m)} + N_1^{(m)}$ or $I^{(m)} + E_{2,m}^{(m)}$ and the others equal to 1, one can show that, for $j - i \geq 2$, the subgroup $\gamma_2 = [\gamma_1, \gamma_1]$ contains all unitriangular matrices with an arbitrary coefficient in position (i, j) and 0 elsewhere. This easily implies that $\gamma_2 \supseteq G_2$. Inductively, we can use the same observation to show the other inclusions $\gamma_i \supseteq G_i$, since $\gamma_i = [\gamma_1, \gamma_{i-1}] = [G_1, G_{i-1}]$, and $I^{(m)} + E_{2,m}^{(m)}$ will always be in G_{i-1} for the given m.

In particular, this computation shows that $\mathrm{UT}(n, R)$ has nilpotency class $n - 1$.

Let now $(Z_i)_{i \geq 0}$ be the upper central series of $\mathrm{UT}(n, R)$. We want to show that $Z_i = G_{n-i} = \gamma_{n-i}$ for $0 \leq i \leq n - 1$.

The inclusion $G_{n-i} = \gamma_{n-i} \subseteq Z_i$ is proved in Lemma 2.12. To show the other inclusion, first observe that for $i = 0$ the statement is true. We now compute $Z_1 = Z(\mathrm{UT}(n, R))$. Let $M \in Z_1$. Then we must have $[I^{(n)} + N_1^{(n)}, M] = I^{(n)}$. But this implies that for each $k = 1, 2, \ldots, n - 1$, the coefficients of M in the positions $(i, i + k)$ for $i = 1, 2, \ldots, n - k$ must all be equal to a common value α_k. Consider now the matrix $I^{(n)} + E_{n-1,n}^{(n)}$. For example

$$I^{(4)} + E_{3,4}^{(4)} = \begin{pmatrix} 1&0&0&0 \\ 0&1&0&0 \\ 0&0&1&1 \\ 0&0&0&1 \end{pmatrix}.$$

Since M is in the center, we have $[I^{(n)} + E_{n-1,n}^{(n)}, M] = I^{(n)}$, which implies that the α_k's must be equal to 0 for $k = 1, 2, \ldots, n - 2$. This gives $\gamma_{n-1} = G_{n-1} \supseteq Z_1$. Inductively, we can argue similarly for the other inclusions $G_{n-i} = \gamma_{n-i} \supseteq Z_i$, recalling that $Z_i/Z_{i-1} = Z(\mathrm{UT}(n, R)/Z_{i-1}) = Z(\mathrm{UT}(n, R)/G_{i-1})$.

Example 2.16. Here we assume that $2 = 1 + 1 \in R$ is invertible in R. For example, this is always the case when R is a field of characteristic different from 2.

Let then $\mathrm{B}(n, R)$ be the group of $n \times n$ upper triangular matrices with diagonal entries from $R^\times$, the group of invertible elements of R.

Let now $(\gamma_i)_{i\geq 1}$ be the lower central series of $\mathrm{B}(n,R)$, and let $(G_i)_{i\geq 1}$ be, as in the previous example, the lower central series of $\mathrm{UT}(n,R)$.

We have the obvious inclusions

$$[\mathrm{B}(n,R),\mathrm{UT}(n,R)] \subseteq [\mathrm{B}(n,R),\mathrm{B}(n,R)] \subseteq \mathrm{UT}(n,R).$$

We want to show that in fact $[\mathrm{B}(n,R),\mathrm{UT}(n,R)] \supseteq \mathrm{UT}(n,R)$, and hence

$$[\mathrm{B}(n,R),\mathrm{UT}(n,R)] = [\mathrm{B}(n,R),\mathrm{B}(n,R)] = \mathrm{UT}(n,R).$$

To see this, in the notation of the previous example, observe that for $m \geq 2$

$$[I^{(m)}+E^{(m)}_{1,m}, I^{(m)}+E^{(m)}_{m,m}] = I^{(m)}+E^{(m)}_{1,m}. \tag{2.8}$$

Remark 2.17. Notice that here we use the assumption that $1+1=2\in R^{\times}$. If 2 is not invertible in R, then the inclusion $[\mathrm{B}(n,R),\mathrm{UT}(n,R)] \supseteq \mathrm{UT}(n,R)$ may not be true. For instance, if $R=\mathbb{Z}$ and $n=2$, then it is an **exercise** to show that $[\mathrm{B}(2,\mathbb{Z}),\mathrm{UT}(2,\mathbb{Z})] = \mathrm{UT}(2,2\mathbb{Z})$.

Using this observation, looking at the block diagonal matrices with one block of the form $I^{(m)}+E^{(m)}_{1,m}$ or $I^{(m)}+E^{(m)}_{m,m}$ and the others equal to 1, we can show that $I^{(n)}+E^{(n)}_{i,j} \in [\mathrm{B}(n,R),\mathrm{UT}(n,R)]$ for all $j>i$, and this easily implies $[\mathrm{B}(n,R),\mathrm{UT}(n,R)] \supseteq \mathrm{UT}(n,R)$.

From this we conclude that the lower central series of $\mathrm{B}(n,R)$ is given by $\gamma_1 = \mathrm{B}(n,R)$, and $\gamma_i = \mathrm{UT}(n,R)$ for all $i\geq 2$.

In particular, this computation shows that $\mathrm{B}(n,R)$ is not a nilpotent group.

Let $(Z_i)_{i\geq 0}$ be the upper central series of $\mathrm{B}(n,R)$. We want to show that for all $i\geq 1$

$$Z_i = Z(\mathrm{B}(n,R)) = R^{\times} I^{(n)}. \tag{2.9}$$

To see this, let $M\in Z_1 = Z(\mathrm{B}(n,R))$. For all $j=1,2,\dots,n$ we must have $[I^{(n)}-2E^{(n)}_{j,j}, M] = I^{(n)}$. But this implies that the coefficients of M in the positions $(j,j+k)$ and $(j+k,j)$ must be equal to zero for $k=1,2,\dots,n-j$. Hence M must be diagonal. But then $[I^{(n)}+N^{(n)}_1, M] = I^{(n)}$ implies that all the diagonal entries must be equal. This shows that $Z_1 \subseteq R^{\times} I^{(n)}$, and hence $Z_1 = R^{\times} I^{(n)}$.

To see now that $Z_2 = Z_1$, just remember that

$$Z_2/Z_1 = Z(\mathrm{B}(n,R)/Z_1) = Z(\mathrm{B}(n,R)/R^{\times} I^{(n)}),$$

and notice that we can apply the arguments that we just used for $\mathrm{B}(n,R)$ to the quotient $\mathrm{B}(n,R)/R^{\times} I^{(n)}$, to deduce that the center of $\mathrm{B}(n,R)/R^{\times} I^{(n)}$ is trivial, and hence $Z_2 = Z_1$. This implies (2.9).

Note that, using Proposition 2.14, this computation gives another way of showing that $\mathrm{B}(n,R)$ is not a nilpotent group.

2.5 Nilpotent Ideals

The term "nilpotency" comes from Ring Theory. Given a unital ring R and an ideal $J \subseteq R$ we denote by $J^n \subseteq R$ the ideal consisting of all finite sums of elements of the form $x_1 x_2 \cdots x_n$ with $x_1, x_2, \ldots, x_n \in J$. We say that J is *nilpotent* provided that there exists an integer $n \geq 1$ such that $J^n = \{0\}$. The minimal n with this property is called the *nilpotency class* of J.

Proposition 2.18. *Let R be a unital ring with unit 1_R. Let $J \subseteq R$ be a nilpotent ideal of nilpotency class n, and consider the set $G := 1_R + J = \{1_R + x : x \in J\} \subseteq R$. Then G is a nilpotent group of nilpotency class at most n.*

Proof. Let us first show that G is a group. Let $x, y \in J$. We have

$$(1_R + x)(1_R + y) = 1_R + (x + y + xy) \in 1_R + J;$$

this shows that G is closed under multiplication. Moreover, 1_R is the identity element 1_G in G. Finally, if n is the class of nilpotency of J, for every $x \in J$ we have

$$(1_R + x)(1_R - x + x^2 - \cdots + (-x)^{n-1}) = 1_R,$$

thus showing that every element of G is invertible. It follows that G is a group.

For $k \geq 1$, let $H_k := 1_R + J^k \subseteq G$. Notice that the ideal J^k is also nilpotent. Indeed $(J^k)^n = J^{nk} = (J^n)^k = \{0\}^k = \{0\}$. Moreover, it follows from the first part of the proof that H_k is a subgroup of G. As $J^k = J \cdot J^{k-1} \subseteq J^{k-1}$, we have $H_k \subseteq H_{k-1}$. Also $H_n = 1_R + J^n = 1_R + \{0\} = \{1_R\} = \{1_G\}$.

Let us prove that the finite descending series $(H_k)_{k \geq 1}$ in G is central. We have to show that if $a \in H_k$ and $b \in H_\ell$, then $[a,b] \in H_{k+\ell} = 1_R + J^{k+\ell}$; equivalently, $[a,b] - 1_R \in J^{k+\ell}$. If $a = 1_R + x$, with $x \in J^k$ and $b = 1_R + y$, with $y \in J^\ell$, we have

$$\begin{aligned}
[a,b] - 1_R &= a^{-1}b^{-1}ab - 1_R \\
&= a^{-1}b^{-1}(ab - ba) \\
&= a^{-1}b^{-1}((1_R + x)(1_R + y) - (1_R + y)(1_R + x)) \\
&= a^{-1}b^{-1}(1_R + x + y + xy - 1_R - y - x - yx) \\
&= a^{-1}b^{-1}(xy - yx) \\
&\in R \cdot (J^k \cdot J^\ell + J^\ell \cdot J^k) \\
&\subseteq J^{k+\ell}.
\end{aligned}$$

This shows that G is nilpotent of nilpotency class $\leq n$. □

2.6 Torsion-Free Finitely Generated Nilpotent Groups

In this section we study finitely generated nilpotent groups which are *torsion-free*, that is, that have no nontrivial elements of finite order.

We start with a lemma.

Lemma 2.19. *Let G be a torsion-free (possibly infinitely generated) nilpotent group with upper central series $(Z_k(G))_{k\geq 0}$. Then all factors $Z_{i+1}(G)/Z_i(G)$, $i \geq 0$, are torsion-free abelian groups.*

Proof. We proceed by induction on the nilpotency class c of G. If $c = 1$, then $Z_1(G)/Z_0(G) = G/\{1_G\} \cong G$ is abelian and, by hypothesis, it is also torsion-free. Suppose that the statement holds true for all nilpotent groups of nilpotency class $c-1$ and let us suppose that G has nilpotency class c. Set $H := G/Z_1(G)$. It follows from the definitions and an immediate induction that $Z_i(H) = Z_{i+1}(G)/Z_1(G)$ for all $i = 1,2,\ldots,c-1$ (**exercise**). In particular, H is nilpotent of class $c-1$. Since $Z_{i+1}(G)/Z_i(G) \cong (Z_{i+1}(G)/Z_1(G))/(Z_i(G)/Z_1(G)) = Z_i(H)/Z_{i-1}(H)$ for all $i = 1,2,\ldots,c-1$, in order to finish the proof it suffices to show that H is torsion-free.

Let $a \in G$ and suppose that there exists an integer $m \geq 1$ such that $a^m \in Z_1(G)$. We need to show that $a \in Z_1(G)$. By contradiction, suppose that $a \notin Z_1(G)$. Since $Z_1(G) = Z(G)$, we can find $b \in G$ which does not commute with a, i.e. such that $[a,b] \neq 1_G$. Let then $i \in \{1,2,\ldots,c-1\}$ be the maximal index for which there exists a $b \in \gamma_i(G)$ such that $[a,b] \neq 1_G$, where $G = \gamma_1(G) \geq \gamma_2(G) \geq \cdots \geq \gamma_{c+1}(G) = \{1_G\}$ is the lower central series of G. It follows from Hall's identities (Lemma 2.1.(2)) that

$$1_G = [a^m,b] = [a \cdot a^{m-1},b] = [a,b]^{a^{m-1}}[a^{m-1},b].$$

Now $[a,b] \in \gamma_{i+1}(G)$, hence, by the maximality of the index i, the element a (and therefore a^{m-1}) commutes with $[a,b]$. This implies $[a,b]^{a^{m-1}} = [a,b]$ so that $1_G = [a^m,b] = [a,b][a^{m-1},b]$. Iterating the above argument, we get

$$1_G = [a,b][a^{m-1},b] = [a,b]^2[a^{m-2},b] = \cdots = [a,b]^m.$$

Since G is torsion-free, this gives $[a,b] = 1_G$, a contradiction. Hence $a \in Z_1(G)$ and therefore $H = G/Z_1(G)$ is torsion-free. □

The remainder of this section is devoted to the proof of the following theorem, due to Malcev, which gives a linear representation of finitely generated torsion-free nilpotent groups.

Theorem 2.20 (Malcev). *Let G be a finitely generated torsion-free nilpotent group. Then there exists an integer $n \geq 1$ and an embedding $G \hookrightarrow \mathrm{UT}(n,\mathbb{Z})$.*

We start with an easy reduction.

Lemma 2.21. *It is sufficient to embed G into $\mathrm{UT}(n,\mathbb{Q})$.*

Proof. Let $\iota\colon G \hookrightarrow \mathrm{UT}(n,\mathbb{Q})$ be an embedding. Consider a finite set S of generators for G. Let N be the least common multiple of the denominators of all the entries of the elements in $\iota(S) \subseteq \mathrm{UT}(n,\mathbb{Q})$. Then, conjugating each element

$$\iota(s) = \begin{pmatrix} 1 & & \alpha_{i,j} \\ & \ddots & \\ 0 & & 1 \end{pmatrix}$$

$s \in S$, by the matrix

$$M_N := \begin{pmatrix} 1 & & & 0 \\ & N & & \\ & & \ddots & \\ 0 & & & N^n \end{pmatrix}$$

we obtain a matrix

$$\begin{pmatrix} 1 & & \beta_{i,j} \\ & \ddots & \\ 0 & & 1 \end{pmatrix}$$

where now $\beta_{i,j} = N^{-i}\alpha_{i,j}N^j \in \mathbb{Z}$ (since $i < j$). Denoting by $\gamma_{M_N} \in \mathrm{Aut}(\mathrm{UT}(n,\mathbb{Q}))$ the conjugation by the matrix M_N, it follows that $\gamma_{M_N} \circ \iota$ yields the desired embedding of G into $\mathrm{UT}(n,\mathbb{Z})$. □

With this lemma we reduced the problem to finding a $\mathbb{Q}$-linear action of G.

We need some notation and terminology.

Definition 2.22. Let G be a group and let $\mathbb{K}$ be a field.

Given a $\mathbb{K}$-vector space V, we denote by $\mathrm{End}_{\mathbb{K}}(V)$ the set (in fact, a $\mathbb{K}$-algebra) of all linear maps $\varphi \colon V \to V$.

A $\mathbb{K}$-vector space V equipped with a linear action of G is called a *G-module*. A *G-endomorphism* of a G-module V is a linear map $\varphi \colon V \to V$ which *centralizes* the action of G, i.e., satisfies $\varphi(gv) = g\varphi(v)$ for all $g \in G$ and $v \in V$.

If V is a G-module, we denote by $\mathrm{End}_G(V)$ the set (in fact, a $\mathbb{K}$-subalgebra of $\mathrm{End}_{\mathbb{K}}(V)$) of all G-endomorphisms of V.

Let V be a G-module. A vector subspace $W \leq V$ which is G-invariant (i.e., $gw \in W$ for all $g \in G$ and $w \in W$) is called a *G-submodule* of V. One then says that V is *irreducible* provided that the only proper G-submodule is the trivial one.

The following is an elementary but extremely useful result in the representation theory of groups and algebras. It is due to Issai Schur.

Lemma 2.23 (Schur). *Let V be an irreducible G-module and suppose that $\varphi \in \mathrm{End}_G(V)$ is a G-endomorphism of V. Suppose that φ admits an eigenvector. Then φ is a scalar multiple of the identity map $\mathrm{id}_V \colon V \to V$.*

Proof. Let $v \in V$ be an eigenvector of φ and let $\alpha \in \mathbb{K}$ denote the corresponding eigenvalue. Since φ centralizes the action of G, the eigenspace $V_\alpha = \{w \in V : \varphi(w) = \alpha w\} \subseteq V$ is a nontrivial (since $v \in V_\alpha$) G-submodule of V. By irreducibility, we necessarily have $V_\alpha = V$. This is equivalent to saying that $\varphi = \alpha\,\mathrm{id}_V$. □

Definition 2.24. Let V be a vector space over a field $\mathbb{K}$. A linear transformation $\varphi \colon V \to V$ is said to be *unipotent* if 1 is its only eigenvalue.

Note that if V is finite-dimensional then, by the Cayley–Hamilton Theorem (cf. [210, Chapter X §2]), $\varphi \in \mathrm{End}_{\mathbb{K}}(V)$ is unipotent if and only if there exists an $n \geq 1$ such that $(\varphi - \mathrm{id}_V)^n = 0$.

The following lemma provides the first step towards proving Malcev's theorem (Theorem 2.20).

Lemma 2.25. *Let V be a finite-dimensional vector space over a field $\mathbb{K}$. Let G be a unipotent nilpotent subgroup of* $\mathrm{End}_{\mathbb{K}}(V)$. *Then there exists a vector basis of V such that $G \subseteq \mathrm{UT}(n,\mathbb{K})$.*

Notice that the statement of the lemma is not a tautology: from the assumption that for every element in G there exists a basis in which the corresponding matrix lies in $\mathrm{UT}(n,\mathbb{K})$, we want to conclude that we can find a single basis in which the matrices of all the elements of G are in $\mathrm{UT}(n,\mathbb{K})$.

Proof of Lemma 2.25. Let us assume first that G acts irreducibly. We show that $G = \{\mathrm{id}_V\}$ and that V is one-dimensional. Suppose by contradiction that G is nontrivial. Then, by nilpotency, we have that its center $Z(G) = Z_1(G)$ is nontrivial as well. Let $z \in Z(G) \setminus \{\mathrm{id}_V\}$ and observe that z centralizes the action of G. By virtue of Schur's lemma (Lemma 2.23), since G acts irreducibly on V, z is a multiple of id_V; but from the unipotency of z, we deduce that $z = \mathrm{id}_V$, a contradiction. Thus $G = \{\mathrm{id}_V\}$. Moreover, since the action is irreducible, V must be one-dimensional.

Let us now drop the irreducibility assumption and denote by

$$\{0\} = V_0 \subsetneqq V_1 \subsetneqq \cdots \subsetneqq V_n = V$$

a maximal chain of G-submodules of V. By maximality, we have that the quotient G-modules V_i/V_{i-1} are irreducible for all $i = 1,2,\ldots,n$. By the first part of the proof, we have $\dim_{\mathbb{K}}(V_i/V_{i-1}) = 1$. Taking $v_i \in V_i \setminus V_{i-1}$ for all $i = 1,2,\ldots,n$ yields a basis of V with respect to which the elements of G are represented by upper-triangular matrices. Indeed, for every $g \in G$ we have $gv_1 = v_1$, $gv_2 \in v_2 + V_1$, $\ldots$, $gv_n = v_n + V_{n-1}$ so that we can find $\alpha_{i,j} = \alpha_{i,j}(g) \in \mathbb{K}$, $1 \le i < j \le n$ such that

$$\begin{cases} gv_1 &= v_1 \\ gv_2 &= v_2 + \alpha_{1,2}v_1 \\ gv_3 &= v_3 + \alpha_{2,3}v_2 + \alpha_{1,3}v_1 \\ \cdots & \\ gv_n &= v_n + \alpha_{n-1,n}v_{n-1} + \cdots \alpha_{1,n}v_1. \end{cases}$$

Setting $\alpha_{i,j} = 0$ for $1 \le j < i \le n$ and $\alpha_{i,i} = 1$ for $i = 1,2,\ldots,n$, we have that the matrix $M(g) = \|\alpha_{i,j}\|_{1\le i,j\le n}$ representing g is upper unitriangular. This shows that $G \subseteq \mathrm{UT}(n,\mathbb{K})$. □

Hence, to complete the proof of Malcev's theorem (Theorem 2.20), we need to find a finite-dimensional $\mathbb{Q}$-vector space V on which G acts unipotently and nilpotently. To construct such a module we look at a special normal series of G.

Lemma 2.26. *Let G be a group and let $(G_i)_{i\ge 1}$ be a central series. Then the map*

$$\begin{aligned} G_i/G_{i+1} \times G_j/G_{j+1} &\to G_{i+j}/G_{i+j+1} \\ (a_iG_{i+1}, b_jG_{j+1}) &\mapsto [a_i,b_j]G_{i+j+1} \end{aligned} \tag{2.10}$$

is well defined and $\mathbb{Z}$-bilinear. Moreover, the quotient groups G_i/G_{i+1} are abelian for all $i \ge 1$.

Proof. Suppose that $a_i = a_i' \mod G_{i+1}$ so that $a_i' = a_ic_{i+1}$, where $c_{i+1} \in G_{i+1}$. By using the Hall identities (Lemma 2.1.(2)) we deduce

$$[a_i', b_j] = [a_i c_{i+1}, b_j] = [a_i, b_j][a_i, b_j, c_{i+1}][c_{i+1}, b_j] = [a_i, b_j] \mod G_{i+j+1},$$

since $[a_i, b_j, c_{i+1}] \in G_{2i+j+1} \subseteq G_{i+j+1}$ and $[c_{i+1}, b_j] \in G_{i+j+1}$. Analogously, if $b_j = b_j' \mod G_{j+1}$ one can show that $[a_i, b_j] = [a_i, b_j'] \mod G_{i+j+1}$. This shows that the map (2.10) is well defined.

To show the $\mathbb{Z}$-bilinearity, let $a_i, a_i' \in G_i$. We have, again by the Hall identities,

$$\begin{aligned} [a_i a_i', b_j] &= [a_i, b_j][a_i, b_j, a_i'][a_i', b_j] \\ &= [a_i, b_j][a_i', b_j][a_i, b_j, a_i']^{[a_i', b_j]} \\ &= [a_i, b_j][a_i', b_j] \mod G_{i+j+1}, \end{aligned} \tag{2.11}$$

since $[a_i, b_j, a_i'] \in G_{2i+j+1} \subseteq G_{i+j+1}$ and G_{i+j+1} is normal in G. This gives the linearity in the first component; the linearity in the second one is proved analogously.

Finally, let $a_i, b_i \in G_i$. Then

$$a_i G_{i+1} b_i G_{i+1} = a_i b_i G_{i+1} = b_i a_i [a_i b_i] G_{i+1} = b_i a_i G_{i+1} = b_i G_{i+1} a_i G_{i+1}$$

since G_{i+1} is a normal subgroup and $[a_i, b_i] \in G_{2i} \subseteq G_{i+1}$. This shows that the groups G_i/G_{i+1}, for $i \geq 1$, are abelian. □

Remark 2.27. From (2.11) we deduce, by an easy inductive argument, that

$$[a_i^m, b_j] = [a_i, b_j]^m \mod G_{i+j+1} \tag{2.12}$$

for all $a_i \in G_i$ and $b_j \in G_j$, and all integers $m \geq 1$.

Proposition 2.28. *Let G be a finitely generated group and let $(\gamma_i)_{i \geq 1}$ denote its lower central series. Let $X \subseteq G$ be a finite generating subset of G. Then for every $i = 1, 2, \ldots$, the group γ_i/γ_{i+1} is abelian and finitely generated. More precisely, it is generated by the elements $[x_1, x_2, \ldots, x_i]\gamma_{i+1}$, where $x_1, x_2, \ldots, x_i \in X$.*

Proof. The groups γ_i/γ_{i+1} are abelian by the previous lemma, since the lower central series is central.

We prove that they are finitely generated arguing by induction on i. First notice that since γ_i is generated by the commutators $[h, k]$ with $h \in \gamma_{i-1}$ and $k \in \gamma_1 = G$, we have that the $\mathbb{Z}$-span of the image of the $\mathbb{Z}$-bilinear map $\varphi_{i,j} \colon \gamma_i/\gamma_{i+1} \times \gamma_j/\gamma_{j+1} \to \gamma_{i+j}/\gamma_{i+j+1}$ given by $\varphi_{i,j}(a_i\gamma_{i+1}, b_j\gamma_{j+1}) := [a_i, b_j]\gamma_{i+j+1}$ is the whole $\gamma_{i+j}/\gamma_{i+j+1}$. Now, it is clear that γ_1/γ_2 is generated by the elements $x\gamma_2$, where $x \in X$. On the other hand, suppose by induction that γ_{i-1}/γ_i is generated by the elements $[x_1, x_2, \ldots, x_{i-1}]\gamma_i$, with $x_1, x_2, \ldots, x_{i-1} \in X$. It follows that γ_i/γ_{i+1} is generated by the elements $\varphi_{i-1,1}([x_1, x_2, \ldots, x_{i-1}]\gamma_i, x\gamma_2) = [[x_1, x_2, \ldots, x_{i-1}], x]\gamma_{i+1} = [x_1, x_2, \ldots, x_{i-1}, x]\gamma_{i+1}$, with $x_1, x_2, \ldots, x_{i-1}, x \in X$. □

We can finally provide the normal series of G that we need.

Corollary 2.29. *Let G be a finitely generated nilpotent group. Then there exists a finite descending normal series in G*

$$G = N_1 \geq N_2 \geq \cdots \geq N_s \geq N_{s+1} = \{1\}$$

such that N_k/N_{k+1} is a cyclic group and $[N_k,G] \subseteq N_{k+1}$ for all $k = 1,2,\dots,s$.

Proof. Let $(\gamma_i)_{i\geq 1}$ be the lower central series of G. Since γ_i/γ_{i+1} is finitely generated and abelian, by the characterization theorem (cf. Corollary 1.30) we can find normal subgroups $N_{i,j}$, $j = 1,2,\dots,t_i$, such that

$$\gamma_i = N_{i,1} \geq N_{i,2} \geq \cdots \geq N_{i,t_i} = \gamma_{i+1}$$

with $N_{i,j}/N_{i,j+1}$ cyclic for all $j = 1,2,\dots,t_i - 1$. Moreover,

$$[N_{i,j},G] \subseteq [\gamma_i,\gamma_1] = \gamma_{i+1} = N_{i,t_i} \subseteq N_{i,j+1}. \tag{2.13}$$

Note that, incidentally, (2.13) implies that the $N_{i,j}$'s are normal in G. To finish, we rename the subgroups $N_{1,1},N_{1,2},\dots,N_{1,t_1} = N_{2,1},N_{2,2},\dots,N_{2,t_2} = N_{3,1},\dots$ as $N_1,N_2,N_3,\dots$ □

With the notation of Corollary 2.29, we now pick elements $a_k \in N_k$ such that $N_k/N_{k+1} = \langle a_k N_{k+1}\rangle$, $k = 1,2,\dots,s$. So every element $g \in G$ can be represented as

$$g = a_1^{\alpha_1} a_2^{\alpha_2} \cdots a_s^{\alpha_s}, \tag{2.14}$$

where $\alpha_i \in \mathbb{Z}$, and such that $0 \leq \alpha_i < |N_i/N_{i+1}|$ whenever the cyclic group N_i/N_{i+1} is finite.

Note that the representation (2.14) is unique (**exercise**).

Consider now the space $F = F(G,\mathbb{Q}) = \mathbb{Q}^G$ consisting of all maps $f\colon G \to \mathbb{Q}$. We define a right action of G on F by setting $gf(h) = f(gh)$ for all $g,h \in G$ and $f \in F$. Also, we define the elements $t_1,t_2\dots,t_s \in F$ by setting $t_i(g) = \alpha_i$ for every element $g \in G$ represented as in (2.14).

Consider the G-module $V := \mathbb{Q}Gt_1 + \mathbb{Q}Gt_2 \cdots + \mathbb{Q}Gt_s$. This is going to be the G-module that we were looking for. At this point, it is not even clear that this module is finite-dimensional. This follows from the next lemma.

Lemma 2.30. *Let G be a finitely generated torsion-free nilpotent group and let V be a G-module over a field $\mathbb{K}$. Suppose that a generating set $A = \{a_1,a_2,\dots,a_s\} \subseteq G$ acts unipotently (i.e., for all $v \in V$ there exists an integer $n(v) \geq 1$ such that $(a_i - \mathrm{id}_V)^{n(v)}v = 0$ for all i's). Then, for all $v \in V$, the G-submodule $\mathbb{K}Gv$ is finite-dimensional.*

Proof. First observe that $\mathbb{K}Gv$ equals the $\mathbb{K}$-span of the vectors $a_1^{\alpha_1}a_2^{\alpha_2}\cdots a_s^{\alpha_s}v$, with $\alpha_1,\alpha_2,\dots,\alpha_s \in \mathbb{Z}$. Notice, however, that since the action is unipotent, there exists a $d_s \in \mathbb{N}$ such that $(a_s - \mathrm{id}_V)^{d_s}v = 0$, so that in fact $a_s^{d_s}v$ belongs to the $\mathbb{K}$-linear span of the vectors $a_s^i v$, $i = 0,1,\dots,d_s - 1$. Also from

$$\begin{aligned}(a_s^{-1} - \mathrm{id}_V)^{d_s}v &= a_s^{-d_s}a_s^{d_s}(a_s^{-1} - \mathrm{id}_V)^{d_s}v \\ &= a_s^{-d_s}(\mathrm{id}_V - a_s)^{d_s}v \\ &= \pm a_s^{-d_s}(a_s - \mathrm{id}_V)^{d_s}v \\ &= 0\end{aligned}$$

we deduce that $a_s^{-d_s}v$ belongs to the $\mathbb{K}$-span of the vectors $a_s^{-i}v$, $i=0,1,\dots,d_s-1$. Altogether, this implies that $\mathbb{K}Gv$ equals the $\mathbb{K}$-linear span of the vectors $a_1^{\alpha_1}a_2^{\alpha_2}\cdots a_s^{\alpha_s}v$ with $\alpha_1,\alpha_2,\dots,\alpha_s\in\mathbb{Z}$ satisfying $|\alpha_s|<d_s$.

Consider the finite set $W_s=\{a_s^{-d_s+1}v,a_s^{-d_s+2}v,\dots,a_s^{-1}v,v,a_sv,a_s^2v,\dots,a_s^{d_s-1}v\}\subseteq V$. Again by unipotency of the G-action, there exists a $d_{s-1}\in\mathbb{N}$ such that $(a_{s-1}^{\pm1}-\mathrm{id}_V)^{d_{s-1}}w=0$ for all $w\in W_s$. As a consequence, the G-module $\mathbb{K}Gv$ equals the $\mathbb{K}$-linear span of the vectors $a_1^{\alpha_1}a_2^{\alpha_2}\cdots a_s^{\alpha_s}v$ with $\alpha_1,\alpha_2,\dots,\alpha_s\in\mathbb{Z}$ satisfying $|\alpha_{s-1}|<d_{s-1}$ and $|\alpha_s|<d_s$.

Continuing in this way, we can find $d_1,d_2,\dots,d_s\in\mathbb{N}$ such that the G-module $\mathbb{K}Gv$ equals the $\mathbb{K}$-linear span of the vectors $a_1^{\alpha_1}a_2^{\alpha_2}\cdots a_s^{\alpha_s}v$ with $\alpha_1,\alpha_2,\dots,\alpha_s\in\mathbb{Z}$ satisfying $|\alpha_i|<d_i$ for $i=1,2,\dots,s$. As all these vectors constitute a finite set, we deduce that $\mathbb{K}Gv$ is finite-dimensional. □

Hence, to finish our proof, we only need to show that we can find a set of generators of G that act unipotently and nilpotently on our module V.

In the following, by the expression $\mathrm{poly}(x_1,x_2,\dots,x_n)$ we mean a (not explicitly determined) polynomial in the n variables $x_1,x_2,\dots,x_n$ with coefficients in $\mathbb{Z}$.

Lemma 2.31. *With the notation we introduced after Corollary* 2.29 (*cf.* (2.14)), *we have*

$$(a_1^{\alpha_1}a_2^{\alpha_2}\cdots a_s^{\alpha_s})\cdot(a_1^{\beta_1}a_2^{\beta_2}\cdots a_s^{\beta_s})=a_1^{f_1}a_2^{f_2}\cdots a_s^{f_s},\tag{2.15}$$

where

$$f_i=\alpha_i+\beta_i+\mathrm{poly}(\alpha_1,\dots,\alpha_{i-1},\beta_1,\dots,\beta_{i-1}).$$

In particular, for all $m\geq 1$,

$$(a_1^{\alpha_1}a_2^{\alpha_2}\cdots a_s^{\alpha_s})^m=a_1^{h_1}a_2^{h_2}\cdots a_s^{h_s}\tag{2.16}$$

where

$$h_i=m\alpha_i+\mathrm{poly}(\alpha_1,\dots,\alpha_{i-1}).$$

Proof. In order to prove (2.15), we use a triple induction on j (with $1\leq j\leq s$) to simultaneously prove that, for all $i=1,2,\dots,j-1$,

$$a_i^{-\alpha_i}a_j^{\alpha_j}a_i^{\alpha_i}=a_j^{\alpha_j}[a_j^{\alpha_j},a_i^{\alpha_i}]=a_j^{\alpha_j}a_{j+1}^{z_{j+1}}a_{j+2}^{z_{j+2}}\cdots a_s^{z_s}\tag{2.17}$$

where

$$z_t=\mathrm{poly}(\alpha_i;\alpha_j)$$

for $t=j+1,j+2,\dots,s$,

$$(a_j^{\alpha_j}a_{j+1}^{\alpha_{j+1}}\cdots a_s^{\alpha_s})\cdot(a_j^{\beta_j}a_{j+1}^{\beta_{j+1}}\cdots a_s^{\beta_s})=a_j^{f_j}a_{j+1}^{f_{j+1}}\cdots a_s^{f_s},\tag{2.18}$$

where, for $t=j,j+1,\dots,s$,

$$f_t=\alpha_t+\beta_t+\mathrm{poly}(\alpha_j,\alpha_{j+1},\dots,\alpha_{t-1};\beta_j,\beta_{j+1},\dots,\beta_{t-1}),$$

and, for all $i=1,2,\dots,j-1$,

$$a_i^{-\alpha_i}(a_j^{\alpha_j}a_{j+1}^{\alpha_{j+1}}\cdots a_s^{\alpha_s})a_i^{\alpha_i}=a_j^{g_j}a_{j+1}^{g_{j+1}}\cdots a_s^{g_s},\tag{2.19}$$

where, for $t = j, j+1, \ldots, s$,

$$g_t = \alpha_t + \mathrm{poly}(\alpha_i; \alpha_j, \alpha_{j+1}, \ldots, \alpha_{t-1}).$$

The base of the induction for (2.18) is

$$a_s^{\alpha_s} \cdot a_s^{\beta_s} = a_s^{\alpha_s + \beta_s}$$

so that $f_s = \alpha_s + \beta_s$.

Similarly, the base of the induction for both (2.17) and (2.19) is

$$a_i^{-\alpha_i} a_s^{\alpha_s} a_i^{\alpha_i} = a_s^{\alpha_s}$$

(recall that a_s is in the center of G), so that $g_s = \alpha_s$.

Assume now that (2.17), (2.18) and (2.19) hold, and let us show that, for all $i = 1, 2, \ldots, j-2$,

$$a_i^{-\alpha_i} a_{j-1}^{\alpha_{j-1}} a_i^{\alpha_i} = a_{j-1}^{\alpha_{j-1}} [a_{j-1}^{\alpha_{j-1}}, a_i^{\alpha_i}] = a_{j-1}^{\alpha_{j-1}} a_j^{z'_j} a_{j+1}^{z'_{j+1}} \cdots a_s^{z'_s} \tag{2.20}$$

where

$$z'_t = \mathrm{poly}(\alpha_i; \alpha_{j-1})$$

for $t = j, j+1, \ldots, s$,

$$(a_{j-1}^{\alpha_{j-1}} a_j^{\alpha_j} \cdots a_s^{\alpha_s}) \cdot (a_{j-1}^{\beta_{j-1}} a_j^{\beta_j} \cdots a_s^{\beta_s}) = a_{j-1}^{f'_{j-1}} a_j^{f'_{j+1}} \cdots a_s^{f'_s}, \tag{2.21}$$

where, for $t = j-1, j, \ldots, s$,

$$f'_t = \alpha_t + \beta_t + \mathrm{poly}(\alpha_{j-1}, \alpha_j, \ldots, \alpha_{t-1}; \beta_{j-1}, \beta_j, \ldots, \beta_{t-1}),$$

and,

$$a_i^{-\alpha_i} (a_{j-1}^{\alpha_{j-1}} a_j^{\alpha_j} \cdots a_s^{\alpha_s}) a_i^{\alpha_i} = a_{j-1}^{g'_{j-1}} a_j^{g'_j} \cdots a_s^{g'_s}, \tag{2.22}$$

where, for $t = j-1, j, \ldots, s$,

$$g'_t = \alpha_t + \mathrm{poly}(\alpha_i; \alpha_{j-1}, \alpha_j, \ldots, \alpha_{t-1}).$$

For $1 \leq i < j \leq s$, and $\varepsilon, \delta \in \{-1, 1\}$, since $[a_j^{\varepsilon}, a_i^{\delta}] \in N_{j+1}$, we fix the notation

$$[a_j^{\varepsilon}, a_i^{\delta}] = a_{j+1}^{k_{j+1}} a_{j+2}^{k_{j+2}} \cdots a_s^{k_s}, \tag{2.23}$$

where $k_t = k_t(i, j, \varepsilon, \delta) \in \mathbb{Z}$ for $t = j+1, j+2, \ldots, s$.

To prove (2.20), we use induction on $|\alpha_i| + |\alpha_{j-1}|$. The base of the induction is simply (2.23), where $|\alpha_i| + |\alpha_{j-1}| = 2$.

Suppose that $|\alpha_i| + |\alpha_{j-1}| \geq 3$. To fix our ideas, we assume that $\alpha_{j-1} \geq 2$ (the other cases are dealt with in a similar way). Using Lemma 2.1.(2), we have

$$\begin{aligned} a_i^{-\alpha_i} a_{j-1}^{\alpha_{j-1}} a_i^{\alpha_i} &= a_{j-1}^{\alpha_{j-1}} [a_{j-1}^{\alpha_{j-1}}, a_i^{\alpha_i}] \\ &= a_{j-1}^{\alpha_{j-1}} [a_{j-1}^{\alpha_{j-1}-1}, a_i^{\alpha_i}]^{a_{j-1}} [a_{j-1}, a_i^{\alpha_i}]. \end{aligned}$$

Using induction for $[a_{j-1}^{\alpha_{j-1}-1}, a_i^{\alpha_i}]$ and $[a_{j-1}, a_i^{\alpha_i}]$, we have

$$[a_{j-1}^{\alpha_{j-1}-1}, a_i^{\alpha_i}] = a_j^{p_j} a_{j+1}^{p_{j+1}} \cdots a_s^{p_s} \text{ and } [a_{j-1}, a_i^{\alpha_i}] = a_j^{q_j} a_{j+1}^{q_{j+1}} \cdots a_s^{q_s},$$

where $p_t = \mathrm{poly}(\alpha_i; \alpha_{j-1} - 1) = \mathrm{poly}(\alpha_i; \alpha_{j-1})$ and $q_t = \mathrm{poly}(\alpha_i)$ for $t = j, j+1, \ldots, s$.

By (2.19),

$$[a_{j-1}^{\alpha_{j-1}-1}, a_i^{\alpha_i}]^{a_{j-1}} = (a_j^{p_j} a_{j+1}^{p_{j+1}} \cdots a_s^{p_s})^{a_{j-1}} = a_j^{g'_j} a_{j+1}^{g'_{j+1}} \cdots a_s^{g'_s}$$

where

$$g'_t = p_t + \mathrm{poly}(\alpha_{j-1}; p_j, p_{j-1}, \ldots, p_{t-1}) = \mathrm{poly}(\alpha_i, \alpha_{j-1}).$$

Using (2.18),

$$\begin{aligned} [a_{j-1}^{\alpha_{j-1}-1}, a_i^{\alpha_i}]^{a_{j-1}} [a_{j-1}, a_i^{\alpha_i}] &= \left(a_j^{g'_j} a_{j+1}^{g'_{j+1}} \cdots a_s^{g'_s} \right) \left(a_j^{q_j} a_{j+1}^{q_{j+1}} \cdots a_s^{q_s} \right) \\ &= a_j^{f''_j} a_{j+1}^{f''_{j+1}} \cdots a_s^{f''_s}, \end{aligned}$$

where

$$f''_t = g'_t + q_t + \mathrm{poly}(g'_j, g'_{j+1}, \ldots, g'_{t-1}; q_j, q_{j+1}, \ldots, q_{t-1}) = \mathrm{poly}(\alpha_i, \alpha_{j-1}).$$

Setting $z'_t := f''_t$ for $t = j, j+1, \ldots, s$, we finally have

$$a_i^{-\alpha_i} a_{j-1}^{\alpha_{j-1}} a_i^{\alpha_i} = a_{j-1}^{\alpha_{j-1}} a_j^{f''_j} a_{j+1}^{f''_{j+1}} \cdots a_s^{f''_s} = a_{j-1}^{\alpha_{j-1}} a_j^{z'_j} a_{j+1}^{z'_{j+1}} \cdots a_s^{z'_s},$$

where $z'_t = \mathrm{poly}(\alpha_i; \alpha_{j-i})$ for $t = j, j+1, \ldots, s$. This proves (2.20).

In order to prove (2.21), we have

$$\begin{aligned} &(a_{j-1}^{\alpha_{j-1}} a_j^{\alpha_j} \cdots a_s^{\alpha_s}) \cdot (a_{j-1}^{\beta_{j-1}} a_j^{\beta_j} \cdots a_s^{\beta_s}) \\ &\qquad = a_{j-1}^{\alpha_{j-1}} a_{j-1}^{\beta_{j-1}} \left(a_{j-1}^{-\beta_{j-1}} (a_j^{\alpha_j} a_{j+1}^{\alpha_{j+1}} \cdots a_s^{\alpha_s}) a_{j-1}^{\beta_{j-1}} \right) \cdot (a_j^{\beta_j} a_{j+1}^{\beta_{j+1}} \cdots a_s^{\beta_s}) \\ &\qquad =_* a_{j-1}^{\alpha_{j-1}+\beta_{j-1}} \cdot \left(a_j^{g_j} a_{j+1}^{g_{j+1}} \cdots a_s^{g_s} \right) \cdot (a_j^{\beta_j} a_{j+1}^{\beta_{j+1}} \cdots a_s^{\beta_s}) \\ &\qquad =_{**} a_{j-1}^{\alpha_{j-1}+\beta_{j-1}} \cdot (a_j^{f'_j} a_{j+1}^{f'_{j+1}} \cdots a_s^{f'_s}), \end{aligned}$$

where in $=_*$ (resp. $=_{**}$) we use the inductive step (2.19) (resp. (2.18)), and, for $t = j-1, j, \ldots, s$,

$$g_t = \alpha_t + \mathrm{poly}(\beta_{j-1}; \alpha_j, \alpha_{j+1}, \ldots, \alpha_{t-1}), \tag{2.24}$$

and

$$\begin{aligned} f'_t &= g_t + \beta_t + \mathrm{poly}(g_j, g_{j+1}, \dots, g_{t-1}; \beta_j, \beta_{j+1}, \dots, \beta_{t-1}) \\ \text{(by (2.24))} &= \alpha_t + \beta_t + \mathrm{poly}(\alpha_{j-1}, \alpha_j, \dots, \alpha_{t-1}; \beta_{j-1}, \beta_j, \dots, \beta_{t-1}). \end{aligned}$$

This shows (2.21).

We now prove (2.22). We have

$$\begin{aligned} a_i^{-\alpha_i}(a_{j-1}^{\alpha_{j-1}} a_j^{\alpha_j} \cdots a_s^{\alpha_s}) a_i^{\alpha_i} &= a_i^{-\alpha_i} a_{j-1}^{\alpha_{j-1}} a_i^{\alpha_i} \cdot a_i^{-\alpha_i}(a_j^{\alpha_j} a_{j+1}^{\alpha_{j+1}} \cdots a_s^{\alpha_s}) a_i^{\alpha_i} \\ &=_* (a_{j-1}^{\alpha_{j-1}} a_j^{z_j} a_{j+1}^{z_{j+1}} \cdots a_s^{z_s}) \cdot (a_j^{g_j} a_{j+1}^{g_{j+1}} \cdots a_s^{g_s}) \\ &=_{**} a_{j-1}^{\alpha_{j-1}} a_j^{f'_j} a_{j+1}^{f'_{j+1}} \cdots a_s^{f'_s}, \end{aligned}$$

where $=_*$ follows from (2.17) and the inductive step (2.19), $=_{**}$ follows from the inductive step (2.18), and, for $t = j, j+1, \dots, s$,

$$f'_t = z_t + g_t + \mathrm{poly}(g_j, g_{j+1}, \dots, g_{t-1}) = \alpha_t + \mathrm{poly}(\alpha_i; \alpha_{j-1}, \dots, \alpha_{t-1}).$$

This proves (2.22) and hence (2.15).

Finally, in order to prove (2.16) we proceed by induction on m. For $m = 1$ we have $h_i = \alpha_i$ for $i = 1, 2, \dots, s$, and the base of the induction is established. Suppose that (2.16) holds. Then

$$\begin{aligned} (a_1^{\alpha_1} a_2^{\alpha_2} \cdots a_s^{\alpha_s})^{m+1} &= (a_1^{\alpha_1} a_2^{\alpha_2} \cdots a_s^{\alpha_s}) \cdot (a_1^{\alpha_1} a_2^{\alpha_2} \cdots a_s^{\alpha_s})^m \\ &=_* (a_1^{\alpha_1} a_2^{\alpha_2} \cdots a_s^{\alpha_s}) \cdot (a_1^{h_1} a_2^{h_2} \cdots a_s^{h_s}) \\ &=_{**} (a_1^{f_1} a_2^{f_2} \cdots a_s^{f_s}) \end{aligned}$$

where, in $=_*$ we used the inductive step and in $=_{**}$ we applied (2.15), and

$$\begin{aligned} f_i &= \alpha_i + h_i + \mathrm{poly}(\alpha_1, \alpha_2 \dots, \alpha_{i-1}, h_1, h_2, \dots, h_{i-1}) \\ &= \alpha_i + m\alpha_i + \mathrm{poly}(\alpha_1, \alpha_2 \dots, \alpha_{i-1}, h_1, h_2, \dots, h_{i-1}) \\ &= (m+1)\alpha_i + \mathrm{poly}(\alpha_1, \alpha_2 \dots, \alpha_{i-1}). \end{aligned}$$

This shows (2.16) and completes the proof. □

End of the proof of Malcev's theorem (Theorem 2.20). We are now in a position to complete the proof of Malcev's theorem. Recall that $t_1, t_2 \dots, t_s \in F$ are defined by $t_i(x) = \gamma_i$, for every $x = a_1^{\eta_1} a_2^{\eta_2} \cdots a_s^{\eta_s} \in G$, so that

$$x = a_1^{t_1(x)} a_2^{t_2(x)} \cdots a_s^{t_s(x)}.$$

Fix $g = a_1^{\alpha_1} a_2^{\alpha_2} \cdots a_s^{\alpha_s} \in G$. From (2.15) we then deduce

$$\begin{aligned} gt_i(x) &= t_i(gx) \\ &= t_i(x) + \alpha_i + \mathrm{poly}(\alpha_1, \alpha_2 \dots, \alpha_{i-1}; t_1(x), t_2(x), \dots, t_{i-1}(x)) \qquad (2.25) \\ &= t_i(x) + \mathrm{poly}(\alpha_1, \alpha_2 \dots, \alpha_{i-1}, \alpha_i; t_1(x), t_2(x), \dots, t_{i-1}(x)). \end{aligned}$$

Let $t = t_1^{\beta_1} t_2^{\beta_2} \cdots t_s^{\beta_s} \in F$ with $\beta_i \geq 0$ for all $i = 1, 2, \ldots, s$. From (2.25) we deduce that

$$gt_i^{\beta_i} = t_i^{\beta_i} + \sum_{k=0}^{\beta_i - 1} \mathrm{poly}(\alpha_1, \alpha_2 \ldots, \alpha_{i-1}, \alpha_i; t_1, t_2, \ldots, t_{i-1}) t_i^k \tag{2.26}$$

We equip the set of all s-tuples $(\beta_1, \beta_2, \ldots, \beta_s) \in \mathbb{N}^s$ with the *right-lexicographic order*. This is defined by induction on s by setting $(\eta_1, \eta_2, \ldots, \eta_s) < (\beta_1, \beta_2, \ldots, \beta_s)$ if either $\eta_s < \beta_s$ or $\eta_s = \beta_s$ and $(\eta_1, \eta_2, \ldots, \eta_{s-1}) < (\beta_1, \beta_2, \ldots, \beta_{s-1})$. From (2.26) we deduce

$$\begin{aligned} gt &= g(t_1^{\beta_1} t_2^{\beta_2} \cdots t_s^{\beta_s}) \\ &= (gt_1)^{\beta_1} (gt_2)^{\beta_2} \cdots (gt_s)^{\beta_s} \\ &= t_1^{\beta_1} t_2^{\beta_2} \cdots t_s^{\beta_s} + \sum_{\eta_1, \eta_2, \ldots, \eta_s} c_{\eta_1, \eta_2, \ldots, \eta_s} t_1^{\eta_1} t_2^{\eta_2} \cdots t_s^{\eta_s}, \end{aligned}$$

equivalently

$$(g - \mathrm{id}_F)t = \sum_{\eta_1, \eta_2, \ldots, \eta_s} c_{\eta_1, \eta_2, \ldots, \eta_s} t_1^{\eta_1} t_2^{\eta_2} \cdots t_s^{\eta_s},$$

where $c_{\eta_1, \eta_2, \ldots, \eta_s} \in \mathbb{Q}$ and the sum runs over all s-tuples

$$(\eta_1, \eta_2, \ldots, \eta_s) < (\beta_1, \beta_2, \ldots, \beta_s).$$

Since there are at most $n := \max_i (\beta_i)^s$ s-tuples $(\eta_1, \eta_2, \ldots, \eta_s)$ preceding the tuple $(\beta_1, \beta_2, \ldots, \beta_s)$ in the given order, it is clear that

$$(g - \mathrm{id}_F)^n t = 0.$$

This shows that G acts unipotently on the G-module $V = \mathbb{Q}Gt_1 + \cdots + \mathbb{Q}Gt_s$ (which is a submodule of the $\mathbb{Q}$-span of the monomials $t_1^{\alpha_1} t_2^{\alpha_2} \cdots t_s^{\alpha_s}$, $\alpha_i \geq 0$). Note that since G is nilpotent, from Lemma 2.30 we deduce that V is finite-dimensional.

By using Lemma 2.25 we can find a homomorphism $G \to \mathrm{UT}(n, \mathbb{Q})$ (with respect to some suitable basis in V), where $n := \dim_{\mathbb{Q}} V$. It only remains to check that this map is injective. Suppose that $g \in G$ is mapped into the identity I_n. Then $t_i(gx) = t_i(x)$ for all $x \in G$ and $i = 1, 2, \ldots, s$. But this in turn implies $gx = x$ so that $g = 1_G$.

It follows that this map $G \hookrightarrow \mathrm{UT}(n, \mathbb{Q})$ in an injection, completing the proof of Malcev's theorem. □

2.7 Finitely Generated Nilpotent Groups with Torsion

We now study the torsion of finitely generated nilpotent groups.

The following lemma does not hold, in general, for non-nilpotent groups (think about the commutator subgroup of the free group F_2 of rank 2).

Lemma 2.32. *Every subgroup of a finitely generated nilpotent group is finitely generated.*

Proof. Let G be a finitely generated nilpotent group and let $G = N_1 \geq N_2 \geq \cdots \geq N_s \geq N_{s+1} = \{1_G\}$ be a normal series with N_i/N_{i+1} cyclic for $i = 1,2,\ldots,s$ (cf. Corollary 2.29). Let us show that every subgroup $H \leq G$ is generated by at most s elements. We proceed by induction on s. If $s = 1$ then G itself is cyclic and the statement is obvious. Suppose the statement holds for all groups with a normal series with cyclic quotients and of length $\leq s-1$. As the group G/N_2 is cyclic, so is its subgroup HN_2/N_2. The latter is isomorphic to $H/(H \cap N_2)$, which is therefore cyclic as well. On the other hand, $H \cap N_2$ is a subgroup of N_2 so that, by induction, it is generated by at most $s-1$ elements. It follows that H is generated by at most $1+(s-1) = s$ elements. □

Definition 2.33. We say that a class $\mathscr{C}$ of groups has the *Burnside property* if it is closed under taking quotients and whenever a group G in $\mathscr{C}$ is generated by finitely many torsion elements it is finite.

The following lemma, combined with Proposition 2.9, shows that the class of nilpotent groups has the Burnside property.

Lemma 2.34. *Let G be a nilpotent group. Suppose that G contains a finite generating subset consisting of torsion elements. Then G is finite.*

Proof. Let $X \subseteq G$ be a finite generating subset consisting of torsion elements, and let $(\gamma_i)_{i=1}^{c}$ be the lower central series of G. We prove the statement by induction on the nilpotency class c of G. If $c = 1$ then G is abelian and $G = \sum_{x \in X} \langle x \rangle$. Since the elements $x \in X$ are torsion, we have $|\langle x \rangle| < \infty$ and therefore G is finite (in fact $|G| \leq \prod_{x \in X} |\langle x \rangle|$).

Suppose that the statement holds for all finitely generated nilpotent groups of nilpotency class $\leq c-1$ and let G have nilpotency class c. Consider the abelian subgroup γ_{c-1}. It follows from Proposition 2.28 that γ_{c-1} is generated by the commutators $[x_1, x_2, \ldots, x_{c-1}]$, with $x_i \in X$ (in particular, it is finitely generated (cf. the previous lemma)). It follows from Lemma 2.26 that the map

$$\gamma_{c-2}/\gamma_{c-1} \times \gamma_1/\gamma_2 \to \gamma_{c-1}/\gamma_c \cong \gamma_{c-1}$$

is bilinear, so that for all $k \in \mathbb{N}$,

$$[x_1, x_2, \ldots, x_{c-2}, x_{c-1}]^k = [x_1, x_2, \ldots, x_{c-2}, x_{c-1}^k].$$

This shows that the commutators $[x_1, x_2, \ldots, x_{c-1}]$ are torsion elements, so that, by the first part of the proof, γ_{c-1} is finite. The quotient group G/γ_{c-1} is nilpotent of class $\leq c-1$ and it is generated by the torsion elements $x\gamma_{c-1}$, with $x \in X$. By the inductive hypothesis, it is finite. It follows that G is finite as well. □

Remark 2.35. The previous result is not true for non-nilpotent groups. For instance, the *modular group* $\mathrm{SL}(2,\mathbb{Z})/\{\pm I_2\}$ is isomorphic to the free product $(\mathbb{Z}/2\mathbb{Z}) * (\mathbb{Z}/3\mathbb{Z})$, i.e. it is generated by an element of order 2 and one of order 3, without any further relations between them. Therefore it is infinite.

Proposition 2.36. *Let G be a nilpotent group and suppose that $a, b \in G$ are torsion elements. Then ab is torsion.*

Proof. The nilpotent subgroup $\langle a,b\rangle \leq G$ is finite by the previous lemma. It follows that its subgroup $\langle ab\rangle$ is finite as well. □

Recalling that the order of a group element is invariant under conjugation, from the above proposition one immediately deduces:

Corollary 2.37. *Let G be a nilpotent group. Then $G_{tor} := \{a \in G : a \text{ is torsion}\}$ is a normal subgroup of G. Moreover, if G is finitely generated, then G_{tor} is finite.* □

Definition 2.38. Let $\mathscr{P}$ be a property of groups. One says that a group G is *virtually* $\mathscr{P}$ if there exists a subgroup $H \leq G$ with $[G:H] < \infty$ such that H satisfies $\mathscr{P}$.

We will give two proofs of the following lemma, which is due to Poincaré.

Lemma 2.39 (Poincaré). *Let G be an arbitrary group. Let $H \leq G$ with $[G:H] < \infty$. Then there exists an $N \leq H$ with $[G:N] < \infty$ and $N \trianglelefteq G$.*

First proof. Let T be a set of representatives for the left cosets of H in G so that $G = \sqcup_{t \in T} Ht$. Thus, for every $g \in G$ there exist unique elements $h \in H$ and $t \in T$ such that $g = ht$ (and therefore $g^{-1} = t^{-1}h^{-1}$). In particular,

$$H^g = g^{-1}Hg = t^{-1}h^{-1}Hht = t^{-1}Ht = H^t.$$

Hence the subgroup $N := \bigcap_{g \in G} H^g \leq G$ equals $\bigcap_{t \in T} H^t$, it is clearly normal and is contained in H. Now if $K, L \leq G$, then $[G:K \cap L] \leq [G:K][G:L]$ (**exercise**). This shows that $[G:N] \leq [G:H]^{[G:H]}$. □

Second proof. For every $x \in G$, the map $\varphi(x)\colon G/H \to G/H$ given by $Hg \mapsto Hgx$ is a permutation of cosets. Moreover, the map $x \mapsto \varphi(x)$ is a homomorphism $\varphi : G \to \mathrm{Sym}(G/H)$ (**exercise**). Then the subgroup $N := \ker(\varphi) \leq G$ is normal and contained in H since it fixes the coset H. Finally, we have $[G:N] = |\varphi(G)| \leq |\mathrm{Sym}(G/H)| = [G:H]! < \infty$. □

Remark 2.40. The second proof yields a better upper bound on the index of N in G.

The following lemma will be useful.

Lemma 2.41. *Let $\mathscr{C}$ be a class of groups with the Burnside property. Let G in $\mathscr{C}$ be finitely generated and let $H \leq G$ be a subgroup of finite index $[G:H] < \infty$. Then there exists a characteristic subgroup $K \leq G$ of finite index in G which is contained in H.*

Proof. By the Poincaré Lemma we know that there exists a normal subgroup N of G of finite index $m := [G:N]$ which is contained in H. Let $\pi\colon G \to G/N$ denote the canonical homomorphism. For every $g \in G$ we have $\pi(g^m) = \pi(g)^m = 1_{G/N}$; equivalently, $g^m \in N$. Consider the subgroup K of G generated by the set $\{g^m : g \in G\}$. Clearly $K \leq N \leq H$, and K is characteristic. By the assumption on $\mathscr{C}$, G/K is in $\mathscr{C}$. Moreover, it is finitely generated and torsion ($x^m = 1_{G/K}$ for all $x \in G/K$), hence, again by the assumption on $\mathscr{C}$, we have that $[G:K] = |G/K|$ is finite. □

Lemma 2.42. *Let G be a finitely generated nilpotent group. Then G is virtually torsion-free. In fact one can find a characteristic subgroup $N \leq G$ of finite index in G and such that N is torsion-free.*

Proof. Let $G = N_1 \geq \cdots \geq N_s \geq N_{s+1} = \{1_G\}$ be a normal series with N_i/N_{i+1} cyclic generated by the element $a_i N_{i+1}$ for $i = 1, 2, \ldots, s$ (cf. Corollary 2.29). We proceed by induction on s. If $s = 1$ then G is cyclic and the statement is obvious.

Suppose that we have proved the statement for all finitely generated nilpotent groups with such a normal series of length $\leq s-1$. Recall that, by Lemma 2.32, in a finitely generated nilpotent group, all subgroups are finitely generated. By induction, we can find a subgroup H of finite index in N_2, such that H is torsion-free and, in addition, H is characteristic in N_2. We distinguish two cases:

Case 1: if the index $[G : N_2]$ of N_2 in G is finite, then the index $[G : H]$ of H in G is also finite. By Lemma 2.41 we can find a subgroup N of H which is characteristic and of finite index in G. Since N is contained in H, it is clearly torsion-free.

Case 2: otherwise, G/N_2 is isomorphic to an infinite cyclic group. Recall that G/N_2 is generated by $a_1 N_2$, where $a_1 \in G \setminus N_2$.

Let K be the subgroup generated by a_1 and H in G. Notice that $K = \langle a_1, H \rangle = \sqcup_{i \in \mathbb{Z}} a_1^i H$. In fact, $Ha_1 = a_1(a_1^{-1} H a_1) = a_1 H$, since H is characteristic in N_2, and the conjugation by a_1 induces an automorphism of N_2, because it is normal in G. (Notice that the normality of H in N_2 would not suffice, since the above mentioned automorphism is not an inner automorphism of N_2, as $a_1 \notin N_2$).

We have $G = \sqcup_{i \in \mathbb{Z}} a_1^i N_2 = \sqcup_{i \in \mathbb{Z}} \sqcup_{t \in T} a_1^i H t$ where T is a (finite) set of representatives for the left cosets of H in N_2. It follows that $G = \bigcup_{t \in T} Kt$, so in particular K has finite index $[G : K] \leq |T|$ in G.

Now, for every $h \in H$ and $i \in \mathbb{Z}$, the element $a_1^i h$ cannot be torsion. Indeed, on the one hand we cannot have $i = 0$ since H was torsion-free. On the other hand, if $i \neq 0$, from $a_1^{-1} H a_1 = H$, we would find an element $h' \in H$ such that $1_G = (a_1^i h)^m = a_1^{i \cdot m} h'$. But this would give $a_1^{im} = 1$ and $h' = 1$. The first equality contradicts the fact that $G/N_2 = \langle a_1 N_2 \rangle$ is infinite cyclic. It follows that K is also torsion-free.

By Lemma 2.41, we can find a subgroup N in K which is both characteristic and of finite index in G. Since N is contained in K, it is also torsion-free. □

The proof of the following lemma is left as an **exercise**.

Lemma 2.43. *Let G be a group and let H be a finite index subgroup of G. If H embeds into* $\mathrm{GL}(m, \mathbb{Z})$ *for some $m \geq 1$, then G embeds into* $\mathrm{GL}(n, \mathbb{Z})$ *for some $n \geq m$.*

Corollary 2.44 (Linearity of finitely-generated nilpotent groups). *Every finitely generated nilpotent group is* linear; *in fact, it is embeddable into* $\mathrm{GL}(m, \mathbb{Z})$ *for some $m \geq 1$.*

Proof. By Lemma 2.43 any finite extension of a subgroup of $\mathrm{UT}(n, \mathbb{Z})$ can be embedded into $\mathrm{GL}(m, \mathbb{Z})$ for some $m \geq n$. Then the statement follows from Lemma 2.42 combined with Malcev's theorem (Theorem 2.20). □

2.8 Notes

The identities in Lemma 2.1 are named after Philip Hall. Formula (5) therein, called the *Hall–Witt identity*, is named also after Ernst Witt.

The term "nilpotent" was coined by Benjamin Peirce in the context of his work on the classification of algebras: given a ring R, an element $x \in R$ is called nilpotent if there exists some positive integer n (in which case the minimal one is called the nilpotency degree of x) such that $x^n = 0$.

Nilpotent groups are then given this name because the "adjoint action" of any element is nilpotent: for a nilpotent group G of nilpotency class n and any element $g \in G$, the map $\mathrm{ad}_g \in \mathrm{End}(G)$ defined by setting $\mathrm{ad}_g(x) := [g,x]$ (where $[g,x] := g^{-1}x^{-1}gx$ is the commutator of g and x) is nilpotent. The above property is not, however, equivalent to nilpotency: groups for which the adjoint maps ad_g, $g \in G$, are nilpotent of degree n, are called n-Engel groups, and need not be nilpotent in general. They are proven to be nilpotent if they have finite order, and are conjectured to be nilpotent as long as they are finitely generated. Abelian groups are precisely those groups for which the adjoint action is not just nilpotent but trivial (1-Engel groups).

In the class of finite groups the following remarkable result holds: *every finite p-group is nilpotent*, where a group G is called a p-group, p a prime number, provided that the order of any element $g \in G$ is a power of p. Conversely, every finite nilpotent group is a direct product of p-groups (for various primes p).

In the setting of *semigroups*, a notion of nilpotency was introduced and developed by Malcev in [229]. A semigroup is said to be nilpotent (of class $\leq n$) provided that it satisfies the verbal identity $X_n = Y_n$, where the words X_n and Y_n are defined inductively as follows: $X_0 = x$, $Y_0 = y$, $X_n = X_{n-1}u_nY_{n-1}$, and $Y_n = Y_{n-1}u_nX_{n-1}$, where x, y and $u_1, u_2, \ldots, u_n$ are variables. A group is nilpotent as a semigroup in the above sense if and only if it is nilpotent in the usual group-theoretical sense. Further significant investigations in this direction are due to Gérard Lallement [208].

Recall that a *Lie group* $\mathfrak{G}$ is a smooth manifold with a group structure whose product and inverse operations are smooth.

A *lattice* in a Lie group $\mathfrak{G}$ is a discrete subgroup $G \subseteq \mathfrak{G}$ such that the quotient $\mathfrak{G}/G$ is compact. Every lattice in a simply-connected nilpotent Lie group $\mathfrak{G}$ is finitely generated, torsion-free, and nilpotent (cf. [286]).

Malcev proved that, conversely, if G is a finitely-generated torsion-free nilpotent group, then G is isomorphic to a lattice of a simply-connected nilpotent Lie group. Such a Lie group, denoted $G^{\mathbb{R}}$, is unique up to isomorphism and is called the *Malcev completion* of G. We present a description of $G^{\mathbb{R}}$. Let $(\gamma_i(G))_{i\geq 1}$ denote the lower central series of G. Suppose that G is nilpotent of class c, so that $\gamma_c(G) \neq \{1_G\}$ and $\gamma_{c+1}(G) = \{1_G\}$. Note that even if G is torsion-free, it may happen that the quotients $\gamma_i(G)/\gamma_{i+1}(G)$ have torsion. The *isolator* of a subgroup H of G, denoted $\sqrt{H}$, is the set

$$\sqrt{H} := \{x \in G : x^k \in H \text{ for some integer } k \geq 1\}.$$

The sequence $(G_i)_{i\geq 1}$, where $G_i := \sqrt{\gamma_i(G)}$ for $i \geq 1$, is a central series with G_i/G_{i+1} (finitely generated and) torsion-free: there exist integers $k_i \geq 1$ such that $G_i/G_{i+1} \cong \mathbb{Z}^{k_i}$ for all $i = 1, 2, \ldots, c$. The quantity $k := \sum_{i=1}^{c} k_i$ is called the Hirsch number of G.

As a consequence, we can find a generating subset

$$\{a_{1,1}, a_{1,2}, \ldots, a_{1,k_1}, a_{2,1}, a_{2,2}, \ldots, a_{2,k_2}, \ldots, a_{c,1}, a_{c,2}, \ldots, a_{c,k_c}\} \tag{2.27}$$

of G such that G_i is generated by $a_{i,1}, a_{i,2}, \ldots, a_{i,k_i}$ and G_{i+1} for all $i = 1, 2, \ldots, c$. One refers to (2.27) as a *Malcev basis* for G. An element $g \in G$ can therefore be uniquely expressed as a product

$$g = a_{1,1}^{z_{1,1}} a_{1,2}^{z_{1,2}} \cdots a_{1,k_1}^{z_{1,k_1}} a_{2,1}^{z_{2,1}} a_{2,2}^{z_{2,2}} \cdots a_{2,k_2}^{z_{2,k_2}} \cdots a_{c,1}^{z_{c,1}} a_{c,2}^{z_{c,2}} \ldots a_{c,k_c}^{z_{c,k_c}}$$

with $z_{i,j} \in \mathbb{Z}$, $1 \leq j \leq k_i$, $1 \leq i \leq c$. This way, we may identify g with the vector

$$\mathbf{z} := (z_{1,1}, z_{1,2}, \ldots, z_{1,k_1}, z_{2,1}, z_{2,2}, \ldots, z_{2,k_2}, \ldots, z_{c,1}, z_{c,2}, \ldots, z_{c,k_c}) \in \mathbb{Z}^k$$

and we shall write $g = g(\mathbf{z})$. Malcev then proved the following (cf. Lemma 2.31):

(1) *there exists a polynomial function* $\mu \colon \mathbb{Z}^k \times \mathbb{Z}^k \to \mathbb{Z}^k$ *satisfying*

$$g(\mathbf{x}) \cdot g(\mathbf{y}) = g(\mu(\mathbf{x}, \mathbf{y})) \tag{2.28}$$

for all $\mathbf{x}, \mathbf{y} \in \mathbb{Z}^k$;

(2) *for any homomorphism* $\alpha \colon G \to G$ *there exists a polynomial function* $\mu_\alpha \colon \mathbb{Z}^k \to \mathbb{Z}^k$ *satisfying*

$$\alpha(g(\mathbf{x})) = g(\mu_\alpha(\mathbf{x})) \tag{2.29}$$

for all $\mathbf{x} \in \mathbb{Z}^k$.

The polynomials μ *and* μ_α *have rational coefficients (in fact in* $\mathbb{Z}[\frac{1}{p_1}, \frac{1}{p_2}, \ldots, \frac{1}{p_s}]$, *where* $p_1, p_2, \ldots, p_s$ *are the primes appearing in the denominators of the coefficients in the Hausdorff–Baker–Campbell formula) but attain integer values at* $\mathbb{Z}^k$.

We may now define $G^{\mathbb{Q}}$ (resp. $G^{\mathbb{R}}$) as the set of all formal products

$$g(\mathbf{q}) := a_{1,1}^{q_{1,1}} a_{1,2}^{q_{1,2}} \cdots a_{1,k_1}^{q_{1,k_1}} a_{2,1}^{q_{2,1}} a_{2,2}^{q_{2,2}} \cdots a_{2,k_2}^{q_{2,k_2}} \cdots a_{c,1}^{q_{c,1}} a_{c,2}^{q_{c,2}} \ldots a_{c,k_c}^{q_{c,k_c}},$$

where $\mathbf{q} := (q_{1,1}, q_{1,2}, \ldots, q_{1,k_1}, q_{2,1}, q_{2,2}, \ldots, q_{2,k_2}, \ldots, q_{c,1}, q_{c,2}, \ldots, q_{c,k_c}) \in \mathbb{Q}^k$ (resp. $\mathbb{R}^k$), and equip it with the product defined by (2.28) in terms of the same polynomials as for G. The group $G^{\mathbb{Q}}$ (resp. $G^{\mathbb{R}}$) is called the *Malcev radical* (resp. *Malcev completion*) of G and is torsion-free nilpotent. Moreover, $G^{\mathbb{Q}}$ is radicable, i.e. for every $h \in G^{\mathbb{Q}}$ and integer $k \geq 1$ there exists (unique, by torsion-freeness) $g \in G^{\mathbb{Q}}$ such that $g^k = h$. Moreover, we have the following characterization: $G^{\mathbb{Q}} = \{g \in G^{\mathbb{R}} : \exists n \geq 1 \text{ s.t. } g^n \in G\}$. Finally, $G^{\mathbb{R}}$ is a nilpotent simply-connected Lie group and G is a lattice.

We have the following:

Every homomorphism $\alpha \colon G \to G$ *extends to a unique homomorphism* $\alpha^{\mathbb{Q}} \colon G^{\mathbb{Q}} \to G^{\mathbb{Q}}$ *(resp. a continuous homomorphism* $\alpha^{\mathbb{R}} \colon G^{\mathbb{R}} \to G^{\mathbb{R}}$*) to the Malcev radical (resp. Malcev completion) of G. In fact, such an extension is obtained by means of the same polynomial function* μ_α *as in* (2.29).

This was proved by Malcev for automorphisms and was extended by Peter Walters in [347, Lemma 1] to homomorphisms.

Note that multiplication in a simply connected nilpotent Lie group $\mathfrak{G}$ is defined by polynomials whose coefficients are referred to as the *structural coefficients* of $\mathfrak{G}$. If the structural coefficients are rational, then the Malcev completion may be reversed and $\mathfrak{G}$ admits a lattice (whose Malcev completion is isomorphic to $\mathfrak{G}$).

For example, the Malcev completion of $\mathbb{Z}^n$ is $\mathbb{R}^n$. Similarly, the Malcev completion of the group $\mathrm{UT}(n,\mathbb{Z})$ of upper unitriangular matrices with integer coefficients is the group $\mathrm{UT}(n,\mathbb{R})$ of upper unitriangular matrices with real coefficients.

Lemma 2.31 as well as the theory of Malcev completion were presented by Malcev in [225]. Theorem 2.20 was proved by Malcev in [226].

Lemma 2.23 was proved by Issai Schur in 1905 [304].

2.9 Exercises

Exercise 2.1. Let G be a group and $H,K \leq G$ be two subgroups. Show that $[H,K] \subseteq K$ (resp. $[H,K] \subseteq H$) if K (resp. H) is normal in G and that $[H,K]$ is normal (resp. characteristic) in G if H and K are both normal (resp. characteristic) in G.

Exercise 2.2. Fill in the details of the proof of the three subgroups lemma (Lemma 2.2).

Exercise 2.3. Let $(\gamma_k(G))_{k\geq 1}$ be the lower central series of a group G. Show that $\gamma_k(G)$ is characteristic (and therefore normal) in G and that $\gamma_{k+1}(G) \subseteq \gamma_k(G)$ for all k.

Exercise 2.4. Show that the set $\mathrm{UT}(n,R)$ consisting of all $n \times n$ upper unitriangular matrices with coefficients in a unital commutative ring R forms a group with the usual multiplication of matrices.

Exercise 2.5. Show that the k-th term $\gamma_k(\mathrm{UT}(n,R))$ of the lower central series of $\mathrm{UT}(n,R)$ consists of all upper unitriangular matrices $M = \|x_{ij}\|_{1\leq i,j\leq n}$ satisfying $x_{ij} = 0$ provided $1 \leq j-i \leq k$, and that, in particular, $\gamma_n(\mathrm{UT}(n,R)) = \{I_n\}$ but $\gamma_{n-1}(\mathrm{UT}(n,R)) \neq \{I_n\}$.

Exercise 2.6. Prove the identity (2.7), and fill in the details of Example 2.15.

Exercise 2.7. Prove the identity (2.8), and fill in the details of Example 2.16.

Exercise 2.8. Show that $[\mathrm{B}(2,\mathbb{Z}),\mathrm{UT}(2,\mathbb{Z})] = \mathrm{UT}(2,2\mathbb{Z})$.

Exercise 2.9. Let G be a nilpotent group of class c with upper central series $(Z_k(G))_{k\geq 0}$ and let $H := G/Z_1(G)$. Show that $Z_i(H) = Z_{i+1}(G)/Z_1(G)$ for all $i = 1,2,\ldots,c-1$.

Exercise 2.10. Prove that the representation (2.14) is unique.

Exercise 2.11. Let $K,L \leq G$ be two subgroups of finite index in G. Show that $[G:K\cap L] \leq [G:K][G:L]$.

Exercise 2.12. For every $x \in G$, let $\varphi(x)$ be the map $G/H \to G/H$ given by $Hg \mapsto Hgx$. Show that the map $x \mapsto \varphi(x)$ is a homomorphism $\varphi\colon G \to \mathrm{Sym}(G/H)$.

Exercise 2.13. Let G be a finitely generated nilpotent group and let G_{tor} denote its (finite normal) subgroup of torsion elements (cf. Corollary 2.37). Show that G/G_{tor} is torsion-free (nilpotent).

Exercise 2.14. Show that any finite extension of a subgroup of $\mathrm{UT}(n,\mathbb{Z})$ is in $\mathrm{GL}(m,\mathbb{Z})$, where m may be larger than n.

Chapter 3
Residual Finiteness and the Zassenhaus Filtration

In this chapter we present an important construction which provides a link between groups and Lie rings. We define the Zassenhaus filtration and the notion of a residually-p (resp. residually-finite, resp. Hopfian) group. In general, given a property (or a class) $\mathscr{P}$ of groups, one says that a group G is residually-$\mathscr{P}$ provided the following holds: given any element $g \in G \setminus \{1_G\}$, there exists a group H satisfying property (or in the class) $\mathscr{P}$ and a group homomorphism $\varphi\colon G \to H$ such that $g \notin \ker(\varphi)$. A group G is Hopfian if every surjective endomorphism $\psi\colon G \to G$ is injective (and therefore an automorphism). We then prove Malcev's theorem (every finitely generated residually finite group is Hopfian (Theorem 3.12)) and the Malcev–Baumslag theorem (the automorphism group of a finitely generated residually finite group is itself residually finite (Theorem 3.13)). We then show that every finitely generated free group F is residually finite and Hopfian as well as its automorphism group $\mathrm{Aut}(F)$ (Corollary 3.15).

3.1 The Lie Ring of a Group

Definition 3.1. A *Lie ring* is a nonassociative ring with an anticommutative multiplication (the *bracket*) satisfying the Jacobi identity. In other words, a Lie ring is an Abelian group L with an operation

$$\begin{array}{ccc} L \times L & \longrightarrow & L \\ (x,y) & \mapsto & [x,y] \end{array}$$

satisfying the following properties:

- $[x+y,z] = [x,z]+[y,z]$ and $[x,y+z] = [x,y]+[x,z]$ for all $x,y,z \in L$ (*bilinearity*);
- $[x,[y,z]]+[y,[z,x]]+[z,[x,y]] = 0$ for all $x,y,z \in L$ (*Jacobi's identity*);
- $[x,x] = 0$ for all $x \in L$ (*anticommutativity*).

The element $[x,y]$ is called the *commutator* of $x,y \in L$.

Notice that the last axiom gives $[x+y,x+y] = 0$, which implies $[x,y] = -[y,x]$ for $x,y \in L$, whence the terminology for the corresponding axiom. Conversely, if L

T. Ceccherini-Silberstein and M. D'Adderio, *Topics in Groups and Geometry*,
Springer Monographs in Mathematics, https://doi.org/10.1007/978-3-030-88109-2_3

is 2-divisible, that is, the map $x \mapsto x^2$ is a group isomorphism, then $[x,x] = -[x,x] \Rightarrow 2[x,x] = 0 \Rightarrow [x,x] = 0$, for all $x \in L$.

For instance, given an associative ring R, we can make it into a Lie ring by defining a bracket by setting $[r,s] := rs - sr$ for all $r,s \in R$ (**exercise**).

Let G be a group with a central series $G = G_1 \geq G_2 \geq \cdots$. Recall that, by definition, we have $[G_i, G_j] \subseteq G_{i+j}$ for all $i,j \geq 1$.

We set

$$L_i := G_i/G_{i+1} \quad \text{and} \quad L(G) := \bigoplus_{i \geq 1} L_i.$$

Then $L(G)$ is a Lie ring with addition supplied by extending linearly the group operation in each homogeneous part

$$a_i G_{i+1} + b_i G_{i+1} := a_i b_i G_{i+1}$$

for all $a_i, b_i \in G_i$, and the bracket operation given by

$$[a_i G_{i+1}, b_j G_{j+1}] := [a_i, b_j] G_{i+j+1}$$

for all $a_i \in G_i$ and $b_j \in G_j$, $i, j \geq 1$.

Note that the centrality of the series ensures that the commutator gives to the bracket operation the appropriate Lie theoretic properties: the anticommutativity of the bracket corresponds to the relation $[a,a] = 1_G$ for all $a \in G$, and Jacobi's identity corresponds to Hall's identity (Lemma 2.1.(6)): in fact we can ignore the conjugations in Hall's identity since if $g_i \in G_i$ and $x \in G$, then $g_i^x G_{i+1} = g_i [g_i, x] G_{i+1} = g_i G_{i+1}$.

3.2 The Zassenhaus Filtration

Let G be a group and let $\mathbb{K}$ be a field. The *group algebra* of G with coefficients in $\mathbb{K}$ is the free vector space $\mathbb{K}G$ over $\mathbb{K}$ with the algebra structure defined by the multiplication in the group. In other words, $\mathbb{K}G$ is the set consisting of all finite formal sums

$$\sum_{g \in G} \alpha_g g, \quad \alpha_g \in \mathbb{K},$$

with the operations defined as follows: the sum

$$\left(\sum_{g \in G} \alpha_g g \right) + \left(\sum_{g \in G} \beta_g g \right) = \sum_{g \in G} (\alpha_g + \beta_g) g,$$

the multiplication by a scalar $c \in \mathbb{K}$

$$c \left(\sum_{g \in G} \alpha_g g \right) = \sum_{g \in G} (c \alpha_g) g,$$

and the product

$$\left(\sum_{g\in G}\alpha_g g\right)\left(\sum_{h\in G}\beta_h h\right)=\sum_{\substack{g,h,k\in G\\ k=gh}}\alpha_g\beta_h k.$$

It is an (**exercise**) to show that the associative and distributive properties of the above operations are satisfied, making indeed $\mathbb{K}G$ into a $\mathbb{K}$-*algebra*.

Definition 3.2. The *augmentation* (or *fundamental*) *ideal* of the group algebra $\mathbb{K}G$ is the ideal $\omega=\omega(\mathbb{K}G)\subseteq\mathbb{K}G$ defined by

$$\omega:=\left\{\sum_{g\in G}\alpha_g g:\sum_{g\in G}\alpha_g=0\right\}.$$

In other words, denoting by $\varepsilon\colon\mathbb{K}G\to\mathbb{K}$ the *augmentation map* defined by

$$\varepsilon\left(\sum_{g\in G}\alpha_g g\right):=\sum_{g\in G}\alpha_g,$$

we have that ε is an algebra homomorphism and $\omega=\ker(\varepsilon)$ (**exercise**). Note that ω is therefore a two-sided ideal in $\mathbb{K}G$. It is generated by the differences $g-g'$ of group elements. Furthermore it is also generated by the differences $g-1_G$ for $g\neq 1_G$, which constitute a basis for ω as a vector space over $\mathbb{K}$.

The proof of the following proposition is left as an **exercise**.

Proposition 3.3. *Let G be a finite p-group (i.e., every element of G has order a power of p), and let $\mathbb{K}$ be a field of characteristic $p>0$ (for instance, $\mathbb{K}=\mathbb{Z}/p\mathbb{Z}$). Then $\omega(\mathbb{K}G)$ is a nilpotent ideal.*

Let now R be a ring with identity $1:=1_R$ and let $I\subseteq R$ be an ideal. We denote by $G:=G(1+I)$ the set consisting of all invertible elements of $1+I\subseteq R$.

Note that this is a group. Indeed, let $1+a\in G$ and $1+b\in G$ with $a,b\in I$. We have $(1+a)(1+b)=1+(a+b+ab)$ where $a+b+ab\in I$ and the product of invertible elements of R is invertible: this shows that G is closed under the multiplication. Moreover, if $x=(1+a)^{-1}\in R$, then, setting $b:=x-1$ we have $x=1+b$ and from $1=x(1+a)=(1+b)(1+a)=1+(b+a+ba)$ we deduce that $b+a+ba=0$, so that $b=-(a+ba)\in I$. This shows that $x\in 1+I$ and therefore $x\in G$. It follows that G is also closed under taking inverses, and it is therefore a group.

Definition 3.4. For $i=1,2,\ldots$, we set

$$G_i:=G(1+\omega(\mathbb{K}G)^i)\cap G=\{g\in G:g=1 \bmod \omega(\mathbb{K}G)^i\}.$$

The series $G=G_1\geq G_2\geq\cdots$ is called the *Zassenhaus filtration*.

Lemma 3.5. *The series $\left(G(1+I^j)\right)_{j\geq1}$ is central.*

Proof. We need to show that

$$[G(1+I^i),G(1+I^j)] \subseteq G(1+I^{i+j}) \text{ for all } i,j \geq 1. \tag{3.1}$$

Let $i,j \geq 1$. Let $x \in G(1+I^i)$ and $y \in G(1+I^j)$, say $x = 1+a$ with $a \in I^i$, and $y = 1+b$ with $b \in I^j$. From

$$yx - xy = (1+b)(1+a) - (1+a)(1+b) = ba - ab \in I^{i+j}$$

we deduce that

$$1 - [x,y] = 1 - x^{-1}y^{-1}xy = x^{-1}y^{-1}(yx - xy) \in I^{i+j}.$$

This proves (3.1). □

Remark 3.6. Let $\mathbb{K}$ be a field of characteristic $p > 0$, and let $G = G_1 \geq G_2 \geq \cdots$ be the Zassenhaus filtration. Consider the Lie ring $L(G) = \bigoplus_{i\geq 1} L_i$, where $L_i = (G_i/G_{i+1})$. Then for every $i \geq 1$ we have $p \cdot L_i = \{0\}$ in $L(G)$.

Indeed, let $i \geq 1$ and $x \in G_i$, say $x = 1+a$ with $a \in \omega(\mathbb{K}G)^i$. Since $\mathbb{K}$ has characteristic p, we have

$$x^p = (1+a)^p = 1 + a^p \in 1 + \omega(\mathbb{K}G)^{pi} \subseteq 1 + \omega(\mathbb{K}G)^{i+1}.$$

This shows that $x^p \in G_{i+1}$.

3.3 Residually-p and Residually Finite Groups

Definition 3.7. Let G be an arbitrary group. We say that a family $(\varphi_j : G \to G_j)_{j\in J}$ of group homomorphisms *approximates* G if $\bigcap_{j\in J} \ker(\varphi_j) = \{1_G\}$ (equivalently, if for every $x,y \in G$ with $x \neq y$ there exists a $j \in J$ such that $\varphi_j(x) \neq \varphi_j(y)$).

We say that G is *residually finite* if there exists an approximating family of group homomorphisms $(\varphi_j : G \to G_j)_{j\in J}$ with G_j finite for all $j \in J$. Note that G is residually finite if and only if

$$\bigcap_{\substack{H \lhd G \\ [G:H]<\infty}} H = \{1_G\}.$$

This is equivalent to saying that given any $g \in G$, $g \neq 1_G$, there exists a finite group H and a group homomorphism $\varphi : G \to H$ such that $\varphi(g) \neq 1_H$.

Given a prime p, we say that G is *residually-p* if there exists an approximating family of group homomorphisms $(\varphi_j : G \to G_j)_{j\in J}$ with G_j a finite p-group for all $j \in J$. Note that G is residually-p if and only if

$$\bigcap_{\substack{H \lhd G \\ [G:H]=p^k,\ k\in\mathbb{N}}} H = \{1_G\}.$$

This is equivalent to saying that given $g \in G$ with $g \neq 1_G$ there exists a finite p-group H and a group homomorphism $\varphi : G \to H$ such that $\varphi(g) \neq 1_H$.

Remark 3.8. Note that every residually-p group is residually finite, but there exist residually finite groups which are not residually-p. Consider, for instance the groups $G_1 = \mathbb{Z}/2\mathbb{Z} \times \mathbb{Z}/3\mathbb{Z}$ and $G_2 = \mathbb{Z} \times \mathbb{Z}/2\mathbb{Z} \times \mathbb{Z}/3\mathbb{Z}$. We have that G_1 and G_2 are both residually finite (indeed G_1 is even finite, while G_2 is a finite product of residually finite groups) but they are not residually-p for any prime p.

Proposition 3.9. *Every subgroup of a residually finite (resp. residually-p) group is residually finite (resp. residually-p).*

Proof. Let G be a residually finite (resp. residually-p) group and let H be a subgroup. Let $h \in H$ such that $h \neq 1_G$. As G is residually finite (resp. residually-p), we can find a finite group (resp. a finite p-group) K and a homomorphism $\varphi\colon G \to K$ such that $\varphi(h) \neq 1_K$. If $\psi\colon H \to K$ denotes the restriction of φ to H, we have $\psi(h) = \varphi(h) \neq 1_K$. Consequently, H is itself residually finite (resp. residually-p). □

Proposition 3.10. *Let G be a finitely generated group, and let $(G_i)_{i\geq 1}$ denote its Zassenhaus filtration with respect to a field $\mathbb{K}$ of characteristic $p > 0$. Then G is residually-p if and only if $\bigcap_{i\geq 1} G_i = \{1_G\}$.*

Proof. Suppose first that $\bigcap_{i\geq 1} G_i = \{1_G\}$.

Let $x \in G$, and write $x = 1 + a$, where $a \in \omega(\mathbb{K}G)$. Since $\mathbb{K}$ has characteristic $p > 0$, we have $x^{p^k} = (1+a)^{p^k} = 1 + a^{p^k}$ for all $k \geq 1$. Hence, given $i \geq 1$, for k such that $p^k \geq i$, since $a^{p^k} \in \omega(\mathbb{K}G)^{p^k}$, we have $x^{p^k} = 1 + a^{p^k} \in 1 + \omega(\mathbb{K}G)^{p^k} \subseteq 1 + \omega(\mathbb{K}G)^i$, showing that G/G_i is a p-group.

Notice also that by Lemma 3.5 the Zassenhaus filtration $(G_i)_{i\geq 1}$ is central, therefore the groups G/G_i are finitely generated nilpotent groups. Since they are also torsion, by Lemma 2.34 the groups G/G_i are finite. Hence they are all finite p-groups.

Then the fact that G is residually-p follows from the assumption $\bigcap_{i\geq 1} G_i = \{1_G\}$ and the fact that each G/G_i is a finite p-group for every i (**exercise**).

Suppose now that G is residually-p, i.e.

$$\bigcap_{\substack{H \lhd G \\ [G:H]=p^k,\ k\in\mathbb{N}}} H = \{1_G\}. \tag{3.2}$$

Let $H \subseteq G$ be a normal subgroup of index $[G:H] = p^k$ for some $k \in \mathbb{N}$. Denote by $\varphi\colon G \to G/H =: \widetilde{G}$ the canonical homomorphism. Observe that φ extends to a surjective homomorphism of $\mathbb{K}$-algebras $\widetilde{\varphi} : \mathbb{K}G \to \mathbb{K}\widetilde{G}$.

Since $\widetilde{G}$ is a finite p-group, by Proposition 3.3 $\omega(\mathbb{K}\widetilde{G}) \cong \omega(\mathbb{K}(G/H))$ is a nilpotent ideal of $\mathbb{K}\widetilde{G}$ (in fact $\omega(\mathbb{K}\widetilde{G})^{p^k} = \{0\}$).

If $g = 1_G + x \in G_i$, so that $x \in \omega(\mathbb{K}G)^i$, then $\varphi(g) = \widetilde{\varphi}(1_G + x) = 1_{\widetilde{G}} + \widetilde{\varphi}(x) \in 1_{\widetilde{G}} + \omega(\mathbb{K}\widetilde{G})^i$, since $\widetilde{\varphi}(\omega(\mathbb{K}G)^i) \subseteq \omega(\mathbb{K}\widetilde{G})^i$. It is now clear that $G_i \subseteq \ker\varphi = H$ for all $i \geq p^k$. By (3.2), this shows that $\bigcap_{i\geq 1} G_i = \{1_G\}$. □

3.4 The Theorems of Malcev and G. Baumslag

Let G be a group. We denote by $\mathrm{Aut}(G)$ the *automorphism group* of G, that is, the group of all bijective homomorphisms $\alpha\colon G \to G$.

Definition 3.11. A group G is called *Hopfian* provided that every surjective homomorphism $\alpha\colon G \to G$ is injective (and therefore $\alpha \in \mathrm{Aut}(G)$).

Theorem 3.12 (Malcev). *Every finitely generated residually finite group is Hopfian.*

Proof. Let G be a finitely generated residually finite group and suppose that $\alpha\colon G \to G$ is a surjective homomorphism. We want to show that α is injective. Let $g \in G \setminus \{1_G\}$. Since G is residually finite, we can find a finite group F and a homomorphism $\rho\colon G \to F$ such that

$$\rho(g) \neq 1_F. \tag{3.3}$$

Consider the set $\mathrm{Hom}(G,F)$ consisting of all group homomorphisms $\varphi\colon G \to F$ and let $\Psi\colon \mathrm{Hom}(G,F) \to \mathrm{Hom}(G,F)$ denote the map defined by

$$\Psi(\varphi) := \varphi \circ \alpha \tag{3.4}$$

for all $\varphi \in \mathrm{Hom}(G,F)$. Since α is surjective, it is an **exercise** to deduce that Ψ is injective. Moreover, since G is finitely generated and F is finite, it is an **exercise** to deduce that $\mathrm{Hom}(G,F)$ is finite. It follows that Ψ is surjective. As a consequence, we can find $\varphi_0 \in \mathrm{Hom}(G,F)$ such that $\varphi_0 \circ \alpha = \Psi(\varphi_0)$ equals ρ. Keeping in mind (3.3), we thus have $\varphi_0(\alpha(g)) = \rho(g) \neq 1_F$, so that, necessarily, $\alpha(g) \neq 1_G$. This shows that $\ker(\alpha) = \{1_G\}$. □

Theorem 3.13 (Malcev, G. Baumslag). *Let G be a finitely generated residually finite group. Then the automorphism group* $\mathrm{Aut}(G)$ *is residually finite.*

Proof. Let $\alpha \in \mathrm{Aut}(G) \setminus \{\mathrm{id}_G\}$ and choose $g_0 \in G$ such that $\alpha(g_0) \neq g_0$. Since G is residually finite, we can find a finite group F and a homomorphism $\rho \in \mathrm{Hom}(G,F)$ such that $\rho(\alpha(g_0)g_0^{-1}) \neq 1_F$, equivalently,

$$\alpha(g_0)g_0^{-1} \notin \ker(\rho). \tag{3.5}$$

Let

$$N := \bigcap_{\varphi \in \mathrm{Hom}(G,F)} \ker(\varphi) \trianglelefteq G. \tag{3.6}$$

and observe (**exercise**) that since (cf. the proof of Theorem 3.12) $\mathrm{Hom}(G,F)$ is finite, N has finite index in G. It follows that G/N and therefore $\mathrm{Aut}(G/N)$ are finite groups. It is an **exercise** to show that if $\beta \in \mathrm{Aut}(G)$ then $\beta(N) = N$ so that the map $\Phi\colon \mathrm{Aut}(G) \to \mathrm{Aut}(G/N)$ given by

$$\Phi(\beta)(gN) = \beta(g)N \tag{3.7}$$

for all $\beta \in \mathrm{Aut}(G)$ and $g \in G$, is well defined and a group homomorphism. By virtue of (3.5) we have $\alpha(g_0)g_0^{-1} \notin N$, equivalently,

$$\Phi(\alpha)(g_0N) = \alpha(g_0)N \neq g_0N.$$

This shows that $\Phi(\alpha) \neq \mathrm{id}_{G/N} = 1_{\mathrm{Aut}(G/N)}$. We deduce that $\mathrm{Aut}(G)$ is residually finite. □

3.5 Residual Finiteness of Free Groups

Theorem 3.14. *Let F be a finitely or countably generated free group. We denote by $(\gamma_i(F))_{i\geq 1}$ its lower central series, and by $(G_i(F))_{i\geq 1}$ its Zassenhaus filtration. Then*

(1) $\bigcap_{i\geq 1} \gamma_i(F) = \{1_F\}$,
(2) $\bigcap_{i\geq 1} G_i(F) = \{1_F\}$.

In particular, F is residually finite.

Proof. We denote by F_m the free group on m free generators, for $2 \leq m \leq \infty$. By virtue of Proposition 3.9 it is sufficient to prove the statement for $F = F_2$, since, by Corollary 1.18, F_2 contains a subgroup isomorphic to F_m for all $2 \leq m \leq \infty$.

As in Example 1.19 we can show that, for every prime p, the matrices $\begin{pmatrix} 1 & p \\ 0 & 1 \end{pmatrix}$ and $\begin{pmatrix} 1 & 0 \\ p & 1 \end{pmatrix}$ generate a group isomorphic to F_2.

In particular, we have $F_2 \subseteq \mathrm{GL}(2,\mathbb{Z})$, so $\mathbb{K}F_2 \subseteq M_2(\mathbb{Z})$ and, in fact, $F_2 = G_1(F_2) \subseteq G(1+pM_2(\mathbb{Z}))$. More generally, $G_i(F_2) \subseteq G(1+M_2(p^i\mathbb{Z}))$.

The groups $G(1+p^iM_2(\mathbb{Z})) = G(1+M_2(p^i\mathbb{Z}))$ for $i \geq 1$ are called the *p-congruence subgroups* of $\mathrm{GL}(2,\mathbb{Z})$.

Now clearly $\bigcap_{i\geq 1} G(1+p^iM_2(\mathbb{Z})) = \{I_2\}$ and this shows (2).

It follows from Lemma 3.5 that $(G_i(F_2))_{i\geq 1}$ is a central series. Therefore, by Lemma 2.7, $\gamma_i(F_2) \subseteq G_i(F_2)$ for all $i \geq 1$, and hence (1) follows from (2). □

From Theorem 3.14, Theorem 3.12, and Theorem 3.13 we immediately deduce the following:

Corollary 3.15. *Let F be a finitely generated free group. Then F is Hopfian and $\mathrm{Aut}(F)$ is residually finite.* □

3.6 Notes

The Jacobi identity in Definition 3.1 is named after Carl Gustav Jakob Jacobi. The Zassenhaus filtration was introduced by Hans Zassenhaus in 1939 [360]. For a survey and applications of the Zassenhaus filtration see the recent survey by Misha Ershov [106]. For a "canonical" text we refer to [93, Chapters 11 and 12].

Recall that a group G is *linear* (see Chapter 8) if there exist an integer $n \geq 1$ and a field K such that G is isomorphic to a subgroup of $\mathrm{GL}(n,K)$. By a fundamental theorem of Malcev [224], every finitely generated linear group is residually finite.

Residual finiteness of free groups was firstly established by Friedrich W. Levi [213]. As F_2, the free group of rank 2, and therefore all finitely generated free groups are linear (cf. Example 1.19 based on Klein's ping-pong lemma (Theorem 1.17)), one may deduce residual finiteness of finitely generated free groups from their linearity and the Malcev theorem we alluded to above. Besides the proof we present here (cf. Theorem 3.14), we also refer to Exercise 3.10. There are other proofs of the residual finiteness of free groups in [293] and [223].

The fact that every finitely generated residually finite group is Hopfian (Theorem 3.12) is due to Malcev [224]. The term *Hopfian* comes from Heinz Hopf who, in 1932, motivated by topological investigations, asked whether or not every finitely generated group satisfies the Hopfian property.

It may be shown that the *Baumslag–Solitar group* $\mathrm{BS}(2,3)$ (named after Gilbert Baumslag and Donald Solitar) which is given by the finite presentation $\langle a,b : ba^2b^{-1} = a^3\rangle$ is not Hopfian (see [22], [223], [218]). As a consequence, the group $\mathrm{BS}(2,3)$ is not residually finite by Theorem 3.12.

Residual finiteness of the automorphism group of a finitely generated residually finite group (Theorem 3.13) was proved by G. Baumslag in [21]. (It was previously shown by A.I. Malcev [224] that every finitely generated residually finite *semi*group has a residually finite monoid of endomorphisms.)

3.7 Exercises

Exercise 3.1. Given an associative ring R, define the bracket as $[r,s] := rs - sr$ for all $r,s \in R$. Show that R with this bracket is a Lie ring.

Exercise 3.2. Prove Proposition 3.3.

Exercise 3.3. Let $\mathbb{K}$ be a field of characteristic $p > 0$, and let $G = G_1 \geq g_2 \geq \cdots$ be the Zassenhaus filtration. Then G_k is generated by the commutators $[g_{i_1}, g_{i_2}, \ldots, g_{i_s}]^{p^\ell}$ such that $g_{i_j} \in G$ for all j, and $s \cdot p^\ell \geq k$.

Exercise 3.4. Fill in the details of the proof of Proposition 3.10.

Exercise 3.5. Let G be a finitely generated group and F a finite group. Show that the set $\mathrm{Hom}(G,F)$ of all group homomorphisms $\varphi\colon G \to F$ is finite.

Exercise 3.6. Show that the map $\Psi\colon \mathrm{Hom}(G,F) \to \mathrm{Hom}(G,F)$ given by (3.4) is injective.

Exercise 3.7. Show that the normal subgroup $N \trianglelefteq G$ defined in (3.6) has finite index in G.

Exercise 3.8. Let $\Phi\colon \mathrm{Aut}(G) \to \mathrm{Aut}(G/N)$ be the map given by (3.7).

(1) Show that Φ is well defined;
(2) show that Φ is a group homomorphism.

Exercise 3.9. Show (without using Theorem 3.14) that the additive group $\mathbb{Z}$ is residually finite.

Exercise 3.10. (1) Show that the group $\mathrm{GL}(n,\mathbb{Z})$ is residually finite for every $n \geq 1$.
(2) Deduce that the free group on two generators is residually finite.

Exercise 3.11. Let $\mathscr{P}$ be a class (or property) of groups. Suppose that $\mathscr{P}$ is closed under finite direct products (i.e., if G_1, G_2 satisfy $\mathscr{P}$ so does their direct product $G_1 \times G_2$). Let G be a group. Show that G is residually-$\mathscr{P}$ if and only if given any finite nonempty subset $F \subseteq G$ there exists a finite index normal subgroup $N \triangleleft G$ such that the canonical quotient map $\varphi\colon G \to G/N$ is injective on F (that is, $\varphi(f_1) = \varphi(f_2)$ implies $f_1 = f_2$, for all $f_1, f_2 \in F$).

Exercise 3.12. Let G be a group. Show that G is residually finite if and only if the following condition is satisfied: for every nonempty finite subset $F \subseteq G$ there exists a finite index subgroup $H \leq H$ such that $(hF)_{h\in H}$ is a disjoint family (i.e. $hF \cap kF = \varnothing$ for all distinct $h, k \in H$).

Exercise 3.13. A group G is termed *divisible* provided that for each $g \in G$ and each integer $n \geq 1$ there exists an $h \in G$ such that $h^n = g$.

(1) Show that the additive groups $\mathbb{Q}, \mathbb{R}$, and $\mathbb{C}$ are divisible.
(2) Show that if G is a divisible group and F is a finite group then there are no nontrivial group homomorphisms $\phi\colon G \to F$.
(3) Deduce that any divisible group is not residually finite.

Exercise 3.14. Let G be a group and denote by N the intersection of all finite index subgroups of G (this is called the *residual subgroup* of G).

(1) Show that N equals the intersection of all finite index *normal* subgroups of G.
(2) Show that N is a normal subgroup of G.
(3) Show that N is residually finite if and only if $N = \{1_G\}$.

Exercise 3.15. Show that every virtually residually finite group is residually finite.

Exercise 3.16. (1) Let $(G_i)_{i\in I}$ be a family of residually finite groups. Show that the direct product $\prod_{i\in I} G_i$ is residually finite.
(2) Deduce that if $(G_i)_{i\in I}$ is a family of residually finite groups, then the direct sum $\bigoplus_{i\in I} G_i$ is residually finite.
(3) Deduce that any finitely generated abelian group is residually finite.
(4) Show that a group G is residually finite if and only if there exists a family $(F_i)_{i\in I}$ of finite groups such that G is isomorphic to a subgroup of $\prod_{i\in I} F_i$.

Exercise 3.17. A *projective systems of groups* consists of the following data: (i) a directed set I; (ii) a family $(G_i)_{i\in I}$ of groups; (iii) for each pair $i, j \in I$ such that $i \leq j$, a group homomorphism $\varphi_{ij}\colon G_j \to G_i$ satisfying the following conditions:

$$\varphi_{ii} = \mathrm{Id}_{G_i} \text{ (the identity map on } G_i\text{) for all } i \in I$$
$$\varphi_{ij} \circ \varphi_{jk} = \varphi_{ik} \text{ for all } i, j, k \in I \text{ such that } i \leq j \leq k.$$

Given a projective system (G_i, φ_{ij}) of groups, set $P = \prod_{i \in I} G_i$. The subgroup

$$G := \{(g_i) \in P : \varphi_{ij}(g_j) = g_i \text{ for all } i, j \in I \text{ such that } i \leq j\}$$

is called the *projective limit* of the projective system (G_i, φ_{ij}).

A group G is called *profinite* if it is the limit of a projective system of finite groups.

(1) Show that the projective limit of a projective system of residually finite groups is itself residually finite.
(2) Deduce that every profinite group is residually finite.
(3) Let p be a prime number. Given integers $0 \leq m \leq n$ let $\varphi_{nm} : \mathbb{Z}/p^m\mathbb{Z} \to \mathbb{Z}/p^n\mathbb{Z}$ denote the reduction modulo p^n. Show that $((\mathbb{Z}/p^n\mathbb{Z}), \varphi_{nm})$ is a projective system of groups (here $I = \mathbb{N}$). The corresponding projective limit, denoted by $\mathbb{Z}_p$, is called the group of *p-adic integers*.

Exercise 3.18. Let R be a ring and suppose that R is finitely generated as a $\mathbb{Z}$-module. Show that the group $\mathrm{GL}(n, R)$ is residually finite for all $n \geq 1$.

Exercise 3.19. (1) Show that an infinite simple group is not residually finite.
(2) Deduce that $\mathrm{Sym}_0(\mathbb{N})$, the group of finitary permutations of $\mathbb{N}$, is not residually finite.
(3) Deduce that the subgroup G_1 of $\mathrm{Sym}(\mathbb{Z})$ generated by the translation $T : n \mapsto n+1$ and the transposition $S = (0\ 1)$ is a finitely generated group which is not residually finite.

Exercise 3.20. Show that if S is a finite simple group, then the wreath product $G := S \wr \mathbb{Z}$ is a Hopfian group.

Chapter 4
Solvable Groups

A group G is solvable if it admits a sequence of subgroups $\{1_G\} = G_0 \leq G_1 \leq \cdots \leq G_n = G$ such that G_{i-1} is normal in G_i and the corresponding quotient group G_i/G_{i-1} is Abelian, for $i = 1, 2, \ldots, n$. Equivalently, G is solvable if its derived series $G = G_{[0]} \geq G_{[1]} \geq G_{[2]} \geq \cdots$, where $G_{[i]}$ is the commutator subgroup $[G_{[i-1]}, G_{[i-1]}]$, $i = 1, 2, \ldots$, eventually reaches the trivial subgroup $\{1_G\}$ of G. Every nilpotent group is solvable, but there are solvable groups that are not nilpotent. The groups $\mathrm{UT}(n,R)$ and $\mathrm{B}(n,R)$ of upper unitriangular and upper triangular $n \times n$ matrices with coefficients in a commutative ring R (with $2 \in R$ invertible) are solvable: in Section 4.2 we compute their derived series. The main result of this chapter is Malcev's theorem (Theorem 4.16) stating that a finitely generated solvable linear group $G \leq \mathrm{GL}(n,\mathbb{K})$, with $\mathbb{K}$ an algebraically closed field, admits a finite index subgroup $H \leq G$ which is triangularizable, that is, there exists an $x \in \mathrm{GL}(n,\mathbb{K})$ such that $x^{-1}Hx \leq \mathrm{B}(n,\mathbb{K})$.

4.1 Solvable Groups: Definitions and Relations with Nilpotent Groups

Definition 4.1. Given a group G, its *derived series* $(G_{[k]})_{k\geq 0}$ is the descending series recursively defined by setting

$$G_{[0]} := G$$

and

$$G_{[n+1]} := \left[G_{[n]}, G_{[n]}\right]$$

for $n \geq 0$. G is said to be *solvable* if there exists an integer n such that $G_{[n]} = \{1\}$. The minimal such n is called the *solvability class* of G.

Notice that the groups $G_{[i]}$ are characteristic, and the quotients $G_{[i]}/G_{[i+1]}$ are abelian.

By induction on n it is easy to see that $G_{[n]} \subseteq \gamma_{2^n}(G)$ (**exercise**), where $(\gamma_k(G))_{k\geq 1}$ denotes the lower central series of G (cf. Definition 2.3). From this we deduce the following:

T. Ceccherini-Silberstein and M. D'Adderio, *Topics in Groups and Geometry*,
Springer Monographs in Mathematics, https://doi.org/10.1007/978-3-030-88109-2_4

Proposition 4.2. *Every nilpotent group is solvable.* □

The converse is false:

Example 4.3. (a) Let R be a commutative ring such that $2 \in R$ is invertible, and let n be a positive integer. Notice (see Section 2.4) that $\mathrm{B}(n,R)_{[1]} = [\mathrm{B}(n,R),\mathrm{B}(n,R)] \subseteq \mathrm{UT}(n,R)$ and hence, by the previous proposition, $\mathrm{B}(n,R)_{[k]} \subseteq \gamma_{2^{k-1}}(\mathrm{UT}(n,R))$ for all $k \geq 1$. Since $\mathrm{UT}(n,R)$ is nilpotent, it follows that $\mathrm{B}(n,R)$ is solvable. But (see again Section 2.4) $Z_k(\mathrm{B}(n,R)) = Z_1(\mathrm{B}(n,R)) = Z(\mathrm{B}(n,R)) = R^\times I_n$ for all $k \geq 1$, where $(Z(G)_k)_{k\geq 0}$ denotes the upper central series of a group G (cf. Definition 2.10) and $R^\times$ is the group of the invertible elements of R. Hence $\mathrm{B}(n,R)$ is not nilpotent.

(b) For the symmetric group $G := S_3$ on 3 elements, $G_{[1]} = \langle(1,2,3)\rangle$, hence G is clearly solvable. But $Z(G) = \{1\}$, hence G is not nilpotent.

The proof of the following proposition is left as an **exercise**.

Proposition 4.4. *The class of solvable groups is closed under the operations of taking subgroups, quotients, and extensions. In particular, given an exact sequence* $\{1\} \to N \to G \to G/N \to \{1\}$ *of solvable groups, the length of the derived series of* G *does not exceed the sum of the lengths of the derived series of* N *and of* G/N.

One the other hand, the class of nilpotent groups is not closed under extensions as the following examples show:

Example 4.5. (a) Let R be a commutative ring such that $2 \in R$ is invertible, and let $G := \mathrm{B}(n,R)$ and $H := \mathrm{UT}(n,R) \trianglelefteq G$. We have that both H and $G/H \cong (R^\times)^n$ are nilpotent (note that G/H is even abelian), but we just saw that G is not (cf. Example 4.3.(a)).

(b) For $G := S_3$ the symmetric group on 3 elements and $H := \langle(1,2,3)\rangle \trianglelefteq G$, both H and $G/H \cong \mathbb{Z}/2\mathbb{Z}$ are nilpotent (note that $\mathbb{Z}/2\mathbb{Z}$ is even abelian), but G is not (cf. Example 4.3.(b)).

Definition 4.6. Let G and H be two groups. A *(right) action* of G on H is a map $\varphi : H \times G \to H$ such that $\varphi(\cdot, g) \in \mathrm{Aut}(H)$ for all $g \in G$ and such that $\varphi(h, g_1g_2) = \varphi(\varphi(h,g_1),g_2)$ for all $g_1, g_2 \in G$ and $h \in H$.

Another way of saying this is that the map $g \mapsto \varphi(\cdot, g)$ is an antihomomorphism from G to $\mathrm{Aut}(H)$.

Remark 4.7. Notice that we are requiring that the elements of the group G act as automorphisms of the group H, and not simply as elements of $\mathrm{Sym}(H)$.

When the action φ is understood we simply write h^g instead of $\varphi(h,g)$ so that we have $h^{g_1g_2} = (h^{g_1})^{g_2}$, for all $g, g_1, g_2 \in G$ and $h \in H$.

As an example, if $N \trianglelefteq G$ is a normal subgroup of G, and H is any subgroup of G, then for every $f \in N$ and $h \in H$ we have that $\varphi(f,h) := f^h = h^{-1}fh$ belongs to N. Then the map $\varphi : N \times H \to N$ is an action of H on N. It is called the *H-conjugation* on N.

Definition 4.8. If G acts on H we define a product on the set $G \times H$ by setting

$$(g_1, h_1) \cdot (g_2, h_2) := (g_1 g_2, h_1^{g_2} h_2), \tag{4.1}$$

for all $g_1, g_2 \in G$ and $h_1, h_2 \in H$. This product endows the set $G \times H$ with a group structure (**exercise**). It is called the *semi-direct product* of G and H and it is denoted by $G \ltimes_\varphi H$ (or simply by $G \ltimes H$, when the action φ is understood).

By identifying H and $\{1_G\} \times H$ (resp. G and $G \times \{1_H\}$), (cf. Exercise 4.3.(c)), we have that $H \trianglelefteq G \ltimes H$ and $G \leq G \ltimes H$. However, in general, G is not normal in $G \ltimes H$, as the following example shows.

Example 4.9. Let $H := \langle (1,2,3) \rangle \trianglelefteq S_3$ and $G := \langle (1,2) \rangle \leq S_3$, and consider the G-conjugation on H. Then $G \ltimes H \cong S_3$ via the isomorphism given by $(1_G, (1,2,3)) \mapsto (1,2,3)$ and $((1,2), 1_H) \mapsto (1,2)$. Here G is not normal in $G \ltimes H$.

In all the examples of solvable groups that we have seen so far, the commutator subgroup was in fact nilpotent. To see that this need not be the case, we introduce an important group-theoretical construction.

Definition 4.10. Let A and B be two groups. We set

$$\mathrm{Fun}(B,A) := A^B = \{f : B \to A\}$$

and

$$\mathrm{fun}(B,A) := \{f \in \mathrm{Fun}(B,A) : \mathrm{supp}(f) \text{ is finite}\},$$

where $\mathrm{supp}(f) := \{b \in B : f(b) \neq 1_A\}$. Notice that

$$\mathrm{Fun}(B,A) = \prod_{b \in B} A \quad \text{and} \quad \mathrm{fun}(B,A) = \bigoplus_{b \in B} A.$$

Then B acts on $\mathrm{Fun}(B,A)$ (resp. $\mathrm{fun}(B,A)$) by $f^b(b') := f(bb')$, for all $b, b' \in B$ and $f \in \mathrm{Fun}(B,A)$ (resp. $f \in \mathrm{fun}(B,A)$).

We call

$$A \,wr\, B = A \wr B := B \ltimes \mathrm{fun}(B,A)$$

the *wreath product* of A and B, and

$$A \,\overline{wr}\, B = A \bar{\wr} B := B \ltimes \mathrm{Fun}(B,A)$$

the *complete wreath product* of A and B.

Proposition 4.11. *Let G and H be two groups. Suppose that G and H are solvable (resp. finitely generated) and that G acts on H. Then $G \ltimes H$ is solvable (resp. finitely generated).*

Proof. This follows from Proposition 4.4 (resp. Proposition 1.12) after observing that we have a short exact sequence $\{1\} \to H \to G \ltimes H \to G \to \{1\}$. $\square$

Corollary 4.12. *Let A and B be two groups. Suppose that A and B are solvable. Then $A \bar{\wr} B$ is solvable. In particular, $A \wr B$ is also solvable.*

Proof. If A is solvable, then $G := \mathrm{Fun}(B,A) = A^B$ is solvable (**exercise**). Thus $A \bar{\wr} B = B \ltimes G$ is solvable by Proposition 4.11. Since $A \wr B$ is a subgroup of $A \bar{\wr} B$, from Proposition 4.4 and the first part of the proof we then deduce that $A \wr B$ is solvable as well. □

Proposition 4.13. *Let A and B be two groups. Suppose that A and B are finitely generated. Then* $A \wr B$ *is also finitely generated.*

Proof. Let $S_A \subseteq A$ and $S_B \subseteq B$ be two finite generating subsets for A and B. Consider the maps $f_a \in \mathrm{fun}(B,A)$, $a \in A$, defined by $f_a(b) := a$ if $b = 1_B$ and $f_a(b) := 1_A$ otherwise. Then given $a \in A$ and $b \in B$ the map $(f_a)^{b^{-1}} \in \mathrm{fun}(B,A)$ satisfies $(f_a)^{b^{-1}}(b') = f_a(b^{-1}b') = a$ if $b' = b$ and $(f_a)^{b^{-1}}(b') = 1_A$ otherwise. From this we easily deduce that $S_B \cup \{f_a : a \in S_A\}$ is a (finite) generating subset of $A \wr B$. □

Remark 4.14. Note that, unless either B is finite (so that $A \bar{\wr} B = A \wr B$) and A is finitely generated, or A is trivial (so that $A \bar{\wr} B = B$) and B is finitely generated, the group $A \bar{\wr} B$ is never finitely generated.

The following proposition will be useful.

Proposition 4.15. *Let A and B be two groups. If A is not nilpotent and* $B \neq \{1_B\}$, *then* $\mathrm{fun}(B,A) \cap [A \wr B, A \wr B]$ *is not nilpotent. In particular, the commutator subgroup* $[A \wr B, A \wr B]$ *of* $A \wr B$ *is not nilpotent either.*

Proof. Let $b \in B \setminus \{1_B\}$ and consider the map

$$\varphi_b \colon \mathrm{fun}(B,A) \cap [A \wr B, A \wr B] \to A$$

given by $\varphi_b(f) = f(b)$ for all $f \in \mathrm{fun}(B,A) \cap [A \wr B, A \wr B]$. Observe that φ_b is a homomorphism.

Claim. *The homomorphism* φ_b *is surjective.*

For $a \in A$, let $f_a : B \to A$ be defined as $f_a(b) := a$ (recall that $b \in B \setminus \{1_B\}$), $f_a(1_B) := a^{-1}$ and $f_a(x) := 1_A$ for all other $x \in B$. We want to show that $f_a \in \mathrm{fun}(B,A) \cap [A \wr B, A \wr B]$, so that $\varphi_b(f_a) = a$, and we are done.

Clearly $f_a \in \mathrm{fun}(B,A)$. To see that $f_a \in [A \wr B, A \wr B]$, consider the element $f \in \mathrm{fun}(B,A)$ defined by $f(b) := a$, and $f(x) := 1_A$ for $x \neq b$. Then $f^b(x) = f(bx)$ equals a if $x = 1_B$, and $f^b(x) = 1_A$ otherwise. Hence

$$[f^{-1}, b](x) = (fb^{-1}f^{-1}b)(x) = (f(f^b)^{-1})(x) = f(x)(f^b(x))^{-1} = \begin{cases} a & \text{if } x = b; \\ a^{-1} & \text{if } x = 1_B; \\ 1_A & \text{otherwise.} \end{cases}$$

So $f_a = [f^{-1}, b] \in [A \wr B, A \wr B]$, as we wanted. This completes the proof of the claim.

Since A is not nilpotent and the class of nilpotent groups is closed under the operation of taking quotients (cf. Proposition 2.9), it follows that $\mathrm{fun}(B,A) \cap [A \wr B, A \wr B]$ is not nilpotent either. Moreover, since the class of nilpotent groups is closed under the operation of taking subgroups (cf. Proposition 2.9), we deduce that $[A \wr B, A \wr B]$ is not nilpotent. □

Let A be a finite solvable group which is not nilpotent. For example, $A = B(n,\mathbb{K})$, where $\mathbb{K}$ is a finite field of characteristic $p \neq 2$, or $A = (\mathbb{Z}/3\mathbb{Z}) \wr (\mathbb{Z}/2\mathbb{Z})$ (**exercise**). Let $B = \mathbb{Z}/2\mathbb{Z}$. Then $G = A \wr B$ is a solvable group whose commutator is not nilpotent.

4.2 Two Important Examples: $\mathrm{UT}(n,R)$ and $\mathrm{B}(n,R)$

In this section we continue our study of two important examples: the groups $\mathrm{UT}(n,R)$ and $\mathrm{B}(n,R)$, where R is a commutative ring such that $2 \in R$ is invertible. In particular, we compute their derived series.

Let $(G_{[i]})_{i\geq 0}$ be the derived series of $\mathrm{UT}(n,R)$. We want to show that $G_{[i]} = G_{2^i}$ for all $i \geq 1$, where $(G_j)_{j\geq 1}$ is the lower central series of $\mathrm{UT}(n,R)$, that we computed in Section 2.4.

We have already seen the inclusions $G_{[i]} \subseteq G_{2^i}$. We need to show the other inclusions $G_{[i]} \supseteq G_{2^i}$.

We introduce some notation: for $j = 1,2,\dots,n-1$ we denote by $N_j^{(n)}$ the $n \times n$ matrix with 1 in the positions $(i,i+j)$ for $i = 1,2,\dots,n-j$ and 0 elsewhere. Then observe that for $m > a \geq 1$, $[I^{(m)} + N_a^{(m)}, I^{(m)} + E_{a+1,m}^{(m)}] = I^{(m)} + E_{1,m}^{(m)}$. In particular for $m \geq 3$ we can use this observation and block diagonal matrices with one block equal to $I^{(m)} + N_a^{(m)}$ or $I^{(m)} + E_{a+1,m}^{(m)}$ and the others equal to 1, to show the inclusions $G_{[i]} \supseteq G_{2^i}$.

For example, in Section 2.4, we already used this observation in the case $a = 1$ and $3 \leq m \leq n$ to show that $G_{[1]} \supseteq [\mathrm{UT}(n,R),\mathrm{UT}(n,R)] = G_2$.

This computation in particular shows that the class of solvability of $\mathrm{UT}(n,R)$ is $\lceil \log_2 n \rceil$, where for $x \in \mathbb{R}$, $\lceil x \rceil$ indicates the minimal integer $\geq x$.

Notice that the derived series of $\mathrm{B}(n,R)$ is just the one of $\mathrm{UT}(n,R)$ shifted by one, since we already observed that $[\mathrm{B}(n,R),\mathrm{B}(n,R)] = \mathrm{UT}(n,R)$. In particular, the class of solvability of $\mathrm{B}(n,R)$ is $\lceil \log_2 n \rceil + 1$.

4.3 Statement of Malcev's Theorem on Solvable Groups

The main goal of this chapter is to prove the following theorem, which is due to Malcev.

Theorem 4.16 (Malcev). *Let $\mathbb{K}$ be an algebraically closed field, and let $G \subseteq \mathrm{GL}(n,\mathbb{K})$ be a finitely generated solvable group. Then there exists an $H \trianglelefteq G$ of finite index which is* triangularizable, *i.e. there exists an $x \in \mathrm{GL}(n,\mathbb{K})$ such that $x^{-1}Hx \subseteq B(n,\mathbb{K})$. Also, there exists a function $f : \mathbb{N} \to \mathbb{N}$ such that $[G : H] \leq f(n)$, i.e. the index has a uniform upper bound which depends only on n, and not on G nor on $\mathbb{K}$.*

Remark 4.17. Notice that Malcev's theorem is "tight": let G be a finite solvable group, with $[G,G]$ not nilpotent (we saw examples of such groups in Section 4.1). Let $\mathbb{K}$ be an arbitrary field and let $n = |G|$. Then $G \hookrightarrow \mathrm{GL}(n,\mathbb{K})$: this is because the map $g \mapsto R_g$, where R_g is the right multiplication by $g \in G$, from G to $\mathrm{GL}(\mathbb{K}G) \cong \mathrm{GL}(n,\mathbb{K})$ is an embedding. But its image is not triangularizable, since $[B(n,\mathbb{K}),B(n,\mathbb{K})]$ is not nilpotent (cf. Example 4.3.(a)).

Notice that in the context of Lie algebras we have the following (cf. [180, Chapter 2, Theorem 4.1 and Corollary A]):

Theorem 4.18 (Lie). *Let $L \subseteq M_n(\mathbb{K})$ be a solvable Lie algebra of $n \times n$ matrices, where $\mathbb{K}$ is an algebraically closed field of characteristic 0. Then there exists an $x \in \mathrm{GL}(n,\mathbb{K})$ such that $x^{-1}Lx \subseteq B(n,\mathbb{K})$.*

Thus, for Lie algebras we can triangularize the whole algebra, but this fails to hold in the group setting.

4.4 Wedderburn Theory

We recall some basic results of Wedderburn theory. For more details, the reader is referred to [181, Chapter IX].

We start with the structural results.

Let $\mathbb{K}$ be a field and let A be a finite-dimensional associative $\mathbb{K}$-algebra. One says that A is *nilpotent* if there exists an $n \geq 1$ such that $A^n = \underbrace{AA\cdots A}_{n} = \{0\}$ (here of course A is not unital).

Example 4.19. Let $\mathbb{K}$ be a field and let $n \geq 2$. A matrix $\|a_{ij}\|_{i,j=1}^n$ in $M_n(\mathbb{K})$ is called *strictly upper triangular* provided that $a_{ij} = 0$ for all $i,j = 1,2,\ldots,n$ such that $j \leq i$. Then the algebra A consisting of all strictly upper triangular $n \times n$ matrices with coefficients in $\mathbb{K}$ is nilpotent.

The following theorem of Wedderburn is fundamental. A proof can be found in [181, Chapter IX].

Recall that a *division algebra* D over a field $\mathbb{K}$ is a $\mathbb{K}$-algebra such that every nonzero element is invertible.

Theorem 4.20 (Wedderburn). *Let A be a finite-dimensional $\mathbb{K}$-algebra. Then there exists a nilpotent ideal $N \trianglelefteq A$, $r \in \mathbb{N}$, positive integers n_i, and division algebras D_i, $i = 1,2,\ldots,r$, such that*

$$A/N \cong M_{n_1}(D_1) \oplus M_{n_2}(D_2) \oplus \cdots \oplus M_{n_r}(D_r).$$

Theorem 4.21. *Let $\mathbb{K}$ be an algebraically closed field and let D be a finite-dimensional division $\mathbb{K}$-algebra. Then $D = \mathbb{K}$.*

Proof. The inclusion $\mathbb{K} \subseteq D$ is obvious. To show the converse, let $x \in D$. Then $\mathrm{span}_{\mathbb{K}}\{1, x, x^2, x^3, \dots\} \subseteq D$ is finite-dimensional over $\mathbb{K}$, so we may assume that it equals $\mathrm{span}_{\mathbb{K}}\{1, x, x^2, x^3, \dots, x^{r-1}\}$ for some $r \geq 1$. Therefore, we can find constants $a_0, a_1, \dots, a_{r-1} \in \mathbb{K}$ such that $x^r = a_0 + a_1 x + a_2 x^2 + \cdots + a_{r-1} x^{r-1}$. Denoting by $P(t)$ the monic polynomial $t^r - a_{r-1} t^{r-1} - \cdots - a_1 t - a_0$ we thus have $P(x) = 0$. Since $\mathbb{K}$ is algebraically closed, we can find $\alpha_1, \alpha_2, \dots, \alpha_r \in \mathbb{K}$ such that $P(t) = (t - \alpha_1)(t - \alpha_2) \cdots (t - \alpha_r)$. Since D is a division algebra and therefore has no zero-divisors, we deduce that $x = \alpha_j \in \mathbb{K}$ for some $1 \leq j \leq n$. This shows that $D \subseteq \mathbb{K}$ and the equality follows. □

As a consequence, in Wedderburn's theorem (Theorem 4.20), if $\mathbb{K}$ is algebraically closed, then all the D_i's will be equal to $\mathbb{K}$.

We now look at the module theory.

Definition 4.22. Let A be a unital associative algebra. Let V be a left A-module (briefly, a module) which is *unital*, i.e. $1_A \cdot v = v$ for all $v \in V$. One says that V is *irreducible* if it contains no nontrivial proper submodules. V is said to be *completely reducible* if one of the following equivalent conditions (**exercise**) holds:

(1) for every submodule $V' \leq V$ there exists a submodule $V'' \leq V$ such that $V = V' \oplus V''$ (every submodule is *complemented*);
(2) V is a direct sum of irreducible submodules;
(3) V is sum of irreducible submodules.

The proof of the following proposition is left as an **exercise**.

Proposition 4.23. *Let A be a unital associative algebra. Then the following holds.*

(1) *A homomorphic image of a completely reducible module is completely reducible.*
(2) *A submodule of a completely reducible module is completely reducible.*
(3) *If $V = \sum_i V_i$ is a submodule and each of the V_i's is completely reducible, then so is V.*

Definition 4.24. Let A be a finite-dimensional algebra over a field $\mathbb{K}$. The *radical* of A, denoted by $N(A)$, is the maximal nilpotent ideal of A. If $N(A) = \{0\}$, then the algebra A is said to be *semisimple*.

Remark 4.25. Observe that if N_1 and N_2 are two nilpotent ideals of A, then their sum $N := N_1 + N_2$ (which is trivially an ideal) is nilpotent as well. For, if n_1 (resp. n_2) is a positive integer such that $(N_1)^{n_1} = \{0\}$ (resp. $(N_2)^{n_2} = \{0\}$), then for $n := n_1 + n_2$ we have $N^n = \{0\}$ by the binomial expansion identity. So, given any family $(N_i)_{i \in I}$ of (nilpotent) ideals, there exist $i_1, i_2, \dots, i_k \in I$ such that $\sum_{i \in I} N_i = N_{i_1} + N_{i_2} + \cdots + N_{i_k}$.

This shows that the radical $N(A)$ equals the sum of all nilpotent ideals of A, proving that it is well defined and unique.

We observe also that for $\varphi \in Aut(A)$, we have $N(A)^{\varphi} = N(A)$.

Notice that by Wedderburn's theorem (Theorem 4.20) a semisimple $\mathbb{K}$-algebra A is a direct sum of full matrix algebras over division algebras. Moreover, for $\mathbb{K}$ algebraically closed, the division algebras are all equal to $\mathbb{K}$.

Example 4.26. Let D be a division algebra over a field $\mathbb{K}$. Also let n be a positive integer, and fix $1 \le i \le n$. In $M_n(D)$ we consider the subset M of matrices of the form $\begin{pmatrix} & * & \\ 0 & \vdots & 0 \\ & * & \end{pmatrix}$, i.e. the set of all matrices whose nonzero entries only occur in the i-th column. This is an irreducible module over $M_n(D)$ (**exercise**).

The following proposition is easily derived from the previous example (**exercise**).

Proposition 4.27. *Let A be a semisimple algebra. Then A, viewed as a module over itself, is completely reducible.*

Theorem 4.28. *If A is a semisimple finite-dimensional algebra, then every A-module is completely reducible.*

Proof. Let V be an A-module. Clearly $V = \sum_{0 \neq v \in V} Av$. Now Av is a homomorphic image of A, viewed as a module over itself, under the map $a \mapsto av$ for $a \in A$ and $v \in V$. It follows from Proposition 4.27 and Proposition 4.23 that V is completely reducible. □

Proposition 4.29. *Let V be a finite-dimensional vector space over $\mathbb{K}$, and let $A \subseteq \mathrm{End}_{\mathbb{K}}(V)$ be a subalgebra. Then V as a left A-module is completely reducible if and only if A is semisimple.*

Proof. If A is semisimple, then it follows from Theorem 4.28 that V is completely reducible.

Conversely, suppose that V is completely reducible. Consider the radical ideal $N := N(A)$ and suppose, by contradiction, that it is not trivial. Let $s \in \mathbb{N}$, $s \geq 2$ be such that $N^s V = \{0\}$ but $N^{s-1} V \neq \{0\}$. By hypothesis, the submodule NV of V is complemented, i.e. there exists a submodule $V' \leq V$ such that $V = NV \oplus V'$. Now $NV' \subseteq NV \cap V' = \{0\}$. Therefore $N^{s-1}V = N^{s-1}(NV \oplus V') = \{0\}$, a contradiction. Hence $N = \{0\}$, i.e. A is semisimple. □

We are interested in the following application of Wedderburn theory.

Theorem 4.30 (Clifford). *Let G be a group and let $H \trianglelefteq G$. Let V be a finite-dimensional completely reducible module over G. Then V is a completely reducible module over H.*

Proof. Since we are interested only in the image of $\mathbb{K}G$ in $\mathrm{End}_{\mathbb{K}}(V)$, we can assume $\mathbb{K}G \subseteq \mathrm{End}_{\mathbb{K}}(V)$, so that $\mathbb{K}G$ is semisimple by Proposition 4.29. Let $N := N(\mathbb{K}H)$ be the radical of the subalgebra $\mathbb{K}H$ of $\mathbb{K}G$. To prove our statement, by Theorem 4.28, it suffices to show that N is trivial.

Suppose by contradiction that $N \neq \{0\}$. Also let s be a positive integer such that $N^s = \{0\}$. Observe that N is not an ideal of $\mathbb{K}G$ in general, but $N' := (\mathbb{K}G)N(\mathbb{K}G)$ is. Let us show that

$$G(NG)^s G = \underbrace{GNGNG \cdots NG}_{N \text{ appears } s \text{ times}} = \{0\}. \tag{4.2}$$

This implies $(N')^s = \{0\}$, that is, N' would be a nilpotent ideal of $\mathbb{K}G$, contradicting the semisimplicity of $\mathbb{K}G$. To prove (4.2) we show that $G(NG)^t G \subseteq GN^t$ for all $t \in \mathbb{N}$. We proceed by induction on t. For $t = 0$ we have $G(NG)^0 G = G^2 \subseteq G = GN^0$. Suppose that the statement is true for an integer $t \geq 0$ and let us show it for $t+1$. We have

$$G(NG)^{t+1}G = GN\left(G(NG)^t G\right) \subseteq GN(GN^t) \subseteq GNN^t = GN^{t+1},$$

where the last inclusion comes from the normality of N: $g_1 n g_2 = g_1 g_2 n^{g_2} \in GN$ for all $g_1, g_2 \in G$ and $n \in N$. Taking $t = s$ and recalling that $N^s = \{0\}$, we deduce (4.2). □

4.5 Proof of Malcev's Theorem on Solvable Groups

We are now ready to start the proof of Malcev's theorem.

Suppose, by contradiction, that the theorem is not true. Let n be the minimal integer for which there is a counterexample, namely a subgroup $G \subseteq \mathrm{GL}(n, \mathbb{K})$ which admits no triangularizable normal subgroup of finite index.

Claim. *We can assume that G acts irreducibly on $V := \mathbb{K}^n$ (we are identifying $\mathrm{GL}(n, \mathbb{K})$ with $\mathrm{GL}(\mathbb{K}^n)$).*

If G does not act irreducibly on V, consider a sequence of submodules

$$\{0\} = V_0 \leq V_1 \leq \cdots \leq V_r = V,$$

such that G acts irreducibly on the quotients V_{i+1}/V_i for $i = 0, 1, \ldots, r-1$. Since $\dim_{\mathbb{K}}(V_{i+1}/V_i) \lneqq n$, for each i there exists a finite index normal subgroup $H_i \trianglelefteq G$ which is triangularizable. More precisely, we can find vectors $v_1^{i+1}, v_2^{i+1}, \ldots, v_{d_{i+1}}^{i+1} \in V_{i+1}$ such that $d_{i+1} = \dim(V_{i+1}/V_i)$ and $V_{i+1} = \oplus_{j=1}^{d_{i+1}} (\mathbb{K}v_j^{i+1} + V_i)$ satisfying, for all $h \in H_i$

$$\begin{cases} hv_1^{i+1} \in \mathbb{K}v_1^{i+1} + V_i \\ hv_2^{i+1} \in \mathbb{K}v_2^{i+1} + \mathbb{K}v_1^{i+1} + V_i \\ \vdots \\ hv_{d_{i+1}}^{i+1} \in \mathbb{K}v_{d_{i+1}}^{i+1} + \mathbb{K}v_{d_{i+1}-1}^{i+1} + \cdots + \mathbb{K}v_1^{i+1} + V_i. \end{cases}$$

But then the intersection $H = \cap_{i=1}^r H_i$ is a finite index normal subgroup of G, which is triangularizable: the basis which triangularizes it is $\{v_j^{i+1} : 1 \leq j \leq d_{i+1}, 0 \leq i \leq r-1\}$. This finishes the proof of the claim.

Hence, from now on we assume that G acts irreducibly on V.

The following remark will be useful.

Remark 4.31. If H is any normal subgroup of finite index of G and G acts irreducibly on V, then H also acts irreducibly on V.

To see this, suppose by contradiction that this is not the case and let $V' \leq V$ be a nontrivial proper H-submodule. Now V, being G-irreducible, is completely reducible for G and therefore, by Clifford's theorem (Theorem 4.30), it is completely reducible for H as well. Thus, we can find an H-submodule $V'' \leq V$ such that $V = V' \oplus V''$. Proceeding in this way, since $\dim_{\mathbb{K}}(V')$ and $\dim_{\mathbb{K}}(V'')$ are both strictly less than $\dim_{\mathbb{K}}(V)$, by applying the same argument in the proof of the above claim, we get a (finite index normal) subgroup $\widetilde{H}$ of H which acts irreducibly. But then H acts irreducibly, since it contains $\widetilde{H}$: a contradiction.

The key idea is in the following lemma.

Lemma 4.32. *Let $H \trianglelefteq G$, and suppose that $[H,H]$ is virtually scalar. Then H is virtually scalar.*

Proof that Lemma 4.32 implies Malcev's theorem (Theorem 4.16). Consider the derived series $G = G_{[0]} \geq G_{[1]} \geq \cdots \geq G_{[r]} \geq \{1_G\}$. Let s be maximal such that $H := G_{[s]}$ is not virtually scalar. Then $G_{[s+1]} = [H,H]$ is virtually scalar, and the lemma implies that $H = G_{[s]}$ is virtually scalar: a contradiction.

Hence G is virtually scalar (and therefore virtually triangularizable), but this contradicts the way we chose G. ☐

Proof of Lemma 4.32. We start with the following observation.

Claim. *H is virtually triangularizable.*

Suppose not. Then H is again a counterexample with the same n as for G. Hence we can take $G := H$, and assume that $[G,G]$ is virtually scalar. Let Z denote a scalar subgroup of finite index in $[G,G]$ and choose a set T of representatives for the right cosets of Z in $[G,G]$ so that $[G,G] = \cup_{t \in T} tZ$.

Fix $a \in [G,G]$. For every $g \in G$ we find unique elements $t = t(a,g) \in T$ and $z = z(a,g) \in Z$ such that $a^g = a[a,g] = atz$. For $z \in Z$, let $\mu_z \in \mathbb{K} \setminus \{0\}$ be the unique coefficient such that $z = \mu_z \mathrm{id}_V$.

Since we are assuming $\mathbb{K}$ is algebraically closed, we can fix an eigenvalue $\lambda \in \mathbb{K} \setminus \{0\}$ of a. Notice that for any $g \in G$ the eigenvalues of a and a^g are the same. So we can find $v \in V$ (which depends on g) such that $\mu_z(at)v = (atz)v = a^g v = \lambda v$. So λ / μ_z is an eigenvalue of at. But, since T is finite, as g varies in G we have finitely many at's, and hence finitely many possibilities for the quantity μ_z. Therefore z assumes only finitely many values. It follows that $a^G := \{a^g \mid g \in G\}$ is finite.

Since G acts on a^G by conjugation, we have a homomorphism $G \to \mathrm{Sym}(a^G)$, whose kernel P_a is the centralizer of a^G, and clearly $[G : P_a] < \infty$. It follows that the centralizer P of $[G,G]$ in G equals the intersection $\cap_{t \in T} P_t$ and therefore it is of finite index in G, since T is finite. Moreover $[P,[G,G]] = \{1_G\}$, which implies, in particular, that $[P,[P,P]] = \{1_G\}$. Hence P is nilpotent. Since G is finitely generated and P is of finite index in G, P is also finitely generated. Therefore by Lemma 2.42 and Theorem 2.20, P is virtually triangularizable. It follows that G is also virtually triangularizable. This ends the proof of the claim.

Let U denote a finite index normal subgroup of H which is triangularizable. Consider $[U,U] \trianglelefteq U \trianglelefteq H \trianglelefteq G$. By three applications of Clifford's theorem (Theorem 4.30), since G acts irreducibly, and hence completely reducibly, we deduce that $[U,U]$ also acts completely reducibly. But, up to a change of basis, $[U,U] \subseteq \mathrm{UT}(n,\mathbb{K})$. It is an **exercise** to show that any subgroup of $\mathrm{UT}(n,\mathbb{K})$ that acts completely reducibly must be trivial. Hence $[U,U] = \{1_G\}$, i.e. U is abelian. It follows that H is virtually abelian.

We need another lemma.

Lemma 4.33. *Let G be a group and suppose that it contains an abelian normal subgroup H of finite index. Then G contains an abelian characteristic subgroup A of finite index.*

Proof. Let T be a set of representatives for the left cosets of H in G so that $G = \sqcup_{t\in T} Ht$ and such that $1_G \in T$. Also let $\widetilde{H}$ denote the (characteristic) subgroup of G generated by $\{\varphi(H) : \varphi \in \mathrm{Aut}(G)\}$ and let us set $A := Z(\widetilde{H})$. Note that A is abelian. Moreover, since the center of a characteristic group is characteristic, A is characteristic as well. It remains to show that it is of finite index in G. We first claim that there exists a $T' \subseteq T$ such that $\widetilde{H} = \sqcup_{t'\in T'} Ht'$. First observe that since $\mathrm{id}_G \in \mathrm{Aut}(G)$ gives $\mathrm{id}_G(H) = H$, we have $H \leq \widetilde{H}$. Observe that if $\varphi(H) \subseteq H$ for all $\varphi \in \mathrm{Aut}(G)$, then $\widetilde{H} = H$ and we take $T' = \{1_G\}$. On the other hand, if there exist $h' \in H$ and $\varphi' \in \mathrm{Aut}(G)$ such that $\varphi'(h') \notin H$, then there exist unique $h \in H$ and $t' \in T \setminus \{1_G\}$ such that $\varphi'(h') = ht'$. Since $h \in H \subseteq \widetilde{H}$, it follows that $t' \in \widetilde{H}$ and therefore $Ht' \subseteq \widetilde{H}$. This proves the claim.

Note that $|T'| \leq |T| = [G:H] < \infty$.

It follows from the preceding argument that we can find $\varphi_{t'} \in \mathrm{Aut}(G)$, $t' \in T'$ such that $\widetilde{H}$ is generated by $\{\varphi_{t'}(H) : t' \in T'\}$. Now, since H is abelian, we have that the subgroups $\varphi_{t'}(H)$ are abelian for all $t' \in T'$. As a consequence, $\bigcap_{t'\in T'} \varphi_{t'}(H) \subseteq Z(\widetilde{H}) = A$ and we have

$$[G:A] \leq [G:\bigcap_{t'\in T'} \varphi_{t'}(H)] \leq |G:H|^{|T'|} < \infty$$

(note that $[G:\varphi(H)] = [G:H]$ for all $\varphi \in \mathrm{Aut}(G)$). □

Applying Lemma 4.33 to our H we find an Abelian subgroup $A \subseteq H$ which is characteristic and of finite index in H, and hence normal in G. We want to show that H' is in fact scalar, concluding the proof of Lemma 4.32.

Notice that since G acts irreducibly and A is Abelian, we have that A is diagonalizable (**exercise**).

Let $h \in A$ and let us show that h is a scalar. For every $g \in G$, we have that $h^g = g^{-1}hg$ has the same eigenvalues of h. Since A is normal in G, $h^G := \{h^g : g \in G\}$ is contained in A. Moreover, $|h^G| < \infty$ since h^G consists of diagonal matrices with the same eigenvalues. Now G acts on h^G by conjugation and this yields a homomorphism $G \to \mathrm{Sym}(h^G)$. If we denote by P the kernel of this homomorphism (P is just the centralizer of h^G), then clearly $[G:P] < \infty$. It follows from Remark 4.31 that P acts irreducibly on V. Since A is abelian, $h \in P$. From the fact that each element of P fixes in particular h, we deduce that h is indeed in $Z(P)$. Then Schur's lemma (Lemma 2.23) implies that h is a scalar. This shows that A is scalar, and hence H is virtually scalar, completing the proof of the Lemma 4.32. □

The proof of Malcev's theorem (Theorem 4.16) is complete.

4.6 Notes

The term “solvable” is related to the attempt to find a formula for the roots of a polynomial

$$p(x) = a_n x^n + a_{n-1} x^{n-1} + \cdots + a_1 x + a_0 \in \mathbb{C}[x]$$

involving the operations of addition, subtraction, multiplication, division, and root extractions, in terms of the coefficients $a_0, a_1, \ldots, a_n$. If the roots of $p(x)$ can be obtained by a formula of this kind, one says that the polynomial equation $p(x) = 0$ is *solvable by radicals*.

For instance, quadratic ($n = 2$) polynomial equations are solvable by radicals: the two roots are expressed by the well-known formula

$$x_{\pm} = \frac{-a_1 \pm \sqrt{a_1^2 - 4a_2 a_0}}{2a_2}.$$

Solvability of the cubic ($n = 3$) and quartic ($n = 4$) polynomial equations by radicals goes back to 15th–16th century mathematicians Scipione del Ferro, Gerolamo Cardano, Niccolò Tartaglia, Rafael Bombelli, and Lodovico Ferrari. In contrast, the *Abel–Ruffini theorem* (named after Paolo Ruffini, who made an incomplete proof in 1799, and Niels Henrik Abel, who provided a proof in 1824) states that the general polynomial equations of degree five or higher are not solvable by radicals. Nowadays, there is an elegant proof of this fundamental result, based on *Galois theory* (Evariste Galois, in the early 19th Century, used permutation groups to describe how the various roots of a given polynomial equation are related to each other. The modern approach, developed by Richard Dedekind, Leopold Kronecker, and Emil Artin, among others, involves the study of automorphism groups of field extensions). Recall that the *Fundamental Theorem of Algebra* states that every polynomial $p(x) \in \mathbb{C}[x]$ of degree $n \geq 0$ has exactly n roots in $\mathbb{C}$ (each counted with its own multiplicity). Denote by $F \subseteq \mathbb{C}$ the smallest field containing all the coefficients $a_0, a_1, \ldots, a_n$ of a given polynomial $p(x) \in \mathbb{C}[x]$. Let $E \subseteq \mathbb{C}$ denote the smallest field containing F and the roots of $p(x)$: this is called the *splitting field* of the polynomial $p(x)$. Let $\mathrm{Aut}(E)$ denote the group of all (field) automorphisms of E and by $\mathrm{Gal}(E,F) = \{\alpha \in \mathrm{Aut}(E) : \alpha(f) = f \text{ for all } f \in F\}$ the subgroup of $\mathrm{Aut}(E)$ fixing (element-wise) all elements in F (this is called the *Galois group* of the extension $F \subseteq E$, or of the polynomial $p(x)$).

The *inverse Galois problem*, posed in the early 19th century and still unsolved, asks whether or not every finite group appears as the Galois group of some Galois extension of the rational numbers $\mathbb{Q}$. More generally, given a finite group G and a field F, one may ask whether or not there is a Galois extension field E such that the Galois group $\mathrm{Gal}(E,F)$ is isomorphic to G: if the answer is positive, one says that G is *realizable* over F. A consistent and reasonably complete survey of the inverse Galois problem is the monograph by Gunter Malle and Heinrich Berndt Matzat [230]. For the original case $F = \mathbb{Q}$, we refer to Serre’s Bourbaki seminar [312]. We only mention that it is easy to show that, in this case, the answer is positive for G abelian; Igor R. Shafarevich [313] has extended this result for G solvable. For non-solvable groups, we mention that there exists a method (see [181, Exercise 14, Section 4, Chapter V]) for constructing a polynomial $p(x) \in \mathbb{Q}[x]$ with Galois group $S_n = \mathrm{Sym}(\{1,2,\ldots,n\})$, the symmetric group of degree n, for $n > 3$ (it is based on the fact that for any finite field F and any integer $n \geq 1$ there exists an irreducible polynomial of degree n in $F[x]$, see, e.g., [181, Corollary 5.9, Chapter V]).

Returning back to our discussion on the solvability of polynomial equations, Galois proved (essentially) the following remarkable result:

A polynomial equation $p(x) = 0$ is solvable by radicals if and only if the Galois group of $p(x)$ is solvable.

As a consequence, while all polynomial equations of degree $n \leq 4$ are solvable by radicals, not all polynomial equations of degree $n \geq 5$ are solvable by radicals, since the symmetric group S_n is not solvable for $n \geq 5$.

The celebrated *Feit–Thompson theorem*, proved by Walter Feit and John Thompson [110], states that every finite group of odd order is solvable.

Theorem 4.16 was proved by Anatoly I. Malcev in [227] (see also [228]).

Theorem 4.20 was proved by Joseph Wedderburn in [349] in 1908. This result was generalized by Emil Artin to (Artinian) semisimple rings in [7] in 1927.

Theorem 4.30 is due to William Kingdon Clifford.

4.7 Exercises

Exercise 4.1. Let G be a group. Let $(G_{[k]})_{k\geq 0}$ and $(\gamma_k(G))_{k\geq 1}$ denote the derived and the lower central series of G, respectively. Show (by induction) that $G_{[k]} \subseteq \gamma_{2^k}(G)$ for all integers $k \geq 0$.

Exercise 4.2. Prove Proposition 4.4.

Exercise 4.3. Let G and H be two groups and suppose that G acts on H.

(a) Show that with the multiplication defined in (4.1) the set $G \times H$ forms a group with neutral element $(1_G, 1_H)$.

(b) Show that $(g,h)^{-1} = (g^{-1}, (h^{g^{-1}})^{-1}$ for all $g \in G$ and $h \in H$.

(c) Show that the map $H \ni h \mapsto (1_G, h) \in G \ltimes H$ (resp. $G \ni g \mapsto (g, 1_H) \in G \ltimes H$) is an injective homomorphism.

(d) Using the identification of G and H with $G \times \{1_H\}$ and $\{1_G\} \times H$ respectively, provided by (c), show that if G and H are generated by the subsets X_G and X_H respectively, then $X = X_H \bigcup X_G$ generates $G \ltimes H$.

(e) Deduce from (d) that if G and H are finitely generated so is $G \ltimes H$.

(f) Show that we have the short exact sequence

$$\{1\} \to H \to G \ltimes H \to G \to \{1\}.$$

Exercise 4.4. (a) Let A be a solvable group and I an index set. Show that $A^I = \prod_{i\in I} A$ is solvable.

(b) Let $(A_i)_{i\in I}$ be a family of solvable groups. Show that, in general, if I is infinite, then $\prod_{i\in I} A_i$ is not solvable.

Exercise 4.5. Show that the group $(\mathbb{Z}/3\mathbb{Z}) \wr (\mathbb{Z}/2\mathbb{Z})$ is solvable, but not nilpotent.

Exercise 4.6. Show that the conditions (i), (ii) and (iii) in Definition 4.22 are equivalent.

Exercise 4.7. Prove Proposition 4.23.

Exercise 4.8. Let D be a division ring over a field $\mathbb{K}$. Also let n be a positive integer, and fix $1 \leq i \leq n$. In $M_n(D)$ we consider the subset M of matrices of the form $\begin{pmatrix} & * & \\ 0 & \vdots & 0 \\ & * & \end{pmatrix}$, i.e. the set of all matrices whose nonzero entries only occur in the i-th column. Show that this is an irreducible module over $M_n(D)$.

Exercise 4.9. Show that any subgroup of $\mathrm{UT}(n,\mathbb{K})$ that acts completely reducibly must be trivial.

Exercise 4.10. Let $\mathbb{K}$ be an algebraically closed field. Suppose that $G \subseteq GL(n,\mathbb{K})$ acts irreducibly and $A \subseteq G$ is an abelian normal subgroup. Show that A is diagonalizable.

Chapter 5
Polycyclic Groups

A group G is called polycyclic (resp. polycyclic-by-finite, resp. poly-infinite-cyclic) if it admits a subnormal series, that is, a sequence of subgroups $\{1_G\} = H_0 \leq H_1 \leq H_2 \leq \cdots \leq H_n = G$ such that H_i is normal in H_{i+1}, with H_{i+1}/H_i a (finite or infinite) cyclic (resp. (infinite) cyclic or finite, resp. infinite cyclic) group for each $0 \leq i \leq n-1$. Clearly, every polycyclic group is solvable. It follows from Corollary 2.29 that every finitely generated nilpotent group is polycyclic. In Section 5.2 we define the Hirsch number of a polycyclic group (the number of infinite factors in its subnormal series) and show that is well defined (Lemma 5.14). In the following sections we prove Malcev's theorem on polycyclic groups (every finitely generated solvable linear group over the integers is polycyclic, cf. Theorem 5.16), the main result of this chapter, as well as Malcev's theorem on polycyclic-by-finite groups (Theorem 5.21). Finally, in the last section we prove the Auslander–Swan theorem (Theorem 5.22) which constitutes a converse to Malcev's theorem on polycyclic groups.

5.1 Polycyclic, Polycyclic-by-Finite, and Poly-Infinite-Cyclic Groups

Definition 5.1. Let G be a group.

We say that a series

$$G = H_s \geq H_{s-1} \geq \cdots \geq H_1 \geq H_0 = \{1_G\} \tag{5.1}$$

of subgroups of G is *subnormal* if $H_i \trianglelefteq H_{i+1}$ for all $i = 1, 2, \ldots, s-1$. The quotients H_{i+1}/H_i are called the *factors* of the series.

If $\mathscr{P}$ is a property groups, we say that G is *poly-$\mathscr{P}$* if there exists a subnormal series as in (5.1) whose factors H_{i+1}/H_i have the property $\mathscr{P}$.

Hence, G is *polycyclic* if there exists a subnormal series in G whose factors are cyclic groups.

T. Ceccherini-Silberstein and M. D'Adderio, *Topics in Groups and Geometry*,
Springer Monographs in Mathematics, https://doi.org/10.1007/978-3-030-88109-2_5

Example 5.2. (1) Every finite solvable group is polycyclic (**exercise**);
(2) Every finitely generated nilpotent group is polycyclic (cf. Corollary 2.29).

Remark 5.3. Observe that every polycyclic group is solvable (**exercise**). The converse is not always true. Before showing an example, we introduce a slightly wider class of groups, which shares many properties with the class of polycyclic groups.

Definition 5.4. A group is *polycyclic-by-finite* if there exists a subnormal series whose factors are cyclic or finite.

It is clear from the definition that every polycyclic group is polycyclic-by-finite. The following result will be widely used in the sequel.

Proposition 5.5. *Let G be polycyclic-by-finite (e.g. polycyclic) group. Then every subgroup of G is finitely generated.*

Proof. Let

$$G = H_s \geq H_{s-1} \geq \cdots \geq H_1 \geq H_0 = \{1_G\}$$

be a subnormal series of G whose factors are cyclic or finite. First observe that the terms H_i, $i = 1, 2, \ldots, s-1$, are polycyclic-by-finite as well. Let H be a subgroup of G. We prove that H is finitely generated by induction on s. For $s = 0$ we have $H = H_0 = \{1_G\}$ and there is nothing to prove. By induction we have that $H \cap H_{s-1} \leq H_{s-1}$ is finitely generated. Now $G/H_{s-1} = H_s/H_{s-1}$ is either $\mathbb{Z}$ or finite; in both cases, it is finitely generated. It follows that $H/(H \cap H_{s-1}) \cong HH_{s-1}/H_{s-1} \leq H_s/H_{s-1}$ is also finitely generated. Since in the exact sequence

$$\{1\} \to H \cap H_{s-1} \to H \to H/(H \cap H_{s-1}) \to \{1\}$$

the lateral terms are finitely generated, we deduce from Proposition 1.12.(2) that the middle term H is finitely generated as well. □

Example 5.6. Let $G := \mathbb{Z} \wr \mathbb{Z} = \mathbb{Z} \ltimes \bigoplus_{j \in \mathbb{Z}} \mathbb{Z}$. This is finitely generated (cf. Proposition 4.13) and solvable (cf. Proposition 4.12). Now $[G, G] \supseteq \bigoplus_{j \in \mathbb{Z}} \mathbb{Z}$, and $\bigoplus_{j \in \mathbb{Z}} \mathbb{Z}$ is a subgroup which is not finitely generated. Thus G is a solvable group which is not polycyclic (by Proposition 5.5).

We think of polycyclic groups as the "good" solvable groups.
The proof of the following proposition is left as an **exercise**.

Proposition 5.7. *Let G be a polycyclic (resp. polycyclic-by-finite) group. Then its subgroups, homomorphic images, or extensions by polycyclic (resp. polycyclic-by-finite) groups, are polycyclic (resp. polycyclic-by-finite) as well.*

Remark 5.8. As we shall see later (cf. Theorem 6.12) the class of polycyclic-by-finite groups has the following property (called the *Burnside property*, see Chapter 6): if G is a finitely generated polycyclic-by-finite torsion group, then G is finite.

According to our definition, a group is *poly-infinite-cyclic* if it admits a subnormal series whose factors are all infinite cyclic (i.e. isomorphic to $\mathbb{Z}$).

Remark 5.9. Notice that every poly-infinite-cyclic group is both polycyclic and torsion free. However the converse is not true: for instance the group

$$\widetilde{G} := \langle x, y, z \mid x^z = x^{-1},\ y^z = y^{-1},\ [x,y] = z^4 \rangle$$

provides a counterexample. In fact, if a group G is poly-infinite-cyclic, then $G/[G,G]$ must be infinite: consider a subnormal series $G = H_s \geq H_{s-1} \geq \cdots$ whose factors are infinite cyclic; then $G/H_{s-1} \cong \mathbb{Z}$, hence $[G,G] \leq H_{s-1}$, so that $|G/[G,G]| \geq |G/H_{s-1}| = \infty$. But notice that $[x,z] = x^{-2}$, $[y,z] = y^{-2}$ and $[x,y] = z^4$ so that $\widetilde{G}/[\widetilde{G},\widetilde{G}]$ is a polycyclic group generated by finitely many torsion elements, hence it is finite by Remark 5.8. This shows that $\widetilde{G}$ cannot be poly-infinite-cyclic.

On the other hand we have the subnormal series

$$\widetilde{G} \geq \langle x, y, z^2 \rangle \geq \langle x, z^2 \rangle \geq \langle x \rangle \geq \{1_{\widetilde{G}}\}.$$

In fact z^2 is in the center of G. Now the first quotient has order two, while the other ones are isomorphic to $\mathbb{Z}$ (**exercise**). Moreover, G is torsion free: in fact any element of G raised to the power two is in $\langle x, y, z^2 \rangle$, which is torsion free.

To see that x and y have infinite order, just quotient by the normal closure of z^2, and observe that the resulting quotient is isomorphic to $\mathbb{Z}^2 \ltimes \mathbb{Z}/2\mathbb{Z}$.

The proof of the following proposition is left as an **exercise**.

Proposition 5.10. *Let G be a poly-infinite-cyclic group. Then every subgroup of G is poly-infinite-cyclic.*

The following lemma will be useful in the sequel.

Lemma 5.11. *Let G be an infinite polycyclic-by-finite group. Then there exists a normal subgroup H of finite index in G which is poly-infinite-cyclic.*

Proof. Let

$$G = H_s \geq H_{s-1} \geq \cdots \geq H_1 \geq H_0 = \{1_G\}$$

be a subnormal series of G whose factors are cyclic or finite. Observe that the terms H_i are also polycyclic-by-finite for $i = 1, 2, \ldots s-1$. We proceed by induction on s. For $s = 1$ we have $G = H_1 \cong H_1/H_0 \cong \mathbb{Z}$, since G is infinite. Thus G itself is poly-infinite-cyclic. By induction on s, suppose that the statement is true for H_{s-1}.

Thus, if H_{s-1} is infinite, we can find $N \trianglelefteq H_{s-1}$ with $[H_{s-1} : N] < \infty$ and N poly-infinite-cyclic, that is, admitting a subnormal series $N = N_r \geq N_{r-1} \geq \cdots \geq N_1 \geq N_0 = \{1_G\}$ with $N_{i+1}/N_i \cong \mathbb{Z}$ for $i = 0, 1, \ldots, r-1$.

If H_{s-1} is finite, we set $N := \{1_G\}$.

We distinguish two cases:

Case 1: $[G : H_{s-1}] < \infty$. In this case H_{s-1} must be infinite. By Lemma 2.41, we can find a characteristic subgroup H of H_{s-1} (and therefore normal in G) such that $H \subseteq N$ and $[H_{s-1}, H] < \infty$ (and therefore $[G : H] < \infty$). Since N is poly-infinite-cyclic, from Proposition 5.10 we deduce that H is poly-infinite-cyclic as well, and this concludes the proof in this case.

Case 2: $[G : H_{s-1}] = \infty$. In this case, we necessarily have $G/H_{s-1} \cong \mathbb{Z}$. Let $a \in G \setminus H_{s-1}$ such that $G/H_{s-1} = \langle aH_{s-1} \rangle$.

Then we set $\widetilde{H} := \langle N, a \rangle$. We claim that this is poly-infinite-cyclic. In fact we have

$$\widetilde{H} \geq N = N_r \geq \cdots \geq N_1 \geq \{1\}.$$

Now a is not torsion modulo N: in fact it is not torsion even modulo H_{s-1} and $N \subseteq H_{s-1}$. So $\widetilde{H}/N \cong \mathbb{Z}$. First we check that $[G : \widetilde{H}] < \infty$. Let $H_{s-1} = \cup_{i=1}^{n} g_i N$ and $G = \cup_{j\in\mathbb{Z}} H_{s-1} a^j$. Then

$$G = \bigcup_{j\in\mathbb{Z}} H_{s-1} a^j = \bigcup_{i=1}^{n} \bigcup_{j\in\mathbb{Z}} g_i N a^j \subseteq \bigcup_{i=1}^{n} g_i \widetilde{H},$$

hence $[G : \widetilde{H}] < \infty$.

In general, $\widetilde{H}$ may not be normal in G. But by Poincaré's lemma (Lemma 2.39) we have a normal subgroup $H \trianglelefteq G$ such that $H \leq \widetilde{H}$ and $[G : H] < \infty$. Also, H is still poly-infinite-cyclic by Proposition 5.10.

This concludes the proof of the theorem. □

5.2 The Hirsch Number

Definition 5.12. If G is polycyclic, we define the *Hirsch number* of G as the number of infinite factors in any subnormal series of G with cyclic factors.

To see that the Hirsch number is well defined, we need one more definition and a lemma.

Definition 5.13. Let G be a group and

$$G = A_n \geq A_{n-1} \geq \cdots \geq A_1 \geq A_0 = \{1_G\} \tag{5.2}$$

and

$$G = B_m \geq B_{m-1} \geq \cdots \geq B_1 \geq B_0 = \{1_G\} \tag{5.3}$$

two series of subgroups of G. We say that (5.3) is a *refinement* of (5.2) if every term of the series $(A_i)_{1\leq i\leq n}$ occurs in the series $(B_j)_{1\leq j\leq m}$.

We also say that (5.2) and (5.3) are *similar* if $n = m$ and $(A_i/A_{i-1})_{1\leq i\leq n}$ is a rearrangement of $(B_j/B_{j-1})_{1\leq j\leq n}$.

The following lemma clearly implies that the Hirsch number is well defined.

Lemma 5.14. *Let G be a group. Then any two subnormal series of G have a similar subnormal refinement.*

Proof. Let

$$G = A_n \geq A_{n-1} \geq \cdots \geq A_1 \geq A_0 = \{1_G\}$$

and

$$G = B_m \geq B_{m-1} \geq \cdots \geq B_1 \geq B_0 = \{1_G\}$$

be two subnormal series of G.

Remark 5.15. Let $K,L \leq G$. Recall that L *normalizes* K if $\ell^{-1}K\ell \subseteq K$ for every $\ell \in L$.

(a) Suppose that L normalizes K. Then $\ell k = \ell k \ell^{-1} \ell \in KL$, hence LK and $KLKL$ are contained in KL, so that KL is a subgroup of G.

(b) Suppose that $L_1, L_2 \leq G$ and $L = L_1 L_2$. Then L normalizes K if and only if both L_1 and L_2 normalize K.

(c) If $K \leq L$, we have that L normalizes K if and only if $K \trianglelefteq L$.

For every $i = 0, 1, \dots, n-1$ we consider the series

$$A_{i+1} = A_i(B_m \cap A_{i+1}) \geq A_i(B_{m-1} \cap A_{i+1}) \geq \cdots \geq A_i(B_1 \cap A_{i+1}) \geq A_i(B_0 \cap A_{i+1}) = A_i.$$

Note that since $A_{i+1} \trianglerighteq A_i$ we have that $(B_j \cap A_{i+1})$ normalizes A_i. Thus, by Remark 5.15.(a), all terms of the above series are indeed subgroups. Collecting together all these series (as i varies) we obtain a series Σ of subgroups of G.

Similarly, for every $j = 0, 1, \dots, m-1$ we consider the series

$$B_{j+1} = B_j(A_n \cap B_{j+1}) \geq B_j(A_{n-1} \cap B_{j+1}) \geq \cdots \geq B_j(A_0 \cap B_{j+1}) = B_j.$$

Collecting together all these series (as j varies) we obtain another series of subgroups of G which has the same length of Σ. We look at the factors.

Let us show that

$$A_i(B_{j+1} \cap A_{i+1}) \trianglerighteq A_i(B_j \cap A_{i+1}).$$

First observe that $(B_{j+1} \cap A_{i+1})$ normalizes both A_i and $B_j \cap A_{i+1}$. Also A_i normalizes $A_i(B_j \cap A_{i+1})$: indeed, since $B_j \cap A_{i+1}$ normalizes A_i, we have

$$a_i^{-1}(a_i' a_{i+1})a_i = a_i^{-1}(a_i'(a_{i+1} a_i a_{i+1}^{-1}))a_{i+1} \in A_i(B_j \cap A_{i+1})$$

for all $a_i, a_i' \in A_i$ and $a_{i+1} \in B_j \cap A_{i+1}$. This shows that $A_i(B_{j+1} \cap A_{i+1})$ normalizes its subgroup $A_i(B_j \cap A_{i+1})$ and, by Remark 5.15.(c), this is equivalent to the statement.

Similarly, we have $B_j(A_{i+1} \cap B_{j+1}) \trianglerighteq B_j(A_i \cap B_{j+1})$.

In order to complete the proof, we are only left to show that

$$\frac{A_i(B_{j+1} \cap A_{i+1})}{A_i(B_j \cap A_{i+1})} \cong \frac{B_j(A_{i+1} \cap B_{j+1})}{B_j(A_i \cap B_{j+1})}. \tag{5.4}$$

Set $A := B_{j+1} \cap A_{i+1}$ and $H := A_i(B_j \cap A_{i+1})$. We have $HA = A_i(B_{j+1} \cap A_{i+1})$ and $H \cap A = A_i(B_j \cap A_{i+1}) \cap B_{j+1} \cap A_{i+1}$. Let us show that

$$H \cap A = B_j(A_i \cap B_{j+1}) \cap (A_{i+1} \cap B_{j+1}). \tag{5.5}$$

Let $a_i \in A_i$ and $b_j \in B_j \cap A_{i+1}$ with $a_i b_j \in B_{j+1} \cap A_{i+1}$. Then we have $a_i b_j = b_j b_j^{-1} a_i b_j \in B_j(A_i \cap B_{j+1})$. Indeed, $a_i b_j \in B_{j+1}$, hence $b_j^{-1} a_i b_j \in B_{j+1}$; also $b_j \in A_{i+1}$, hence $b_j^{-1} a_i b_j \in A_i$. This shows that $H \cap A \subseteq B_j(A_i \cap B_{j+1}) \cap (A_{i+1} \cap B_{j+1})$. The other inclusion is proved in a similar way, and (5.5) follows.

Symmetrically, set $\widetilde{A} := A_{i+1} \cap B_{j+1}$ (notice that $\widetilde{A} = A$) and $\widetilde{H} := B_j(A_i \cap B_{j+1})$ we have $\widetilde{H}\widetilde{A} = B_j(A_{i+1} \cap B_{j+1})$ and (by the definitions)

$$\widetilde{H} \cap \widetilde{A} = B_j(A_i \cap B_{j+1}) \cap A_{i+1} \cap B_{j+1}. \tag{5.6}$$

We then deduce:

$$\begin{aligned}
\frac{A_i(B_{j+1} \cap A_{i+1})}{A_i(B_j \cap A_{i+1})} &= \frac{HA}{H} \\
&\cong \frac{A}{H \cap A} \\
\text{(by (5.5))} \quad &= \frac{B_{j+1} \cap A_{i+1}}{B_j(A_i \cap B_{j+1}) \cap A_{i+1} \cap B_{j+1}} \\
\text{(by (5.6))} \quad &= \frac{\widetilde{A}}{\widetilde{H} \cap \widetilde{A}} \\
&\cong \frac{\widetilde{H}\widetilde{A}}{\widetilde{H}} \\
&= \frac{B_j(A_{i+1} \cap B_{j+1})}{B_j(A_i \cap B_{j+1})}.
\end{aligned}$$

This shows (5.4) and completes the proof of the lemma. □

5.3 Malcev's Theorem on Polycyclic Groups

The main theorem of this chapter is again due to Malcev.

Theorem 5.16 (Malcev). *Let $G \leq \mathrm{GL}(n,\mathbb{Z})$ be finitely generated and solvable. Then G is polycyclic.*

In Theorem 5.22 we will see that, conversely, every polycyclic group is embeddable in $\mathrm{GL}(n,\mathbb{Z})$.

Definition 5.17. A finite extension K of $\mathbb{Q}$ is called a *field of algebraic numbers*.

A number $\alpha \in \mathbb{C}$ is called an *algebraic integer* if there exists a monic polynomial $f(t) \in \mathbb{Z}[t]$ such that $f(\alpha) = 0$.

Remark 5.18. Suppose that K is a field of algebraic numbers and that $\alpha \in K$ is an algebraic integer. Then $\sum_{i \geq 0} \alpha^i \mathbb{Z} \subseteq K$ is a finitely generated abelian group which is a subring.

For the proof of the following lemma we refer to [74, Section 4.9].

Lemma 5.19 (Dirichlet). *If α and β are algebraic integers, then so are $\alpha \pm \beta$ and $\alpha\beta$. The algebraic integers for a fixed field of algebraic numbers K form a subring R. If $\alpha \in K$, then there exists an $m \in \mathbb{Z}$ such that $m\alpha \in R$. Moreover, $(R,+)$ is finitely generated, and the multiplicative group of R is finitely generated as well.*

Proof of Theorem 5.16. Let $\overline{\mathbb{Q}}$ be the algebraic closure of $\mathbb{Q}$. Consider the inclusions $G \leq \mathrm{GL}(n,\mathbb{Z}) \leq \mathrm{GL}(n,\overline{\mathbb{Q}})$. By Theorem 4.16, G is virtually triangularizable over $\overline{\mathbb{Q}}$. Let $H \trianglelefteq G$ be such that $[G:H] < \infty$ and H is triangularizable.

By Proposition 5.7 it is sufficient to show that H is polycyclic, since we already know that G/H, being finite and solvable (cf. Proposition 4.4), is polycyclic (cf. Example 5.2).

Hence, without loss of generality, we can assume that G is triangularizable, that is, up to conjugation, we have $G \subseteq B(n, \overline{\mathbb{Q}})$. As G is finitely generated, we can find a finite generating subset $S \subseteq G$. Let $X \subseteq \overline{\mathbb{Q}}$ denote the (finite) set consisting of all matrix entries of the elements in S. Let K be the extension of $\mathbb{Q}$ generated by X, and let R be the subring of algebraic integers of K. By Dirichlet's lemma (Lemma 5.19), there exists an integer $m \geq 1$ such that $mX \subseteq R$. Consider the matrix

$$x := \begin{pmatrix} 1 & 0 & 0 & \dots & 0 \\ 0 & m & 0 & \dots & 0 \\ 0 & 0 & m^2 & \dots & 0 \\ \vdots & \vdots & \vdots & \ddots & \vdots \\ 0 & 0 & 0 & \dots & m^{n-1} \end{pmatrix}.$$

Then $x^{-1}Gx \subseteq B(n,K)$ and above the diagonal all the coefficients are in R. By the Cayley–Hamilton theorem the diagonal entries of each element g of $x^{-1}Gx \subseteq B(n,K)$ are roots of the characteristic polynomial $p_g(t)$ of g. Since $p_g(t)$ is invariant under conjugation of g, and G was originally in $\mathrm{GL}(n,\mathbb{Z})$, we have that indeed $p_g(t)$ has coefficients in $\mathbb{Z}$. As a consequence, the diagonal entries of g are algebraic integers. This shows that $x^{-1}Gx \subseteq B(n,R)$.

Claim. *The group $B(n,R)$ is polycyclic.*

Consider the group epimorphism

$$B(n,R) = \begin{pmatrix} R^\times & & R \\ & \ddots & \\ 0 & & R^\times \end{pmatrix} \to \begin{pmatrix} R^\times & & 0 \\ & \ddots & \\ 0 & & R^\times \end{pmatrix}$$

given by replacing by 0 all the entries out of the main diagonal. The image, being abelian and finitely generated, is polycyclic (cf. Example 5.2), hence, by Proposition 5.7, it is enough to show that the kernel, which is $\mathrm{UT}(n,R)$, is polycyclic.

Recall (cf. Example 2.15) that for $t = 1,2,\dots,n-1$, the t-th term $\gamma_t(\mathrm{UT}(n,R))$ of the lower central series is given by

$$\gamma_t(\mathrm{UT}(n,R)) := \{x = \|x_{i,j}\|_{i,j} \in \mathrm{UT}(n,R) : x_{i,j} = 0 \text{ for } 1 \leq j-i \leq t-1\}.$$

Each factor of the lower central series is a finitely generated (torsion-free) abelian group and therefore, by Proposition 5.7, polycyclic. By applying again Proposition 5.7, we deduce that $\mathrm{UT}(n,R)$, being an iterated extension of polycyclic groups, is polycyclic as well. This shows that the multiplicative group of $B(n,R)$ is polycyclic and finishes the proof of the claim.

We are now in position to complete the proof of Malcev's theorem. Indeed, since we already showed that $x^{-1}Gx \subseteq B(n,R)$, by virtue of Proposition 5.7 and the last claim we deduce that G itself is polycyclic, concluding the proof of the theorem. □

5.4 Malcev's Theorem on Polycyclic-by-Finite Groups

We start with a useful lemma.

Lemma 5.20. *If G is polycyclic-by-finite, then we have a normal series*

$$G \geq K_t \geq \cdots \geq K_1 \geq K_0 = \{1_G\}$$

such that G/K_t is finite and each K_i/K_{i-1} is free abelian.

Proof. Since G is polycyclic-by-finite, by Lemma 5.11 we can find a subnormal series

$$G = H_s \geq H_{s-1} \geq \cdots \geq H_1 \geq H_0 = \{1_G\}$$

whose factors H_i/H_{i-1} are infinite cyclic for $i = 1,2,\ldots,s-1$, and G/H_{s-1} is finite.

Since the factors H_i/H_{i-1}, $i = 1,2,\ldots,s-1$, are abelian, we have that H_{s-1} is solvable. Note that H_{s-1} is also torsion-free. Let K_1 denote the last but one term of its derived series. It is abelian, torsion-free and finitely generated by Proposition 5.5, equivalently, it is free abelian. Moreover, K_1 is characteristic in H_{s-1}, hence it is normal in G.

Then we repeat this procedure with G replaced by $\widetilde{G} := G/K_1$ (which is still polycyclic-by-finite by Proposition 5.7). We thus find a subnormal series

$$\widetilde{G} = \widetilde{H}_r \geq \widetilde{H}_{r-1} \geq \cdots \geq \widetilde{H}_1 \geq \widetilde{H}_0 = \{1_{\widetilde{G}}\},$$

where $r \leq s$, with $\widetilde{H}_i/\widetilde{H}_{i-1}$ infinite cyclic for $i = 1,2,\ldots,r-1$, and $[G : \widetilde{H}_{r-1}] < \infty$. Let $\widetilde{K}_1$ denote the last but one term of the derived series of the solvable group $\widetilde{H}_{r-1}$. As before, it is free abelian and we define the subgroup $K_2 \subseteq G$ by $\widetilde{K}_1 = K_2/K_1$. Again $\widetilde{K}_1 = K_2/K_1$ is normal in G/K_1, hence K_2 is normal in G.

Continuing this way, we construct a chain of subgroups of G

$$\{1_G\} \leq K_1 \leq K_2 \leq \cdots \leq G.$$

By construction each K_i is normal in G and the quotient K_i/K_{i-1} is free abelian. Clearly, the process stops in finitely many steps at some K_t which will be of finite index in G, so that the resulting series is the one that we were looking for. □

Theorem 5.21 (Malcev). *Let G be a polycyclic-by-finite group. Then G has a normal subgroup $N \subseteq G$ such that N is torsion-free nilpotent, and G/N is abelian-by-finite, i.e., there is an abelian normal subgroup of G/N of finite index.*

Proof. By Lemma 5.20 we have a normal series

$$G \geq K_t \geq \cdots \geq K_1 \geq K_0 = \{1_G\}$$

such that G/K_t is finite and each factor K_i/K_{i-1} is free abelian, say of rank n_i. Consider the centralizer $C_{K_t}(K_i/K_{i-1})$ of the action of K_t on K_i/K_{i-1} by conjugation. This gives an embedding of $K_t/C_{K_t}(K_i/K_{i-1})$ into $\mathrm{Aut}(K_i/K_{i-1}) \cong \mathrm{GL}(n_i,\mathbb{Z}) \subseteq \mathrm{GL}(n_i,\mathbb{C})$.

Notice that K_t is finitely generated, polycyclic and hence solvable. Therefore its quotient $K_t/C_{K_t}(K_i/K_{i-1}) \subseteq \mathrm{GL}(n_i,\mathbb{C})$ is also solvable and finitely generated. By Malcev's theorem on solvable linear groups (Theorem 4.16), $K_t/C_{K_t}(K_i/K_{i-1})$ has a triangularizable normal subgroup of finite index, whose preimage T_i in K_t is normal of finite index. Since in a suitable basis, T_i acts as a subgroup of $B(n,\mathbb{C})$, then in the same basis its commutator $[T_i,T_i]$ acts as a subgroup of $\mathrm{UT}(n,\mathbb{C})$. Hence the action of $[T_i,T_i]$ *stabilizes* a series in K_i/K_{i-1} of length n_i, i.e., we can find a series

$$K_i/K_{i-1} = L_{n_i} \geq L_{n_i-1} \geq \cdots \geq L_1 \geq L_0 = \{1\}$$

such that for all $x \in L_j$, $t \in [T_i,T_i]$, and $j = 1,2,\ldots,n_i$,

$$(xL_{j-1})^t = xL_{j-1}, \tag{5.7}$$

in particular $L_j^t = L_j$.

Now $T := T_1 \cap T_2 \cap \cdots \cap T_t \subseteq K_t$ is a torsion-free (since K_t is) normal subgroup of K_t of finite index and its commutator $H := [T,T]$ stabilizes a series in K_t.

Claim. *H is nilpotent.*

Let $(L_i)_{i=0}^m$ be the series stabilized by H. Hence in particular $L_0 = \{1\}$ and $L_m = K_t$. We will show that

$$[L_i, \gamma_j(H)] \leq L_{i-j} \tag{5.8}$$

for all $i \geq 1$ and $j \geq 1$.

If we can prove this, then for $i = j = m$ we have

$$\gamma_{m+1}(H) = [H, \gamma_m(H)] \leq [K_t, \gamma_m(H)] = [L_m, \gamma_m(H)] \leq L_0 = \{1\},$$

showing that H is nilpotent.

We prove (5.8) by induction on j. For $j = 1$, $\gamma_1(H) = H$, and this is the definition of stability (5.7): for $x \in L_i$, $\ell \in L_{i-1}$ and $t \in H$, $(x\ell)^t = x\widetilde{\ell}$ for some $\widetilde{\ell} \in L_{i-1}$, hence

$$\widetilde{\ell} = x^{-1}(x\ell)^t = x^{-1}t^{-1}x\ell t = (x^{-1}t^{-1}xt)t^{-1}\ell t = [x,t]\ell^t,$$

therefore

$$[x,t] = \widetilde{\ell}(\ell^t)^{-1} \in L_{i-1}.$$

Assume now that (5.8) is true for all i and some fixed $j \geq 1$. Then we have

$$[[H,L_i], \gamma_j(H)] \leq [L_{i-1}, \gamma_j(H)] \leq L_{i-j-1}$$

and

$$[[L_i, \gamma_j(H)], H] \leq [L_{i-j}, H] \leq L_{i-j-1}.$$

Therefore from the three subgroups lemma (Lemma 2.2) and the normality of L_{i-j-1} in K_t we deduce

$$[\gamma_{j+1}(H), L_i] = [[\gamma_j(H), H], L_i] \leq L_{i-j-1},$$

as we wanted. This ends the proof the claim.

Hence H is torsion-free (since T is) and nilpotent. By Proposition 5.5, H is also finitely generated, and it has finite index in K_t and therefore in G.

Using Poincaré's lemma (Lemma 2.39), we can find a subgroup $\widetilde{T}$ of T which is normal in G and of finite index. Let us set $N := [\widetilde{T}, \widetilde{T}] \subseteq [T,T] = H$. Now N is nilpotent, torsion-free, finitely generated and normal in G (since $\widetilde{T}$ is). Hence we have $\widetilde{T}/N \trianglelefteq G/N$, $[G/N : \widetilde{T}/N] = [G : \widetilde{T}] < \infty$ and $\widetilde{T}/N$ is abelian.

This finishes the proof of the theorem. □

5.5 The Auslander–Swan Theorem

The goal of this section is to prove the following theorem which constitutes a converse to Malcev's theorem on polycyclic groups (Theorem 5.16).

Theorem 5.22 (Auslander, Swan). *Let G be a polycyclic-by-finite group. Then G embeds into* $\mathrm{GL}(n,\mathbb{Z})$ *for some $n \geq 1$.*

Proof. By Malcev's theorem (Theorem 5.21), we can find a normal subgroup N of G which is torsion-free, nilpotent, and finitely generated, such that G/N is abelian-by-finite. The last condition means that there exists a finite index subgroup T of G such that $N \subseteq T$ and T/N is (finitely generated) abelian. So we can also find a finite index subgroup $\widetilde{T}$ of T such that $\widetilde{T}/N$ is (finitely generated) free abelian.

Applying Lemma 2.43, it is enough to show that $\widetilde{T}$ is linear. Hence, up to replacing G by $\widetilde{T}$, from now on we can assume that G/N is free abelian.

We argue by induction on the rank of the free abelian group G/N. If this rank is 0, then $G = N$, and by Malcev's theorem (Theorem 2.20) this is linear.

Suppose now that the rank r of G/N is greater than 0. Then we can choose a normal subgroup H of G such that $N \subseteq H$, $G/H \cong \mathbb{Z}$, and H/N is free abelian of rank $r-1$. Notice that in this case $[H,G] \subseteq [G,G] \subseteq N$.

By induction, there exists an embedding $\iota : H \hookrightarrow \mathrm{GL}(m,\mathbb{Z})$ for some integer $m \geq 1$. Let $a \in G$ be such that $G/H = \langle Ha \rangle \cong \mathbb{Z}$ and observe that $G = HA$, where $A := \langle a \rangle$ is infinite cyclic. We want to lift ι to an embedding of the whole group G into $\mathrm{GL}(n,\mathbb{Z})$ for a suitable $n \geq m$. We start by extending the homomorphism ι to a homomorphism of rings $\bar{\iota} : \mathbb{Z}H \to M_m(\mathbb{Z})$ ($\cong \mathbb{Z}^{m^2}$ as $\mathbb{Z}$-modules) in the obvious way, and we call $K \subseteq \mathbb{Z}H$ the kernel of $\bar{\iota}$.

We define a right action of G on $\mathbb{Z}H$ as follows. Since $G = HA$, any element $g \in G$ can be uniquely expressed as $g = hb$ with $h \in H$ and $b \in A$. Then, given $x = \sum_i \alpha_i h_i \in \mathbb{Z}H$ and $g = hb$ we set

$$x \cdot g = x \cdot hb := (xh)^b = \sum_i \alpha_i b^{-1} h_i hb.$$

In other words, H acts by right multiplication on $\mathbb{Z}H$, while A acts by conjugation.

We would like G to act on the quotient $\mathbb{Z}H/K$. If we could do so, then $\mathbb{Z}H/K$ would be a G-module. Unfortunately, it is not clear whether K is globally A-invariant so that the action of A passes to the quotient $\mathbb{Z}H/K$ or not. Hence we need to find a different quotient of this ring.

We observe that if $h \in H$ and $b \in A$ then

$$h \cdot b = h^b = h[h,b] = h + h([h,b] - 1). \tag{5.9}$$

Since $[H,G] \subseteq N$, equation (5.9) shows that h^b is equal to h modulo the left ideal I of $\mathbb{Z}H$ generated by the elements $x-1$ where $x \in N$.

Notice that if $h \in H$, then using the normality of N we have $h(x-1) = (x^{h^{-1}} - 1)h \in I$. Thus I is in fact a two sided ideal, and $K+I$ is now preserved by the action of A and hence of the whole G. Therefore $\mathbb{Z}H/(K+I)$ is a G-module. But we want a G-module which is faithful as an H-module.

Hence for any $k \geq 1$ we consider the ideal I^k: it is generated, as a left ideal, by the products $(x_1-1)(x_2-1)\cdots(x_k-1)$, where $x_i \in N$ for all $i = 1,2,\ldots,k$. Since N is unitriangularizable, these product vanish for $k \geq m$. Therefore $I^m \subseteq K = \ker(\overline{\iota})$, and so $(K+I)^m \subseteq K$. Moreover, if $b \in A$, from (5.9) we deduce that

$$(K+I)^m \cdot b = ((K+I)^m)^b = ((K+I)^b)^m \subseteq (K+I)^m.$$

Thus $(K+I)^m$ is a G-submodule of $\mathbb{Z}H$. Hence $\mathbb{Z}H/(K+I)^m$ is a G-module, and since $(K+I)^m \subseteq K$, it is faithful as an H-module.

Consider then the quotient $\mathbb{Z}H/(K+I)^m$. We are left with two problems: we need to check that this is finitely generated as a $\mathbb{Z}$-module, and we need to take care of the torsion.

Notice that $\mathbb{Z}H/(K+I)$ is a quotient of $\mathbb{Z}H/K$ which is finitely generated as a $\mathbb{Z}$-module (indeed $\mathbb{Z}H/K \hookrightarrow M_m(\mathbb{Z}) \cong \mathbb{Z}^{m^2}$). Hence $\mathbb{Z}H/(K+I)$ is finitely generated as a $\mathbb{Z}$-module. We need a lemma.

Lemma 5.23. *Let G be a finitely generated group, and let I a two-sided ideal of $\mathbb{Z}G$ such that $\mathbb{Z}G/I$ is finitely generated as a $\mathbb{Z}$-module. Then I is a finitely generated ideal and $\mathbb{Z}G/I^r$ is a finitely generated $\mathbb{Z}$-module for all $r \geq 1$.*

Before proving it, we show how to conclude the proof of the Auslander–Swan theorem assuming the lemma.

The lemma implies that the quotient $V := \mathbb{Z}H/(K+I)^m$ is finitely generated as a $\mathbb{Z}$-module.

We are left with the problem of the torsion. Let T be the torsion part of V, say $T = J/(K+I)^m$. Then T is fully invariant in V, i.e., it is invariant under all endomorphisms of V. Also, since $\mathbb{Z}H/K \hookrightarrow M_m(\mathbb{Z}) \cong \mathbb{Z}^{m^2}$, we have that $\mathbb{Z}H/K$ is torsion free as a $\mathbb{Z}$-module. Hence we necessarily have $T = J/(K+I)^m \subseteq K/(K+I)^m$. Thus $U := V/T \cong \mathbb{Z}H/J$ is a G-module (as it is a quotient of G-modules), and it is a free $\mathbb{Z}$-module of finite rank. Moreover, it is a faithful H-module since $J \subseteq K$ and $\mathbb{Z}H/K$ is H-faithful. However U is not necessarily faithful as a G-module.

To make it faithful, it is enough to add to it a free abelian group W of finite rank on which A acts faithfully and H trivially (we can choose A itself acting by multiplication), i.e., consider $U \oplus W$. This yields a faithful $\mathbb{Z}G$-module (**exercise**), and the proof of the theorem is complete. □

Proof of Lemma 5.23. Let $X \subseteq G$ be a finite and symmetric generating subset of G. Let also $u_1 := 1, u_2, \ldots, u_t \in \mathbb{Z}G$ generate $\mathbb{Z}G$ modulo I as a $\mathbb{Z}$-module and let M be the $\mathbb{Z}$-submodule of $\mathbb{Z}G$ generated by the u_i's and X. Thus, $I + M = \mathbb{Z}G$ as $\mathbb{Z}$-modules.

Therefore, given m_1 and m_2 in M, we have $m_1 m_2 = x + m$ with $x \in I$ and $m \in M$. In fact $x = m_1 m_2 - m \in M^2 + M$, so that indeed $x \in I \cap (M + M^2)$.

Let J be the ideal in $\mathbb{Z}G$ generated by $I \cap (M + M^2)$. Notice that $I \cap (M + M^2)$ is a $\mathbb{Z}$-submodule of the finitely generated $\mathbb{Z}$-module $M + M^2$, hence it is finitely generated as well. This implies that J is finitely generated as an ideal.

Also $J + M$ is, by construction, multiplicatively closed: in fact $(J + M)(J + M) \subseteq J^2 + MJ + JM + M^2 \subseteq J + M^2 \subseteq J + M$, where the last inclusion follows from the fact that for $m_1, m_2 \in M$ we have $m_1 m_2 = x + m \in I \cap (M + M^2) + M \subseteq J + M$. Moreover, $J + M$ contains generators of G, hence, being multiplicatively closed, it equals $\mathbb{Z}G$. We deduce that $\mathbb{Z}G/J = (J + M)/J \cong M/(J \cap M)$ is finitely generated as a $\mathbb{Z}$-module, since M is.

Note that, by construction, $J \subseteq I$. Now I/J is a $\mathbb{Z}$-submodule of the finitely generated $\mathbb{Z}$-module $\mathbb{Z}G/J$, hence it is finitely generated as a $\mathbb{Z}$-module as well.

All this implies that I is finitely generated as an ideal (because both J and I/J are). This shows the first part of the statement of the lemma.

Let then $v_1, v_2 \ldots, v_s \in I$ denote some generators of I as an ideal. Then the finite set $\{u_i v_j u_k : i, k = 1, 2, \ldots, t; j = 1, 2, \ldots, s\} \subseteq I$ generates I mod I^2 as a $\mathbb{Z}$-module (recall that $u_1 = 1$) (**exercise**).

Remark 5.24. A more conceptual way of seeing this is to observe that $\mathbb{Z}G/I$ acts both on the left and on the right on I/I^2, and the two actions commute, so that I/I^2 is a bimodule over $\mathbb{Z}G/I$, i.e., it is a (right) module over the ring

$$R := (\mathbb{Z}G/I)^\circ \otimes_{\mathbb{Z}} \mathbb{Z}G/I.$$

Here with $\otimes_{\mathbb{Z}}$ we denote the tensor product of $\mathbb{Z}$-modules, and for a ring S, we denote by S° the opposite ring, i.e., S° is equal to S as an abelian group, but the product "$*$" of two elements x and y is defined to be $x * y := y \cdot x$, where "$\cdot$" indicates the original product in S.

In fact, I/I^2 is finitely generated as an R-module: just pick the generators of I as an ideal. But since $\mathbb{Z}G/I$ is a finitely generated $\mathbb{Z}$-module, R is finitely generated too. So I/I^2 is a finitely generated $\mathbb{Z}$-module.

Since $\mathbb{Z}G/I \cong (\mathbb{Z}G/I^2)/(I/I^2)$, we have that $\mathbb{Z}G/I^2$ is an extension of finitely generated $\mathbb{Z}$-modules, hence it is a finitely generated $\mathbb{Z}$-module as well.

By induction, we conclude that $\mathbb{Z}G/I^r$ is a finitely generated $\mathbb{Z}$-module for every integer $r \geq 1$, completing the proof of the lemma. □

5.6 Notes

The study of polycyclic groups was initiated by Kurt August Hirsch in 1938 in [175, 176], where, in particular, it is shown that a group G is polycyclic if and only if it is solvable and each subgroup is finitely generated (G satisfies the maximal condition). Theorem 5.16 (every solvable group isomorphic to a matrix group over the integers is polycyclic) was proved by Anatoly I. Malcev in [228]. The theorem that every polycyclic-by-finite group is isomorphic to a matrix group over the integers was first proved by Louis Auslander [10] (the proof involves considerable knowledge of the theory of Lie groups) and later by Richard G. Swan [324] (a purely algebraic proof). It follows in particular that every polycyclic group is residually finite. This last result is due to Hirsch [177] (see also [293, p. 154]). For a detailed study of polycyclic groups we refer the interested reader to the book by Dan Segal [308] (see also the two volumes book by Derek J.S. Robinson [292]).

5.7 Exercises

Exercise 5.1. Show that every finite solvable group is polycyclic.

Exercise 5.2. Show that every polycyclic group is solvable.

Exercise 5.3. Let G be a group. (a) Show that if G is polycyclic, then every subgroup, homomorphic image, or extension by a polycyclic group of G is polycyclic.

(b) Show that if G is polycyclic-by-finite, then every subgroup, homomorphic image or extension by a polycyclic group is polycyclic-by-finite.

(c) Show that if G is polycyclic of special type, then every subgroup of G is polycyclic of special type.

Exercise 5.4. Show that the group G in the example considered in Remark 5.9, the two quotients $\langle x,y,z^2\rangle/\langle x,z^2\rangle$ and $\langle x,z^2\rangle/\langle x\rangle$ are both infinite cyclic.

Exercise 5.5. Show that every subgroup of a poly-infinite-cyclic group is poly-infinite-cyclic.

Exercise 5.6. Show that the module $U\oplus W$ at the end of the proof of Theorem 5.22 is a faithful $\mathbb{Z}G$-module.

Exercise 5.7. Show that the finite set $\{u_iv_ju_k : i,k=1,2,\ldots,t; j=1,2,\ldots,s\}\subseteq I$ in the proof of Lemma 5.23 generates I mod I^2 as a $\mathbb{Z}$-module.

Chapter 6
The Burnside Problem

In this chapter we discuss some instances of the Burnside Problem. There are three versions of this problem, the first one being the

General Burnside Problem: *Is it true that if a group G is finitely generated and torsion, then it is finite?*

We discuss the General Burnside problem for locally finite groups (Section 6.2), for polycyclic-by-finite and solvable groups (Section 6.3), as well as its bounded version for linear groups (Section 6.4). Finally, in Section 6.5 we discuss the Kurosh–Levitzky problem (on nil algebras) and explain the construction of Golod and Shafarevich yielding a negative answer to the Kurosh–Levitzky problem and therefore to the General Burnside Problem.

6.1 Formulation of the Burnside Problems

The *General Burnside Problem*, posed by William Burnside in 1902 [46] – one of the oldest and most influential questions in group theory – asks whether or not a finitely generated group in which every element has finite order is necessarily finite.

Problem 6.1 (General Burnside Problem). Is it true that if a group G is finitely generated and torsion, then it is finite?

Sometimes, the word *periodic* is used instead of "torsion". In order to approach the study of the General Burnside Problem, we introduce the following useful notion.

Definition 6.2 (Burnside property). A class $\mathscr{C}$ of groups satisfies the *Burnside property* if for every torsion group G in $\mathscr{C}$ the following holds: every finitely generated subgroup of G is finite.

A weaker formulation of the General Burnside Problem is the following. First recall that a group G is said to be *periodic with bounded exponent*, or just a group

T. Ceccherini-Silberstein and M. D'Adderio, *Topics in Groups and Geometry*,
Springer Monographs in Mathematics, https://doi.org/10.1007/978-3-030-88109-2_6

with *bounded exponent*, if there exists an integer $n \geq 1$ such that $g^n = 1_G$ for all $g \in G$.

Problem 6.3 (Bounded Burnside Problem). Is it true that if a group G is finitely generated and of bounded exponent, then G is finite?

There is also a restricted version of the Burnside problem.

Problem 6.4 (Restricted Burnside Problem). Is it true that for every $m, n \in \mathbb{N}$ there are finitely many (up to isomorphism) finite groups G with m generators and of bounded exponent n?

6.2 Locally Finite Groups and the General Burnside Problem

Definition 6.5. Let $\mathscr{P}$ be a property of groups (e.g., being finite). We say that a group G is *locally* $\mathscr{P}$ if every finitely generated subgroup of G satisfies $\mathscr{P}$.

Example 6.6. Every abelian torsion group is locally finite. This immediately follows from the structure theorem of finitely generated abelian groups (see Corollary 1.30).

We can rephrase the General Burnside Problem in the following way.

Problem 6.7 (Reformulation of the General Burnside Problem). Is every torsion group locally finite?

Notice that Lemma 2.34 says that the General Burnside Problem has a positive solution for nilpotent groups, equivalently, the class of nilpotent groups has the Burnside property.

The class of locally finite groups is clearly closed under taking subgroups, homomorphic images and finite direct products (**exercise**). Next we show that it is also closed under extensions.

Lemma 6.8. *Let G be a group. Let $H \leq G$ be a normal subgroup and suppose that both H and G/H are locally finite. Then G is locally finite.*

Proof. Let $K \leq G$ be a finitely generated subgroup of G. The image of K in G/H is $KH/H \cong K/(K \cap H)$, which is finite by assumption. Hence $[K : K \cap H] < \infty$, which implies that $K \cap H$ is finitely generated (cf. Corollary 1.11). Hence $K \cap H \leq H$ is finite by the assumption on H. We deduce that K is also finite. $\square$

In fact, we can show that every group contains a largest locally finite subgroup. Here, "largest" means that it contains all the other locally finite subgroups: this is stronger than "maximal". In order to do so, the following two propositions will be useful.

Proposition 6.9. *Let $K, L \trianglelefteq G$, and let both K and L be locally finite. Then KL is locally finite.*

Proof. $KL/K \cong L/(K \cap L)$ is locally finite, hence KL is an extension of locally finite groups (K and KL/K). Therefore, by the previous lemma, KL itself is locally finite. □

Let I be a *directed set*, that is, a set equipped with an order $\preceq$ such that for every $i_1, i_2 \in I$ there exists an $i \in I$ with $i_1 \preceq i$ and $i_2 \preceq i$. Let G be a group. A family $(H_i)_{i \in I}$ of subgroups of G is said to be *increasing* if $H_i \leq H_j$ for all $i, j \in I$ such that $i \preceq j$. If in addition we have $\bigcup_{i \in I} H_i = G$, then we say that the family $(H_i)_{i \in I}$ *exhausts* G.

Proposition 6.10. *Let G be a group. Let $(H_i)_{i \in I}$ be an increasing and exhausting family of subgroups of G. Suppose that H_i is locally finite for every $i \in I$. Then G itself is locally finite.*

Proof. Let $X \subseteq G$ be a finite subset. Since $(H_i)_{i \in I}$ is increasing and exhausting, we can find $i = i(X)$ such that $X \subseteq H_i$. Since H_i is locally finite, it follows that the subgroup generated by X is finite. This shows that G is locally finite. □

Corollary 6.11. *Every group G contains a largest locally finite normal subgroup $L(G)$ such that $G/L(G)$ does not contain nontrivial locally finite normal subgroups.*

Proof. Let I denote the set of all locally finite normal subgroups of G. Equip I with the order given by inclusion and observe that, by Proposition 6.9, it is a directed set. It follows from Proposition 6.10 that the subgroup $L(G) := \bigcup_{H \in I} H$ is locally finite. Moreover, since conjugation by elements in G preserves local-finiteness of subgroups, we have that $L(G) \trianglelefteq G$. It is clear from the construction that every locally finite normal subgroup of G is contained in $L(G)$.

On the other hand, if $H/L(G) \leq G/L(G)$ is a locally finite normal subgroup, then $H \leq G$ is locally finite by Lemma 6.8. But we have just seen that H must be contained in $L(G)$, hence $H/L(G)$ is the trivial subgroup. □

6.3 The General Burnside Problem for Polycyclic-by-Finite and Solvable Groups

The following theorem gives a positive solution to the General Burnside Problem for polycyclic-by-finite and solvable groups.

Theorem 6.12 (General Burnside Problem for polycyclic-by-finite and solvable groups). *Let G be a torsion group. Then*

(1) *if G is solvable, then it is locally finite;*

(2) *if G is polycyclic-by-finite (e.g polycyclic), then it is finite.*

Proof. If G is solvable, then consider its derived series. The quotients of this series are abelian and torsion, and therefore locally finite (cf. Example 6.6). Then G is locally finite by recursively applying Lemma 6.8.

If G is polycyclic-by-finite, consider any finite subnormal series with cyclic quotients. Since G is torsion, these quotients are necessarily finite. Since extensions of finite groups by finite groups are finite, it follows that G is finite. □

6.4 The Bounded Burnside Problem for Linear Groups

This section is devoted to the proof of the following theorem, which is due to Burnside.

Theorem 6.13 (Bounded Burnside Problem for linear groups). *Let G be a subgroup of* $\mathrm{GL}(n,\mathbb{K})$, *where $\mathbb{K}$ is a field. Suppose that G is finitely generated and of bounded exponent. Then G is finite.*

Proof. Notice that if $\overline{\mathbb{K}}$ is the algebraic closure of $\mathbb{K}$, then $G \leq \mathrm{GL}(n,\mathbb{K}) \leq \mathrm{GL}(n,\overline{\mathbb{K}})$. Hence, without loss of generality, we can assume that $\mathbb{K}$ is algebraically closed.

We can also assume that G acts irreducibly on $V = \mathbb{K}^n$. Indeed, we can always find a chain of subspaces

$$\{0\} \leq V_1 \leq V_2 \leq \cdots \leq V_s = V,$$

such that G acts irreducibly on each V_i/V_{i-1}, $i = 1,2,\ldots,s$ (cf. the proof of the first Claim in Section 4.5). Taking a basis for each factor, consider the basis of V obtained by taking the union of these bases. Then, in this basis, G will be in block upper triangular form:

$$g = \begin{pmatrix} M_1(g) & * & \cdots & * \\ 0 & M_2(g) & \cdots & * \\ \vdots & \vdots & \ddots & \vdots \\ 0 & 0 & \cdots & M_s(g) \end{pmatrix},$$

for every $g \in G$, where $M_i(g) \in \mathrm{GL}(V_i/V_{i-1})$, $i = 1,2,\ldots,s$. Consider now the homomorphism

$$g \mapsto \varphi(g) := \begin{pmatrix} M_1(g) & 0 & \cdots & 0 \\ 0 & M_2(g) & \cdots & 0 \\ \vdots & \vdots & \ddots & \vdots \\ 0 & 0 & \cdots & M_s(g) \end{pmatrix}.$$

The kernel of this map consists of matrices of the form

$$\begin{pmatrix} I_{n_1} & * & \cdots & * \\ 0 & I_{n_2} & \cdots & * \\ \vdots & \vdots & \ddots & \vdots \\ 0 & 0 & \cdots & I_{n_s} \end{pmatrix}$$

where $I_{n_i} \in \mathrm{GL}(n_i,\mathbb{K})$ denotes the identity matrix and $n_i = \dim_{\mathbb{K}}(V_i/V_{i-1})$, $i = 1,2,\ldots,s$. Thus $\ker(\varphi)$, being a subgroup of $\mathrm{UT}(n,\mathbb{K})$, is a nilpotent group (cf. Example 2.5.(b)). Since every nilpotent group is solvable, we deduce from Theorem 6.12 and the hypothesis that G is torsion, that $\ker(\varphi)$ is locally finite. On the other hand, the image of φ is a finite direct product of linear groups acting irreducibly, by construction. Hence if we know the result for G acting irreducibly, then $\varphi(G)$, being the finite direct product of locally finite groups, is locally finite by Lemma 6.8. We then deduce that G, being an extension of locally finite groups, is locally finite by applying once more Lemma 6.8. Since G is finitely generated, it is in fact finite.

To finish the proof, we need some more Wedderburn theory (cf. Section 4.4).

Let A be an algebra, $A \subseteq \mathrm{End}_{\mathbb{K}}(V)$ with $n := \dim_{\mathbb{K}} V < \infty$ and suppose that A acts irreducibly. Then A acts completely reducibly on V and therefore, by Proposition 4.29, its radical vanishes: $N(A) = \{0\}$. Then, since $\mathbb{K}$ is algebraically closed, Wedderburn's Theorem (Theorem 4.20) guarantees that $A \cong A/N(A) \cong M_{n_1}(\mathbb{K}) \oplus \cdots \oplus M_{n_r}(\mathbb{K})$. Let us show that A is in fact simple, that is, that $r = 1$. Suppose, by contradiction, that $r \geq 2$. Let e_1 denote the identity of $M_{n_1}(\mathbb{K}) \subseteq A$ (here we are identifying $M_{n_1}(\mathbb{K})$ with the corresponding ideal of A). Then $e_1 V$ is a proper A-submodule of V, contradicting the irreducibility of the action of A on V. Thus $A \cong M_s(\mathbb{K}) \subseteq \mathrm{End}_{\mathbb{K}}(V) \cong M_n(\mathbb{K})$. In particular, $s \leq n$. Let us show that, in fact, $s = n$, so that $A = \mathrm{End}_{\mathbb{K}}(V) \cong M_n(\mathbb{K})$.

Choosing a suitable basis of V, we can assume that $E_{1,1}$, the matrix that has 1 in the $(1,1)$ position and zero elsewhere, belongs to A. Since $E_{1,1}V \neq \{0\}$, we can find a vector $v \in V$ such that $E_{1,1}v \neq 0$. Now, $AE_{11}v$ is an A-submodule of V, therefore $AE_{1,1}v = V$, since V is A-irreducible. Moreover, $AE_{1,1}$ is the set of matrices with possibly nonzero entries only in the first column, i.e. of the form

$$\begin{pmatrix} * & 0 & \cdots & 0 \\ * & 0 & \cdots & 0 \\ \vdots & \vdots & \ddots & \vdots \\ * & 0 & \cdots & 0 \end{pmatrix} \in M_s(\mathbb{K}).$$

Now $\dim_{\mathbb{K}} V = \dim_{\mathbb{K}} Ae_{11}v \leq s$, so that $n \leq s$. Hence $n = s$, and $A = \mathrm{End}_{\mathbb{K}}(V)$. We just proved:

Theorem 6.14 (Wedderburn). *Let $\mathbb{K}$ be an algebraically closed field and let V be a finite-dimensional vector space over $\mathbb{K}$. Suppose that $A \subseteq \mathrm{End}_{\mathbb{K}}(V)$ acts irreducibly on V. Then $A = \mathrm{End}_{\mathbb{K}}(V)$.* □

Corollary 6.15 (Burnside). *Let $\mathbb{K}$ be an algebraically closed field and let V be a finite-dimensional vector space over $\mathbb{K}$. If $G \subseteq \mathrm{GL}(V)$ and G acts irreducibly on V, then* $\mathrm{span}_{\mathbb{K}}(G) = \mathrm{End}_{\mathbb{K}}(V)$. □

We now finish the proof of the Burnside Theorem for linear groups.

Recall that $G \leq \mathrm{GL}(n, \mathbb{K})$, G is finitely generated and there exists a positive integer d such that $g^d = 1$ for all $g \in G$. Also, by the preceding arguments, we may assume that G acts irreducibly. Then, by Corollary 6.15, G spans $M_n(\mathbb{K})$. Hence there exist $g_1, g_2, \ldots, g_{n^2} \in G$ which form a basis for $M_n(\mathbb{K})$. Let $g \in G$. Observe that since $\mathbb{K}$ is algebraically closed g is triangularizable, that is, g is similar to a matrix of the form

$$\begin{pmatrix} \alpha_1 & * & \cdots & * \\ 0 & \alpha_2 & \cdots & * \\ \vdots & \vdots & \ddots & \vdots \\ 0 & 0 & \ldots & \alpha_n \end{pmatrix}$$

where $\alpha_1, \alpha_2, \ldots, \alpha_n \in \mathbb{K}$. Since $g^d = 1$, we have $\alpha_i^d = 1$ for all i. It follows that we have at most d possible values for each α_i. This implies that, as g varies in G, the distinct values of the traces $\mathrm{tr}(g)$ are at most d^n. Hence the number of distinct n^2-tuples $(\mathrm{tr}(g_1 g), \mathrm{tr}(g_2 g), \ldots, \mathrm{tr}(g_{n^2} g))$ as g varies in G is at most $(d^n)^{n^2} = d^{n^3}$.

Claim. *Given $g \in G$, the tuple $(\mathrm{tr}(g_1 g), \mathrm{tr}(g_2 g), \ldots, \mathrm{tr}(g_{n^2} g))$ determines g uniquely.*

We first observe that the bilinear form $f\colon M_n(\mathbb{K}) \times M_n(\mathbb{K}) \to \mathbb{K}$ defined by $f(a,b) := \mathrm{tr}(ab)$ for all $a, b \in M_n(\mathbb{K})$ satisfies the condition

$$f(ab, c) = f(a, bc)$$

for all $a, b, c \in M_n(\mathbb{K})$. It follows that the set $I = \{a \in M_n(\mathbb{K}) : f(a,b) = 0 \text{ for all } b \in M_n(\mathbb{K})\}$ is an ideal of $M_n(\mathbb{K})$. As the algebra $M_n(\mathbb{K})$ is simple (**exercise**), we have that $I = \{0\}$, equivalently, f is non-degenerate.

Let then $g', g'' \in G$ and suppose that $\mathrm{tr}(g_i g') = \mathrm{tr}(g_i g'')$ for all $i = 1, 2, \ldots, n^2$. By subtracting we get $\mathrm{tr}(g_i(g' - g'')) = 0$ for $i = 1, \ldots, n^2$. Since the g_i's span $M_n(\mathbb{K})$, this implies $f(a, (g' - g'')) = \mathrm{tr}(a(g' - g'')) = 0$ for all $a \in M_n(\mathbb{K})$. It follows that $g' - g'' = 0$ by the non-degeneracy of f. The claim follows.

All this shows that there are finitely many possibilities for $g \in G$. Hence G is finite, and this finishes the proof of the Burnside theorem for linear groups. □

6.5 The Golod–Shafarevich Construction

In this section we describe the negative solution to the General Burnside Problem provided by Golod and Shafarevich.

It turns out that their solution goes through the negative solution to a problem in associative algebras strictly connected to the General Burnside Problem.

Let $\mathbb{K}$ be a field and let A be an associative algebra over $\mathbb{K}$. For two subsets S and T of such an algebra we set $ST := \mathrm{span}_{\mathbb{K}}\{st \mid s \in S, t \in T\}$. For a positive integer n, we denote by S^n the product $SS \cdots S$ of S with itself n times: by associativity, this is well defined.

Definition 6.16. An element $a \in A$ is called *nilpotent* if there exists an integer $n \geq 1$ such that $a^n = 0$. We say that A is a *nil algebra* if every $a \in A$ is nilpotent. We say that A is *nilpotent* if there exists an integer $n \geq 1$ such that $A^n = \{0\}$.

Note that every nilpotent algebra is nil. Conversely, we have the following theorem, whose proof is left as an **exercise**.

Theorem 6.17 (Wedderburn). *A finite-dimensional nil algebra is nilpotent.*

Remark 6.18. In fact, Wedderburn proved even more, namely that a finite-dimensional algebra which admits a linear basis consisting of nilpotent elements is nilpotent.

The following problem is related to the General Burnside Problem.

Problem 6.19 (Kurosh–Levitzky). Let A be a finitely generated nil algebra. Does this imply that A is nilpotent (and hence finite-dimensional)?

An algebra in which any finitely-generated subalgebra is nilpotent is called *locally nilpotent*. The following proposition is somewhat similar to Lemma 6.8. Its proof is left as an **exercise**.

Proposition 6.20. *Let A be a algebra. Let $I \leq A$ be an ideal and suppose that both A/I and I are locally nilpotent. Then A is locally nilpotent.*

So for any associative algebra A there exists a largest locally nilpotent ideal which is called the *Levitzky radical*.

Remark 6.21. In analogy with the reformulation of the General Burnside Problem, the Kurosh–Levitzky Problem asks whether nil and local nilpotence are equivalent conditions.

The answer to the Kurosh–Levitzky Problem is negative. To see this, we fix some notation.

Let $\mathbb{K}$ be a field. Denote by $\mathbb{K}\langle x_1, x_2, \ldots, x_n\rangle = \mathbb{K}\langle X\rangle$ the free associative algebra with coefficients in $\mathbb{K}$ freely generated by $X = \{x_1, x_2, \ldots, x_n\}$. We simply call it the *free algebra generated by X*.

Let $R \subseteq \mathbb{K}\langle X\rangle$ be any subset. We denote by (R) the (two-sided) ideal generated by R, i.e. the set consisting of all finite sums $\sum_i a_i r_i b_i$ where $r_i \in R$ and $a_i, b_i \in \mathbb{K}\langle X\rangle$. We then say that the algebra $\mathbb{K}\langle X\rangle/(R)$ has the *presentation* $\langle X \mid R\rangle$ and that the elements of X (resp. R) are the corresponding *generators* (resp. *relators*).

A unital algebra A is said to be *graded* if it has a direct sum decomposition into $\mathbb{K}$-subspaces

$$A = A_0 \oplus A_1 \oplus A_2 \oplus \cdots = \bigoplus_{i \in \mathbb{N}} A_i \tag{6.1}$$

where $A_0 := \mathbb{K}1_A$ and $A_i A_j \subseteq A_{i+j}$ for all $i, j = 0, 1, \ldots$ We say that the elements of A_i are the *homogeneous elements of degree i*. An ideal I of a graded algebra A is said to be *homogeneous* provided that for every element $a \in I$, the homogeneous parts of a are also contained in I. If I is a homogeneous ideal of a graded algebra A, then A/I is also a graded algebra, and it has decomposition

$$A/I = \bigoplus_{i \in \mathbb{N}} (A_i + I)/I.$$

Example 6.22. (1) Let X be a set. Then the free algebra $\mathbb{K}\langle X\rangle$ generated by X is graded. Indeed, the homogeneous elements of degree i are the homogeneous (non-commutative) polynomials of degree i together with the 0 polynomial.

(2) The algebra $A := \mathbb{K}[x_1, \ldots, x_n]$ of (commutative) polynomials with coefficients in $\mathbb{K}$ is also graded. Here, the homogeneous elements of degree i are the homogeneous (commutative) polynomials of degree i together with the 0 polynomial.

Let A be a graded algebra as in (6.1) and suppose that $\dim_{\mathbb{K}} A_i < \infty$ for all i. Then the associated *Hilbert series* is the formal power series

$$H_A(t) := \sum_{i \geq 0} \dim_{\mathbb{K}}(A_i)t^i.$$

Given two formal powers series $\sum_{i \geq 0} a_i t^i$ and $\sum_{i \geq 0} b_i t^i$, we write

$$\sum_{i \geq 0} a_i t^i \preceq \sum_{i \geq 0} b_i t^i$$

provided that $a_i \leq b_i$ for all $i \geq 0$, and define their product as

$$\sum_{k \geq 0} c_k t^k := \left(\sum_{i \geq 0} a_i t^i \right) \left(\sum_{j \geq 0} b_j t^i \right)$$

where

$$c_k := a_0 b_k + a_1 b_{k-1} + \cdots + a_k b_0$$

for all $k \geq 0$.

Suppose that $R \subseteq \mathbb{K}\langle X \rangle$ is a subset consisting of homogeneous linearly independent elements of degree ≥ 2, and let r_i denote the number of elements of degree i in R for all $i \geq 2$. We set

$$H_R(t) := r_2 t^2 + r_3 t^3 + \cdots.$$

Then the ideal (R) generated by R is a graded ideal and the algebra $A = \langle X \mid R \rangle := \mathbb{K}\langle X \rangle / (R)$ is a graded algebra.

The following theorem constitutes the key ingredient of the Golod–Shafarevich construction.

Theorem 6.23 (Golod–Shafarevich). *With the above notation we have*

$$H_A(t)(1 - nt + H_R(t)) \succeq 1. \tag{6.2}$$

Before proving the theorem, let us show how we can derive from it a negative answer to the Kurosh–Levitzky Problem.

Suppose that we manage to find a real number $0 < t_0 < 1$ such that

(1) $H_R(t)$ converges at t_0 and
(2) $1 - nt_0 + H_R(t_0) < 0$.

Then $H_A(t)$ does not converge at t_0. In fact if it converges, then necessarily $H_A(t_0) \geq 0$, since $t_0 > 0$, which, together with (2), contradicts (6.2). This implies that A is infinite-dimensional: in fact for a finite-dimensional algebra A, the power series $H_A(t)$ is a polynomial, which converges everywhere.

Remark 6.24. This argument can be used in several ways to conclude that an algebra with a given presentation is infinite-dimensional.

For example, suppose that we are given a finite subset $R \subseteq \mathbb{K}\langle X \rangle_2$ of quadratic relators such that $r := |R| < n^2/4$ (recall that $n = |X|$). Then for $t = 2/n$, one has $1 - nt + H_R(t) = 1 - nt + rt^2 < 0$, and therefore the algebra $A = \langle X \mid R \rangle$ is infinite-dimensional.

Let $\mathbb{K}$ be countable, and observe that $\mathbb{K}\langle X\rangle$ is also countable. Denote by $\mathbb{K}_0\langle X\rangle$ the ideal of $\mathbb{K}\langle X\rangle$ consisting of all elements with 0 constant term and let $a_1, a_2, \ldots$ be an enumeration of all the elements of $\mathbb{K}_0\langle X\rangle$. Finally, recursively define $R \subseteq \mathbb{K}_0\langle X\rangle$ as follows. Let $R_1 \subseteq \mathbb{K}_0\langle X\rangle$ be the set of all homogeneous components of $(a_1)^2$ and let $n_2 \in \mathbb{N}$ be greater than the degrees of all elements in R_1.

Example 6.25. If $a_1 = x_1 + x_2x_3$, then $a_1^2 = x_1^2 + x_1x_2x_3 + x_2x_3x_1 + x_2x_3x_2x_3$, so $R_1 = \{x_1^2, x_1x_2x_3 + x_2x_3x_1, x_2x_3x_2x_3\}$. Here we can take $n_2 \geq 5$.

Suppose we have defined $R_i \subseteq \mathbb{K}_0\langle X\rangle$ and $n_{i+1} \in \mathbb{N}$. Then we define $R_{i+1} \subseteq \mathbb{K}_0\langle X\rangle$ as the set of all homogeneous components of $(a_1)^2, (a_2)^{n_2}, \ldots, (a_{i+1})^{n_{i+1}}$ and we choose $n_{i+2} \in \mathbb{N}$ greater than the degrees of all elements in R_{i+1}. Note that $R_i \subseteq R_{i+1}$. We then set $R := \bigcup_{i\geq 1} R_i$.

Remark 6.26. The choice of starting with a_1^2 is made in order to ensure that all the elements of R have degree at least 2 (recall that the a_i's have zero constant term).

Set

$$B := \mathbb{K}_0\langle X\rangle/(R).$$

We first notice that B is nil. Indeed, R contains all the homogeneous components of $(a_i)^{n_i}$ for all $i \geq 1$, so that every element of B is nilpotent. Note that B is clearly finitely generated, so in order to prove that B is a counterexample to the Kurosh–Levitzky problem, we only need to show that it is infinite-dimensional.

Consider now the (graded) algebra $A := \mathbb{K}\langle X\rangle/(R)$ and observe that $A \cong \mathbb{K}1_A \oplus B$ as vector spaces, so that B is infinite-dimensional if and only if A is. Now, to prove that $\dim_{\mathbb{K}} A$ is infinite, it will be enough to show that there exists a t_0 such that $1 - nt_0 + \sum_{i\geq 2} r_i t_0^i < 0$. Recall that r_i is the number of homogeneous elements of degree i in R. By construction, the r_i's are either 0 or 1, hence we can assume that $r_i = 1$ for all i, since $\sum_{i\geq 2} r_i t^i \preceq \sum_{i\geq 2} t^i$.

Now $1 - nt + \sum_{i\geq 2} t^i$ converges for all $0 < t < 1$ and for these values of t we have

$$1 - nt + \sum_{i\geq 2} t^i = 1 - nt + \frac{t^2}{1-t} = \frac{(n+1)t^2 - (n+1)t + 1}{1-t}.$$

Consider the inequality

$$\frac{(n+1)t^2 - (n+1)t + 1}{1-t} < 0. \tag{6.3}$$

The discriminant of the quadratic polynomial at the numerator of (6.3) is positive for $n \geq 4$ and, in this case, the corresponding roots are

$$\alpha^{\pm} := \frac{n+1 \pm \sqrt{(n+1)^2 - 4(n+1)}}{2(n+1)}.$$

Note that $0 < \alpha^- < \alpha^+ < 1$ and that (6.3) is satisfied for every $\alpha^- < t < \alpha^+$. It follows that we can find $0 < t_0 < 1$ such that $1 - nt_0 + H_R(t_0) < 0$.

This completes the proof that A is infinite-dimensional over $\mathbb{K}$, so that B is a counterexample to the Kurosh–Levitzky problem.

Proof of the Golod–Shafarevich theorem (Theorem 6.23). It is straightforward to check the inequalities implicit in (6.2) for the constant term and the coefficient of t. Indeed for the constant term it reduces to $1 \geq 1$, while for the coefficient of t, it reduces to $\dim_{\mathbb{K}} A_1 - n = n - n \geq 0$.

To check it for the other coefficients, we proceed as follows. Let $\mathbb{K}\langle X\rangle = \bigoplus_{i\geq 0} \mathbb{K}\langle X\rangle_i$ be the decomposition in homogeneous components, so that we have $\dim_{\mathbb{K}} \mathbb{K}\langle X\rangle_i = n^i$. Set $I := (R)$, and denote by $I = \bigoplus_{i\geq 2} I_i$ the decomposition of the graded ideal I into homogeneous components (note that $I_i = I \cap \mathbb{K}\langle X\rangle_i$ for all i). Also set $a_i := \dim_{\mathbb{K}} A_i$ and observe that we have $\dim_{\mathbb{K}} I_i = n^i - a_i$. Moreover, set $R_i := R \cap \mathbb{K}\langle X\rangle_i$ and $r_i = |R_i|$ for all i. For every i we choose a subspace $\widetilde{A}_i$ such that $\mathbb{K}\langle X\rangle_i = I_i \oplus \widetilde{A}_i$, so that $\dim_{\mathbb{K}} \widetilde{A}_i = a_i$.

We clearly have $I = \mathbb{K}\langle X\rangle R + IX$. It follows that

$$I_s = \sum_{i=2}^{s} \mathbb{K}\langle X\rangle_{s-i} R_i + I_{s-1} X$$

for all $s \geq 2$. In fact, we have

$$I_s = \sum_{i=2}^{s} \widetilde{A}_{s-i} R_i + I_{s-1} X$$

for $s \geq 2$. To see this, it is enough to observe that $I_{s-i} R_i \subseteq I_{s-1} X$ for $i \geq 2$, so that

$$\begin{aligned} I_s &= \sum_{i=2}^{s} \mathbb{K}\langle X\rangle_{s-i} R_i + I_{s-1} X = \sum_{i=2}^{s} (I_{s-i} \oplus \widetilde{A}_{s-i}) R_i + I_{s-1} X \\ &= \sum_{i=2}^{s} \widetilde{A}_{s-i} R_i + \sum_{i=2}^{s} I_{s-i} R_i + I_{s-1} X = \sum_{i=2}^{s} \widetilde{A}_{s-i} R_i + I_{s-1} X. \end{aligned}$$

Hence, looking at the dimension over $\mathbb{K}$ of these subspaces, we deduce that

$$n^s - a_s \leq \sum_{i=2}^{s} a_{s-i} r_i + (n^{s-1} - a_{s-1}) n,$$

that is,

$$a_s + \sum_{i=2}^{s} a_{s-i} r_i - n a_{s-1} \geq 0.$$

It remains only to notice that, for $s \geq 1$, the coefficient of t^s in $H_A(t)(1 - nt + H_R(t))$ is exactly $a_s - n a_{s-1} + \sum_{i=2}^{s} a_{s-i} r_i$. □

We are now going to use our counterexample to the Kurosh–Levitzky problem to produce a counterexample to the General Burnside Problem.

Let $\mathbb{K}$ be a countable or finite field of characteristic $ch(\mathbb{K}) = p > 0$. Consider the algebra A that we just constructed. We have $A = \mathbb{K}1 \oplus B$ as vector spaces, where $1 = 1_A$ is the unit of the algebra A, and $B = \mathbb{K}_0\langle X\rangle/(R)$ is an infinite dimensional nil algebra. Consider the set $A^{\times}$ of all invertible elements of A.

Now for every $b \in B$, the element $1+b \in A$ is invertible, and in fact has finite order. Indeed, since B is nil, there exists an $\ell \geq 1$ such that $b^{p^\ell} = 0$; hence $(1+b)^{p^\ell} = 1+b^{p^\ell} = 1$ (here we are using $ch(\mathbb{K}) = p > 0$). It follows that $1+B \subseteq A^\times$.

Let us denote again by x_i the cosets $x_i + (R)$ in $A \equiv \mathbb{K}\langle X\rangle/(R)$, where $x_i \in X$. Then the elements $1+x_i$, $i = 1,2,\ldots,n$ are invertible and torsion. Consider the subgroup $G \subseteq 1+B \subseteq A^\times$ generated by $\{1+x_1, 1+x_2, \ldots, 1+x_n\}$. This is clearly a finitely generated torsion group.

Theorem 6.27. *G is infinite.*

Proof. Suppose not and assume $|G| = d$. Then every element g of G can be expressed as

$$g = (1+x_{i_1})(1+x_{i_2})\cdots(1+x_{i_r})$$

with $r < d$ (we don't need the inverses, since $(1+x_j)^{k_j} = 1$ for some k_j). Indeed, suppose that $g = (1+x_{j_1})\cdots(1+x_{j_\ell})$ with ℓ minimal and $\ell \geq d$. Then at least two of the $d+1$ elements

$$1, 1+x_{j_1}, (1+x_{j_1})(1+x_{j_2}), \ldots, (1+x_{j_1})(1+x_{j_2})\cdots(1+x_{j_d})$$

must be equal, say $(1+x_{j_1})(1+x_{j_2})\cdots(1+x_{j_h})$ and $(1+x_{j_1})(1+x_{j_2})\cdots(1+x_{j_k})$, where $0 \leq h < k \leq d$. But then

$$g = (1+x_{j_1})(1+x_{j_2})\cdots(1+x_{j_h})(1+x_{j_{k+1}})(1+x_{j_{k+2}})\cdots(1+x_{j_\ell})$$

is a product of $h+(\ell-k) < \ell$ generators, contradicting the minimality of ℓ. So $\ell < d$.

Let us show that the set of products $\{x_{j_1}x_{j_2}\cdots x_{j_r} : 1 \leq j_i \leq n, 1 \leq i \leq r < d\}$ spans B. To do so, it is sufficient to prove that every word $w = x_{i_1}x_{i_2}\cdots x_{i_d}$ of length d is a linear combination of shorter words.

We have

$$(1+x_{i_1})(1+x_{i_2})\cdots(1+x_{i_d}) = (1+x_{j_1})(1+x_{j_2})\cdots(1+x_{j_r})$$

with $r < d$ as we have just shown. Keeping the factor $x_{i_1}x_{i_2}\cdots x_{i_d}$ on the left-hand side, and bringing everything else to the right-hand side, gives the desired expression of w as a linear combination of shorter words.

It follows that $\dim_{\mathbb{K}} A = 1 + \dim_{\mathbb{K}} B \leq 1 + (1+n+\cdots+n^{d-1})$, contradicting the fact that A is infinite-dimensional. Therefore G must be infinite. □

6.6 Notes

William Burnside [47] solved the Bounded Burnside Problem for linear groups. Issai Schur [305] proved the General Burnside Problem for linear groups. The Bounded Burnside Problem has been checked for exponent $n = 2$ (trivial: abelian groups), $n = 3$ (Burnside [46]), $n = 4$ (Ivan N. Sanov [302]) and $n = 6$ (Marshall Hall [153]).

In 1964 Evgenii S. Golod and Igor R. Shafarevich [125] constructed a 2-generated infinite p-group, thus providing a counterexample to the General Burnside Problem.

In 1980 Rostislav I. Grigorchuk [129] constructed his renowned group of intermediate growth which, among other most important properties, provides a negative solution to the General Burnside Problem. See the Notes to Chapter 7 for more on the Grigorchuk group.

In 1968 Pëtr S. Novikov and Sergei I. Adyan [258] found a counterexample to the Bounded Burnside Problem for all odd exponents $n \geq 4381$.

In 1992 both Sergei V. Ivanov and Igor Lysënok announced a counterexample to the Bounded Burnside Problem for all but finitely many exponents: Ivanov [188] for $n \geq 2^{48}$ and Lysënok [220] for $n \geq 8000$.

In 1980 Alexander Yu. Olshanskii [260] constructed the so-called Tarski monsters. A *Tarski monster* is an infinite group G such that every proper subgroup H of G, other than the identity subgroup, is a cyclic group of order a fixed prime number p. Such a group G is necessarily finitely generated. In fact it is clearly generated by every two non-commuting elements. Then Olshanskii showed that there is a Tarski p-group for every prime $p > 10^{75}$.

In 1991 Efim I. Zelmanov [361, 362] gave a positive solution to the Restricted Burnside Problem.

The Kurosh–Levitzky problem goes back to Alexander G. Kurosh [206] and Jakob Levitzky [214] in the early 1940s.

For a comprehensive relatively recent account on the Burnside problem, we also refer to Adyan's survey [3].

6.7 Exercises

Exercise 6.1. Show that the class of locally finite groups is closed under taking subgroups, homomorphic images, and finite direct products.

Exercise 6.2. Let $\mathbb{K}$ be a field. Show that the algebra $M_n(\mathbb{K})$ is simple.

Exercise 6.3. Show that a finitely-dimensional algebra which admits a linear basis consisting of nilpotent elements is nilpotent. This proves Wedderburn theorem (Theorem 6.17).

Part II
Geometric Theory

Chapter 7
Finitely Generated Groups and their Growth Functions

This chapter is devoted to the growth of finitely generated groups. The choice of a finite symmetric generating subset $X \subseteq G$ for a finitely generated group G defines a word metric d_X on the group and a labelled graph $\mathrm{Cay}(G,X)$, which is called the Cayley graph of G with respect to X. The associated growth function $b_X(n)$ counts the number of group elements in a ball of radius n with respect to the word metric. We define a notion of equivalence for such growth functions and observe that the growth functions of a given group G associated with different finite symmetric generating subsets are in the same equivalence class (Corollary 7.10). This equivalence class, denoted by b^G, is called the growth type of the group G. The important notions of (sub-)polynomial, sub-exponential, and exponential growth are introduced in Section 7.4. In the following sections we define the growth rate $\beta_X \in [1,+\infty)$ of G with respect to X and study the growth of subgroups and quotients. In Section 7.7 we show that a group of linear growth is virtually infinite cyclic: this is Justin's theorem (Corollary 7.27). In Section 7.8 we study the growth of finitely generated nilpotent groups: these have polynomial growth. We then present the Bass–Guivarc'h formula for the corresponding degree of polynomial growth (Theorem 7.29). Finally, in Section 7.9 we present the theorems of Milnor (Theorem 7.36) and Wolf (Theorem 7.37) and deduce that finitely generated solvable groups of sub-exponential growth are virtually nilpotent, and hence have polynomial growth (Corollary 7.41).

7.1 The Word Metric

Let G be a group. Recall that a subset $X \subseteq G$ is said to *generate* G, and we write $\langle X \rangle = G$, provided that every element $g \in G$ can be expressed as a product of elements in $X \cup X^{-1}$, where $X^{-1} = \{x^{-1} : x \in X\}$, that is, there exist $n \geq 0$, $x_1, x_2, \dots, x_n \in X$ and $\varepsilon_1, \varepsilon_2, \dots, \varepsilon_n \in \{1,-1\}$ such that

$$g = x_1^{\varepsilon_1} x_2^{\varepsilon_2} \cdots x_n^{\varepsilon_n}. \tag{7.1}$$

A subset $X \subseteq G$ is called *symmetric* if $X = X^{-1}$, that is, if $x^{-1} \in X$ for all $x \in X$.

T. Ceccherini-Silberstein and M. D'Adderio, *Topics in Groups and Geometry*,
Springer Monographs in Mathematics, https://doi.org/10.1007/978-3-030-88109-2_7

Moreover, one says that G is *finitely generated* if it admits a finite generating subset. Note that if G is finitely generated then it admits a finite generating subset which is symmetric.

Let G be a finitely generated group and let X be a finite symmetric generating subset of G. The *word length* $\ell_X(g)$ of an element $g \in G$ with respect to X is the minimal integer $n \geq 0$ such that g can be expressed as a product of n elements in X, that is,

$$\ell_X(g) := \min\{n \geq 0 : g = x_1 x_2 \cdots x_n,\ x_i \in X, 1 \leq i \leq n\}. \tag{7.2}$$

Consider the map $d_X : G \times G \to \mathbb{N}$ defined by

$$d_X(g,h) := \ell_X(g^{-1}h) \tag{7.3}$$

for all $g, h \in G$.

The proof of the following lemma is left as an **exercise**.

Lemma 7.1. *The map d_X is a left-invariant metric on G. In other words:*

(i) $d_X(g,h) = 0$ *if and only if* $g = h$;
(ii) $d_X(g,h) = d_X(h,g)$ *(symmetry)*;
(iii) $d_X(g,h) \leq d_X(g,k) + d_X(k,h)$ *(triangular inequality)*;
(iv) $d_X(kg,kh) = d_X(g,h)$ *(invariance by left multiplication)*,

for all $g, h, k \in G$.

The metric d_X is called the *word metric* on G associated with the finite symmetric generating subset X.

As a consequence of the left-invariance of d_X, for every $k \in G$ the map $L_k \colon G \to G$ defined by $L_k(g) = kg$ for all $g \in G$ is an isometry of the metric space (G, d_X). Note that the map $L \colon G \to \mathrm{Isom}(G, d_X)$, $k \mapsto L_k$, where $\mathrm{Isom}(G, d_X)$ denotes the group of all isometries of (G, d_X), is injective.

For $g \in G$ and $n \in \mathbb{N}$, we denote by

$$B_X(g,n) := \{h \in G : d_X(g,h) \leq n\}$$

the *ball of radius n* in G centered at the element $g \in G$ with respect to X.

Note that $B_X(1_G, n) = \{h \in G : \ell_X(h) \leq n\} = \{x_1 x_2 \cdots x_m : x_i \in X, 0 \leq m \leq n\}$ so that, in particular, $B_X(1_G, 0) = \{1_G\}$, and

$$B_X(1_G,0) \subseteq B_X(1_G,1) \subseteq B_X(1_G,2) \subseteq \cdots \tag{7.4}$$

Moreover, the map L_g induces, by restriction, an isometry from $B_X(1_G, n)$ onto $B_X(g,n)$ for all $g \in G$ so that, in particular, $|B_X(1_G,n)| = |B_X(g,n)|$ for all $n \in \mathbb{N}$ and $g \in G$.

For $n \in \mathbb{N}$, we set $b_X(n) := |B_X(1_G, n)|$. Observe that $b_X(0) = 1$ and

$$b_X(0) \leq b_X(1) \leq b_X(2) \leq \cdots, \tag{7.5}$$

which immediately follows from (7.4).

Also note that the group G is finite if and only if the sequence $(B_X(1_G,n))_{n\in\mathbb{N}}$ (resp. $(b_X(n))_{n\in\mathbb{N}}$) eventually stabilizes, i.e., there exists an $n_0 \in \mathbb{N}$ such that $B_X(1_G,n) = B_X(1_G,n_0)$ (resp. $b_X(n) = b_X(n_0)$) for all $n \geq n_0$.

7.2 Cayley Graphs

In this section we rephrase in a more geometric and pictorial way the material presented in the previous section.

Let G be a finitely generated group and let X be a finite symmetric generating subset of G.

Definition 7.2. The *Cayley graph* of G with respect to X is the X-labelled graph $\mathrm{Cay}(G,X) = (V,E)$, where the set of *vertices* is $V := G$ and the set of (X*-labelled*) *edges* is $E := \{(g,x,gx) : g \in G \text{ and } x \in X\} \subseteq G \times X \times G$.

Let α (resp. ω, resp. λ) denote the map $E \to V$ (resp. $E \to V$, resp. $E \to X$) defined for all $e = (g,x,h) \in E$ by $\alpha(e) := g$ (resp. $\omega(e) := h$, resp. $\lambda(e) := x$). We then call $\alpha(e)$ (resp. $\omega(e)$, resp. $\lambda(e)$) the *initial vertex* (resp. the *terminal vertex*, resp. the *label*) of $e \in E$.

Note that, since X is symmetric, the inversion map $x \mapsto x^{-1}$ is an involution on X. Moreover, if $e = (g,x,h) \in E$ (so that $h = gx$), then, since $g = (gx)x^{-1} = hx^{-1}$, we have that the *inverse edge* $e^{-1} := (h,x^{-1},g)$ also belongs to E. One then says that the Cayley graph $\mathrm{Cay}(G,X)$ is *edge-symmetric* with respect to the inversion map on X. We say that $g,h \in G$ are *neighbors*, and we write $g \sim h$, provided there exists an $e \in E$ such that $\alpha(e) = g$ and $\omega(e) = h$; since one always has $\omega(e) = \alpha(e^{-1})$ and $\alpha(e) = \omega(e^{-1})$, it is clear that $\sim$ is a symmetric relation.

Moreover, $\mathrm{Cay}(G,X)$ is *connected*, that is, given $g,h \in G$ there exists a finite sequence $\pi := (e_i)_{i=1}^n$ of edges in E such that $\alpha(e_1) = g$, $\omega(e_i) = \alpha(e_{i+1})$ for $i = 1,2,\ldots,n-1$, and $\omega(e_n) = h$; such a sequence π of edges is then called a *path* connecting g to h and the integer $\ell(\pi) := n$ is called its *length*. Indeed, as X generates G, we can find a nonnegative integer n and $x_1,x_2,\ldots,x_n \in X$ such that $g^{-1}h = x_1x_2\cdots x_n$. Then the path $\pi = (e_1,e_2,\ldots,e_n)$, where

$$e_i = (gx_1x_2\cdots x_{i-1}, x_i, gx_1x_2\cdots x_{i-1}x_i),$$

$i = 1,2,\ldots,n$, connects g to $h = gx_1x_2\cdots x_n$.

Also $\mathrm{Cay}(G,X)$ is *regular*, that is, for every $g \in G$ the number of $h \in G$ such that $g \sim h$, is constant (does not depend on g). This number, which clearly equals $|X|$, is called the *degree* of $\mathrm{Cay}(G,X)$.

Given two vertices $g,h \in V = G$ of the Cayley graph $\mathscr{G} := \mathrm{Cay}(G,X)$, we set

$$d_{\mathscr{G}}(g,h) := \min\{\ell(\pi) : \pi \text{ a path connecting } g \text{ to } h\}. \tag{7.6}$$

A path π of minimal length connecting g to h, that is, such that $\ell(\pi) = d_{\mathscr{G}}(g,h)$, is called a *geodesic path* from g to h.

We leave it as an **exercise** to show that the map $d_{\mathcal{G}} : V \times V \to \mathbb{N}$ defines a metric on $V = G$ called the *geodesic metric* on $\mathcal{G}$. In fact, we have the following proposition whose proof is left as an **exercise**.

Proposition 7.3. *Let G be a finitely generated group. Let X be a finite symmetric generating subset of G. Then, the word distance of two group elements with respect to X equals the geodesic distance of the same elements viewed as vertices in the associated Cayley graph $\mathcal{G} = \mathrm{Cay}(G,X)$. In other words,*

$$d_X(g,h) = d_{\mathcal{G}}(g,h) \tag{7.7}$$

for all $g,h \in G$.

Example 7.4. We graphically represent Cayley graphs by connecting two neighboring vertices $g,h \in G$ by a *single* directed arc e labelled by $\lambda(e) \in X$ (see Figure 7.1). Thus, one should think of the inverse edge e^{-1} as the oppositely directed arc with label $\lambda(e^{-1}) = \lambda(e)^{-1}$.

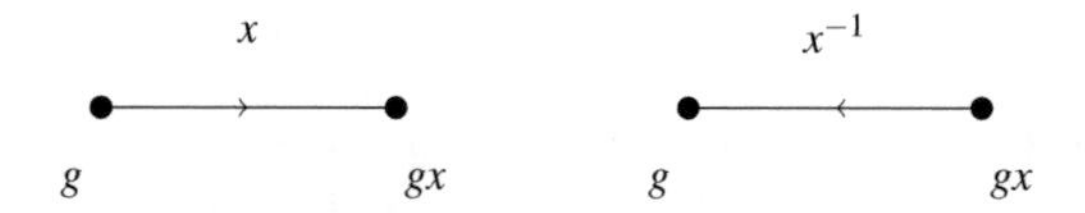

Fig. 7.1 The edges $e = (g,x,gx)$ (with $\alpha(e) = g$, $\omega(e) = gx$, and $\lambda(e) = x$) and $e^{-1} = (gx,x^{-1},g)$ (with $\alpha(e) = gx$, $\omega(e) = x$, and $\lambda(e^{-1}) = x^{-1}$).

(a) Let $G = \mathbb{Z}$ and take $X = \{1,-1\}$ as a finite symmetric generating subset of G. Then the Cayley graph $\mathrm{Cay}(\mathbb{Z},X)$ is represented in Figure 7.2.

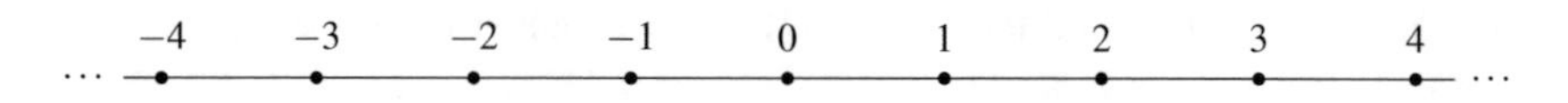

Fig. 7.2 The Cayley graph of $G = \mathbb{Z}$ for $X = \{1,-1\}$

(a') Let $G = \mathbb{Z}$ and take $X = \{1,0,-1\}$ as a finite symmetric generating subset of G. Then the Cayley graph $\mathrm{Cay}(\mathbb{Z},X)$ is represented in Figure 7.3. Note that as $0 = 1_{\mathbb{Z}} \in X$, we have a loop at each vertex in $\mathrm{Cay}(\mathbb{Z},X)$.

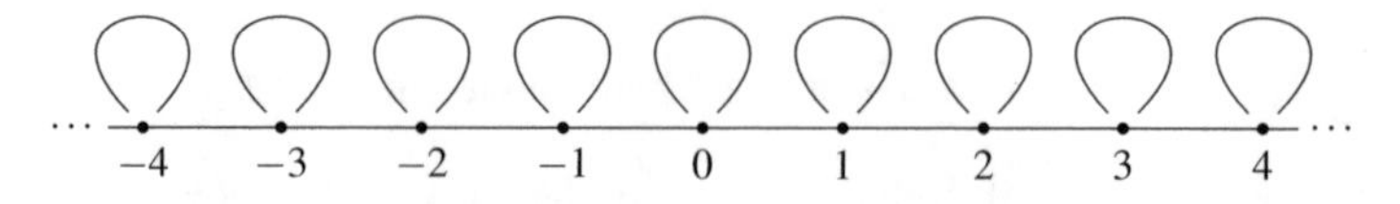

Fig. 7.3 The Cayley graph of $G = \mathbb{Z}$ for $X = \{1,0,-1\}$

(a") Let $G = \mathbb{Z}$ and $X = \{2, -2, 3, -3\}$. Note that X generates G (**exercise**). Then, the corresponding Cayley graph $\mathrm{Cay}(\mathbb{Z}, X)$ is represented in Figure 7.4.

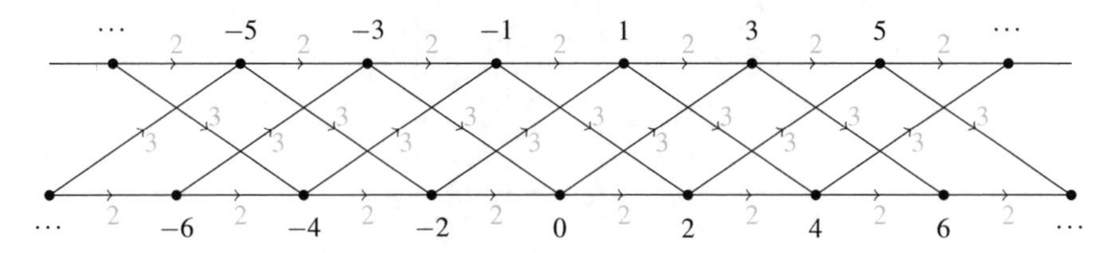

Fig. 7.4 The Cayley graph of $G = \mathbb{Z}$ for $X = \{2, -2, 3, -3\}$.

(b) Let $G = \mathbb{Z}^2$ and consider the finite and symmetric generating subset $X = \{(1,0), (-1,0), (0,1), (0,-1)\}$. Then, the corresponding Cayley graph $\mathrm{Cay}(\mathbb{Z}^2, X)$ is represented in Figure 7.5.

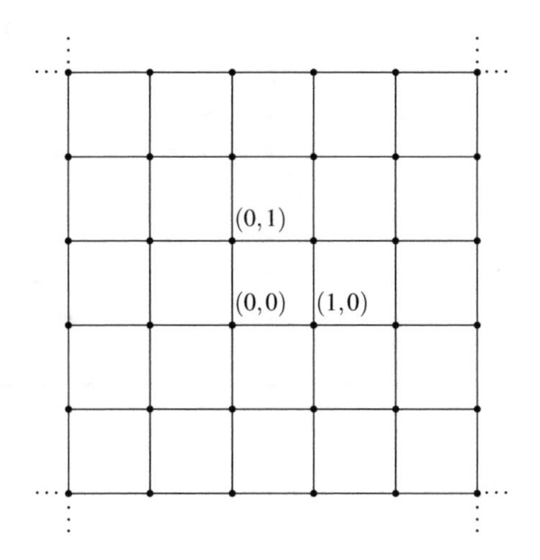

Fig. 7.5 The Cayley graph of $G = \mathbb{Z}^2$ for $X = \{(1,0), (-1,0), (0,1), (0,-1)\}$

(b') Let $G = \mathbb{Z}^2$ and take

$$X = \{(1,0), (-1,0), (0,1), (0,-1), (1,1), (-1,1), (1,-1), (-1,-1)\}.$$

Then, the corresponding Cayley graph $\mathrm{Cay}(\mathbb{Z}^2, X)$ is as in Figure 7.6.

(c) Let $G = F_k$ be the free group of rank $k \geq 1$. Let $\{a_1, a_2, \ldots, a_k\}$ be a free basis and set $X = \{a_1, a_1^{-1}, a_2, a_2^{-1}, \ldots, a_k, a_k^{-1}\}$. Then, the Cayley graph $\mathrm{Cay}(F_k, X)$ is a regular tree of degree $2k$ (see Figure 7.7).

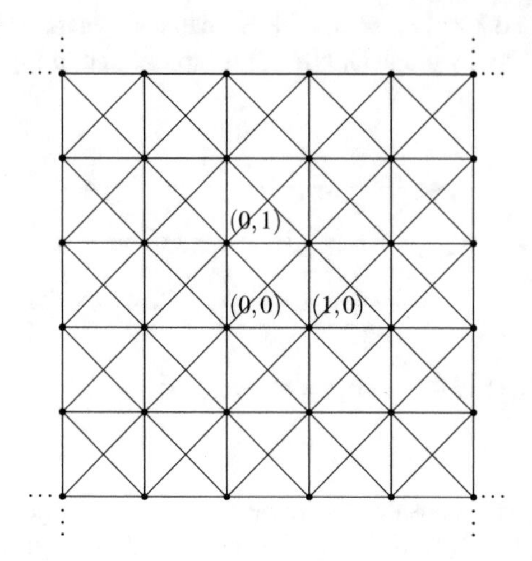

Fig. 7.6 The Cayley graph of $G = \mathbb{Z}^2$ for $X = \{(1,0), (-1,0), (0,1), (0,-1), (1,1), (-1,1), (1,-1), (-1,-1)\}$

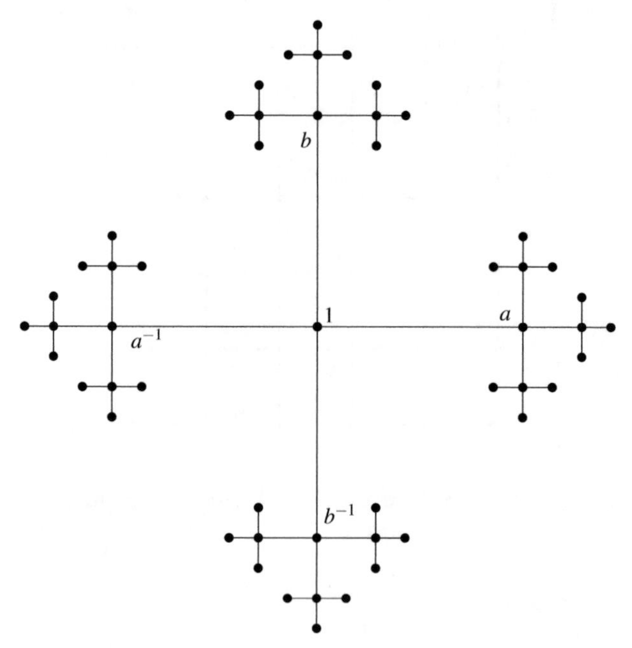

Fig. 7.7 The Cayley graph of $G = F_2$ for $X = \{a, a^{-1}, b, b^{-1}\}$.

7.3 Growth Functions

Let G be a finitely generated group and let X be a finite and symmetric generating subset of G. The *growth function* of G relative to X is the function $b_X \colon \mathbb{N} \to \mathbb{N}$ defined by

$$b_X(n) = |B_X(1_G, n)| = |\{g \in G : \ell_X(g) \leq n\}| \tag{7.8}$$

for all $n \in \mathbb{N}$.

Recall that $b_X(0) = |B_X(1_G, 0)| = |\{1_G\}| = 1$ and that $b_X(n) \leq b_X(n+1)$ for all $n \in \mathbb{N}$. Also, as the map $(x_1, x_2, \ldots, x_n) \mapsto x_1 x_2 \cdots x_n$ is a surjection from $(X \cup \{1_G\})^{\times n}$ onto $B_X(n)$, one has

$$b_X(n) \leq |X \cup \{1_G\}|^n \tag{7.9}$$

for all $n \geq 2$.

Example 7.5. (a) Let $G = \mathbb{Z}$ and consider the finite symmetric generating subset $X = \{1, -1\}$. Then the ball $B_X(0, n)$ of radius n centered at the identity element 0 is the interval $[-n, n] = \{-n, -n+1, \ldots, -1, 0, 1, \ldots, n-1, n\}$, see Figure 7.8. We thus have $b_X(n) = 2n+1$.

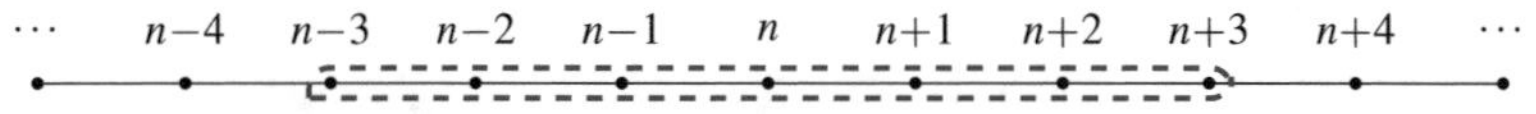

Fig. 7.8 The ball $B_X(n, 2) \subset \mathbb{Z}$ with $X = \{1, -1\}$.

(b) Let $G = \mathbb{Z}^2$ and let $X = \{(1,0), (-1,0), (0,1), (0,-1)\}$. Then the ball of radius n centered at the identity element $(0,0)$ is $B_X((0,0), n) = \{(i,j) \in G : |i| + |j| \leq n\}$, that is, the diagonal square with vertices $(r,0), (-r,0), (0,-r), (0,r)$, see Figure 7.9. We have $b_X(n) = 1 + \sum_{k=1}^{n} 4k = 2n^2 + 2n + 1$.

(c) Let $G = \mathbb{Z}^2$ and consider now the new finite symmetric generating subset

$$X' = \{(1,0), (-1,0), (0,1), (0,-1), (1,1), (-1,1), (1,-1), (-1,-1)\}.$$

Then the ball of radius n centered at the identity element $(0,0)$ is $B_{X'}((0,0), n) = \{(i,j) \in G : |i| \leq n, |j| \leq n\}$, that is, the square $[-r, r] \times [-r, r]$, see Figure 7.10. We have $b_{X'}(n) = (2n+1)^2 = 4n^2 + 4n + 1$.

(d) For $G = \mathbb{Z} \times \mathbb{Z}_2$ and $X = \{(1, \bar{0}), (-1, \bar{0}), (0, \bar{1})\}$ the ball of center $(n, \bar{0})$ and radius 4 is shown in Figure 7.11.

(e) Let $G = F_k$ be the free group of rank $k \geq 2$. Let $\{a_1, a_2, \ldots, a_k\}$ be a free basis and set $X = \{a_1, a_1^{-1}, a_2, a_2^{-1}, \ldots, a_k, a_k^{-1}\}$. Then the ball of radius r centered at the element $g \in G$ is the finite tree rooted at g of depth r (see Figure 7.12). We have

$$b_X(n) = 1 + 2k \sum_{j=0}^{n-1} (2k-1)^j = \frac{k(2k-1)^n - 1}{k-1}.$$

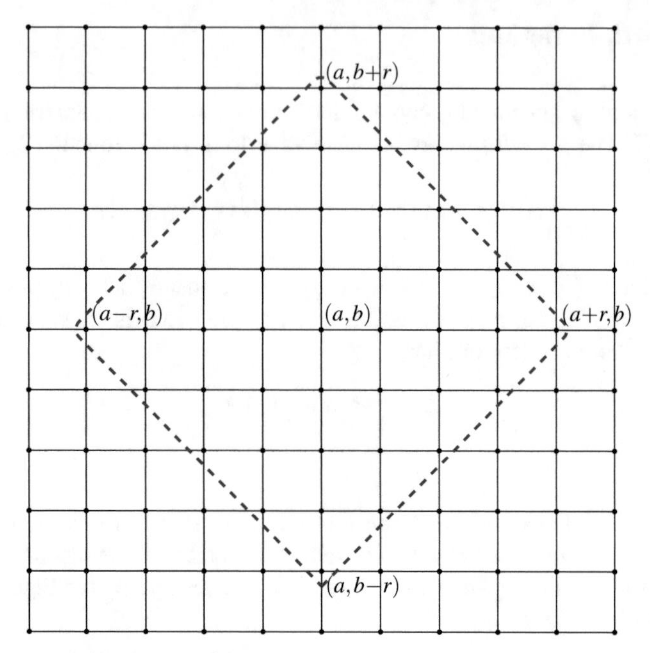

Fig. 7.9 The ball $B_X((a,b),r) \subset \mathbb{Z}^2$ with $X = \{(1,0),(-1,0),(0,-1),(0,1)\}$.

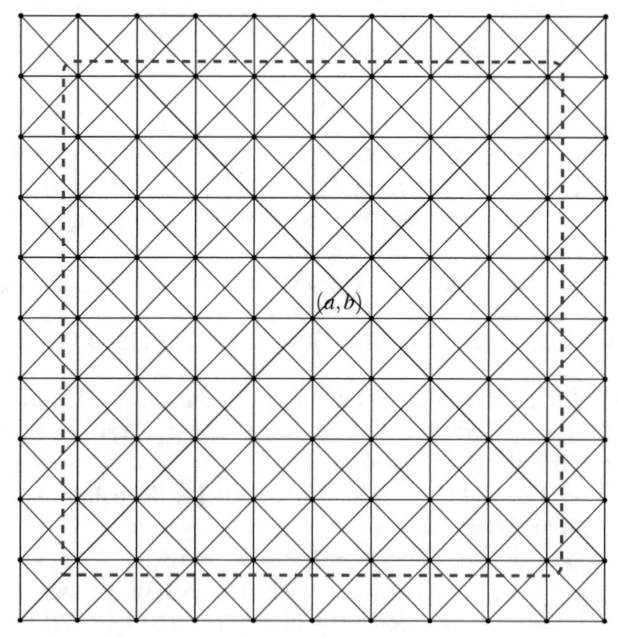

Fig. 7.10 The ball $B_{X'}((a,b),r) \subset \mathbb{Z}^2$ with $X' = \{(1,0),\ (-1,0),\ (0,-1),\ (0,1),\ (1,1),\ (-1,1),\ (1,-1),\ (-1,-1)\}$.

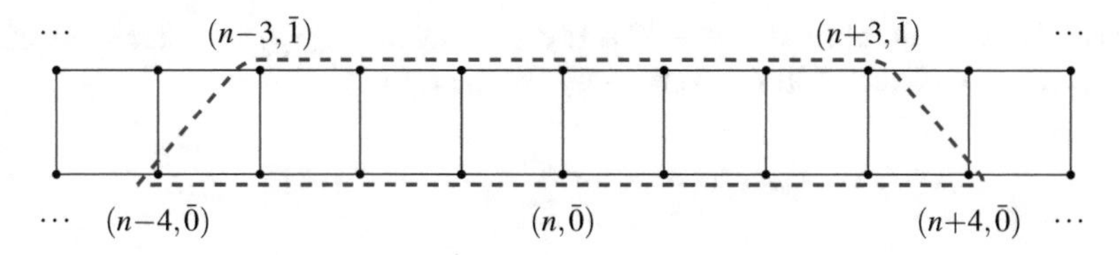

Fig. 7.11 The ball $B_X((n,\bar{0}),4) \subset \mathbb{Z} \times \mathbb{Z}_2$ with $X = \{(1,\bar{0}),(-1,\bar{0}),(0,\bar{1})\}$.

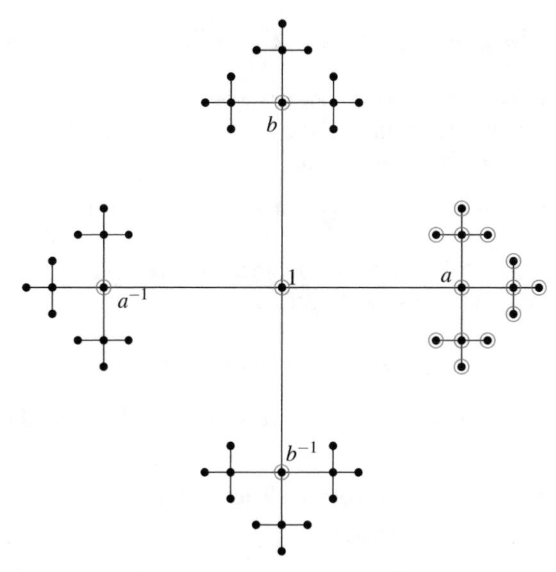

Fig. 7.12 The ball $B_X(a,2) \subset F_2$ with $X = \{a,a^{-1}b,b^{-1}\}$.

7.4 Growth Types

Recall that two metrics d and d' on a set Z are said to be *Lipschitz-equivalent* if there exist constants $c, C > 0$ such that

$$cd(z_1,z_2) \leq d'(z_1,z_2) \leq Cd(z_1,z_2)$$

for all $z_1, z_2 \in Z$.

Proposition 7.6. *Let G be a finitely generated group and let X and X' be two finite symmetric generating subsets of G. Then, there exist two real numbers c and C with $0 < c \leq 1 \leq C$ satisfying*

$$cd_X(g_1,g_2) \leq d_{X'}(g_1,g_2) \leq Cd_X(g_1,g_2)$$

for all $g_1, g_2 \in G$. In particular, d_X and $d_{X'}$ are Lipschitz-equivalent.

Proof. Set $C:=\max\{\ell_X(x') : x' \in X'\}$. If $g \in G$ satisfies $\ell_X(g) = n$ then there exist $x_1, x_2, \ldots, x_n \in X$ such that $g = x_1 x_2 \cdots x_n$. We then get

$$\ell_{X'}(g) = \ell_{X'}(x_1 x_2 \cdots x_n) \leq \sum_{i=1}^{n} \ell_{X'}(x_i) \leq Cn = C\ell_X(g).$$

This shows that $\ell_{X'}(g) \leq C\ell_X(g)$ for all $g \in G$. As a consequence, for all $g, h \in G$ we have that

$$d_{X'}(g,h) = \ell_{X'}(g^{-1}h) \leq C\ell_X(g^{-1}h) = Cd_X(g,h).$$

Setting $c:=1/\max\{\ell_{X'}(x) : x \in X\}$ and exchanging the roles of X and X' in the previous argument one shows that $cd_X(g,h) \leq d_{X'}(g,h)$ for all $g, h \in G$. It follows that d_X and $d_{X'}$ are Lipschitz-equivalent. □

Definition 7.7. A non-decreasing function $f\colon \mathbb{N} \to [0,+\infty)$ is called a *growth function.*

Let $f, f'\colon \mathbb{N} \to [0,+\infty)$ be two growth functions. We say that f' *dominates* f, and we write $f \preceq f'$, if there exists an integer $c \geq 1$ such that

$$f(n) \leq cf'(cn) \text{ for all } n \geq 1.$$

We say that f and f' are *equivalent* and we write $f \sim f'$ provided that $f \preceq f'$ and $f' \preceq f$.

The proof of the following proposition is left as an **exercise**.

Proposition 7.8. *The following hold:*

(i) $\preceq$ *is reflexive and transitive;*
(ii) $\sim$ *is an equivalence relation;*
(iii) *let* $f_1, f_2, f_1', f_2'\colon \mathbb{N} \to [0,+\infty)$ *be growth functions. Suppose that* $f_1 \sim f_1'$, $f_2 \sim f_2'$ *and that* $f_1 \preceq f_2$. *Then* $f_1' \preceq f_2'$.

Let $f\colon \mathbb{N} \to [0,+\infty)$ be a growth function. We denote by $[f]$ the $\sim$-equivalence class of f and we refer to it as to the *asymptotic behavior* of f. By abuse of notation, we shall also write $[f] \sim f(n)$. If f_1 and f_2 are two growth functions, we write $[f_1] \preceq [f_2]$ if $f_1 \preceq f_2$. This definition makes sense by virtue of Proposition 7.8.(iii). Note that, this way, $\preceq$ becomes a partial ordering on the set of equivalence classes of growth functions.

Example 7.9. (a) Let α and β be nonnegative real numbers. Then $n^\alpha \preceq n^\beta$ if and only if $\alpha \leq \beta$, and $n^\alpha \sim n^\beta$ if and only if $\alpha = \beta$.

(b) Let $f\colon \mathbb{N} \to [0,+\infty)$ be a growth function. Suppose that f is a polynomial of degree d for some $d \geq 0$. Then one has $f(n) \sim n^d$.

(c) Let $a, b \in (1,+\infty)$. Then

$$a^n \sim b^n. \tag{7.10}$$

We leave this as an **exercise**. In particular, we have $a^n \sim \exp(n)$ for all $a \in (1,+\infty)$.

(d) Let $d \geq 0$ be an integer. Then $n^d \preceq \exp(n)$ and $n^d \not\sim \exp(n)$ (**exercise**). As a consequence, if $f\colon \mathbb{N} \to [0,+\infty)$ is a growth function such that $f(n) \preceq n^d$, then $f \preceq \exp(n)$ and $f \not\sim \exp(n)$.

Corollary 7.10. *Let G be a finitely generated group and let X and X' be two finite symmetric generating subsets of G. Then, the growth functions associated with X and X' are equivalent, that is, $b_X \sim b_{X'}$. Moreover, $b_X(n) \preceq \exp(n)$.*

Proof. From Proposition 7.6, there exists a $C \geq 1$ such that

$$\ell_{X'}(g) = d_{X'}(1_G, g) \leq C d_X(1_G, g) = C\ell_X(g),$$

hence

$$B_X(1_G, n) = \{g \in G : \ell_X(g) \leq n\} \subseteq \{g \in G : \ell_{X'}(g) \leq Cn\} = B_{X'}(1_G, Cn).$$

This shows that $b_X(n) \leq b_{X'}(Cn) \leq C b_{X'}(Cn)$ for all $n \geq 1$, yielding $b_X \preceq b_{X'}$. By symmetry, we also have $b_{X'} \preceq b_X$, hence $b_X \sim b_{X'}$.

The last statement follows from (7.9). □

Let G be a finitely generated group. By the previous corollary, all growth functions associated with finite symmetric generating subsets of G are in the same equivalence class, which is called the *growth type* of G. We denote it by b^G.

One says that G has *exponential* (resp. *sub-exponential*) *growth* if $b^G(n) \sim \exp(n)$ (resp. $b^G(n) \not\sim \exp(n)$).

One says that G has *sub-polynomial growth* (resp. *polynomial growth*) if there exists an integer $d \geq 0$ such that $b^G(n) \preceq n^d$ (resp. $b^G(n) \sim n^d$).

Remark 7.11. In the literature, in relation to the growth of groups, one often finds that no distinction is made between the notions of "sub-polynomial" and "polynomial" growth. Although, *a priori*, the two notions are distinct (and, in fact, are not all equivalent outside the setting of the growth of groups: this is the case, for instance, for the growth of finitely-generated algebras, where the so-called *Gelfand–Kirillov dimension* plays the role of the present polynomial degree of growth) it turns out, as a consequence of a deep theorem of Gromov (Theorem 12.1; cf. Corollary 12.24) that for finitely generated groups the two notions coincide. For this reason, in the study of the growth of groups, one often directly refers to "polynomial growth" having in mind "sub-polynomial growth". However, for the sake of clarity, we shall try to keep these two notions distinct.

The proof of the following proposition is left as an **exercise**.

Proposition 7.12. (1) *Let $b \colon \mathbb{N} \to [0, +\infty)$ be a growth function with $b(0) > 0$. Then $b \sim 1$ if and only if b is bounded.*

(2) *Let G be a finitely generated group. Then $b^G \sim 1$ if and only if G is finite. As a consequence, all finite groups have the same growth type.*

(3) *Let G be a finitely generated group. Then G is infinite if and only if $n \preceq b^G(n)$.*

(4) *Every finitely generated group of sub-polynomial growth (in particular, of polynomial growth) has sub-exponential growth.*

7.5 The Growth Rate

In this section we give an analytic characterization of the exponential growth. It relies on the following classical result in undergraduate-level analysis.

Lemma 7.13 (Fekete [111]). *Let $(a_n)_{n\geq 1}$ be a sequence of positive real numbers such that $a_{n+m} \leq a_n a_m$ for all $n,m \geq 1$. Then the limit*

$$\lim_{n\to\infty} \sqrt[n]{a_n}$$

exists and equals $\inf_{n\geq 1} \sqrt[n]{a_n}$.

Proof. Fix an integer $t \geq 1$. Then for all $n \geq 1$ there exist $q_n, r_n \in \mathbb{N}$ such that $n = tq_n + r_n$ and $0 \leq r_n < t$. Then

$$a_n \leq a_{tq_n} a_{r_n} \leq a_t^{q_n} a_{r_n}$$

so that, taking the n-th roots, $a_n^{1/n} \leq a_t^{q_n/n} a_{r_n}^{1/n}$.

As $0 \leq r_n/n < t/n$ and $\lim_{n\to\infty} t/n = 0$, we have that $\lim_{n\to\infty} r_n/n = 0$ and

$$\lim_{n\to\infty} \frac{q_n}{n} = \lim_{n\to\infty} \left(\frac{1}{t} - \frac{r_n}{tn}\right) = \frac{1}{t}.$$

In particular, $\lim_{n\to\infty} a_t^{q_n/n} a_{r_n}^{1/n} = a_t^{1/t}$. Thus,

$$\limsup_{n\to\infty} a_n^{1/n} \leq \limsup_{n\to\infty} a_t^{q_n/n} a_{r_n}^{1/n} = \lim_{n\to\infty} a_t^{q_n/n} a_{r_n}^{1/n} = a_t^{1/t}.$$

As t was arbitrary, it follows that

$$\limsup_{n\to\infty} \sqrt[n]{a_n} \leq \inf_{t\geq 1} \sqrt[t]{a_t} \leq \liminf_{t\to\infty} \sqrt[t]{a_t},$$

completing the proof. □

Proposition 7.14. *Let G be a finitely generated group and let X be a finite symmetric generating subset of G. Then, the limit*

$$\beta_X := \lim_{n\to\infty} \sqrt[n]{b_X(n)} \tag{7.11}$$

exists and $\beta_X \in [1, +\infty)$.

Proof. Clearly $B_X(1_G, n+m) \subseteq B_X(1_G, n) B_X(1_G, m)$ and therefore

$$\begin{aligned} b_X(n+m) &= |B_X(1_G, n+m)| \leq |B_X(1_G, n) B_X(1_G, m)| \\ &\leq |B_X(1_G, n)| \cdot |B_X(1_G, m)| = b_X(n) b_X(m). \end{aligned}$$

Thus, the sequence $(b_X(n))_{n\geq 1}$ satisfies the hypotheses of the previous lemma and β_X exists and is finite. As $b_X(n) \geq 1$ for all $n \in \mathbb{N}$ we also have $\beta_X \geq 1$. □

Definition 7.15. The number β_X in (7.11) is called the *growth rate* of G with respect to X.

The proof of the proposition is left as an **exercise**.

Proposition 7.16. *Let G be a finitely generated group and let X be a finite symmetric generating subset of G. Then $\beta_X > 1$ (resp. $\beta_X = 1$) if and only if G has exponential (resp. sub-exponential) growth. In particular, the condition $\beta_X > 1$ (resp. $\beta_X = 1$) is independent of X.*

Note that if $X_1, X_2 \subseteq G$ are two finite symmetric generating subsets of G then β_{X_1} may differ from β_{X_2}. The above proposition states that either $\beta_{X_1} > 1$ and $\beta_{X_2} > 1$, or $\beta_{X_1} = 1 = \beta_{X_2}$.

7.6 Growth of Subgroups and Quotients

In the following, in order to avoid confusion with the ambient group, given a finitely generated group G and a finite generating subset $X \subseteq G$, we denote by $b_X^G = b_X$ the corresponding growth function.

Proposition 7.17. *Let G be a finitely generated group and let H be a finitely generated subgroup of G. Then $b^H \preceq b^G$.*

Proof. Let X_G (resp. X_H) be a finite symmetric generating subset of G (resp. H). Then the set $X := X_H \cup X_G$ is a finite symmetric generating subset of G. As $X_H \subseteq X$ we have $B_{X_H}(1_H, n) \subseteq B_X(1_G, n)$ and therefore $b_{X_H}^H(n) \leq b_X^G(n)$ for all $n \in \mathbb{N}$. Thus, $b^H \preceq b^G$. □

Corollary 7.18. *Every finitely generated group which contains a finitely generated subgroup of exponential growth (e.g. a subgroup isomorphic to the free group F_2) has exponential growth.* □

Example 7.19. Consider the groups $\mathrm{SL}(n, \mathbb{Z}) = \{A \in M_n(\mathbb{Z}) : \det A = 1\}$, $n \geq 2$. For $n = 2$ it is straightforward to check that the matrices

$$x_1 = \begin{pmatrix} 1 & 1 \\ 0 & 1 \end{pmatrix} \quad \text{and} \quad x_2 = \begin{pmatrix} 1 & 0 \\ 1 & 1 \end{pmatrix}$$

generate $\mathrm{SL}(2, \mathbb{Z})$. Moreover, it follows from Example 1.19 that $\mathrm{SL}(2, \mathbb{Z})$ contains a subgroup isomorphic to the free group F_2. Hence $\mathrm{SL}(2, \mathbb{Z})$ has exponential growth. We leave it as an **exercise** to show that for $n \geq 3$ the group $\mathrm{SL}(n, \mathbb{Z})$ is also finitely generated and of exponential growth.

We now show that finitely generated groups and their finite index subgroups (which are finitely generated as well, by Corollary 1.11) have the same growth type.

Proposition 7.20. *Let G be a finitely generated group and let H be a finite index subgroup of G. Then $b^H = b^G$.*

Proof. From Proposition 7.17 we have that $b^H \preceq b^G$.

Let now X be a finite symmetric generating subset of G and $S \subseteq G$ a transversal of H in G. It follows from Proposition 1.9 that the (finite) set $Y = \{sx(\overline{sx})^{-1} : s \in S,\ x \in X\}$ is a generating subset of H. Since X is symmetric, it follows from (1.11) in the proof of Proposition 1.9 that Y is symmetric as well.

Let then $g \in B_X(1_G, n)$ and write $g = x_1 x_2 \cdots x_n$, where $x_1, x_2, \ldots, x_n \in X$. As in the proof of Proposition 1.9 we may find $s_0 = 1_G, s_1, \ldots, s_n \in S$ such that

$$\begin{aligned} g &= x_1 x_2 \cdots x_n \\ &= (1_G x_1 s_1^{-1})(s_1 x_2 s_2^{-1}) \cdots (s_{n-2} x_{n-1} s_{n-1}^{-1})(s_{n-1} x_n s_n^{-1}) s_n \\ &= h_1 h_2 \cdots h_{n-1} h_n s_n, \end{aligned}$$

where $s_{i+1} = \overline{s_i x_{i+1}}$ so that $h_i := s_{i-1} x_i s_i^{-1} \in Y \subseteq H$, for $i = 0, 1, \ldots, n-1$. It follows that $B_X(1_G, n) \subseteq B_Y(1_H, n)S$ so that, taking cardinalities, and setting $c := |S| = [G : H]$,

$$b_X^G(n) = |B_X(1_G, n)| \leq |B_Y(1_H, n)| \cdot |S| = c b_Y^H(n) \leq c b_Y^H(cn)$$

This shows that $b^G \preceq b^H$. It follows from the beginning of the proof that $b^G = b^H$. □

Proposition 7.21. *Let G be a finitely generated group and let N be a normal subgroup of G. Then the quotient group G/N is finitely generated and one has $b^{G/N} \preceq b^G$. If in addition N is finite, then $b^{G/N} = b^G$.*

Proof. Let X be a finite symmetric generating subset of G and let $\pi \colon G \to G/N$ denote the quotient homomorphism. Then $X' := \pi(X)$ is a finite symmetric generating subset of G/N. Thus, for all $n \in \mathbb{N}$ one has (**exercise**)

$$B_{X'}(1_{G/N}, n) = \pi(B_X(1_G, n)) \tag{7.12}$$

and therefore

$$b_{X'}^{G/N}(n) = |B_{X'}(1_{G/N}, n)| = |\pi(B_X(1_G, n))| \leq |B_X(1_G, n)| = b_X^G(n).$$

This shows that $b^{G/N} \preceq b^G$.

Suppose now that N is finite. Then the map π is $|N|$-to-one. Moreover, from (7.12) we deduce that $B_X(1_G, n) \subseteq \pi^{-1}(B_{X'}(1_{G/N}, n))$. Thus, setting $c := |N|$ we have

$$\begin{aligned} b_X^G(n) &= |B_X(1_G, n)| \leq |\pi^{-1}(B_{X'}(1_{G/N}, n))| \\ &= c|B_{X'}(1_{G/N}, n)| = c b_{X'}^{G/N}(n) \leq c b_{X'}^{G/N}(cn). \end{aligned}$$

This shows that $b^G \preceq b^{G/N}$. From the first part of the statement it follows that $b^{G/N} = b^G$. □

The proof of the following lemma is left as an **exercise**.

Lemma 7.22. *Let $f_1, f_2, f_1', f_2' \colon \mathbb{N} \to [0, +\infty)$ be growth functions. Suppose that $f_i \preceq f_i'$, $i = 1, 2$. Then the products $f_1 f_2, f_1' f_2' \colon \mathbb{N} \to [0, \infty)$ are also growth functions and one has $f_1 f_2 \preceq f_1' f_2'$.*

Given two growth functions f_1 and f_2 we set $[f_1] \cdot [f_2] := [f_1 f_2]$. This is well defined by virtue of Lemma 7.22.

The proof of the following proposition is left as an **exercise**.

Proposition 7.23. *Let G_1 and G_2 be two finitely generated groups. Then the direct product $G_1 \times G_2$ is also finitely generated and $b^{G_1 \times G_2} = b^{G_1} b^{G_2}$.*

From Example 7.5.(a) and Proposition 7.23 one immediately deduces the following:

Corollary 7.24. $b^{\mathbb{Z}^d}(n) \sim n^d$. □

From the structure theorem of finitely generated abelian groups (Corollary 1.30), Corollary 7.24 and Proposition 7.20 we deduce the following:

Corollary 7.25. *Every finitely generated abelian group has polynomial growth.* □

7.7 Groups of Linear Growth

Let G be a finitely generated group. One says that G has *sub-linear growth* (resp. *linear growth*) provided that $b^G(n) \preceq n$ (resp. $b^G(n) \sim n$). In other words, let X be a finite symmetric generating subset of G, then G has sub-linear growth if there exists a constant c such that $b_X(n) \leq cn$ for all $n \geq 1$. Thus a group of sub-linear growth has sub-polynomial (and therefore sub-exponential) growth.

It follows from Proposition 7.12.(2) that all finite groups have sub-linear growth. Also, it follows from Example 7.5.(a) and Proposition 7.20 that $\mathbb{Z}$ and all finite extensions of $\mathbb{Z}$ have linear growth. The main result of this section tells us that there are no other examples of groups of sub-linear growth and, as a by-product, for infinite finitely generated groups the notions of "sub-linear" and "linear" growth coincide.

Theorem 7.26 (Justin). *Every infinite finitely generated group of sub-linear growth contains an element of infinite order.*

Proof. Let G be a finitely generated group and let $X = \{x_1, x_2, \ldots, x_r\}$ be a finite symmetric generating subset of G. Let also $Y = \{y_1, y_2, \cdots, y_r\}$ be a finite set (of the same cardinality r of X). As in Section 1.1, we denote by Y^* the free monoid generated by Y, with the empty word ε as a neutral element, and by $\ell(w)$ the length of a word $w \in Y^*$.

Let then $\pi \colon Y^* \to G$ denote the monoid homomorphism defined by $\pi(y_i) := x_i$ for all $i = 1, 2, \ldots, r$. We equip Y^* with a total ordering $\preceq$ by lexicographically extending the order $\varepsilon \preceq y_1 \preceq y_2 \preceq \cdots \preceq y_r$ on Y. We then recursively construct a subset $W \in Y^*$ by setting:

- $W_1 := Y \cup \{\varepsilon\} = \{\varepsilon, y_1, y_2, \cdots, y_r\}$;
- assume that W_n has been obtained for $n \geq 1$ and consists of words w of length $\ell(w) \leq n$. Consider $Y^{\times(n+1)}$ and, starting from the $\preceq$-minimal element, remove all those elements w such that there exists a $w' \in W_n \cup Y^{\times(n+1)}$ satisfying $w' \prec w$ and $\pi(w') = \pi(w)$. Define W_{n+1} as the union of W_n and the set of words in $Y^{\times(n+1)}$ that remain after this process has been completed;
- $W := \bigcup_{n=1}^{\infty} W_n$.

Let then $Y^\infty := \{f : \mathbb{N} \to Y\}$ be the set of all *infinite words* on Y, and denote by $W_\infty := \{f \in Y^\infty : f(1)f(2)\cdots f(n) \in W \text{ for all } n \in \mathbb{N}\}$ the set of infinite words on Y whose prefixes are all in W. We call the elements of W_∞ *chains*.

We leave it as an **exercise** to check the following facts:

(1) every subword of a word in W belongs to W;
(2) the map π induces a bijection between W and G such that $\ell(w) = \ell_X(\pi(w))$ for all $w \in W$;
(3) G is infinite if and only if W_∞ is nonempty;
(4) if $w \in W_\infty$ then $w[h,k] := w(h)w(h+1)\cdots w(k) \in W$ for all $h,k \in \mathbb{N}$ such that $h \leq k$.

Given $f \in Y^\infty$ and $h \in \mathbb{N}$, we denote by $f_h \in Y^\infty$ the shifted infinite word defined by $f_h(k) = f(h+k)$, for all $k \in \mathbb{N}$. For all $w \in W_\infty$ and $h \in \mathbb{N}$, we have $w_h \in W_\infty$, i.e. W_∞ is shift-invariant. We then say that a chain w is *periodic* if there exist $h \in \mathbb{N}$ (the *preperiod*) and $k \in \mathbb{N} \setminus \{0\}$ (the *period*) such that $w_h = w_{h+k}$. In other words, a periodic chain is a right-infinite word w of the form

$$(y_0 y_1 \cdots y_{h-1}) \cdot (y_h y_{h+1} \cdots y_{h+k-1}) \cdot (y_h y_{h+1} \cdots y_{h+k-1}) \cdot \ldots \cdot (y_h y_{h+1} \cdots y_{h+k-1}) \cdot \ldots,$$

where $y_0, y_1, \ldots, y_{h+k-1} \in Y$.

Suppose now that G has sub-linear growth and let c be a positive integer such that $b_X(n) \leq cn$ for all $n \geq 1$.

Suppose that G is infinite. By (3) there exists a chain $w \in W_\infty$. We claim that indeed w is periodic. If not, for all $h,k \in \mathbb{N}$ with $k \geq 1$, we have $w_h \neq w_{h+k}$, and therefore there exists an $M(h,k) \in \mathbb{N}$ such that $w_h[0,m] \neq w_{h+k}[0,m]$ for all $m \geq M(h,k)$.

Set $M := \max\{M(h,k) : 1 \leq h+k \leq c+1\}$ and, for $n \geq M$, define

$$W_c(n) := \{w_j[0,m] : 1 \leq j \leq c+1,\ M \leq m \leq n\}.$$

By construction, the elements in $W_c(n)$ are all distinct and by (4) they all belong to W_n so that

$$b_X(n) = |W_n| \geq |W_c(n)| = (c+1)(n-M+1),$$

which, for large values of n, invalidates the definition of c.

Let then w be a periodic chain of period k and preperiod h. Let $u := w_h[0,k-1] \in W_k$ and set $g := \pi(u) \in G$. Periodicity and (4) imply that for all integers $m \geq 1$ we have $u^m = w_h[0,km-1] \in W_{km}$ so that, by (2), all the powers g^m are distinct. We deduce that g has infinite order, completing the proof of the theorem. □

Corollary 7.27 (Justin). *A finitely generated group of sub-linear growth is virtually cyclic.*

Proof. Let G be a finitely generated group of sub-linear growth. If G is finite, the statement is obvious.

Suppose that G is infinite. By Theorem 7.26 there exist elements in G of infinite order. Since G has sub-linear growth, we have that

$(*)$ given two elements $g_1, g_2 \in G$ of infinite order, there exist $s,t \in \mathbb{Z} \setminus \{0\}$ such that $g_1^s = g_2^t$,

otherwise the subgroup $\langle g_1, g_2 \rangle$ would have at least quadratic growth (**exercise**): a contradiction.

Let H denote the subgroup generated by all elements of infinite order: it is clearly characteristic. Moreover, it is of finite index in G. Otherwise G/H would be again a group of linear growth (cf. Proposition 7.21) and, by Theorem 7.26, it would contain an element g_2H of infinite order. Lifting g_2H to an element $g_2 \in G$ (which *a fortiori* has infinite order), for any $g_1 \in H$ of infinite order we would have $g_1^s \neq g_2^t$ for all $s,t \in \mathbb{Z} \setminus \{0\}$, contradicting $(*)$.

As $[G : H] < \infty$, the subgroup H is finitely generated, say by $h_1, h_2, \ldots, h_s$ (we may suppose that these generators all have infinite order). By $(*)$ there exist $m_1, m_2, \ldots, m_s \in \mathbb{Z}$ and $h \in H$ such that $h_1^{m_1} = h_2^{m_2} = \ldots = h_k^{m_k} = h$. Note that h has infinite order. Moreover, as h is centralized by all the h_i's, the subgroup $\langle h \rangle$ is normal in H and, by the same argument given before, of finite index in H. Thus $\mathbb{Z} \cong \langle h \rangle$ has finite index also in G. This completes the proof. □

Remark 7.28. It follows from Justin's theorem (Corollary 7.27) that a finitely generated group G of sub-linear growth is either finite or has (exactly) linear growth, i.e., either $b^G \sim 1$ or $b^G(n) \sim n$. Thus, a sub-linear function such as $\sqrt{n}$ cannot occur as a growth type of a finitely generated group (cf. Remark 7.11). We shall see later, as a consequence of Gromov's theorem (Theorem 12.1) that, similarly, a finitely generated group G of sub-polynomial growth has (exactly) polynomial growth (Corollary 12.24).

7.8 The Growth of Nilpotent Groups and the Bass–Guivarc'h Formula

The goal of this section is to prove that finitely generated nilpotent groups have polynomial growth and present the Bass–Guivarc'h formula for the corresponding degree of polynomial growth.

Theorem 7.29 (Bass–Guivarc'h). *Let G be a finitely generated nilpotent group, and let $G = \gamma_1(G) \geq \gamma_2(G) \geq \cdots \geq \gamma_c(G) \geq \gamma_{c+1}(G) = \{1_G\}$ be its lower central series. Denote by r_i the torsion-free rank of $\gamma_i(G)/\gamma_{i+1}(G)$ and set $d := \sum_i i r_i$. Then one has*

$$b^G(n) \sim n^d.$$

We will focus first on the upper bound.

A first reduction

The first observation is that we can reduce ourselves to the torsion-free case. In order to keep track of the ranks of the factors of the lower central series, rather than passing to a finite index torsion-free subgroup of G (cf. Lemma 2.42), we will mod out the torsion subgroup of G.

Recall that in a nilpotent group G all torsion elements form a subgroup G_{tor}, which is finite if G is finitely generated (cf. Corollary 2.37).

For $i = 1, 2, \ldots, c+1$, let us simply write γ_i in place of $\gamma_i(G)$ and then denote by $\sqrt{\gamma_i}$ the set all elements of G which are torsion modulo γ_i, that is,

$$\sqrt{\gamma_i} := \{g \in G : \exists n \geq 1 \text{ such that } g^n \in \gamma_i\}.$$

In other words, $\sqrt{\gamma_i}$ is the preimage of the torsion subgroup $\sqrt{\gamma_i}/\gamma_i$ of the nilpotent group G/γ_i. Hence the $\sqrt{\gamma_i}$'s are clearly normal subgroups of G, and we have the inclusions

$$G = \sqrt{\gamma_1} \geq \sqrt{\gamma_2} \geq \cdots \geq \sqrt{\gamma_c} \geq \sqrt{\gamma_{c+1}} = \sqrt{\{1\}}.$$

Note that $\sqrt{\{1\}} = G_{tor}$, the subgroup of torsion elements of G. Moreover, by Corollary 2.37, the torsion subgroup $\sqrt{\gamma_i}/\gamma_i$ of G/γ_i is finite, i.e. $[\sqrt{\gamma_i} : \gamma_i] < \infty$.

Let $1 \leq i, j \leq c+1$. We claim that $[\sqrt{\gamma_i}, \sqrt{\gamma_j}] \subseteq \sqrt{\gamma_{i+j}}$. Let $a \in \sqrt{\gamma_i}$ and $b \in \sqrt{\gamma_j}$. Then we can find $n, m \in \mathbb{N}$ such that $a^n \in \gamma_i$ and $b^m \in \gamma_j$. Now recall that by virtue of Lemma 2.26 the map $(\gamma_i/\gamma_{i+1}) \times (\gamma_j/\gamma_{j+1}) \to \gamma_{i+j}/\gamma_{i+j+1}$, defined by $(a_i\gamma_{i+1}, b_j\gamma_{j+1}) \mapsto [a_i, b_j]\gamma_{i+j+1}$ for all $a_i \in \gamma_i$ and $b_j \in \gamma_j$, is bilinear. We then have $[a^n, b^m]\gamma_{i+j+1} = [a, b]^{nm}\gamma_{i+j+1}$. As $[a^n, b^m] \in [\gamma_i, \gamma_j] \subseteq \gamma_{i+j}$, where the last inclusion follows from centrality of the lower central series, we deduce that $[a, b]^{nm} \in \gamma_{i+j}$ as well. This shows that $[a, b] \in \sqrt{\gamma_{i+j}}$, and the claim is proved.

Now the quotient group $G/\sqrt{\{1\}} = G/G_{tor}$ is torsion-free nilpotent, and has the same growth of G by Proposition 7.21. Moreover, its series $\left(\sqrt{\gamma_i}/\sqrt{\{1\}}\right)_{i=1}^{c}$ is central by virtue of the claim above, and the corresponding factors

$$\frac{\sqrt{\gamma_i}/\sqrt{\{1\}}}{\sqrt{\gamma_{i+1}}/\sqrt{\{1\}}} \cong \sqrt{\gamma_i}/\sqrt{\gamma_{i+1}}$$

are torsion-free (since $G/\sqrt{\gamma_{i+1}}$ is torsion-free) and free-abelian of the same free-abelian rank of γ_i/γ_{i+1} (since $[\sqrt{\gamma_i}/\gamma_{i+1} : \gamma_i/\gamma_{i+1}] = [\sqrt{\gamma_i} : \gamma_i] < \infty$).

Remark 7.30. Notice that it was not enough to observe that G has a normal subgroup of finite index which is torsion free, since we don't know, a priori, the ranks of its lower central series. We actually needed to find an appropriate series, as we did.

Thus, up to replacing G by $G/\sqrt{\{1\}}$ if necessary, we can suppose that G is torsion-free. Moreover, setting $G_i := \sqrt{\gamma_i}/\sqrt{\{1\}}$, we have that the series

$$G = G_1 \geq G_2 \geq \cdots \geq G_c \geq \{1\}$$

is central (that is, $[G_i, G_j] \subseteq G_{i+j}$), the factors G_i/G_{i+1} are torsion-free abelian, and $\mathrm{rk}(G_i/G_{i+1}) = r_i$. This is all we need to prove the upper bound in the Bass–Guivarc'h formula.

An example: $\mathrm{UT}(m, \mathbb{Z})$

The ideas behind the Bass–Guivarc'h formula are better understood by analyzing the motivating example $G := \mathrm{UT}(m, \mathbb{Z})$ of upper triangular $m \times m$ matrices with integer coefficients. Note that for $m = 3$, the group $\mathrm{UT}(3, \mathbb{Z})$ is the so-called *Heisenberg group*.

Let us fix $m \geq 1$.

Definition 7.31. A *transvection* is an $m \times m$ matrix of the form $t_{i,j}(a) := I^{(m)} + E_{i,j}^{(m)}(a)$, where $I^{(m)}$ is the identity matrix of size m, and for fixed $1 \leq i \neq j \leq m$, $E_{i,j}^{(m)}(a)$ is the square matrix of size m with $a \in \mathbb{Z}$ in the (i, j)-th position, and 0 elsewhere.

Notice that $t_{i,j}(a)^{-1} = t_{i,j}(-a)$ for all $a \in \mathbb{Z}$.

Consider the finite symmetric generating subset $X := \{t_{i,j}(\pm 1) : j > i\}$ of the group $G := \mathrm{UT}(m, \mathbb{Z})$.

Lemma 7.32. *There exists a $C > 0$ such that every product of elements of X of length $\leq n$ is of the form* $\begin{pmatrix} 1 & & a_{i,j} \\ & \ddots & \\ 0 & & 1 \end{pmatrix}$ *with $|a_{i,j}| \leq C \cdot n^{j-i}$.*

Assuming the lemma, we count these words of length $\leq n$. Since the matrices have size m, looking at the upper diagonals (which have length $m-1, m-2, \ldots, 2, 1$), we have

$$\begin{aligned} b_X(n) &\leq (2Cn+1)^{m-1} \cdot (2Cn^2+1)^{m-2} \cdot \ldots \cdot (2Cn^{m-1}+1) \\ &\leq C' n^{(m-1)+2(m-2)+\cdots+(m-1)}, \end{aligned} \tag{7.13}$$

where C' is a constant. The last term is clearly a polynomial in n. Notice also that its degree can be thought of as $\sum_{i=1}^{m} i r_i$, where $r_i = m - i$ equals the rank of the factor $\gamma_i(\mathrm{UT}(m, \mathbb{Z}))/\gamma_{i+1}(\mathrm{UT}(m, \mathbb{Z}))$ (see Section 2.4).

Proof of Lemma 7.32. We rename each element $t_{j,k}(\pm 1)$ of X by $x_i^{\pm 1}$, and we set $x_i^{\pm 1} =: 1 \pm a_i$ (here on the right-hand side 1 denotes the identity matrix), where a_i is the matrix $\pm E_{j,k}^{m}$.

Consider the ring $R = \mathrm{B}(m, \mathbb{Z})$ of upper triangular matrices of size m and observe that the ideal of strictly upper triangular matrices

$$I := \left\{ \begin{pmatrix} 0 & & * \\ & \ddots & \\ 0 & & 0 \end{pmatrix} \right\}$$

is nilpotent as $I^m = \{0\}$. Notice that the nilpotency class of the group $\mathrm{UT}(m,\mathbb{Z})$ is $m-1$ (see Section 2.4).

Let $M \subseteq R$ denote the (finite) set consisting of all products of the form $a_{i_1}a_{i_2}\cdots a_{i_k}$, where $1 \leq k \leq n$, and set

$$c := \max\{|v_{i,j}| : 1 \leq i < j \leq m, v = \|v_{i,j}\| \in M\}.$$

We then deduce an estimate for the (p,q)-entry of a product $x_{i_1}x_{i_2}\cdots x_{i_k}$, where $1 \leq p < q \leq m$ and $1 \leq k \leq n$. The key observation is that if $a_{j_1}a_{j_2}\cdots a_{j_r}$ has a nonzero (p,q)-entry, then necessarily $r \leq q-p$: this is clear after looking at how matrices multiply.

We have

$$\begin{aligned} x_{i_1}x_{i_2}\cdots x_{i_k} &= (1+a_{i_1})(1+a_{i_2})\cdots(1+a_{i_k}) \\ &= 1+(a_{i_1}+a_{i_2}+\cdots+a_{i_k})+\sum_{\substack{j,j'=1\\ j<j'}}^{k} a_{i_j}a_{i_{j'}} \\ &\quad + \sum_{\substack{j,j',j''=1\\ j<j'<j''}}^{k} a_{i_j}a_{i_{j'}}a_{i_{j''}}+\cdots+a_{i_1}a_{i_2}\cdots a_{i_k}. \end{aligned} \tag{7.14}$$

It follows that the absolute value $|(x_{i_1}x_{i_2}\cdots x_{i_k})_{p,q}|$ of the (p,q)-entry in (7.14) is bounded above by c times the number of all possible summands of monomials $a_{i_{j_1}}a_{i_{j_2}}\cdots a_{i_{j_\ell}}$ of length ℓ not exceeding $q-p$. In other words,

$$|(x_{i_1}x_{i_2}\cdots x_{i_k})_{p,q}| \leq c\left(n+\binom{n}{2}+\cdots+\binom{n}{q-p}\right) \leq C\cdot n^{q-p},$$

where C is a constant. □

This shows how to get the upper bound (cf. equation (7.13)) for the Bass–Guivarc'h formula in the case $G = \mathrm{UT}(m,\mathbb{Z})$.

Recall that by Malcev's theorem (Theorem 2.20) any finitely generated torsion free nilpotent group is embeddable into $\mathrm{UT}(m,\mathbb{Z})$ for some $m \geq 1$. Thus, from Proposition 7.17 we immediately deduce the following:

Corollary 7.33. *Every finitely generated nilpotent group has sub-polynomial growth.* □

However, we still don't know if the growth is (exactly) polynomial, nor do we know the corresponding degree of polynomial growth.

The upper bound

Recall that we can assume that G has a series

$$G = G_1 \geq G_2 \geq \cdots \geq G_c \geq \{1\}$$

that is central (i.e. $[G_i, G_j] \subseteq G_{i+j}$), the factors G_i/G_{i+1} are torsion-free abelian, and $\mathrm{rk}(G_i/G_{i+1}) = r_i := \mathrm{rk}(\gamma_i(G)/\gamma_{i+1}(G))$. For each $i = 1, 2, \ldots, c$, let $x_{ij}G_{i+1}$, $j = 1, 2, \ldots, r_i$, be free generators of G_i/G_{i+1}. Notice that with this choice the subgroup G_ℓ is generated by the x_{ij}'s with $i \geq \ell$, for all $\ell = 1, 2, \ldots, c$. In particular, the symmetric subset $X := \{x_{ij}^{\pm 1} : 1 \leq i \leq c, 1 \leq j \leq r_i\}$ generates G, and every element $g \in G$ can be uniquely written in the *canonical form*

$$g = x_{11}^{e_{11}} x_{12}^{e_{12}} \cdots x_{1r_1}^{e_{1r_1}} \cdots x_{c1}^{e_{c1}} \cdots x_{cr_c}^{e_{cr_c}},$$

where $e_{ij} \in \mathbb{Z}$ for all i, j.

Given an element $g \in G$ written as a word w in the elements of X of length $\ell_X(g) = n$, we describe a procedure to express g in its canonical form. We start by looking at the occurrences of $x_{11}^{\pm 1}$ in w, from left to right. We want to move them to the leftmost positions. In order to do this, for every $y \in X$, $y \neq x_{11}^{\pm 1}$, we replace iteratively every possible occurrence of $yx_{11}^{\pm 1}$ with $x_{11}^{\pm 1} y [y, x_{11}^{\pm 1}]$. Notice that, after all these replacements, the element g is expressed as a word $w' = x_{11}^{e_{11}} w_1$, with $e_{11} \in \mathbb{Z}$, the number $|e_{11}|$ is than or equal to the number of occurrences of $x_{11}^{\pm 1}$ in w, and w_1 is a word in $X \setminus \{x_{11}^{\pm 1}\}$ and simple commutators of weight at least 2 in the elements of X.

Example 7.34. Consider the word $w = x_{11}x_{21}x_{11}x_{21}^{-1}x_{12}x_{11}^{-1}x_{22}$. Moving the occurrences of $x_{11}^{\pm 1}$ to the left, we get

$$x_{11}x_{21}x_{11}x_{21}^{-1}x_{12}x_{11}^{-1}x_{22} \rightsquigarrow$$

$$\begin{aligned}
&\rightsquigarrow x_{11}x_{11}x_{21}[x_{21}, x_{11}]x_{21}^{-1}x_{12}x_{11}^{-1}x_{22} \\
&\rightsquigarrow x_{11}x_{11}x_{21}[x_{21}, x_{11}]x_{21}^{-1}x_{11}^{-1}x_{12}[x_{12}, x_{11}^{-1}]x_{22} \\
&\rightsquigarrow x_{11}x_{11}x_{21}[x_{21}, x_{11}]x_{11}^{-1}x_{21}^{-1}[x_{21}^{-1}, x_{11}^{-1}]x_{12}[x_{12}, x_{11}^{-1}]x_{22} \\
&\rightsquigarrow x_{11}x_{11}x_{21}x_{11}^{-1}[x_{21}, x_{11}][x_{21}, x_{11}, x_{11}^{-1}]x_{21}^{-1}[x_{21}^{-1}, x_{11}^{-1}]x_{12}[x_{12}, x_{11}^{-1}]x_{22} \\
&\rightsquigarrow x_{11}x_{11}x_{11}^{-1}x_{21}[x_{21}, x_{11}^{-1}][x_{21}, x_{11}][x_{21}, x_{11}, x_{11}^{-1}]x_{21}^{-1}[x_{21}^{-1}, x_{11}^{-1}]x_{12}[x_{12}, x_{11}^{-1}]x_{22}.
\end{aligned}$$

Hence the word w has been rewritten as $w' = x_{11}w_1$, where

$$w_1 = x_{21}[x_{21}, x_{11}^{-1}][x_{21}, x_{11}][x_{21}, x_{11}, x_{11}^{-1}]x_{21}^{-1}[x_{21}^{-1}, x_{11}^{-1}]x_{12}[x_{12}, x_{11}^{-1}]x_{22}.$$

Observe that every commutator occurring in w_1 is an element of G_2, hence we can express it as a word in $X \setminus \{x_{11}^{\pm 1}, x_{12}^{\pm 1}, \ldots, x_{1r_1}^{\pm 1}\}$. Therefore we can express w_1 as a word w_1' in $X \setminus \{x_{11}^{\pm 1}\}$.

Now we can work on w_1', and repeat the same argument iteratively for all the generators $x_{12}, x_{13}, \ldots, x_{1r_1}, \ldots, x_{cr_c}$, proceeding in lexicographic order. Since G_c is in the center of G, this procedure brings g to its canonical form in finitely many steps.

Carefully estimating the number of operations of this procedure will provide us with the upper bound that we are looking for.

Let $X_i := \{x_{ij}^{\pm 1} : 1 \leq j \leq r_i\}$ denote the set of *generators of weight* i, for $i = 1, 2, \ldots, c$. Notice that $X = \cup_i X_i$. We also set

$$A := \sup\{\ell_X([y_1, y_2, \ldots, y_s]) : 1 \leq s \leq c, y_i \in X \text{ for all } i\}.$$

After moving all the elements of X_1 occurring in w to the left, we will have rewritten w as

$$x_{11}^{e_{11}} x_{12}^{e_{12}} \cdots x_{1r_1}^{e_{1r_1}} w^{(1)},$$

where $w^{(1)}$ is expressed as a word in $X \setminus X_1$.

For each $s = 2, 3, \ldots, c$, we want to estimate the number of generators of X_s occurring in $w^{(1)}$.

Notice that, during our rewriting process, we will get commutators of the form

$$[x_{ij}^{\pm 1}, y_1, \ldots, y_t],$$

where $y_1, y_2, \ldots, y_t \in X_1$ and $0 \leq t < c$ (for $t = 0$ these are simply the generators $x_{ij}^{\pm 1}$'s). Notice that these commutators are in G_{i+t}.

The crucial observation is that in our rewriting process, when moving each single generator of X to the left, we get precisely one extra simple commutator of weight $i+1$ from each simple commutator of weight i that gets passed by.

After moving all the elements of X_1 to the left, in what remains on the right, there are obviously at most n simple commutators of weight 1 (i.e. the generators in X); by our observation above, since the simple commutators of weight 2 can only come from an x_{1j} passing a simple commutator of weight 1, there are at most $n \cdot n = n^2$ of them; inductively, since again the simple commutators of weight $i+1$ can only come from an x_{1j} passing a simple commutator of weight i, there are at most $n \cdot n^i = n^{i+1}$ of them.

Now each commutator has length at most A, hence the number of generators in X_s occurring in $w^{(1)}$ is at most

$$An + An^2 + \cdots + An^s \leq cAn^s.$$

By induction, given $k \in \{1, 2, \ldots, c-1\}$, after moving all the elements of $X_1 \cup X_2 \cup \cdots \cup X_k$ to the left, so that we have rewritten w as

$$x_{11}^{e_{11}} \cdots x_{kr_k}^{e_{kr_k}} w^{(k)},$$

where $w^{(k)}$ is expressed as a word in $X \setminus (X_1 \cup X_2 \cup \cdots \cup X_k)$, we can assume that there exists a constant $B > 0$ such that the number of elements of X_s occurring in $w^{(k)}$ is bounded by Bn^s, for every $s = k+1, k+2, \ldots, c$.

After moving all the elements of X_{k+1} to the left, we rewrite w as

$$x_{11}^{e_{11}} \cdots x_{(k+1)r_{k+1}}^{e_{(k+1)r_{k+1}}} w^{(k+1)},$$

where $w^{(k+1)}$ is expressed as a word in $X \setminus (X_1 \cup X_2 \cup \cdots \cup X_{k+1})$. We want to estimate the number of elements of X_s occurring in $w^{(k+1)}$ for $s = k+2, k+3, \ldots, c$. Arguing as before, in the rewriting process we get simple commutators of the form

$$[x_{ij}^{\pm 1}, y_1, \ldots, y_t],$$

where $y_1, y_2, \ldots, y_t \in X_{k+1}$, $i \geq k+1$ and $0 \leq t < c$ (for $t = 0$ these are simply the generators $x_{ij}^{\pm 1}$'s). Notice that these commutators are in $G_{i+t(k+1)}$. Fixing i and t, by the inductive hypothesis, the number of such commutators is bounded by

$$(Bn^i) \cdot (Bn^{k+1})^t = B^{t+1} n^{i+t(k+1)}.$$

Given $p \geq k+1$, varying i and t so that $p = i + t(k+1)$, we deduce that there are at most

$$Bn^p + B^2 n^p + \cdots + B^{\lfloor p/(k+1) \rfloor + 1} n^p \leq Bn^p + B^2 n^p + \cdots + B^{c+1} n^p \leq (c+1) B^{c+1} n^p$$

simple commutators of weight p. Therefore the number of elements of X_s occurring in $w^{(k+1)}$ is bounded by

$$(c+1) A B^{c+1} n^s$$

for $s = k+2, k+3, \ldots, c$. So, by induction, we proved that there exists a constant $D > 0$ such that the number of generators of weight s in the canonical form of w is bounded by Dn^s for every $s = 1, 2, \ldots, c$.

Finally, if the canonical form of w is

$$x_{11}^{e_{11}} \cdots x_{1r_1}^{e_{1r_1}} x_{21}^{e_{21}} \cdots x_{2r_2}^{e_{2r_2}} \cdots x_{c1}^{e_{c1}} \cdots x_{cr_c}^{e_{cr_c}},$$

then we showed that

$$\sum_{j=1}^{r_i} |e_{ij}| \leq Dn^i \quad \text{for all } i = 1, 2, \ldots, c.$$

Now, in the abelian group $G_i/G_{i+1} \cong \mathbb{Z}^{r_i}$, we know that the number of distinct elements of length at most m with respect to the canonical generators is a polynomial $f_i(m)$ in m of degree r_i. Hence the number of possible values of g of length $\ell_X(g) = n$ is at most

$$\prod_{i=1}^{c} f_i(Dn^i) \leq Cn^d$$

for a suitable constant $C > 0$, where $d = \sum_{i=1}^{c} i r_i$. This completes the proof of the upper bound.

The lower bound

It remains to prove the lower bound for the Bass–Guivarc'h formula, which is due to Wolf.

We need a lemma. Recall that G is a finitely generated nilpotent group of nilpotency class c. Let also X be a finite symmetric generating subset of G.

Lemma 7.35. *Let $z := [x_1, x_2, \ldots, x_c]$, where $x_i \in X$. Then there exists a constant $B > 0$ such that $\ell_X(z^n) \le Bn^{\frac{1}{c}}$ for all $n \in \mathbb{N}$.*

Proof. Recall (cf. Lemma 2.26) that the commutator yields a bilinear map

$$\gamma_i/\gamma_{i+1} \times \gamma_j/\gamma_{j+1} \to \gamma_{i+j}/\gamma_{i+j+1}$$

defined by $(a_i\gamma_{i+1}, b_j\gamma_{j+1}) \mapsto [a_i, b_j]\gamma_{i+j+1}$ for all $a_i \in \gamma_i$ and $b_j \in \gamma_j$. More generally, the c-fold (left-normed) commutator gives a multilinear map

$$\underbrace{\gamma_1/\gamma_2 \times \gamma_1/\gamma_2 \times \cdots \times \gamma_1/\gamma_2}_{c} \to \gamma_c$$

defined by $(a_1\gamma_2, a_2\gamma_2, \ldots, a_c\gamma_2) \mapsto [a_1, a_2, \ldots, a_c]$, for all $a_1, a_2, \ldots, a_c \in \gamma_1$ (we consider only terms of the form γ_i/γ_{i+1} with $i = 1$ because if $i \ge 2$, the corresponding image is always 0).

Observe that, since $n \ge 1$, there exists an integer m such that $n \le m^c \le 2^c n$. Hence there exist $q, r \in \mathbb{N}$ such that $n = qm^{c-1} + r$ with $0 \le q \le m$ and $0 \le r < m^{c-1}$.

For $n = m^c$, by multilinearity, we have $[x_1, x_2, \ldots, x_c]^{m^c} = [x_1^m, x_2^m, \ldots, x_c^m]$. Thus

$$\ell_X(z^n) = \ell_X(z^{m^c}) = \ell_X([x_1, x_2, \ldots, x_c]^{m^c}) = \ell_X([x_1^m, x_2^m, \ldots, x_c^m]) \le 2^{c+1}m = Bn^{\frac{1}{c}},$$

where $B = 2^{c+1}$ is clearly independent of n.

If $n < m^c$, we have $n = qm^{c-1} + r$ with and $1 \le q < m$ and $0 \le r < m^{c-1}$.

Again by multilinearity we have

$$\begin{aligned}[x_1, x_2, \ldots, x_c]^n &= [x_1, \ldots, x_c]^{qm^{c-1}+r} \\ &= [[x_1, x_2, \ldots, x_{c-1}]^{m^{c-1}}, x_c^q] \cdot [[x_1, x_2, \ldots, x_{c-1}]^r, x_c].\end{aligned}$$

By induction on c we have that in the group G/γ_c (which is nilpotent of class $c-1$) the length of $[x_1, x_2, \ldots, x_{c-1}]^{m^{c-1}}$ modulo γ_c is bounded above by $B_1 m$, where $B_1 > 0$ is a constant independent of n. Also, since $\gamma_c \subseteq Z(G)$ (and $[az, b] = [a, b]$ if $z \in Z(G)$), if $y \in \gamma_c$ is such that

$$[x_1, \ldots, x_{c-1}]^{m^{c-1}} = [x_1^m, \ldots, x_{c-1}^m]y,$$

we have

$$\ell_X([[x_1, \ldots, x_{c-1}]^{m^{c-1}}, x_c^q]) = \ell_X([[x_1^m, \ldots, x_{c-1}^m]y, x_c^q]) = \ell_X([[x_1^m, \ldots, x_{c-1}^m], x_c^q]).$$

Since $\ell_X(x_c^q) = q < m$, we have that the length of $[[x_1,\ldots,x_{c-1}]^{m^{c-1}}, x_c^q]$ is bounded by a linear function of m, indeed

$$\ell_X([[x_1,\ldots,x_{c-1}]^{m^{c-1}}, x_c^q]) \leq (2B_1 + 2)m.$$

For the other term $[[x_1,x_2,\ldots,x_{c-1}]^r, x_c]$, we argue in the same way by induction on c and deduce that the length of $[x_1,x_2,\ldots,x_{c-1}]^r$ modulo γ_c is bounded above by $B_2 r^{1/(c-1)}$, where B_2 is a constant independent of r. Since $r < m^{c-1}$, this yields the upper bound $B_2 m$. Arguing as before, the length of $[[x_1,x_2,\ldots,x_{c-1}]^r, x_c]$ is bounded above by $2B_2 m + 2$.

All together

$$\ell_X(z^n) \leq (2B_1 + 2 + 2B_2)m + 2 < C'm < Cn^{1/c},$$

where $C' = 2B_1 + 2 + 2B_2 + 2$ and $C = 2C'$ are positive constants independent of n.

□

Consider now γ_c. It has free abelian rank r_c and we have $\gamma_c = \langle [x_1,x_2,\ldots,x_c] : x_i \in X \rangle$. So we can choose $z_1, z_2, \ldots, z_{r_c}$, where each z_i is a c-fold (left-normed) commutator of elements of X, and they generate a free abelian group of rank r_c.

By induction on c, there exists a constant $C' > 0$ such that G/γ_c contains at least $C' n^{\sum_{i=1}^{c-1} i r_i}$ distinct elements of length $\leq n$. Let W be a maximal set of words on X of length $\leq n$, distinct modulo γ_c. Consider the products $w z_1^{\alpha_1} z_2^{\alpha_2} \cdots z_{r_c}^{\alpha_{r_c}}$ with $w \in W$ and $0 \leq \alpha_j < n^c$. The number of such words is $\geq C'(n^{\sum_{i=1}^{c-1} i \cdot r_i})(n^c)^{r_c} = C' n^{\sum_{i=1}^{c} i \cdot r_i}$. By the previous lemma, their length is $\leq B'n$ for some constant $B' > 0$. It follows that $b_X(B'n) \geq C'n^d$, where $d = \sum_{i=1}^{c} i r_i$. This gives $b_X(n) \succeq n^d$ as we wanted, and finishes the proof of Theorem 7.29.

7.9 The Theorems of Milnor and Wolf

The goal of this section is to present the theorems of Milnor (Theorem 7.36) and Wolf (Theorem 7.37) which together yield the fact that finitely generated solvable group of sub-exponential growth are virtually nilpotent, and hence have polynomial growth (Corollary 7.41).

Theorem 7.36 (Milnor). *A finitely generated solvable group of sub-exponential growth is polycyclic.*

Proof. We claim that it suffices to show that $[G,G]$ is finitely generated. In fact, if this is the case, then $[G,G]$ is of sub-exponential growth (by Proposition 7.17) and, by induction on the derived length, it is polycyclic. On the other hand, the quotient group $G/[G,G]$ is finitely generated and abelian, hence polycyclic. Thus G, being an extension of polycyclic groups, is itself polycyclic (cf. Proposition 5.7).

Let us then show that $[G,G]$ is finitely generated. Since $G/[G,G]$ is finitely generated and abelian, by virtue of the structure theorem for finitely generated abelian groups (Corollary 1.30), we can find a series

$$G \geq H_s \geq \cdots \geq H_1 \geq H_0 = [G,G],$$

where $[G : H_s]$ is finite and H_i/H_{i-1} is infinite cyclic. Note that H_s is finitely generated, by virtue of Proposition 1.9. So the result will follow by an iterative application of the following fact.

Claim. *Let G be finitely generated of sub-exponential growth and suppose that H is a normal subgroup of G such that $G/H \cong \mathbb{Z}$. Then H is finitely generated.*

So let $a \in G$ be such that $G/H = \langle aH \rangle$, and let $X \subseteq G$ be a finite symmetric generating subset of G, which we may also suppose to contain a (and therefore a^{-1}). For every $x \in X$ we can find $n \in \mathbb{Z}$ and $h \in H$ such that $x = a^n h$. It follows that, up to replacing every element $x \neq a^{\pm 1}$ by the corresponding h, we may suppose that $X = \{a^{\pm 1}, h_1^{\pm 1}, h_2^{\pm 1}, \ldots, h_\ell^{\pm 1}\}$ where $h_i \in H$ for all $i = 1, 2, \ldots, \ell$.

Let $H^{(m)} \subseteq H$ denote the subgroup generated by the elements $a^j h_i^{\pm 1} a^{-j}$ with $i = 1, 2, \ldots, \ell$ and $j = 0, 1, \ldots, m$. Note that $H^{(0)} \subseteq H^{(1)} \subseteq H^{(2)} \subseteq \ldots$ and set $H^+ = \bigcup_{m=0}^{\infty} H^{(m)}$. Let us show that H^+ is finitely generated by proving that $H^+ = H^{(m)}$ for some $m \geq 1$. If not, for every $m \geq 0$ we can find $j_m \in \{1, 2, \ldots, \ell\}$ such that $k_m := a^m h_{j_m} a^{-m} \in H^{(m)} \setminus H^{(m-1)}$. Now, for $m \in \mathbb{N}$, consider the products

$$k_0^{\varepsilon_0} k_1^{\varepsilon_1} \cdots k_m^{\varepsilon_m},$$

where $\varepsilon_i \in \{0,1\}$. We have 2^{m+1} words of this type which represent distinct group elements of length $\leq 3m+1$ with respect to the generating subset X. Indeed the maximal length is attained when $\varepsilon_i = 1$ for all i so that the corresponding element is $k_0 k_1 \cdots k_m = h_{j_0} a h_{j_1} a h_{j_2} \cdots a h_{j_m} a^{-m}$ (in the product of two consecutive terms $k_i = a^i h_{j_i} a^{-i}$ and $k_{i+1} = a^{i+1} h_{j_{i+1}} a^{-i-1}$ we have a cancellation between a^{-i} and a^{i+1}).

This would yield $b_X(3m+1) \geq 2^{m+1}$ and therefore $b^G(n) \succeq 2^n$ contradicting the hypothesis that G had sub-exponential growth.

As a consequence, we have $H^+ = H^{(m')}$ for some $m' \geq 1$ and H^+ is finitely generated. Similarly, exchanging the roles of a and a^{-1} we can show that the subgroup $H^- = \bigcup_{m=0}^{\infty} H^{(-m)}$, where $H^{(-m)}$ is the subgroup of G generated by the elements $a^{-j} h_i^{\pm 1} a^j$ with $i = 1, 2, \ldots, \ell$ and $j = 0, 1, \ldots, m$, is also finitely generated, say $H^- = H^{(-m'')}$. It follows that $H = \langle \bigcup_{j=-\infty}^{\infty} H^{(j)} \rangle = \langle \bigcup_{j=-m''}^{m'} H^{(j)} \rangle$ is finitely generated as well. This ends the proof of the claim and the theorem follows. □

Theorem 7.37 (Wolf). *Every polycyclic group of sub-exponential growth is virtually nilpotent.*

Proof. Let G be a polycyclic group. By definition, we have a series

$$G = H_s \geq H_{s-1} \geq \cdots \geq H_1 \geq H_0 = \{1_G\},$$

where the quotients H_i/H_{i-1} are all cyclic. By induction on the length of the series, we assume the result for H_{s-1}. If $[G : H_{s-1}] < \infty$ we are done. Hence we may assume that $G/H_{s-1} \cong \mathbb{Z}$ and we have a nilpotent normal subgroup $\widetilde{N} \subseteq H_{s-1}$ with $[H_{s-1} : \widetilde{N}] < \infty$. By applying the Poincaré lemma (cf. Lemma 2.39) we can find a subgroup N of $\widetilde{N}$ which is characteristic in H_{s-1} and such that $[H_{s-1} : N] < \infty$. So N is normal in G and nilpotent (by Proposition 2.9).

Let $a \in G$ be such that $G/H_{s-1} = \langle aH_{s-1} \rangle$. Consider the subgroup $\langle a, N \rangle$: this has finite index in G, since $G = \bigcup_{i \in \mathbb{Z}} a^i H_{s-1}$ and there exist $h_1, h_2, \ldots, h_k \in H_{s-1}$, $k = [H_{s-1} : N]$, such that $H_{s-1} = \cup_{j=1}^{k} N h_j$, so that

$$G = \bigcup_{i \in \mathbb{Z}, 1 \leq j \leq k} a^i N h_j = \langle a, N \rangle h_1 \cup \langle a, N \rangle h_2 \cup \cdots \cup \langle a, N \rangle h_k.$$

Clearly $\langle a, N \rangle / N \cong \mathbb{Z}$, since a has infinite order in G. So, up to replacing G by $\langle a, N \rangle$, we may reduce to the case where G has a normal nilpotent subgroup N such that $G/N \cong \mathbb{Z}$. Let then $x \in G$ be such that $G/N = \langle xN \rangle$, and consider the lower central series of N:

$$N = N_1 \geq N_2 \geq \cdots \geq N_m = \{1_G\}.$$

Recall that the subgroups N_i are characteristic in N, hence normal in G. Also recall (cf. Lemma 2.6) that the series is central so that, in particular,

$$[N_i, N] \subseteq N_{i+1} \tag{7.15}$$

for all i. Moreover, if $M \trianglelefteq G$ and $N_i \leq M \leq N_{i+1}$, then the series

$$N = N_1 \geq \cdots \geq N_i \geq M \geq N_{i+1} \geq \cdots \geq N_m = \{1_G\}$$

has still the property (7.15). Thus for each inclusion $N_i \geq N_{i+1}$ we insert M so that $N_i \geq M \geq N_{i+1}$ and M/N_{i+1} is the torsion subgroup of the finitely generated (cf. Lemma 2.32) abelian group N_i/N_{i+1}. Observe that since the torsion subgroup is characteristic, M is characteristic in N. Continuing this way, we end up with a new series (which by simplicity we denote in the same way)

$$N = N_1 \geq N_2 \geq \cdots \geq N_m = \{1_G\}$$

such that $N_i \trianglelefteq G$ and N_i/N_{i+1} is either finite or torsion free for all i. After possibly inserting further intermediate subgroups in the above series, we may also assume that such a series has the maximal number of infinite factors (which we will show to be the Hirsch number).

Claim. *Let $x \in G$ be such that $G/N = \langle xN \rangle$. For each $i = 1, 2, \ldots, m$ there exists $k = k(i) \geq 1$ such that x^k centralizes N_i/N_{i+1}, i.e. $[N_i, x^k] \subseteq N_i$.*

Assuming the claim we are in position to end the proof of the theorem: in fact we can now find a $k \geq 1$ (in fact $k = \prod_{i=1}^{s-1} k(i)$ would work) such that x^k centralizes all factors. Then $N_0 := \langle x^k, N \rangle$ is nilpotent: consider the series

$$N_0 \geq N_1 \geq N_2 \geq \cdots \geq N_m = \{1_G\}.$$

Notice that for all $i = 1, 2, ...m+1$, we have $\gamma_i(N_0) \subseteq N_{i-1}$. Indeed, by induction,

$$\gamma_{i+1}(N_0) = [N_0, \gamma_i(N_0)] \subseteq [\langle x^k, N \rangle, N_{i-1}] \subseteq [N, N_{i-1}] \subseteq N_i.$$

Moreover, $N_0 = \langle x^k, N\rangle$ is of finite index in $G = \langle x, N\rangle$.

Thus, in order to end the proof of the theorem we are only left to prove the claim.

If N_i/N_{i+1} is finite, then x acts by permutations, hence x^k acts as the identity for some k. Hence we can assume that N_i/N_{i+1} is torsion free.

Set $V := \mathbb{Q} \otimes_{\mathbb{Z}} (N_i/N_{i+1})$. Let us show that $\widehat{x}$, the conjugation by x, acts irreducibly on V. Indeed, suppose that there exists a nontrivial proper $\mathbb{Q}$-subspace $W \leq V$ which is invariant under $\widehat{x}$. Then $W \cap (N_i/N_{i+1}) \neq \{0\}$: for, given $(a_1, a_2, \ldots, a_n) \in \mathbb{Q}^n$, multiplying by the product of the denominators of the a_i's we obtain an element in $\mathbb{Z}^n$, hence in N_i/N_{i+1}. Now $W \cap (N_i/N_{i+1})$ is invariant under $\widehat{x}$ since $N_i \trianglelefteq G$. Let M be the subgroup of G such that $M/N_{i+1} = W \cap (N_i/N_{i+1})$, and let us insert M in the series: $N_i \geq M \geq N_{i+1}$. Now M is normal in G (recall that $G = \langle x, N\rangle$), so by our assumptions on the series $(N_j)_{j=1}^m$ one of the two quotients N_i/M and M/N_{i+1} has to be finite. If N_i/M is finite, then there exists an $n \geq 1$ such that $nN_i \subseteq M$ and hence $nV \subseteq W$. But this implies $V = W$, a contradiction. On the other hand M/N_{i+1} cannot be finite since we assumed N_i/N_{i+1} to be torsion-free. This shows that $\hat{x}$ acts irreducibly on V.

Let now $n \in \mathbb{N}$ and $v \in N_i/N_{i+1}$ and let $a \in N_i$ be such that $v = aN_{i+1}$. Consider the elements

$$a^{\varepsilon_0}(a^x)^{\varepsilon_1} \cdots (a^{x^{n-1}})^{\varepsilon_{n-1}}$$

where $\varepsilon_i \in \{0,1\}$. There are 2^n such words, and their lengths in the a^{x^j}'s are $\leq n$. Since G has sub-exponential growth, we can argue as we did in the proof of Milnor's theorem and find two equal such words

$$a^{\varepsilon_0}(a^x)^{\varepsilon_1} \cdots (a^{x^{n-1}})^{\varepsilon_{n-1}} = a^{\delta_0}(a^x)^{\delta_1} \cdots (a^{x^{n-1}})^{\delta_{n-1}}$$

where the vectors $(\varepsilon_0, \varepsilon_1, \ldots, \varepsilon_{n-1})$ and $(\delta_0, \delta_1, \ldots, \delta_{n-1})$ are distinct. So, modulo N_{i+1}, we have

$$\varepsilon_0 v + \varepsilon_1 \widehat{x}(v) + \cdots + \varepsilon_{n-1}\widehat{x}^{n-1}(v) = \delta_0 v + \delta_1 \widehat{x}(v) + \cdots + \delta_{n-1}\widehat{x}^{n-1}(v).$$

Hence there exists a polynomial $f_v(t) = t^n + \alpha_{n-1}t^{n-1} + \cdots + \alpha_0$, $\alpha_j \in \{-1,0,1\}$, $j = 0,1,2,\ldots,n-1$, such that $f_v(\widehat{x})v = 0$: in fact, $f_v(t)$ is the difference $\varepsilon_0 + \varepsilon_1 t + \cdots + \varepsilon_{n-1}t^{n-1} - (\delta_0 + \delta_1 t + \cdots + \delta_{n-1}t^{n-1})$.

Let $N_i/N_{i+1} = \oplus_{k=1}^r v_k\mathbb{Z}$, and consider $f(t) := \prod_{k=1}^r f_{v_k}(t)$. Then $f(\hat{x})V = \{0\}$. Now $v_1, v_2, \ldots, v_r$ constitute a basis of $\mathbb{C} \otimes_{\mathbb{Z}} (N_i/N_{i+1}) \cong \mathbb{C}^r$, and, since $\hat{x}(v_k) \in N_i/N_{i+1}$ for all $k = 1,2,\ldots,r$, the conjugation operator $\hat{x}$ has an integer matrix representation with respect to this basis.

Lemma 7.38. *$\hat{x}$ does not have eigenvalues with absolute value ≥ 2.*

Proof. Let $v \in V$ be an eigenvector of $\hat{x}$ and let λ be the corresponding eigenvalue. We already observed that $f_v(\hat{x})v = 0$, hence $f_v(\lambda) = 0$. Suppose $|\lambda| \geq 2$. Then, since $\alpha_j \in \{-1,0,1\}$ for all j,

$$|\lambda^n| \leq |\alpha_{n-1}\lambda^{n-1} + \cdots + \alpha_1\lambda + \alpha_0| \leq |\lambda|^{n-1} + \cdots + |\lambda| + 1 = \frac{|\lambda|^n - 1}{|\lambda| - 1} \lneqq |\lambda|^n,$$

a contradiction. ☐

Lemma 7.39. *Every eigenvalue of $\hat{x}$ has absolute value* 1.

Proof. Let λ be an eigenvalue of $\hat{x}$. Since $\hat{x}$ is invertible, $\lambda \neq 0$. If $|\lambda| \neq 1$, then we can find $k \in \mathbb{Z}$ such that $|\lambda|^k \geq 2$. But then, setting $y := x^k$, we would have that $\mu := \lambda^k$ is an eigenvalue of $\hat{y}$ and $|\mu| \geq 2$, contradicting the previous lemma. □

Lemma 7.40. *Every eigenvalue of $\hat{x}$ is a root of* 1.

Proof. $\hat{x}$ corresponds to an integer $r \times r$ matrix, and its characteristic polynomial $p_{\hat{x}}(t) \in \mathbb{Z}[t]$ has degree r, say $p_{\hat{x}}(t) = t^r + k_{r-1}t^{r-1} + \cdots + k_0$. Since $|\lambda| = 1$, $|k_i| \leq C_r$ where $C_r > 0$ is a constant depending on r but not on x: indeed each coefficient k_j is an elementary symmetric polynomial in the roots of $p_{\hat{x}}(t)$, which have absolute value 1. Since the k_i's are all integers, there are finitely many such polynomials. Hence there are finitely many possibilities for the eigenvalues of $\hat{x}^k$, for $k \in \mathbb{Z}$. It follows that if λ is an eigenvalue of $\hat{x}$, then the set $\{\lambda^k : k \in \mathbb{Z}\}$ is finite, i.e. λ is a root of unity. □

It follows from the preceding lemma that we can find an integer k such that $\lambda^k = 1$ for all eigenvalues λ of $\hat{x}$. Thus $\hat{x}^k$ can be expressed as the sum of the identity matrix and a nilpotent matrix: $\hat{x}^k = \mathrm{id}_V + T$, with $T^r = 0$ for some $r \geq 0$. Then $TV \subsetneqq V$, since $T^r = 0$, and TV is invariant with respect to $\hat{x}$:

$$\hat{x}TV = \hat{x}(\mathrm{id}_V - \hat{x}^k)V = (\mathrm{id}_V - \hat{x}^k)\hat{x}V = T\hat{x}V \subseteq TV.$$

Since TV is invariant under $\hat{x}$ and $\hat{x}$ acts irreducibly on V, we necessarily have $T = 0$, equivalently $\hat{x}^k = \mathrm{id}_V$. Thus $\hat{x}^k$ centralizes N_i/N_{i+1}, as we wanted. This ends the proof of the claim.

The proof of the theorem of Wolf is now completed.

We immediately deduce the following:

Corollary 7.41 (Milnor–Wolf). *Every finitely generated solvable group of sub-exponential growth is virtually nilpotent, and hence has polynomial growth.* □

7.10 Notes

Cayley graphs are named after Arthur Cayley who, in a finite groups setting, first considered such graphs in 1878 [57]. Max Dehn, in his unpublished lectures on group theory from 1909–10, reintroduced Cayley graphs under the name "Gruppen-bilder" (group diagrams) and therefore one also refers to them as "Dehn's Gruppen-bilder", see Pierre de la Harpe's survey [162, Section 4] and the references therein for more historical information and interesting examples of Cayley graphs coming from topology. As a generalization of a Cayley graph, given a finitely generated group G, a finite symmetric generating subset $X \subseteq G$, and a subgroup $H \leq G$, the *Schreier coset graph* (or, simply, *Schreier graph*) $\mathrm{Sch}(G,H,X) = (V,E)$ is the X-labelled graph whose set of vertices is $V := H\backslash G := \{Hg : g \in G\}$, the set of all right cosets of H in G, and the set of (X-labelled) edges is $E := \{(Hg, x, Hgx) : g \in G$ and $x \in X\} \subseteq (H\backslash G) \times X \times (H\backslash G)$. Schreier graphs are named after Otto Schreier who, in his 1928 paper [303], used the term "Nebengruppenbilder" (subgroup diagrams).

The notion of growth of a finitely generated group arose in group theory in relation to volume growth in Riemannian manifolds in the 1950s. This line of study was initiated by Vadim A. Efremovich [103] and Albert S. Schwarz [307] in the USSR and, slightly later and completely independently, by John Milnor [237] and Joseph A. Wolf [357] in the USA. In [237] Milnor proved that fundamental groups of closed Riemannian manifolds with negative sectional curvature have exponential growth. Wolf [357] proved that a polycyclic group has sub-exponential growth if it contains a nilpotent subgroup of finite index (Theorem 7.37). Then, Milnor [238] proved that every finitely generated non-polycyclic solvable group has exponential growth (Theorem 7.36). Combining the above two results we then have that finitely generated solvable groups of sub-exponential growth have polynomial growth (Corollary 7.41). Finally, in 1972 Yves Guivarc'h [145] and independently Hyman Bass [19] and Brian Hartley [164] (with different proofs) computed the exact order of polynomial growth (we refer to [158, page 201] for more information on the history and prehistory of these results).

The results of Milnor and Bass–Guivarc'h–Hartley imply that a finitely generated solvable group has either polynomial or exponential growth. It was shown by Jacques Tits [333] (see Chapter 8) that every finitely generated linear group either is virtually solvable or contains a free subgroup of rank two. This last result, which is known as the *Tits alternative* for linear groups, together with the results of Milnor and Wolf, implies that every finitely generated linear group has either polynomial growth or exponential growth.

The problem of the characterization of finitely generated groups with (sub-) polynomial growth originally conjectured by Milnor [239] was solved by Gromov [138] who proved that a finitely generated group of polynomial growth contains a nilpotent subgroup of finite index (see Chapter 12).

It follows from Gromov's theorem (Theorem 12.1), the above mentioned result of Bass–Guivarc'h (Theorem 7.29), and Proposition 7.20 that a group G of sub-polynomial growth has, in fact, (exactly) polynomial growth: its growth function $b_X(n)$ (relative to some, equivalently, any symmetric generating subset $X \subseteq G$) is equivalent to a polynomial.

Corollary 7.27 is due to Jacques Justin [192]. In [95], Lou van den Dries and Alex Wilkie gave the following bound on the index of a cyclic subgroup in an infinite group G of linear growth: *let $X \subset G$ be a finite symmetric generating subset and suppose that there exists an integer $m \geq 1$ such that $b_X(m) - b_X(m-1) \leq m$. Then G has an infinite cyclic subgroup of index not exceeding $m^4/2$.*

In 1980, Rostislav I. Grigorchuk [129] constructed what is now called the *Grigorchuk group* (also known as the *first Grigorchuk group*) yielding a new example of a finitely generated infinite periodic group (other than the Golod–Shafarevich groups, see Section 6.5), thus providing another counterexample to the general Burnside problem (cf. Chapter 6). Explicitly the Grigorchuk group G can be described as follows. Let $\Sigma = \{0, 1\}$ and denote by Σ^* the monoid consisting of all words over Σ (ε denotes the empty word). Then G is the subgroup of $\mathrm{Sym}(\Sigma^*)$, the permutation group of Σ^*, generated by the elements $a,b,c,d \in \mathrm{Sym}(\Sigma^*)$ recursively defined by setting

- $a(\varepsilon):=\varepsilon$ and $a(0w):=1w$, $a(1w):=0w$
- $b(\varepsilon):=\varepsilon$ and $b(0w):=0a(w)$, $b(1w):=1c(w)$
- $c(\varepsilon):=\varepsilon$ and $c(0w):=0a(w)$, $c(1w):=1d(w)$
- $d(\varepsilon):=\varepsilon$ and $d(0w):=0w$, $d(1w):=1b(w)$

for all $w \in \Sigma^*$. Later, in 1984, Grigorchuk [131] proved that his group has *intermediate growth* (this was announced by Grigorchuk in 1983 [130]), thus providing a positive answer to the *Milnor problem*, posed by Milnor in 1968, about the existence of finitely generated groups of intermediate growth. More precisely, in [131] Grigorchuk proved, among other things, that his group G is such that $\exp(\sqrt{n}) \preceq b^G(n) \preceq \exp(n^s)$, where $s = \log_{32}(31) \approx 0.991$. A sharp asymptotic was recently obtained by Anna Erschler and Tianyi Zheng [105] who proved that $b^G(n) \sim \exp(n^{\alpha_0+o(1)})$, where $\alpha_0 := \log 2/\log \lambda_0 \approx 0.7674$, where λ_0 is the positive root of the polynomial $x^3 - x^2 - 2x - 4$.

For more on the Grigorchuk group we refer to [158, 133, 59].

The theory of growth of groups has continued to develop and present interesting and remarkable results. For instance, the following solves a problem posed by Gromov. Let G be a finitely generated group. Recall that one says that G is of exponential growth if for some (equivalently any) finite symmetric generating subset $X \subset G$, the growth rate $\beta_X \in [1,+\infty)$ (cf. Definition 7.15) satisfies $\beta_X > 1$ (cf. Proposition 7.16). One says that G has *uniformly exponential growth* if $\inf_X \beta_X > 1$, where the infimum is taken over all finite symmetric generating subsets X of G. In 1981 Gromov asked whether groups of exponential growth necessarily have uniformly exponential growth. Positive answers were given for wide classes of groups including: free groups, and, more generally, hyperbolic groups (a result by Malik Koubi [203]), linear groups (a result by Alex Eskin, Shahar Mozes, and Hee Oh [108, 109]), Golod–Shafarevich groups from Section 6.5 (as observed by Laurent Bartholdi and Grigorchuk [18], see [160, Section 5]), one-relator groups (a result by Grigorchuk and Pierre de la Harpe [134]): see the research-expository paper by de la Harpe [160]. Eventually, John S. Wilson [351] gave a negative answer to Gromov's question: by using certain permutational wreath products involving the alternating group A_{31}, he constructed a group G with nonabelian free subgroups (so that it has exponential growth) and admitting generating subsets X_n (consisting of two elements, one of order 2 and the other of order 3) such that $\lim_{n\to\infty} \beta_{X_n} = 1$. The construction by Laurent Bartholdi [15] is, according to the author, "somewhat shorter and more specific" than Wilson's example.

Among the most recent results we mention the following. For a hyperbolic group (see Section 11.16), the set of growth rates (relative to all (finite) symmetric generating subsets) is well-ordered. Moreover, given a positive real number, there are at most finitely many – up to the action of the automorphism group of the given hyperbolic group – (finite) symmetric generating sets with this real number as a growth rate. This is due to Koji Fujiwara and Zlil Sela [120].

A parallel branch of *asymptotic group theory* is that of *subgroup growth*, started as early as 1891, in a paper by Adolf Hurwitz, who wanted to count the number of covering spaces of surfaces by counting the subgroups of finite index in the fundamental group of the surface. Let G be a finitely generated group. It follows from a well-known result by Marshall Hall [152] that G contains only a finite number of subgroups of index n. In fact, Hall also gave an explicit recurrence formula for computing the number of subgroups of index n in a free group of finite rank r. Thus, for each integer n define $a_n = a_n(G)$ (resp. $m_n = m_n(G)$, resp. $s_n = s_n(G)$) to be the number of subgroups (resp. maximal subgroups, resp. normal subgroups) of index n in G. Then, subgroup growth studies these functions, their interplay, and the characterization of group-theoretical properties in terms of these functions. The 2003 monograph [217] by Alex Lubotzky and Dan Segal is entirely devoted to subgroup growth, yet another active area of current research.

Another variant of growth in groups is *conjugacy growth*. Given a finitely generated group G together with a finite symmetric generating subset X, one denotes by $\gamma_c^X(n)$ the number of conjugacy classes of G intersecting the ball $B_X^G(n)$ of radius n in G (with respect to X). The function γ_c^X is called the *conjugacy growth function* of G with respect to X. The study of conjugacy growth functions was motivated by counting closed geodesics (up to free homotopy) on complete Riemannian manifolds. Suppose that M is a complete Riemannian manifold admitting a negative upper bound for the sectional curvature. Then there is only one closed geodesic in each free homotopy class. Grigorii Margulis [232], improving on a result by Yakov Sinai [317], proved that for compact manifolds of pinched negative curvature and exponential volume growth $\exp(ht)$, $h > 0$, the number of primitive closed geodesics of period not exceeding t is approximately $\exp(ht)/ht$. From the group theoretical point of view, this implies that the number of primitive conjugacy classes intersecting the ball of radius n in the Cayley graph of $\pi_1(M)$, the fundamental group of M, (with respect to some finite symmetric generating subset) is $\approx \exp(hn)/hn$. Conjugacy growth is always dominated by group growth, but there are groups for which these two notions of growth dramatically differ, as shown by Sergei Ivanov, who constructed finitely generated groups of exponential growth having a finite number of conjugacy classes (see [261]). However, exponential growth should imply exponential conjugacy growth for "ordinary" finitely generated groups. This was shown to be true for hyperbolic groups by Michel Coornaert and Gerhard Knieper in [80], where an upper bound for the growth rate of primitive conjugacy classes in torsion-free hyperbolic groups is given, for solvable groups by Emmanuel Breuillard and Yves de Cornulier [36], and for linear groups by Breuillard, de Cornulier, Lubotzky, and Chen Meiri [37]. In the research-expository paper [144] by Victor Guba and Mark Sapir, estimates for the conjugacy growth of several interesting classes of finitely generated groups, including Baumslag–Solitar groups, the Heisenberg group, and diagram groups are provided. Also, several conjectures, examples, and statements showing that in "normal" cases, groups of exponential growth also have exponential conjugacy growth functions, are provided.

The interested reader my find several other results related to the growth of groups in the very informative and delightful monograph [231] by Avinoam Mann.

We end this historical survey by quoting the recent note [163] by Pierre de la Harpe, where one may find an interesting discussion on a few articles showing that, before 1968 (the year of publication of the papers by Milnor [237] and Wolf [357]) and at least retrospectively, the notion of growth has already played a significant role in various subjects in Mathematics.

7.11 Exercises

Exercise 7.1. Prove Lemma 7.1.

Exercise 7.2. Show that the map $d_{\mathscr{G}} : V \times V \to \mathbb{N}$ defined by (7.6) is a metric on V.

Exercise 7.3. Let $G = \mathbb{Z}$. Show that $X = \{2, -2, 3, -3\}$ generates G.
More generally, find the conditions for a finite symmetric subset $Y \subseteq \mathbb{Z}$ to generate G.

Exercise 7.4. Prove Proposition 7.3.

Exercise 7.5. Prove Proposition 7.8.

Exercise 7.6. Let $a, b \in (1, +\infty)$. Show that $a^n \sim b^n$.

Exercise 7.7. Let $d \geq 0$ be an integer. Show that $n^d \preceq \exp(n)$ and $n^d \not\sim \exp(n)$.

Exercise 7.8. Prove Proposition 7.12.

Exercise 7.9. Prove Proposition 7.16.

Exercise 7.10. Use Proposition 7.16 to give an alternative prove of the fact that every finitely generated group of polynomial growth has sub-exponential growth (cf. Proposition 7.12.(4)).

Exercise 7.11. Let G be a finitely generated group and let $X \subset G$ be a finite symmetric generating subset. Suppose that there exists an integer $m \geq 1$ such that $b_X(m) \leq m$. Show that G is finite (in fact $|G| \leq b_X(m)$).

Exercise 7.12. Prove the statement of Example 7.19 for $n \geq 3$.

Exercise 7.13. Two groups G_1 and G_2 are said to be *commensurable* if there exist finite index subgroups $H_1 \subseteq G_1$ and $H_2 \subseteq G_2$ such that H_1 and H_2 are isomorphic. Show that if G_1 and G_2 are commensurable groups and G_2 is finitely generated, then G_1 is finitely generated, and one has $b^{G_1} = b^{G_2}$.

Exercise 7.14. Prove Formula (7.12).

Exercise 7.15. Prove Lemma 7.22.

Exercise 7.16. Prove Proposition 7.23.

Exercise 7.17. With the notation from the proof of Theorem 7.26, prove the following:

(1) every subword of a word in W belongs to W;
(2) π induces a bijection between W and G and $\ell(w) = \ell_X(\pi(w))$ for all $w \in W$;
(3) G is infinite if and only if W_∞ is non empty;
(4) if $w \in W_\infty$ then $w[h,k] := w(h)w(h+1)\cdots w(k) \in W_{k-h+1}$.

Exercise 7.18. Suppose that a finitely generated group G has linear growth and that $g_1, g_2 \in G$ have infinite order. Then there exist $s, t \in \mathbb{Z} \setminus \{0\}$ such that $g_1^s = g_2^t$.

Chapter 8
Hyperbolic Plane Geometry and the Tits Alternative

This chapter, partially based on the expository paper by Pierre de la Harpe [158], is an introduction to the following deep result due to Jacques Tits [333], originally conjectured by Bass and Jean-Pierre Serre:

Theorem 8.1 (Tits alternative). *A finitely generated linear group Γ over a field $\mathbb{K}$ of characteristic* 0 *either is virtually solvable, that is, it contains a solvable subgroup of finite index, or it contains a nonabelian free subgroup.*

Recall that a *linear group* over a field $\mathbb{K}$ is a group which admits a faithful finite-dimensional representation over $\mathbb{K}$, equivalently, a group isomorphic to a subgroup of $\mathrm{GL}(n,\mathbb{K})$ for some n. If Γ is a finitely generated subgroup of $\mathrm{GL}(n,\mathbb{K})$, and $S \subset \Gamma$ is a finite generating subset, then, denoting by $\mathbb{K}_0$ the subfield of $\mathbb{K}$ generated by the entries of the elements in S, one has that $\Gamma \leq \mathrm{GL}(n,\mathbb{K}_0)$. If $\mathbb{K}$ is of characteristic zero, then, $\mathbb{K}_0$ is a finitely generated extension of $\mathbb{Q}$, and hence there exists an embedding of $\mathbb{K}_0$ in $\mathbb{C}$. One may thus assume that Γ lies in $\mathrm{GL}(n,\mathbb{C})$. As a consequence, it is sufficient to prove the theorem for $\mathbb{K} = \mathbb{C}$, in fact for $\mathbb{K} = \mathbb{R}$, since $\mathrm{GL}(n,\mathbb{C})$ is a subgroup of $\mathrm{GL}(2n,\mathbb{R})$. We thus have the following reformulation of Theorem 8.1:

Theorem 8.2 (Tits alternative (reformulated)). *Let n be a positive integer and let Γ be a finitely generated subgroup of* $\mathrm{GL}(n,\mathbb{R})$. *Then either Γ is virtually solvable or Γ contains a nonabelian free subgroup.*

Actually, this apparent simplification is misleading, because the proof does indeed require fields other than $\mathbb{C}$ or $\mathbb{R}$. In fact, the proof has two important ingredients, namely, Klein's Ping-Pong lemma (Theorem 1.17) and the theory of affine algebraic groups over various fields (not necessarily algebraically closed, not necessarily subfields of $\mathbb{C}$).

Here, to give a taste of some of the ideas underlying the proof of the Tits alternative, we limit ourselves to prove a special case of the theorem (namely $n = 2$ in Theorem 8.2), for which only the first ingredient is essentially sufficient together with some basic notions of geometry of the hyperbolic plane and its isometries.

T. Ceccherini-Silberstein and M. D'Adderio, *Topics in Groups and Geometry*,
Springer Monographs in Mathematics, https://doi.org/10.1007/978-3-030-88109-2_8

8.1 Möbius Transformations

Recall that a *Möbius transformation* is a rational function of one complex variable z of the form

$$f(z) = \frac{az+b}{cz+d} \tag{8.1}$$

where the coefficients $a,b,c,d \in \mathbb{C}$ satisfy $ad-bc \neq 0$.

If $\widehat{\mathbb{C}} = \mathbb{C} \cup \{\infty\}$ denotes the one-point compactification of the complex plane $\mathbb{C}$ (also called the *extended complex plane*, or *Riemann sphere*), so that ∞ is the *point at infinity*, then (8.1) extends to a map $f\colon \widehat{\mathbb{C}} \to \widehat{\mathbb{C}}$ as follows:

- if $c \neq 0$,
$$f\left(\frac{-d}{c}\right) := \infty \quad \text{and} \quad f(\infty) := \frac{a}{c}$$
- if $c = 0$,
$$f(\infty) := \infty.$$

Thus, any Möbius transformation yields a bijective holomorphic function from the extended complex plane to itself.

The set of all Möbius transformations forms a group under composition, called the *Möbius group*. It can be shown that the Möbius group is isomorphic to the *projective linear group* $\mathrm{PGL}(2,\mathbb{C})$. This group can be given the structure of a complex manifold (turning it into a complex Lie group) in such a way that composition and inversion are holomorphic maps. It is isomorphic to the automorphism group of the extended complex plane.

Given a circle $\mathscr{C}_0$ in the complex plane $\mathbb{C}$, with center $z_0 \in C$ and radius $r > 0$, the *inverse* of a point $z \in \mathbb{C}$ with respect to $\mathscr{C}_0$ is the point $z' \in \mathbb{C}$ lying on the ray from z_0 through z such that $|z-z_0| \cdot |z'-z_0| = r^2$. The map $z \mapsto z'$ is called *circle inversion* with *reference circle* $\mathscr{C}_0$ (see Figure 8.1).

It is an **exercise** to check that circle inversion (with respect to $\mathscr{C}_0$) maps z' back to z, that is, the map $z \mapsto z'$ is involutive. Extending circle inversion to the extended complex plane, one has that $z_0 \mapsto \infty$ and $\infty \mapsto z_0$.

It follows from the definition that the inversion of any point inside the reference circle must lie outside it, and vice versa, with the center and the point at infinity switching positions, whilst any point on the circle is a fixed-point: in other words, the nearer z to the center, the further away its inverse z', and vice versa.

When $\mathscr{C}_0 = \{z \in \mathbb{C} : |z| = 1\}$ is the unit circle, we simply call the circle inversion with respect to $\mathscr{C}_0$ *inversion*: clearly, it is given by $z \mapsto z/|z|^2$.

Example 8.3. We present *simple Möbius transformations* and show that every Möbius transformation is a composition of these.

- $f(z) = z+b$ ($a=d=1$ and $d=0$): this is a *translation* (by b);
- $f(z) = az$ ($a \neq 0$, $d=1$, and $b=c=0$):
  if $a \in \mathbb{R}$ this is a *homothety* (of *factor* a);
  if $a \in \mathbb{C}$ with $|a|=1$, say $a = e^{i\theta}$, with $\theta \in [0,2\pi)$, it is a *rotation* (of *angle* θ);
  in general, writing $a = |a|e^{i\theta}$, it is a composition of a rotation (by angle θ) and a homothety (of factor $|a|$);

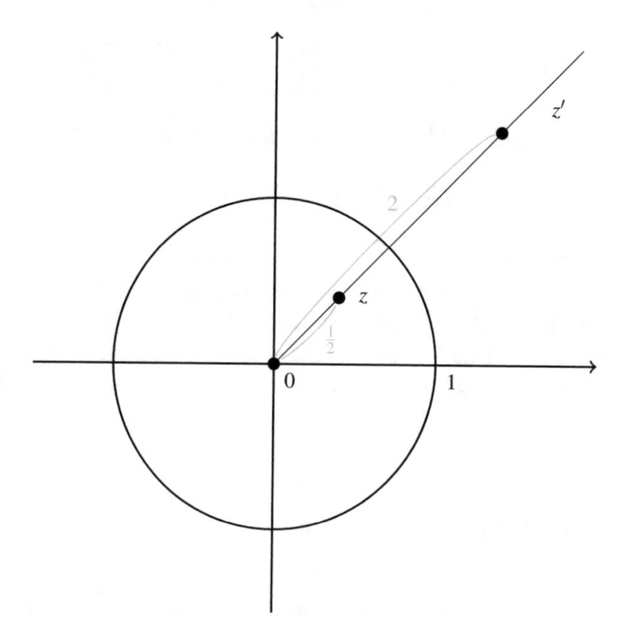

Fig. 8.1 Inversion: the circle inversion $z \mapsto z'$ with reference circle $\mathscr{C}_0 = \{z \in \mathbb{C} : |z| = 1\}$.

- $f(z) = 1/z$ ($a = d = 0$ and $b = c = 1$): this is an inversion and reflection with respect to the real axis.

  Suppose $c \neq 0$. Then setting

- $f_1(z) := z + d/c$ (translation by d/c);
- $f_2(z) := 1/z$ (inversion and reflection with respect to the real axis);
- $f_3(a) := \left((bc - ad)/c^2\right) z$ (composition of a rotation and a homothety);
- $f_4(z) := z + a/c$ (translation by a/c),

it is an **exercise** to show that

$$\frac{az+b}{cz+d} = f_4 \circ f_3 \circ f_2 \circ f_1(z). \tag{8.2}$$

From the decomposition (8.2), we see that Möbius transformations carry over all nontrivial properties of circle inversion. For example, the preservation of angles is reduced to proving that circle inversion preserves angles since the other types of transformations are dilation and isometries (translation, reflection, rotation), which trivially preserve angles. We leave the details as an **exercise**.

The *cross-ratio* of a 4-uple (z_1, z_2, z_3, z_4) of distinct points in $\mathbb{C}$ is defined as

$$(z_1, z_2; z_3, z_4) := \frac{(z_3 - z_1)(z_4 - z_2)}{(z_3 - z_2)(z_4 - z_1)}. \tag{8.3}$$

This notion can be extended to the case when one of the points is $\infty \in \widehat{\mathbb{C}}$. For instance, if $z_1 = \infty$, one has

$$(\infty, z_2; z_3, z_4) := \frac{(z_3 - \infty)(z_4 - z_2)}{(z_3 - z_2)(z_4 - \infty)} = \frac{(z_4 - z_2)}{(z_3 - z_2)}.$$

It is an **exercise** to check that Möbius transformations preserve cross-ratios, that is,

$$(f(z_1), f(z_2); f(z_3), f(z_4)) = (z_1, z_2; z_3, z_4)$$

for any Möbius transformation $f\colon \widehat{\mathbb{C}} \to \widehat{\mathbb{C}}$ and all distinct points $z_1, z_2, z_3, z_4 \in \widehat{\mathbb{C}}$.

A *generalized circle* is either a circle or a line, the latter being considered as a circle through ∞, the point at infinity. It is an **exercise** to check that Möbius transformations map generalized circles to generalized circles. Note, however, that a Möbius transformation does not necessarily map circles to circles and lines to lines: it can mix the two.

8.2 Hyperbolic (Plane) Geometry

In order to understand the nature of hyperbolic geometry and its motivations we start with some historical background. Near the beginning of the first book of the *Elements*, Euclid gives five postulates (axioms) for plane geometry. The last one, known as the *parallel axiom*, states that through a given point outside a given line there passes a unique line which does not intersect the given line.

It was a long standing problem to determine whether or not the last axiom was independent of the first ones. The main investigations were due to Janos Bolyai, Carl Friedrich Gauss, and Nicolaj I. Lobachevsky who eventually showed that the fifth postulate is indeed independent of the others, in the sense that there exists a plane, the *hyperbolic plane*, which satisfies the first four axioms but not the parallel axiom.

In the following, we present two equivalent models of the hyperbolic plane, namely, the Lobachevsky–Poincaré half-plane and the Poincaré disc.

8.3 The Lobachevsky–Poincaré Half-Plane

The *upper half-plane* is the set

$$X := \mathbb{R} \times \mathbb{R}_{>0} \equiv \{z = x + iy \in \mathbb{C} : y > 0\}$$

of complex numbers with positive imaginary part. The real axis

$$\ell_\infty := \mathbb{R} \times \{0\} \equiv \{z = x + iy \in \mathbb{C} : y = 0\}$$

is called the *boundary line*.

The *hyperbolic distance* of two points $z_1 = x_1 + iy_1$ and $z_2 = x_2 + iy_2$ in X is given by

$$d(z_1,z_2) = \operatorname{arcosh}\left(1+\frac{(x_2-x_1)^2+(y_2-y_1)^2}{2y_1y_2}\right). \tag{8.4}$$

Some special cases can be simplified:
(i) if z_1 and z_2 lie on a vertical line, say $z_1 = x+y_1$ and $z_2 = x+y_2$, then (**exercise**)

$$d(z_1,z_2) = \left|\ln\frac{y_1}{y_2}\right| = |\ln(y_2)-\ln(y_1)|;$$

(ii) if $z_1 = x_1+iy$ and $z_2 = x_2+iy$, then (**exercise**)

$$d(z_1,z_2) = \operatorname{arsinh}\left(\frac{|x_2-x_1|}{2y}\right);$$

(iii) if z_1 and z_2 do not lie on a vertical line, then they lie on a unique semicircle ℓ with center in the boundary line. This uniquely determines two points z_1^∞ and z_2^∞ in the boundary line (in other words $\{z_1^\infty, z_2^\infty\} := \ell \cap \ell_\infty$). Arranging the four points in cyclic order (on ℓ), say $z_1, z_2, z_2^\infty, z_1^\infty$, see Figure 8.2,

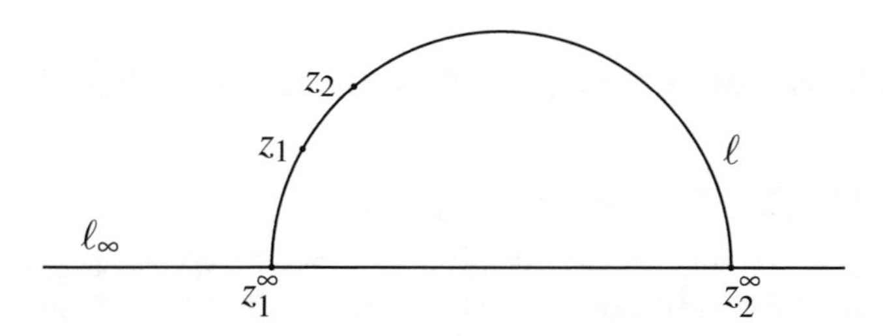

Fig. 8.2 The four points $z_1, z_2, z_2^\infty, z_1^\infty$ on ℓ.

the hyperbolic distance between z_1 and z_2 is then given in terms of the cross-ratio of Euclidean distances (**exercise**):

$$d(z_1,z_2) := \log\left(\frac{|z_1-z_2^\infty|\cdot|z_2-z_1^\infty|}{|z_1-z_1^\infty|\cdot|z_2-z_2^\infty|}\right). \tag{8.5}$$

A *path* from z_1 to z_2, for $z_1, z_2 \in X$, is a differentiable map $\gamma\colon [0,1] \to X \subset \mathbb{R}^2$ such that $\gamma(0) = z_1$ and $\gamma(1) = z_2$. Writing $\gamma(t) = (x(t), y(t))$, the *length* $L(\gamma)$ of a path is

$$L(\gamma) := \int_0^1 \frac{|\gamma'(t)|}{y(t)}\,\mathrm{d}t.$$

It is an **exercise** to show that the hyperbolic distance between z_1 and z_2 equals the infimum of the lengths of all paths from z_1 to z_2, in formulæ,

$$d(z_1,z_2) = \inf\{L(\gamma) : \gamma \text{ a path from } z_1 \text{ to } z_2\}.$$

A *geodesic path*, or simply, a *geodesic*, is a path γ such that $L(\gamma) = d(\gamma(0),\gamma(1))$. The geodesics, also called *hyperbolic lines*, are traces of Euclidean half-circles (with centers in the boundary line) and Euclidean lines which orthogonally intersect the boundary line (see Figure 8.3).

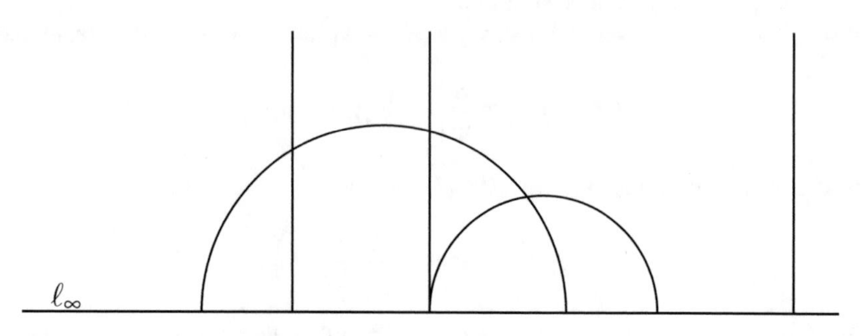

Fig. 8.3 Geodesics in the hyperbolic half-plane.

8.4 Isometries of the Lobachevsky–Poincaré Half-Plane

Recall that $\mathrm{GL}(2,\mathbb{R})$ is the group of 2-by-2 matrices $g = \begin{pmatrix} a & b \\ c & d \end{pmatrix}$ with real coefficients which are invertible (that is, $\det(g) = ad - bc \neq 0$). We denote by $\mathrm{GL}_+(2,\mathbb{R})$ the index-two subgroup consisting of all elements in $\mathrm{GL}(2,\mathbb{R})$ with positive determinant. The complement $\mathrm{GL}_-(2,\mathbb{R}) := \mathrm{GL}(2,\mathbb{R}) \setminus \mathrm{GL}_+(2,\mathbb{R})$ is the nontrivial coset of $\mathrm{GL}_+(2,\mathbb{R})$ in $\mathrm{GL}(2,\mathbb{R})$. We define an action of $\mathrm{GL}(2,\mathbb{R})$ on X as follows.

Given $g := \begin{pmatrix} a & b \\ c & d \end{pmatrix} \in \mathrm{GL}(2,\mathbb{R})$ and $z = x + iy \in X$, we set

$$gz := \frac{az+b}{cz+d} \quad \text{if } g \in \mathrm{GL}_+(2,\mathbb{R}) \tag{8.6}$$

and

$$gz := \frac{a\bar{z}+b}{c\bar{z}+d} \quad \text{if } g \in \mathrm{GL}_-(2,\mathbb{R}), \tag{8.7}$$

where $\bar{z} = x - iy$.

Thus, any $g \in \mathrm{GL}(2,\mathbb{R})$ yields either a Möbius transformation (if $g \in \mathrm{GL}_+(2,\mathbb{R})$) or the composition of the conjugation map $z \mapsto \bar{z}$ and a Möbius transformation (if $g \in \mathrm{GL}_-(2,\mathbb{R})$).

It is an **exercise** to show that $gz \in X$ for all $g \in \mathrm{GL}(2,\mathbb{R})$ and $z \in X$, and that the map $(g,z) \mapsto gz$ defines an action of $\mathrm{GL}(2,\mathbb{R})$ on X.

We leave it as an **exercise** to check that the center $Z = Z(\mathrm{GL}(2,\mathbb{R}))$ of $\mathrm{GL}(2,\mathbb{R})$ consists exactly of the nonzero multiples of the identity matrix I, that is, $Z = \mathbb{R}^* I = \{kI : k \in \mathbb{R}^*\}$. Note that if $k \in \mathbb{R}^*$ and $g \in \mathrm{GL}_\pm(2,\mathbb{R})$ then $kg \in \mathrm{GL}_\pm(2,\mathbb{R})$ and $(kg)z = gz$ for all $z \in X$. As a consequence, setting $\mathrm{PGL}(2,\mathbb{R}) := \mathrm{GL}(2,\mathbb{R})/\mathbb{R}^* I$, called the *projective real linear group*, and denoting by $[g] \in \mathrm{PGL}(2,\mathbb{R})$ the class of

$g = \begin{pmatrix} a & b \\ c & d \end{pmatrix} \in \mathrm{GL}(2,\mathbb{R}) \bmod \mathbb{R}^* I$, the quantity

$$[g]z := gz$$

is well defined for all $z \in X$, yielding an action of $\mathrm{PGL}(2,\mathbb{R})$ on X. In the following, unless explicitly indicated, in order to avoid a heavy notation, we shall simply write g in place of $[g]$ and make no distinction between $g \in \mathrm{GL}(2,\mathbb{R})$ or $g \in \mathrm{PGL}(2,\mathbb{R})$.

As usual, we denote by $\mathrm{Isom}(X)$ the *isometry group* of (X,d).

It is an **exercise** to check that

$$d(gz_1, gz_2) = d(z_1, z_2)$$

for all $g \in \mathrm{GL}(2,\mathbb{R})$ (resp. $g \in \mathrm{PGL}(2,\mathbb{R})$) and $z_1, z_2 \in X$. In other words, $\mathrm{GL}(2,\mathbb{R})$ (resp. $\mathrm{PGL}(2,\mathbb{R})$) acts by isometries on X.

Example 8.4. Here we list some basic examples of isometries of X: every other isometry can be expressed as a suitable composition of these.

(1) For $s \in \mathbb{R}$, the *translation* by s is given by the matrix $T_s := \begin{pmatrix} 1 & s \\ 0 & 1 \end{pmatrix} \in \mathrm{GL}_+(2,\mathbb{R})$, so that

$$T_s z = z + s = (x+s) + iy$$

for all $z = x + iy \in X$.

(2) For $\lambda > 0$, the *dilation* by λ is given by the matrix $D_\lambda := \begin{pmatrix} \lambda & 0 \\ 0 & 1 \end{pmatrix} \in \mathrm{GL}_+(2,\mathbb{R})$, so that

$$D_\lambda z = \lambda z = \lambda x + i\lambda y$$

for all $z = x + iy \in X$.

(3) Given a hyperbolic line ℓ in X, the *reflection about* ℓ is the unique nontrivial isometry of X which fixes all points in ℓ. For example:

(3.1) The *reflection about the y-axis* is given by the matrix $r := \begin{pmatrix} -1 & 0 \\ 0 & 1 \end{pmatrix} \in \mathrm{GL}_-(2,\mathbb{R})$, so that

$$rz = -\bar{z} = -x + iy$$

for all $z = x + iy \in X$. By conjugating r by the translation T_s, one obtains the reflection about the axis $x = s$. Note that such a reflection is given by a matrix in $\mathrm{GL}_-(2,\mathbb{R})$.

(3.2) The *reflection about the unit circle* is given by the matrix $R := \begin{pmatrix} 0 & 1 \\ 1 & 0 \end{pmatrix} \in \mathrm{GL}_-(2,\mathbb{R})$, so that

$$Rz = \frac{1}{\bar{z}} = \frac{z}{|z|^2} = \frac{x+iy}{x^2+y^2}$$

for all $z = x + iy \in X$. By suitably conjugating R by a dilation and a translation, one obtains the reflection about any circle with center in ℓ_∞. Note that such a reflection is given by a matrix in $\mathrm{GL}_-(2,\mathbb{R})$.

(4) The composition of the reflections about two intersecting hyperbolic lines in X is called a *rotation*. It can be shown (cf. Example 8.14 below) that the rotation about the point $i \in X$ of angle $\theta \in [0, 2\pi)$ corresponds to the matrix $\rho_\theta \in \mathrm{GL}_+(2, \mathbb{R})$ given by

$$\rho_\theta := \begin{pmatrix} \cos(\theta/2) & \sin(\theta/2) \\ -\sin(\theta/2) & \cos(\theta/2) \end{pmatrix}.$$

Clearly, any other rotation is conjugate to ρ_θ for a suitable $\theta \in [0, 2\pi)$ and therefore corresponds to a matrix in $\mathrm{GL}_+(2, \mathbb{R})$.

Let ℓ be a hyperbolic line in X. A pair (z_1, z_2) of distinct points in ℓ defines an *orientation* of the line by declaring that z_1 is met before z_2 when traveling along ℓ. Clearly, the symmetric pair (z_2, z_1) gives the only other possible orientation. The line ℓ divides X into two connected components and we denote by X_ℓ^+ (resp. X_ℓ^-) the one on the right (resp. left) of a traveler along ℓ in the direction given by the orientation (see Figure 8.4).

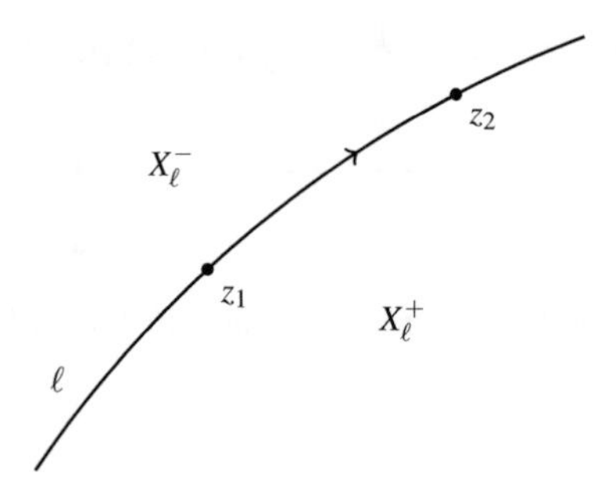

Fig. 8.4 The line ℓ divides X into two connected components X_ℓ^+ and X_ℓ^-.

Definition 8.5 (Orientation-preserving isometry). Let φ be an isometry of X. We say that φ is *orientation-preserving* if given a pair (z_1, z_2) of distinct points in X and denoting by ℓ the hyperbolic line through z_1 and z_2, oriented by the pair, one has

$$\varphi(X_\ell^+) = X_{\varphi(\ell)}^+$$

where the hyperbolic line $\varphi(\ell)$ is orientated by the pair $(\varphi(z_1), \varphi(z_2))$ (see Figure 8.5).

We leave it as an **exercise** to check that the definition above does not depend on the chosen pair (z_1, z_2) of distinct points in X. For instance, translations, dilations, and rotations are orientation-preserving, while reflections are not. We leave it as an **exercise** to show that orientation-preserving isometries form an index-two subgroup, denoted by $\mathrm{Isom}_+(X)$, in $\mathrm{Isom}(X)$.

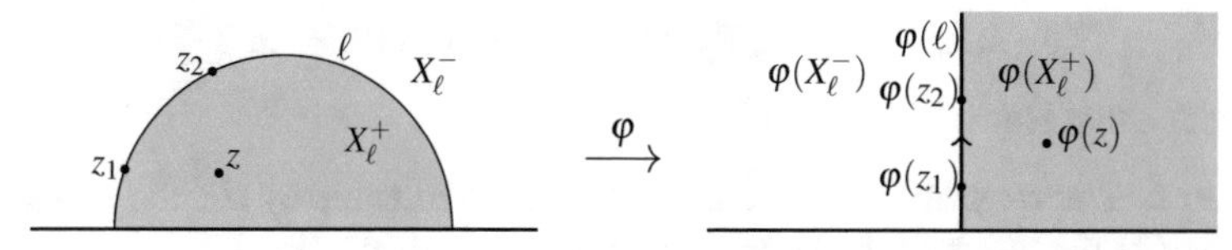

Fig. 8.5 The isometry $\varphi\colon X \to X$, satisfying $\varphi(X_\ell^+) = X_{\varphi(\ell)}^+$, is orientation-preserving.

Theorem 8.6. *Any orientation-preserving isometry of X is either a translation, a dilation, or a rotation, or some composition of these. As a consequence,* $\mathrm{Isom}_+(X) \cong \mathrm{PSL}(2,\mathbb{R})$.

Any isometry of X is either a translation, a dilation, a rotation, or a reflection, or some composition of these. As a consequence, $\mathrm{Isom}(X) \cong \mathrm{PGL}(2,\mathbb{R})$.

Sketch of the proof. Let $\phi \in \mathrm{Isom}(X)$. Let also $P_1, P_2, P \in X$ be distinct points, where P does not lie in the hyperbolic line ℓ through P_1 and P_2.

We claim that ϕ is uniquely determined by the values $\phi(P_1)$, $\phi(P_2)$, and $\phi(P)$. Suppose that $\phi(P_1) = P_1$ and $\phi(P_2) = P_2$. Observe that all points in ℓ are fixed by ϕ. Denote by $P_\ell \in \ell$ the unique point such that $d(P,P_\ell) = d(P,\ell)$. Let $P' \in X \setminus \{P\}$ be the unique point such that $d(P',P_\ell) = d(P',\ell)$. Since $\phi(P_\ell) = P_\ell$, we have two possibilities: either $\phi(P) = P$ (and $\phi(P') = P'$) or $\phi(P) = P'$ and $\phi(P') = P$. By continuity, the same holds for every point Q in the connected component of $X \setminus \ell$ containing P. In the first case ϕ is the identity map and in the second case ϕ is the reflection about ℓ. This proves the claim if $\phi(P_1) = P_1$ and $\phi(P_2) = P_2$. The general case is handled as follows: if $\phi_1, \phi_2 \in \mathrm{Isom}(X)$ attain the same values on P_1, P_2 and P, then $\phi := \phi_1 \circ \phi_2^{-1}$ fixes P_1, P_2 and P and by the previous case, ϕ must be the identity. This proves the claim.

Let us show that ϕ can be expressed as a suitable composition of a translation, a dilation, a reflection, and a rotation. Clearly there exists a composition of a dilation and a translation $\phi_1(z) = az + b$ with $a, b \in \mathbb{R}$ and $a > 0$ such that $\phi_1(P_1) = \phi(P_1)$. Now there exists a rotation ϕ_2 centered at $\phi(P_1)$ such that $(\phi_2 \circ \phi_1)(P_2) = \phi(P_2)$. Therefore $\psi := \phi \circ \phi_1^{-1} \circ \phi_2^{-1} \in \mathrm{Isom}(X)$ fixes both P_1 and P_2. By the argument above, ψ is either the identity or the reflection about ℓ.

We leave it as an **exercise** to show that the matrices T_s, $s \in \mathbb{R}$, D_λ, $\lambda > 0$, and the rotations ρ_θ, $\theta \in [0, 2\pi)$, generate the group $\mathrm{GL}_+(2,\mathbb{R})$ so that $[T_s]$, $s \in \mathbb{R}$, $[D_\lambda]$, $\lambda > 0$, and $[\rho_\theta]$, $\theta \in [0,2\pi)$, generate $\mathrm{PSL}(2,\mathbb{R})$). This shows that $\mathrm{Isom}_+(X) \cong \mathrm{PSL}(2,\mathbb{R})$.

The reflection ϕ about the y-axis, corresponding to the matrix $r \in \mathrm{GL}_-(2,\mathbb{R})$ has order 2. Thus, on the one hand,

$$\mathrm{Isom}(X) = \mathrm{Isom}_+(X) \rtimes \{1, \phi\} \cong \mathrm{Isom}_+(X) \rtimes \mathbb{Z}/2\mathbb{Z}$$

and, on the other hand,

$$\mathrm{GL}(2,\mathbb{R}) = \mathrm{GL}_+(2,\mathbb{R}) \rtimes \{I, r\} \cong \mathrm{GL}_+(2,\mathbb{R}) \rtimes \mathbb{Z}/2\mathbb{Z}$$

so that

$$\mathrm{PGL}(2,\mathbb{R}) \cong \mathrm{PSL}(2,\mathbb{R}) \rtimes \{[I],[r]\} \cong \mathrm{PSL}(2,\mathbb{R}) \rtimes \mathbb{Z}/2\mathbb{Z}.$$

We deduce that $\mathrm{Isom}(X) \cong \mathrm{PGL}(2,\mathbb{R})$. □

Let ∞ be a new symbol called the *ideal point*. We denote by ∂X the *boundary* of X defined as the union of the boundary line and the ideal point:

$$\partial X := \ell_\infty \cup \{\infty\} \equiv \{z = x+iy \in \mathbb{C} : y = 0\} \cup \{\infty\}.$$

Then $\overline{X} := X \cup \partial X$ is the *closed upper half-plane*. Topologically, $\overline{X}$ is the one-point compactification of $\{z = x+iy \in \mathbb{C} : y \geq 0\}$.

It is an **exercise** to show that every isometry $\phi \in \mathrm{Isom}(X)$ extends to a unique homeomorphism of the closed upper half-plane.

8.5 The Poincaré Disc

Let

$$D := \{z \in \mathbb{C} : |z| < 1\} \equiv \{(x,y) \in \mathbb{R}^2 : x^2 + y^2 < 1\}$$

denote the *open unit disc* in the complex plane $\mathbb{C}$. We shall refer to it as to the *Poincaré disc*. Its *boundary* is the circle

$$\partial D := \{(x,y) \in \mathbb{R}^2 : x^2 + y^2 = 1\}.$$

We equip D with a distance, called the *hyperbolic distance*, by setting for two points $z_1 = x_1 + iy_1$ and $z_2 = x_2 + iy_2$ in D

$$d(z_1,z_2) := \operatorname{arcosh}\left(1 + 2\frac{(x_2-x_1)^2 + (y_2-y_1)^2}{(1-x_1^2-x_2^2)\,(1-y_1^2-y_2^2)}\right). \tag{8.8}$$

Equivalently, given z_1 and z_2 in D either they lie on the same diameter, say ℓ or, they lie on a (unique) Euclidean circle, say ℓ, orthogonally intersecting the boundary ∂D. In either case, z_1 and z_2 uniquely determine two points z_1^∞ and z_2^∞ in the boundary ∂D (in other words $\{z_1^\infty, z_2^\infty\} := \ell \cap \partial D$). Arranging the four points in cyclic order (on ℓ), say $z_1, z_2, z_2^\infty, z_1^\infty$, see Figure 8.6,
the hyperbolic distance between z_1 and z_2 is then given in terms of the cross-ratio of Euclidean distances as in (8.5) (**exercise**).

The following special cases can be simplified:
(i) if z_1 and z_2 lie on the same radius so that, by using polar coordinates, $z_1 = r_1 e^{i\theta}$ and $z_2 = r_2 e^{i\theta}$, with $0 \leq r_2 < r_1 < 1$ then (**exercise**)

$$d(z_1,z_2) = \ln\left(\frac{1+r_1}{1-r_1} \cdot \frac{1-r_2}{1+r_2}\right) = 2\,(\operatorname{arctanh} r_1 - \operatorname{arctanh} r_2),$$

where arctanh is the inverse hyperbolic function of the hyperbolic tangent;
(ii) in particular, when one of the two points is the origin, we have (**exercise**)

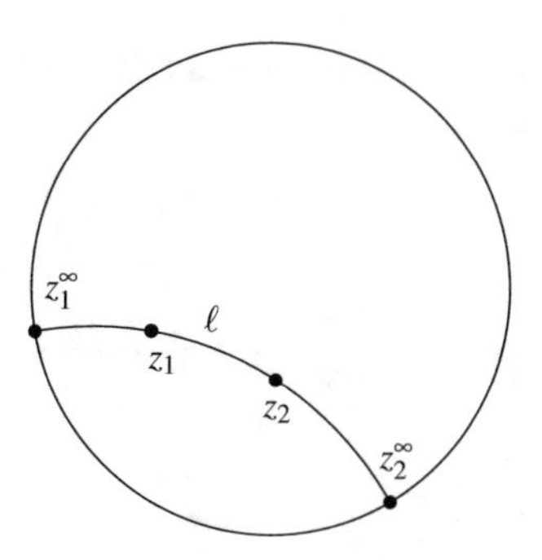

Fig. 8.6 The four points $z_1, z_2, z_2^\infty, z_1^\infty$ on ℓ.

$$d(z,0) = \ln\left(\frac{1+r}{1-r}\right) = 2\,\mathrm{arctanh}\, r.$$

The geodesics, also called *hyperbolic lines*, are traces of Euclidean circles and Euclidean lines which orthogonally intersect the boundary ∂D (see Figure 8.7).

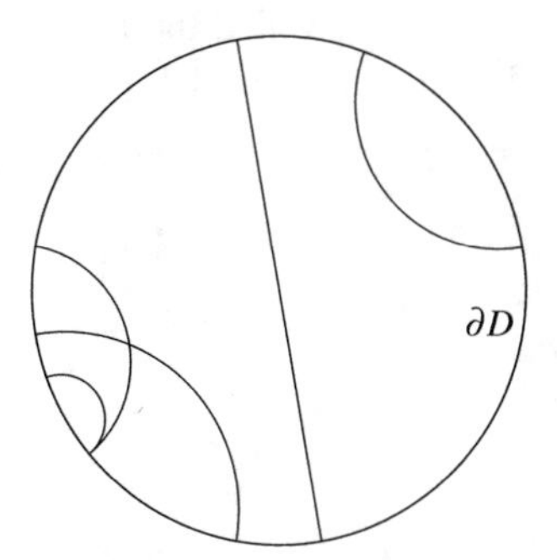

Fig. 8.7 Geodesics in the Poincaré disc.

8.6 Isometries of the Poincaré Disc

We denote by $\mathrm{U}(1,1)$ the group of 2-by-2 complex matrices of the form

$$\begin{pmatrix} u & v \\ \overline{v} & \overline{u} \end{pmatrix}$$

such that $u\overline{u} - v\overline{v} \neq 0$. It is an **exercise** to show that $\mathrm{U}(1,1)$ is indeed a group: it is called the *pseudo-unitary group*. We denote by $\mathrm{U}_+(1,1)$ the index-two subgroup consisting of all elements in $\mathrm{U}(1,1)$ with $u\overline{u} - v\overline{v} > 0$. The complement

$U_-(1,1) := U(1,1) \setminus U_+(1,1)$ is the nontrivial coset of $U_+(1,1)$ in $U(1,1)$. We define an action of $U(1,1)$ on D as follows.

Given $g = \begin{pmatrix} u & v \\ \overline{v} & \overline{u} \end{pmatrix} \in U(1,1)$ and $z \in D$ we set

$$gz := \frac{uz+v}{\overline{v}z+\overline{u}} \quad \text{if } g \in U_+(1,1)$$

and

$$gz := \frac{u+v\overline{z}}{\overline{v}+\overline{u}\overline{z}} \quad \text{if } g \in U_-(1,1).$$

Thus, any $g \in U(1,1)$ yields either a Möbius transformation (if $g \in U_+(1,1)$) or the composition of the conjugation map $z \mapsto \overline{z}$ and a Möbius transformation (if $g \in U_-(1,1)$).

It is an **exercise** to show that $gz \in D$ for all $g \in U(1,1)$ and $z \in D$, and that the map $(g,z) \mapsto gz$ defines an action of $U(1,1)$ on D.

It is an **exercise** to show that the center $Z(U(1,1))$ of $U(1,1)$ consists exactly of the nonzero real multiples of the identity matrix, that is, $Z(U(1,1)) = \{aI : a \in \mathbb{R}^*\}$. Note that $Z(U(1,1)) \leq U_+(1,1)$ and that if $g \in U(1,1)$ and $a \in \mathbb{R}^*$ then $(ag)z = gz$ for all $z \in D$. As a consequence, there is an action of the quotient group $PU(1,1) := U(1,1)/Z(U(1,1))$, called the *projective pseudo-unitary group*. Also, we denote by $SU(1,1)$ the subgroup of $U_+(1,1)$ consisting of all matrices $g = \begin{pmatrix} u & v \\ \overline{v} & \overline{u} \end{pmatrix}$ such that $u\overline{u} - v\overline{v} = 1$. This is called the *special pseudo-unitary group*. Finally, we denote by $PSU(1,1) := U_+(1,1)/Z(U_+(1,1)) \cong SU(1,1)/\{I,-I\}$ the *projective special pseudo-unitary group*.

It is an **exercise** to check that

$$d(gz_1, gz_2) = d(z_1, z_2)$$

for all $g \in U(1,1)$ and $z_1, z_2 \in D$. In other words, $g \in \mathrm{Isom}(D)$ for all $g \in U(1,1)$ (resp. in $PU(1,1)$), so that $U(1,1)$ (resp. $PU(1,1)$) acts by isometries on D.

Example 8.7. Here we list some basic examples of isometries of D (cf. Example 8.14). For more examples, see Example 8.16 and Example 8.17.

(1) For $\theta \in [0, 2\pi)$, the *rotation by angle* θ is given by the matrix

$$R_\theta := \begin{pmatrix} e^{i\theta/2} & 0 \\ 0 & e^{-i\theta/2} \end{pmatrix} \in U_+(1,1),$$

so that

$$R_\theta z = e^{i\theta} z$$

for all $z \in D$.

(2) Given a diameter ℓ, the *reflection about* ℓ is the unique nontrivial isometry of D which fixes all points in ℓ. For example:

(2.1) the *reflection about the x-axis* corresponds to the matrix $r_x := \begin{pmatrix} 0 & 1 \\ 1 & 0 \end{pmatrix} \in \mathrm{U}_-(1,1)$ so that

$$r_x z = \overline{z}$$

for all $z \in D$, as expected.

(2.2) the *reflection about the y-axis* corresponds to the matrix $r_y := \begin{pmatrix} 0 & i \\ -i & 0 \end{pmatrix} \in \mathrm{U}_-(1,1)$ so that

$$r_x z = -\overline{z}$$

for all $z \in D$, as expected.

(2.3) the *reflection about the line* $x = y$ corresponds to the matrix

$$r_{x,y} := \begin{pmatrix} 0 & e^{i\pi/4} \\ e^{-i\pi/4} & 0 \end{pmatrix} \in \mathrm{U}_-(1,1)$$

so that

$$r_{x,y} z = i\overline{z}$$

for all $z \in D$, as expected.

The definition of an orientation-preserving isometry of X given in Definition 8.5 holds verbatim for isometries of D. We are now in a position to state the following result describing the groups $\mathrm{Isom}(D)$ and $\mathrm{Isom}_+(D)$ of all isometries (resp. all orientation-preserving isometries of D): the proof will be given in the next section.

Theorem 8.8. $\mathrm{Isom}(D) \cong \mathrm{PU}(1,1)$ *and* $\mathrm{Isom}_+(D) \cong \mathrm{PSU}(1,1)$.

Finally note (**exercise**) that every $g \in \mathrm{Isom}(D)$ uniquely extends to a homeomorphism of the *closed unit disc*

$$\overline{D} := D \cup \partial D = \{z \in \mathbb{C} : |z| \leq 1\} = \{(x,y) \in \mathbb{R}^2 : x^2 + y^2 \leq 1\}.$$

8.7 The Cayley Transform and the Definition of $\mathbb{H}$

Definition 8.9. The *Cayley transform* is the Möbius transformation

$$f(z) := \frac{z - i}{z + i}.$$

Like every Möbius transformation, the Cayley transform extends to a homeomorphism $f : \widehat{\mathbb{C}} \to \widehat{\mathbb{C}}$ of the extended complex plane. It is an easy **exercise** to check that (see Figure 8.8)

(i) $f(\infty) = 1$
(ii) $f(1) = -i$
(iii) $f(-1) = i$
(iv) $f(i) = 0$

(v) $f(-i) = \infty$
(vi) $f(0) = -1$.

Since Möbius transformations permute the generalized circles in the complex plane, it follows from (i)–(iii) that f maps the real line onto the unit circle. Furthermore, by (iv) and continuity of f, the upper half-plane is mapped onto the unit disc.

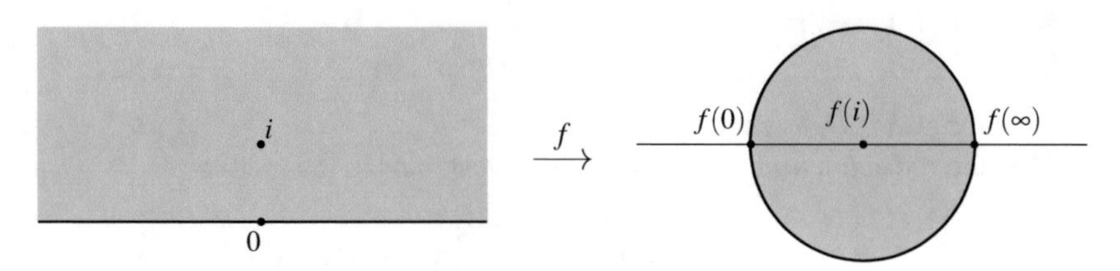

Fig. 8.8 The Cayley transform.

In fact, we have the following:

Theorem 8.10. *The Cayley transform yields an isometry of the Lobachevsky–Poincaré half-plane (X,d) onto the Poincaré disc (D,d).*

Proof. The Cayley transform yields a bijection from the closed upper half-plane $\overline{X} = X \cup \partial X$ onto the closed unit disc $\overline{D} = D \cup \partial D$ mapping ∂X onto ∂D and therefore X onto D. We leave it as an **exercise**, by using (8.4) and (8.8), to check that the Cayley transform indeed yields an isometry $(X,d) \to (D,d)$. □

It is an **exercise** to check that the map

$$\begin{pmatrix} u & v \\ \overline{v} & \overline{u} \end{pmatrix} \mapsto \begin{pmatrix} a+d & b+c \\ c-b & a-d \end{pmatrix}$$

for all $u = a+ib$ and $v = c+id$, with $a,b,c,d \in \mathbb{R}$, yields an isomorphism of $U(1,1)$ (resp. $U_+(1,1)$) onto $\mathrm{GL}(2,\mathbb{R})$ (resp. $\mathrm{GL}_+(2,\mathbb{R})$). As a consequence, the following isomorphisms hold:

$$\mathrm{PU}(1,1) \cong \mathrm{PGL}(2,\mathbb{R}) \quad \text{and} \quad \mathrm{PSU}(1,1) \cong \mathrm{PSL}(2,\mathbb{R}). \tag{8.9}$$

Proof of Theorem 8.8. This follows immediately from Theorem 8.10, yielding $\mathrm{Isom}(D) \cong \mathrm{Isom}(X)$ (resp. $\mathrm{Isom}_+(D) \cong \mathrm{Isom}_+(X)$), Theorem 8.6, and (8.9). □

We denote by $\mathbb{H}$ the isometry class of the Lobachevsky–Poincaré half-plane and the Poincaré disc, and call it the *hyperbolic plane*. We then denote by $\partial\mathbb{H}$ the *boundary* of the hyperbolic plane.

From Theorem 8.6 and Theorem 8.8 we immediately deduce the following:

Theorem 8.11. *The group $\mathrm{Isom}(\mathbb{H})$ (resp. $\mathrm{Isom}_+(\mathbb{H})$) of all isometries (respectively orientation-preserving isometries) of $\mathbb{H}$ is isomorphic to $\mathrm{PGL}(2,\mathbb{R}) \cong \mathrm{PU}(1,1)$ (resp. $\mathrm{PSL}(2,\mathbb{R}) \cong \mathrm{PSU}(1,1)$).* □

8.8 Classification of the Orientation-Preserving Isometries of $\mathbb{H}$

Definition 8.12. Let $g \in \mathrm{Isom}_+(\mathbb{H})$ be an orientation-preserving isometry of $\mathbb{H}$. One says that g is:

- *elliptic* if there is some point in $\mathbb{H}$ fixed by g;
- *parabolic* if in $\partial\mathbb{H}$ there is exactly one point fixed by g;
- *hyperbolic* if in $\partial\mathbb{H}$ there are exactly two points fixed by g.

Remark 8.13. It is clear that the elliptic (resp. parabolic, resp. hyperbolic) type of an orientation-preserving isometry of $\mathbb{H}$ is invariant under conjugation by elements in $\mathrm{Isom}(\mathbb{H})$.

It is an **exercise** to check that an orientation-preserving isometry of $\mathbb{H}$ is hyperbolic if and only if there is a line in $\mathbb{H}$ invariant by g on which g has no fixed-points (namely, the unique line connecting the two fixed-points of g in $\partial\mathbb{H}$).

It follows from the proof of Theorem 8.6 that if an orientation-preserving isometry g of $\mathbb{H}$ has two fixed-points in $\mathbb{H}$, then g is the identity. As a consequence, the point in $\mathbb{H}$ fixed by a nontrivial elliptic isometry is unique.

Example 8.14 (Elliptic isometry). In the Poincaré disc model D of $\mathbb{H}$, the counterclockwise rotation $R^D_\theta(0)$ around the origin of angle $\theta \in [0, 2\pi)$, given by the map $z \mapsto e^{i\theta}z$, and corresponding to the matrix $\begin{pmatrix} e^{i\theta/2} & 0 \\ 0 & e^{-i\theta/2} \end{pmatrix} \in \mathrm{SU}(1,1)$, fixes the origin and therefore is an elliptic isometry. Note that for $\theta = \pi$, the rotation $R^D_\pi(0)$ is given by the map $z \mapsto -z$ with corresponding matrix $\begin{pmatrix} i & 0 \\ 0 & -i \end{pmatrix} \in \mathrm{SU}(1,1)$.

Conjugating $R^D_\theta(0)$ by the Cayley transform, one obtains the rotation $R^X_\theta(i)$ of center i of angle θ in the Lobachevsky–Poincaré half-plane model X of $\mathbb{H}$. We leave it as an **exercise** to check that for $\theta \in [0, 2\pi)$ the corresponding matrix is $\begin{pmatrix} \cos(\theta/2) & \sin(\theta/2) \\ -\sin(\theta/2) & \cos(\theta/2) \end{pmatrix} \in \mathrm{GL}_+(2,\mathbb{R})$. Note that for $\theta = \pi$, the rotation $R^X_\pi(i)$ is expressed by the matrix $\begin{pmatrix} 0 & 1 \\ -1 & 0 \end{pmatrix} \in \mathrm{GL}_+(2,\mathbb{R})$ and therefore it is given by the map $z \mapsto -z^{-1}$.

It follows from the proof of Theorem 8.6 that every elliptic isometry of X (resp. of D) is conjugate to a rotation $R^D_\theta(0)$ around the origin (resp. $R^X_\theta(i)$ around i) of angle $\theta \in [0, 2\pi)$, and therefore it is a rotation around its fixed-point.

Remark 8.15. The counterclockwise rotations $R^D_\theta(0)$ (resp. $R^X_\theta(i)$) around the origin $0 \in D$ (resp. $i \in X$) act transitively on the boundary ∂D (resp. ∂X).

Example 8.16 (Parabolic isometry). In the Lobachevsky–Poincaré half-plane model X of $\mathbb{H}$, given $b \in \mathbb{R}^*$, the *translation* T^X_b by b, given by the map $z \mapsto z + b$, and corresponding to the matrix $\begin{pmatrix} 1 & b \\ 0 & 1 \end{pmatrix} \in \mathrm{SL}(2,\mathbb{R})$, fixes only the point $\infty \in \partial X$ and therefore is a parabolic isometry.

It follows from the isomorphism $\mathrm{Isom}_+(X) \cong \mathrm{PSL}(2,\mathbb{R})$ (cf. Theorem 8.6) that the only orientation-preserving isometries fixing $\infty \in \partial X$ are the T_b^Xs. By using Remark 8.15, we deduce that every parabolic isometry of X is conjugate to a translation T_b^X for some $b \in \mathbb{R}^*$.

Conjugating T_b^X by the Cayley transform f, one obtains the translation T_b^D in the Poincaré model D of $\mathbb{H}$. We leave it as an **exercise** to check that the corresponding matrix is $\begin{pmatrix} 2+ib & -ib \\ ib & 2-ib \end{pmatrix} \in \mathrm{U}_+(1,1)$. Note that the only fixed-point in ∂D is $1 = f(\infty)$. As above, every parabolic isometry of D is conjugate to a translation T_b^D for some $b \in \mathbb{R}^*$.

Example 8.17 (Hyperbolic isometry). In the Lobachevsky–Poincaré half-plane model X of $\mathbb{H}$, given $0 < \lambda \neq 1$ the *dilation* D_λ^X, given by the map $z \mapsto \lambda z$, and corresponding to the matrix $\begin{pmatrix} \lambda & 0 \\ 0 & 1 \end{pmatrix} \in \mathrm{GL}_+(2,\mathbb{R})$, fixes only the points 0 and ∞ in ∂X and therefore is a hyperbolic isometry. Using the fact that translations T_b^X, $b \in \mathbb{R}$, act transitively on the boundary line ℓ_∞ (**exercise**) one can show (**exercise**) that every hyperbolic isometry of X is conjugate to a dilation D_λ^X for some $0 < \lambda \neq 1$.

Conjugating D_λ^X by the Cayley transform f, one obtains the dilation D_λ^D in the Poincaré model D of $\mathbb{H}$. We leave it as an **exercise** to check that the corresponding matrix is $\begin{pmatrix} \lambda+1 & \lambda-1 \\ \lambda-1 & \lambda+1 \end{pmatrix} \in \mathrm{U}_+(1,1)$. Note that the only fixed-points in ∂D are $-1 = f(0)$ and $1 = f(\infty)$. As above, every hyperbolic isometry of D is conjugate to a dilation D_λ^D for some $0 < \lambda \neq 1$.

Theorem 8.18. *Elliptic, parabolic, and hyperbolic isometries define a partition of* $\mathrm{Isom}_+(\mathbb{H})$ *into three disjoint classes.*

Proof. We first check that the three classes do not overlap.

Let $g \in \mathrm{Isom}_+(\mathbb{H})$. Suppose first that g is hyperbolic and let ℓ denote the invariant line. Since g has two fixed-points in $\partial\mathbb{H}$ it is not parabolic. But g is not elliptic either, because if $P \in \mathbb{H}$ was fixed by g, then the foot $P_\ell \in \ell$ of the perpendicular from P to ℓ would also be fixed (**exercise**), and the line ℓ itself should be pointwise fixed (recall that g is orientation-preserving), contradicting the definition of hyperbolicity.

Suppose now that g is at the same time elliptic, with fixed-point $P \in \mathbb{H}$, and parabolic with fixed-point $Q \in \partial\mathbb{H}$. Then the line $\ell_{P,Q}$ from P to Q should be invariant. Thus, if Q' is the other point in $\ell \cap \partial\mathbb{H}$, then g should also fix $Q' \in \partial\mathbb{H}$, contradicting the definition of parabolicity.

This shows that the three classes are disjoint.

In order to prove that they exhaust the whole of $\mathrm{Isom}_+(\mathbb{H})$, we need the following celebrated theorem of Luitzen Egbertus Jan Brouwer (see [44]):

Theorem 8.19 (Brouwer fixed-point theorem). *Let $B^n \subseteq \mathbb{R}^n$ denote the n-dimensional closed unit ball. Then every continuous map $f\colon B^n \to B^n$ has a fixed-point.*

Let $g \in \mathrm{Isom}(\mathbb{H})$. By Theorem 8.19 applied to $B^2 \equiv \overline{D} \cong \overline{\mathbb{H}}$ and $f = g$, we deduce that g has at least one fixed-point P in $\overline{\mathbb{H}}$. If $P \in \mathbb{H}$, then g is elliptic. If $P \in \partial\mathbb{H}$ and it is unique with this property, then g is parabolic. Otherwise, if $P' \in \partial\mathbb{H}$ is another fixed-point in the boundary, then the line $\ell = \ell_{P,P'}$ is g-invariant, and g is hyperbolic. The proof of Theorem 8.18 is complete. □

8.9 Characterizations of Orientation-Preserving Isometries of $\mathbb{H}$

Given $g \in \mathrm{Isom}_+(\mathbb{H})$, by Theorem 8.11 there exists a unique $m(g) \in \mathrm{PSL}(2,\mathbb{R})$ which represents g. Let $M(g) \in \mathrm{SL}(2,\mathbb{R})$ such that $m(g) = [M(g)]$. Such a matrix $M(g)$ is uniquely defined up to sign: if $M' \in \mathrm{SL}(2,\mathbb{R})$ satisfies $m(g) = [M']$, then $M' = \pm M(g)$. We then set $\tau(g) := |\mathrm{tr}(M(g))|$. Thus, $\tau(g)$ equals the absolute value of the trace of a determinant-one matrix corresponding to g. Note that this definition is independent of the model X or D for $\mathbb{H}$ and of the choice of the matrix (**exercise**).

The proof of the following proposition is left as an **exercise**.

Proposition 8.20 (Characterization of elliptic isometries). *Let $g \in \mathrm{Isom}_+(\mathbb{H})$. Then the following conditions are equivalent:*

(a) *there exists a $P \in \mathbb{H}$ such that $g(P) = P$, i.e., g is elliptic;*
(b) $\inf_{P\in\mathbb{H}} d(P, g(P)) = 0$ *and the infimum is attained;*
(c) $g = \mathrm{Id}_{\mathbb{H}}$ *or* $\tau(g) < 2$.

Proposition 8.21 (Characterization of parabolic isometries). *Let $g \in \mathrm{Isom}_+(\mathbb{H})$. Then the following conditions are equivalent:*

(a) *there exists a unique $P \in \partial\mathbb{H}$ such that $g(P) = P$, i.e., g is parabolic;*
(b) $\inf_{P\in\mathbb{H}} d(P, g(P)) = 0$ *and the infimum is* not *attained;*
(c) $\tau(g) = 2$.

Moreover, if one of the above conditions is satisfied, denoting by $P \in \partial\mathbb{H}$ the unique fixed-point in $\mathbb{H} \cup \partial\mathbb{H}$ the following holds. For every neighborhood $U \subseteq \partial\mathbb{H}$ of P in $\partial\mathbb{H}$ and every compact subset $K \subset \partial\mathbb{H} \setminus \{P\}$, there exists an $n_0 = n_0(g;U,K) \in \mathbb{N}$ such that $g^n(K) \subseteq U$ for all $n \in \mathbb{Z}$ such that $|n| \geq n_0$.

Proof. Given $x \in \partial X$, the rotation R^X_θ by $\theta = \theta(x) := \arcsin(2x/(1+x^2))$ (cf. Example 8.14) maps x to $\infty \in \partial X$ (**exercise**). Thus, given a parabolic isometry of X fixing $x \in \partial X$, conjugation by $R^X_{\theta(x)}$ yields a parabolic isometry of X fixing ∞. We leave it as an **exercise** to check (by looking at the expressions (8.6) and (8.7)) that the only isometries of X only fixing ∞ on ∂X are indeed the translations T^X_b with $b \in \mathbb{R}^*$. It follows that every parabolic isometry of $\mathbb{H}$ is conjugate to a translation of $\mathbb{H}$. It is also clear that $\tau(T^X_b) = 2$ for all $b \in \mathbb{R}$.

In the Lobachevsky–Poincaré model X of $\mathbb{H}$, given a neighborhood U of $\infty \in \partial X$, we can find $M \geq 0$ such that $U \supset U_M := \{z \in X : |z| > M\} \cup \{\infty\}$. Also, given a compact subset $K \subset \partial X \setminus \{\infty\}$, there exists an $m \in \mathbb{R}$ such that $K \subseteq \{z = x + iy \in \mathbb{C} : |x| \leq m, y = 0\}$, and setting $n_0 := (M+m)/|b|$ one has that $(T^X_b)^n(K) \subseteq U$ for all $n \in \mathbb{Z}$ such that $|n| \geq n_0$. □

Proposition 8.22 (Characterization of hyperbolic isometries). *Let $g \in \mathrm{Isom}_+(\mathbb{H})$. Then the following conditions are equivalent:*

(a) *there exists two distinct points P_+ and P_- in $\partial\mathbb{H}$ such that $g(P_\pm) = P_\pm$, i.e., g is hyperbolic;*
(b) *there exists a line $\ell \subseteq \mathbb{H}$ which is g-invariant and fixed-point-free;*
(c) $\inf_{P\in\mathbb{H}} d(P, g(P)) > 0$ *(the infimum is attained on the invariant line ℓ);*
(d) $\tau(g) > 2$.

Moreover, if one of the above conditions is satisfied, then, up to exchanging P_+ and P_-, the following holds: for every neighborhood $U_- \subseteq \partial\mathbb{H}$ of P_- (resp. $U_+ \subseteq \partial\mathbb{H}$ of P_+) in $\partial\mathbb{H}$ and every compact subset $K_- \subseteq \partial\mathbb{H}\setminus\{P_+\}$ (resp. $K_+ \subseteq \partial\mathbb{H}\setminus\{P_-\}$) there exists an $n_0 \in \mathbb{N}$ such that $g^n(K_-) \subseteq U_-$ for all integers $n > n_0$ (resp. $g^n(K_+) \subseteq U_+$ for all integers $n < -n_0$).

Proof. We leave the equivalences (a)–(d) as an **exercise**. Just note that in the Lobachevsky–Poincaré model X of $\mathbb{H}$, the dilation $g = D^X_\lambda$, $0 < \lambda \neq 1$, satisfies $d(P, g(P)) = |\ln\lambda|$ for all $P \in \ell = \{iy : y > 0\}$. To prove the dynamical characterization of hyperbolicity, we again use the Lobachevsky–Poincaré model X of $\mathbb{H}$ and consider a dilation $g = D^X_\lambda$, with say $\lambda > 1$. In this case $P_- = \infty$ and $P_+ = 0$. A neighborhood of P_- in ∂X contains $U_R := \{x+iy : |x| > R, y = 0\} \cup \{\infty\}$ for some $R > 0$ and a neighborhood of P_+ in ∂X contains $U_{\varepsilon'} := \{x+iy : |x| < \varepsilon', y = 0\} \cup \{0\}$ for some $\varepsilon' > 0$. Also any compact subset in $\partial X \setminus \{P_+\}$ (resp. in $\partial X \setminus \{P_-\}$) is contained in $K_\varepsilon := \{x+iy : \varepsilon < |x|, y = 0\} \cup \{\infty\}$ for some $\varepsilon > 0$ (resp. in $K_{R'} := \{x+iy : |x| < R', y = 0\} \cup \{0\}$ for some $R' > 0$). We leave it as an **exercise** to check that any $n_0 > \max\{\log_\lambda(R/\varepsilon), \log_\lambda(R'/\varepsilon')\}$ has the required properties. □

Remark 8.23. Suppose that $g \in \mathrm{Isom}_+(\mathbb{H})$ is hyperbolic. Then the invariant line is unique. For, suppose that ℓ, ℓ' are two distinct invariant lines. If $\ell \cap \ell' \neq \varnothing$, then the (unique) point $P \in \mathbb{H}$ such that $\ell \cap \ell' = \{P\}$ would be fixed by g, and g would be elliptic (contradicting Theorem 8.18). If $\ell \cap \ell' = \varnothing$ and ℓ and ℓ' have no common point in $\partial\mathbb{H}$, then there exists a unique line $\ell^\perp$ perpendicular to both of them. Note that $\ell^\perp$ is g-invariant. We deduce that the (unique) point P in $\ell \cap \ell^\perp$ is fixed by g, and again g would be elliptic (contradicting Theorem 8.18). Finally, suppose that $\ell \cap \ell' = \varnothing$ and ℓ and ℓ' have a common point P in $\partial\mathbb{H}$. Let $\rho > 0$ and denote by C_ρ the set of points in $\mathbb{H}$ at distance ρ from ℓ. It is an **exercise** to show that C_ρ is g-invariant and $C_\rho \cap \ell' \neq \varnothing$. We deduce that the (unique) point P in $C_\rho \cap \ell'$ is fixed by g, again a contradiction.

8.10 The Tits Alternative for $\mathrm{GL}(2,\mathbb{R})$

Proposition 8.24. *Let $g, h \in \mathrm{Isom}_+(\mathbb{H}) \setminus \{1\}$ and suppose that they have no common fixed-points in $\mathbb{H} \cup \partial\mathbb{H}$. Then the group $\Gamma := \langle g, h\rangle$ contains a subgroup isomorphic to a nonabelian free group, up to the following two exceptional cases:*

(i) $g^2 = h^2 = 1$;
(ii) *(up to exchanging g and h)* $g^2 = 1$, *h is hyperbolic, and g exchanges the two fixed-points of h in* $\partial\mathbb{H}$.

In these cases, Γ is isomorphic to the infinite dihedral group D_∞ (which is solvable).

Proof. The proof consists of a case-by-case analysis.

Case 1. One element, say g, is parabolic with fixed-point $P_1 \in \partial\mathbb{H}$. Set $g_1 := g$. Then, the element $g_2 := hg_1h^{-1}$ is also parabolic and its fixed-point in $\partial\mathbb{H}$ is $P_2 := h(P_1)$ which is distinct from P_1 by our hypotheses on g and h. Let K_1 (resp. K_2) be a compact neighborhood of P_1 (resp. P_2) in $\partial\mathbb{H}$ with $K_1 \cap K_2 = \varnothing$. Then, by the dynamical description of parabolicity (cf. Proposition 8.21) there exists a positive integer n_0 such that $g_1^n(K_2) \subseteq K_1$ and $g_2^n(K_1) \subseteq K_2$ for all $n \in \mathbb{Z}$ with $|n| \geq n_0$. It follows from Klein's ping-pong lemma (Theorem 1.17) that $g_1^{n_0}$ and $g_2^{n_0}$ generate a free subgroup in Γ.

Case 2. Both g and h are hyperbolic. Let P_+ and P_- (resp. Q_+ and Q_-) be the fixed-points of g (resp. of h) in $\partial\mathbb{H}$. Note that, by our assumptions on g and h, these points are distinct. Let then K_+ and K_- (resp. H_+ and H_-) be pairwise disjoint compact neighborhoods of P_+ and P_- (resp. of Q_+ and Q_-) in $\partial\mathbb{H}$. Then, setting $K := K_+ \cup K_-$ (resp. $H := H_+ \cup H_-$) we get a compact neighborhood of both P_+ and P_- (resp. of Q_+ and Q_-) in $\partial\mathbb{H}$ such that $K \cap H = \varnothing$. By the dynamical description of hyperbolicity (cf. Proposition 8.22), there exists an $n_0 \in \mathbb{N}$ such that $g^n(H) \subseteq K_-$ and $g^{-n}(H) \subseteq K_+$ (resp. $h^n(K) \subseteq H_-$ and $h^{-n}(K) \subseteq H_+$) for all $n \geq n_0$. In other words, for $|n| \geq n_0$, $g^n(H) \subseteq K$ and $h^n(K) \subseteq H$. Then, as in Case 1, from Klein's ping-pong lemma we deduce that g^{n_0} and h^{n_0} generate a free subgroup in Γ.

Case 3. One of the elements, say h, is hyperbolic with fixed-points P, Q in $\partial\mathbb{H}$, and g does not exchange them. Note that $g(Q) \neq Q$ and $g(P) \neq P$, by our assumptions on g and h. If $g(P) \neq Q$ and $g(Q) \neq P$, then h and ghg^{-1} have no common fixed-points in $\partial\mathbb{H}$ and therefore we are as in Case 2. If not, say $g(P) = Q$. By assumption, $g(Q) \neq P$.

If $g^2(Q) \neq P$, then h and g^2hg^{-2} have no common fixed-points in $\partial\mathbb{H}$ and therefore we are again in Case 2.

Thus we may assume that $g^2(Q) = P$. Consider the isometries $h' := g^{-1}hg$ and

$$h'' := ghg^{-1}hgh^{-1}g^{-1} \equiv h^{gh^{-1}g^{-1}}.$$

These are both hyperbolic with fixed-points in $\partial\mathbb{H}$ given by $R := g^{-1}(P) = g(Q)$ and $g^{-1}(Q) = P$, and $ghg^{-1}(Q) = Q$ and $S := ghg^{-1}(P)$, respectively. We already know that P, Q and R are distinct.

We claim that S is distinct from P, Q, and R.

Since the only fixed-points in $\partial\mathbb{H}$ of h are P and Q, we have $h(R) \neq R$, so that $S = ghg^{-1}(P) = gh(R) \neq g(R) = P$. Clearly, $S = ghg^{-1}(P) \neq ghg^{-1}(Q) = Q$, since $P \neq Q$ and ghg^{-1} is injective. Finally, since $R = g(Q) \neq Q = h^{-1}(Q)$, we have $h(R) \neq Q$ and thus $S = gh(R) \neq g(Q) = R$.

Consequently, h' and h'' are again as in Case 2.

Case 4. Both g and h are elliptic with $g^2 \neq 1$. Note that $k := hgh^{-1} \neq g$, since g and h have no common fixed-points.

Claim: *There exists a $P \in \partial D$ such that $k(P) = g(P)$.*

In order to prove the Claim, we recall some notions and results from the basic theory of Dynamical Systems (for details, we refer, for instance, to [195, Chapter 11] and [42, Chapter 7]). We denote by $\mathbb{S}^1 := \{z \in \mathbb{C} : |z| = 1\}$ the unit circle and by $p\colon \mathbb{R} \to \mathbb{S}^1$ the universal covering map given by $p(x) = e^{2\pi i x}$ for all $x \in \mathbb{R}$. Given an orientation-preserving homeomorphism $f\colon \mathbb{S}^1 \to \mathbb{S}^1$, a continuous map $\widetilde{f}\colon \mathbb{R} \to \mathbb{R}$ such that $p \circ \widetilde{f} = f \circ p$ is called a *lifting* of f. It is an **exercise** to show that such liftings exist and, moreover, they are uniquely determined by the value $\widetilde{f}(0)$. In fact, for two liftings $\widetilde{f_1}$ and $\widetilde{f_2}$ of f there exists an $n = n(\widetilde{f_1}, \widetilde{f_2}) \in \mathbb{Z}$ such that $\widetilde{f_1}(x) - \widetilde{f_2}(x) = n$ for all $x \in \mathbb{R}$.

Theorem 8.25 (Poincaré). *Let $f\colon \mathbb{S}^1 \to \mathbb{S}^1$ be an orientation-preserving homeomorphism of the circle and let $\widetilde{f}\colon \mathbb{R} \to \mathbb{R}$ be the lifting such that $0 \leq \widetilde{f}(0) < 1$. Then the limit*

$$\rho(f) := \lim_{n\to\infty} \frac{\widetilde{f}^n(x) - x}{n} \tag{8.10}$$

exists and is independent of $x \in \mathbb{R}$.

Moreover, if $g\colon \mathbb{S}^1 \to \mathbb{S}^1$ is a homeomorphism, then $\rho(g^{-1}fg) = \rho(f)$.

The number $\rho(f)$ defined by (8.10) is called the *rotation number* of f.

For example, it is an **exercise** to show that if R^D_θ is the rotation by angle $\theta \in [0, 2\pi)$ around the origin in D yielding, by continuity, an orientation-preserving homeomorphism f_θ of $\mathbb{S}^1 \equiv \partial D$, then

$$\rho(f_\theta) = \frac{\theta}{2\pi}. \tag{8.11}$$

It is an **exercise** to show that

$$\min_{x\in\mathbb{R}} \left(\widetilde{f}(x) - x\right) \leq \rho(f) \leq \max_{x\in\mathbb{R}} \left(\widetilde{f}(x) - x\right). \tag{8.12}$$

We are now in a position to give a proof of the claim. Using the Poincaré disc model D of $\mathbb{H}$, and modulo a conjugation within $\mathrm{Isom}_+(D)$, one may assume that $g = r_\alpha$ is the rotation around the origin O of the disc D by some angle $\alpha \in (0, \pi) \cup (\pi, 2\pi)$. Moreover, $k = hgh^{-1}$ induces an orientation-preserving homeomorphism $k\colon : \mathbb{S}^1 \to \mathbb{S}^1$ with rotation number $\rho(k) = \rho(hgh^{-1}) = \rho(g) = \alpha/(2\pi)$, cf. (8.11). By (8.12), there exists an $x \in \mathbb{R}$ such that $\widetilde{k}(x) - x = \alpha/(2\pi)$, where $\widetilde{k}\colon \mathbb{R} \to \mathbb{R}$ is the lifting of k such that $\widetilde{k}(0) \in [0, 1)$. Setting $P := \pi(x) \in \partial D$, we have

$$k(P) = k(p(x)) = p(\widetilde{k}(x)) = p(x + \alpha/(2\pi)) = p(x)p(\alpha/(2\pi)) = e^{i\alpha}P = g(P).$$

This shows that there exists a $P \in \partial D$ such that $k(P) = g(P)$. The claim follows.

By the claim, the element $g' := g^{-1}k = g^{-1}hgh^{-1}$ has a fixed-point P in $\partial\mathbb{H}$. Let us show that h and g' have no common fixed-points in $\mathbb{H} \cup \partial\mathbb{H}$. Suppose $Q \in \mathbb{H} \cup \partial\mathbb{H}$ is fixed by both h and g'. We leave it as an **exercise** to check that $g^{-1}(Q)$ is fixed by h. If $Q \in \partial\mathbb{H}$, then $g^{-1}(Q) \in \partial\mathbb{H}$, contradicting the hypothesis that h is elliptic. If $Q \in \mathbb{H}$, then $g^{-1}(Q) \in \mathbb{H}$ and either $Q = g^{-1}(Q)$, that is, $g(Q) = Q$ and this is not possible because h and g have no common fixed-points, or $Q \neq g^{-1}(Q)$ so that h has two fixed-points in $\mathbb{H}$. This is not possible either since $h \neq 1$.

Now, if the g'-fixed-point $P \in \partial\mathbb{H}$ is unique, then g' is parabolic, and one reduces to Case 1 (with g replaced by g'). Otherwise, g' is hyperbolic and we denote by Q the other element in $\partial\mathbb{H}$ such that $g'(Q) = Q$. If h does not exchange P and Q, then we reduce to Case 3 (with g replaced by h and h replaced by g'). We are only left with the following:

Exceptional case (i). If $g^2 = h^2 = 1$, then both g and h are elliptic. Since g and h have no common fixed-point, gh cannot be elliptic (**exercise**) and therefore it generates an infinite cyclic subgroup of index 2 in Γ. Thus Γ is isomorphic to the infinite dihedral group D_∞ (and therefore it is solvable).

Exceptional case (ii). If h is hyperbolic and g exchanges the fixed-points of h, then $ghg^{-1} = h^{-1}$. Since $g^2 = 1$ we also have $(gh)^2 = 1$. Moreover (**exercise**) g and gh have no common fixed-points in $\partial\mathbb{H}$, so that $\Gamma = \langle g, h\rangle = \langle g, gh\rangle$ is as in the exceptional case (i).

The proof of Proposition 8.24 is complete. □

Lemma 8.26. *Let $g, h \in \mathrm{Isom}_+(\mathbb{H})$ and suppose that they have a common fixed-point in $\mathbb{H} \cup \partial\mathbb{H}$. Then the subgroup Γ generated by a and b is solvable.*

Proof. Let $P \in \mathbb{H} \cup \partial\mathbb{H}$ denote the common fixed-point of g and h. If $P \in \mathbb{H}$ then both g and h are elliptic, in fact rotations around P (cf. Example 8.14). It is then clear that Γ is Abelian and therefore solvable. If $P \in \partial\mathbb{H}$, then, up to conjugation, we can suppose that $P = \infty \in \partial X$. The stabilizer of ∞ is isomorphic to the group $\Gamma_0 \leq \mathrm{PSL}(2,\mathbb{R})$ consisting of all $[g]$ with $g = \begin{pmatrix} a & b \\ 0 & d \end{pmatrix}$ in $\mathrm{SL}(2,\mathbb{R})$ (**exercise**). Thus Γ_0 is isomorphic to a quotient of the subgroup $\mathrm{B}(2,\mathbb{R}) \cap \mathrm{SL}(2,\mathbb{R})$ of $\mathrm{B}(2,\mathbb{R})$ (cf. Example 4.3.(a)). By Proposition 4.4, Γ_0 and therefore its subgroup Γ are solvable. The proof of Lemma 8.26 is complete. □

Theorem 8.27 (The Tits alternative for $\mathrm{GL}(2,\mathbb{R})$). *A subgroup of $\mathrm{GL}(2,\mathbb{R})$ which is not solvable contains a nonabelian free subgroup.*

Proof. We start by showing that a subgroup Γ of $\mathrm{Isom}_+(\mathbb{H})$ which is not solvable contains a nonabelian free subgroup. Suppose first that Γ contains a parabolic isometry, say g. Let $P \in \partial\mathbb{H}$ denote the (unique) fixed-point of g and let Γ_P denote the stabilizer of P in $\mathrm{Isom}_+(\mathbb{H})$. Since Γ_P is solvable (cf. the proof of Lemma 8.26), by Proposition 4.4 and our assumptions, $\Gamma \not\leq \Gamma_P$. If $h \in \Gamma \setminus \Gamma_P$, then g and h satisfy the hypotheses of Proposition 8.24 and therefore, by Case 1 therein, $\langle g,h\rangle \leq \Gamma$ contains a nonabelian free subgroup.

Suppose now that Γ contains a hyperbolic isometry, say h, and denote by $\mathrm{Fix}(h) \subset \partial\mathbb{H}$ the set of its fixed-points. Let $g \in \Gamma \setminus \{h\}$ and suppose that it is also hyperbolic, and denote by $\mathrm{Fix}(g) \subset \partial\mathbb{H}$ the set of its fixed-points. If $\mathrm{Fix}(h) \cap \mathrm{Fix}(g) = \varnothing$, then by Proposition 8.24 $\langle g,h\rangle \leq \Gamma$ contains a nonabelian free subgroup. Suppose now that $\mathrm{Fix}(h) \cap \mathrm{Fix}(g) = \{P\}$. We observe that in this case g and h do not commute, that is, their commutator $[g,h]$ is nontrivial. Up to conjugation, we can suppose that $P = \infty \in \partial X$ so that the action of g (resp. of h) is given by a matrix in $\mathrm{B}(2,\mathbb{R}) \cap \mathrm{SL}(2,\mathbb{R})$. It follows that $[g,h]$ is represented by a nontrivial matrix in $\mathrm{UT}(2,\mathbb{R})$ (cf. Example 4.3.(a)). Thus $[g,h]$ is parabolic and one reduces to the previous case.

Suppose now that $\mathrm{Fix}(h) = \mathrm{Fix}(g)$ for all hyperbolic isometries $g \in \Gamma$. As a consequence, the subgroup Γ_{hyp} generated by all hyperbolic elements in Γ is Abelian and therefore solvable. Thus, by our assumptions, Γ_{hyp} does not exhaust the whole of Γ and therefore there exist nonhyperbolic isometries in Γ. By virtue of the first case above, we are only left to consider the case when $\Gamma \setminus \Gamma_{\mathrm{hyp}}$ consists only of elliptic elements. Let $g \in \Gamma \setminus \Gamma_{\mathrm{hyp}}$ and suppose that it is elliptic. If g does not exchange the two elements in $\mathrm{Fix}(h)$, then it follows from Proposition 8.24 that $\langle g,h\rangle \leq \Gamma$ contains a nonabelian free subgroup. Otherwise, all $g \in \Gamma \setminus \Gamma_{\mathrm{hyp}}$ are elliptic and exchange the two elements in $\mathrm{Fix}(h)$: the product of two distinct such elements fixes the elements in $\mathrm{Fix}(h)$ and therefore belongs to Γ_{hyp}. Thus $[\Gamma : \Gamma_{\mathrm{hyp}}] = 2$ and Γ would be solvable, contradicting our assumptions.

Finally, suppose that Γ is elliptic. If $\mathrm{Fix}(g) = \mathrm{Fix}(h)$ for all $g,h \in \Gamma$ (resp. $g^2 = 1$ for all $g \in \Gamma$) then Γ is Abelian and therefore solvable, contradicting our assumptions. Thus we can find $g,h \in \Gamma$ such that $g^2 \neq 1$ and $\mathrm{Fix}(g) \neq \mathrm{Fix}(h)$. Then $\langle g,h\rangle \leq \Gamma$ contains a nonabelian free subgroup by Proposition 8.24.

This proves the Tits alternative for $\mathrm{PSL}(2,\mathbb{R}) \cong \mathrm{Isom}_+(\mathbb{H})$.

Suppose now that $\widetilde{\Gamma} \leq \mathrm{PGL}(2,\mathbb{R})$ is not solvable. Set $\Gamma := \widetilde{\Gamma} \cap \mathrm{PSL}(2,\mathbb{R})$. Since $[\mathrm{PGL}(2,\mathbb{R}) : \mathrm{PSL}(2,\mathbb{R})] = 2$, either $\Gamma = \widetilde{\Gamma}$ or $[\widetilde{\Gamma} : \Gamma] = 2$. In either case, Γ is not solvable (**exercise**). By the Tits alternative for $\mathrm{PSL}(2,\mathbb{R})$, we deduce that Γ and therefore $\widetilde{\Gamma}$ contain a nonabelian free subgroup. This proves the Tits alternative for $\mathrm{PGL}(2,\mathbb{R})$.

Finally, suppose that $\overline{\Gamma} \leq \mathrm{GL}(2,\mathbb{R})$ is not solvable. Denote by $\pi\colon \mathrm{GL}(2,\mathbb{R}) \to \mathrm{PGL}(2,\mathbb{R})$ the quotient map and set $\widetilde{\Gamma} := \pi(\overline{\Gamma}) \leq \mathrm{PGL}(2,\mathbb{R})$. We claim that $\widetilde{\Gamma}$ is not solvable. Otherwise, as $\widetilde{\Gamma} = (\overline{\Gamma}\,\mathbb{R}^*I)/\mathbb{R}^*I \cong \overline{\Gamma}/(\overline{\Gamma} \cap \mathbb{R}^*I)$, from exactness of

$$1 \to \overline{\Gamma} \cap \mathbb{R}^*I \to \overline{\Gamma} \to \overline{\Gamma}/(\overline{\Gamma} \cap \mathbb{R}^*I) \to 1$$

and Proposition 4.4 we get solvability of $\overline{\Gamma}$, contradicting our assumptions. By virtue of the Tits alternative for $\mathrm{PGL}(2,\mathbb{R})$, we deduce that $\widetilde{\Gamma}$ contains a nonabelian free subgroup F. It is clear that $\pi^{-1}(F) \leq \overline{\Gamma}$ contains a free subgroup as well. This proves the Tits alternative for $\mathrm{GL}(2,\mathbb{R})$. $\square$

8.11 Growth of Finitely Generated Linear Groups

From the Tits alternative theorem (Theorem 8.1) and the Milnor–Wolf theorem (Corollary 7.41) we deduce the following:

Corollary 8.28 (Growth of finitely generated linear groups). *Let G be a finitely generated linear group. Then the growth of G is either exponential or polynomial. In the latter case, G is virtually nilpotent.* □

As a consequence of the above corollary, a finitely generated group of intermediate growth (e.g., the Grigorchuk group) cannot be linear.

8.12 Notes

Möbius transformations are named after August Ferdinand Möbius; they are also variously called homographies, homographic transformations, linear fractional transformations, bilinear transformations, or fractional linear transformations. The Möbius group is isomorphic to the projective linear group $\mathrm{PGL}(2,\mathbb{C})$ (the group of all projective transformations of the complex projective line $\widehat{\mathbb{C}}$).

The Tits alternative theorem was proved by Jacques Tits in [333]. Its original proof consists in looking at the Zariski closure of the $\mathbb{K}$-linear group G in $\mathrm{GL}(n,\mathbb{K})$. If it is solvable, then the group is solvable. Otherwise, one looks at the image of G in the Levi component. If it is noncompact then a ping-pong argument yields a nonabelian free subgroup. If it is compact then, either all eigenvalues of elements in the image of G are roots of unity and then the image is finite, otherwise one applies again the ping-pong strategy yielding nonabelian free subgroups. For a complete and comprehensive proof of the Tits alternative, a part the original paper [333] by Tits, we refer to the monograph [96] by Cornelia Druţu and Misha Kapovich as well as to the notes [334] by Matthew C.H. Tointon.

In geometric group theory, a group G is said to satisfy the Tits alternative if for every subgroup H of G either H is virtually solvable or H contains a nonabelian free subgroup (in some versions of the definition, this condition is only required to be satisfied for all *finitely generated* subgroups of G).

Examples of groups satisfying the Tits alternative which are either not linear, or at least not known to be linear, are, without intending to be exhaustive:

- hyperbolic groups [139, 8.2.F] (see also [122, Chapter 8]);
- mapping class groups [186], [139], [157], [300], [264], [234];
- $\mathrm{Out}(F_n) := \mathrm{Aut}(F_n)/\mathrm{Inn}(F_n)$ the group of *outer* automorphisms of the free group F_n [28, 29];
- fundamental groups of certain 3-manifolds [269];
- certain groups of birational transformations of algebraic surfaces [49];
- one-relator groups [194, Theorem 3].

Examples of groups *not* satisfying the Tits alternative are:

- automorphism groups of locally finite trees; e.g., the Grigorchuk group [130] and the Gupta–Sidki groups [148] (see also [248] by Claudio Nebbia and [271] by Isabelle Pays and Alain Valette, for a variant of the Tits alternative (involving amenability) in this context and, more generally, for fixed-point properties versus existence of free groups in groups of homeomorphisms acting on metric spaces (and on their *end compactification*) see [353] and the reference therein, by Wolfgang Woess);
- the Richard Thompson group F [48];
- most *big* mapping class groups [211], and [6] where the Grigorchuk group is used;
- the group $\mathrm{Aut}(\Sigma^{\mathbb{Z}})$ of all automorphisms (i.e., invertible cellular automata [59, 215]) of a full shift (over a finite alphabet Σ and the group $\mathbb{Z}$) [301] by Ville Salo.

Also, the original result has been improved by quantifying the depth in G of the free subgroup with respect to some fixed generating set S of G. Emmanuel Breuillard proved that there exists a universal function $N(d)$ such that for any finite subset $S \subseteq \mathrm{GL}(d,K)$ either the group G generated by S is virtually solvable or there exist two words a,b on S of S-length less than $N(d)$ which generate a non-abelian free group (see [34, 35]), and previous quantifications in this direction [38, 109]).

Similar forms of quantification of the Tits Alternative for Gromov hyperbolic groups were proved by Thomas Delzant [92], by Malik Koubi [203] (for Gromov hyperbolic groups with torsion), and by Goulnara Arzhantsheva and Igor Lysënok [8] (for subgroups of a given hyperbolic group), for a constant N depending however always on the group G under consideration.

A weaker form of the alternative, generally easier to establish, the *weak Tits alternative*, asks for the existence of free subsemigroups in G instead of free subgroups, provided that G is not virtually solvable. Notice that, in this weaker form, the Tits alternative is no longer a dichotomy for linear groups, since it is well known that there exist solvable groups of $\mathrm{GL}(n,\mathbb{R})$ which also contain free semigroups (and actually, any finitely generated solvable group which is not virtually nilpotent contains a free semigroup on two generators [295]). It remains a dichotomy for those classes of groups for which virtual solvability implies sub-exponential growth, e.g. hyperbolic groups, groups acting geometrically on CAT(0)-spaces, etc. Quantitative results on the weak Tits alternative can be found in [72].

Note that, in order to prove Gromov's theorem (Theorem 12.1), the weak form of the Tits alternative suffices. We thank Mark Sapir for pointing this out to us.

Chapter 9
Topological Groups, Lie Groups, and Hilbert's Fifth Problem

This chapter is mainly expositive. We review the notions of a topological group and, in particular, of a locally compact group, also presenting a few examples. The latter admit a left-invariant regular Borel measure, called a Haar measure (Theorem 9.10): we present a complete (and self-contained) proof of this result in Section 9.3. We then briefly discuss locally compact Abelian groups and the celebrated Pontryagin duality. In the last sections we briefly review the notion of a Lie group, Hilbert's fifth problem and its solutions: in particular, we focus on the Theorem of Gleason and Montgomery–Zippin (Theorem 9.17) which will play a significant role in the proof of Gromov's theorem (Theorem 12.1) in Section 12.3.

9.1 Topological Groups

A topological group is a group G together with a Hausdorff topology on G such that both the product map and the inverse map are continuous functions with respect to the given topology.

Definition 9.1. Let G be a group that is also a Hausdorff topological space. Suppose that the topology on G is *compatible with the group operations*, that is, the product $G \times G \to G, (g,h) \mapsto gh$ (here $G \times G$ is equipped with the product topology) and the inversion $G \to G, g \mapsto g^{-1}$ are continuous maps. Then the group G is called a *topological group* and the topology is called a *group topology*.

Note that a topology on a group G is Hausdorff if and only if the trivial subgroup $\{1_G\} \leq G$ is closed (**exercise**). Moreover, the product map is continuous if and only if for any $g,h \in G$ and any neighborhood W of gh in G, there exist neighborhoods U of g and V of h in G such that $U \cdot V := \{uv : u \in U, v \in V\} \subseteq W$. Similarly, the inversion map is continuous if and only if for any $g \in G$ and any neighborhood V of g^{-1} in G, there exists a neighborhood U of g in G such that $U^{-1} := \{u^{-1} : u \in U\} \subseteq V$. Now, in order to show that a topology is compatible with the group operations, it suffices to check that the map $G \times G \to G, (g,h) \mapsto gh^{-1}$ is continuous. Explicitly, this means that for any $g,h \in G$ and any neighborhood W of gh^{-1} in G, there exist

T. Ceccherini-Silberstein and M. D'Adderio, *Topics in Groups and Geometry*, Springer Monographs in Mathematics, https://doi.org/10.1007/978-3-030-88109-2_9

neighborhoods U of g and V of h in G such that $U \cdot V^{-1} := \{uv^{-1} : u \in U, v \in V\} \subseteq W$.

A *homomorphism* of topological groups is a continuous group homomorphism $G \to H$. This way, topological groups, together with their homomorphisms, form a category.

An *isomorphism* of topological groups is a group isomorphism that is also a homeomorphism of the underlying topological spaces. Note that this condition is stronger than simply requiring a continuous group isomorphism: indeed, its inverse must also be continuous. There are examples of topological groups that are isomorphic as ordinary groups but not as topological groups. For instance, any non-discrete topological group is also a topological group when considered with the discrete topology: then the underlying groups are the same, however, as topological groups, they are not isomorphic.

Example 9.2. (a) Every group can be trivially made into a topological group by equipping it with the discrete topology. Such groups are then called *discrete groups*. In this sense, the theory of discrete topological groups reduces to that of ordinary groups. The *indiscrete topology* (i.e. the trivial topology) also makes every group into a topological group.

(b) An *Abelian topological group* is a topological group such that the group operation is commutative. The additive group of real numbers $(\mathbb{R}, +)$ with the usual Euclidean topology is an Abelian topological group. More generally, $\mathbb{R}^n$ is also an Abelian topological group under addition. In fact, every topological vector space V forms (with respect to vector addition) an Abelian topological group. Some other examples of Abelian topological groups include the *circle group* $\mathbb{S}^1 := \{z \in \mathbb{C} : |z| = 1\}$ (equipped with multiplication) which is isomorphic to $\mathbb{T} := \mathbb{R}/\mathbb{Z}$ (equipped with addition mod 1 and the quotient topology), and, more generally, the *n-torus* $\mathbb{T}^n$ for any natural number $n \geq 1$.

(c) The *classical groups* contain important examples of non-Abelian topological groups. For instance, the *general linear group* $\mathrm{GL}(n, \mathbb{R})$ of all invertible n-by-n matrices with real entries can be viewed as a topological group with the topology defined by viewing it as a subspace of $\mathbb{R}^{n^2}$. Another classical group is the *orthogonal group* $\mathrm{O}(n)$, the group of all linear isometries $\mathbb{R}^n \to \mathbb{R}^n$. The orthogonal group is compact as a topological space. These classical groups are *Lie groups*, meaning that they are smooth manifolds in such a way that the group operations are smooth, not just continuous. See Section 9.5 for more on this.

(d) An example of a topological group that is not a Lie group is the additive group $(\mathbb{Q}, +)$ of rational numbers, with the topology inherited from $\mathbb{R}$. This is a countable space, and it does not have the discrete topology.

(e) Another important example, from number theory, is the group $\mathbb{Z}_p$ of p-adic integers (here p is a prime number). It can be defined as the inverse limit of the finite groups $\mathbb{Z}/p^n\mathbb{Z}$ as $n \to \infty$. This is a compact group (in fact, it is homeomorphic to the Cantor set), but it differs from (real) Lie groups in that it is totally disconnected. The group $\mathbb{Z}_p$ is a *pro-finite group*: it is isomorphic to a subgroup of the product $\prod_{n \geq 1} \mathbb{Z}/p^n\mathbb{Z}$ with the *prodiscrete topology* (i.e., it is induced by the product topology, where the finite groups $\mathbb{Z}/\mathbb{Z}p^n$ are given the discrete topology). More generally, there is a theory of p-adic Lie groups, including compact groups such as

$\mathrm{GL}(n,\mathbb{Z}_p)$ as well as locally compact groups such as $\mathrm{GL}(n,\mathbb{Q}_p)$, where $\mathbb{Q}_p$ is the locally compact field of p-adic numbers (see, for instance [290]).

(e) A *solenoid* is a compact connected Abelian topological group obtained as the inverse limit of an inverse system of topological groups and continuous homomorphisms $(S_i, f_i)_{i\in\mathbb{N}}$, where $S_i := \mathbb{S}^1$ and $f_i\colon S_{i+1} \to S_i$ is the map that uniformly wraps $n_i \geq 2$ times S_{i+1} around S_i, for all $i \in \mathbb{N}$. In the simple case when $n_i = n$ for all $i \in \mathbb{N}$, one then has $f_i(z) = z^n$ for all $z \in S_i$. These were introduced by Leopold Vietoris for $n = 2$ [343] and by van Dantzig for an arbitrary n [87]. In the theory of hyperbolic dynamical systems, such solenoids arise as one-dimensional expanding attractors, also called *Smale–Williams attractors* [320], named after Stephen Smale and Robert F. Williams

(f) Some topological groups can be viewed as *infinite-dimensional* Lie groups. For example, a topological vector space, such as a Banach space or a Hilbert space, is an Abelian topological group under addition. Also, in every unital infinite-dimensional Banach algebra, the set of invertible elements forms a (non-Abelian) topological group under multiplication: this is the case, for example, for the group of invertible bounded operators on an infinite-dimensional Hilbert space.

Let H be a subgroup of a topological group G. Then H is itself a topological group when equipped with the subspace topology. If H is open in G, then it is also closed in G, since the complement of H is the open set given by the union of the open cosets gH with $g \in G \setminus H$. Also, the closure $\overline{H}$ of H is also a subgroup of G; if, in addition, H is normal in G, then $\overline{H}$ is normal in G.

The set of left cosets G/H with the quotient topology is called a *homogeneous space* for G: it is Hausdorff if and only if H is closed in G. The quotient map $\pi\colon G \to G/H$ is always open. For example, for a positive integer n, the *n-sphere* $\mathbb{S}^n := \{(x_1, x_2, \ldots, x_{n+1}) \in \mathbb{R}^{n+1} : \sum_{i=1}^{n+1} x_i^2 = 1\}$ is a homogeneous space for the *rotation group* $\mathrm{SO}(n+1) := \{g \in \mathrm{O}(n+1) : \det(g) = 1\}$ and its subgroup $\mathrm{SO}(n)$, viewed as the stabilizer of $(1,0,0,\ldots,0) \in \mathbb{R}^{n+1}$, so that $\mathbb{S}^n = \mathrm{SO}(n+1)/\mathrm{SO}(n)$. If H is a closed *normal* subgroup of G, then the quotient group G/H becomes itself a topological group when given the quotient topology. For example, as mentioned above, the quotient group $\mathbb{R}/\mathbb{Z}$ is isomorphic to the circle group $\mathbb{S}^1$.

In any topological group G, the connected component containing 1_G (this is called the *identity component*) is a closed normal subgroup. If C is the identity component and $g \in G$, then the left coset gC is the connected component of G containing g. So, the collection of all left (or right) cosets of C in G is equal to the collection of all connected components of G. As a consequence, the quotient group G/C is totally disconnected.

We end this section with the *Birkhoff–Kakutani theorem* (named after Garrett Birkhoff and Shizuo Kakutani) which characterizes metrizable topological groups (recall that in our setting, topological groups are, by definition, Hausdorff spaces).

Theorem 9.3 (Birkhoff–Kakutani). *Let G be a topological group. The following conditions are equivalent:*

- *G is* first-countable, *that is, there exists a countable base of neighborhoods for 1_G;*
- *G is* metrizable *(as a topological space);*

- *there exists a* left-invariant metric *on G that induces the given topology on G.*

Recall that a metric $d\colon G \times G \to [0, +\infty)$ on a group G is called left-invariant if left multiplication by g, that is, the map $h \mapsto gh$, is an isometry.

9.2 Locally Compact Groups

A locally compact group is a topological group for which the underlying topology is locally compact. Locally compact groups are important because many examples of groups that arise throughout mathematics and physics are locally compact. Moreover such groups have a natural measure called the *Haar measure*: this allows one to define integrals of Borel measurable functions on the group and to generalize classical analysis notions such as the Fourier transform and L^p spaces.

Definition 9.4. A topological group G is called *locally compact* provided it is locally compact as a topological space, that is, every element $g \in G$ admits a compact neighborhood.

Recall that in a topological space X, a neighborhood of a point $x \in X$ is a subset $V \subset X$ containing an open set U such that $x \in U$. Also recall that in this book a topological group is, by definition, Hausdorff. Thus, in a locally compact group G any element $g \in G$ in fact admits a base of compact neighborhoods.

Example 9.5. (a) Every *compact* topological group is, trivially, locally compact. In particular, the circle group $\mathbb{S}^1 := \{z \in \mathbb{C} : |z| = 1\}$ under multiplication is compact.

(b) The additive groups R^n (and in particular R) are locally compact as a consequence of the *Heine–Borel theorem* (a subset $X \subset \mathbb{R}^n$ is compact if and only if it is closed and bounded). More generally, Lie groups, which are locally Euclidean, are all locally compact groups (see Section 9.5).

(c) All discrete topological groups are locally compact. A discrete topological group is compact if and only if it is finite.

(d) The additive group of a Hausdorff topological vector space V is locally compact if and only if V is finite-dimensional.

(e) All open (resp. closed) subgroups of a locally compact group are locally compact in the subspace topology.

(f) The additive group $(\mathbb{Q}, +)$ of rational numbers is not locally compact if given the relative topology as a subset of the real numbers. It is locally compact if given the discrete topology.

(e) The additive topological group $(\mathbb{Q}_p, +)$ of p-adic numbers is locally compact for any prime number p. In fact, it is homeomorphic to the Cantor set minus one point.

By homogeneity, local compactness for a topological group G need only be checked at the identity element 1_G. That is, a topological group G is locally compact if and only if the identity element 1_G admits a compact neighborhood. Indeed, if $V \subset G$ is such a compact neighborhood of 1_G, then $gV \subset G$ serves as a compact neighborhood of g, for any $g \in G$.

A subgroup of a locally compact group is locally compact if and only if it is closed. Every quotient G/N of a locally compact group G by a closed normal subgroup $N \leq G$ is locally compact. The product of a family of locally compact groups is locally compact if and only if all but a finite number of factors are actually compact.

Every locally compact group which is second-countable is metrizable as a topological group (cf. Theorem 9.3) and complete.

Below, we present a few preliminary technical results on locally compact groups, of some interest on their own, which we shall need later.

Lemma 9.6. *Let G be a locally compact group. Then the following holds.*

(1) *Let V be a symmetric neighborhood of 1_G. Then $V^\infty := \bigcup_{n\geq 1} V^n$ is a clopen subgroup of G.*
(2) *Let $A \subseteq G$. Then the closure of A equals $\overline{A} = \bigcap_V AV$, where V ranges over all symmetric neighborhoods of 1_G.*
(3) *Let $K \subseteq U \subset G$ with K compact and U open. Then there exists a symmetric neighborhood V of 1_G such that $KV \cup VK \subseteq U$.*

Proof. (1) First observe that since $1_G \in V$, one has $V \subseteq V^2 \subseteq \cdots \subseteq V^n \subseteq V^{n+1} \subseteq \cdots$. Also, since $V = V^{-1}$ one has $(V^n)^{-1} = V^n$, that is, the V^ns are also symmetric. Thus, given $g,h \in V^\infty$, we can find $m,n \in N$ such that $g \in V^m$ and $h \in V^n$. We deduce that $gh^{-1} \in V^{m+n} \subset V^\infty$. This shows that V^∞ is a subgroup. It is open since if $g \in V^\infty$ then gV is an open neighborhood of g in G and $gV \subseteq V^\infty V = V^\infty$. Moreover, for any $g \in G$ the right coset $V^\infty g$ is homeomorphic to V^∞ and therefore is open. We deduce that $V^\infty = G \setminus \bigcup_{g\in G\setminus V^\infty} V^\infty g$ is also closed.

(2) Let $g \in \overline{A}$ and let V be a symmetric neighborhood of 1_G. Then gV is a neighborhood of g and $gV \cap A \neq \varnothing$. Let $h \in gV \cap A$ so that there exists a $v \in V$ such that $g = hv^{-1} \in AV^{-1} = AV$. This shows that $\overline{A} \subseteq \bigcap_V AV$. Conversely, suppose that $g \in AV$, for every symmetric neighborhood V of 1_G. This means that for every such V one has $gV \cap A = gV^{-1} \cap A \neq \varnothing$. As $(gV)_V$, where V ranges over all symmetric neighborhoods of 1_G, is a base of neighborhoods of g, this shows that $g \in \overline{A}$.

(3) The set $O := \{(g,h) : gh \in U\} \cap \{(g,h) : hg \in U\} \subseteq G \times G$ is open in $G \times G$ and $(1_G, u), (u, 1_G) \in O$ for all $u \in U$ (**exercise**). As a consequence, given $h \in K$, there exist neighborhoods V_h of 1_G and W_h of h such that $V_h \times W_h \subseteq O$. As $K \subset U$, the family $(W_h)_{h\in K}$ covers K. Since K is, by our assumptions, compact, we can find $h_1, h_2, \ldots, h_n \in K$ such that $\bigcup_{i=1}^n W_{h_i} \supseteq K$. Let V be a symmetric neighborhood contained in $\bigcap_{i=1}^n V_{h_i}$. Now, if $g \in V$ and $h \in K$, say $h \in W_{h_i}$, we have $(g,h) \in V \times W_{h_i} \subset V_{h_i} \times W_{h_i} \subseteq O$. We deduce that $gh, hg \in U$. This shows that $KV \cup VK \subseteq U$. □

Definition 9.7. Let G be a group. Given $g \in G$ we denote by $\lambda(g) \colon \mathbb{C}^G \to \mathbb{C}^G$ (resp. $\rho(g) \colon \mathbb{C}^G \to \mathbb{C}^G$) the linear map defined by setting

$$[\lambda(g)f](h) := f(g^{-1}h) \text{ (resp. } [\rho(g)f](h) := f(hg))$$

for all $f \in \mathbb{C}^G$ and $h \in G$. This is called the *left* (resp. *right*) *translation* by g.

Lemma 9.8. *Let G be a locally compact group. Let $f\colon G \to \mathbb{C}$ be a continuous function with compact support. Then, given $\varepsilon > 0$ there exists a neighborhood V of 1_G such that if $g,h \in G$ satisfy $g^{-1}h \in V$, then*

$$|[\lambda(g)f](k) - [\lambda(h)f](k)| < \varepsilon$$

for all $k \in G$.

Proof. We first observe that $\lambda(g^{-1})(\lambda(g)f - \lambda(h)f) = f - \lambda(g^{-1}h)f$ (**exercise**). Thus, it suffices to show that if $g' := g^{-1}h$ belongs to a suitable neighborhood V of 1_G, then $|f(k) - f((g')^{-1}k)| = |f(k) - [\lambda(g')f](k)| < \varepsilon$ for all $k \in G$.

Let $K := \mathrm{supp}(f)$ and let U be a compact and symmetric neighborhood of 1_G. The set $\{(g,h) : |f(h) - f(g^{-1}h)| < \varepsilon\} \subseteq G \times G$ is open (**exercise**) and contains $(1,h)$ for all $h \in H$. As a consequence, for each $h \in G$ we can find neighborhoods V_h of 1_G and W_h of h such that $(1,h) \in V_h \times W_h \subseteq \{(g,h) : |f(h) - f(g^{-1}h)| < \varepsilon\}$. As $(W_h)_{h \in UK}$ is an open cover of UK and UK is compact (**exercise**), we can find $h_1, h_2 \dots h_n \in UK$ such that $\bigcup_{i=1}^n W_{h_i} \supseteq UK$. Then $V := \left(\bigcap_{i=1}^n V_{h_i}\right) \cap U$ is an open neighborhood of 1_G. Let us show that V is indeed the sought neighborhood. Suppose that $g' \in V$ and $k \in G$. If $k \in UK$, say $k \in W_{h_i}$, then $(g',k) \in V_{h_i} \times W_{h_i} \subset \{(g,h) : |f(h) - f(g^{-1}h)| < \varepsilon\}$, and we have $|f(h) - f((g')^{-1}h)| < \varepsilon$, as required. If $k \notin UK$, then on the one hand, since $K \subseteq UK$ we have $k \notin K = \mathrm{supp}(f)$ so that $f(k) = 0$, and on the other hand, recalling that $g \in V \subseteq U$, we have $(g')^{-1}k \notin K = \mathrm{supp}(f)$, $f((g')^{-1}k) = 0$. We deduce that $|f(k) - f((g')^{-1}k)| = 0 < \varepsilon$. □

In the following we denote by $C_c(G)$ (resp. $C_0(G)$) the space of all continuous functions $f\colon G \to \mathbb{C}$ with compact support (which vanish at infinity: $\lim_{g\to\infty} f(g) = 0$. This means that for all $\varepsilon > 0$ there exists a compact subset $K \subset G$ such that $|f(g)| < \varepsilon$ for all $g \in G \setminus K$). Note that $C_c(G) \subseteq C_0(G)$. We define the *sup-norm* $\|\cdot\|_\infty$ on $C_0(G)$ by setting

$$\|f\|_\infty := \sup_{g \in G} |f(g)|. \tag{9.1}$$

It follows from the preceding lemma that, given $f \in C_c(G)$ the map $g \mapsto \lambda(g)f$ is a continuous function. It is easy to see (**exercise**) that $C_c(G)$ is dense in $C_0(G)$ in the norm $\|\cdot\|_\infty$. We deduce that given $f \in C_0(G)$, the map $g \mapsto \lambda(g)f$ is a continuous function for all $g \in G$.

9.3 The Haar Measure

Let G be a locally compact topological group. Recall that a *σ-algebra* of subsets of G is a collection Σ of subsets of G such that: (i) $G \in \Sigma$, (ii) Σ is closed under complementation (if $S \in \Sigma$ then $(G \setminus S) \in \Sigma$), and (iii) Σ is closed under countable unions (if $(S_n)_{n\in\mathbb{N}}$ is a sequence of subsets $S_n \in \Sigma$, then $(\bigcup_{n\in\mathbb{N}} S_n) \in \Sigma$). Note that a σ-algebra also includes the empty subset ($\varnothing \in \Sigma$) and it is closed under countable intersections (if $(S_n)_{n\in\mathbb{N}}$ is a sequence of subsets $S_n \in \Sigma$, then $(\bigcap_{n\in\mathbb{N}} S_n) \in \Sigma$).

The σ-algebra $\mathscr{B}(G)$ generated by (that is, the smallest σ-algebra of subsets of G containing) all open subsets of G is called the *Borel algebra* and its elements are called the *Borel sets* of G. Given $g \in G$ and a subset $S \subseteq G$, the *left* and *right translates* of S by g are the sets $gS := \{gs : s \in S\}$ and $Sg := \{sg : s \in S\}$, respectively. The Borel algebra $\mathscr{B}(G)$ is clearly invariant under left and right translates.

Definition 9.9. A *Borel measure* on G is a map $\mu : \mathscr{B}(G) \to [0,+\infty) \cup \{+\infty\}$ such that

- $\mu(\varnothing) = 0$ (*null empty set condition*);
- $\mu\left(\bigcup_{n\in\mathbb{N}} S_n\right) = \sum_{n\in\mathbb{N}} \mu(S_n)$ for any countable sequence $(S_n)_{n\in\mathbb{N}}$ of pairwise *disjoint* ($S_i \cap S_j = \varnothing$ for all $i \neq j$) Borel sets of G (*countable additivity*).

Let μ be a Borel measure on G. One says that μ is:

- *nontrivial* provided that there exists an $S \in \mathscr{B}(G)$ such that $\mu(S) > 0$;
- *locally compact* provided that it is finite on compact sets ($\mu(K) < \infty$ for every compact subset $K \subset G$);
- *outer regular* on Borel sets if $\mu(S) = \inf\{\mu(U) : S \subset U,\ U \text{ open}\}$ for all $S \in \mathscr{B}(G)$;
- *inner regular* on open sets if $\mu(U) = \sup\{\mu(K) : K \subset U,\ K \text{ compact}\}$ for all $U \subset G$ open;
- *left-* (resp. *right-*) *invariant* if $\mu(gS) = \mu(S)$ (resp. $\mu(Sg) = \mu(S)$ for all Borel subsets $S \subset G$ and all $g \in G$.

We are now in a position to state Haar's Theorem (named after Alfred Haar).

Theorem 9.10 (Haar). *Let G be a locally compact group. Then there is, up to a positive multiplicative constant, a unique nontrivial Borel measure μ on G satisfying the following properties:*

- *μ is left-invariant;*
- *μ is locally compact;*
- *μ is outer regular on Borel sets;*
- *μ is inner regular on open sets.*

Such a measure on G is called a left Haar measure.

It can be shown that a left Haar measure satisfies that $\mu(U) > 0$ for every nonempty open subset $U \subset G$. Moreover, if G is compact then $\mu(G)$ is finite, so we can uniquely specify a left Haar measure on G by adding the normalization condition $\mu(G) = 1$; in this case μ is also right-invariant. Locally compact groups for which the left Haar measures are also right-invariant (and therefore are *bi-invariant*) are called *unimodular*: this is the case, besides the compact case we alluded to above, for Abelian groups (e.g., the additive group $(\mathbb{R},+)$, where the Haar measures are nothing but positive multiples of the Lebesgue measure), and for discrete groups (here the Haar measures are positive multiples of the *counting measure*: $\mu(g) = 1$ for all $g \in G$). Other examples of unimodular groups include *semisimple Lie groups* and *connected nilpotent Lie groups*. An example of a *non-unimodular group* is the *Affine group*

$$\mathrm{Aff}(\mathbb{R}) := \left\{ \begin{pmatrix} a & b \\ 0 & 1 \end{pmatrix} : a,b \in \mathbb{R}, a \neq 0 \right\} \leq \mathrm{GL}(2,\mathbb{R}),$$

the group of affine transformations of the real line (a *solvable Lie group*) given by $x \mapsto ax+b$ for all $x \in \mathbb{R}$.

Haar established Theorem 9.10 in the special case of second countable locally compact groups in 1933 [149]. Existence and uniqueness (up to scaling) of a left Haar measure in full generality was first proven (by using the axiom of choice) by André Weil [348]. Henri Cartan gave a proof (which establishes existence and uniqueness simultaneously) which avoided the axiom of choice [53].

The remainder of this section is devoted to the proof of Theorem 9.10. Our exposition is based on the notes [113] by Alessandro Figà-Talamanca. By virtue of the Riesz–Markov–Kakutani representation theorem, instead of directly defining the measure (the Haar measure), we shall prove the existence of an *integral* on the space $C_c(G)$ of all compactly supported complex functions on G, that is, of a nontrivial continuous linear map (a *functional*) $I\colon C_c(G) \to \mathbb{C}$ such that

(i) $I(f) \geq 0$ if $f \in C_c(G)$ is nonnegative ($f(g) \geq 0$ for all $g \in G$);
(ii) $I(\lambda(g)f) = I(f)$ for all $f \in C_c(G)$ and $g \in G$.

Moreover, we shall show that if $J\colon C_c(G) \to \mathbb{C}$ is another nontrivial continuous linear map satisfying (i) and (ii), then there exists a constant $k > 0$ such that $J(f) = kI(f)$ for all $f \in C_c(G)$.

We denote by $C_c^+(G)$ the set of all nonnegative functions in $C_c(G)$. Given $f_1, f_2 \in C_c^+(G)$ we set

$$(f_1;f_2) := \inf_c \{\sum_{j=1}^n c_j : \exists g_1,\ldots,g_n \in G \text{ s.t. } f_1(g) \leq \sum_{j=1}^n c_j[\lambda(g_j)f_2](g), \forall g \in G\}$$

where the infimum is take over all $c = (c_1,c_2,\ldots,c_n) \in \mathbb{R}_+^n$, $n \in \mathbb{N}$. Note that such linear combinations of translates exist since $\mathrm{supp}(f_1)$ is compact and there exists an open subset $U \subseteq G$ and $\delta > 0$ such that $f_2(g) \geq \delta$ for all $g \in U$ (**exercise**).

The number $(f_1;f_2)$ satisfies the following properties:

(1) $(f_1;f_2) > 0$;
(2) $(\lambda(g)f_1;f_2) = (f_1;f_2)$ for all $g \in G$;
(3) $(f_1 + f_1';f_2) \leq (f_1;f_2) + (f_1';f_2)$;
(4) $(cf_1;f_2) = c(f_1;f_2)$ for all constants $c > 0$;
(5) $(f_1;f_2) \leq (f_1';f_2)$ if $f_1 \leq f_1'$;
(6) $(f_1;f_3) \leq (f_1;f_2)(f_2;f_3)$,

for all $f_1, f_1', f_2, f_3 \in C_c^+(G)$. We leave the proof of properties (1)–(5) as an easy **exercise**. Let us show (6). Suppose that $f_1 \leq \sum_{j=1}^n c_j[\lambda(g_j)f_2]$ (respectively $f_2 \leq \sum_{i=1}^m d_i[\lambda(h_i)f_3]$ for suitable $(c_1,g_1),\ldots,(c_n,g_n) \in \mathbb{C} \times G$ (respectively $(d_1,h_1),\ldots,(d_m,h_m) \in \mathbb{C} \times G$). Then,

$$f_1 \leq \sum_{j=1}^n \sum_{i=1}^m c_j d_i \lambda(g_j h_i) f_3.$$

This shows that $(f_1;f_3) \leq (f_1;f_2)(f_2;f_3)$.

Let us now fix, once and for all, a nontrivial element $f_0 \in C_c^+(G)$ and, for $\phi \in C_c^+(G)$ define $I_\phi : C_c^+(G) \to \mathbb{R}$ by setting

$$I_\phi(f) := \frac{(f;\phi)}{(f_0;\phi)} \tag{9.2}$$

for all $f \in C_c^+(G)$. Note that, by property (1), $(f_0;\phi) \neq 0$. It is easy to verify (**exercise**) that I_ϕ satisfies the following properties:

(1') $I_\phi(f) > 0$;
(2') $I_\phi(\lambda(g)f) = I_\phi(f)$ for all $g \in G$;
(3') $I_\phi(f_1 + f_2) \leq I_\phi(f_1) + I_\phi(f_2)$;
(4') $I_\phi(cf) = cI_\phi(f)$ for all constants $c > 0$;
(5') $I_\phi(f_1) \leq I_\phi(f_2)$ if $f_1 \leq f_2$.

Claim 1. *Let $f_1, f_2 \in C_c^+(G)$ and $\varepsilon > 0$. Then there exists a neighborhood V of 1_G such that*

$$I_\phi(f_1) + I_\phi(f_2) < I_\phi(f_1) + I_\phi(f_2) + \varepsilon,$$

for all $\phi \in C_c^+(G)$ vanishing in $G \setminus V$.

Let $u \in C_c^+(G)$ such that $u(t) = 1$ for all $t \in \mathrm{supp}(f_1 + f_2)$. Let $\delta, \eta > 0$ such that

$$2\eta(f_1 + f_2; f_0) + \delta(1 + 2\eta)(u; f_0) < \varepsilon. \tag{9.3}$$

Set $f = f_\delta := f_1 + f_2 + \delta u \in C_c^+(G)$. Also set $h_1 = h_{1,\delta} := f_1/f$ (resp. $h_2 = h_{2,\delta} := f_2/f$) with the convention that if $f_i(t) = 0$, then $h_i(t) = 0$, $i = 1,2$. Then $h_1, h_2 \in C_c^+(G)$ and we can find a symmetric neighborhood $V = V_{\delta,\eta}$ of 1_G such that $|h_1(g) - h_1(g')| \leq \eta$ and $|h_2(g) - h_2(g')| \leq \eta$ for all $g, g' \in G$ such that $g^{-1}g' \in V$. Suppose now that $\phi \in C_c^+(G)$ vanishes in $G \setminus V$ and let $c = (c_1, c_2, \ldots, c_n) \in \mathbb{R}_+^n$ and $g_1, g_2, \ldots, g_n \in G$ such that $f(g) \leq \sum_{j=1}^n c_j[\lambda(g_j)]\phi(g)$ for all $g \in G$.

Observe that $[\lambda(g_j)\phi](g) \neq 0$ implies $g_j^{-1}g \in V$ and thus $|h_i(g) - h_i(g_j)| < \eta$, for $i = 1,2$. We deduce

$$\begin{aligned} f_i(g) = f(g)h_i(g) &\leq \sum_{j=1}^n c_j h_i(g)[\lambda(g_j)\varphi](g) \\ &\leq \sum_{j=1}^n c_j (h_i(g_j) + \eta)\, \varphi(g_j^{-1}g) \end{aligned}$$

for all $i = 1,2$, and $g \in G$, that is, $(f_i;\phi) \leq \sum_{j=1}^n c_j(h_i(g_j) + \eta)$. As a consequence,

$$(f_1;\phi) + (f_2;\phi) \leq \sum_{j=1}^n c_j (h_1(g_j) + h_1(g_j) + 2\eta) = \sum_{j=1}^n c_j(1 + 2\eta).$$

Taking the inf over all $c = (c_1, c_2, \ldots, c_n) \in \mathbb{R}_+^n$ and $g_1, g_2, \ldots, g_n \in G$, we deduce

$$I_\phi(f_1) + I_\phi(f_2) \leq (1 + 2\eta)I_\phi(f) \leq (1 + 2\eta)(I_\phi(f_1 + f_2) + \delta I_\phi(u)).$$

It follows from (9.3) that $I_\phi(f_1) + I_\phi(f_2) \leq I_\phi(f_1 + f_2) + \varepsilon$. This ends the proof of the claim.

Consider now the compact space

$$S := \prod_{f \in C_c^+(G)} [1/(f_0;f),\ (f;f_0)]$$

and observe that by virtue of (6), one has $I_\phi(f) \in [1/(f_0;f),\ (f;f_0)]$. As a consequence, $I_\phi \equiv \big(I_\phi(f)\big)_{f \in C_c^+(G)} \in S$. For a neighborhood V of 1_G set

$$C_V := \overline{\{I_\phi : \phi \in C_c^+(G) \text{ vanishing in } G \setminus V\}} \subseteq S.$$

Observe that $(C_V)_V$ is a family of closed subsets of S with the finite intersection property. Indeed, if $V_1, V_2, \ldots, V_n$ are neighborhoods of 1_G, then $V := V_1 \cap V_2 \cap \cdots \cap V_n$ is a neighborhood of 1_G and $C_{V_1} \cap C_{V_2} \cap \cdots \cap C_{V_n} = C_V \neq \varnothing$. By compactness of S, there exists an $I \in \bigcap_V C_V$. Note that $I \colon C_c^+(G) \to \mathbb{R}_+$ and, in fact, $I(f) \in [1/(f_0;f),\ (f;f_0)]$ for all $f \in C_c^+(G)$. Moreover, given any (arbitrarily small) neighborhood V of 1_G, any functions $f_1, f_2, \ldots, f_n \in C_c^+(G)$, and $\varepsilon > 0$, there exists a $\phi \in C_c^+(G)$ vanishing in $G \setminus V$ such that $|I_\phi(f_i) - I(f_i)| < \varepsilon$ for all $i = 1, 2, \ldots, n$. It follows that I satisfies properties (1')–(5') and, in addition, $I(f_1 + f_2) = I(f_1) + I(f_2)$. We extend $I \colon C_c(G) \to \mathbb{R}$ by setting

$$I(f_1 - f_2) := I(f_1) + I(f_2)$$

for all $f_1, f_2 \in C_c^+(G)$: this is well defined (**exercise**). By linearity, we indeed extend I to a linear functional on the space of all *complex* compactly-supported functions on G. This proves the *existence* of the left-invariant integral. We are only left to show *uniqueness* up to positive multiplicative constants. This is immediately deduced from the following:

Claim 2. *Let J be another linear functional on the space of all complex compactly-supported functions on G. Then*

$$\frac{I(f_1)}{J(f_1)} = \frac{I(f_2)}{J(f_2)} \tag{9.4}$$

for all $f_1, f_2 \in C_c^+(G)$.

Let $f_1, f_2 \in C_c^+(G)$. Let $K \subseteq G$ be a compact subset containing the support of f_1. Let $U \subseteq G$ be an open subset such that $\overline{U}$ is compact and contains K. Finally, let $u \in C_c^+(G)$ such that $u(g) = 1$ for all $g \in U$. Fix $\varepsilon > 0$. Then we can find a symmetric neighborhood V of 1_G such that $VK \cup KV \subset U$ and, moreover, $|f_1(hg) - f_1(gk)| < \varepsilon$ for $h, k \in V$ and $g \in G$ (**exercise**). Then, if $h \in V$, one has $f_1(gh) = f_1(gh)u(g)$ and $f_1(hg) = f_1(hg)u(g)$ for all $g \in G$ (**exercise**). Let now $\phi \in C_c^+(G)$ and suppose that ϕ is symmetric ($\phi(g^{-1}) = \phi(g)$ for all $g \in G$). By *Fubini's theorem* we have

$$I(\phi)J(f_1) = I_\phi(J_g(\phi(h)f_1(g))) = I_h(J_g(\phi(h)f_1(hg))) = J_g(I_h(\phi(h)f_1(hg))),$$

where J_g (resp. I_h) denote integration with respect to the variable g (resp. h). Analogously, since $\phi(h^{-1}g) = \phi(g^{-1}h)$,

$$I(f_1)J(\phi) = J_g(I_h(\phi(h^{-1}g)f_1(h))) = J_g(I_h(\phi(g^{-1}h)f_1(h))) = J_g(I_h(\phi(h)f_1(gh))).$$

As a consequence,

$$\begin{aligned}|I(\phi)J(f_1) - I(f_1)J(\phi)| &\le J_g(I_h(\phi(h)|f_1(hg) - f_1(gh)|)) \\ &\le \varepsilon J_g(I_h(\phi(h)u(g))) = \varepsilon I(\phi)J(u).\end{aligned}$$

Analogously, choosing a suitable function $v \in C_c^+(G)$ such that $u(g) = 1$ for all g in a neighborhood of $\operatorname{supp}(f_2)$, we obtain, with the same function ϕ,

$$|I(\phi)J(f_2) - I(f_2)J(\phi)| \le \varepsilon I(\phi)J(v).$$

We deduce that

$$\left|\frac{J(f_1)}{I(f_1)} - \frac{J(\phi)}{I(\phi)}\right| \le \varepsilon \frac{J(u)}{I(f_1)} \quad \text{and} \quad \left|\frac{J(f_2)}{I(f_2)} - \frac{J(\phi)}{I(\phi)}\right| \le \varepsilon \frac{J(v)}{I(f_2)}$$

and therefore

$$\left|\frac{J(f_1)}{I(f_1)} - \frac{J(f_2)}{I(f_2)}\right| \le \varepsilon \left(\frac{J(u)}{I(f_1)} + \frac{J(v)}{I(f_2)}\right).$$

Since $\varepsilon > 0$ was arbitrary, (9.4) follows.

The claim is proved and the proof of Theorem 9.10 is complete.

9.4 Locally Compact Abelian Groups and Pontryagin Duality

Recall that a topological group G is called locally compact if the underlying topological space is locally compact (and Hausdorff). The topological group G is called Abelian if the underlying group is Abelian. Examples of locally compact Abelian groups include:

- finite abelian groups (with the discrete topology),
- the additive group $(\mathbb{Z}, +)$ of the integers, or, more generally, $(\mathbb{Z}^d, +)$ (with the discrete topology, but also with any metric as a finitely generated group),
- the additive group $(\mathbb{R}, +)$ of real numbers, or, more generally, $(\mathbb{R}^d, +)$ (with the discrete topology, but also with the usual Euclidean metric),
- the additive group $(\mathbb{Q}, +)$ of rational numbers (with the discrete topology),
- the circle group $\mathbb{T} = \mathbb{R}/\mathbb{Z} \cong \mathbb{S}^1$ (with its usual metric topology),
- the additive group $(\mathbb{Z}_p, +)$ of p-adic integers (with the usual p-adic topology, induced by the non-Archimedean norm $\|\cdot\|_p$),
- the additive group $(\mathbb{Q}_p, +)$ of p-adic numbers (with the usual p-adic topology, induced by the non-Archimedean norm $\|\cdot\|_p$).

Definition 9.11. Let G be a locally compact Abelian group. A *character* of G is a continuous homomorphism $\chi \colon G \to \mathbb{T}$ of G to the circle group (the latter equipped with the usual metric topology). The set $\widehat{G} := \operatorname{Hom}(G, \mathbb{T})$ consisting of all characters of G equipped with pointwise multiplication and the *compact-open topology* is called the *dual group* or *Pontryagin dual* of G.

Pointwise multiplication is defined as follows: given $\chi_1,\chi_2 \in \widehat{G}$ one defines $\chi_1\chi_2\colon G \to \mathbb{T}$ by setting $(\chi_1\chi_2)(g) := \chi_1(g)\chi_2(g) \in \mathbb{T}$ for all $g \in G$. Then one easily checks that $\chi_1\chi_2 \in \widehat{G}$, and that all group axioms are satisfied. In particular, the *trivial character* $\chi_0 \in \widehat{G}$, defined by setting $\chi_0(g) = 1$ for all $g \in G$, is the identity element $1_{\widehat{G}}$ of $\widehat{G}$, and, for any element $\chi \in \widehat{G}$, the inverse $\chi^{-1} \in \widehat{G}$ is defined by setting $\chi^{-1}(g) := \chi(g)^{-1} = \overline{\chi(g)}$ for all $g \in G$.

Moreover, the *compact-open topology* is the topology given by uniform convergence on compact sets, viewing $\widehat{G}$ as a subset of the space of all continuous functions from G to $\mathbb{T}$.

Theorem 9.12. *Let G be a locally compact Abelian group. Then $\widehat{G}$ is a locally compact Abelian group. Moreover, if G is a compact (resp. discrete) group then $\widehat{G}$ is a discrete (resp. compact) group.*

For example we have:

- the dual $\widehat{F}$ of any finite abelian group F (with the discrete topology) is isomorphic with the group F itself;
- the dual $\widehat{\mathbb{Z}}$ of the additive group $(\mathbb{Z},+)$ of the integers (with the discrete topology, but also with any metric as a finitely generated group) is isomorphic to the circle group $\mathbb{T}$. More generally, the dual $\widehat{\mathbb{Z}^d}$ of $\mathbb{Z}^d$ is isomorphic to the d-dimensional torus $\mathbb{T}^d$;
- the dual $\widehat{\mathbb{T}}$ of the circle group $\mathbb{T}$ (with its usual metric topology) is isomorphic to the additive group $(\mathbb{Z},+)$ of the integers. More generally, the dual $\widehat{\mathbb{T}^d}$ of the d-dimensional torus $\mathbb{T}^d$ is isomorphic to the additive group $(\mathbb{Z}^d,+)$;
- the dual $\widehat{\mathbb{R}}$ of the additive group $(\mathbb{R},+)$ of real numbers (with the usual Euclidean metric) is isomorphic to $(\mathbb{R},+)$ itself. More generally, the dual $\widehat{\mathbb{R}^d}$ of $(\mathbb{R}^d,+)$ is isomorphic to $(\mathbb{R}^d,+)$.
- the dual $\widehat{\mathbb{Q}}$ of the additive group $(\mathbb{Q},+)$ of rational numbers (with the discrete topology) is isomorphic to the quotient $\mathbb{A}_{\mathbb{Q}}/\mathbb{Q}$, where $\mathbb{A}_{\mathbb{Q}}$ is the Abelian group (in fact a commutative ring) of *rational adèles*. This is based on the famous theorem of Ostrowski, stating that every nontrivial evaluation on the rational numbers $\mathbb{Q}$ is equivalent to either the usual real absolute value or a p-adic absolute value [265]. See [23, 77] for a gentle and clear exposition of this duality result;
- the dual $\widehat{\mathbb{Z}_p}$ of the additive group $(\mathbb{Z}_p,+)$ of p-adic integers (with the induced p-adic topology, which is induced by an ultrametric) is isomorphic to the *Prüfer p-group* $\mathbb{Z}(p^\infty) := \mathbb{Z}[p^{-1}]/\mathbb{Z}$ (equivalently, to the quotient group $\mathbb{Q}_p/\mathbb{Z}_p$);
- the dual $\widehat{\mathbb{Q}/\mathbb{Z}}$ of the quotient group $\mathbb{Q}/\mathbb{Z}$ (equipped with the discrete topology) is isomorphic to the product $\prod_p \mathbb{Z}_p$ (which can be regarded as the profinite-completion of $\mathbb{Z}$). See, e.g., [23];
- the dual $\widehat{\mathbb{Q}_p}$ of the additive group $(\mathbb{Q}_p,+)$ of p-adic numbers (with the usual p-adic topology) is isomorphic to $(\mathbb{Q}_p,+)$ itself. See, e.g., [23].

Let G be a locally compact Abelian group. The dual group $\widehat{\widehat{G}}$ of the dual group $\widehat{G}$ of G is called the *bi-dual* of G. The *evaluation map* $\mathrm{ev}_G\colon G \to \widehat{\widehat{G}}$ is defined by setting

$$\mathrm{ev}_G(g)(\chi) := \chi(g)$$

for all $g \in G$ and $\chi \in \widehat{G}$.

We are now in a position to state the *Pontryagin duality* theorem (see [244] for a nice and comprehensive treatment).

Theorem 9.13 (Pontryagin). *Let G be a locally compact Abelian group. Then the evaluation map* $\mathrm{ev}_G \colon G \to \widehat{\widehat{G}}$ *establishes a canonical isomorphism (of locally compact topological groups) of G onto its bi-dual* $\widehat{\widehat{G}}$.

The above result is a generalization of the canonical isomorphism between a finite-dimensional vector space V and its double dual V^{**} (recall that the additive group $(V,+)$ is an Abelian group). Note that any finite-dimensional vector space V is isomorphic to its dual V^*, but there is no canonical isomorphism. The same situation occurs when G is a finite Abelian group: then, as we mentioned above, $G \cong \widehat{G}$, but this isomorphism is not canonical.

Pontryagin duality has a clear categorical formulation. Let LCA denote the category of locally compact abelian groups and continuous group homomorphisms. The dual group construction of $G \mapsto \widehat{G}$ is a contravariant functor LCA $\to$ LCA. In particular, the double dual functor $G \mapsto \widehat{\widehat{G}}$ is covariant. Pontryagin duality can therefore be stated as follows: the natural transformation between the identity functor on LCA and the double dual functor is an isomorphism.

9.5 Lie Groups

In order to give the definition of a Lie group, we first review the notion of a smooth manifold.

Definition 9.14 (Smooth manifold). Let $d \in \mathbb{N}$. A d-dimensional *topological manifold* is a Hausdorff topological space M which is *locally Euclidean*, that is, every point in M has a neighborhood which is homeomorphic to (an open subset of) $\mathbb{R}^d$.

A *smooth atlas* on a d-dimensional topological manifold M is a family $(\phi_\alpha : U_\alpha \to V_\alpha)_{\alpha \in A}$ of homeomorphisms, where $(U_\alpha)_{\alpha \in A}$ is an open cover of M and $V_\alpha \subset \mathbb{R}^d$ are open for all $\alpha \in A$ such that, for all $\alpha, \beta \in A$ the map

$$\phi_\beta \circ \phi_\alpha^{-1}|_{\phi_\alpha(U_\alpha \cap U_\beta)} \colon \phi_\alpha(U_\alpha \cap U_\beta) \to \phi_\beta(U_\alpha \cap U_\beta)$$

is *smooth*, that is, it is infinitely differentiable.

Two smooth atlases on M are said to be *equivalent* if their union is a smooth atlas. An equivalence class of smooth atlases is called a *smooth structure* on M.

A *smooth manifold* is a topological manifold equipped with a smooth structure.

Given two smooth manifolds M and M', a map $\Psi : M \to M'$ is *smooth* if the maps

$$\phi'_\alpha \circ \Psi \circ \phi_\beta^{-1} \colon \phi_\beta(U_\beta \cap \Psi^{-1}(U'_\alpha)) \to \phi'_\alpha(\Psi(U_\beta) \cap U'_\alpha)$$

are smooth for all $\alpha \in A$ and $\beta \in B$, where $(\phi_\beta : U_\beta \to V_\beta)_{\beta\in B}$ (resp. $(\phi'_\alpha : U'_\alpha \to V'_\alpha)_{\alpha\in A}$) is a smooth atlas on M (resp. on M').

Smooth manifolds together with smooth maps form a category.

Cartesian product of two manifolds.

Definition 9.15 (Lie group). A *Lie group* is a group which is a smooth manifold such that the product $G\times G \to G, (g,h) \mapsto gh$ (here $G\times G$ is equipped with the natural product smooth structure) and the inversion $G \to G, g \mapsto g^{-1}$ are smooth maps.

Given two Lie groups G and G' a group homomorphism $\Psi : G \to G$ which is also a smooth map is called a *Lie homomorphism*.

Lie groups together with Lie homomorphisms form a category.

As every smooth map is continuous and $\mathbb{R}^d$ is locally compact, we have that every Lie group is a locally compact topological group.

Example 9.16. (a) Any group G equipped with the discrete topology is a Lie group (it is a 0-dimensional manifold).

(b) For any integer $n \geq 1$, $(\mathbb{R}^n,+)$ is a Lie group (with the smooth structure given by the atlas $(\mathrm{id}_{\mathbb{R}^n} : \mathbb{R}^n \to \mathbb{R}^n)$ consisting of one single chart).

(c) For any integer $n \geq 1$, $(\mathbb{T}^n,+)$, the nth-dimensional torus, can be given the structure of a Lie group with underlying smooth manifold of dimension n (**exercise**).

(d) For any integer $n \geq 1$, the *general linear group* $\mathrm{GL}(n,\mathbb{R})$ consisting of invertible $n\times n$ matrices, can be given the structure of a Lie group with underlying smooth manifold of dimension n^2 (**exercise**).

(e) For any integer $n \geq 1$, the *special orthogonal group* $\mathrm{SO}(n,\mathbb{R})$ consisting of $n\times n$ orthogonal matrices with determinant one, can be given the structure of a Lie group with underlying smooth manifold of dimension $n(n-1)/2$ (**exercise**).

(f) For any integer $n \geq 1$, the *special linear group* $\mathrm{SL}(n,\mathbb{R})$ consisting of $n\times n$ matrices with determinant one, can be given the structure of a Lie group with underlying smooth manifold of dimension n^2-1 (**exercise**).

(g) For any integer $n \geq 1$, the *projective general linear group* (resp. *projective special linear group*)

$$\mathrm{PGL}(n,\mathbb{R}) \cong \mathrm{GL}(n,\mathbb{R})/\mathbb{R}^* I \text{ (resp. } \mathrm{PSL}(n,\mathbb{R}) = \mathrm{SL}(n,\mathbb{R})/\{\pm I\})$$

can be given the structure of a Lie group with underlying smooth manifold of dimension n^2-1 (**exercise**).

(h) The *real Heisenberg group* $\mathrm{UT}(3,\mathbb{R})$ consisting of all 3×3 upper unitriangular matrices with real coefficients can be given the structure of a Lie group with underlying smooth manifold of dimension 3 (**exercise**).

(i) For any integer $n \geq 1$, the *affine group* $\mathrm{Aff}(n,\mathbb{R}) := \mathbb{R}^n \rtimes \mathrm{GL}(n,\mathbb{R})$, where the natural action of $\mathrm{GL}(n,\mathbb{R})$ on $\mathbb{R}^n$ is matrix multiplication of a vector, can be given the structure of a Lie group with underlying smooth manifold of dimension n^2+n (**exercise**). Note that any element of $\mathrm{Aff}(n,\mathbb{R})$ can be represented in $(n+1)\times(n+1)$ matrix form

$$\begin{pmatrix} M & v \\ 0 & 1 \end{pmatrix}$$

with $M \in \mathrm{GL}(n,\mathbb{R})$ and $v \in \mathbb{R}^n$ a column vector.

(j) The Cartesian product of two Lie groups is a Lie group (**exercise**).

(k) Any topologically closed subgroup of a Lie group is a Lie group. This is known as the Closed subgroup theorem or Cartan's theorem.

(l) The quotient of a Lie group by a closed normal subgroup is a Lie group (**exercise**).

(m) The universal cover of a connected Lie group is a Lie group. For example, the group $\mathbb{R}$ is the universal cover of $\mathbb{T}$. In fact, any covering of a differentiable manifold is also a differentiable manifold, but in the case of the universal cover, one indeed has a group structure (compatible with its other structures).

9.6 Hilbert's Fifth Problem

Hilbert's fifth problem entitled "Lie's concept of a continuous group of transformations without the assumptions of the differentiability of the functions defining of the group" is one of the twenty-three mathematical problems presented by David Hilbert [172] at the Paris conference of the International Congress of Mathematicians in 1900.

Roughly speaking, it asks whether, in the definition of a Lie group, the smoothness condition is redundant, that is, if replacing "differentiable manifold" by "topological manifold" can yield new examples. The answer to this question turned out to be negative: in 1952, Andrew Mattei Gleason, Deane Montgomery and Leo Zippin showed that if G is a topological manifold with continuous group operations, then there exists exactly one differentiable (in fact analytic) structure on G which turns it into a Lie group.

As remarked above, every Lie group is a (locally compact) topological group. The converse is clearly false: an example is provided, for instance, by the infinite-dimensional torus $\mathbb{T}^{\mathbb{N}} = (\mathbb{R}/\mathbb{Z})^{\mathbb{N}}$ equipped with the product topology. This is a compact (and therefore locally compact) topological group which is not locally Euclidean and therefore is not a Lie group.

We are interested in the following application of the solution of Hilbert's fifth problem.

Theorem 9.17 (Gleason and Montgomery–Zippin). *Let X be a finite-dimensional, locally compact, connected and locally connected, homogeneous metric space. Then* $\mathrm{Isom}(X)$, *the group of isometries of X, can be given the structure of a Lie group with finitely many components.*

As an illustration of the above result, we observe that the Lie group $\mathrm{SO}(n,\mathbb{R})$, the special orthogonal group of degree n, is the group of orientation-preserving isometries of $\mathbb{R}^n$ fixing the origin. Thus, it is a subgroup of the isometry group $\mathrm{Isom}(\mathbb{R}^n)$ of $\mathbb{R}^n$ equipped with its natural Euclidean structure. Similarly, the projective real linear group $\mathrm{PGL}(2,\mathbb{R})$ (resp. the projective special linear group $\mathrm{PSL}(2,\mathbb{R})$) is the group $\mathrm{Isom}(\mathbb{H})$ (resp. $\mathrm{Isom}_+(\mathbb{H})$) of all isometries (resp. orientation-preserving isometries) of $\mathbb{H}$, the hyperbolic plane (cf. Theorem 8.11).

The interested reader may find a much richer discussion on the structure theory of locally compact groups and related topics connected to Hilbert's fifth problem in Tao's monograph [327] which indeed constitutes a beautiful and accessible account of all these fascinating topics as well as of recent important developments (such as additive combinatorics [329] and the theory of approximate groups [335]).

9.7 Exercises

Exercise 9.1. Show that every open subgroup of a topological group is closed (and therefore clopen).

Exercise 9.2. Let G be a locally compact group.

(i) Show that the map $g \mapsto g^{-1}$ is a homeomorphism of G onto G.
(ii) Fix $g_0 \in G$. Show that the maps $g \mapsto gg_0$ and $g \mapsto g_0 g$ are homeomorphisms of G onto G.
(iii) Show that every neighborhood V of $g \in G$ is of the form $V = gU$ (resp. $V = Ug$), where U is a neighborhood of 1_G.
(iv) Show that if U is a neighborhood of 1_G, so is U^{-1}.
(v) Show that every neighborhood of 1_G contains a *symmetric* neighborhood W of 1_G.
(vi) Show that if $K_1, K_2 \subset G$ are compact subsets, so is their product K_1K_2.

Exercise 9.3. Let G be a locally compact group. Show that G has a clopen subgroup which is the union of a countable family of compact subsets.

Exercise 9.4. Let G be a group. Let $\lambda(g)\colon \mathbb{C}^G \to \mathbb{C}^G$ (resp. $\rho(g)\colon \mathbb{C}^G \to \mathbb{C}^G$) be the left (resp. right) translation by $g \in G$ (cf. Definition 9.7). Show that:
(i) $\lambda(g)$ (resp. $\rho(g)$) is linear for all $g \in G$.
(ii) $\lambda(g_1g_2) = \lambda(g_1)\lambda(g_2)$ and $\rho(g_1g_2) = \rho(g_1)\rho(g_2)$ for all $g_1, g_2 \in G$.

Exercise 9.5. Let G be a locally compact group. Show that $C_c(G)$, the space of all functions $f\colon G \to \mathbb{C}$ with compact support, is dense in $C_0(G)$, the space of all functions $f\colon G \to \mathbb{C}$ vanishing at infinity, in the norm $\|\cdot\|_\infty$ defined by (9.1).

Exercise 9.6. Let G be a locally compact group. Use Lemma 9.8 and the preceding exercise to show that given $f \in C_0(G)$, the map $g \mapsto \lambda(g)f$ is a continuous function.

Exercise 9.7. Let G be a locally compact group. Let $f_1, f_2 \in C_c^+(G)$. Show that the set of $c = (c_1, c_2, \ldots, c_n) \in \mathbb{C}^n$, $n \in \mathbb{N}$, such that there exists a linear combination $\sum_{j=1}^n c_j[\lambda(g_j)f_2]$ of left-translates of f_2 majorizing f_1 is nonempty.
Hint. Use the fact that (i) $\mathrm{supp}(f_1)$ is compact and (ii) there exists an open subset $U \subseteq G$ and $\delta > 0$ such that $f_2(g) \geq \delta$ for all $g \in U$.

Exercise 9.8. Let $G = (\mathbb{R}, +)$. For any integer $n \geq 1$ let $\phi_n \in C_c^+(\mathbb{R})$ be the piecewise-linear function defined by setting

$$\phi_n(t) := \begin{cases} 0 & \text{if } t \in (-\infty, -1/n] \cup [1/n, +\infty) \\ 2nt + 2 & \text{if } t \in [-1/n, -1/(2n)] \\ 1 & \text{if } t \in [-1/(2n), 1/(2n)] \\ -2nt - 2 & \text{if } t \in [1/(2n), 1/n] \end{cases}$$

for all $t \in \mathbb{R}$. Let also $f_0 \in C_c^+(\mathbb{R})$ satisfy $\int_{-\infty}^{\infty} f_0(t)\mathrm{d}t = 1$. Define I_{ϕ_n} as in (9.2). Show that

$$\lim_{n\to\infty} I_{\phi_n}(f) = \int_{-\infty}^{\infty} f(t)\mathrm{d}t$$

for all $f \in C_c^+(\mathbb{R})$.

Chapter 10
Dimension Theory

In this chapter, we present an introduction to dimension theory. We based our exposition on Chapters II, III, IV, and VII of the monograph by Witold Hurewicz and Henry Wallman [184], an authentic masterpiece in clarity and elegance of writing. We closely followed this source, and our slight modifications, far from improving anything, are only aimed at getting a more modern presentation, closer to the style of our book.

The topological dimension we present here is often called *inductive dimension* (to distinguish it from the equivalent notion of *covering dimension*: see the Notes) and, as in the monograph by Hurewicz and Wallman, we limit ourselves to separable metrizable spaces. We thus define topological dimension by first defining the 0-dimensional spaces as those admitting a base of clopen subsets (Definition 10.3) and then by inductively defining spaces of dimension $\leq n$ as those admitting a base of open subsets whose boundaries have dimension $\leq n-1$ and declaring a space to be of dimension n if it is of dimension $\leq n$ but not of dimension $\leq n-1$ (Definition 10.16).

One of the main nontrivial examples of a 0-dimensional space is provided by the Cantor set, which we define and study in Section 10.1. 0-dimensional spaces are defined and studied in Section 10.2. Then in Section 10.3 we define and study spaces of dimension n. A characterization of n-dimensionality is given in terms of "separation" of points and/or closed subsets by closed subsets of dimension $n-1$ (cf. Theorem 10.9 and Theorem 10.33). The fact that the dimension of $[0,1]^n$ and $\mathbb{R}^n$ is n is established in Section 10.4.

Finally, in Section 10.5 we define and study p-dimensional measure, $p \in [0,+\infty)$, and Hausdorff dimension. The Hausdorff dimension of $[0,1]^n$ and $\mathbb{R}^n$ coincides with the topological dimension n (Theorem 10.50). We also establish the result, important for its role in the proof of Gromov's theorem (see Chapter 12), that a space of finite Hausdorff dimension is finite-dimensional (Corollary 10.47).

In this chapter, unless otherwise specified, all (topological) spaces are *separable and metrizable*. For the convenience of the reader and the sake of completeness, we review some basic properties satisfied by such spaces. Recall that a topological space is *separable* if it admits a countable dense subset and it is *metrizable* if it admits a metric inducing its topology. A separable and metrizable space X is:

T. Ceccherini-Silberstein and M. D'Adderio, *Topics in Groups and Geometry*,
Springer Monographs in Mathematics, https://doi.org/10.1007/978-3-030-88109-2_10

(SM-1) *Hausdorff*, i.e., any two distinct points P_1 and P_2 admit disjoint neighborhoods, that is, there exist disjoint open subsets V_1 and V_2 such that $P_1 \in V_1$ and $P_2 \in V_2$ (**exercise**);

(SM-2) *normal*, i.e, for any two disjoint closed subsets C_1 and C_2 there exist disjoint open subsets V_1 and V_2 such that $C_1 \subseteq V_1$ and $C_2 \subseteq V_2$ (**exercise**);

(SM-3) *completely normal*, i.e., every subspace is normal, (**exercise**). Moreover, given two disjoint subsets X_1 and X_2, neither containing a cluster point of the other, there exist disjoint open subsets W_1 and W_2 such that $X_1 \subseteq W_1$ and $X_2 \subseteq W_2$ (**exercise**);

(SM-4) *first countable*, i.e., each point admits a countable neighbourhood base (**exercise**);

(SM-5) *second countable*, i.e., it admits a countable base (**exercise**).

10.1 The Cantor Set

In this section, as a prelude to the general theory of dimension, we present the Cantor set, which will serve as a fundamental example of a 0-dimensional space (cf. Example 10.4.(c)) as well as a space of non-integral Hausdorff dimension (cf. Theorem 10.51).

Given $a, b \in \mathbb{R}$ with $a < b$, the open interval $(a + (b-a)/3, b - (b-a)/3)$ is called the *open middle third* of the interval $[a,b]$.

Definition 10.1. The *Cantor set* is the set

$$K := \bigcap_{j \in \mathbb{N}} K_j \subseteq [0,1]$$

where

$$\begin{aligned} K_0 &:= [0,1] \\ K_1 &:= [0,1/3] \cup [2/3,1] \subseteq K_0 \\ K_2 &:= [0,1/9] \cup [2/9,1/3] \cup [2/3,7/9] \cup [8/9,1] \subseteq K_1, \end{aligned}$$

and, more generally, $K_{j+1} \subseteq K_j$ is obtained by removing the open middle third of each interval constituting K_j. See Figure 10.1.

We have the following expressions of the Cantor set (**exercise**):

$$K = \bigcap_{m=1}^{\infty} \bigcap_{k=0}^{3^{m-1}-1} \left(\left[0, \frac{3k+1}{3^m}\right] \cup \left[\frac{3k+2}{3^m}, 1\right] \right) \tag{10.1}$$

and

$$K = [0,1] \setminus \bigcup_{m=1}^{\infty} \bigcup_{k=0}^{3^{m-1}-1} \left(\frac{3k+1}{3^m}, \frac{3k+2}{3^m} \right). \tag{10.2}$$

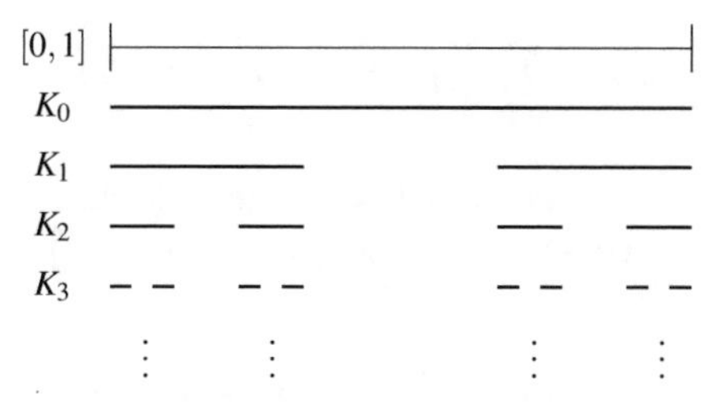

Fig. 10.1 The construction of the Cantor set.

We summarize the properties of the Cantor set in the following:

Theorem 10.2. *The Cantor set K is:*

(1) *compact;*
(2) *with empty interior (and therefore nowhere dense: a set is nowhere dense if its closure has empty interior);*
(3) *homeomorphic to $\{0,1\}^{\mathbb{N}}$ equipped with the prodiscrete topology;*
(4) *totally disconnected (i.e., the connected component of each point is a singleton);*
(5) *uncountable;*
(6) *perfect (i.e., a closed set with no isolated points);*
(7) *fractal (it is equal to two copies of itself, where each copy is shrunk by a factor of 3 and translated).*

Proof. (1) The sets K_n are closed in $[0,1]$. It follows that K is closed in $[0,1]$ and therefore it is compact.

(2) Let I be an interval such that $I \subset K$. As I is connected, for each $n \in \mathbb{N}$ the set I must be contained in one of the connected components of K_n. But then, the length $\ell(I)$ of I cannot exceed the length $1/3^n$ of such connected components of K_n for all $n \in \mathbb{N}$. It follows that $\ell(I) = 0$, that is, I is either the empty set or it is a singleton. This shows that the interior of K is empty. Since K is closed, it is nowhere dense.

(3) Recall that any number $z \in [0,1]$ can be expressed as

$$z = 0.x_0x_1x_2\cdots x_nx_{n+1}\cdots := \sum_{n\in\mathbb{N}} \frac{x_n}{3^{n+1}},$$

where $(x_n)_{n\in\mathbb{N}} \in \{0,1,2\}^{\mathbb{N}}$. The right-hand side is called the *ternary expansion* of the number z. Unless $z \in [0,1]$ is a *triadic rational number*, that is, $z = m/3^n$ with $m,n \in \mathbb{N}$ such that $1 \le m \le 3^n - 1$, its ternary expansion is unique. For example $1/2 = 0.111\cdots 11\cdots$. On the other hand, any triadic rational number admits two ternary expansions, one whose terms are eventually 0 (this is called *proper*) and one whose terms are eventually 2 (this is called *improper*). For example, $1/3 = 0.100\cdots 0\cdots = 0.022\cdots 22\cdots$.

Now, the set K_n consists of all numbers $z \in [0,1]$ that admit a ternary expansion $0.x_0x_1x_2\cdots x_ix_{i+1}\cdots$ such that $x_i \in \{0,2\}$ for all $0 \le i \le n-1$. We deduce that the map $\varphi\colon \{0,1\}^{\mathbb{N}} \to K$ defined by setting

$$\varphi(x) := \sum_{n\in\mathbb{N}} \frac{2x_n}{3^{n+1}} = 0.x_0x_1x_2\cdots x_nx_{n+1}\cdots$$

for all $x = (x_n)_{n\in\mathbb{N}} \in \{0,1\}^{\mathbb{N}}$, is well defined and a bijection of $\{0,1\}^{\mathbb{N}}$ onto K.

Let us show that φ is continuous. Let $x = (x_n)_{n\in\mathbb{N}} \in \{0,1\}^{\mathbb{N}}$. Let $\varepsilon > 0$. For $m \in \mathbb{N}$ the set

$$V_m(x) := \{y = (y_n)_{n\in\mathbb{N}} \in \{0,1\}^{\mathbb{N}} : y_n = x_n \text{ for } n = 0,1,\ldots,m\}$$

is an open neighborhood of x in the prodiscrete topology of $\{0,1\}^{\mathbb{N}}$. If $m \geq \log_3(\varepsilon^{-1})$ one has, for all $y \in V_m(x)$,

$$|\varphi(x) - \varphi(y)| \leq \sum_{n=m+1}^{\infty} \frac{2}{3^{n+1}} = \frac{1}{3^{m+1}} < \varepsilon,$$

and continuity of φ follows. Since $\{0,1\}^{\mathbb{N}}$ is compact by Tychonov's theorem, we deduce that, in fact, φ is a homeomorphism.

(4) Since $\{0,1\}^{\mathbb{N}}$ is totally disconnected (**exercise**), we deduce from (3) that K is totally disconnected as well.

(5) Since $\{0,1\}^{\mathbb{N}}$ is uncountable (**exercise**: use the Cantor diagonal argument), we deduce from (3) that K is uncountable as well.

(6) Let $x \in \{0,1\}^{\mathbb{N}}$. First observe that the sets $V_m(x)$, $m \in \mathbb{N}$, constitute a base of (clopen) neighborhoods of x for the prodiscrete topology. Since each $V_m(x)$ is infinite, we deduce that x is not isolated, equivalently, $\{0,1\}$ is perfect. It follows from (3) that K is also perfect.

(7) This is obvious. □

10.2 0-Dimensional Spaces

Recall that, unless otherwise specified, all (topological) spaces are *separable and metrizable*.

Definition 10.3. A nonempty space X has *dimension* 0, and we shall write $\dim(X) = 0$, if it admits a base of the topology made up of clopen subsets.

A topological space (not necessarily separable nor metrizable) admitting a base consisting of clopen subsets is called *scattered*. Thus, a separable metric space is 0-dimensional if and only if it is scattered.

It is clear that 0-dimensionality is a topological invariant.

Example 10.4. (a) Every nonempty finite or countable space is 0-dimensional. Indeed, let $(x_n)_{n\in\mathbb{N}}$ be an enumeration of X. Let also $x \in X$ and $\varepsilon > 0$, and set $U := B_\varepsilon(x)$, the open ball of radius ε centered at x. Then we can find $\varepsilon' > 0$ such that $\varepsilon' < \varepsilon$ and $\varepsilon' \neq d(x_n, x)$ for all $n \in \mathbb{N}$. Then $V := B_{\varepsilon'}(x)$ satisfies $\partial V = \varnothing$, i.e., V is clopen, and $V \subseteq U$.

In particular, the set $\mathbb{Q}$ of rational numbers (with the topology induced by $\mathbb{R}$) is 0-dimensional.

(b) The set $\mathscr{I} := \mathbb{R} \setminus \mathbb{Q}$ of irrational numbers (with the topology induced by $\mathbb{R}$) is 0-dimensional. Indeed, let $x \in \mathscr{I}$ and let U be a neighborhood of x. Then we can find $p, q \in \mathbb{Q}$ such that $p < x < q$ and $V := [p,q] \cap \mathscr{I}$ is a clopen neighborhood of x satisfying $V \subseteq U$.

More generally, the set $\mathscr{I}^n \subseteq \mathbb{R}^n$ of completely-irrational points in $\mathbb{R}^n$ is 0-dimensional for all $n \geq 1$ (**exercise**).

(c) The *Cantor set* K (cf. Section 10.1) is 0-dimensional. Use Theorem 10.2.(3) and observe that for each $x \in \{0,1\}^{\mathbb{N}}$ the sets $C(x;n) := \{y \in \{0,1\}^{\mathbb{N}} : y(i) = x(i)$ for $i = 0, 1, \ldots, n\}$ constitute a base of clopen neighborhoods of x. We leave it as an **exercise** to fill in the details. See also (d) or, more generally, Theorem 10.38.

(d) Any set X of real numbers with empty interior is 0-dimensional (**exercise**).

(e) The set $(\mathscr{I} \times \mathbb{Q}) \cup (\mathbb{Q} \times \mathscr{I}) \subseteq \mathbb{R}^2$ is 0-dimensional (**exercise**).

(f) The *Hilbert cube* is the space $I^\infty := \prod_{n \geq 1} [0, 1/n]$. Then the space $I^\infty_{\mathbb{Q}} \subseteq I^\infty$ consisting of points all of whose coordinates are rational is (uncountable and) 0-dimensional. Indeed, let $x = (x_n)_{n \geq 1} \in I^\infty_{\mathbb{Q}}$ and let U be a neighborhood of x in $I^\infty_{\mathbb{Q}}$. Then (**exercise**) we can find $n_0 \in \mathbb{N}$ (large enough) and $p_i, q_i \in \mathscr{I}$, $i = 1, 2, \ldots, n_0$, such that $p_i < x_i < q_i$ (and $q_i - p_i$ small enough) such that $V := \{y = (y_n)_{n \geq 1} \in I^\infty_{\mathbb{Q}} : p_i < y_i < q_i, i = 1, 2, \ldots, n_0\}$ is a neighborhood of x satisfying $V \subseteq U$. Now if y is in the boundary of V in I^∞, then at least one of its first n_0 coordinates must be irrational (**exercise**). This shows that ∂V in $I^\infty_{\mathbb{Q}}$ is empty, that is, V is clopen in $I^\infty_{\mathbb{Q}}$.

Similarly, the space $I^\infty_{\mathscr{I}} \subseteq I^\infty$ consisting of points all of whose coordinates are irrational is 0-dimensional (**exercise**).

(g) Let $\ell^2(\mathbb{N}) := \{(x_n)_{n \in \mathbb{N}} : x_n \in \mathbb{R}, \sum_{n \in \mathbb{N}} |x_n|^2 < \infty\}$ denote the Hilbert space of all square summable real sequences. The set $X \subseteq \ell^2(\mathbb{N})$ all of whose coordinates are rational is (uncountable and) *not* 0-dimensional. Indeed, we now show that any bounded neighborhood U in X of the origin $(0)_{n \in \mathbb{N}}$ has a nonempty boundary. Consider the subspace $Y_1 := \{x \in X : x_1 = x_2 = \cdots = 0\} \subseteq X$ (this is homeomorphic to $\mathbb{R}$). Then $Y_1 \cap U \neq \varnothing$ and $Y_1 \cap (X \setminus U) \neq \varnothing$. Thus we can find $q_1 \in \mathbb{Q}$ such that $x^1 := (q_1, 0, 0, \ldots)$ is in U and $d(x^1, X \setminus U) < 1$. Similarly, by considering the subspace $Y_2 := \{x \in X : x_1 = q_1, x_3 = x_4 = \cdots = 0\} \subseteq X$ (this is again homeomorphic to $\mathbb{R}$), we find $q_2 \in \mathbb{Q}$ such that $x^2 := (q_1, q_2, 0, 0 \ldots)$ belongs to U and $d(x^2, X \setminus U) < 1/2$. Proceeding inductively, we determine a sequence $(q_n)_{n \in \mathbb{N}}$ in $\mathbb{Q}$ and points $x^n := (q_1, q_2, \ldots, q_n, 0, 0, \ldots)$ in U such that $d(x^n, X \setminus U) < 1/n$. Then the point $x := (q_1, q_2, \ldots, q_n, q_{n+1}, \ldots)$ is in X (**exercise**) and, moreover, is a boundary point of U.

(h) The unit interval $[0,1]$ and $\mathbb{R}$ with the usual Euclidean topology are *not* 0-dimensional (**exercise**). In fact, as we expect and indeed we shall see (cf. Example 10.18.(a) and, more generally, Corollary 10.36), $[0,1]$ and $\mathbb{R}$ are one-dimensional.

Proposition 10.5. *A nonempty subset of a* 0*-dimensional space is* 0*-dimensional.*

Proof. Let X be 0-dimensional and let $X' \subseteq X$ be a nonempty subset. Let $x \in X'$ and let U' be a neighborhood of x in X'. Then we can find a neighborhood U of x in X' such that $U' = U \cap X'$. Since $\dim X = 0$, there exists a clopen V in X such that $x \in V \subseteq U$. It then follows that $V' := V \cap X'$ is a clopen neighborhood of x such that $V' \subseteq U'$. This shows that $\dim(X') = 0$. ☐

Definition 10.6. Let X be a separable metrizable space. Two (disjoint) subsets A_1 and A_2 are *separated* provided that there exist (cl)open subsets A_1' and A_2' in X such that (i) $A_1' \cup A_2' = X$, (ii) $A_1' \cap A_2' = \varnothing$, and (iii) $A_i \subseteq A_i'$ for $i = 1,2$.

By abuse of notation, we say that a point $x \in X$ and a subset $A \subseteq X$ (resp. another point $y \in X$) are *separated* provided the subsets $\{x\}$ and A (resp. $\{x\}$ and $\{y\}$) are separated.

Remark 10.7. A_1 and A_2 are separated if and only if there exists a clopen subset $A_1' \subseteq X$ such that $A_1 \subseteq A_1'$ and $A_1' \cap A_2 = \varnothing$: indeed, one may take $A_2' := X \setminus A_1'$ (**exercise**).

Proposition 10.8. *A nonempty space X is 0-dimensional if and only if any point $x \in X$ and any closed subset $C \subseteq X$ not containing x can be separated.*

Proof. Suppose first that X is 0-dimensional and let $x \in X$ and C be a closed subset of X such that $x \notin C$. Then the set $U := X \setminus C$ is an open neighborhood of x. Since $\dim(X) = 0$, there exists a clopen V in X such that $x \in V \subseteq U$. Since $V \cap C = \varnothing$, it follows from Remark 10.7 that x and C are separated.

Conversely, suppose that any point $x \in X$ and any closed subset of X not containing x can be separated. Let U be an open neighborhood of x and set $C := X \setminus U$. Then C is a closed subset of X. By our assumptions and Remark 10.7, there exists a clopen subset $V \subseteq X$ such that $x \in V$ and $V \subseteq U$ (since $V \cap C = \varnothing$). This shows that $\dim(X) = 0$. □

It follows immediately from the above proposition that a nonempty space such that any two disjoint closed subsets can be separated is 0-dimensional. It turns out that the converse is also true:

Theorem 10.9. *A nonempty space X is 0-dimensional if and only if any two disjoint closed subsets can be separated.*

Proof. As observed, we only need to show that 0-dimensionality is a sufficient condition. So, let X be 0-dimensional and let C and K be two disjoint closed subsets of X. Given any point $x \in X$ we necessarily have that at least one of the conditions $x \notin C$ or $x \notin K$ holds true. Since X is metrizable, and therefore normal (and therefore regular), for every $x \in X$ we can find a neighborhood U_x of x such that at least one of the conditions $U_x \cap C = \varnothing$ or $U_x \cap K = \varnothing$ holds true. Since X is 0-dimensional, we may suppose that U_x is clopen. Since X is separable and metrizable, it is second countable (that is, it admits a countable base of open subsets). Thus we can find a sequence $(x_n)_{n\in\mathbb{N}}$ such that $\bigcup_{n\in\mathbb{N}} U_{x_n} = X$. Let us set $V_0 := U_{x_0}$ and, for $n \geq 1$,

$$V_n := U_{x_n} \setminus \bigcup_{i=1}^{n} U_{x_i}.$$

We then have

(i) $\bigcup_{n\in\mathbb{N}} V_n = X$,
(ii) $V_m \cap V_n = \varnothing$ if $m \neq n$,

(iii) V_n is open,
(iv) at least one of the conditions $V_n \cap C = \varnothing$ or $V_n \cap K = \varnothing$ holds true.

Let us set

$$C' := \bigcup_{V_n \cap K = \varnothing} V_n \quad \text{and} \quad K' := \bigcup_{V_n \cap K \neq \varnothing} V_n.$$

We then have $X = C' \cup K'$ (by (i)), $C' \cap K' = \varnothing$ (by (ii)), C' and K' are open (by (iii)) and in fact clopen by the two preceding properties, and $C' \cap K = \varnothing = K' \cap C$ (by (iv)). The last condition implies that $C \subseteq C'$ and $K \subseteq K'$. This shows that C and K are separated. □

Corollary 10.10. *A* 0*-dimensional space is totally disconnected. In particular, a connected* 0*-dimensional space is a singleton.*

Proof. The fact that a connected 0-dimensional space is a singleton follows immediately from Proposition 10.8. That a 0-dimensional space is totally disconnected follows from the previous fact and Proposition 10.5. □

It follows from Examples 10.4.(a), (b), and (g) that the disjoint union of two 0-dimensional spaces may fail to be 0-dimensional. However we have the following:

Theorem 10.11. *A space which is a* countable *union of* 0*-dimensional* closed *subsets is* 0*-dimensional.*

Proof. Let

$$X = \bigcup_{n \geq 1} C_n,$$

where each C_n is closed and 0-dimensional.

Let K and L be two disjoint closed subsets in X and let us show that they are separated.

The sets $K \cap C_1$ and $L \cap C_1$ are closed and disjoint subsets of C_1. Since C_1 is 0-dimensional, it follows from Theorem 10.9 that there exist disjoint subsets A_1 and B_1 of C_1, closed in C_1 (and therefore in X), such that $K \cap C_1 \subseteq A_1$, $L \cap C_1 \subseteq B_1$, and $A_1 \cup B_1 = C_1$. Now, the sets $K \cup A_1$ and $L \cup B_1$ are closed and disjoint in X. Since X is normal, we can find open subsets $G_1, H_1 \subseteq X$ such that $K \cup A_1 \subseteq G_1$, $L \cup B_1 \subseteq H_1$, and $\overline{G}_1 \cap \overline{H}_1 = \varnothing$. Note that, in particular, $C_1 \subset G_1 \cup H_1$, $K \subset G_1$, and $L \subset H_1$. Repeating the above construction by replacing K, L, and C_1 by $\overline{G}_1$, $\overline{H}_1$, and C_2, respectively, we obtain open subsets $G_2, H_2 \subseteq X$ such that $C_2 \subset G_2 \cup H_2$, $\overline{G}_1 \subset G_2$, $\overline{H}_1 \subset H_2$, and $\overline{G}_2 \cap \overline{H}_2 = \varnothing$. By induction, we construct two sequences $(G_n)_{n \geq 1}$ and $(H_n)_{n \geq 1}$ of open sets in X for which $C_n \subset G_n \cup H_n$, $\overline{G}_{n-1} \subset G_n$, $\overline{H}_{n-1} \subset H_n$, and $\overline{G}_n \cap \overline{H}_n = \varnothing$. Then the sets $G := \bigcup_{n \geq 1} G_n$ and $H := \bigcup_{n \geq 1} H_n$ are disjoint open subsets of X such that $G \cup H = X$ and, moreover, $K \subset G$ and $L \subset H$. This shows that K and L are separated. It follows from Proposition 10.8 that X is 0-dimensional. □

Example 10.12. Let $m, n \in \mathbb{N}$, with $1 \leq m \leq n$ and set

$$R^n_m := \{x = (x_1, x_2, \ldots, x_n) \in \mathbb{R}^n : \text{ exactly } m \text{ of the } x_i\text{s are in } \mathbb{Q}\}. \tag{10.3}$$

For each choice of $\mathbf{i} = (i_1, i_2, \ldots, i_m)$, where $0 \leq i_1 < i_2 < \ldots < i_m \leq n$, and of $\mathbf{r} = (r_1, r_2, \ldots, r_m) \in \mathbb{Q}^m$, we have an $(n-m)$-dimensional vector subspace

$V_{\mathbf{i},\mathbf{r}}$ of $\mathbb{R}^n$ determined by the system of equations $x_{i_k} = r_k$, $k = 1,2,\ldots,m$. Then $C_{\mathbf{i},\mathbf{r}} := \{x = (x_1,x_2,\ldots,x_n) \in V_{\mathbf{i},\mathbf{r}} : x_j \notin \mathbb{Q}$ for all $j \neq i_k, k = 1,2,\ldots,m\}$ is congruent to $\mathscr{I}^{n-m}$, the 0-dimensional subset of completely-irrational points in $\mathbb{R}^{n-m}$ (cf. Example 10.4.(b)). Thus each $C_{\mathbf{i},\mathbf{r}}$ satisfies that $\dim C_{\mathbf{i},\mathbf{r}} = 0$, it is closed in R^n_m and $\bigcup_{\mathbf{i},\mathbf{r}} C_{\mathbf{i},\mathbf{r}} = R^n_m$. Since this union is countable, we deduce from Theorem 10.11 that $\dim R^n_m = 0$.

Lemma 10.13. *Let X be a compact space. Let $x \in X$ and let $C \subset X$ be a closed subset. Suppose that x and any $c \in C$ can be separated. Then x and C can be separated.*

Note that under our assumptions $x \notin C$.

Proof. For each $c \in C$ we can find disjoint clopen subsets U_c and V_c of X, with $x \in U_c$ and $c \in V_c$. Note that $(V_c)_{c \in C}$ is a (cl)open cover of C. Since C is a closed subset of a compact space, it is itself compact, and therefore we can find $c_1, c_2, \ldots, c_n \in C$ such that $V := \bigcup_{i=1}^n V_{c_1} \supseteq C$. Setting $U := \bigcap_{i=1}^n U_{c_1}$, we have that $x \in U$. Moreover, U and V are both clopen. This shows that x and C are separated. □

Lemma 10.14. *Let X be a compact space. Let $x \in X$ and denote by $M(x)$ the set of all points of X which cannot be separated from x. Then $M(x)$ is connected.*

Proof. We first claim that $M(x)$ is closed. Let $y \in X \setminus M(x)$. By the hypotheses, x and y can be separated, that is, there exist disjoint clopen neighborhoods U and V of x and y, respectively, such that $U \cup V = X$. As all elements in V can be separated from x, we have that $V \subset X \setminus M(x)$. It follows that $X \setminus M(x)$ is open, and the claim follows.

Suppose, by contradiction, that $M(x)$ is disconnected, say we have a partition $M(x) = C \cup K$ with C and K nonempty disjoint subsets in $M(x)$. Since a point cannot be separated from itself, we have $x \in M(x)$. To fix our ideas, let us suppose that $x \in C$. Since X is normal, we can find an open set $U \subset X$ such that $C \subseteq U$ and $\overline{U} \cap K = \varnothing$. We have $\partial U \cap M(x) = (\overline{U} \setminus U) \cap (C \cup K) = \varnothing$, that is, every $y \in \partial U$ is separated from x. By Lemma 10.13, x and ∂U can be separated, that is, there exists a clopen V not containing x such that $\partial U \subset V$. Note that since $x \in C \subset U$, then $x \in W := \overline{U} \setminus V = U \setminus V$. Moreover, W is clearly clopen and $W \cap K = \varnothing$. It follows that x is separated from (all points of) $K \subset M(x)$, a contradiction. □

Theorem 10.15. *Let X be a compact space. Then the following conditions are equivalent.*

(a) *X is totally disconnected;*
(b) *any two distinct points in X can be separated;*
(c) *any point $x \in X$ and any closed subset $C \subset X$ not containing x can be separated;*
(d) *any two disjoint closed subsets can be separated;*
(e) *X is 0-dimensional.*

Proof. We already know that (e) $\iff$ (d) by Theorem 10.9 and (e) $\iff$ (c) by Proposition 10.8: in both cases, compactness of X is not needed. The implications (c) $\Rightarrow$ (b) $\Rightarrow$ (a) are obvious. (b) $\Rightarrow$ (c) follows from Lemma 10.13. Finally, suppose (a). Let $x \in X$. We have that $M(x)$ is connected by Lemma 10.14, and therefore, since X is totally disconnected, $M(x) = \{x\}$. Thus x is the only point which cannot be separated by itself. This shows that any two points in X can be separated, and (a) $\Rightarrow$ (b) follows. □

10.3 n-Dimensional Spaces

Definition 10.16. Let X be a separable metrizable space.

(i) $\dim(\varnothing) := -1$ and $\dim(X) \geq 0$ if $X \neq \varnothing$;
(ii) X has dimension $\leq n$, $n \in \mathbb{N}$, if it admits a base of the topology made up of open sets whose boundaries have dimension $\leq n-1$;
(iii) $\dim(X) = n$ if X has dimension $\leq n$ but X has *not* dimension $\leq n-1$;
(iv) $\dim(X) = \infty$ if X has *not* dimension $\leq n$ for all $n \in \mathbb{N}$.

Note that $\dim(X) \leq n$ is equivalent to the following: *every $x \in X$ admits a base of neighborhoods whose boundaries have dimension $\leq n-1$.*

For $n = 0$, we recover Definition 10.3.

It is also clear that n-dimensionality (resp. ∞-dimensionality) is a topological invariant. Note that, however, it is not an invariant by continuous transformations: the projection $[0,1]^2 \ni (x,y) \mapsto x \in [0,1]$ lowers the dimension, while the *Peano curve* (cf. Exercise 10.13) which maps $[0,1]$ continuously onto $[0,1]^2$ raises the dimension.

Proposition 10.17. *Suppose that* $\dim(X) = n$, $n \in \mathbb{N}$. *Then for every* $-1 \leq m \leq n-1$ *there exists a subset* $Y \subset X$ *satisfying* $\dim(Y) = m$.

Proof. Since $\dim(X) > n-1$, there exists a (basic) open set U_0 such that $\dim(\partial U_0) = n-1$. Setting $X_0 := X$, and repeating the above argument, we determine a sequence

$$X = X_0 \supseteq U_0 \supseteq X_1 := \partial U_0 \supseteq U_1 \supseteq X_2 \supseteq \cdots \supseteq X_n := \partial U_{n-1} \supseteq U_n \supseteq X_{n+1} := \partial U_n = \varnothing,$$

where $\dim(X_k) = n-k$ for all $k = 0, 1, \ldots, n+1$. □

Example 10.18. (a) $[0,1]$, $\mathbb{R}$, and (the boundary of) a polygon have dimension 1 (**exercise**).

(b) $[0,1]^n$ and $\mathbb{R}^n$ have dimension $\leq n$ (**exercise**). The proof that $\dim([0,1]^n) = \dim(\mathbb{R}^n)$ is *exactly* n is not trivial and will be given in the next section (cf. Corollary 10.36).

(c) The set X of points in $\ell^2(\mathbb{N})$ all of whose coordinates are rational is one-dimensional. Indeed, we already know that $\dim(X) \geq 1$ (cf. Example 10.4.(g)). Since the balls $B_\varepsilon := \{x \in X : \|x\|_2 < \varepsilon\}$, $\varepsilon > 0$, constitute a base of neighborhoods of $0 \in X$, it suffices to show that $\dim(B_\varepsilon) \leq 1$. Let $\varepsilon < 1$ and set $S := \{x \in \ell^2(\mathbb{N}) : \|x\|_2 = \varepsilon\}$. Note that $Y := S \cap X = \partial B_\varepsilon$. Let us show that $\dim(Y) = 0$. Given $x = (x_n)_{n \geq 1} \in \ell^2(\mathbb{N})$ we set $x' := \left(\frac{x_n}{n}\right)_{n \geq 1} \in \ell^2(\mathbb{N})$. Note that if $x \in S$, then x' belongs to the Hilbert cube I^∞. Thus the map $x \mapsto x'$ yields an embedding $\varphi \colon S \to I^\infty$ (**exercise**). In particular, $\varphi(Y) \subset I^\infty_{\mathbb{Q}}$. As $\dim(I^\infty_{\mathbb{Q}}) = 0$ (cf. Example 10.4.(f)), we deduce from Proposition 10.5 that $\dim(Y) = 0$, as desired.

Proposition 10.19. *A subset of an n-dimensional space has dimension $\leq n$.*

Proof. We proceed by induction on n. For $n = -1$ (resp. $n = 0$) the statement is obvious (resp. follows from Proposition 10.5). Assume the statement is true for $n-1$. Suppose $\dim(X) = n$ and let $X' \subset X$. Let $x \in X'$ and let U' be a neighborhood of x' in X'. Then there exists a neighborhood U of x in X such that $U' = U \cap X'$. Since $\dim(X) \le n$, there exists a neighborhood V of x in X such that $V \subset U$ and $\dim(\partial^X V) \le n-1$. Then $V' := V \cap X'$ is a neighborhood of x in X' and $V' \subset U'$. Setting $B := \partial^X V = \overline{V} \setminus V$ and $B' := \partial^{X'} V' = (\overline{V'} \cap X') \setminus V' = (\overline{V'} \cap X') \setminus V$, one has $B' \subseteq B \cap X'$. Recalling that $\dim(B) \le n-1$, induction yields $\dim(B') \le n-1$. $\square$

Definition 10.20. Let X be a separable metrizable space. Suppose that A_1, A_2 and B are mutually disjoint subsets of X. One says that B *separates* A_1 and A_2 *in* X provided that there exist open sets $U_1, U_2 \subseteq X$ such that $A_i' := U_i \cap (X \setminus B)$ satisfy: (i) $A_1' \cup A_2' = X \setminus B$, (ii) $A_1' \cap A_2' = \varnothing$, and (iii) $A_i \subseteq A_i'$, $i = 1, 2$ (see Figure 10.2).

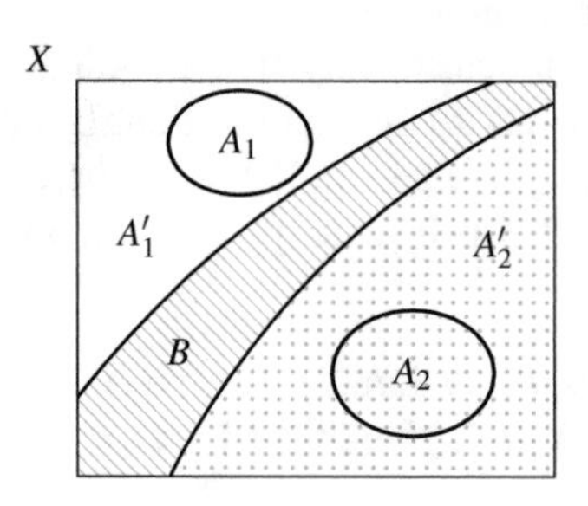

Fig. 10.2 B separates A_1 and A_2 in X.

By abuse of notation, we say that a point $x \in X$ and a subset $A \subseteq X$ (resp. another point $y \in X$) are *separated* by B provided that the subsets $\{x\}$ and A (resp. $\{x\}$ and $\{y\}$) are separated by B.

Note that if $B = \varnothing$, then Definition 10.20 reduces to Definition 10.6, that is, A_1 and A_2 are separated by $\varnothing$ if and only if they are separated.

The following generalizes Proposition 10.8.

Proposition 10.21. *X has dimension $\le n$ if and only if any point $x \in X$ and any closed subset $C \subseteq X$ not containing x can be separated by a closed set of dimension $\le n-1$.*

Proof. Suppose $\dim(X) \le n$ and let $x \in X$ and let $C \subseteq X$ be a closed subset not containing x. The set $U := X \setminus C$ is a neighborhood of x. Since X is metrizable, it is regular, and therefore we can find a neighborhood V of x such that $\overline{V} \subset U$. Since $\dim(X) \le n$, we can find another neighborhood W of x such that $W \subset V$ and $B := \partial W$ satisfies $\dim(B) \le n-1$. Moreover, B separates x and C (**exercise**).

Conversely, suppose that any point $x \in X$ and any closed subset of X not containing x can be separated by a closed set of dimension $\le n-1$. Let $x \in X$ and let U be a neighborhood of x. Then $C := X \setminus U$ is a closed subset not containing x, and, by our assumptions, it may be separated from x by a closed set B of dimension $\le n-1$. In other words, there exist open sets $U_1, U_2 \subseteq X$ such that $A_i' := U_i \cap (X \setminus B)$ satisfy that (i) $A_1' \cup A_2' = X \setminus B$, (ii) $A_1' \cap A_2' = \varnothing$, and (iii) $x \in A_1'$ and $C \subset A_2'$. Note that A_1' and A_2' are open. Now, A_1' is a neighborhood of x contained in $X \setminus C = U$ and its boundary

$\partial A'_1$ is contained in B (**exercise**). Since $\dim(B) \leq n-1$, it follows from Proposition 10.19 that $\dim(\partial A'_1) \leq n-1$. Since $x \in A'_1 \subset U$, this shows that $\dim(X) \leq n-1$. □

Proposition 10.22. *Let X' be a subset of X. Then the following conditions are equivalent:*

(a) $\dim(X') \leq n$;
(b) *every $x \in X'$ admits a base of neighborhoods in X whose boundaries have intersection with X' of dimension $\leq n-1$.*

Proof. Suppose (a) Let $x \in X'$ and let U be a neighborhood of x in X. Then $U' := U \cap X'$ is a neighborhood of x in X'. Since $\dim(X') \leq n$, we can find a neighborhood V' of x in X' such that $V' \subset U'$ and $\dim(B') \leq n-1$, where $B' := \partial^{X'} V' = (\overline{V'} \cap X') \setminus V'$. Since every metric space is *completely normal* (i.e., every subspace is normal, **exercise**: see [339]), we can find an open set W of X such that $V' \subset W \subset U$ and $\overline{W} \cap (X' \setminus \overline{V}') = \varnothing$. Now $\partial W \cap V' = \varnothing = \partial W \cap (X' \setminus \overline{V}')$. As a consequence, $\partial W \cap X' \subseteq B'$ and therefore, by Proposition 10.19, it has dimension $\leq n-1$. This shows (a) $\Rightarrow$ (b).

Conversely, suppose (b). Let $x \in X'$ and let U' be a neighborhood of x in X'. Then there exists a neighborhood U of x in X such that $U' = U \cap X'$. By our assumptions there exists a neighborhood V of x in X such that $V \subset U$ and $\dim(X' \cap \partial V) \leq n-1$. Now $V' := V \cap X'$ is a neighborhood of x in X' and $V' \subset U'$. Setting $B := \partial_X V = \overline{V} \setminus V$ and $B' := \partial_{X'} V' = (\overline{V'} \setminus V') \cap X'$, we have $B' \subset B \cap X'$. It follows from Proposition 10.19 that $\dim(B') \leq n-1$. This shows that $\dim(X') \leq n$. This shows (b) $\Rightarrow$ (a). □

Corollary 10.23. *Let A and B be two subsets of X. Then*

$$\dim(A \cup B) \leq 1 + \dim(A) + \dim(B). \tag{10.4}$$

First note that (10.4) is optimal since, taking $A = \varnothing = B$ one has $\dim(A \cup B) = \dim(\varnothing) = -1 = 1 + (-1) + (-1) = 1 + \dim(\varnothing) + \dim(\varnothing) = 1 + \dim(A) + \dim(B)$.

Proof of Corollary 10.23. We proceed by a double induction on the dimensions of A and B. As remarked above, (10.4) holds true if $\dim(A) = -1 = \dim(B)$. Assume that $\dim(A) = m$ and $\dim(B) = n$ and that (10.4) holds whenever $\dim(A) \leq m$ and $\dim(B) \leq n-1$, as well as whenever $\dim(A) \leq m-1$ and $\dim(B) \leq n$. Let $x \in A \cup B$. To fix our ideas, we suppose that $x \in A$. Let U be a neighborhood of x in X. By Proposition 10.22, we can find an open set V such that $x \in V \subset U$ and $\dim(A \cap \partial V) \leq m-1$. By Proposition 10.19, $\dim(B \cap \partial V) \leq \dim(B) \leq n$. By the induction hypotheses we have

$$\begin{aligned}\dim((A \cup B) \cap \partial V) &= \dim((A \cap \partial V) \cup (B \cap \partial V)) \\ &\leq 1 + \dim(A \cap \partial V) + \dim(B \cap \partial V) \\ &\leq 1 + (m-1) + n = m + n.\end{aligned}$$

Then Proposition 10.22 guarantees us that $\dim(A \cup B) \leq 1 + (m+n) = 1 + \dim(A) + \dim(B)$. □

Corollary 10.24. *The union of $n+1$ subsets of dimension ≤ 0 has dimension $\leq n$.* □

Example 10.25. (a) Recall that the sets $\mathbb{Q}$ and $\mathscr{I} := \mathbb{R} \setminus \mathbb{Q}$ of rational and irrational numbers are 0-dimensional (cf. Example 10.4.(a) and (b)). Their union is $\mathbb{R}$, which has dimension $\dim(\mathbb{R}) = 1$ (cf. Example 10.18.(a)).

(b) For $0 \leq m \leq n$ let us set

$$M_m^n := \{x = (x_1, x_2, \ldots, x_n) \in \mathbb{R}^n : \text{ at most } m \text{ of the } x_i\text{'s are in } \mathbb{Q}\}$$

and

$$N_m^n := \{x = (x_1, x_2, \ldots, x_n) \in \mathbb{R}^n : \text{ at least } m \text{ of the } x_i\text{'s are in } \mathbb{Q}\}.$$

We clearly have $M_m^n = \bigcup_{i=0}^m R_i^n$ and $N_m^n = \bigcup_{i=m}^n R_i^n$, where the R_i^n's are as in (10.3). Since $\dim R_i^n = 0$ for all $i = 0, 1, \ldots, n$ (cf. Example 10.12), it follows from Corollary 10.24 that $\dim M_m^n \leq m$ and $\dim N_m^n \leq n - m$ (in fact, as we shall see in Example 10.37, one has $\dim M_m^n = m$ and $\dim N_m^n = n - m$).

The following is a generalization of Theorem 10.11.

Theorem 10.26. *A space which is the countable union of closed subsets of dimension $\leq n$ has dimension $\leq n$.*

Proof. We proceed by induction on n. The statement is obvious for $n = -1$ and it follows from Theorem 10.11 for $n = 0$. Suppose the statement holds for $n-1$. Recall from general topology that an F_σ is a countable union of closed sets.

Claim. *Any set of dimension $\leq n$ is the union of an F_σ of dimension $\leq n-1$ and a subspace of dimension ≤ 0.*

Let X be a space of dimension $\leq n$. Since X is separable, we can find a countable base $\{U_i : i \in \mathbb{N}\}$ of open sets of X whose boundaries $B_i := \partial U_i$ have dimension $\leq n-1$. By the inductive hypothesis, the F_σ set $B := \bigcup_{i\in\mathbb{N}} B_i$ has dimension $\leq n-1$. It is an **exercise** to show that $\dim(X \setminus B) \leq 0$. Then the claim follows as $X = B \cup (X \setminus B)$.

Let then $X = \bigcup_{i\in\mathbb{N}} C_i$, with C_i closed and $\dim(C_i) \leq n$ for all $i \in \mathbb{N}$. Set $K_0 := C_0$ and, for $i = 1, 2, \ldots$ define

$$K_i := C_i \setminus \bigcup_{j\leq i-1} C_j = C_i \cap \left(X \setminus \bigcup_{j\leq i-1} C_j \right).$$

Observe that K_i is an F_σ in X: indeed, the intersection of a closed set with an F_σ is also an F_σ, and moreover, in a metric space, any open set is an F_σ (**exercise**). We then have (i) $X = \bigcup_{i\in\mathbb{N}} K_i$, (ii) $K_i \cap K_j = \varnothing$ for $i \neq j$, (iii) each K_i is an F_σ, and (iv) $\dim(K_i) \leq \dim(C_i) \leq n$ (cf. Proposition 10.19).

By applying the claim to each K_i, we can find subsets M_i and N_i with $\dim(M_i) \leq n-1$ and $\dim(N_i) \leq 0$ such that $K_i = M_i \cup N_i$. Setting $M := \bigcup_{i\in\mathbb{N}} M_i$ and $N := \bigcup_{i\in\mathbb{N}} N_i$ we have $X = M \cup N$. Now, each M_i is an F_σ in M: indeed $M_i = M_i \cap K_i = M \cap K_i$ with K_i an F_σ in X. Therefore, by the induction hypothesis, $\dim(M) \leq n-1$. Similarly, each N_i is an F_σ in N and therefore $\dim(N) \le 0$. As $X = M \cup N$, we deduce from Corollary 10.23 that $\dim(X) \le n$, and the proof is complete. □

Corollary 10.27. *Any set of dimension $\le n$, $n \in \mathbb{N}$, is the union of a subspace of dimension $\le n-1$ and a subspace of dimension ≤ 0.*

Proof. This is the claim in the proof of the preceding theorem (recall that its proof was subject to the inductive hypothesis for the proof (still incomplete) of the theorem). □

From Corollary 10.24 and by repeated applications of Corollary 10.27 we deduce (**exercise**):

Corollary 10.28. *A space X is of dimension $\le n$, $n \in \mathbb{N}$, if and only if it is the union of $n+1$ subspaces of dimension ≤ 0.*

Moreover, if $\dim(X) = n$ and n_1, n_2 are integers ≥ -1 such that $n = 1 + n_1 + n_2$, then there exist subspaces X_1 and X_2 of X such that $\dim(X_i) = n_i$, $i = 1,2$, and $X = X_1 \cup X_2$. □

Theorem 10.29. *Let A and B be two spaces and suppose that $A \neq \varnothing$ or $B \neq \varnothing$. Then $\dim(A \times B) \le \dim(A) + \dim(B)$.*

Proof. We proceed by a double induction on the dimensions of A and B. If either $A = \varnothing$ or $B = \varnothing$ we have $\dim(A \times B) = \dim(\varnothing) = -1 \le \dim(A) + \dim(B)$, where the last equality follows since exactly one of $\dim(A)$ and $\dim(B)$ equals -1 and the other is ≥ 0.

Let $m, n \ge 0$ and suppose that the statement is true whenever $\dim(A) \le m$ and $\dim(B) \le n-1$ as well as $\dim(A) \le m$ and $\dim(B) \le n-1$. Let $\dim(A) = m$ and $\dim(B) = n$. Let $x = (a,b) \in A \times B$ and suppose that W is a neighborhood of x in $A \times B$. Then, we can find neighborhoods U and V of a in A and b in B, respectively, such that $U \times V \subset W$. Moreover, up to taking smaller neighborhoods $U' \subset U$ and $V' \subset V$ of a and b, respectively, we may suppose that $\dim(\partial U) \le m-1$ and $\dim(\partial V) \le n-1$. We have (**exercise**)

$$\partial(U \times V) = \left(\overline{U} \times \partial V\right) \cup \left(\partial U \times \overline{V}\right),$$

where each of the two terms in the right-hand side is closed and, by the induction hypothesis, has dimension $\le m+n-1$. It follows from Theorem 10.26 that $\dim(\partial(U \times V)) \le m+n-1$. This shows that $\dim(A \times B) \le m+n = \dim(A) + \dim(B)$. □

Corollary 10.30. *Let A and B be two spaces and suppose that $\dim(B) = 0$. Then $\dim(A \times B) = \dim(A) + \dim(B)$.*

Proof. It follows from the preceding theorem that $\dim(A \times B) \le \dim(A) + \dim(B)$. To show the reverse inequality observe that since $B \neq \varnothing$ there is an embedding $\varphi\colon A \to A \times B$. It follows from Proposition 10.19 that $\dim(A \times B) \ge \dim(\varphi(A)) = \dim(A) = \dim(A) + \dim(B)$. □

Lemma 10.31. *Let C_1 and C_2 be two disjoint closed subsets of a space X and let $A \subset X$ be a 0-dimensional subset. Then there exists a closed set $B \subset X$ separating C_1 and C_2 such that $A \cap B = \varnothing$.*

Proof. Since X is normal, we can find two open subsets U_1 and U_2 such that $C_i \subset U_i$, $i = 1,2$ and $\overline{U}_1 \cap \overline{U}_2 = \varnothing$. The sets $\overline{U}_1 \cap A$ and $\overline{U}_2 \cap A$ are closed and disjoint in A and since $\dim(A) = 0$, by Theorem 10.9 they can be separated in A. In other words, there exist (cl)open disjoint subsets C_1' and C_2' in A such that $\overline{U}_i \cap A \subset C_i'$, $i = 1,2$, and $C_1' \cup C_2' = A$. It is an **exercise** to show that the disjoint subsets $B_1 := C_1 \cup C_1'$ and $B_2 := C_2 \cup C_2'$ satisfy $\overline{B}_1 \cap B_2 = \varnothing = B_1 \cap \overline{B}_2$. By complete normality of X, we can find an open subset $W \subset X$ such that $B_1 \subseteq W$ and $\overline{W} \cap B = \varnothing$. Then $B := \partial W = \overline{W} \setminus W$ is a closed set separating C_1 and C_2 and it is disjoint from $C_1' \cup C_2' = A$. □

Theorem 10.32. *Let X be a space and let $A \subset X$ be a subset of dimension $\leq n$. Then any two disjoint closed subsets in X can be separated by a closed subset B such that $\dim(A \cap B) \leq n-1$.*

Proof. We proceed by induction on n. Suppose first that $n = 0$. Then either $A = \varnothing$ and the statement follows after taking $B = \varnothing$, or $\dim(A) = 0$, and then the statement follows from Theorem 10.9. Suppose that $n > 0$. By Corollary 10.27 we can write $A = M \cup N$ with $\dim(M) \leq n-1$ and $\dim(N) \leq 0$. By applying Lemma 10.31 to $N \subseteq A$, we can find a closed subset $B \subseteq A \subseteq X$ which separates the two given closed subsets in X and such that $N \cap B = \varnothing$. Thus $A \cap B \subset M$ and, by Proposition 10.19, we have $\dim(A \cap B) \leq \dim(M) \leq n-1$. □

The following is a generalization of Theorem 10.15.

Theorem 10.33. *Let X be a compact space. Then the following conditions are equivalent:*

(a) *any two distinct points in X can be separated by a closed set of dimension $\leq n-1$;*
(b) *any point $x \in X$ and any closed subset $C \subset X$ not containing x can be separated by a closed set of dimension $\leq n-1$;*
(c) *any two disjoint closed sets can be separated by a closed set of dimension $\leq n-1$;*
(d) *X has dimension $\leq n$.*

Proof. We already know that (b) $\iff$ (d) by Proposition 10.21 and moreover, the implications (c) $\Rightarrow$ (b) $\Rightarrow$ (a) are obvious (in either case, compactness of X is not needed). The implication (d) $\Rightarrow$ (c) follows from Theorem 10.32 (by taking $A = X$).

Finally, suppose (a). Let $x \in X$ and let $C \subset X$ be a closed subset not containing x. For each $c \in C$ we can find an open neighborhood U_c of c in X such that $x \notin \overline{U}_c$ and $\dim(\partial U) \leq n-1$. Note that $\{U_c : c \in C\}$ is an open cover of C. Since C is closed and X is compact, C is also compact. As a consequence, we can find $c_1, c_2, \ldots, c_k \in C$ such that $C \subseteq U := \bigcup_{i=1}^k U_{c_i}$. Then $B := \partial U \subseteq \bigcup_{i=1}^k \partial U_{c_i}$ and therefore (since the ∂U_{c_i}'s are closed) we deduce from Theorem 10.26 that $\dim(B) \leq n-1$. As $x \notin \overline{U}$, this shows that x is separated from C by the closed set B of dimension $\leq n-1$. This shows (a) $\Rightarrow$ (b). □

10.4 The Dimension of $\mathbb{R}^n$

Recall (cf. Example 10.18.(b) and the remark therein) that $\dim(\mathbb{R}^n) \leq n$. In this section we prove that $\dim(\mathbb{R}^n) \geq n$, so that, in fact, $\dim(\mathbb{R}^n) = n$.

We need the following result from Algebraic Topology and Fixed Point Theory: we refer to [184, Chapter IV] as well as to [98, Theorem page 95] for the proof. Recall that a topological space X is said to be *contractible* provided that there exist a continuous function $f\colon X \times [0,1] \to X$ and $x_0 \in X$ such that $f(\cdot,0) = \mathrm{Id}_X$ and $f(\cdot,1) = x_0$ (this condition may be rephrased by saying that the identity map Id_X is *homotopic* to a constant map). Also, given two spaces X and Y with $Y \subset X$, a *retraction* of X onto Y is a continuous map $r\colon X \to Y$ such that $r|_Y = \mathrm{Id}_Y$.

We denote by

$$\mathbb{S}^n := \{x \in \mathbb{R}^{n+1} : d(x,0) = 1\} \text{ (resp. } B^{n+1} := \{x \in \mathbb{R}^{n+1} : d(x,0) \leq 1\})$$

the *n-sphere* (resp. the *closed* $(n+1)$*-ball*), so that $\mathbb{S}^n = \partial B^{n+1}$.

Theorem 10.34. , *The following equivalent conditions hold true:*

(a) *The n-sphere* $\mathbb{S}^n$ *is* not *contractible.*
(b) *Every continuous map* $f\colon B^{n+1} \to \mathbb{R}^{n+1}$ *satisfies (at least) one of the following properties:*

(i) f *has a fixed-point (i.e., there exists an* $x \in B^{n+1} \subset \mathbb{R}^{n+1}$ *such that* $f(x) = x$*);*
(ii) *there are* $x \in \mathbb{S}^n$ *and* $\lambda \in (0,1)$ *such that* $x = \lambda f(x)$*.*

(c) *Every continuous map* $f\colon B^{n+1} \to B^{n+1}$ *has at least one fixed-point.*
(d) *There exists* no *retraction* $r\colon B^{n+1} \to \mathbb{S}^n$.

Let $I_n := \{x = (x_1, x_2, \ldots, x_n) \in \mathbb{R}^n : |x_i| \leq 1, i = 1,2,\ldots,n\} \subset \mathbb{R}^n$. Since I_n is homeomorphic to I^n, the unit n-cube in $\mathbb{R}^n$, we have that $\dim(I_n) \leq n$ (cf. Example 10.18.(b)). Also observe that I_n is homeomorphic to B^n the closed n-ball. For $1 \leq i \leq n$ we denote by $C_i := \{x \in I_n : x_i = 1\}$ and $C_i' := \{x \in I^n : x_i = -1\}$ the pairs of opposite faces of I_n.

Corollary 10.35. *For* $1 \leq i \leq n$ *let* B_i *be a closed set separating* C_i *and* C_i' *in* I_n*. Then*

$$\bigcap_{i=1}^{n} B_i \neq \varnothing. \tag{10.5}$$

Proof. Recall (cf. Definition 10.20) that by our hypothesis on B_i we can find subsets $U_i, U_i' \subseteq I_n$, open in $I_n \setminus B_i$ (and therefore in I_n) such that (i) $U_i \cup U_i' = I_n \setminus B_i$, (ii) $U_i \cap U_i' = \varnothing$, and (iii) $C_i \subseteq U_i$ and $C_i' \subseteq U_i'$. For each $x \in I_n$, let $v(x)$ denote the vector whose ith component has the value $\pm d(x, B_i)$, the sign being $+$ if $x \in U_i'$ and $-$ if $x \in U_i$. Let us set $f(x) := x + v(x)$ for all $x \in I_n$ (see Figure 10.3). It is an **exercise** to check that $f(x) \in I_n$.

Moreover, the map $f\colon I_n \to I_n$ is continuous (**exercise**). It follows from Theorem 10.34.(c) that there exists a fixed-point $x^0 \in I_n$ for f, that is, $d(x^0, B_i) = 0$ for all $i = 1,2,\ldots,n$. This means that $x^0 \in \bigcap_{i=1}^n B_i$. □

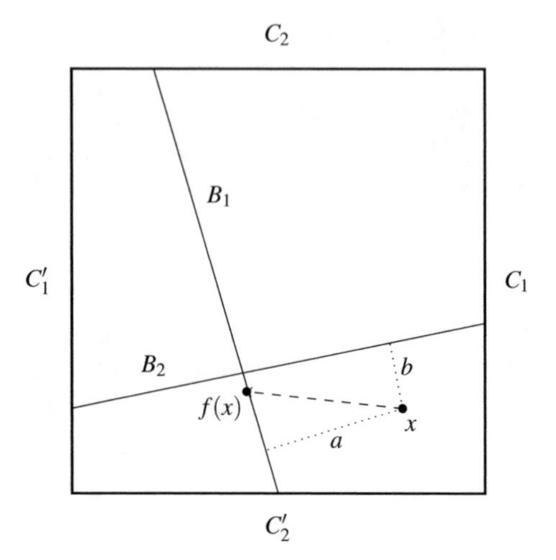

Fig. 10.3 Definition of $f(x)$.

Corollary 10.36 (Dimension of $\mathbb{R}^n$). *We have* $\dim I^n = \dim \mathbb{R}^n = n$.

Proof. We first show that $\dim I^n \geq n$. Suppose, by contradiction, that $\dim I^n \leq n-1$. Then by Theorem 10.32 (with $A = X$) there exists a closed subset $B_1 \subset I^n$ separating C_1 and C'_1 of dimension $\leq n-2$. Again by Theorem 10.32 (with $A = B_1$) there exists a closed subset $B_2 \subseteq I^n$ separating C_2 and C'_2 such that $\dim(B_1 \cap B_2) \leq n-3$. Iterating the argument, we get a finite sequence $B_1, B_2, \ldots, B_i, \ldots$ of closed subsets $B_i \subseteq I_n$ separating C_i and C'_i such that $\dim(B_1 \cap B_2 \cap \cdots \cap B_i) \leq n-i-1$. For $i = n$ we get $\dim(B_1 \cap B_2 \cap \cdots \cap B_n) = -1$, that is, $\bigcap_{i=1}^n B_i = \varnothing$, contradicting Corollary 10.35. This shows that $\dim I^n \geq n$. Since $I^n \subset \mathbb{R}^n$ we deduce from Proposition 10.19 that $\dim \mathbb{R}^n \geq n$. It then follows from Example 10.18.(b) that $\dim I^n = \dim \mathbb{R}^n = n$. □

Example 10.37. With the notation from Example 10.25.(b), we have $\dim M^n_m = m$ and $\dim N^n_m = n - m$. Indeed, we have $\mathbb{R}^n = M^n_m \cup N^n_{m+1}$, with $\dim M^n_m \leq m$ and $\dim N^n_{m+1} \leq n-m-1$. If at least one of the inequalities were strict, we would deduce from Corollary 10.23 that $\dim \mathbb{R}^n < 1+m+(n-m-1) = n$, a contradiction.

Theorem 10.38. *Let $N \subseteq \mathbb{R}^n$. Then* $\dim(N) = n$ *if and only if N contains a nonempty open subset.*

Proof. We first observe that, by Proposition 10.19, $\dim(N) \leq n$.

The condition is clearly sufficient: if N contains a nonempty open subset, then there is a point $x \in N$ and $\varepsilon > 0$ such that $B_\varepsilon(x) \subseteq N$. Since $B_\varepsilon(x)$ is homeomorphic to I^n, we deduce that $\dim B_\varepsilon(x) = \dim I^n = n$, so that $\dim(N) \geq n$, again by Proposition 10.19.

Conversely, let us show that if $M := \mathbb{R}^n \setminus N$ is dense in $\mathbb{R}^n$, then $\dim N \leq n-1$.

It is not restrictive to suppose that M is countable: indeed, M (being a subset of a separable space) contains a countable dense subset $A \subset M$, and $\dim(\mathbb{R}^n \setminus A) \leq n-1$ implies $\dim(\mathbb{R}^n \setminus M) \leq n-1$, by Proposition 10.19.

The statement clearly holds true if $M := N_n^n$, the countable set of points of $\mathbb{R}^n$ all of whose coordinates are rational (cf. Example 10.25). Then, $N = \mathbb{R}^n \setminus M = M_{n-1}^n$ has dimension $n-1$ (cf. Example 10.37).

Claim. *Let A and B be two countable dense subsets of* $\mathbb{R}^n$. *Then there exists a homeomorphism* $\varphi\colon \mathbb{R}^n \to \mathbb{R}^n$ *such that* $\varphi(A) = B$.

We first introduce some notation. We say that (x^1, x^2) and (y^1, y^2) in $\mathbb{R}^n \times \mathbb{R}^n$ are *similarly placed* if the vectors $x^1 - x^2$ and $y^1 - y^2$ are in the same n-dimensional quadrant (that is, the signs of $x_i^1 - x_i^2$ and $y_i^1 - y_i^2$ are the same for all $i = 1, 2, \dots, n$). Let then $X = \{x^1, x^2, \dots\}$ and $Y = \{y^1, y^2, \dots\}$ be two countable sets in $\mathbb{R}^n$. Since X and Y are countable, it is always possible to choose a coordinate system for $\mathbb{R}^n$ so that the axes are in general position (that is, no parallel to a coordinate hyperplane contains more than one point x^i or one point y^j). We then say that X and Y are *similarly placed* if (x^{i_1}, x^{i_2}) and (y^{i_1}, y^{i_2}) are similarly placed for each index pair (i_1, i_2). This ends the preliminaries.

Let $a^1, a^2, \dots$ (resp. $b^1, b^2, \dots$) be an enumeration of A (resp. B). We inductively rearrange the enumerations of A and B in a such way that the resulting sequences $c^1, c^2, \dots$ and $d^1, d^2, \dots$ are similarly placed. We start by setting $c^1 := a^1$ and $d^1 := b^1$. Then we take $d^2 := b^2$ and $c^2 = a^{i_2}$, where i_2 is the least integer i such that (c^1, a^i) and (d^1, d^2) are similarly placed: note that such an i exists (in fact there are infinitely many such) since A is dense in $\mathbb{R}^n$. Suppose that we have constructed $c^1, c^2, \dots, c^{2j}$ and $d^1, d^2, \dots, d^{2j}$, for some integer $j \geq 1$, which are similarly placed. We then set c^{2j+1} to be the first $a^i \neq c^1, c^2, \dots, c^{2j}$ and then set d^{2j+1} to be the first $b^k \neq d^1, d^2, \dots, d^{2j}$ such that $\{c^1, c^2, \dots, c^{2j}, c^{2j+1}\}$ and $\{d^1, d^2, \dots, d^{2j}, d^{2j+1}\}$ are similarly placed. Note that such a b^k exists (in fact there are infinitely many such) since B is dense in $\mathbb{R}^n$. We then denote by d^{2j+2} the first $b^{k'} \neq d^1, d^2, \dots, d^{2j}, d^{2j+1}$ and choose c^{2j+2} to be the first $a^{i'} \neq c^1, c^2, \dots, c^{2j+1}$ such that $\{c^1, c^2, \dots, c^{2j+2}\}$ and $\{d^1, d^2, \dots, d^{2j+2}\}$ are similarly placed. This completes the inductive step. It is then clear that the resulting rearrangements are the ones we were looking for.

To complete the proof of the claim, we only need to show that the bijective map $\varphi\colon A \to B$ defined by setting $\varphi(c_i) = d_i$ for all $i = 1, 2, \dots$ extends to a (unique) homeomorphism of $\mathbb{R}^n$. Let $x = (x_1, x_2, \dots, x_n) \in \mathbb{R}^n \setminus A$. Fix $k \in \{1, 2, \dots, n\}$. We partition A into the two subsets $A_k^- := \{y = (y_1, y_2, \dots, y_n) \in A : y_k \leq x_k\}$ and $A_k^+ := \{y = (y_1, y_2, \dots, y_n) \in A : y_k > x_k\}$. We then set $B_k^- := \varphi(A_k^-)$ and $B_k^+ := \varphi(A_k^+)$, so that $B = B_k^- \sqcup B_k^+$. Consider the projection of B on the k-th coordinate $K := \{y_k : y = (y_1, y_2, \dots, y_n) \in B\} \subset \mathbb{R}$.

The partition $B = B_k^- \sqcup B_k^+$ induces a partition $K = K_- \sqcup K_+$. It is an **exercise** to check that either every element of K_- is equal or less than every element of K_+, or, vice versa, every element of K_+ is equal or less than every element of K_-. Since K is dense in $\mathbb{R}$, this Dedekind cut uniquely defines a separating real number, which we denote by z_k. Repeating this argument for each $k = 1, 2, \dots, n$, this defines $z := (z_1, z_2, \dots, z_n) \in \mathbb{R}^n$. It is an **exercise** to check that setting $\varphi(x) := z$ for all $x \in \mathbb{R}^n \setminus A$ yields a homeomorphism. This completes the proof of the claim.

The proof of the theorem then immediately follows from the fact that the statement holds true when M equals the countable set N_n^n (cf. the initial part of the proof), from the claim, and topological invariance of dimension. □

10.5 Dimension and Measure

Definition 10.39. Let (X,d) be a separable metric space and let $p \in [0,+\infty)$.

Let $\varepsilon > 0$. We say that a countable cover $\mathscr{A} = \{A_i : i \in \mathbb{N}\}$ of X is an *ε-cover* of X provided that each subset $A_i \in \mathscr{A}$ has *diameter* $\delta(A_i) := \sup\{d(x,y) : x,y \in A_i\} \leq \varepsilon$. We then set

$$m_p^\varepsilon(X) := \inf_{\mathscr{A}} \sum_{i=1}^{\infty} \delta(A_i)^p, \tag{10.6}$$

where the infimum is taken over all ε-covers $\mathscr{A} = \{A_i : i \in \mathbb{N}\}$ of X. By convention, we set $\delta(\varnothing)^0 := 0$ and $\delta(X)^0 := 1$ if X is nonempty.

Then

$$m_p(X) := \sup_{\varepsilon > 0} m_p^\varepsilon(X) \tag{10.7}$$

is called the *p-dimensional* measure of X.

Remark 10.40. Let (X,d), $p \in [0,+\infty)$, and $\varepsilon > 0$ be as in Definition 10.39. We say that a countable cover $\mathscr{B} = \{B_i : i \in \mathbb{N}\}$ of X is an *ε-b-cover* of X provided that each subset $B_i \in \mathscr{B}$ is a closed ball of diameter $\delta(B_i) \leq \varepsilon$. We then set

$$\widetilde{m}_p^\varepsilon(X) := \inf_{\mathscr{B}} \sum_{i=1}^{\infty} \delta(B_i)^p,$$

where the infimum is taken over all ε-b-covers $\mathscr{B} = \{B_i : i \in \mathbb{N}\}$ of X, and

$$\widetilde{m}_p(X) := \sup_{\varepsilon > 0} \widetilde{m}_p^\varepsilon(X). \tag{10.8}$$

Given an ε-cover $\mathscr{A} = \{A_i : i \in \mathbb{N}\}$ of X, it is an **exercise** to show that for each $i \in \mathbb{N}$ there exists a closed ball B_i such that $\overline{A_i} \subseteq B_i$ and $\delta(B_i) = 2\delta(A_i)$. It is clear that $\mathscr{B} = \{B_i : i \in \mathbb{N}\}$ is a 2ε-b-cover of X such that

$$\sum_{i=1}^{\infty} \delta(B_i)^p = 2^p \sum_{i=1}^{\infty} \delta(A_i)^p.$$

Since every 2ε-b-cover is a 2ε-cover, we deduce that

$$m_p^{2\varepsilon}(X) \leq \widetilde{m}_p^{2\varepsilon}(X) \leq 2^p m_p^\varepsilon(X)$$

so that

$$m_p(X) \leq \widetilde{m}_p(X) \leq 2^p m_p(X). \tag{10.9}$$

Example 10.41. Let (X,d) be a separable metric space.

(a) Suppose that $p=0$. Then (i) $m_0(X)=0$ if and only if $X=\varnothing$ (**exercise**); (ii) $m_0(X)=n$, $n\in\mathbb{N}$, if and only if $|X|=n$ (**exercise**); (iii) $m_0(X)=\infty$ if and only if X is infinite (**exercise**).

(b) Suppose that $0\le p<q$. Then $m_p(X)\ge m_q(X)$, so that if $m_p(X)<\infty$ one also has $m_q(X)<\infty$ (**exercise**).

(c) Let X be the n-dimensional cube I^n. Then $m_n(X)<\infty$ and $m_p(X)=0$ for all $p>n$ (**exercise**).

Proposition 10.42. *Let C be a compact metric space and let $p\in[0,+\infty)$. Then $m_p(X)=0$ if and only if given $\varepsilon>0$ there exists a* finite *ε-cover $\mathscr{A}=\{A_i:i=1,2,\dots,k\}$ such that $\sum_{i=1}^k\delta(A_i)^p<\varepsilon$.*

Proof. Suppose that $m_p(X)=0$ and let $\varepsilon>0$. Then $m_p^\varepsilon(X)=0$ and therefore there exists an ε-cover $\mathscr{A}'=\{A_i':i\in\mathbb{N}\}$ of X such that $\sum_{i=1}^\infty\delta(A_i')^p<\varepsilon/2$. We may enlarge each A_i' to an open set A_i in a such a way that $\delta(A_i)<\varepsilon$ and $\delta(A_i)^p<\delta(A_i')^p+\varepsilon/2^{i+1}$ (**exercise**: recall that X is second countable). Since C is compact, we can find $k\in\mathbb{N}$ such that $\mathscr{A}:=\{A_i:i=1,2,\dots,k\}$ covers C. Note that $\mathscr{A}$ is an ε-cover and $\sum_{i=1}^k\delta(A_i)^p<\varepsilon$.

The converse being obvious, this completes the proof. □

For the next result, we need some preliminaries from functional analysis.

Recall that a sequence $(f_j)_{j\in\mathbb{N}}$ of measurable functions $f_j\colon I\to\mathbb{R}$, where $I\subseteq\mathbb{R}$ is an interval, is said to *converge in mean* to a (measurable) function $f\colon I\to\mathbb{R}$ *in I* provided that $\lim_{j\to\infty}\int_I|f_j(t)-f(t)|dt=0$. The following is a classical result (cf. [332, Section 12.5], [298, Exercise], [40]).

Theorem 10.43. *Suppose that a sequence $(f_j)_{j\in\mathbb{N}}$ of measurable functions converges in mean to f in I. Then there exists a subsequence $(j_k)_{k\in\mathbb{N}}$ such that $\lim_{k\to\infty}f_{j_k}(t)=f(t)$ for almost all (with respect to Lebesgue measure) $t\in I$.*

The next theorem relates n-dimensionality and the *n-dimensional* measure.

Theorem 10.44. *Suppose that* $\dim(X)=n$. *Then $m_n(X)>0$.*

Proof. Keeping in mind Example 10.41.(b), the statement is equivalent (**exercise**) to the implication: $m_{n+1}(X)=0\Rightarrow\dim(X)\le n$. So, suppose that $m_{n+1}(X)=0$ and let us show that $\dim(X)\le n$. Let $x\in X$ and for $r>0$ denote by $S_r:=\{y\in X:d(x,y)=r\}$ the sphere of radius r centered at x.

Claim. *$m_n(S_r)=0$ for almost all (with respect to Lebesgue measure) $r\in(0,\infty)$.*

Let $Y\subset X$ be a subset and set

$$r_1:=d(x,Y)=\inf\{d(x,y):y\in Y\}\text{ and }r_2:=\sup\{d(x,y):y\in Y\}.$$

It follows from the triangle inequality that $r_2-r_1\le\delta(Y)$. We then have

$$\int_0^\infty\delta(S_r\cap Y)^n\mathrm{d}r=\int_{r_1}^{r_2}\delta(S_r\cap Y)^n\mathrm{d}r\le\delta(Y)^n\int_{r_1}^{r_2}\mathrm{d}r\le\delta(Y)^{n+1}.\qquad(10.10)$$

Since $m_{n+1}(X) = 0$, we can find a family $(\mathscr{A}_j)_{j\in\mathbb{N}}$ of $\frac{1}{j+1}$-covers of X such that

$$\lim_{j\to\infty} \sum_{i=1}^{\infty} \delta(A_{i,j})^{n+1} = 0.$$

Using (10.10), we deduce that

$$\lim_{j\to\infty} \sum_{i=1}^{\infty} \int_0^{\infty} \delta(S_r \cap A_{i,j})^n \mathrm{d}r = 0.$$

As each of the integrands is nonnegative, we may exchange integration and summation, obtaining

$$\lim_{j\to\infty} \int_0^{\infty} \sum_{i=1}^{\infty} \delta(S_r \cap A_{i,j})^n \mathrm{d}r = 0,$$

so that, in mean,

$$\lim_{j\to\infty} \sum_{i=1}^{\infty} \delta(S_r \cap A_{i,j})^n = 0.$$

By Theorem 10.43 there exists a subsequence $(j_k)_{k\in\mathbb{N}}$ such that

$$\lim_{k\to\infty} \sum_{i=1}^{\infty} \delta(S_r \cap A_{i,j_k})^n = 0 \text{ for almost all } r,$$

and this shows that $m_n(S_r) = 0$ for almost all r. This ends the proof of the claim.

To complete the proof of Theorem 10.44, we proceed by induction on n. Let $n = 0$ and suppose that $m_1(X) = 0$. Then, by the Claim, $m_0(S_r) = 0$, equivalently (cf. Example 10.41.(a)) $S_r = \varnothing$, for almost all r. Since the balls $B_r := \{y \in X : d(x,y) < r\}$, $r > 0$, constitute a base of neighborhoods of x and $S_r = \partial B_r$, we deduce that $\dim(X) = 0$. Now, suppose by induction that the implication $m_n(Z) = 0 \Rightarrow \dim(Z) \leq n-1$ holds for all spaces Z. Suppose that $m_{n+1}(X) = 0$. Then the claim yields $m_n(S_r) = 0$ for almost all r and, by the inductive hypothesis, we have $\dim(\partial B_r) = \dim(S_r) \leq n-1$ for almost all r, and we deduce that $\dim(X) \leq n$. □

Remark 10.45. From the proof of the above theorem we deduce that, in fact, if $m_{n+1}(X) = 0$, not only every point $x \in X$ has arbitrarily small neighborhoods with boundaries of dimension $\leq n-1$ (that is, $\dim(X) \leq n$), but indeed, for every point $x \in X$, *almost all* spherical neighborhoods of x have boundaries of dimension $\leq n-1$.

10.6 Hausdorff Dimension

Definition 10.46 (Hausdorff dimension). The *Hausdorff dimension* of a metric space (X,d) is the nonnegative number $\mathrm{Hdim}(X)$ defined by setting

$$\mathrm{Hdim}(X) := \sup\{p \geq 0 : m_p(X) > 0\}.$$

It follows from Remark 10.40 (in particular, from (10.9)) that

$$\mathrm{Hdim}(X) = \sup\{p \geq 0 : \widetilde{m}_p(X) > 0\},$$

where $\widetilde{m}_p(X)$ is as in (10.8). In other words, in the computation of the Hausdorff dimension of a space X, we can limit ourselves to the use of ε-b-covers of X.

From the above definition and Theorem 10.44 we immediately have the following.

Corollary 10.47. *Let X be a metric space. Then*

$$\mathrm{Hdim}(X) \geq \dim(X).$$

In particular, a metric space of finite Hausdorff dimension is finite-dimensional. □

Remark 10.48. Note that if $Y_1 \subseteq Y_2 \subseteq X$ then $\mathrm{Hdim}(Y_1) \leq \mathrm{Hdim}(Y_2)$ since any cover of Y_2 is also a cover of Y_1.

Proposition 10.49. *Countable metric spaces have Hausdorff dimension* 0.

Proof. Let X be a countable metric space and denote by $x_0, x_1, \ldots$ its elements. Let $p, \varepsilon > 0$. For $i = 0, 1, \ldots$ set

$$B_i := B(x_i, (\varepsilon/2^{i+2})^{\frac{1}{p}}/2).$$

Then $(B_i)_{i\in\mathbb{N}}$ is a b-cover of X and $\sum_{i=1}^{\infty} \delta(B_i)^p = \sum_{i=0}^{\infty} \varepsilon/2^{i+2} = \varepsilon/2 < \varepsilon$. This shows that $\widetilde{m}_p(X) = 0$. Since p was arbitrary, we deduce that $\mathrm{Hdim}(X) = 0$. □

Theorem 10.50. $\mathrm{Hdim}(\mathbb{R}^n) = n$.

Proof. Let $X = \mathbb{R}^n$. We claim that it is enough to show that $\mathrm{Hdim}(B) = n$, where B is any ball of radius 1. Suppose that $m_p(B) = 0$ for some $p \geq 0$, and let $\varepsilon > 0$. Let $(Y_j)_{j\in\mathbb{N}}$ denote an enumeration of the balls of radius 1 centered at the points $(x/2, y/2)$ with $x, y \in \mathbb{Z}$. It is clear that $(Y_j)_{j\in\mathbb{N}}$ covers X. For $j \in \mathbb{N}$ we set $\varepsilon_j = \varepsilon/2^{j+1}$ and, in accordance with our hypothesis on p (so that $m_p(Y_j) = 0$ for all $j \in \mathbb{N}$), we denote by $(B_{i,j})_{i\in\mathbb{N}}$ a sequence of balls covering Y_j and such that $\sum_{i\in\mathbb{N}} \delta(B_{i,j})^p < \varepsilon_j$. Then the sequence $(B_{i,j})_{i,j\in\mathbb{N}}$ of balls covers X and satisfies $\sum_{i,j\in\mathbb{N}} \delta(B_{i,j})^p < \sum_{j\in\mathbb{N}} \varepsilon_j = \varepsilon$, thus showing that $m_p(X) = 0$. This shows that $\mathrm{Hdim}(X) \leq \mathrm{Hdim}(B)$. Since $B \subseteq X$ the previous remark gives $\mathrm{Hdim}(B) \leq \mathrm{Hdim}(X)$, and therefore $\mathrm{Hdim}(X) = \mathrm{Hdim}(B)$. This proves the claim.

Recall that in $X = \mathbb{R}^n$, the volume of any ball $B(r)$ of radius r is given by

$$\mathrm{Vol}(B(r)) = \mu r^n, \tag{10.11}$$

where $\mu = \mu(n)$ is a constant (independent of r). In particular, $\mathrm{Vol}(B) = \mathrm{Vol}(B(1)) = \mu$.

Let $(B_i)_{i\in\mathbb{N}}$ be a cover of B by balls. Using (10.11) we have $\delta(B_i)^n = 2^n \mathrm{Vol}(B_i)/\mu$ for all $i \in \mathbb{N}$ so that

$$\sum_{i\in\mathbb{N}} \delta(B_i)^n = \frac{2^n}{\mu}\sum_{i\in\mathbb{N}} \mathrm{Vol}(B_i) \geq \frac{2^n}{\mu}\mathrm{Vol}(B) = 2^n.$$

This shows that the LHS cannot be made arbitrarily small, thus showing that $m_n(B) > 0$.

On the other hand, for every integer $N \geq 1$, let $(c_i(N))_{i=1}^{(2N+1)^n}$ be an enumeration of the points

$$(i_1/N, i_2/N, \ldots, i_n/N) \in \mathbb{R}^n, \ \ i_j \in \mathbb{Z} \text{ such that } |i_j| \leq N \text{ for all } j = 1,2,\ldots,n.$$

For $i = 1,2,\ldots,(2N+1)^n$ set $B_i(N) := B(c_i(N); 3/(2N+1))$. It is an **exercise** to check that

$$B = B((0,0,\ldots,0);1) \subseteq \bigcup_{i=1}^{(2N+1)^n} B_i(N).$$

For every $\varepsilon > 0$ we have

$$\sum_{i=1}^{(2N+1)^n} \delta(B_i(N))^{n+\varepsilon} = (2N+1)^n \cdot \frac{6^{n+\varepsilon}}{(2N+1)^{n+\varepsilon}} = \frac{6^{n+\varepsilon}}{(2N+1)^{\varepsilon}} \to 0$$

as $N \to \infty$. This shows that $m_{n+\varepsilon}(B) = 0$. As ε was arbitrary, this shows that $\mathrm{Hdim}(B) \leq n$. We conclude that $\mathrm{Hdim}(\mathbb{R}^n) = \mathrm{Hdim}(B) = n$. □

Theorem 10.51 (Hausdorff dimension of the Cantor set). $\mathrm{Hdim}(K) = \log 2/\log 3$.

Proof. We start by observing that every set K_j consists of 2^j intervals, each of length 3^{-j}. Thus, since $K_0 \supseteq K_1 \supseteq \cdots \supseteq K_j \supseteq \cdots \supseteq K$, the Cantor set may be covered, for each $j \in \mathbb{N}$, by the finite 3^{-j}-b-cover $\mathscr{B}_j$ consisting of all 2^j balls B of K of diameter $\delta(B) = 3^{-j}$.

Given $p > 0$ we have $\sum_{B\in\mathscr{B}_j} \delta(B)^p = 2^j 3^{-jp}$. Thus, in particular,

$$\sum_{B\in\mathscr{B}_j} \delta(B)^{\log 2/\log 3} = 2^j 3^{-j\log 2/\log 3} = 2^j 2^{-j} = 1. \tag{10.12}$$

For $\varepsilon > 0$ we set $p_\varepsilon := \log(2+\varepsilon)/\log 3$. We then have

$$\sum_{B\in\mathscr{B}_j} \delta(B)^{p_\varepsilon} = 2^j 3^{-jp_\varepsilon} = 2^j(2+\varepsilon)^{-j} = \left(\frac{2}{2+\varepsilon}\right)^j$$

which tends to 0 as $j \to \infty$. This shows that $\mathrm{Hdim}(K) \leq \log 2/\log 3$.

To prove that $\mathrm{Hdim}(K) \geq \log 2/\log 3$, let us show that

$$\sum_{B\in\mathscr{B}} \delta(B)^{\log 2/\log 3} \geq 1 \tag{10.13}$$

for all collections $\mathscr{B}$ of balls of K covering K.

Suppose, by contradiction, that there exists a collection $\mathscr{B}$ of balls of K covering K such that (10.13) fails to hold. Then, expanding each ball $B \in \mathscr{B}$ slightly in order to make it an open ball, we can still have the left-hand side of (10.13) strictly less than 1. By compactness of K, we can then extract a finite subcover $\mathscr{B}'$ of $\mathscr{B}$ which then, a fortiori, satisfies $\sum_{B\in\mathscr{B}'}\delta(B)^{\log 2/\log 3}<1$. Moreover, up to replacing every ball B in $\mathscr{B}'$ with its closure, we may suppose that all balls in $\mathscr{B}'$ are closed.

Let $B\in\mathscr{B}'$ and denote by I_B the smallest interval in $[0,1]$ containing B (see Figure 10.4). Note that $\delta(I_B)=\delta(B)$.

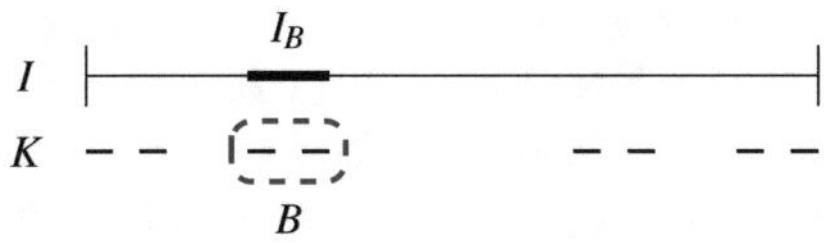

Fig. 10.4 I_B is the smallest interval in $[0,1]$ containing B.

Let J_B denote the largest (open) subinterval of I_B contained in $[0,1]\setminus K$. Finally, let L_B and R_B denote the left and right components of $I_B\setminus J_B$ so that $I_B=L_B\sqcup J_B\sqcup R_B$. Note that $\delta(L_B),\delta(R_B)\le\delta(J_B)$ so that, in particular, $\delta(J_B)\ge\frac{1}{2}(\delta(L_B)+\delta(R_B))$.

Using the concavity of the function $f(x)=x^{\log 2/\log 3}$ and the fact that $3^{\log 2/\log 3}=2$, we deduce

$$\begin{aligned}
\delta(B)^{\log 2/\log 3} &= \delta(I_B)^{\log 2/\log 3}\\
&=(\delta(L_B)+\delta(J_B)+\delta(R_B))^{\log 2/\log 3}\\
&\ge\left(\frac{3}{2}(\delta(L_B)+\delta(R_B))\right)^{\log 2/\log 3}\\
&=3^{\log 2/\log 3}\left(\frac{1}{2}\delta(L_B)+\frac{1}{2}\delta(R_B)\right)^{\log 2/\log 3}\\
&\ge 2\left(\frac{1}{2}\delta(L_B)^{\log 2/\log 3}+\frac{1}{2}\delta(R_B)^{\log 2/\log 3}\right)\\
&=\delta(L_B)^{\log 2/\log 3}+\delta(R_B)^{\log 2/\log 3}\\
&\ge\delta(B_L)^{\log 2/\log 3}+\delta(B_R)^{\log 2/\log 3},
\end{aligned}$$

where $B_L:=B\cap L_B$ and $B_R:=B\cap R_B$.

It follows that the family

$$\mathscr{B}'':=(\mathscr{B}'\setminus\{B\})\cup\{B_L,B_R\}$$

still covers K and satisfies $\sum_{A\in\mathscr{B}''}\delta(A)^{\log 2/\log 3}\le\sum_{A\in\mathscr{B}'}\delta(A)^{\log 2/\log 3}<1$. Continuing this way, after a finite number of steps, we reach a covering $\mathscr{B}'''$ of K consisting of balls all of the same diameter, say 3^{-j} for a suitable $j\in\mathbb{N}$. It follows that these are exactly the intervals constituting K_j, that is, $\mathscr{B}'''=\mathscr{B}_j$. From (10.12) we then deduce $1=\sum_{B\in\mathscr{B}'''}\delta(B)^{\log 2/\log 3}<1$, a contradiction. This shows that $\mathrm{Hdim}(K)\ge\log 2/\log 3$. ☐

10.7 Notes

According to Witold Hurewicz and Henry Wallman [184, Introduction], the first precise and topologically invariant definition of dimension is due to Luitzen Egbertus Jan (Bertus) Brouwer [45] in 1913, based on a first intuition due to Poincaré [282, 283]. The definition of dimension that we used here, namely the *inductive dimension* (cf. Definitions 10.3 and 10.16) is due to Karl Menger and Pavel Urysohn who, in 1922, independently of each other and of Brouwer, rediscovered and improved on Brouwer's notion. Note that in 1911 Brouwer gave the first proof of the fact that $\mathbb{R}^n$ and $\mathbb{R}^m$ are not homeomorphic if $n \neq m$. In his 1913 paper he introduced the notion of "Dimensionsgrad", an integer-valued topological invariant, and showed that the "Dimensionsgrad" of $\mathbb{R}^n$ is n, thus yielding a new proof of his 1911 result we just alluded to.

Example 10.4.(f) is due to Paul Erdös [104]. Wacław Sierpiński [316] found an example of a subset of $\mathbb{R}^2$ which is totally disconnected but admitting distinct points that cannot be separated (cf. conditions (a) and (b) in Theorem 10.15).

Proposition 10.17 cannot be extended to infinite-dimensional spaces, as shown, under the continuum hypothesis, by Hurewicz [182].

The inequality $\dim(A \times B) \leq \dim(A) + \dim(B)$ in Theorem 10.29 may be strict, as shown by taking $A = B$ to be the 1-dimensional set $X \subseteq \ell^2(\mathbb{N})$ all of whose coordinates are rational (cf. Example 10.18.(c)), which is clearly homeomorphic to its own square $X \times X$. Recall that, however, equality holds if one of the two spaces is 0-dimensional (cf. Corollary 10.30). An example due to Lev S. Pontryagin [285] shows that the above inequality may be strict even if both A and B are compact.

James Dugundji, the author of the monograph [98], was a pupil of Hurewicz. Theorem 10.34.(b) is due to Piers Bohl [30]. Theorem 10.34.(c) is the celebrated *Brouwer fixed-point theorem* of Brouwer [44]. Theorem 10.34.(d), known as *Borsuk non-retraction theorem*, is due to Karol Borsuk [31].

Corollary 10.36 was first proved by Brouwer [45].

The *Peano curve* (cf. Exercise 10.13), the first example of a *space-filling curve*, was constructed by Giuseppe Peano in 1890 [272], motivated by an earlier counterintuitive result of Georg Cantor, namely, that the unit interval and the unit square have the same cardinality. A year later, David Hilbert published in the same journal a variation of Peano's construction [171]. A space-filling curve is often called a *Peano–Hilbert curve*. Exercise 10.13 is inspired by the beautiful treatment in [150, Chapter IV.2].

Theorem (Menger–Nöbeling). *Let X be a space of dimension $\leq n$ for some $n \in \mathbb{N}$. Then X is homeomorphic to a subset of $\mathbb{R}^{2n+1}$.*

Karl Menger proved the above theorem under the hypothesis of compactness of X and $\dim(X) = 1$ in [235] and in [236] for arbitrary (finite) dimensions. Georg August Nöbeling proved the general statement (dropping the compactness assumption) in [256] by combining Menger's result with a result by Hurewicz [183] stating that any space is a (topological) subspace of a compact space of the same dimension. We refer to [184, Theorem V.2, Theorem V.3, and Theorem V.7] for a proof of these results. An example due to Antonio Flores [114] (see also [184, Example V.3]) shows that it is not possible, in general, to imbed an n-dimensional space in $\mathbb{R}^{2n}$, thus showing that $2n+1$ in the Menger–Nöbeling theorem is optimal.

There is another approach, due to Henri Lebesgue [212], in terms of the *(topological) covering dimension* (also called the *Čech–Lebesgue covering dimension*). Roughly speaking, Lebesgue observed that the square $I^2 = [0,1] \times [0,1]$ can be covered by arbitrarily small "bricks" in such a way that no point in I^2 is contained in more than three of these bricks; moreover, if these bricks are sufficiently small, at least three have a common point. He then extended the same argument to the n-cube I^n, where the number of bricks becomes $n+1$, and conjectured that this number $n+1$ cannot be reduced. The conjecture was proved by Brouwer in [45].

We now give the definition of topological covering dimension introduced by Eduard Čech [68, 69, 70], heavily relying on Lebesgue's work. This equivalent approach to dimension is presented, with extreme clarity and elegance, in Part I of the monograph by Michel Coornaert [78], with the aim of developing the necessary preparatory tools for the theory of *mean topological dimension*, a conjugacy invariant of topological dynamical systems due to Gromov [142], which was used by Elon Lindenstrauss and Benjy Weiss [216] to solve in the negative a long-standing open problem on the embeddability of dynamical systems into shifts (see [78, Part II]).

Let X be a topological space. Recall that a *cover* of X is a family $\alpha = (A_i)_{i \in I}$ of subsets $A_i \subseteq X$ such that $\bigcup_{i \in I} A_i = X$. If A_i is open for all $i \in I$, one says that α is an *open cover* of X. Given two covers $\alpha = (A_i)_{i \in I}$ and $\beta = (B_j)_{j \in J}$, one says that β is *finer* that α, and one writes $\alpha \preceq \beta$, if for every $j \in J$ there exists an $i \in I$ such that $B_j \subseteq A_i$.

Let α be a cover of X. The *order* of α is

$$\operatorname{ord}(\alpha) := \sup_{x \in X} \operatorname{ord}_x(\alpha),$$

where, for $x \in X$, $\operatorname{ord}_x(\alpha) := -1 + |\{i \in I : x \in A_i\}| \in \mathbb{N} \cup \{+\infty\}$ is the order of α at x, with the convention that if $X = \varnothing$, then $\operatorname{ord}(\alpha) = -1$. If α is finite, we set

$$D(\alpha) := \min_{\beta} \operatorname{ord}(\beta),$$

where β runs over all *finite* open covers of X such that $\alpha \preceq \beta$. We are now in a position to give the definition of Čech–Lebesgue covering dimension.

Definition (Topological covering dimension). Let X be a topological space. The *topological covering dimension* of X is the quantity

$$\dim(X) := \sup_{\alpha} D(\alpha) \in \{-1\} \cup \mathbb{N} \cup \{+\infty\},$$

where α runs over all finite open covers of X.

The Lebesgue method, leading to the definition of covering dimension, coincides with that of Brouwer, Menger, and Urysohn [44] (see [184, Chapter V]). Another significant approach is due to Pavel Alexandroff [5] (cf. [184, Chapters VI and

VII]). Another approach to the notion of dimension, from a purely algebraic and combinatorial viewpoint, comes from homology theory. Once again, the homology-dimension turns out to be the same as the topological dimension (see [184, Chapter VIII]).

Intuitively, with the topological notion of one-dimensionality (resp. two-dimensionality, resp. three-dimensionality) one associates the metrical notion of length, or linear measure (resp. of area, or two-dimensional measure, resp. of volume, or three-dimensional measure). For metric spaces this intuitive feeling was made precise by Edward Szpilrajn [325] who established the connections between topological dimension and measure, in particular with the Hausdorff dimension such as Theorem 10.47 (see [184, Chapter VII]).

For $X \subset \mathbb{R}$, the 1-dimensional measure of X coincides with Lebesgue outer measure. However, in general, for $n \geq 2$, the n-measure of a set $X \subset \mathbb{R}^n$ may differ numerically from its Lebesgue outer measure. However, the two measures are absolutely continuous with respect to each other (cf. [184, Chapter VII, Section 1.(E)]).

The following result (cf. [184, Theorem VII.4]) provides an interesting refinement of the Menger–Nöbeling theorem in terms of n-measure.

Theorem. *Let X be a space of dimension $\leq n$ for some $n \in \mathbb{N}$. Then X is homeomorphic to a subset of $\mathbb{R}^{2n+1}$ of $(n+1)$-measure zero.*

Hausdorff dimension was introduced by Felix Hausdorff in [167] in 1919. As we have seen, (cf. Theorem 10.50) one has $\mathrm{Hdim}(\mathbb{R}^n) = n$ for all $n \in \mathbb{N}$. More generally, it can be shown that, for sets of points that define a smooth shape or a shape that has a small number of corners, the Hausdorff dimension is an integer agreeing with the topological dimension. However, Hausdorff dimension may take non-integer values: for instance, the Hausdorff dimension of the Cantor set K is $\mathrm{Hdim}(K) = \log 2/\log 3$ (cf. Theorem 10.51). Significant technical advances, allowing computation of the Hausdorff dimensions of more general highly irregular or "rough" sets, were made by Abram S. Besicovitch [25, 26].

Szpilrajn [325] (cf. [184, Theorem VII.5 and Chapter VII, Section 4]) proved that given an arbitrary separable metric space X,

$$\dim(X) = \inf_{(Y,d)} \mathrm{Hdim}(Y),$$

where (Y,d) runs over all metric spaces homeomorphic to X.

In our treatment, we have only considered separable metric spaces. There are spaces of a more general nature which proved to be very interesting from the dimension theory point of view (cf. [184, Appendix] for a discussion on a "general dimension theory" and "dimension functions"). Part I of the monograph by Michel Coornaert [78] is devoted to dimension theory in the covering-dimension approach in a much more general setting than separable metric spaces. There are plenty of fascinating examples providing counterintuitive results in this more general setting. In the Notes to each of the chapters one may find a careful historical account on these.

10.8 Exercises

Exercise 10.1. Let X be a nonempty space. Show that $\dim(X)=0$ if and only if the following holds: for every $x \in X$ and every neighborhood U of x there exists a neighborhood V of x such that $V \subseteq U$ and $\partial V = \varnothing$.

Exercise 10.2. Show that the set $\mathbb{Q}^n$ of rational points in Euclidean space $\mathbb{R}^n$ is 0-dimensional for all $n \geq 1$.

Exercise 10.3. Let $I^\infty := \prod_{n\geq 1}[0,1/n]$ be the Hilbert cube (cf. Example 10.4.(e)). Observe that I^∞ is contained in the real Hilbert space $\mathscr{H} := \ell^2(\mathbb{N})$ of square summable real sequences and show that it is compact and contains no nonempty open subsets of $\mathscr{H}$.

Exercise 10.4. Let A and C be two 0-dimensional subsets of a space X. Suppose that C is closed. Show that $A \cup C$ is 0-dimensional. Deduce that a 0-dimensional space remains 0-dimensional after the adjunction of finitely many points. (Cf. Exercise 10.9).

Exercise 10.5. Let $0 \leq m \leq n$ and denote by X the set of points in $\mathbb{R}^n$ that have exactly m coordinates in $\mathbb{Q}$. Show that X is 0-dimensional.

Exercise 10.6. Let $0 \leq m$ and denote by X the set of points in the Hilbert cube I^∞ that have exactly m coordinates in $\mathbb{Q}$. Show that X is 0-dimensional.

Exercise 10.7. The set $X \subseteq \ell^2(\mathbb{N})$ all of whose coordinates are rational is (uncountable and) not 0-dimensional (cf. Example 10.4.(f)). Show that any two distinct points in X can be separated (cf. conditions (b) and (e) in Theorem 10.15).

Exercise 10.8. Recall that a topological (not necessarily separable nor metrizable) space is scattered if it admits a base of clopen subsets.

(1) Show that every set equipped with the discrete topology is scattered;
(2) show that a connected space is scattered if and only if the topology is the trivial one (only $\varnothing$ and the whole space are open);
(3) suppose that a space X is scattered and every point is closed. Show that X is Hausdorff;
(4) show that every subspace of a scattered space is itself scattered;
(5) show that the product of scattered spaces is itself scattered;
(6) show that a subset of $\mathbb{R}$ is scattered if and only if it has empty interior.

Exercise 10.9. Let A and C be two subsets of a space X. Suppose that both $\dim(A)$, $\dim(C) \leq n$ and that C is closed. Show that $\dim(A \cup C) \leq n$. Deduce that the dimension of a nonempty space cannot be increased by the adjunction of finitely many points (cf. Exercise 10.4).

Exercise 10.10. Show that the unit circle $\mathbb{S}^1$ has Hausdorff dimension 1.

Exercise 10.11. Let M be an n-dimensional manifold (so that each point $P \in M$ admits a neighborhood homeomorphic to $\mathbb{R}^n$). Show that $\dim(X) = n = \mathrm{Hdim}(X)$.

Exercise 10.12. Let $X = \{0,1\}^{\mathbb{N}}$. For $q \in (0,1)$ define the function $d_q\colon X \times X \to [0,1]$ by setting

$$d_q(x,y) := \begin{cases} 0 & \text{if } x = y \\ 2^{-n/q} & \text{if } x(i) = y(i) \text{ for } i = 0,1,\dots,n-1 \text{ and } x(n) \neq y(n). \end{cases}$$

(a) Show that d_q is a metric on X and that the topology it induces is the prodiscrete topology.

(b) Show that $m_p(X,d_q) = 0$ for $p > q$ and deduce that $\mathrm{Hdim}(X,d_q) \leq q$.

(c) Verify that $\dim(K) = \inf \mathrm{Hdim}(Y,d)$, where (Y,d) runs over all metric spaces homeomorphic to the Cantor set K.

Exercise 10.13 (The Peano curve). Let $\varphi\colon [0,1] \to [0,1] \times [0,1]$ be a continuous function such that $\varphi(0) = (0,0)$ and $\varphi(1) = (1,0)$. Define a new function $F(\phi)\colon [0,1] \to [0,1] \times [0,1]$ by setting

$$F(\phi) := \begin{cases} \frac{1}{2}(y(4t), x(4t)) & \text{if } 0 \leq t \leq \frac{1}{4} \\ \frac{1}{2}(x(4t-1), 1 + y(4t-1)) & \text{if } \frac{1}{4} \leq t \leq \frac{1}{2} \\ \frac{1}{2}(1 + x(4t-2), 1 + y(4t-2)) & \text{if } \frac{1}{2} \leq t \leq \frac{3}{4} \\ \frac{1}{2}(2 - y(4t-3), 1 - x(4t-3)) & \text{if } \frac{3}{4} \leq t \leq 1, \end{cases}$$

where $\varphi(t) = (x(t), y(t))$, for all $t \in [0,1]$.

(a) Show that $F(\phi)$ is continuous and $F(\phi)(0) = (0,0)$ and $F(\phi)(1) = (1,0)$.

(b) For any continuous function $f\colon [0,1] \to [0,1] \times [0,1]$, with $f(t) = (x(t), y(t))$ for all $t \in [0,1]$, the nonnegative quantity

$$\|f\|_\infty := \sup\{(x(t)^2 + y(t)^2)^{1/2} : t \in [0,1]\} = \max\{(x(t)^2 + y(t)^2)^{1/2} : t \in [0,1]\}$$

denotes the sup-norm of f. Show that if $\psi\colon [0,1] \to [0,1] \times [0,1]$ is another continuous function such that $\psi(0) = (0,0)$ and $\psi(1) = (1,0)$, then (cf. Figure 10.5)

$$\|F(\varphi) - F(\psi)\|_\infty \leq \frac{1}{2} \|\varphi - \psi\|_\infty.$$

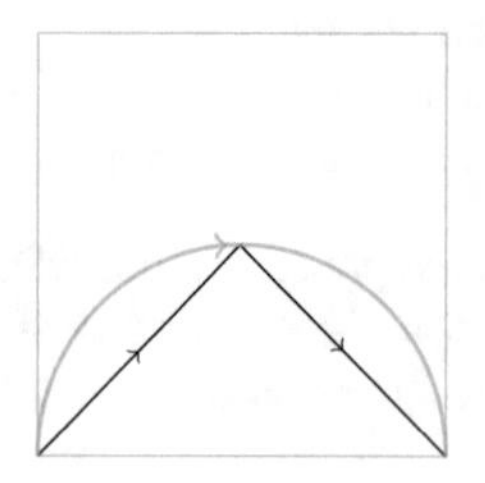
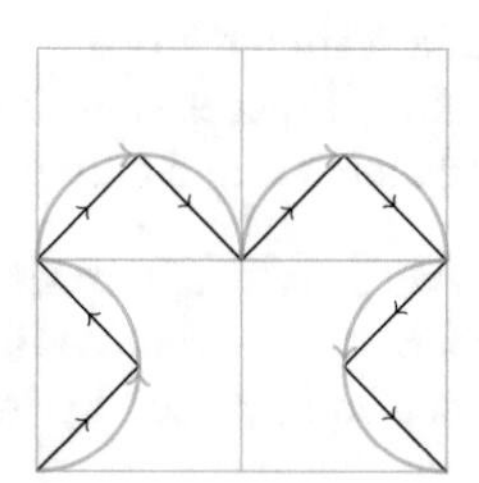
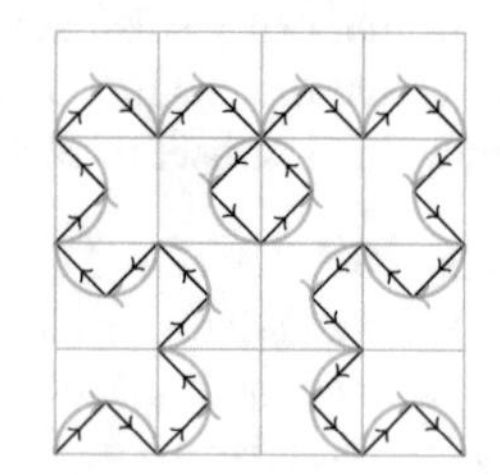

Fig. 10.5 Two maps $\varphi, \psi\colon [0,1] \to [0,1] \times [0,1]$, $F(\varphi)$, $F(\psi)$, and $F^2(\varphi)$, $F^2(\psi)$.

(c) Set $\varphi_n := F^n(\varphi)$ for all $n \in \mathbb{N}$, so that $\varphi_0 := \varphi$ and $\varphi_1 := F(\varphi)$. Show that the sequence $(\varphi_n)_{n\in\mathbb{N}}$ is uniformly convergent, that is, there exists a function $\varphi_\infty \colon [0,1] \to [0,1] \times [0,1]$ (necessarily continuous and such that $\varphi_\infty(0) = (0,0)$ and $\varphi_\infty(1) = (1,0)$) such that

$$\lim_{n\to\infty} \|\varphi_n - \varphi_\infty\|_\infty = 0.$$

(d) Show that $\varphi_\infty([0,1]) = [0,1] \times [0,1]$, that is, φ_∞ (continuously) maps the interval $[0,1]$ *onto* the square $[0,1] \times [0,1]$.

(e) Can φ_∞ be injective? Why?

Exercise 10.14. Show that the Hilbert space $\ell^2(\mathbb{N})$ has infinite Hausdorff dimension.

Chapter 11
Ultrafilters, Ultraproducts, Ultrapowers, and Asymptotic Cones

This chapter is devoted to the study of filters and ultrafilters on the naturals and ultraproducts and asymptotic cones of metric spaces. A filter on $\mathbb{N}$ is a nonempty subset $\mathscr{F}$ of the set $\mathscr{P}(\mathbb{N})$ of all subsets of $\mathbb{N}$ such that: $\varnothing \notin \mathscr{F}$; $\bigcap_{i=1}^n A_i \in \mathscr{F}$ for all $A_1, A_2, \ldots, A_n \in \mathscr{F}$; $B \in \mathscr{F}$ for all $B \supseteq A$ such that $A \in \mathscr{F}$. For example, the set of all cofinite subsets of $\mathbb{N}$ is a filter, called the Fréchet filter (Example 11.2.(c)). Also, if $\varnothing \neq A_0 \subseteq \mathbb{N}$ then the set of all subsets containing A_0 is a filter, called the principal filter based at A_0 (Example 11.2.(b)). A filter which is maximal (with respect to set-theoretical inclusion) is called an ultrafilter. If an ultrafilter is principal based, say, at A_0, then A_0 is singleton. A non-principal ultrafilter is also called a free ultrafilter. A characterization of free ultrafilters is given in Theorem 11.12; in particular, an ultrafilter is free if and only if it contains the Fréchet filter (Corollary 11.13). Given a metric space (X,d) and a filter (in particular, an ultrafilter) $\mathscr{F}$ on $\mathbb{N}$, a notion of convergence for sequences $(x_n)_{n\in\mathbb{N}}$ in X along $\mathscr{F}$ is introduced and studied in Section 11.4: a sequence that is convergent (to some limit $x \in X$) in the usual sense also converges along any free ultrafilter to the same limit x (and vice versa). If X is compact and ω is an ultrafilter then any sequence is convergent along ω (Theorem 11.19, a Bolzano–Weierstass type theorem). A characterization of compact metric spaces in terms of ultrafilters is derived (Theorem 11.21). As a consequence, given an ultrafilter ω on $\mathbb{N}$, every bounded sequence $x = (x_n)_{n\in\mathbb{N}}$ of real numbers is convergent along ω, moreover the map $x \mapsto \lim_\omega x_n$ is linear, continuous (i.e., bounded, with respect to the norm $\|\cdot\|_\infty$ on the Banach space $\ell^\infty(\mathbb{R})$), and monotone (Theorem 11.22). The set $\beta\mathbb{N}$ of all ultrafilters on $\mathbb{N}$ can be given a topology making it into a Hausdorff compact space and one has an injection $\mathbb{N} \hookrightarrow \beta\mathbb{N}$ (by setting $n \mapsto \omega_n$, the principal filter based at $\{n\}$) with dense image (Proposition 11.23 and Theorem 11.25): it is called the Stone–Čech compactification of $\mathbb{N}$.

In Section 11.7 (resp. Section 11.8) we introduce and study the ultrapower (resp. the ultraproduct) of a metric space (resp. of a sequence of pointed metric spaces). The case where the metric space (resp. the sequence of metric spaces) is a metric group (resp. a sequence of metric groups) such as the symmetric group equipped with the Hamming distance or the unitary group equipped with the Hilbert–Schmidt norm is of particular interest: the resulting construction of an ultrapower or ultraproduct is a metric group as well (Theorem 11.37). In Section 11.10 a similar con-

T. Ceccherini-Silberstein and M. D'Adderio, *Topics in Groups and Geometry*,
Springer Monographs in Mathematics, https://doi.org/10.1007/978-3-030-88109-2_11

struction is applied to define the ultrapower of a field, yielding a new field called an ultrafield (Theorem 11.39). The ω-ultrapower $\mathrm{GL}(n,\mathbb{F})_\omega$ of the general linear group $\mathrm{GL}(n,\mathbb{F})$ of all invertible $n \times n$ matrices with coefficients in a field $\mathbb{F}$ is canonically isomorphic to the general linear group $\mathrm{GL}(n,\mathbb{F}_\omega)$ of all invertible $n \times n$ matrices with coefficients in the ω-ultrafield $\mathbb{F}_\omega$ (Theorem 11.40).

In Section 11.12 we define and study the asymptotic cone $(K_\omega(X,d),d_\omega)$ of a metric space (X,d) relative to a free ultrafilter ω: it is just the ultraproduct of the sequence (X_n,d_n,x_n^0) of pointed metric spaces, where $X_n = X$, $d_n(x,y) := d(x,y)/n$, and $x_n^0 := x_0 \in X$ is a fixed base point. In Section 11.13 we introduce the notions of a quasi-isometry and of a bi-Lipschitz map between metric spaces as well as the induced equivalences. We then show that given a finitely generated group G, the bi-Lipschitz class of the asymptotic cone $K_\omega(G,d_S)$ is independent of the finite symmetric generating subset $S \subset G$ (Corollary 11.49). In Section 11.14 we present some properties of asymptotic cones (associated with a free ultrafilter): the asymptotic cone of a complete metric space is complete (Theorem 11.51); the asymptotic cone of a finitely generated group (with respect to a finite symmetric generating subset) is homogeneous, arcwise connected (and therefore connected) and in fact geodesic, locally arcwise connected (and therefore locally connected), and complete (Theorem 11.52). In the subsequent section we describe some examples of asymptotic cones (associated with a free ultrafilter ω on $\mathbb{N}$): the asymptotic cone of a bounded metric space reduces to a point (Proposition 11.54). The asymptotic cone $K_\omega(\mathbb{Z},d_S)$, where $S = \{-1,1\} \subset \mathbb{Z}$, is homeomorphic to $(\mathbb{R},d)$, where d denotes the Euclidean distance on $\mathbb{R}$ (i.e., $d(x,y) = |x-y|$ for all $x,y \in \mathbb{R}$) (Proposition 11.56). As the asymptotic cone of a Cartesian product is naturally homeomorphic to the Cartesian product of the asymptotic cones (Proposition 11.57), we deduce that the bi-Lipschitz class of the asymptotic cone of $\mathbb{Z}^n$ is the same as that of $(\mathbb{R}^n,d_n)$, where d_n is the ℓ^1-distance ($d_n(x,y) = \sum_{i=1}^n |x_i - y_i|$ for all $x = (x_1,\ldots,x_n), y = (y_1,\ldots,y_n) \in \mathbb{R}^n$) (Corollary 11.58).

The final part of the chapter is devoted to the introduction and study of hyperbolic metric spaces. In Section 11.16 we define hyperbolicity, presenting a few equivalent conditions (Proposition 11.63), and illustrate several examples. We remark that hyperbolicity is a quasi-isometry invariant (Theorem 11.67). In Section 11.17 we then define $\mathbb{R}$-trees (essentially, an $\mathbb{R}$-tree is the same thing as a 0-hyperbolic metric space, by virtue of Proposition 11.72) and characterize hyperbolic metric spaces in terms of their asymptotic cones being $\mathbb{R}$-trees (Theorem 11.75).

11.1 Filters

Let $\mathbb{N} = \{0,1,2,\ldots\}$ be the set of naturals and denote by $\mathscr{P}(\mathbb{N}) = \{A : A \subseteq \mathbb{N}\}$ the set of all subsets of $\mathbb{N}$.

Definition 11.1. A *filter* on $\mathbb{N}$ (briefly, a filter) is a nonempty set $\mathscr{F} \subseteq \mathscr{P}(\mathbb{N})$ satisfying the following conditions:

(F1) $\varnothing \notin \mathscr{F}$;
(F2) if $A_1, A_2, \ldots, A_n \in \mathscr{F}$, $n \geq 1$, then $\bigcap_{i=1}^n A_i \in \mathscr{F}$;
(F3) if $A \in \mathscr{F}$ and $A \subseteq B \subseteq \mathbb{N}$, then $B \in \mathscr{F}$.

Example 11.2. (a) $\mathscr{F} = \{\mathbb{N}\}$ is clearly a filter.

(b) Let $A_0 \in \mathscr{P}(\mathbb{N})$. Then the set $\mathscr{F}_{A_0} := \{A \in \mathscr{P}(\mathbb{N}) : A_0 \subseteq A\}$ is a filter. It is called the *principal filter based* at A_0. A filter $\mathscr{F}$ on $\mathbb{N}$ is said to be *principal* if there exists an $A_0 \subseteq \mathbb{N}$ such that $\mathscr{F} = \mathscr{F}_{A_0}$.

(c) The set $\mathscr{F} = \{A \in \mathscr{P}(\mathbb{N}) : \mathbb{N} \setminus A \text{ is finite}\}$, consisting of all *cofinite* subsets of $\mathbb{N}$, is a filter. It is called the *Fréchet filter*.

Definition 11.3. A *base of a filter* is a set $\mathscr{B} \subseteq \mathscr{P}(\mathbb{N})$ satisfying the following conditions:

(B1) $\varnothing \notin \mathscr{B}$;
(B2) if $B_1, B_2 \in \mathscr{B}$, then there exists a $B \in \mathscr{B}$ such that $B \subseteq B_1 \cap B_2$.

It follows immediately from (B1)–(B2) and (F1)–(F3) that if $\mathscr{B}$ is a base of a filter then the set

$$\mathscr{F} = \{A \in \mathscr{P}(\mathbb{N}) : A \text{ contains an element of } \mathscr{B}\}$$

is a filter. It is called the *filter generated* by $\mathscr{B}$.

Definition 11.4. A subset $\Omega \subseteq \mathscr{P}(\mathbb{N})$ is said to be *saturated* provided that it has the *finite intersection property*, that is, for any $A_1, A_2, \ldots, A_n \in \Omega$, $n \geq 1$, one has $\bigcap_{i=1}^n A_i \neq \varnothing$.

Note that if Ω is saturated then $\varnothing \notin \Omega$.

Lemma 11.5. *A nonempty subset $\Omega \subseteq \mathscr{P}(\mathbb{N})$ is embeddable in a filter if and only if it is saturated.*

Proof. Suppose that Ω is saturated. Then the set

$$\mathscr{B}(\Omega) := \left\{ \bigcap_{i=1}^n A_i : A_1, A_2, \ldots, A_n \in \Omega, n \geq 1 \right\} \tag{11.1}$$

is a base of a filter. The filter generated by $\mathscr{B}(\Omega)$ is

$$\mathscr{F}(\Omega) := \{A \in \mathscr{P}(X) : \exists A_1, A_2, \ldots, A_n \in \Omega \text{ such that } \bigcap_{i=1}^n A_i \subseteq A\} \tag{11.2}$$

and one clearly has $\Omega \subseteq \mathscr{F}(\Omega)$.

The converse follows immediately from (F1) and (F2). $\square$

If $\Omega \subseteq \mathscr{P}(\mathbb{N})$ is saturated then we say that $\mathscr{F}(\Omega)$ as in (11.2) is the filter *generated* by Ω.

Note that any filter $\mathscr{F} \subseteq \mathscr{P}(\mathbb{N})$ satisfies the following properties:

(F4) if $A \subseteq \mathbb{N}$, then A and $\mathbb{N} \setminus A$ cannot both belong to $\mathscr{F}$;
(F5) $\mathbb{N} \in \mathscr{F}$.

Indeed, (F4) follows from (F1) and (F2) since $A \cap (\mathbb{N} \setminus A) = \varnothing$, and (F5) is an immediate consequence of the fact that $\mathscr{F} \neq \varnothing$ and (F3).

11.2 Ultrafilters

Let $\mathscr{F} \subseteq \mathscr{P}(\mathbb{N})$ be a filter. Consider the set $\Phi_{\mathscr{F}}$ consisting of all filters containing $\mathscr{F}$. This is a nonempty set (as $\mathscr{F} \in \Phi_{\mathscr{F}}$), partially ordered by inclusion. Let $\Phi = (\mathscr{F}_i)_{i \in I}$ be a totally ordered subset of $\Phi_{\mathscr{F}}$. We claim that $\widetilde{\mathscr{F}} := \cup_{i \in I} \mathscr{F}_i$ is an upper bound for Φ. We only have to show that $\widetilde{\mathscr{F}}$ belongs to $\Phi_{\mathscr{F}}$. As $\varnothing \notin \mathscr{F}_i$ for all $i \in I$ we also have $\varnothing \notin \widetilde{\mathscr{F}}$ so that condition (F1) is satisfied. Suppose now that $A_1, A_2, \dots, A_n \in \widetilde{\mathscr{F}}$. Then there exist $i_1, i_2, \dots, i_n \in I$ such that $A_j \in \mathscr{F}_{i_j}$ for all $j = 1, 2, \dots, n$. Let $k \in I$ be such that $k \geq i_1, i_2, \dots, i_n$. We then have $A_1, A_2, \dots, A_n \in \mathscr{F}_k$ and therefore $\bigcap_{j=1}^n A_j \in \mathscr{F}_k$, by (F2). It follows that $\bigcap_{j=1}^n A_j \in \widetilde{\mathscr{F}}$. This shows that $\mathscr{F}$ satisfies condition (F2) as well. Finally, let $A \in \widetilde{\mathscr{F}}$ and $B \subseteq \mathbb{N}$ be such that $A \subseteq B$. Then there exists an $i \in I$ such that $A \in \mathscr{F}_i$ and therefore $B \in \mathscr{F}_i$, by (F3). It then follows that $B \in \widetilde{\mathscr{F}}$ so that $\widetilde{\mathscr{F}}$ also satisfies condition (F3). Thus $\widetilde{\mathscr{F}}$ is a filter. It is also clear that $\widetilde{\mathscr{F}}$ contains $\mathscr{F}$. In other words, $\widetilde{\mathscr{F}} \in \Phi_{\mathscr{F}}$. This shows that $\Phi_{\mathscr{F}}$ is inductive.

By Zorn's lemma, $\Phi_{\mathscr{F}}$ contains a maximal element.

Definition 11.6. A maximal filter $\omega \subseteq \mathscr{P}(\mathbb{N})$ is called an *ultrafilter.*

As an immediate consequence of the above discussion and of Lemma 11.5, we have the following.

Corollary 11.7. *Every filter is contained in some ultrafilter. Moreover a nonempty subset $\Omega \subseteq \mathscr{P}(\mathbb{N})$ is embeddable in an ultrafilter if (and only if) it is saturated.* □

Theorem 11.8. *Let $\omega \in \mathscr{P}(\mathbb{N})$ be a filter on X. Then the following conditions are equivalent:*

(a) *ω is an ultrafilter;*
(b) *for every $A \in \mathscr{P}(\mathbb{N})$ one has either $A \in \omega$ or $(\mathbb{N} \setminus A) \in \omega$.*

Proof. To prove (a) $\Rightarrow$ (b) assume that ω is an ultrafilter. Let $A \in \mathscr{P}(\mathbb{N})$ and suppose that $A' := (\mathbb{N} \setminus A) \notin \omega$. Then, by virtue of Lemma 11.5 and the maximality of ω, the set $\omega \cup \{A'\} \subseteq \mathscr{P}(\mathbb{N})$ is not saturated. Therefore there exist $A_1, A_2, \dots, A_n \in \omega$ such that $(\bigcap_{i=1}^n A_i) \cap A' = \varnothing$, equivalently, $\bigcap_{i=1}^n A_i \subseteq \mathbb{N} \setminus A' = A$. By (F2) we have $\bigcap_{i=1}^n A_i \in \omega$ and by (F3) this yields $A \in \omega$.

Conversely, assume (b) and suppose by contradiction that ω is not maximal. Then we can find a filter $\mathscr{F}$ properly containing ω. Let $A \in \mathscr{F} \setminus \omega$. By (b) one has $(\mathbb{N} \setminus A) \in \omega \subseteq \mathscr{F}$. Since $A \in \mathscr{F}$ and $(\mathbb{N} \setminus A) \in \mathscr{F}$, this contradicts (F4). It follows that ω is maximal, that is, ω is an ultrafilter. This shows (b) $\Rightarrow$ (a). □

Example 11.9 (Principal ultrafilters). Let $n_0 \in \mathbb{N}$. Then the principal filter based at the singleton $\{n_0\}$, which we denote by $\omega_{n_0} := \mathscr{F}_{\{n_0\}}$, is an ultrafilter. Indeed if ω is an ultrafilter containing ω_{n_0} and we had $A \in \omega \setminus \omega_{n_0}$ then necessarily $A \not\ni n_0$. Since $\{n_0\} \in \omega_{n_0} \subseteq \omega$ and $A \cap \{n_0\} = \varnothing$, by (F2) this would contradict (F1) for ω. The element n_0 is called the *principal element* of ω_{n_0} and one says that ω_{n_0} is the *principal ultrafilter based* at n_0. Conversely, suppose that $A \subseteq \mathbb{N}$ is a nonempty set such that the principal filter $\mathscr{F}_A$ is an ultrafilter. Then there exists an $n_0 \in \mathbb{N}$ such that $A = \{n_0\}$. Indeed if we had distinct elements $n_0, n_1 \in A$ then necessarily $\{n_0\} \notin \mathscr{F}_A$ (since $\{n_0\}$ does not contain n_1) and $\mathbb{N} \setminus \{n_0\} \notin \mathscr{F}_A$ (since $\mathbb{N} \setminus \{n_0\}$ does not contain n_0) contradicting the fact that $\mathscr{F}_A$ is an ultrafilter (cf. Theorem 11.8.(b)).

11.3 Free Ultrafilters

Definition 11.10. An ultrafilter which is not principal is called *free* (or *nonprincipal*).

Lemma 11.11. *Let ω be an ultrafilter on $\mathbb{N}$. Let $A_1, A_2, \dots, A_n \in \mathscr{P}(\mathbb{N})$ and suppose that $A := \bigcup_{i=1}^n A_i \in \omega$. Then there exists $1 \leq i \leq n$ such that $A_i \in \omega$.*

Proof. Suppose by contradiction that $A_i' := \mathbb{N} \setminus A_i \in \omega$ for all $i = 1, 2, \dots, n$. Then by (F2) we have $\mathbb{N} \setminus A = \bigcap_{i=1}^n A_i' \in \omega$. Since $A \in \omega$, this contradicts (F4). □

Theorem 11.12. *Let ω be an ultrafilter. The following conditions are equivalent:*

(a) *ω is free;*
(b) *$A \notin \omega$ for all finite subsets $A \subseteq \mathbb{N}$;*
(c) $\bigcap_{A \in \omega} A = \varnothing$.

Proof. Suppose that there exists a finite subset $A = \{a_1, a_2, \dots, a_n\}$ of $\mathbb{N}$ such that $A \in \omega$. As $A = \cup_{i=1}^n \{a_i\}$, by Lemma 11.11 there exists a (unique) $1 \leq i \leq n$ such that $\{a_i\} \in \omega$. Hence, by (F3), $\omega \supseteq \mathscr{F}_{\{a_i\}}$. By maximality of $\mathscr{F}_{\{a_i\}}$ we deduce that $\omega = \mathscr{F}_{\{a_i\}}$, i.e., ω is principal. This shows (a) $\Rightarrow$ (b).

Assume now (b) and set $B := \bigcap_{A \in \omega} A$. Let us show that $B = \varnothing$. If not, let $b \in B$. By hypothesis we have $\{b\} \notin \omega$. It follows from Theorem 11.8 that $\mathbb{N} \setminus \{b\} \in \omega$. Thus B, being the intersection of all subsets $A \subseteq \mathbb{N}$ which belong to ω, satisfies $B \subseteq \mathbb{N} \setminus \{b\}$. But this contradicts the fact that $b \in B$. This shows (b) $\Rightarrow$ (c).

Finally, suppose that ω is principal, say based at $n_0 \in \mathbb{N}$. We then have $\bigcap_{A \in \omega} A = \{n_0\} \neq \varnothing$. This shows (c) $\Rightarrow$ (a). □

Corollary 11.13. *An ultrafilter is free if and only if it contains the Fréchet filter.*

Proof. Let ω be an ultrafilter and let $\mathscr{F}$ denote the Fréchet filter (cf. Example 11.2.(c)). Suppose that ω is free. Then by virtue of Theorem 11.12.(b) and Theorem 11.8.(b) we have $(\mathbb{N} \setminus A) \in \omega$ for all finite subsets $A \subseteq \mathbb{N}$, that is, $\mathscr{F} \subseteq \omega$.

Conversely, suppose that ω is principal so that there exists an $n_0 \in \mathbb{N}$ such that $\omega = \omega_{n_0}$ (cf. Example 11.9), in particular $\{n_0\} \in \omega$. It follows from (F4) that the cofinite set $\mathbb{N} \setminus \{n_0\}$ does not belong to ω. Thus $\mathscr{F} \not\subseteq \omega$. □

Corollary 11.14. *Free ultrafilters exist.*

Proof. Consider the Fréchet filter $\mathscr{F}$. By Corollary 11.7 there exists an ultrafilter ω containing $\mathscr{F}$. By Corollary 11.13 ω is free. □

11.4 Limits along Filters in Metric Spaces

Let (X,d) be a metric space and let $\mathscr{F} \subseteq \mathscr{P}(\mathbb{N})$ be a filter.

Definition 11.15. A sequence $(x_n)_{n\in\mathbb{N}}$ of points in X is said to be *convergent along* $\mathscr{F}$ provided there exists a point $x \in X$ such that the following holds: for every $\varepsilon > 0$ one has $\{n \in \mathbb{N} : d(x_n,x) < \varepsilon\} \in \mathscr{F}$. If this is the case we say that x is a *limit* of the sequence $(x_n)_{n\in\mathbb{N}}$ and we write

$$\lim_{\mathscr{F}} x_n = x$$

(we shall also write $x_n \underset{\mathscr{F}}{\longrightarrow} x$).

Remark 11.16. Let $\mathscr{F}' \subseteq \mathscr{P}(\mathbb{N})$ be another filter and suppose that $\mathscr{F} \subseteq \mathscr{F}'$. Then every sequence $(x_n)_{n\in\mathbb{N}}$ of points in X which is convergent along $\mathscr{F}$ is also convergent along $\mathscr{F}'$ and the two limits are the same, in formulæ

$$\lim_{\mathscr{F}} x_n = x \implies \lim_{\mathscr{F}'} x_n = x.$$

Example 11.17. Let $(x_n)_{n\in\mathbb{N}}$ be a sequence of points in X and let $x \in X$.

(a) Let $n_0 \in \mathbb{N}$ and consider the principal ultrafilter ω_{n_0} based at n_0. Then one has

$$\lim_{\omega_{n_0}} x_n = x \iff x = x_{n_0}.$$

In particular, $(x_n)_{n\in\mathbb{N}}$ is always convergent along ω_{n_0}.

(b) Suppose that $(x_n)_{n\in\mathbb{N}}$ is convergent to x in the usual sense, i.e. for every $\varepsilon > 0$ there exists an $n(\varepsilon) \in \mathbb{N}$ such that $d(x_n,x) < \varepsilon$ for all $n \geq n(\varepsilon)$. Then $(x_n)_{n\in\mathbb{N}}$ is convergent to x along the Fréchet filter. It follows from Remark 11.16 and Corollary 11.13 that $(x_n)_{n\in\mathbb{N}}$ is convergent to x along every free ultrafilter.

(c) Conversely, suppose that $(x_n)_{n\in\mathbb{N}}$ is convergent to x along every free ultrafilter. Let us show that $\lim_{n\to\infty} x_n = x$ (in the usual sense). If this is not the case, then we can find $\varepsilon_0 > 0$ such that for every $n_0 \in \mathbb{N}$ there exists an $n = n(n_0) \geq n_0$ satisfying $d(x_n,x) \geq \varepsilon_0$. The set $A := \{n(n_0) : n_0 \in \mathbb{N}\} \subseteq \mathbb{N}$ is clearly infinite. Now the set $\Omega = \{A \setminus \{n\} : n \in \mathbb{N}\} \subseteq \mathscr{P}(\mathbb{N})$ is saturated and by Corollary 11.7 there exists an ultrafilter ω containing Ω. Note that ω is necessarily free. By (F3) $A \in \omega$. On the other hand $\{n \in \mathbb{N} : d(x_n,x) < \varepsilon_0\} \subseteq \mathbb{N} \setminus A$. Hence by (F1) and (F2) $\{n \in \mathbb{N} : d(x_n,x) < \varepsilon_0\} \notin \omega$. It follows that $x_n \underset{\omega}{\not\to} x$, a contradiction.

Lemma 11.18. *The limit of a converging sequence is unique.*

Proof. Suppose that $(x_n)_{n\in\mathbb{N}}$ is a sequence of points in X converging along a filter $\mathscr{F}$ both to the points x and y in X. Let $\varepsilon > 0$. Then the sets $A := \{n \in \mathbb{N} : d(x_n,x) <$

$\varepsilon\}$ and $B := \{n \in \mathbb{N} : d(x_n, y) < \varepsilon\}$ both belong to $\mathscr{F}$ and therefore, by (F2), their intersection is nonempty. By the triangular inequality (applied to x, y and x_n with $n \in A \cap B$) we then have $d(x,y) < 2\varepsilon$. Since ε was arbitrary, we deduce that $x = y$.

□

Theorem 11.19. *Let (X,d) be a metric space and let $\omega \subseteq \mathscr{P}(\mathbb{N})$ be an ultrafilter. Suppose that X is compact. Then every sequence $(x_n)_{n\in\mathbb{N}}$ of points in X converges along ω.*

By the Bolzano–Weierstass theorem, in the hypotheses of Theorem 11.19, we can (only) deduce the existence of a subsequence $(x_{n_k})_{k\in\mathbb{N}}$ converging in the usual sense, which, by Example 11.17.(b), will also converge along ω. Theorem 11.19 states that, for the convergence along ω of the original sequence, there is no need to extract a subsequence. However, the proof of Theorem 11.19 is essentially the same as that of the Bolzano–Weierstass theorem.

To emphasize the analogies with the Bolzano–Weierstrass theorem, before giving the general proof of the above theorem, we present the special case of compact intervals in $\mathbb{R}$.

Proof of Theorem 11.19 (for compact intervals in $\mathbb{R}$). We consider the special case where $X = [a,b] \subseteq \mathbb{R}$ and $d(x,y) = |x-y|$ for all $x, y \in X$.

Set $a_0 := a$ and $b_0 := b$ and partition the interval $C_0 := [a_0, b_0]$ into the two subintervals $A_0 := [a_0, (a_0+b_0)/2]$ and $B_0 := [(a_0+b_0)/2, b_0]$. Let also $\alpha_0 := \{n \in \mathbb{N} : x_n \in A_0\}$ and $\beta_0 := \{n \in \mathbb{N} : x_n \in B_0\}$, so that $\gamma_0 := \mathbb{N} = \alpha_0 \cup \beta_0$. By virtue of Lemma 11.11 and (F4), at least one of the two subsets α_0 and β_0 belongs to ω. We pick one that is in ω and call it γ_1. We set $C_1 := [a_1, b_1]$ where $a_1 := a_0$ and $b_1 := (a_0+b_0)/2$ if $\gamma_1 = \alpha_0$, while $a_1 := (a_0+b_0)/2$ and $b_1 := b_0$ if $\gamma_1 = \beta_0$. Let then $A_1 := [a_1, (a_1+b_1)/2]$ and $B_1 := [(a_1+b_1)/2, b_1]$ and set $\alpha_1 := \{n \in \mathbb{N} : x_n \in A_1\}$ and $\beta_1 := \{n \in \mathbb{N} : x_n \in B_1\}$ so that $\gamma_1 = \alpha_1 \cup \beta_1$. From Lemma 11.11 and (F4) we again deduce that at least one of the two subsets α_1 and β_1 belongs to ω, we pick one of them and call it γ_2. Continuing in this way, we construct a sequence $C_0 \supseteq C_1 \supseteq \cdots$ of intervals with length $\ell(C_i) = (b-a)2^{-i}$ for all $i \in \mathbb{N}$ and a sequence $\gamma_0 \supseteq \gamma_1 \supseteq \gamma_2 \supseteq \cdots$ of elements in ω such that $\gamma_i := \{n \in \mathbb{N} : x_n \in C_i\}$ for all $i \in \mathbb{N}$. Let x denote the unique element in $\cap_{i\in\mathbb{N}} C_i$ (here we are using the compactness of $X = [a,b]$). Let us show that $(x_n)_{n\in\mathbb{N}}$ converges to x along ω. Fix $\varepsilon > 0$. Then we can find $n_\varepsilon \in \mathbb{N}$ such that $(x-\varepsilon, x+\varepsilon) \supseteq C_{n_\varepsilon}$. It follows that the set $\{n \in \mathbb{N} : x_n \in (x-\varepsilon, x+\varepsilon)\}$ contains $\{n \in \mathbb{N} : x_n \in C_{n_\varepsilon}\} = \gamma_{n_\varepsilon} \in \omega$ so that, by (F3), $\{n \in \mathbb{N} : d(x_n, x) < \varepsilon\} = \{n \in \mathbb{N} : x_n \in (x-\varepsilon, x+\varepsilon)\}$ also belongs to ω. This shows that $\lim_\omega x_n = x$. □

Proof of Theorem 11.19. For every $x \in X$ and $\varepsilon > 0$ let $B(x,\varepsilon) = \{y \in X : d(x,y) < \varepsilon\}$ (resp. $\overline{B(x,\varepsilon)} = \{y \in X : d(x,y) \leq \varepsilon\}$) denote the open (resp. closed) ball of radius ε centered at x.

We recursively construct a decreasing sequence $(X_n)_{n\in\mathbb{N}}$ of compact subsets of X such that $\operatorname{diam}(X_n) \leq 2^{-n+1}$ for all $n \in N$, and such that the sets $A_n := \{k \in \mathbb{N} : x_k \in X_n\}$ for $n \in \mathbb{N}$ form a decreasing sequence of elements of ω, as follows.

We start by setting $X_0 := X$, so that $A_0 := \mathbb{N}$, which is clearly in ω.

Suppose that we have constructed X_n and A_n with $n \geq 0$.

Since X_n is compact, we can find finite $Y_n \subseteq X_n$ such that $X_n \subseteq \cup_{y\in Y_n} B(y, 2^{-n-1})$. Since $A_n = \cup_{y\in Y_n}\{k \in \mathbb{N} : x_k \in \overline{B(y,2^{-n-1})} \cap X_n\}$, by Lemma 11.11 there exists $y_n \in Y_n$ such that $\{k \in \mathbb{N} : x_k \in \overline{B(y_n,2^{-n-1})} \cap X_n\} \in \omega$. Hence we set

$$X_{n+1} := \overline{B(y_n, 2^{-n-1})} \cap X_n,$$

so that $A_{n+1} = \{k \in \mathbb{N} : x_k \in \overline{B(y_n,2^{-n-1})} \cap X_n\} \in \omega$.

Clearly $\operatorname{diam}(X_{n+1}) \leq 2^{-n}$ and $X_{n+1} \subseteq X_n$, so that X_{n+1} is compact. Moreover $A_{n+1} \subseteq A_n$.

Let y be the unique element in $\bigcap_{n\in\mathbb{N}} X_n$, which exists by compactness of X and is unique since $\operatorname{diam}(X_n) \to 0$ as $n \to \infty$.

Fix $\varepsilon > 0$. Then we can find $n_\varepsilon \in \mathbb{N}$ such that $B(y,\varepsilon) \supseteq X_{n_\varepsilon}$. It follows that the set $\{n \in \mathbb{N} : x_n \in B(y,\varepsilon)\}$ contains A_{n_ε}, which is an element of ω. So by (F3) also $\{n \in \mathbb{N} : x_n \in B(y,\varepsilon)\}$ is in ω. Thus $\lim_\omega x_n = y$, and therefore $(x_n)_{n\in\mathbb{N}}$ is convergent along the ultrafilter ω. □

Corollary 11.20. *Let ω be an ultrafilter. Then every bounded sequence $(x_n)_{n\in\mathbb{N}}$ of real numbers is convergent along ω and its limit is unique.* □

Theorem 11.21. *Let (X,d) be a metric space. Then the following conditions are equivalent*

(a) *X is compact;*
(b) *for every free ultrafilter ω one has that every sequence $(x_n)_{n\in\mathbb{N}}$ in X is convergent along ω;*
(c) *there exists a free ultrafilter ω on $\mathbb{N}$ such that every sequence $(x_n)_{n\in\mathbb{N}}$ in X is convergent along ω.*

Proof. The implication (a) $\Rightarrow$ (b) follows immediately from Theorem 11.19. The implication (b) $\Rightarrow$ (c) is trivial. To complete the proof, let us show (c) $\Rightarrow$ (a) by contradiction. Suppose X is not compact. Then either X is not totally bounded, that is, there exists no $\varepsilon > 0$ such that X is covered by a finite collection of open balls of radius ε, or X is not complete. In the first case, for every $\varepsilon > 0$ there exists an infinite sequence $(x_n)_{n\in\mathbb{N}}$ in X such that $d(x_n,x_m) \geq \varepsilon$ for all $n \neq m$. Suppose now that there exists a free ultrafilter ω on which $(x_n)_{n\in\mathbb{N}}$ converges, and set $x := \lim_\omega x_n$. Then, by the triangular inequality, there exists (at most) one $n_0 \in \mathbb{N}$ such that $d(x,x_{n_0}) < \varepsilon/2$. This implies that either $\varnothing \in \omega$ or $\{n_0\} \in \omega$, both contradicting the fact that ω is a free ultrafilter.

Suppose instead that X is not complete, and let $(x_n)_{n\in\mathbb{N}}$ be a Cauchy sequence in X which is not convergent in X. Let us show that if ω is a free ultrafilter on $\mathbb{N}$ then $(x_n)_{n\in\mathbb{N}}$ does not converge along ω. Assume by contradiction that there exists an $x \in X$ such that $\lim_\omega x_n = x$. Thus for every $t \in \mathbb{N}$ the set $A_t := \{n \in \mathbb{N} : d(x_n,x) < 1/t\}$ is in ω, in particular it is infinite, since ω is free, and clearly $A_t \supseteq A_{t+1}$. Note that if the decreasing sequence $(A_t)_{t\in\mathbb{N}}$ eventually stabilizes then the sequence $(x_n)_{n\in\mathbb{N}}$ admits a constant and therefore convergent subsequence in the usual sense. Being Cauchy, it must also be convergent in X, a contradiction.

So there exists a sequence of natural numbers $(t_k)_{k\in\mathbb{N}}$ such that $A_{t_k} \supsetneq A_{t_{k+1}}$ for all $k \in \mathbb{N}$. Thus, if $n_k \in A_{t_k} \setminus A_{t_{k+1}}$ for all $k \in \mathbb{N}$, then we have that the subsequence

$(x_{n_k})_{k\in\mathbb{N}}$ converges (in the usual sense) to x. Again, since $(x_n)_{n\in\mathbb{N}}$ is Cauchy, this means that $(x_n)_{n\in\mathbb{N}}$ itself converges to $x \in X$, a contradiction. □

The set $\ell^\infty(\mathbb{R})$ consisting of all bounded real sequences has the natural structure of a vector space over $\mathbb{R}$ in which addition and scalar multiplication are given by

$$(x+y)_n = x_n + y_n \quad \text{and} \quad (\lambda x)_n = \lambda x_n$$

for all $x = (x_n)_{n\in\mathbb{N}}$ and $y = (y_n)_{n\in\mathbb{N}}$ in $\ell^\infty(\mathbb{R})$, and $\lambda \in \mathbb{R}$. The map

$$x \mapsto \|x\|_\infty := \sup_{n\in\mathbb{N}} |x_n|$$

defines a *norm* on $\ell^\infty(\mathbb{R})$. With this norm, $\ell^\infty(\mathbb{R})$ becomes a real *Banach space*.

It follows from Corollary 11.20 that if ω is an ultrafilter then for every $x = (x_n)_{n\in\mathbb{N}} \in \ell^\infty(\mathbb{R})$ there exists the limit

$$m_\omega(x) := \lim_\omega x_n \in \mathbb{R} \tag{11.3}$$

in fact,

$$m_\omega(x) \in [-\|x\|_\infty, \|x\|_\infty]. \tag{11.4}$$

Theorem 11.22. *Let ω be an ultrafilter. Then the map $m_\omega \colon \ell^\infty(\mathbb{R}) \to \mathbb{R}$ defined by $m_\omega(x) := \lim_\omega x_n \in \mathbb{R}$ is linear, continuous (i.e., bounded), and monotone. In other words we have:*

(i) $\lim_\omega(\lambda x)_n = \lambda \lim_\omega x_n$;
(ii) $\lim_\omega(x+y)_n = \lim_\omega x_n + \lim_\omega y_n$;
(iii) $|\lim_\omega x_n| \leq \|x\|_\infty$;
(iv) $\lim_\omega x_n \leq \lim_\omega z_n$,

for all $\lambda \in \mathbb{R}$ and $x, y, z \in \ell^\infty(\mathbb{R})$ with $x \leq z$ (i.e. $x_n \leq z_n$ for all $n \in \mathbb{N}$).

Proof. Let $x = (x_n)_{n\in\mathbb{N}}, y = (y_n)_{n\in\mathbb{N}}, z = (z_n)_{n\in\mathbb{N}} \in \ell^\infty(\mathbb{R})$. Let also $\varepsilon > 0$.

(i) Let $\lambda \in \mathbb{R}$. We have $\{n \in \mathbb{N} : |\lambda x_n - \lambda m_\omega(x)| < \varepsilon\} = \{n \in \mathbb{N} : |x_n - m_\omega(x)| < \varepsilon/|\lambda|\} \in \omega$, if $\lambda \neq 0$, and $\{n \in \mathbb{N} : |\lambda x_n - \lambda m_\omega(x)| < \varepsilon\} = \mathbb{N} \in \omega$ if $\lambda = 0$. This shows that in either case $m_\omega(\lambda x) = \lim_\omega(\lambda x)_n = \lambda \lim_\omega x_n = \lambda m_\omega(x)$.

(ii) By virtue of the triangle inequality we have that the set

$$A := \{n \in \mathbb{N} : |(x_n + y_n) - (m_\omega(x) + m_\omega(y))| < \varepsilon\}$$

contains

$$\{n \in \mathbb{N} : |x_n - m_\omega(x)| < \varepsilon/2\} \cap \{n \in \mathbb{N} : |y_n - m_\omega(y)| < \varepsilon/2\}$$

so that $A \in \omega$ by virtue of (F3) and (F2). This shows that $m_\omega(x+y) = \lim_\omega(x+y)_n = \lim_\omega x_n + \lim_\omega y_n = m_\omega(x) + m_\omega(y)$.

(iii) This follows immediately from (11.4).

(iv) Suppose $x \leq z$, i.e. $x_n \leq z_n$ for all $n \in \mathbb{N}$, and assume by contradiction that $m_\omega(x) > m_\omega(z)$. Take $\varepsilon := \frac{1}{2}(m_\omega(x) - m_\omega(z))$ and set $A_x := \{n \in \mathbb{N} : |x_n - m_\omega(x)| <$

$\varepsilon\}$ and $A_z := \{n \in \mathbb{N} : |z_n - m_\omega(z)| < \varepsilon\}$. Since $m_\omega(x) = \lim_\omega x_n$ (resp. $m_\omega(z) = \lim_\omega z_n$) we have that both A_x and A_y belong to ω. On the other hand, if $n \in A_x \cap A_y$ we have

$$z_n < m_\omega(z) + \varepsilon = m_\omega(x) - \varepsilon < x_n$$

so that our assumptions force $A_x \cap A_y = \varnothing$, contradicting (F2) and (F1). It follows that $m_\omega(x) = \lim_\omega x_n \leq \lim_\omega z_n = m_\omega(z)$. □

11.5 The Stone–Čech Compactification

We denote by $\beta\mathbb{N}$ the set of all ultrafilters and by $\mathscr{P}^*(\mathbb{N}) := \mathscr{P}(\mathbb{N}) \setminus \{\varnothing\}$ the set of all nonempty subsets of $\mathbb{N}$.

The map

$$\begin{aligned} \mathbb{N} &\to \beta\mathbb{N} \\ n &\mapsto \omega_n \end{aligned}$$

where $\omega_n = \{A \subseteq \mathbb{N} : n \in A\}$ is the principal ultrafilter based at n (cf. Example 11.9), is injective (**exercise**). In other words, we have an embedding

$$\mathbb{N} \hookrightarrow \beta\mathbb{N}. \tag{11.5}$$

Given a nonempty subset $A \subseteq \mathbb{N}$ we denote by

$$V(A) = \{\omega \in \beta\mathbb{N} : A \in \omega\}$$

the set of all ultrafilters containing A.

Recall that given a nonempty set X and $\mathfrak{B} \subseteq \mathscr{P}(X)$, the set of all arbitrary unions of elements of $\mathfrak{B}$ is a topology on X if and only if the following conditions hold:

(B1) $X = \cup_{B \in \mathfrak{B}} B$;
(B2) for all B_1, B_2 in $\mathfrak{B}$ and for each $x \in B_1 \cap B_2$ there exists a $B_3 \in \mathfrak{B}$ such that $x \in B_3 \subseteq B_1 \cap B_2$.

In this case $\mathfrak{B}$ is called a *base* for that topology.

Proposition 11.23. *The collection*

$$\mathscr{V} := \{V(A) : A \in \mathscr{P}^*(\mathbb{N})\} \tag{11.6}$$

constitutes a base for a topology on $\beta\mathbb{N}$.

Proof. First of all, since $\mathbb{N} \in \omega$ for every $\omega \in \beta\mathbb{N}$, we have $V(\mathbb{N}) = \beta\mathbb{N}$, so that $\cup_{A \in \mathscr{P}^*(\mathbb{N})} V(A) = \beta\mathbb{N}$. This shows that (B1) is satisfied by $\mathscr{V}$. Let now $A, B \subseteq \mathbb{N}$ and suppose that $V(A) \cap V(B) \neq \varnothing$. Let then $\omega \in V(A) \cap V(B)$. Let us show that

$$\omega \in V(A \cap B) \subseteq V(A) \cap V(B). \tag{11.7}$$

From our assumptions we deduce that ω contains both A and B, so that by (F2) ω contains $A \cap B$ (note that this implies $A \cap B \neq \varnothing$ by (F1)), equivalently $\omega \in V(A \cap B)$.

This shows the first inclusion in (11.7). Suppose now that $\omega' \in V(A\cap B)$, that is, $A\cap B \in \omega'$. Since both A and B contain $A\cap B$, it follows from (F3) that $A,B \in \omega'$, equivalently $\omega' \in V(A)\cap V(B)$. This shows the second inclusion in (11.7). Thus (B2) is also satisfied. Hence $\mathscr{V}$ is a base. □

Remark 11.24. Note that in the proof of Proposition 11.23, since $\omega \in V(A)\cap V(B)$ was arbitrary, we actually have $V(A\cap B) = V(A)\cap V(B)$, and this condition itself guarantees that $\mathscr{V}$ is a base for a topology on $\beta\mathbb{N}$.

Theorem 11.25. *Let $\mathscr{T}$ denote the topology on $\beta\mathbb{N}$ generated by the base $\mathscr{V}$ defined in* (11.6). *Then the following properties hold.*

(i) *$\mathscr{T}$ induces the discrete topology on its subspace $\mathbb{N}$ (cf. (11.5));*
(ii) *$\mathscr{T}$ is Hausdorff;*
(iii) *$(\beta\mathbb{N},\mathscr{T})$ is compact;*
(iv) *$\mathbb{N}$ is $\mathscr{T}$-dense in $\beta\mathbb{N}$;*
(v) *for every map $f\colon \mathbb{N}\to X$, where (X,d) is a compact metric space, there exists a unique continuous map $\widetilde{f}\colon \beta\mathbb{N}\to X$ extending f.*

Proof. (i) Let $n\in\mathbb{N}$ and let $\omega_n \in \beta\mathbb{N}$ denote the principal ultrafilter based at n. Observe that if $\omega\in\beta\mathbb{N}$ contains $\{n\}$ then, by maximality of ultrafilters, $\omega=\omega_n$. This shows that $V(\{n\})=\{\omega_n\}$. As a consequence, the singleton $\{\omega_n\}$ is open in $\beta\mathbb{N}$ and therefore it is open in the subspace $\mathbb{N}$.

(ii) Let $\omega,\omega'\in\beta\mathbb{N}$ and suppose that $\omega\neq\omega'$. Then there exists an $A\in\mathscr{P}^*(\mathbb{N})$ such that $A\in\omega$ but $A\notin\omega'$ (hence $(\mathbb{N}\setminus A)\in\omega'$). Then the sets $V(A)$ and $V(\mathbb{N}\setminus A)$ constitute open neighborhoods of ω and ω', respectively. Moreover, they are disjoint: if $\omega\in V(A)\cap V(\mathbb{N}\setminus A)$ then ω contains both A and its complement $\mathbb{N}\setminus A$, contradicting (F4).

(iii) Let $\mathscr{U}$ be an open cover of $\beta\mathbb{N}$ and let us show that there exists a finite subcover. Since $\mathscr{V}$ is a base of the topology,

$$\mathscr{U}' := \{V(A) : V(A)\subseteq V \text{ for some } V\in\mathscr{U}\} \tag{11.8}$$

is also an open cover of $\beta\mathbb{N}$. Let us show that $\mathscr{U}'$ admits a finite subcover. First observe that the set $\mathscr{A} := \{A\in\mathscr{P}^*(\mathbb{N}) : V(A)\in\mathscr{U}'\}$ is a cover of $\mathbb{N}$. Indeed if $n\in\mathbb{N}$ then we can find $V\in\mathscr{U}$ containing ω_n and if $A\in\mathscr{P}^*(\mathbb{N})$ is such that $\omega_n\in V(A)\subseteq V$, so that $A\in\mathscr{A}$, then necessarily $n\in A$ (in fact $\omega_n\in V(A)\Leftrightarrow A\in\omega_n\Leftrightarrow n\in A$). Let us show that $\mathscr{A}$ admits a finite subcover. Suppose by contradiction that $\mathscr{A}$ admits no finite subcover and observe (**exercise**) that this is equivalent to saying that the set

$$\Omega := \{\mathbb{N}\setminus A : A\in\mathscr{A}\}\subseteq\mathscr{P}^*(\mathbb{N})$$

is saturated. By Corollary 11.7, we can then find an ultrafilter ω containing Ω. This means that $(\mathbb{N}\setminus A)\in\omega$, equivalently $A\notin\omega$, for all $A\in\mathscr{A}$. This in turn may be expressed by saying that $\omega\notin V(A)$ for all $A\in\mathscr{A}$, that is, $\omega\notin\cup_{A\in\mathscr{A}}V(A) = \cup_{V'\in\mathscr{U}'}V'$, contradicting the fact that $\mathscr{U}'$ covers $\beta\mathbb{N}$. We deduce that there exist $n\in\mathbb{N}$ and $A_1,A_2,\ldots,A_n\in\mathscr{A}$ such that $A_1\cup A_2\cup\cdots\cup A_n=\mathbb{N}$.

Let now $\omega\in\beta\mathbb{N}$. By virtue of Lemma 11.11 there exists $1\leq i\leq n$ such that $A_i\in\omega$, equivalently $\omega\in V(A_i)$. Thus

$$V(A_1) \cup V(A_2) \cup \cdots \cup V(A_n) = \beta\mathbb{N}, \tag{11.9}$$

showing that $\mathscr{U}'$ admits a finite subcover. To show that $\mathscr{U}$ itself admits a finite subcover and therefore to end the proof, it suffices to observe that, by definition, for every $i = 1, 2, \ldots, n$ there exists a $V_i \in \mathscr{U}$ such that $V(A_i) \subseteq V_i$ so that, by (11.9), $V_1 \cup V_2 \cup \cdots \cup V_n = \beta\mathbb{N}$.

(iv) Let $U \in \mathscr{T}$ be nonempty and let us show that there exists an $n \in \mathbb{N}$ such that $\{\omega_n\} \subseteq U$. Since $\mathscr{V}$ is a base, we can find a nonempty subset $A \subseteq \mathbb{N}$ such that $V(A) \subseteq U$. Taking any $n \in A$ we have $A \in \omega_n$ by (F3) and therefore $\{\omega_n\} = V(\{n\}) \subseteq V(A) \subseteq U$.

(v) For every $\omega \in \beta\mathbb{N}$ let us set

$$\widetilde{f}(\omega) := \lim_{\omega} f(n). \tag{11.10}$$

Note that (11.10) exists and is well defined since X is compact (cf. Lemma 11.18 and Theorem 11.19). Note that if $n_0 \in \mathbb{N}$ then by Example 11.17.(a) we have

$$\widetilde{f}(\omega_{n_0}) = \lim_{\omega_{n_0}} f(n) = f(n_0)$$

thus showing that $\widetilde{f}$ extends f. We are only left to prove continuity of $\widetilde{f}$. Let $\overline{\omega} \in \beta\mathbb{N}$ and set $x_0 := \widetilde{f}(\overline{\omega}) \in X$. Let also $\varepsilon > 0$ and consider the open neighborhood $B(x_0, \varepsilon)$ of x_0. By definition of limit along $\overline{\omega}$, the set $A := \{n \in \mathbb{N} : f(n) \in B(x_0, \varepsilon/2)\}$ belongs to $\overline{\omega}$. Consider the basic open set $V(A) \subseteq \beta\mathbb{N}$ and observe that $\overline{\omega} \in V(A)$. Given $\omega \in V(A)$ we have $A \in \omega$ and therefore $\widetilde{f}(\omega) = \lim_{\omega} f(n) \in \overline{B(x_0, \varepsilon/2)} \subseteq B(x_0, \varepsilon)$ (**exercise**: let (X,d) be a metric space, and let $(x_n)_{n\in\mathbb{N}}$ be a sequence in X converging along an ultrafilter ω. Suppose that there exists a ball $B \subseteq X$ such that $\{n \in \mathbb{N} : x_n \in B\} \in \omega$. Show that $\lim_{\omega} x_n \in \overline{B}$.) This shows that $\widetilde{f}(V(A)) \subseteq B(x_0, \varepsilon)$. Thus the continuity of the extension $\widetilde{f}$ of f follows. Uniqueness of $\widetilde{f}$ is left as an **exercise.** □

By Theorem 11.25.(iii) the topological space $(\beta\mathbb{N}, \mathscr{T})$ is compact. It is called the *Stone–Čech compactification* of $\mathbb{N}$.

11.6 The Completion of a Metric Space

Let (X,d) be a metric space.

Recall that a sequence $(x_n)_{n\in\mathbb{N}}$ in X is *bounded* provided it satisfies one of the following equivalent (**exercise**) conditions:

(i) $\sup_{n,m\in\mathbb{N}} d(x_n, x_m) < \infty$;
(ii) there exists an $x^0 \in X$ such that $\sup_{n\in\mathbb{N}} d(x^0, x_n) < \infty$;
(iii) for every $x^0 \in X$ one has $\sup_{n\in\mathbb{N}} d(x^0, x_n) < \infty$.

Also recall that (X,d) is said to be *totally bounded* provided that for every $\varepsilon > 0$, there exists a finite collection of open balls in X of radius ε whose union contains X. Clearly a totally bounded space is bounded (as the union of finitely many bounded sets is bounded), but the converse is not true in general. For example, an infinite set equipped with the discrete metric is bounded but not totally bounded. On the other hand, for $(\mathbb{R}^n, d_n)$, where d_n is the Euclidean distance, a subset (with the induced distance) is totally bounded if and only if it is bounded.

We denote by $\ell^\infty(X) = \ell^\infty(X,d) \subseteq X^{\mathbb{N}}$ the set of all bounded sequences in X. We also denote by $\mathrm{Cau}(X) = \mathrm{Cau}(X,d) \subseteq \ell^\infty(X)$ the set of all Cauchy sequences in X.

Given $x = (x_n)_{n\in\mathbb{N}}$ and $y = (y_n)_{n\in\mathbb{N}}$ in $\mathrm{Cau}(X)$ we write $x \sim y$ provided that $\lim_{n\to\infty} d(x_n, y_n) = 0$. It is easy to see (**exercise**) that $\sim$ is an equivalence relation on $\mathrm{Cau}(X)$ and that, denoting by $[x] := \{y \in \mathrm{Cau}(X) : y \sim x\}$ the equivalence class of $x \in \mathrm{Cau}(X)$, the quantity

$$\tilde{d}([x],[y]) := \lim_{n\to\infty} d(x_n, y_n)$$

is well defined (note that the limit exists since $(d(x_n,y_n))_{n\in\mathbb{N}}$ is a Cauchy sequence in $\mathbb{R}$ (**exercise**) and $\mathbb{R}$ is complete) and yields a metric on the corresponding quotient space $\tilde{X} := \mathrm{Cau}(X)/\sim$. The metric space $(\tilde{X}, \tilde{d})$ is called the *metric completion* (or *Cauchy completion*) of (X,d).

The proof of the following proposition is left as an **exercise**.

Proposition 11.26. *Let (X,d) be a metric space and denote by $(\tilde{X},\tilde{d})$ its metric completion. Then the following holds:*

(i) *the map $x \mapsto [(x_n)_{n\in\mathbb{N}}]$, where $x_n = x$ for all $n \in \mathbb{N}$, yields an isometric embedding of X into $\tilde{X}$;*
(ii) *X is dense in $\tilde{X}$;*
(iii) *X is totally bounded if and only if $\tilde{X}$ is totally bounded.*

Moreover, a subset of a complete metric space is totally bounded if and only if it is pre-compact (**exercise**): for this reason one also calls a totally bounded metric space *pre-compact*.

Proposition 11.27. *Let (X,d) be a metric space. Then the following conditions are equivalent:*

(a) *(X,d) isometrically embeds into a metric space $(\hat{X},\hat{d})$ in which every bounded sequence converges along every free ultrafilter;*
(b) *every ball in X is totally bounded.*

In this case, one may take $\hat{X} = \tilde{X}$, the metric completion of X.

Proof. Suppose that every ball B in X is totally bounded. Then every ball $\tilde{B}$ in the metric completion $\tilde{X}$ of X is totally bounded and therefore compact (cf. Proposition 11.26). Let $x = (x_n)_{n\in\mathbb{N}}$ be a bounded sequence in $\tilde{X}$, then there exists a ball $\tilde{B}$ such that $x_n \in \tilde{B}$ for all $n \in \mathbb{N}$. Then by virtue of Theorem 11.21 the sequence x converges in $\tilde{B}$ (and therefore in $\tilde{X}$) along every free ultrafilter.

Conversely, suppose there exists a ball B in X which is not totally bounded, so that there exists $\varepsilon_0 > 0$ and a sequence $(x_n)_{n\in\mathbb{N}}$ contained in B, and therefore bounded, such that

$$d(x_n, x_m) \geq \varepsilon_0 \text{ for all distinct } n,m \in \mathbb{N}. \tag{11.11}$$

Let $(\hat{X}, \hat{d})$ be a metric completion of X, so that we have an isometric embedding $X \hookrightarrow \hat{X}$. Then we may regard $(x_n)_{n \in \mathbb{N}}$ as a sequence in $\hat{X}$. It immediately follows from (11.11) and the fact that $\hat{d}(x,y) = d(x,y)$ for all $x,y \in X$ that $(x_n)_{n \in \mathbb{N}}$ does not converge in $\hat{X}$ along any free ultrafilter (**exercise**). □

11.7 Ultrapowers of Metric Spaces

Let (X,d) be a metric space.

In view of Proposition 11.27, we cannot, in general, find a metric extension $\widetilde{X}$ of X such that, for every bounded sequence $(x_n)_{n \in \mathbb{N}}$ in X there exists an ultralimit $\lim_\omega x_n$ in $\widetilde{X}$. The following construction, however, enables us to associate with any bounded sequence in X an ideal new point which will be the limit (in the usual sense, and therefore the ultralimit along any ultrafilter) in the case of a Cauchy sequence.

We equip the set $\ell^\infty(X)$ consisting of all bounded sequences in X with the metric d_∞ defined by

$$d_\infty(x,y) := \sup_{n \in \mathbb{N}} d(x_n, y_n) \tag{11.12}$$

for all $x = (x_n)_{n \in \mathbb{N}}$ and $y = (y_n)_{n \in \mathbb{N}}$ in $\ell^\infty(X)$. Note that the quantity in (11.12) is finite by virtue of the triangle inequality (**exercise**).

It can be shown that $(\ell^\infty(X), d_\infty)$ is complete if and only if (X,d) is complete (**exercise**).

Let now ω be an ultrafilter. For $x = (x_n)_{n \in \mathbb{N}}$ and $y = (y_n)_{n \in \mathbb{N}}$ in $\ell^\infty(X)$ we write

$$x \sim_\omega y \quad \text{provided} \quad \lim_\omega d(x_n, y_n) = 0. \tag{11.13}$$

The relation $\sim_\omega$ defined by (11.13) is an equivalence relation on $\ell^\infty(X)$ (**exercise**). We then denote by

$$[x]_\omega := \{y \in \ell^\infty(X) : y \sim_\omega x\}$$

the equivalence class of x, and by $X_\omega := \ell^\infty(X)/\sim_\omega$ the corresponding quotient space. Also, the quantity

$$d_\omega([x]_\omega, [y]_\omega) := \lim_\omega d(x_n, y_n) \tag{11.14}$$

is well defined: if $x' = (x'_n)_{n \in \mathbb{N}}$ and $y' = (y'_n)_{n \in \mathbb{N}}$ in $\ell^\infty(X)$ satisfy $x' \sim_\omega x$ and $y' \sim_\omega y$, then $\lim_\omega d(x_n, y_n) = \lim_\omega d(x'_n, y'_n)$ (**exercise**). Moreover, $d_\omega([x]_\omega, [y]_\omega) = \inf_{\substack{x' \sim_\omega x \\ y' \sim_\omega y}} d_\infty(x', y')$, and d_ω defines a metric on X_ω (**exercise**).

Definition 11.28. The metric space (X_ω, d_ω) is called the *ultrapower* of (X,d) with respect to the ultrafilter ω.

It is easy to show that the map

$$\begin{aligned} \iota_\omega : X &\to X_\omega \\ x &\mapsto [(x,x,\dots)]_\omega \end{aligned}$$

is an isometric embedding of (X,d) into (X_ω, d_ω). Moreover, ι_ω is surjective if and only if X is compact (**exercise**).

Remark 11.29. Suppose that the ultrafilter ω is principal. Then there is an element $n_0 \in \mathbb{N}$ such that ω consists of all subsets of $\mathbb{N}$ containing n_0. Thus for all $x = (x_n)_{n\in\mathbb{N}}$ and $y = (y_n)_{n\in\mathbb{N}}$ in $\ell^\infty(X)$ we have $x \sim_\omega y$ if and only if $x_{n_0} = y_{n_0}$. This implies that the map ι_ω is also surjective (given $[x]_\omega \in X_\omega$, we have $[x]_\omega = \iota_\omega(x_{n_0})$). It follows that the metric spaces X_ω and X are canonically isometric.

Note that given a bounded sequence $(x_n)_{n\in\mathbb{N}}$ in X, the associated "ideal new point" we alluded to above is nothing but $[(x_n)_{n\in\mathbb{N}}]_\omega \in X_\omega$. Moreover, this will be the limit in case of a Cauchy sequence, as the proof of the following theorem shows.

Theorem 11.30. *Suppose that (X,d) is a complete metric space. Then the ultrapower (X_ω, d_ω) is also complete.*

Proof. We start by showing the following useful result which we shall recursively apply in the proof.

Claim. *Let $x, y \in X_\omega$ and let $A > 0$ be such that $D := d_\omega(x,y) < A$. Then we can find $(x_n)_{n\in\mathbb{N}}$ and $(y_n)_{n\in\mathbb{N}}$ in $\ell^\infty(X)$ such that $x = [(x_n)_{n\in\mathbb{N}}]_\omega$, $y = [(y_n)_{n\in\mathbb{N}}]_\omega$ and $d(x_n, y_n) < A$ for all $n \in \mathbb{N}$.*

Indeed, let $(x_n)_{n\in\mathbb{N}}$ and $(y'_n)_{n\in\mathbb{N}}$ in $\ell^\infty(X)$ be such that $x = [(x_n)_{n\in\mathbb{N}}]_\omega$, $y = [(y'_n)_{n\in\mathbb{N}}]_\omega$ and set $S := \{n \in \mathbb{N} : d(x_n, y'_n) < A\}$. We claim that $S \in \omega$. Indeed, if $\varepsilon := (A-D)/2$ we have, by definition, $S_\varepsilon := \{n \in \mathbb{N} : D - \varepsilon < d(x_n, y'_n) < D + \varepsilon\} \in \omega$. But since $D + \varepsilon < A$ we have $S_\varepsilon \subseteq \{n \in \mathbb{N} : d(x_n, y'_n) < D + \varepsilon\} \subseteq S$ so that by (F3) we have $S \in \omega$. Thus setting $y_n := y'_n$ for $n \in S$ and $y_n := x_n$ for $n \notin S$ we have that $[(y_n)_{n\in\mathbb{N}}]_\omega = y$ and $d(x_n, y_n) < A$ holds for all $n \in \mathbb{N}$. The claim follows.

Let now $(x^i)_{i\in\mathbb{N}}$ be a Cauchy sequence in X_ω. Possibly disregarding a finite number of initial terms, it is not restrictive to suppose that $d_\omega(x^j, x^i) < 1$ for all $i, j \in \mathbb{N}$. For all $i \in \mathbb{N}$, by applying the claim with $x = x^1$, $y = x^i$ and $A = 1$, we can find $(x^i_n)_{n\in\mathbb{N}} \in \ell^\infty(X)$ such that $x^i = [(x^i_n)_{n\in\mathbb{N}}]_\omega$ and $d(x^1_n, x^i_n) < 1$ for all $n \in \mathbb{N}$. By the triangle inequality, we therefore have $d(x^j_n, x^i_n) < 2$ for all $i, j, n \in \mathbb{N}$.

Let now ℓ be the first index $\ell > 1$ such that $d_\omega(x^i, x^j) < 1/2$ for all $i, j \geq \ell$. By applying the claim with $x = x^\ell$, $y = x^i$ for $i \geq \ell$, and $A = 1/2$, we can suppose that the sequences $(x^i_n)_{n\in\mathbb{N}} \in \ell^\infty(X)$ satisfy, in addition, $d(x^\ell_n, x^i_n) < 1/2$ for all $i, n \in \mathbb{N}$ such that $i \geq \ell$ (note that since in the process indicated by the claim we may have just replaced some x^i_n by x^1_n, the inequalities from the previous paragraph remain true). Again, by the triangle inequality, we therefore have $d(x^j_n, x^i_n) < 1$ for all $i, j, n \in \mathbb{N}$ such that $i, j \geq \ell$.

Continuing in this way, it is therefore possible to choose the sequences $(x^i_n)_{n\in\mathbb{N}} \in \ell^\infty(X)$, $i \in \mathbb{N}$, in a such a way that for every $\varepsilon > 0$ there exists an $\ell_\varepsilon \in \mathbb{N}$ such that

$$d(x^j_n, x^i_n) < \varepsilon \quad \text{for all } i, j, n \in \mathbb{N} \text{ such that } i, j \geq \ell_\varepsilon. \tag{11.15}$$

Fix $n \in \mathbb{N}$. It follows immediately from (11.15) that the sequence $(x^i_n)_{i\in\mathbb{N}}$ is Cauchy in X. Since (X,d) is complete, there exists $x_n := \lim_{i\to\infty} x^i_n \in X$.

Consider the sequence $x := (x_n)_{n\in\mathbb{N}}$ and let us show that $x \in \ell^\infty(X)$ and $[x]_\omega = \lim_{i\to\infty} x^i$ in X_ω.

Fix $\varepsilon > 0$ and let ℓ_ε as above. Letting $j \to \infty$ in (11.15) we obtain

$$d(x_n, x_n^i) < \varepsilon \text{ for all } i, n \in \mathbb{N} \text{ such that } i \geq \ell_\varepsilon. \tag{11.16}$$

In order to prove the first assertion, fix $i \geq \ell_\varepsilon$ and let $A > 0$ be such that $d(x_0, x_n^i) < A$ for all $n \in \mathbb{N}$. Then, using the triangle inequality, from (11.16) we deduce $d(x_0, x_n) \leq d(x_0, x_n^i) + d(x_n, x_n^i) < (A + \varepsilon)$ for all $n \in \mathbb{N}$, thus showing that the sequence $(x_n)_{n\in\mathbb{N}}$ is in $\ell^\infty(X)$.

On the other hand, still from (11.16) we deduce that

$$d_\omega([x]_\omega, x^i) = \lim_\omega d(x_n, x_n^i) < \varepsilon$$

for all $i \geq \ell_\varepsilon$. Since ε was arbitrary, this shows that $[x]_\omega = \lim_{i\to\infty} x^i$. Thus the Cauchy sequence $(x^i)_{i\in\mathbb{N}}$ converges. It follows that (X_ω, d_ω) is complete. □

11.8 Ultraproducts of Sequences of Pointed Metric Spaces

A *pointed metric space* is a triple (X, x^0, d) where (X, d) is a metric space and $x^0 \in X$ is a fixed point, called a *base point*.

Let $\mathbf{X} = ((X_n, x_n^0, d_n))_{n\in\mathbb{N}}$ be a sequence of pointed metric spaces and let $\omega \in \beta\mathbb{N}$ be an ultrafilter.

Consider the set $\ell^\infty(\mathbf{X})$ consisting of all sequences $(x_n)_{n\in\mathbb{N}}$ such that $x_n \in X_n$ for all $n \in \mathbb{N}$ and $d_\infty((x_n)_{n\in\mathbb{N}}, (x_n^0)_{n\in\mathbb{N}}) := \sup_{n\in\mathbb{N}} d_n(x_n, x_n^0) < \infty$. For instance, if (X, x^0, d) is a pointed metric space and $(X_n, x_n^0, d_n) = (X, x^0, d)$ for all $n \in \mathbb{N}$, then we clearly have $\ell^\infty(\mathbf{X}) = \ell^\infty(X)$.

As in the previous section, with $\ell^\infty(\mathbf{X})$ playing the role of $\ell^\infty(X)$, we define an equivalence relation $\sim_\omega$ on $\ell^\infty(\mathbf{X})$ and a distance function d_ω on

$$\mathbf{X}_\omega := \ell^\infty(\mathbf{X}) / \sim_\omega$$

as follows: for $x = (x_n)_{n\in\mathbb{N}}$ and $y = (y_n)_{n\in\mathbb{N}}$ in $\ell^\infty(\mathbf{X})$, we write $x \sim_\omega y$ provided $\lim_\omega d_n(x_n, y_n) = 0$ and we set $d_\omega([x]_\omega, [y]_\omega) := \lim_\omega d_n(x_n, y_n)$, where $[x]_\omega := \{x' \in \ell^\infty(\mathbf{X}) : x' \sim_\omega x\} \in \mathbf{X}_\omega$ is the $\sim_\omega$ equivalence class of x.

Definition 11.31. The metric space $(\mathbf{X}_\omega, d_\omega)$ is called the *ultraproduct* of the sequence $\mathbf{X} = (X_n, x_n^0, d_n)_{n\in\mathbb{N}}$ of pointed metric spaces, with respect to the ultrafilter ω.

Example 11.32. Let (X, x^0, d) be a pointed metric space and $(X_n, x_n^0, d_n) = (X, x^0, d)$ for all $n \in \mathbb{N}$. It is clear that $\mathbf{X}_\omega$ coincides with X_ω, the ultrapower of X.

Given a set X, recall that the *discrete metric* is the metric d^* defined by $d^*(x, y) = 0$ if $x = y$ and $d^*(x, y) = 1$ otherwise, for all $x, y \in X$. Note that $\ell^\infty(X, d^*) = X^\mathbb{N}$.

Example 11.33. Let $(X_n)_{n\in\mathbb{N}}$ be a sequence of sets. Equip each X_n with the discrete metric d_n^* and pick a fixed point x_n^0 in each X_n. Observe that the sequence $\mathbf{X} = ((X_n, x_n^0, d_n^*))_{n\in\mathbb{N}}$ of pointed metric spaces has uniformly bounded diameter (≤ 1), so clearly $\ell^\infty(\mathbf{X}) = \prod_{n\in\mathbb{N}} X_n$. Moreover, given $x = (x_n)_{n\in\mathbb{N}}$ and $y = (y_n)_{n\in\mathbb{N}}$ in $\prod_{n\in\mathbb{N}} X_n$ we have $x \sim_\omega y$ if and only if there exists an $A \in \omega$ such that $x_n = y_n$ for all $n \in A$ (**exercise**). Moreover, d_∞^* (resp. d_ω^*) coincides with the discrete metric on $\prod_{n\in\mathbb{N}} X_n$ (resp. $\mathbf{X}_\omega$) (**exercise**).

11.9 Ultraproducts of Groups

Let G be a group and let $d\colon G \times G \to [0, +\infty)$ be a metric on G. One says that the metric d is *left-* (resp.*right-*) *invariant* provided that $d(hg_1, hg_1) = d(g_1, g_2)$ (resp. $d(g_1h, g_1h) = d(g_1, g_2)$) for all $h, g_1, g_2 \in G$; one then refers to the pair (G, d) as to a *left* (resp. *right*) *metric group*. A metric d which is both left and right invariant is termed *bi-invariant* and in this case one refers to (G, d) simply as a *metric group*.

Note that in a left or right metric group (G, d) one has

$$d(1_G, g^{-1}) = d(1_G, g) \tag{11.17}$$

for all $g \in G$, while if d is bi-invariant, then

$$d(g, h) = d(g^{-1}, h^{-1}) \tag{11.18}$$

and

$$d(hg_1h^{-1}, hg_2h^{-1}) = d(g_1, g_2) \tag{11.19}$$

for all $g, h, g_1, g_2 \in G$ (**exercise**).

Example 11.34. (a) Let G be a finitely generated group and let $S \subseteq G$ be a finite generating subset. Then (G, d_S) (cf. (7.3)) is a left metric group. In general, d_S is not right invariant
(b) Let G be any group and denote by d^* the *discrete metric*. Then (G, d^*) is a metric group (**exercise**).
(c) Let $G = S_n$ denote the *symmetric group* on n letters: this is the group of all *permutations* (= bijective self-maps) of a set X of cardinality n. Also denote by $d_n^H\colon S_n \times S_n \to [0, 1]$ the (normalized) *Hamming distance*. This is defined by

$$d_n^H(g, h) := \frac{n - |\{x \in X : g(x) = h(x)\}|}{n} \equiv \frac{|\{x \in X : g(x) \neq h(x)\}|}{n} \tag{11.20}$$

for all $g, h \in S_n$. Then (S_n, d_n^H) is a metric group (**exercise**).
(d) Let $U_n \subset M_n(\mathbb{C})$ denote the *unitary group* consisting of all $n \times n$ *unitary* matrices. Recall that a matrix $A \in M_n(\mathbb{C})$ is called *unitary* provided $A^*A = I = AA^*$, where $A^* = \overline{A}^T$ is the *conjugate-transpose* (or *adjoint*) of A and I denotes the identity matrix. The *Hilbert–Schmidt norm* of a matrix $A = (a_{i,j})_{i,j=1}^n \in M_n(\mathbb{C})$ is defined as the non-negative number

$$\|A\|_{HS} := \sqrt{\sum_{i,j=1}^{n} |a_{i,j}|^2}.$$

For $A, B \in U_n$ we set $d_n^{HS}(A,B) := \|A - B\|_{HS}$. Then (U_n, d_n^{HS}) is a metric group (**exercise**).

(e) Let $\mathbb{F}$ be a field and let $\mathrm{GL}(n,\mathbb{F}) \subseteq M_n(\mathbb{F})$ denote the *general linear group* consisting of all $n \times n$ invertible matrices with coefficients in $\mathbb{F}$. The (normalized) *rank* of a matrix $A \in M_n(\mathbb{F})$ is defined as the non-negative number

$$\rho_n(A) := \frac{\dim \operatorname{coker}(A)}{n} = 1 - \frac{\dim \ker(A)}{n}.$$

For $A, B \in M_n(\mathbb{F})$ we set $d_n(A,B) := \rho_n(A - B)$. Then $(\mathrm{GL}(n,\mathbb{F}), d_n)$ is a metric group. (**exercise**).

Given a metric group (G,d) we implicitly choose as a base point the identity element $1_G \in G$.

Thus, given a sequence $((G_n, d_n))_{n\in\mathbb{N}}$ of metric groups we use the notation $\mathbf{G} = ((G_n, 1_{G_n}, d_n))_{n\in\mathbb{N}}$. Given an ultrafilter $\omega \in \beta\mathbb{N}$ we may construct the ultraproduct $\mathbf{G}_\omega$ along the lines of the previous section. In the following, we show that $(\mathbf{G}_\omega, d_\omega)$ is a metric group.

We first observe that $\ell^\infty(\mathbf{G})$ has a natural group structure defined as follows. Given $x = (x_n)_{n\in\mathbb{N}}$ and $y = (y_n)_{n\in\mathbb{N}}$ in $\ell^\infty(\mathbf{G})$ we define $z := xy \in \ell^\infty(\mathbf{G})$ by setting $z_n = x_n y_n \in G_n$ for all $n \in \mathbb{N}$. Indeed,

$$\begin{aligned} d_n(z_n, 1_{G_n}) &= d_n(x_n y_n, 1_{G_n}) \\ \text{(by left invariance)} &= d_n(y_n, x_n^{-1}) \\ \text{(by the triangular inequality)} &\le d_n(y_n, 1_{G_n}) + d_n(1_{G_n}, x_n^{-1}) \\ \text{(by (11.17))} &= d_n(y_n, 1_{G_n}) + d_n(x_n, 1_{G_n}), \end{aligned}$$

thus showing that $z \in \ell^\infty(\mathbf{G})$. Moreover, $x^{-1} := (x_n^{-1})_{n\in\mathbb{N}}$ is the inverse of x and $1 := (1_{G_n})_{n\in\mathbb{N}}$ is the neutral element.

Lemma 11.35. *Let $x, x', y, y' \in \ell^\infty(\mathbf{G})$. Then one has*

$$x \sim_\omega x' \text{ and } y \sim_\omega y' \implies xy \sim_\omega x'y' \tag{11.21}$$

and

$$x \sim_\omega y \Rightarrow x^{-1} \sim_\omega y^{-1}. \tag{11.22}$$

Consider the subset $N_\omega \subseteq \ell^\infty(\mathbf{G})$ defined by

$$N_\omega := \{x \in \ell^\infty(\mathbf{G}) : x \sim_\omega 1\}.$$

The proof of the following proposition is left as an **exercise**.

Proposition 11.36. *The set N_ω is a normal subgroup of $\ell^\infty(\mathbf{G})$.*

Observe that, given x and y in $\ell^\infty(\mathbf{G})$, one has

$$xN_\omega = yN_\omega \iff x \sim_\omega y. \tag{11.23}$$

Indeed, one has $xN_\omega = yN_\omega$ if and only if $xy^{-1} \in N_\omega$, that is, if and only if $xy^{-1} \sim_\omega 1$. This is equivalent to $x \sim_\omega y$ by (11.21). As a consequence, the ultraproduct $\mathbf{G}_\omega$ of the sequence of metric groups $(G_n, d_n)_{n\in\mathbb{N}}$ is itself a group, called the *ultraproduct group*, and one has

$$\mathbf{G}_\omega = \ell^\infty(\mathbf{G})/N_\omega.$$

Moreover, the metric d_ω on $\mathbf{G}_\omega$ given by

$$d_\omega(xN_\omega, yN_\omega) := \lim_\omega d_n(x_n, y_n) \tag{11.24}$$

for all $x = (x_n)_{n\in\mathbb{N}}$ and $y = (y_n)_{n\in\mathbb{N}}$ in $\ell^\infty(\mathbf{G})$, is bi-invariant (**exercise**).

From the discussion above we deduce the following:

Theorem 11.37. *Let* $\mathbf{G} = ((G_n, 1_{G_n}, d_n))_{n\in\mathbb{N}}$ *be a sequence of (pointed) metric groups. Then the ultraproduct* $(\mathbf{G}_\omega, d_\omega)$ *is a metric group.* □

Example 11.38. Let $\omega \in \beta\mathbb{N}$ be an ultrafilter.
(a) Let $\mathbf{G} := (S_n, 1, d_n^H)_{n\in\mathbb{N}}$ (cf. Example 11.34.(c)). It follows from Theorem 11.37 that the ultraproduct $(\mathbf{G}_\omega, d_\omega)$ is a metric group. It is called the *universal sofic group*.
(b) Let $\mathbf{G} := (U_n, 1, d_n^{HS})_{n\in\mathbb{N}}$ (cf. Example 11.34.(d)). It follows from Theorem 11.37 that the ultraproduct $(\mathbf{G}_\omega, d_\omega)$ is a metric group. It is called the *universal hyperlinear group*.
(c) Let $\mathbf{G} := (\mathrm{GL}(n, \mathbb{F}), I, \rho_n)_{n\in\mathbb{N}}$ (cf. Example 11.34.(e)). It follows from Theorem 11.37 that the ultraproduct $(\mathbf{G}_\omega, d_\omega)$ is a metric group. It is called the *universal linearly-sofic group*.

11.10 Ultrafields

Let $\mathbb{F}$ be a field and let $\mathscr{A}$ be a unital $\mathbb{F}$-algebra.

Let $\mathscr{A}^\mathbb{N} := \{(x_n)_{n\in\mathbb{N}} : x_n \in \mathscr{A} \text{ for all } n \in \mathbb{N}\}$ denote the space of all sequences in $\mathscr{A}$. Given $x = (x_n)_{n\in\mathbb{N}}$ and $y = (y_n)_{n\in\mathbb{N}}$ in $\mathscr{A}^\mathbb{N}$, and $a \in \mathscr{A}$, we define the elements $x+y$, xy, and ax in $\mathscr{A}^\mathbb{N}$ by setting

$$x+y := (x_n + y_n)_{n\in\mathbb{N}}, \quad xy := (x_n y_n)_{n\in\mathbb{N}} \quad \text{and} \quad ax = (ax_n)_{n\in\mathbb{N}}. \tag{11.25}$$

It is easy to check (**exercise**) that, with the operations in (11.25), $\mathscr{A}^\mathbb{N}$ is a unital $\mathbb{F}$-algebra. For example, $\mathbb{F}^\mathbb{N}$ is a unital commutative $\mathbb{F}$-algebra. We denote by $0 := (0_\mathscr{A})_{n\in\mathbb{N}}$ (resp. $1 := (1_\mathscr{A})_{n\in\mathbb{N}}$) the corresponding zero (resp. unit) element.

Let now ω be an ultrafilter. As usual, given $x = (x_n)_{n\in\mathbb{N}}$ and $y = (y_n)_{n\in\mathbb{N}}$ in $\mathscr{A}^\mathbb{N}$, we write $x \sim_\omega y$ provided that $\{n \in \mathbb{N} : x_n = y_n\} \in \omega$, and denote by $[x]_\omega = \{x' \in \mathscr{A}^\mathbb{N} : x' \sim_\omega x\} \subseteq \mathscr{A}^\mathbb{N}$ the class of x with respect to the equivalence relation $\sim_\omega$. Finally, denote by $\mathscr{A}_\omega := \mathscr{A}^\mathbb{N}/\sim_\omega$ the corresponding ultrapower.

We set

$$[x]_\omega + [y]_\omega := [x+y]_\omega, \quad [x]_\omega [y]_\omega := [xy]_\omega \quad \text{and} \quad a[x]_\omega := [ax]_\omega \tag{11.26}$$

for all $x, y \in \mathscr{A}^{\mathbb{N}}$ and $a \in \mathbb{F}$. It is easy to check (**exercise**) that the operations in (11.26) are well defined and that the ultrapower $\mathscr{A}_\omega := \mathscr{A}^{\mathbb{N}}/\sim_\omega$ is then endowed with the structure of a unital $\mathbb{F}$-algebra with zero element $[0]_\omega$ and unit element $[1]_\omega$.

Theorem 11.39. *The ultrapower* $\mathbb{F}_\omega := \mathbb{F}^{\mathbb{N}}/\sim_\omega$ *is a field.*

Proof. We first observe that since the operations (11.26) are commutative in $\mathbb{F}_\omega$, we only need to show that every nonzero element in $\mathbb{F}_\omega$ is invertible. Let $x = (x_n)_{n\in\mathbb{N}} \in \mathbb{F}^{\mathbb{N}}$ and suppose that $[x]_\omega \neq [0]_\omega$. This means that the set $\Omega_x := \{n \in \mathbb{N} : x_n \neq 0\}$ belongs to ω. Let then $x' = (x'_n)_{n\in\mathbb{N}} \in \mathbb{F}^{\mathbb{N}}$ be defined by $x'_n := {x_n}^{-1}$ for $n \in \Omega_x$ and $x'_n := 1_{\mathbb{F}}$ otherwise. Since $x_n x'_n = 1_{\mathbb{F}}$ for all $n \in \Omega_x \in \omega$, we deduce that $[x]_\omega [x']_\omega = [1]_\omega$, that is, $[x']_\omega$ is the inverse of $[x]_\omega$ in $\mathbb{F}_\omega$. It follows that $\mathbb{F}_\omega$ is a field. □

Note that $\mathscr{A}$ can be viewed as a subalgebra of $\mathscr{A}_\omega$ via the map $x \mapsto [(x)_{n\in\mathbb{N}}]_\omega$, which is an injective algebra homomorphism from $\mathscr{A}$ into $\mathscr{A}_\omega$. Moreover, if ω is principal, then $\mathscr{A}_\omega \cong \mathscr{A}$. The field $\mathbb{F}_\omega$ is called the *ultrafield* associated with $\mathbb{F}$ and the ultrafilter ω. When $\mathbb{F}$ is the field $\mathbb{R}$ of real numbers, the ultraproduct $\mathbb{R}_\omega$ is called the field of *hyperreal numbers* (also called *nonstandard reals*).

11.11 Ultrapowers of General Linear Groups

Let $\mathbb{F}$ be a field and let $\mathscr{A}$ be an $\mathbb{F}$-algebra. Let also ω be an ultrafilter. For and integer $k \geq 1$ we denote by $M_k(\mathscr{A})$ the $\mathbb{F}$-algebra of all $k \times k$ matrices with coefficients in $\mathscr{A}$ and by $I_{M_k(\mathscr{A})} = (\delta_{i,j} 1_{\mathscr{A}})_{i,j=1}^k \in M_k(\mathscr{A})$ the identity matrix.

Consider the map $\Phi \colon M_k(\mathbb{F})^{\mathbb{N}} \to M_k(\mathbb{F}^{\mathbb{N}})$ defined by

$$\Phi\left((A_n)_{n\in\mathbb{N}}\right) = \left((a_n(i,j))_{n\in\mathbb{N}}\right)_{i,j=1}^k$$

where $A_n = (a_n(i,j))_{i,j=1}^k \in M_k(\mathbb{F})$ for all $n \in \mathbb{N}$. It is easy to see (**exercise**) that Φ is a unital isomorphism of $\mathbb{F}$-algebras and that the induced map $\overline{\Phi} \colon M_k(\mathbb{F})_\omega \to M_k(\mathbb{F}_\omega)$ defined by

$$\overline{\Phi}\left([A]_\omega\right) = \left([(a_n(i,j))_{n\in\mathbb{N}}]_\omega\right)_{i,j=1}^k$$

for all $A = (A_n)_{n\in\mathbb{N}} \in M(k,\mathbb{F})^{\mathbb{N}}$ is also a unital isomorphism of $\mathbb{F}$-algebras. In other words, we have

$$M_k(\mathbb{F})_\omega \cong M_k(\mathbb{F}_\omega). \tag{11.27}$$

Let now $\mathrm{GL}(k,\mathscr{A}) \subseteq M_k(\mathscr{A})$ denote the *general linear group* consisting of all invertible $k \times k$ matrices with coefficients in $\mathscr{A}$. We observe that since the ultrapower construction preserves the inclusion, we have

$$\mathrm{GL}(k,\mathscr{A})_\omega \subseteq M_k(\mathscr{A})_\omega. \tag{11.28}$$

Our next task is to show that the algebra isomorphism (11.27) restricts, by virtue of the inclusion (11.28), to a group isomorphism

$$\mathrm{GL}(k,\mathbb{F})_\omega \cong \mathrm{GL}(k,\mathbb{F}_\omega). \tag{11.29}$$

Thus, let $A = (A_n)_{n\in\mathbb{N}} \in \mathrm{GL}(k,\mathbb{F})^{\mathbb{N}}$ and denote by $A^{-1} := (A_n^{-1})_{n\in\mathbb{N}} \in \mathrm{GL}(k,\mathbb{F})^{\mathbb{N}}$ the corresponding sequence of inverse matrices. We have

$$\begin{aligned}\overline{\Phi}([A]_\omega)\,\overline{\Phi}([A^{-1}]_\omega) &= \overline{\Phi}([A]_\omega[A^{-1}]_\omega)\\ &= \overline{\Phi}([AA^{-1}]_\omega)\\ &= ([(\delta_{i,j}1_{\mathbb{F}})_{n\in\mathbb{N}}]_\omega)_{i,j=1}^k\\ &= (\delta_{i,j}1_{\mathbb{F}_\omega})_{i,j=1}^k\\ &= I_{M_k(\mathbb{F}_\omega)}.\end{aligned}$$

This shows that $\overline{\Phi}(\mathrm{GL}(k,\mathbb{F})_\omega) \subseteq \mathrm{GL}(k,\mathbb{F}_\omega)$. In order to show the reverse inclusion, let $\overline{B} = ([(b_n(i,j))_{n\in\mathbb{N}}]_\omega)_{i,k=1}^k \in \mathrm{GL}(k,\mathbb{F}_\omega)$. This means that there exists a matrix $\overline{C} = ([(c_n(i,j))_{n\in\mathbb{N}}]_\omega)_{i,j=1}^k \in \mathrm{GL}(k,\mathbb{F}_\omega)$ such that $\overline{B}\,\overline{C} = \overline{C}\,\overline{B} = I_{M(k,\mathbb{F}_\omega)}$. This implies that the set $\Omega := \{n \in \mathbb{N} : \sum_{h=1}^n b_n(i,h)c_n(h,j) = \delta_{i,j}1_{\mathbb{F}} = \sum_{\ell=1}^n c_n(i,\ell)b_n(\ell,j)\}$ belongs to ω. It follows that for all $n \in \Omega$, the matrix $B'_n := (b_n(i,j))_{i,j=1}^k \in M_k(\mathbb{F})$ is in fact invertible, that is, $B'_n \in \mathrm{GL}(k,\mathbb{F})$. Thus, if we denote by $B = (B_n)_{n\in\mathbb{N}} \in M_k(\mathbb{F})^{\mathbb{N}}$ the sequence defined by $B_n := B'_n$ for $n \in \Omega$ and $B_n := I_{M_k(\mathbb{F})}$ otherwise, we have $B \in \mathrm{GL}(k,\mathbb{F})^{\mathbb{N}}$ and, by construction, $\overline{\Phi}([B]_\omega) = \overline{B}$. This shows that $\overline{\Phi}(\mathrm{GL}(k,\mathbb{F})_\omega) = \mathrm{GL}(k,\mathbb{F}_\omega)$ and (11.29) follows.

We can rephrase this result as follows.

Theorem 11.40. *Let $\mathbb{F}$ be a field, let $k \geq 1$ an integer, and let ω be an ultrafilter. Then the ultrapower $\mathrm{GL}(k,\mathbb{F})_\omega$ of the general linear group with coefficients in $\mathbb{F}$ is canonically isomorphic to the general linear group $\mathrm{GL}(k,\mathbb{F}_\omega)$ with coefficients in the ultrafield $\mathbb{F}_\omega$.* □

Corollary 11.41. *Let $\mathbb{F}$ be a field and let ω be an ultrafilter. Let also $\mathbf{G} = (G_n, d_n)_{n\in\mathbb{N}}$ be a sequence of metric groups which are $\mathbb{F}$-linear of uniformly bounded rank (i.e. there exists an integer $k = k(\mathbf{G}) \geq 1$ and an injective group homomorphism $\varphi_n \colon G \to \mathrm{GL}(k,\mathbb{F})$ for every $n \in \mathbb{N}$). Then the ultraproduct $\mathbf{G}_\omega$ is $\mathbb{F}_\omega$-linear.*

Proof. The product map $\prod_{n\in\mathbb{N}} \varphi_n \colon \mathbf{G}_\omega \to \mathrm{GL}(k,\mathbb{F})_\omega$ defined by

$$\left(\prod_{n\in\mathbb{N}} \varphi_n\right)([g]_\omega) := [(\varphi_n(g_n))_{n\in\mathbb{N}}]_\omega$$

for all $g = (g_n)_{n\in\mathbb{N}} \in \ell^\infty(\mathbf{G})$ is well defined and injective (**exercise**). The statement then follows from Theorem 11.40. □

11.12 Asymptotic Cones

Let (X,d) be a metric space and fix a *base point* $x^0 \in X$. Let also ω be a free ultrafilter on $\mathbb{N}$.

Definition 11.42. A sequence $x = (x_n)_{n\in\mathbb{N}}$ in X is said to be *moderate* provided that there exists a constant $A = A_x > 0$ such that

$$d(x_n, x^0) \leq An$$

for all $n \in \mathbb{N}$, $n \geq 1$. We denote by $\mathcal{M}(X,d)$ the set of all moderate sequences in X.

Note that if we change the basepoint, the set $\mathcal{M}(X,d)$ of moderate sequences remains the same (**exercise**).

Let $x = (x_n)_{n\in\mathbb{N}}$ and $y = (y_n)_{n\in\mathbb{N}}$ be two moderate sequences in (X,d). Then by the triangle inequality one has $d(x_n,y_n) \leq d(x_n,x^0) + d(x^0,y_n) \leq (A_x + A_y)n$. It follows that the sequence $(d(x_n,y_n)/n)_{n\geq 1}$ is bounded. Then Corollary 11.20 ensures that the limit

$$\lim_{\omega} \frac{d(x_n,y_n)}{n}$$

exists in $[0,+\infty)$. We then write $x \sim_\omega y$ provided that $\lim_\omega d(x_n,y_n)/n = 0$. One immediately checks that $\sim_\omega$ is an equivalence relation on $\mathcal{M}(X,d)$ (**exercise**). We then denote by

$$K_\omega(X,d) := \mathcal{M}(X,d)/\sim_\omega$$

the corresponding quotient space. Also, for $x = (x_n)_{n\in\mathbb{N}} \in \mathcal{M}(X,d)$ we denote by $[x]_\omega = [(x_n)_{n\in\mathbb{N}}]_\omega \in K_\omega(X,d)$ the corresponding equivalence class. Given $[x]_\omega, [y]_\omega \in K_\omega(X,d)$ one immediately checks that the nonnegative real number

$$d_\omega([x]_\omega, [y]_\omega) := \lim_{\omega} \frac{d(x_n,y_n)}{n}$$

is well defined, that is, it does not depend on the particular choice of the representatives $x, y \in \mathcal{M}(X,d)$ of the classes $[x]_\omega$ and $[y]_\omega$. Moreover,

$$d_\omega \colon K_\omega(X,d) \times K_\omega(X,d) \to [0,+\infty)$$

is a distance function (**exercise**).

Definition 11.43. The metric space $(K_\omega(X,d), d_\omega)$ is called the *asymptotic cone* of the metric space (X,d) relative to the free ultrafilter ω.

Example 11.44. Let (G,d) be a left metric group (for instance a group G generated by a finite symmetric subset $S \subseteq G$ with the metric $d = d_S$). We also choose as basepoint the identity element 1_G of G. We then equip $\mathcal{M}(G,d)$ with a group structure given by pointwise multiplication: if $g = (g_n)_{n\in\mathbb{N}}$ and $h = (h_n)_{n\in\mathbb{N}}$ are moderate sequences, then the sequence $gh = (g_nh_n)_{n\in\mathbb{N}}$ is also moderate. Indeed by the triangle inequality and left-invariance of the metric d we have $d(g_nh_n, 1_G) \leq d(h_n, 1_G) + d(g_n, 1_G) \leq (A_g + A_h)n$ for all $n \in \mathbb{N}$. It is clear that the constant sequence $1 := (1_G)_{n\in\mathbb{N}}$ is the neutral element and that $g^{-1} := (g_n^{-1})_{n\in\mathbb{N}} \in \mathcal{M}(G,d)$ is the inverse of $g = (g_n)_{n\in\mathbb{N}} \in \mathcal{M}(G,d)$. Let ω be an ultrafilter and set

$$N_\omega := \{g = (g_n)_{n\in\mathbb{N}} \in \mathcal{M}(G,d) : \lim_{\omega} \frac{d(g_n, 1_G)}{n} = 0\}.$$

It is easy to see (**exercise**) that N_ω is a subgroup of $\mathcal{M}(G,d)$ and that for $g,g',h,h' \in \mathcal{M}(G,d)$

$$g \sim_\omega g' \text{ and } h \sim_\omega h' \text{ implies } gh \sim_\omega g'h'. \tag{11.30}$$

Moreover, $g \sim_\omega g'$ if and only if $gN_\omega = g'N_\omega$ for all $g,g' \in \mathcal{M}(G,d)$. It follows that we have a bijection between the asymptotic cone and the set of the right cosets of N_ω in $\mathcal{M}(G,d)$

$$K_\omega(G,d) \simeq \mathcal{M}(G,d)/N_\omega.$$

If the metric d is bi-invariant, then the subgroup N_ω is also normal and $(K_\omega(G,d),d_\omega)$ is also a metric group (**exercise**).

Remark 11.45. Let (X,d) be a metric space and fix a base point $x^0 \in X$. Let $d_0 := d$ and for $n \geq 1$ denote by d_n the metric obtained by dividing d by n, that is, $d_n(x,y) := d(x,y)/n$ for all $x,y \in X$. It is clear that a sequence $x = (x_n)_{n\in\mathbb{N}}$ in X is moderate if and only if $\sup_{n\in\mathbb{N}} d_n(x_n,x^0) < \infty$. In other words, we have

$$\mathcal{M}(X,d) = \ell^\infty(\mathbf{X}),$$

where $\mathbf{X} = ((X_n,d_n,x_n^0))_{n\in\mathbb{N}}$, with $X_n = X$ and $x_n^0 = x^0$ for all $n \in \mathbb{N}$. Let now ω be an ultrafilter. It is clear that the equivalence relations $\sim_\omega$ defined in $\mathcal{M}(X,d)$ and $\ell^\infty(\mathbf{X})$ also coincide. It follows that the asymptotic cone is just the ultraproduct of this sequence of pointed metric spaces:

$$K_\omega(X,d) = \mathbf{X}_\omega.$$

11.13 Asymptotic Cones and Quasi-Isometries

Definition 11.46. Let (X,d_X) and (Y,d_Y) be two metric spaces. A map $\Phi\colon X \to Y$ is called a *quasi-isometry* provided that there exist constants $\lambda \geq 1$ and $\mu,\nu \geq 0$, called the *parameters* of Φ, such that the following conditions hold:

(Q1-1) for all $x_1,x_2 \in X$

$$\frac{1}{\lambda}d_X(x_1,x_2) - \mu \leq d_Y(\Phi(x_1),\Phi(x_2)) \leq \lambda d_X(x_1,x_2) + \mu;$$

(Q2-2) for every $y \in Y$ there exists an $x \in X$ such that

$$d_Y(\Phi(x),y) \leq \nu.$$

If such a quasi-isometry exists one then says that the metric space (X,d_X) is *quasi-isometric* to (Y,d_Y). It is an **exercise** to show that this yields an equivalence relation.

When $\mu = 0$, that is, if

$$\frac{1}{\lambda}d_X(x_1,x_2) \leq d_Y(\Phi(x_1),\Phi(x_2)) \leq \lambda d_X(x_1,x_2)$$

for all $x_1,x_2 \in X$, then Φ is called a *bi-Lipschitz* mapping.

Note that a surjective bi-Lipschitz mapping (that is, a quasi-isometry with parameters $\mu = 0 = \nu$) is a metric space homeomorphism (called a *bi-Lipschitz homeomorphism*).

We leave the proof of the following proposition as an **exercise**.

Proposition 11.47. *Let G be a finitely generated group, and H be a finite index subgroup of G. Then G and H with any word metric are quasi-isometric.*

Proposition 11.48. *Let (X,d_X) and (Y,d_Y) be metric spaces and let ω be a free ultrafilter on $\mathbb{N}$. Then every quasi-isometry $\Phi\colon X \to Y$ induces a bi-Lipschitz homeomorphism*

$$\Phi_\omega : K_\omega(X,d_X) \to K_\omega(Y,d_Y).$$

Proof. Let (λ,μ,ν) be the parameters of Φ. Let $x^0 \in X$ be a base point of X and choose $y^0 := \Phi(x^0)$ to be the base point of Y. Let $x := (x_n)_{n\in\mathbb{N}}$ be a moderate sequence in X. We have

$$d_Y(\Phi(x_n),\Phi(x^0)) \leq \lambda d_X(x_n,x^0) + \mu \leq (\lambda A_x + \mu)n$$

for all $n \in \mathbb{N}$ showing that $(\Phi(x_n))_{n\in\mathbb{N}}$ is a moderate sequence in Y. Given another moderate sequence $x' = (x'_n)_{n\in\mathbb{N}}$ in X, we have

$$\lim_\omega \frac{d_Y(\Phi(x_n),\Phi(x'_n))}{n} \leq \lim_\omega \frac{\lambda d_X(x_n,x'_n)+\mu}{n} = \lambda d_{X,\omega}([x]_\omega,[x']_\omega).$$

We deduce that the map $\Phi_\omega\colon K_\omega(X,d_X) \to K_\omega(Y,d_Y)$ given by

$$\Phi_\omega([x]_\omega) := [(\Phi(x_n))_{n\in\mathbb{N}}]_\omega$$

for all $x = [(x_n)_{n\in\mathbb{N}}] \in K_\omega(X,d_X)$ is well defined and that

$$d_{Y,\omega}(\Phi_\omega([x]_\omega),\Phi_\omega([x']_\omega) \leq \lambda d_{X,\omega}([x]_\omega,[x']_\omega)$$

for all $x,x' \in K_\omega(X,d_X)$. This proves one half of the bi-Lipschitz condition for Φ_ω. The other half is proved similarly (**exercise**).

Let us show that Φ_ω is surjective. Let $y = (y_n)_{n\in\mathbb{N}}$ be a moderate sequence in Y. Then, for every $n \in \mathbb{N}$ we can find $x_n \in X$ such that $d_Y(\Phi(x_n),y_n) \leq \nu$. Let us show that the sequence $x := (x_n)_{n\in\mathbb{N}}$ in X is moderate. We have

$$\begin{aligned} d_X(x_n,x^0) &\leq \lambda d_Y(\Phi(x_n),\Phi(x^0)) \\ &\leq \lambda\left(d_Y(\Phi(x_n),y_n) + d_Y(y_n,\Phi(x^0))\right) \\ &\leq \lambda(\nu + A_y n) \\ &\leq \lambda(\nu + A_y)n. \end{aligned}$$

Thus $x \in \mathscr{M}(X,d_X)$ and we have

$$d_{Y,\omega}(\Phi_\omega([x]_\omega),[y]_\omega) = \lim_\omega \frac{d_Y(\Phi(x_n),y_n)}{n} \leq \lim_\omega \frac{\nu}{n} = 0,$$

thus showing that $\Phi_\omega([x]_\omega) = [y]_\omega$. □

Corollary 11.49. *Let G be a finitely generated group. Let $S \subseteq G$ be a finite symmetric generating subset. Then the bi-Lipschitz class of the asymptotic cone $K_\omega(G, d_S)$ is independent of the generating subset S.* □

For a finitely generated group G we denote by $K_\omega(G)$ the *bi-Lipschitz class* of the asymptotic cone $K_\omega(G, d_S)$, where S is any finite symmetric generating subset of G.

Proposition 11.50. *Let G be a finitely generated group and let H be a finite index subgroup of G. Then $K_\omega(G) = K_\omega(H)$.*

Proof. This follows immediately from Proposition 11.48 after observing that G and its finite index subgroup H (which is finitely generated by Corollary 1.11) are quasi-isometric by Proposition 11.47. □

11.14 Properties of Asymptotic Cones

Before stating the main properties of the asymptotic cone of a finitely generated group, we review a few concepts.

A metric space (X, d) is called *homogeneous* provided that the group $\mathrm{Isom}(X)$ of all isometries of X acts transitively on X: for all $x, y \in X$ there exists an $\alpha \in \mathrm{Isom}(X)$ such that $\alpha(x) = y$. Also, (X, d) is *arcwise connected* if for any $x, y \in X$ there exists a continuous map $\alpha\colon [0,1] \to X$ such that $\alpha(0) = x$ and $\alpha(1) = y$; we shall call such a map α a *continuous path* connecting x to y. Note that if (X, d) is arcwise connected then it is also connected. One also says that (X, d) is *locally connected* (resp. *locally arcwise connected*) provided that every $x \in X$ admits connected (resp. arcwise connected) neighboring balls of arbitrarily small radii. Note that local arcwise connectedness implies local connectedness. A *geodesic* in a metric space is an isometric embedding $f\colon [0, d] \to X$: one then says that f connects the points $f(0)$ and $f(d)$ in X. A metric space is called *geodesic* if every two points are connected by a geodesic. Note that every geodesic metric space is arcwise connected.

Theorem 11.51. *Let (X, d) be a complete metric space and ω an ultrafilter. Then the asymptotic cone $(K_\omega(X, d), d_\omega)$ is complete.*

Proof. The proof is essentially the same, but slightly more involved than the proof of the analogous result for the ultrapower (X_ω, d_ω) (Theorem 11.30). We include the detailed proof for the sake of completeness and the convenience of the reader.

Claim. *Let $h, k \in K_\omega(X, d)$ and $A > 0$ be such that $D := d_\omega(h, k) < A$. Then we can find $(x_n)_{n\in\mathbb{N}}$ and $(y_n)_{n\in\mathbb{N}}$ in $\mathscr{M}(X, d)$ such that $h = [(x_n)_{n\in\mathbb{N}}]_\omega$, $k = [(y_n)_{n\in\mathbb{N}}]_\omega$ and $d(x_n, y_n) < An$ for all $n \in \mathbb{N}$.*

The proof is, again, basically the same as the proof of the claim in Theorem 11.30. Indeed, let $(x_n)_{n\in\mathbb{N}}$ and $(y'_n)_{n\in\mathbb{N}}$ be in $\mathscr{M}(X, d)$ such that $h = [(x_n)_{n\in\mathbb{N}}]_\omega$, $k = [(y'_n)_{n\in\mathbb{N}}]_\omega$ and set $S := \{n \in \mathbb{N} : d(x_n, y'_n) < An\}$. We claim that $S \in \omega$. Indeed, if $\varepsilon := (A - D)/2$ we have, by definition, $S_\varepsilon := \{n \in \mathbb{N} : D - \varepsilon < d(x_n, y'_n)/n < D + \varepsilon\} \in$

ω. But since $D+\varepsilon < A$ we have $S_\varepsilon \subseteq \{n \in \mathbb{N} : d(x_n, y'_n)/n < D+\varepsilon\} \subseteq S$ so that by (F3) we have $S \in \omega$. Thus setting $y_n := y'_n$ for $n \in S$ and $y_n := x_n$ for $n \notin S$ we have that $[(y_n)_{n\in\mathbb{N}}]_\omega = k$ and $d(x_n, y_n) < An$ holds for all $n \in \mathbb{N}$. This proves the claim.

Let now $(k^i)_{i\in\mathbb{N}}$ be a Cauchy sequence in $K_\omega(X,d)$. Possibly disregarding a finite number of initial terms, it is not restrictive to suppose that $d_\omega(k^j, k^i) < 1$ for all $i, j \in \mathbb{N}$. For all $i \in \mathbb{N}$, by applying the previous claim with $h = k^1$, $k = k^i$ and $A = 1$, we can find $(x^i_n)_{n\in\mathbb{N}} \in \mathcal{M}(X,d)$ such that $k^i = [(x^i_n)_{n\in\mathbb{N}}]_\omega$ and $d(x^1_n, x^i_n) < n$ for all $n \in \mathbb{N}$. By the triangle inequality we therefore have $d(x^j_n, x^i_n) < 2n$ for all $i, j, n \in \mathbb{N}$.

Let now ℓ be the first index $\ell > 1$ such that $d_\omega(k^i, k^j) < 1/2$ for all $i, j \geq \ell$. By applying the claim with $h = k^\ell$, $k = k^i$ for $i \geq \ell$, and $A = 1/2$, we can suppose that the sequences $(x^i_n)_{n\in\mathbb{N}} \in \mathcal{M}(X,d)$ satisfy, in addition, $d(x^\ell_n, x^i_n) < n/2$ for all $i, n \in \mathbb{N}$ such that $i \geq \ell$ (note that since in the process indicated by the claim we may have just replaced some x^i_n by x^1_n, the inequalities from the previous paragraph remain true). Again, by the triangle inequality we therefore have $d(x^j_n, x^i_n) < n$ for all $i, j, n \in \mathbb{N}$ such that $i, j \geq \ell$.

Continuing in this way, it is therefore possible to choose the sequences $(x^i_n)_{n\in\mathbb{N}} \in \mathcal{M}(X,d)$, $i \in \mathbb{N}$, in a such a way that for every $\varepsilon' > 0$ there exists an $\ell_{\varepsilon'} \in \mathbb{N}$ such that

$$d(x^j_n, x^i_n) < \varepsilon' n \text{ for all } i, j, n \in \mathbb{N} \text{ such that } i, j \geq \ell_{\varepsilon'}. \tag{11.31}$$

Fix $n \in \mathbb{N}$ and let us show that the sequence $(x^i_n)_{i\in\mathbb{N}}$ is Cauchy. Indeed, let $\varepsilon > 0$. Then if $\varepsilon' := \varepsilon/n$ and $i_\varepsilon := \ell_{\varepsilon'}$, from (11.31) we have $d(x^i_n, x^j_n) < \varepsilon$ for all $i, j \geq i_\varepsilon$. Since (X,d) is complete, there exists $x_n := \lim_{i\to\infty} x^i_n \in X$.

Consider the sequence $x := (x_n)_{n\in\mathbb{N}}$ and let us show that $x \in \mathcal{M}(X,d)$ and $[x]_\omega = \lim_{i\to\infty} k^i$.

Fix $\varepsilon' > 0$ and let $\ell_{\varepsilon'}$ as above. Letting $j \to \infty$ in (11.31) we obtain

$$d(x_n, x^i_n) < \varepsilon' n \text{ for all } i, n \in \mathbb{N} \text{ such that } i \geq \ell_{\varepsilon'}. \tag{11.32}$$

In order to prove the first assertion, fix a point x^0 in X, fix $i \geq \ell_{\varepsilon'}$ and let $A > 0$ be such that $d(x^0, x^i_n) < An$ for all $n \in \mathbb{N}$. Then, using the triangle inequality, from (11.32) we deduce $d(x^0, x_n) \leq d(x^0, x^i_n) + d(x_n, x^i_n) < (A+\varepsilon')n$ for all $n \in \mathbb{N}$, thus showing that the sequence $(x_n)_{n\in\mathbb{N}}$ is moderate.

On the other hand, still from (11.32) we deduce that

$$d_\omega([x]_\omega, k^i) = \lim_\omega d(x_n, x^i_n)/n < \varepsilon'$$

for all $i \geq \ell_{\varepsilon'}$. Since ε' was arbitrary, this shows that $[x]_\omega = \lim_{i\to\infty} k^i$. Thus the Cauchy sequence $(k^i)_{i\in\mathbb{N}}$ is converging. It follows that the asymptotic cone $(K_\omega(X,d), d_\omega)$ is complete. □

Theorem 11.52. *Let G be a finitely generated group, let $S \subseteq G$ be a finite symmetric generating subset, and let ω be an ultrafilter. Then the associated asymptotic cone $K_\omega(G, d_S)$ is:*

(i) *homogeneous;*
(ii) *arcwise connected (and therefore connected) and in fact geodesic;*

(iii) *locally arcwise connected (and therefore locally connected);*
(iv) *complete.*

Proof. Set $K := K_\omega(G, d_S)$.

(i) Recall that for $g \in G$ we denote by $L_g \colon G \to G$ the left-multiplication by g (cf. Section 7.1). We can generalize this in the following way.

Let $g = (g_n)_{n\in\mathbb{N}} \in \mathscr{M}(G, d_S)$. Then, recalling that $\mathscr{M}(G, d_S)$ is a group (cf. Example 11.44) the map $L_g \colon \mathscr{M}(G, d_S) \to \mathscr{M}(G, d_S)$ is well defined. Explicitly, $L_g(g') = (g_n g'_n)_{n\in\mathbb{N}}$, for all moderate sequences $g' = (g'_n)_{n\in\mathbb{N}} \in \mathscr{M}(G, d_S)$. It follows from (11.30) that if $g' = (g'_n)_{n\in\mathbb{N}} \sim_\omega g'' = (g''_n)_{n\in\mathbb{N}}$ then $L_g(g') \sim_\omega L_g(g'')$. Thus L_g induces a map $\Lambda_g \colon K \to K$ defined by $\Lambda_g([g']_\omega) = [L_g(g')]_\omega = [gg']_\omega$. Let us show that Λ_g is an isometry: indeed,

$$\begin{aligned} d_\omega(\Lambda_g([g']_\omega), \Lambda_g([g'']_\omega)) &= d_\omega([gg']_\omega, [gg'']_\omega) \\ &= \lim_\omega d_S(g_n g'_n, g_n g''_n) \\ &= \lim_\omega d_S(g'_n, g''_n) \\ &= d_\omega([g']_\omega, [g'']_\omega) \end{aligned}$$

for all $g', g'' \in \mathscr{M}(G, d_S)$. It follows that $\Lambda_g \in \mathrm{Isom}(K)$ for all $g \in \mathscr{M}(G, d_S)$.

Let $g = (g_n)_{n\in\mathbb{N}}, g' = (g'_n)_{n\in\mathbb{N}} \in \mathscr{M}(G, d_S)$ and consider the element $h = g'g^{-1} \in \mathscr{M}(G, d_S)$. Then $\Lambda_h([g]_\omega) = [g']_\omega$. This shows that K is homogeneous.

(ii) In order to show that K is arcwise connected, we need a simple technical lemma which will also be useful in the sequel. Given a real number a we denote by $\lfloor a \rfloor \in \mathbb{N}$ the *integer part* of a, that is, the largest integer $n \le a$.

Lemma 11.53. *Let $a, b \in \mathbb{R}$ and suppose that $a > b$. Then*

$$(a-b) - 1 \le \lfloor a \rfloor - \lfloor b \rfloor \le (a-b) + 1. \tag{11.33}$$

Proof. Let $n \in \mathbb{N}$ and $\varepsilon \in [0, 1)$ be such that $a = b + n + \varepsilon$. We have, on the one hand

$$\lfloor a \rfloor = \lfloor b + n + \varepsilon \rfloor = \lfloor b + \varepsilon \rfloor + n \le \lfloor b \rfloor + 1 + n$$

so that

$$\lfloor a \rfloor - \lfloor b \rfloor \le n + 1 \le (n + \varepsilon) + 1 = (a - b) + 1. \tag{11.34}$$

On the other hand

$$\lfloor a \rfloor = \lfloor b + n + \varepsilon \rfloor \ge \lfloor b + n \rfloor = \lfloor b \rfloor + n$$

so that

$$\lfloor a \rfloor - \lfloor b \rfloor \ge n \ge (n-1) + \varepsilon = (n + \varepsilon) - 1 = (a - b) - 1. \tag{11.35}$$

From (11.34) and (11.35) one immediately deduces (11.33). □

Let then $g = (g_n)_{n\in\mathbb{N}} \in \mathscr{M}(G, d_S)$ and let us show that there exists a continuous path connecting $1_K := [(1_G, 1_G, \dots)]_\omega$ to $[g]_\omega$ in K. Set $d := d_\omega(1_K, [g]_\omega)$.

Let also $w_n = x_{n,1}x_{n,2}\cdots x_{n,r_n}$, where $x_{n,j} \in S$ and $r_n = d_S(1_G, g_n) = \ell_S(g_n)$, be a minimal word representing g_n for all $n \in \mathbb{N}$.

For $t \in [0,1]$ and any word $w \in S^*$ we denote by $w(t)$ the prefix of w of length $\lfloor t\ell_S(w)\rfloor$. In other words, if $w = x_1x_2\cdots x_\ell$ then $w(t) = x_1x_2\cdots x_{\lfloor t\ell\rfloor}$. Note that the sequence $(w_n(t))_{n\in\mathbb{N}}$ is moderate, since $d_S(1_G, w_n(t)) \le d_S(1_G, w_n) = d_S(1_G, g_n)$ for all $n \in \mathbb{N}$, and $(g_n)_{n\in\mathbb{N}}$ is moderate. Let us set $\alpha(t) := [(w_n(t))_{n\in\mathbb{N}}]_\omega \in K$.

It is clear that $\alpha(0) = 1_K$ and $\alpha(1) = [g]_\omega$. Moreover, for $0 \le t \le t' \le 1$ and w_n as above, from (11.33) we deduce

$$(t'-t)\ell_S(w_n) - 1 \le \ell_S(w_n(t')) - \ell_S(w_n(t)) \le (t'-t)\ell_S(w_n) + 1.$$

Dividing by n and taking the limit along ω, this gives

$$d_\omega(\alpha(t'), \alpha(t)) = \lim_\omega \frac{d_S(w_n(t'), w_n(t))}{n} = (t'-t)d. \tag{11.36}$$

Thus α is a continuous path connecting 1_K to $[g]_\omega$. This shows that K is arcwise connected.

By defining $f : [0,d] \to K$ as $f(t) := \alpha(t/d)$ for all $t \in [0,d]$, equation (11.36) implies that f is an isometric embedding. This shows that K is geodesic.

(iii) Note that in the above argument, one has $d_\omega(\alpha(t), 1_K) \le d_\omega([g]_\omega, 1_K)$, for all $t \in [0,1]$. Thus, the path connecting 1_K to $[g]_\omega$ entirely lies inside the ball of radius $r := d_\omega([g]_\omega, 1_K)$ centered at 1_K in K. As a consequence, each ball around 1_K is arcwise connected and therefore connected. By homogeneity, the same holds for all balls in K. This shows that K is locally connected as well.

(iv) In order to show that K is complete, it suffices to observe that $d_S(g,h) \ge 1$ for all distinct $g, h \in G$, so (G, d_S) is complete; then we can apply Theorem 11.51.

The proof of Theorem 11.52 is complete. □

11.15 Examples of Asymptotic Cones

Proposition 11.54. *Let (X,d) be a bounded metric space and let ω be a free ultrafilter. Then $K_\omega(X,d)$ reduces to a point.*

Proof. Denote by D the finite diameter $\mathrm{diam}(X) := \sup_{x,y\in X} d(x,y)$ of X. Observe that, in this case, all sequences in X are moderate, that is $\mathscr{M}(X,d) = X^{\mathbb{N}}$. Moreover, given $x = (x_n)_{n\in\mathbb{N}}$ and $y = (y_n)_{n\in\mathbb{N}}$ in $\mathscr{M}(X,d)$ we clearly have $\lim_\omega d(x_n,y_n)/n \le \lim_\omega D/n = 0$, yielding $x \sim_\omega y$, i.e. $[x]_\omega = [y]_\omega$. Thus $K_\omega(X,d) = \mathscr{M}(X,d)/\sim_\omega$ consists of a single class. □

Remark 11.55. Since any asymptotic cone is an arcwise connected metric space, as soon as it has at least two distinct points it is uncountable (**exercise**).

Proposition 11.56. *Let $G = \mathbb{Z}$ with any word metric. Then $K_\omega(\mathbb{Z})$ is the bi-Lipschitz class of $\mathbb{R}$ with the Euclidean distance.*

Proof. By Proposition 11.50, we can pick any finite symmetric generated subset of $\mathbb{Z}$ with the respective word metric. For simplicity we choose $S := \{1, -1\}$, so that $d_S(n,m) = |n-m|$ for all $n, m \in \mathbb{Z}$. For $x \in \mathbb{R}$ we set

$$f(x) := (\lfloor xn \rfloor)_{n \in \mathbb{N}}$$

and observe that $|\lfloor xn \rfloor / n| \leq |x| + 1$ for all $n \geq 1$, so that $f(x) \in \mathcal{M}(\mathbb{Z}, d_S)$. Consider the map $F \colon \mathbb{R} \to K_\omega(\mathbb{Z}, d_S)$ defined by setting $F(x) := [f(x)]_\omega$ and let us show that F is an isometric homeomorphism. Indeed, if $x, y \in \mathbb{R}$, from (11.33) we deduce

$$|y-x| - \frac{1}{n} \leq \frac{|\lfloor yn \rfloor - \lfloor xn \rfloor|}{n} \leq |y-x| + \frac{1}{n}. \tag{11.37}$$

Taking the ultralimit and observing that

$$d_\omega(F(y), F(x)) = \lim_\omega \frac{|\lfloor yn \rfloor - \lfloor xn \rfloor|}{n},$$

from (11.37) we deduce

$$d_\omega(F(y), F(x)) = |y-x|.$$

This shows that F is an isometry, in particular F is injective. In order to show that F is surjective, let $(g_n)_{n \in \mathbb{N}} \in \mathcal{M}(\mathbb{Z}, d_S)$. Then there exists an $A > 0$ such that $|g_n/n| < A$ for all $n \geq 1$. Let us show that $r := \lim_\omega g_n/n \in \mathbb{R}$ satisfies $F(r) = [(g_n)_{n \in \mathbb{N}}]_\omega$. Indeed we have

$$\begin{aligned} d_\omega(F(r), [(g_n)_{n \in \mathbb{N}}]_\omega) &= \lim_\omega \frac{|\lfloor rn \rfloor - g_n|}{n} \\ \text{(by (11.33))} &\leq \lim_\omega \left(\frac{|rn - g_n|}{n} + \frac{1}{n} \right) \\ &= \lim_\omega \frac{|rn - g_n|}{n} \\ &= \lim_\omega \left| r - \frac{g_n}{n} \right| = 0. \end{aligned}$$

As a consequence,

$$F(r) = [(g_n)_{n \in \mathbb{N}}]_\omega$$

showing that F is also surjective. □

In order to analyze the asymptotic cone of the Cartesian product of two metric spaces we introduce some notation. Let (X, d_X) and (Y, d_Y) be two metric spaces. We define their *Cartesian product* as the metric space $(X \times Y, d_X \times d_Y)$ where $d := d_X \times d_Y$ is given by $d((x_1, y_1), (x_2, y_2)) := d_X(x_1, x_2) + d_Y(y_1, y_2)$ for all $x_1, x_2 \in X$ and $y_1, y_2 \in Y$.

Proposition 11.57. *Let (X, d_X) and (Y, d_Y) be two metric spaces. Let ω be an ultrafilter. Then the asymptotic cone of the Cartesian product can be identified with the Cartesian product of the asymptotic cones, in formulae*

$$(K_\omega(X \times Y, d_X \times d_Y), d_\omega) \cong (K_\omega(X, d_X) \times K_\omega(Y, d_Y), d_{X,\omega} \times d_{Y,\omega}).$$

Proof. Let $\mathcal{M}$ (resp. $\mathcal{M}_X$, resp. $\mathcal{M}_Y$) denote the set of all moderate sequences in $X \times Y$ (resp. X, resp. Y) and observe that we have a bijective map $f\colon \mathcal{M}_X \times \mathcal{M}_Y \to \mathcal{M}$ given by $f((x_n)_{n\in\mathbb{N}}, (y_n)_{n\in\mathbb{N}}) = ((x_n, y_n))_{n\in\mathbb{N}}$. Let $x = (x_n)_{n\in\mathbb{N}}, x' = (x'_n)_{n\in\mathbb{N}} \in \mathcal{M}_X$ and $y = (y_n)_{n\in\mathbb{N}}), y' = (y'_n)_{n\in\mathbb{N}} \in \mathcal{M}_Y$. In the following we then identify $(x,y) \in \mathcal{M}_X \times \mathcal{M}_Y$ and $f(x,y) \in \mathcal{M}$. For simplicity of notation we also denote $d_X \times d_Y$ by d. We then have

$$\lim_\omega \frac{d((x_n,y_n),(x'_n,y'_n))}{n} = 0 \iff \frac{d_X(x_n,x'_n)}{n} = 0 \text{ and } \frac{d_Y(y_n,y'_n)}{n} = 0$$

equivalently,

$$(x,y) \sim_\omega (x',y') \iff x \sim_\omega x' \text{ and } y \sim_\omega y'.$$

As a consequence, f preserves $\sim_\omega$ and induces a bijection

$$(\mathcal{M}_X/\sim_\omega) \times (\mathcal{M}_Y/\sim_\omega) \to \mathcal{M}/\sim_\omega.$$

Moreover,

$$\begin{aligned} d_\omega([(x,y)]_\omega, [(x',y')]_\omega) &= \lim_\omega \frac{d((x_n,y_n),(x'_n,y'_n))}{n} \\ &= \lim_\omega \frac{d_X(x_n,x'_n) + d_Y(y_n,y'_n)}{n} \\ &= \lim_\omega \frac{d_X(x_n,x'_n)}{n} + \lim_\omega \frac{d_Y(y_n,y'_n)}{n} \\ &= d_{X,\omega}([x]_\omega,[x']_\omega) + d_{Y,\omega}([y]_\omega,[y']_\omega) \end{aligned}$$

and this shows that $d_\omega = d_{X,\omega} \times d_{Y,\omega}$. □

For $k \geq 1$ we denote by d_k the metric on $\mathbb{R}^k$ defined by $d_k(r,s) := \sum_{i=1}^k |r_i - s_i|$ for all $r = (r_i)_{i=1}^k$ and $s = (s_i)_{i=1}^k \in \mathbb{R}^k$. This is the so-called ℓ^1 distance on $\mathbb{R}^k$. (Note that the bi-Lipschitz class of $(\mathbb{R}^k, d_k)$ is the same as that of $(\mathbb{R}^k, d'_k)$, where d'_k is the Euclidean distance, defined by setting $d'_k(r,s) := \left(\sum_{i=1}^k |r_i - s_i|^2\right)^{1/2}$ for all $r = (r_i)_{i=1}^k$ and $s = (s_i)_{i=1}^k \in \mathbb{R}^k$.) From Proposition 11.56 and Proposition 11.57 we deduce:

Corollary 11.58. *$K_\omega(\mathbb{Z}^k)$ is the bi-Lipschitz class of $(\mathbb{R}^k, d_k)$.* □

More generally, we have:

Corollary 11.59. *Let G be a finitely generated abelian group. Then $K_\omega(G)$ is the bi-Lipschitz class of $(\mathbb{R}^k, d_k)$, where k denotes the free abelian rank of G.*

Proof. This follows from the classification of finitely generated abelian groups (Corollary 1.30), Proposition 11.54, and Corollary 11.58. □

Corollary 11.60. *The asymptotic cone $K_\omega(\mathbb{R}^k, d_k)$ is bi-Lipschitz homeomorphic to $(\mathbb{R}^k, d_k)$.*

Proof. This follows from the fact that $(\mathbb{R}^k, d_k)$ is quasi-isometric to $(\mathbb{Z}^k, d_k)$ combined with Proposition 11.48 and Corollary 11.58. □

11.16 Hyperbolic Metric Spaces

Given a metric space (X,d) and two points $x,y \in X$, an *arc* from x to y is a continuous map $f\colon [a,b] \to X$, where $a,b \in \mathbb{R}$, such that $f(a) = x$ and $f(b) = y$. If, in addition, f is an isometric embedding, that is, $d(f(z), f(t)) = |z-t|$ for all $z,t \in [a,b]$, such an arc is called a *geodesic segment*. For a geodesic segment f from x to y we denote by $[x,y] = [x,y]_f := \{f(c) : c \in [a,b]\}$ its image in X.

Recall that a *geodesic metric space* is a metric space (X,d) such that for all $x,y \in X$ there exists a geodesic segment $f\colon [a,b] \to X$ from x to y. Hence in particular every geodesic metric space is arcwise-connected.

Let (X,d) be a geodesic metric space. Given three points $x,y,z \in X$, we choose for each pair of such points a geodesic segment, and we denote by $[x,y]$, $[y,z]$, and $[z,x]$ their images. Hence we call the set

$$\Delta(x,y,z) := [x,y] \cup [y,z] \cup [z,x] \subseteq X$$

a *geodesic triangle* with vertices x, y, and z.

Definition 11.61. Given a geodesic triangle $\Delta = \Delta(x,y,z)$, with $x,y,z \in X$, and a $\delta \geq 0$, we say that Δ is δ*-thin* if (see Figure 11.1)

$$\begin{aligned} d(s, [y,z] \cup [z,x]) &\leq \delta \quad \text{for all } s \in [x,y] \\ d(t, [z,x] \cup [x,y]) &\leq \delta \quad \text{for all } t \in [y,z] \\ d(u, [x,y] \cup [y,z]) &\leq \delta \quad \text{for all } u \in [z,x]. \end{aligned} \tag{11.38}$$

In other words, a geodesic triangle Δ is δ-thin if each side of Δ is contained in the δ-neighborhood of the other two sides.

For instance, if T is a simplicial tree (see the next section for the definition of a simplicial tree) and d is the geodesic distance, then every triangle Δ in (T,d) is 0-thin (see Figure 11.2).

A *tripod* is a metric simplicial tree (see the next section for the definition of a simplicial tree) with at most three edges and at most one vertex of degree ≥ 2. Given three positive real numbers $A,B,C \in \mathbb{R}$ there exists a unique tripod $T(A,B,C)$ up to isometries whose edges have length A,B,C (see Figure 11.3)

Given a nondegenerate geodesic triangle $\Delta = \Delta(x,y,z)$, with $x,y,z \in X$, we define T_Δ to be the unique tripod $T(A,B,C)$ such that $A+B = d(x,y)$, $A+C = d(x,z)$, and $B+C = d(y,z)$. We thus have $A = (y \mid z)_x$, $B = (x \mid z)_y$, and $C = (x \mid y)_z$, where

$$(p \mid q)_r := \frac{1}{2}(d(p,r) + d(q,r) - d(p,q)) \quad \text{for all } p,q,r \in X$$

is called the *Gromov product* of p and q relative to r. We denote by v_x, v_y, v_z the vertices of degree 1 of T_Δ such that the distance between v_p and v_q in T_Δ equals

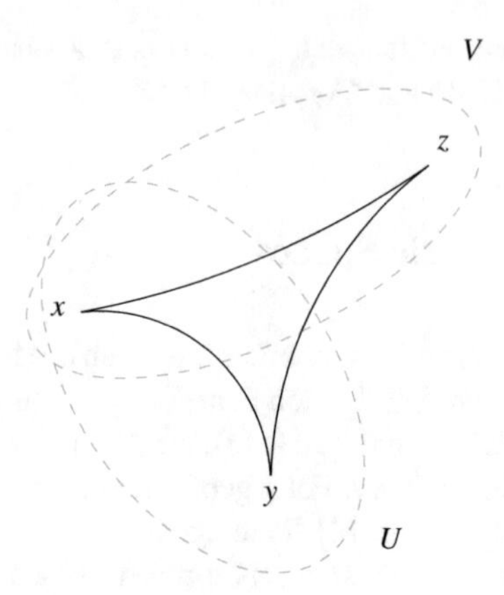

Fig. 11.1 The geodesic triangle $\Delta(x,y,z)$ is δ-thin: $U:=\{p : d(p,[x,y]) \leq \delta\}$ and $V:=\{p : d(p,[x,z]) \leq \delta\}$ satisfy $[y,z] \subseteq U \cup V$.

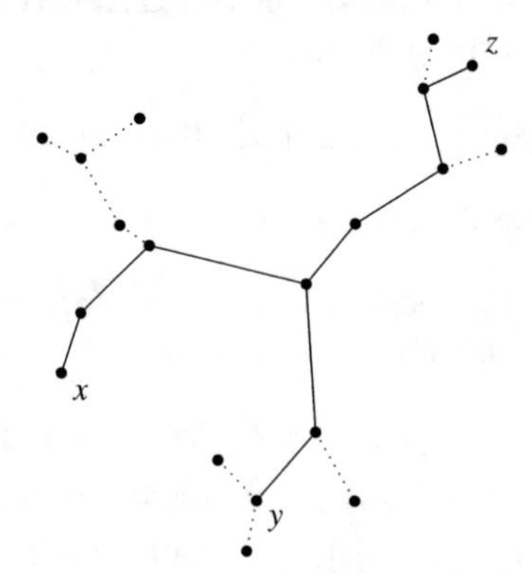

Fig. 11.2 A 0-thin triangle in a simplicial tree: in this case, $[y,z] \subseteq [x,y] \cup [x,z]$.

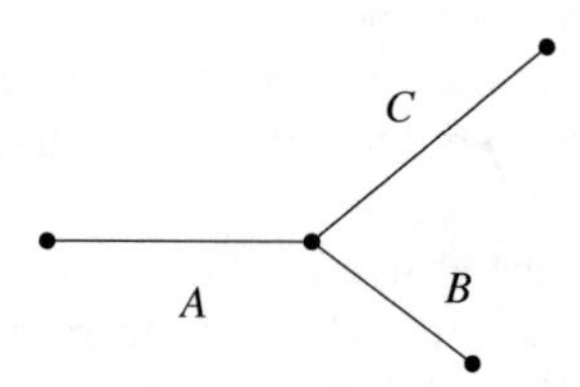

Fig. 11.3 A tripod $T(A,B,C)$.

$d(p,q)$ for all $p,q \in \{x,y,z\}$, and by v_Δ the vertex of degree ≥ 2 of T_Δ. There is a unique map $f_\Delta : \Delta \to T_\Delta$ such that $f_\Delta(p) = v_p$ for all $p \in \{x,y,z\}$ and f_Δ restricted to each edge of Δ is an isometry.

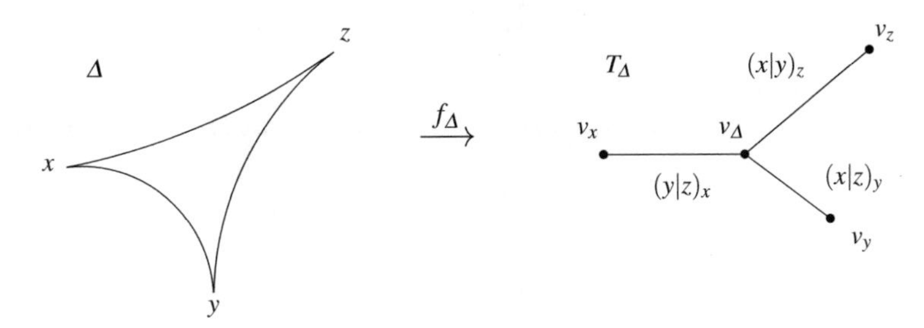

Fig. 11.4 The map $f_\Delta : \Delta \to T_\Delta$.

Definition 11.62. Given a geodesic triangle $\Delta = \Delta(x,y,z)$, with $x,y,z \in X$, and a $\delta \geq 0$, we say that Δ is δ-*slim* if for all $r \in T_\Delta$ and $p,q \in f_\Delta^{-1}(r)$ we have $d(p,q) \leq \delta$.

Proposition 11.63. *Let (X,d) be a geodesic metric space. The following conditions are equivalent:*

(1) *there exists a $\delta_1 \geq 0$ such that every geodesic triangle in X is δ_1-thin;*
(2) *there exists a $\delta_2 \geq 0$ such that every geodesic triangle in X is δ_2-slim;*
(3) *there exists a $\delta_3 \geq 0$ such that for every $x,y,z,w \in X$ we have*

$$(x \mid z)_w \geq \min\{(x \mid y)_w, (y \mid z)_w\} - \delta_3;$$

(4) *there exists a $\delta_4 \geq 0$ such that for every $x,y,z,w \in X$ we have*

$$d(x,z) + d(y,w) \leq \max\{d(x,y) + d(z,w), d(x,w) + d(y,z)\} + \delta_4.$$

Proof. (1) $\iff$ (2). It is obvious that (2) implies (1) by taking $\delta_1 = \delta_2$. To show the converse, we suppose that (1) holds, and we show that (2) holds with $\delta_2 = 2\delta_1$. Let $\Delta = \Delta(x,y,z)$ be a geodesic triangle in X with vertices $x,y,z \in X$, and let $f_\Delta^{-1}(v_\Delta) = \{i_x, i_y, i_z\}$ with $i_p \in [q,r]$ with $\{p,q,r\} = \{x,y,z\}$. By assumption, there exists a $u \in [x,y] \cup [x,z]$, say $u \in [x,y]$ (otherwise the argument is the same), such that $d(i_x,u) \leq \delta_1$. Hence

$$d(y,u) - d(i_x,y) \leq d(i_x,u) \leq \delta_1.$$

Since $d(i_x,y) = d(i_z,y) = (x \mid z)_y$, it follows that $d(i_z,y) \leq \delta_1$, so that $d(i_x,i_z) \leq 2\delta_1$. Similarly, we get $d(i_p,i_q) \leq 2\delta_1$ for any $p,q \in \{x,y,z\}$.

Now given $u \in [y,z]$ with $d(y,u) < d(y,i_x)$ (otherwise take z instead of y), let $v \in [y,x]$ be such that $d(y,u) = d(y,v)$, so that $f_\Delta(u) = f_\Delta(v)$.

We can find a point z' on $[y,z]$ such that $\Delta' := \Delta(x,y,z')$ is such that $f_{\Delta'}(u) = f_{\Delta'}(v) = v_{\Delta'}$ (**exercise**). From the first part of the proof it follows that $d(u,v) \leq 2\delta_1$. This shows that Δ is $2\delta_1$-thin.

(2) $\iff$ (3). Consider four points $x,y,z,w \in X$, and let

$$t := \min\{(x \mid y)_w, (y \mid z)_w\}.$$

We will show that (2) implies $(x \mid z)_w \geq t - 2\delta_2$.

If $t \leq (x \mid z)_w$ then the inequality is obvious. So we can assume $t > (x \mid z)_w$. Hence

$$\begin{aligned} t > (x \mid z)_w &= \frac{1}{2}(d(x,w) + d(z,w) - d(x,z)) \\ &\geq \frac{1}{2}(d(x,w) + (d(x,w) - d(x,z)) - d(x,z)) \\ &= d(x,w) - d(x,z), \end{aligned}$$

therefore $d(w,x) - t < d(x,z)$. Similarly, $d(w,z) - t < d(x,z)$. So there exist $x'' \in [x,z]$ with $d(x,x'') = d(x,w) - t$ and $z'' \in [x,z]$ with $d(z,z'') = d(z,w) - t$. Moreover,

$$d(x,x'') = d(x,x') = d(w,x) - t < d(w,x) - (x \mid z)_w = (w \mid z)_x,$$

so $f_{\Delta(w,x,z)}(x') = f_{\Delta(w,x,z)}(x'')$, therefore, by hypothesis, $d(x',x'') \leq \delta_2$. Similarly $d(z',z'') \leq \delta_2$. Therefore

$$d(x'',z'') \leq d(x'',x') + d(x',z') + d(z',z'') \leq d(x',z') + 2\delta_2.$$

Since clearly $(x \mid y)_w \leq d(x,w)$, $(x \mid y)_w \leq d(y,w)$, $(z \mid y)_w \leq d(z,w)$ and $(z \mid y)_w \leq d(y,w)$, we have $t \leq \min\{d(w,x), d(w,z), d(w,y)\}$. Therefore we can find $x' \in [w,x]$, $y' \in [w,y]$, and $z' \in [w,z]$ such that $d(w,x') = d(w,y') = d(w,z') = t$. Observe that $t = d(w,x') = d(w,y') \leq (x \mid y)_w$, hence $f_{\Delta(w,x,y)}(x') = f_{\Delta(w,x,y)}(y')$, and therefore, by hypothesis, $d(x',y') \leq \delta_2$. Similarly, $t = d(w,z') = d(w,y') \leq (z \mid y)_w$, hence $f_{\Delta(w,z,y)}(z') = f_{\Delta(w,z,y)}(y')$, and therefore $d(z',y') \leq \delta_2$. We deduce that $d(x',z') \leq d(x',y') + d(y',z') \leq 2\delta_2$.

All this implies

$$\begin{aligned} 2\delta_2 &\geq d(x',z') \\ &\geq d(x'',z'') - 2\delta_2 \\ &\geq d(x,z) - d(x,x'') - d(z,z'') - 2\delta_2 \\ &= d(x,z) - d(x,x') - d(z,z') - 2\delta_2 \\ &= d(x,z) - (d(w,x) - d(w,x')) - (d(w,z) - d(w,z')) - 2\delta_2 \\ &= (d(w,x') + d(w,z')) - (d(x,w) + d(z,w) - d(x,z)) - 2\delta_2 \\ &= 2t - 2(x \mid z)_w - 2\delta_2. \end{aligned}$$

We conclude that $(x \mid z)_w \geq t - 2\delta_2$.

Conversely, let $\Delta = \Delta(x,y,z)$, with $x,y,z \in X$, be a geodesic triangle, and let $u \in [x,y]$ and $v \in [z,x]$ such that $f_\Delta(u) = f_\Delta(v)$. We want to show that (3) implies (2) with $\delta_2 = 4\delta_3$, that is $d(u,v) \le 4\delta_3$. We know that $t := d(x,u) = d(x,v) \le (y \mid z)_x$, and clearly $(u \mid y)_x = (v \mid z)_x = t$. By hypothesis

$$(u \mid v)_x \ge \min\{(u \mid y)_x, (y \mid v)_x\} - \delta_3$$

that is

$$(u \mid v)_x \ge (u \mid y)_x - \delta_3 \ge (u \mid y)_x - 2\delta_3$$

or

$$(u \mid v)_x \ge (y \mid v)_x - \delta_3 \ge \min\{(y \mid z)_x, (z \mid v)_x\} - 2\delta_3$$

which implies

$$(u \mid v)_x \ge \min\{(u \mid y)_x, (y \mid z)_x, (z \mid v)_x\} - 2\delta_3 = t - 2\delta_3.$$

As $d(u,x) = d(v,x)$, we have

$$(u \mid v)_x = t - \frac{1}{2} d(u,v),$$

so that

$$d(u,v) = 2t - 2(u \mid v)_x \le 2t - 2(t - 2\delta_3) = 4\delta_3.$$

(3) $\iff$ (4). Finally, for $\delta \ge 0$ we have

$$\begin{aligned}
&(x \mid z)_w \ge \min\{(x \mid y)_w, (y \mid z)_w\} - \delta \\
&\Leftrightarrow [(x \mid z)_w \ge (x \mid y)_w - \delta] \text{ or } [(x \mid z)_w \ge (y \mid z)_w - \delta] \\
&\Leftrightarrow [d(x,w) + d(z,w) - d(z,x) \ge d(x,w) + d(y,w) - d(y,x) - 2\delta] \\
&\qquad \text{or } [d(x,w) + d(z,w) - d(z,x) \ge d(y,w) + d(z,w) - d(z,y) - 2\delta] \\
&\Leftrightarrow [d(z,w) + d(y,x) \ge d(y,w) + d(z,x) - 2\delta] \\
&\qquad \text{or } [d(x,w) + d(z,y) \ge d(y,w) + d(z,x) - 2\delta] \\
&\Leftrightarrow \max\{d(z,w) + d(y,x), d(x,w) + d(z,y)\} \ge d(y,w) + d(z,x) - 2\delta \\
&\Leftrightarrow d(y,w) + d(z,x) \le \max\{d(z,w) + d(y,x), d(x,w) + d(z,y)\} + 2\delta,
\end{aligned}$$

which shows that (3) implies (4) with $\delta_4 = 2\delta_3$ and (4) implies (3) with $\delta_3 = \delta_4/2$.

The proof of Proposition 11.63 is complete. □

Definition 11.64. Let (X,d) be a geodesic metric space. We say that (X,d) is δ-*hyperbolic* if it satisfies the condition (1) in Proposition 11.63 with $\delta = \delta_1$. We say that (X,d) is *hyperbolic* if it is δ-hyperbolic for some $\delta \ge 0$, and therefore if it satisfies any of the equivalent conditions of Proposition 11.63.

Note that a δ-hyperbolic metric space is δ'-hyperbolic for every $\delta' \ge \delta$.

Remark 11.65. It follows from the proof of Proposition 11.63 that the following conditions for a geodesic metric space (X,d) are equivalent:

(1) X is 0-hyperbolic (i.e., every geodesic triangle in X is 0-thin);
(2) every geodesic triangle in X is 0-slim;

(3) for every $x,y,z,w \in X$ we have

$$(x \mid z)_w \geq \min\{(x \mid y)_w, (y \mid z)_w\};$$

(4) for every $x,y,z,w \in X$ we have

$$d(x,z) + d(y,w) \leq \max\{d(x,y) + d(z,w), d(x,w) + d(y,z)\}.$$

Example 11.66. (a) Every bounded geodesic metric space (X,d) is δ-hyperbolic for $\delta := \mathrm{diam}(X) = \sup\{d(x,y) : x,y \in X\}$.

(b) The real line $(\mathbb{R},d)$, where d is the Euclidean distance, is 0-hyperbolic (**exercise**).

(c) Any *simplicial tree* (T,d) (see the next section for the definition of a simplicial tree) is 0-hyperbolic (**exercise**).

(d) The *hyperbolic plane* $(\mathbb{H}, ds)$ (see Chapter 8 for the definition of the hyperbolic plane) is hyperbolic (in fact, 2-hyperbolic (**exercise**)).

(e) The real plane $(\mathbb{R}^2, d_2)$ with the Euclidean metric d_2 is *not* hyperbolic (**exercise**).

A proof of the following theorem can be found in [79, Theorème 2.2], [122, Chapitre 5, Theorème 12], and [41, H. Theorem 1.9] (recall Definition 11.46).

Theorem 11.67. *Hyperbolicity for geodesic metric spaces is a quasi-isometric invariant.*

A connected graph $\mathscr{G} = (V,E)$ with its graph distance is embedded isometrically into a (unique up to isometry) minimal geodesic metric space $\overline{\mathscr{G}}$, the same way as $(\mathbb{Z}, d_1)$ embeds into $(\mathbb{R}, d_1)$, where d_1 is, as usual, the Euclidean distance. (We shall describe the construction of $\overline{\mathscr{G}}$ in detail in the case where $\mathscr{G}$ is a tree, in the next section). It is clear that $\mathscr{G}$ and $\overline{\mathscr{G}}$ are quasi-isometric. Thus, by virtue of Theorem 11.67, it is meaningful to extend the notion of hyperbolicity to connected graphs (equipped with their graph distance). This way, we can give the following:

Definition 11.68. Let G be a finitely generated group and let $X \subseteq G$ be a finite and symmetric generating subset. One says that G is *hyperbolic* provided the Cayley graph $\mathrm{Cay}(G,X)$ with the graph distance ($= d_X$) is hyperbolic.

It follows from Proposition 11.47 and Theorem 11.67 that the above definition is well posed, that is, it does not depend on the particular choice of the (finite) generating subset X of G.

Example 11.69. (a) Every finite group is hyperbolic.

(b) The infinite cyclic group $\mathbb{Z}$ is hyperbolic. Note that if one takes $X := \{1, -1\}$ as a generating system, then $(\mathbb{Z}, d_X)$ is 0-hyperbolic, while for the generating system $X' := \{2, -2, 3, -3\}$ then $(\mathbb{Z}, d_{X'})$ is 1-hyperbolic (**exercise**). It follows from Proposition 11.47 and Theorem 11.67 that every group which is virtually cyclic, e.g. the infinite dihedral group $D_\infty = \langle a,b : a^2, b^2 \rangle = \langle r,s : s^2, srsr \rangle$, is hyperbolic.

(c) More generally, the free group F_r of rank $r \geq 1$ is 0-hyperbolic (**exercise**). It follows from Proposition 11.47 and Theorem 11.67 that every group which is virtually free, e.g. the *modular group* $G := \langle a,b : a^2, b^3 \rangle \cong \mathrm{SL}(2,\mathbb{Z})$, is hyperbolic.

(d) The *surface group* (i.e., the fundamental group of a closed orientable surface Σ_g of *genus* $g \geq 1$)

$$G := \pi_1(\Sigma_g) = \left\langle a_1, b_1, a_2, b_2, \ldots, a_g, b_g : \prod_{i=1}^{g} [a_i, b_i] \right\rangle$$

is hyperbolic exactly if $g \geq 2$.

(e) The free abelian group $\mathbb{Z}^d$ of rank $d \geq 2$ is *not* hyperbolic (**exercise**).

11.17 $\mathbb{R}$-trees and Asymptotic Cones of Hyperbolic Metric Spaces

A *simple undirected graph* is a pair $\mathscr{G} = (V, E)$, where V is a nonempty set of *vertices* and E, called the set of *edges*, is a subset of the set $\binom{V}{2}$ of all 2-subsets of V.

Let $\mathscr{G} = (V, E)$ be a simple undirected graph.

We say that two vertices $u, v \in V$ are *neighbors*, and we write $u \sim v$, if $\{u, v\} \in E$.

An *orientation* in $\mathscr{G}$ is a function $\phi \colon E \to V^2$ such that $\phi(\{u,v\}) \in \{(u,v),(v,u)\}$. In other words, we choose an order for each edge $\{u, v\}$. For an element $e = (u, v)$ of $\phi(E)$, we will use the notation $e^+ := u$ and $e^- := v$. The elements of $\phi(E)$ are called *oriented edges*.

$\mathscr{G}$ is said to be *locally finite* if for every $v \in V$ the number $k(v) := |\{u \in V : u \sim v\}|$ of its neighbors is finite. If, in addition, $k = k(u) = k(v)$ for all $u, v \in V$, we say that $\mathscr{G}$ is *regular* of *degree k*.

A finite (resp. infinite) *path* in $\mathscr{G}$ is a sequence $\pi = (v_0, v_1, \ldots, v_n)$ (resp. an infinite sequence $\pi = (v_0, v_1, \ldots)$) of vertices in V such that $v_i \sim v_{i+1}$ for all $i = 0, 1, \ldots n-1$ (resp. for all $i \in \mathbb{N}$). Given a finite path $\pi = (v_0, v_1, \ldots, v_n)$ we call $\ell(\pi) := n \in \mathbb{N}$ the *length* of π and we say that π connects its *initial vertex* $\pi^- := v_0$ and its *terminal vertex* $\pi^+ := v_n$. A path of length 0 is called *trivial*. The graph $\mathscr{G}$ is said to be *connected* if for every $u, v \in V$ there exists a path in $\mathscr{G}$ which connects them. If this is the case, the minimal length of a path connecting u, v is called the *graph distance* of the vertices u and v, denoted $d_{\mathscr{G}}(u, v)$, and a path realizing this minimum is called a *geodesic path* of $\mathscr{G}$. It is easy to see (**exercise**) that the map $d_{\mathscr{G}} \colon V \times V \to \mathbb{N}$ is a distance function: it is called the *graph metric* on $\mathscr{G}$. A path $\pi = (v_0, v_1, \ldots, v_n)$ in $\mathscr{G}$ is said to be *closed* (resp. *proper*) if $v_0 = v_n$ (resp. $v_i \neq v_{i+2}$ for all $i = 0, 1, \ldots, n-2$).

A *tree* is a connected simple undirected graph with no nontrivial closed proper paths. Observe that in a tree for every pair of vertices there exist a unique geodesic path connecting them.

Let $\mathscr{T} = (V, E)$ be a tree.

We now associate with the "discrete" metric space $(\mathscr{T}, d_{\mathscr{T}})$ a "continuous" metric space $(X(\mathscr{T}), d)$ in which V embeds isometrically, the same way as $(\mathbb{Z}, d_1)$ embeds into $(\mathbb{R}, d_1)$, where d_1 is, as usual, the Euclidean distance.

We fix a vertex $w \in V$ and refer to it as to a *base point*. Then there is a natural orientation $\phi = \phi_w$ in $\mathscr{T}$: given $e = \{u,v\} \in E$ one sets $\phi(\{u,v\}) = (u,v)$ if and only if $d_{\mathscr{T}}(u,w) < d_{\mathscr{T}}(v,w)$, and therefore $d(v,w) = d(u,w) + 1$ (notice that we cannot have $d_{\mathscr{T}}(u,w) = d_{\mathscr{T}}(v,w)$ since $\mathscr{T}$ is a tree).

We also set $X' := V \sqcup \bigsqcup_{e \in \phi(E)} I_e$, where the I_e's are disjoint copies of the unit interval $[0,1]$, and then *identify* the extreme points 0 and 1 of each interval I_e with the vertices e^- and e^+ in V, respectively. We denote by $X = X(\mathscr{T})$ the corresponding quotient space. Note that V embeds in X, and also I_e embeds in X for every $e \in E$.

Let $x \in X$. We say that x is *extremal* if there exists a vertex $v_x \in V$ (necessarily unique) such that x is identified with v_x. Otherwise, we say that x is *internal*. In this case we denote by $e_x \in E$ the unique edge e such that $x \in I_e$.

Now let $x \in X$ and $e \in E$ such that $x \notin I_e$. Then we denote by $x(e)$ the element of e which minimizes the graph distance from e_x (which is unique since $\mathscr{T}$ is a tree). So for example $w(e_x) = \phi(e_x)^+$.

For $x, y \in X$ we set

$$d(x,y) := \begin{cases} d_{\mathscr{T}}(x,y) & \text{if } x = v_x \text{ and } y = v_y \\ d_{\mathscr{T}}(x(e_y),x) + |y - x(e_y)| & \text{if } x = v_x,\ y \text{ is internal and } x \notin e_y \\ d_{\mathscr{T}}(y,y(e_x)) + |x - y(e_x)| & \text{if } y = v_y,\ x \text{ is internal and } y \notin e_x \\ d_{\mathscr{T}}(x(e_y),y(e_x)) + |y - x(e_y)| + |x - y(e_x)| & \text{if both } x \text{ and } y \text{ are internal and } e_x \neq e_y \\ |x - y| & \text{if } e_x = e_y, \end{cases}$$

where $|\cdot|$ is the Euclidean distance in $I_e = [0,1]$ for all $e \in E^+$. One easily shows (**exercise**) that $d\colon X \times X \to [0,+\infty)$ is a distance on X making $(X(\mathscr{T}),d)$ into a geodesic metric space. We call it the *simplicial tree* associated with the tree T.

Let $x, y \in X$. Up to exchanging x and y, only the following two cases may occur (see Figure 11.5):

(1) there exists a $v \in T$ such that $[w,x] \cap [w,y] = [w,v]$: then $d(x,y) = d(x,v) + d(v,y)$;

(2) $x \in [w,y]$: then $d(x,y) = d(y,w) - d(x,w)$.

Moreover, if $\pi = (w = v_0, v_1, \ldots, v_n)$ (resp. $\pi = (w = v_0, v_1, \ldots))$ is a finite (resp. infinite) geodesic path in $\mathscr{T}$ starting at the base point, then setting $e_i = (v_i, v_{i+1})$ we have $e_i \in \phi(E)$ and $\bigcup_{i=1}^n I_{e_i}$ (resp. $\bigcup_{i=1}^\infty I_{e_i}$) is isometrically homeomorphic to the interval $[0,n]$ (resp. $[0,+\infty)$). Analogously, given $x \in X$ the arc $[w,x]$ has length $d(w,\phi(e_x)^+) + |x - \phi(e_x)^+|$.

Definition 11.70. A *real tree*, or simply an $\mathbb{R}$-*tree*, is a geodesic metric space (X,d) such that for all $x, y \in X$ (i) there exists an arc from x to y whose image $[x,y]$ is uniquely determined and (ii) this arc can be chosen to be a geodesic segment.

Let (X,d) be an $\mathbb{R}$-tree. A point $x \in X$ is called *ordinary* if $X \setminus \{x\}$ has exactly two connected components. The points which are not ordinary are called *singular*. The $\mathbb{R}$-tree (X,d) is said to be *simplicial* provided the set of singular points is discrete and closed.

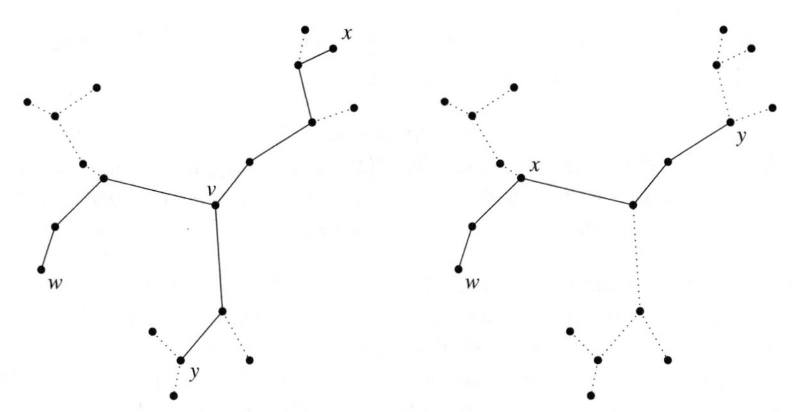

Fig. 11.5 On the left, case (1): there exists a $v \in T$ such that $[w,x] \cap [w,y] = [w,v]$ so that $d(x,y) = d(x,v) + d(v,y)$. On the right, case (2): $x \in [w,y]$ so that $d(x,y) = d(y,w) - d(x,w)$.

Condition (ii) means that, among all continuous maps $f\colon [0,D] \to X$ such that $D = d(x,y)$ and $f(0) = x$ and $f(D) = y$, there exists one (necessarily unique) which is isometric.

Example 11.71. (1) The real line $(\mathbb{R}, d)$, where d is the Euclidean distance, is a real tree.

(2) Let T be a tree. It is an **exercise** to show that the associated simplicial tree $(X(\mathscr{T}), d)$ is a simplicial real tree whose singular points are exactly the extremal points.

Proposition 11.72. *Let (X,d) be a geodesic metric space. The following conditions are equivalent.*

(a) *X is a real tree;*
(b) *X is 0-hyperbolic.*

Proof. Suppose that X is a real tree. Recall that in this case given any two points $x, y \in X$, there exists a geodesic arc from x to y which is unique. Let now $x, y, z \in X$, let $s \in [z,y]$, and consider the (images of the) arcs $[s,x]$, $[s,y] \cup [y,x]$ and $[s,z] \cup [z,x]$ connecting s and x. By uniqueness of the arc we necessarily have $[s,x] = [s,y] \cup [y,x] = [s,z] \cup [z,x]$. This implies that s must belong to $[y,x] \cup [z,x]$ and therefore $d(s, [y,x] \cup [z,x]) = 0$. It follows that the geodesic triangle $T(x,y,z)$ is 0-thin. Therefore X is 0-hyperbolic.

Conversely, suppose that X is geodesic and 0-hyperbolic. Let $x, y \in X$. Suppose that there exist two geodesic segments connecting them, say $f\colon [0,D] \to X$ and $f'\colon [0,D] \to X$ such that $f(0) = f'(0) = x$ and $f(D) = f'(D) = y$, where $D = d(x,y)$. Let us denote by $[x,y]$ (resp. $[x,y]'$) the image of f (resp. f'). For every $t \in (0,D)$ consider the geodesic triangle $T(f(t),x,y) := [f(t),x] \cup [f(t),y] \cup [x,y]'$, where $[f(t),x]$ is the image of the restriction of f to $[0,t]$ and $[f(t),y]$ is the image of the restriction of f to $[t,D]$. Since this is 0-thin, any point $z \in [x,y]'$ is at 0-distance from $[f(t),x] \cup [f(t),y] = [x,y]$, equivalently $z \in [x,y]$. This shows that $[x,y]' \subseteq [x,y]$.

By exchanging the roles of f and f' we deduce that $[x,y] \subseteq [x,y]'$. It follows that $[x,y] = [x,y]'$. We deduce that X is a real tree. □

Theorem 11.73. *Let $\mathscr{T} = (V,E)$ be a locally-finite infinite tree and let ω be a free ultrafilter. Then the asymptotic cone $K_\omega := (K_\omega(V, d_\mathscr{T}), d_\omega)$ is isometrically homeomorphic to the simplicial tree $(X(\mathscr{T}), d)$ associated with $\mathscr{T}$. In particular, K_ω is isometrically homeomorphic to a simplicial $\mathbb{R}$-tree.*

Proof. Let $w \in V$ denote, as usual, a base-point. For $v \in V$ we denote by $[w,v] \subseteq X(\mathscr{T})$ the geodesic segment connecting the base point with v, and by $V^+(v) = \{u \in V : v \in [w,u]\}$ the *future cone* based at v. Note that $v \in V^+(v)$.

Recall that $\mathscr{T}$ is locally finite. So for $v \in V$ we denote by $v^+(1), v^+(2), \ldots, v^+(k)$, $k := k(v) - 1$, the neighbors of v such that $v^+(i) \in V^+(v)$ for $i = 1, 2, \ldots, k$. In this way, $V^+(v) = \{v\} \bigcup \left(\sqcup_{i=1}^k V(v^+(i))\right)$ (recall that $\mathscr{T}$ is a tree).

Let now $v = (v_n)_{n \in \mathbb{N}}$ be an unbounded moderate sequence in $(V, d_\mathscr{T})$.

Set $w_0 := w$ and consider the forward cones $V^+(w_0^+(i))$, $i = 1, 2, \ldots, k(w) - 1$. Then there exists (here we are using that ω is free) $1 \leq i_1 \leq k(w) - 1$ such that setting $w_1 := w^+(i_1)$ and $V_1 := \{w_0, w_1\} \cup V^+(w_1)$ one has $I_1 := \{n \in \mathbb{N} : v_n \in V_1\} \in \omega$. For $n \in \mathbb{N} \setminus I_1$ we set $u_n := w \in V_1$, and denote by u the moderate sequence obtained from v by replacing v_n by u_n for all $n \in \mathbb{N} \setminus I_1$. It is clear that $[v]_\omega = [u]_\omega$ in K_ω. Thus, up to passing from v to u, we may suppose that $v_n \in V_1$ for all $n \in \mathbb{N}$. Continuing in this way, we construct an infinite geodesic path $\pi = (w = w_0, w_1, w_2 \ldots)$ such that for every $m \in \mathbb{N}$ we clearly have $d(w, w_m) = m$ and, up to changing the values of v on the complement of a set in ω, we also have $v_n \in V_m := \{w_0, w_1, \ldots, w_m\} \cup V^+(w_m)$ for all $n, m \in \mathbb{N}$. We have shown that for every moderate sequence v in T we can find an infinite geodesic $\overline{\pi(v)}$ in $X(\mathscr{T})$ and a moderate sequence $\overline{v} = (\overline{v}_n)_{n \in \mathbb{N}}$ in V such that $[\overline{v}]_\omega = [v]_\omega$ and $\overline{v}_n \in \overline{\pi(v)}$ for all $n \in \mathbb{N}$. We call $\overline{v}$ a *geodesic-representative* of v. It is also clear that if $v' \sim_\omega v$ then $\overline{\pi(v)} = \overline{\pi(v')}$.

For every $x \in X(\mathscr{T})$ and $\alpha \in [0,1]$ we denote by $\alpha x \in X(\mathscr{T})$ the unique element in the geodesic segment $[w,x] \subseteq X(\mathscr{T})$ such that $d(w, \alpha x) = \alpha d(w,x)$. Note that, in particular, $\alpha \overline{v}_n \in \overline{\pi(v)}$ for all $n \in \mathbb{N}$.

We then define $F \colon K_\omega \to X(\mathscr{T})$ by setting $F(1_K) = w$ and, for v unbounded,

$$F([v]_\omega) = \lim_\omega \frac{1}{n} \overline{v}_n \in \overline{\pi(v)}.$$

Let us show that F is isometric. Let v and u be two moderate sequences and denote by $\overline{\pi(v)}$ and $\overline{\pi(u)}$ (resp. $\overline{v}$ and $\overline{u}$) their associated infinite geodesic in $X(\mathscr{T})$ (resp. their geodesic-representatives in $[v]_\omega$ and $[u]_\omega$, respectively). We distinguish two cases.

(1) There exists a $w' \in V$ such that $[w, F([v]_\omega)] \cap [w, F([u]_\omega)] = [w, w']$. We then have

$$\begin{aligned} d_\omega([v]_\omega, [u]_\omega) &= \lim_\omega \frac{d_\mathscr{T}(\overline{v}_n, \overline{u}_n)}{n} \\ &= \lim_\omega \frac{d_\mathscr{T}(\overline{v}_n, w') + d_\mathscr{T}(w', \overline{u}_n)}{n} \\ &= d(F([v]_\omega), w') + d_X(w', F([u]_\omega)) \end{aligned}$$

$$= d(F([v]_\omega), F([u]_\omega)).$$

(2) Up to exchanging the roles of v and u we have $F([v]_\omega) \in [w_0, F([u]_\omega)]$). Then

$$\begin{aligned} d_\omega([v]_\omega, [u]_\omega) &= \lim_\omega \frac{d_{\mathscr{T}}(\overline{v}_n, \overline{u}_n)}{n} \\ &= \lim_\omega \frac{d_{\mathscr{T}}(\overline{v}_n, w) - d_{\mathscr{T}}(\overline{u}_n, w)}{n} \\ &= d(F([v]_\omega), w) + d(F([u]_\omega), w) \\ &= d(F([v]_\omega), F([u]_\omega)). \end{aligned}$$

It follows that F is isometric. Let us show that F is also surjective.

For every $v \in V$ we arbitrarily choose an infinite geodesic path $\underline{\pi(v)} = (v_0 = w, v_1, \ldots, v_{m-1}, v_m = v, v_{m+1}, \ldots)$ (thus $m = d(w,v)$) and denote by $\overline{\pi(v)} = \sqcup_{i=0}^{\infty} I_{(v_i, v_{i+1})}$ the corresponding infinite geodesic in $X(\mathscr{T})$. For $x \in X(\mathscr{T})$ we set $\overline{\pi(x)} := \overline{\pi(\phi(e_x)^+)}$. For any real $\alpha \geq 0$ we denote by αx the unique element in $\overline{\pi(x)}$ such that $d(w, \alpha x) = \alpha d(w,x)$ and set $\lfloor x \rfloor := \phi(e_x)^+ \in V$.

For $x \in X(\mathscr{T})$ we then set

$$v(x) = (v_n(x))_{n\in\mathbb{N}} := (\lfloor nx \rfloor)_{n\in\mathbb{N}}$$

and observe that

$$d(x,w) - \frac{1}{n} = \frac{d(nx,w) - 1}{n} \leq \frac{d_{\mathscr{T}}(\lfloor nx \rfloor, w)}{n} \leq \frac{d(nx,w)}{n} = d(x,w) \tag{11.39}$$

for all $n \geq 1$, so that in particular $v(x) \in \mathscr{M}(V, d_{\mathscr{T}})$, i.e. $v(x)$ is a moderate sequence. Moreover, $\underline{\pi(v(x))} = \underline{\pi(x)}$ and $\overline{v(x)} \sim_\omega v(x)$. Finally from (11.39) we deduce

$$F([v(x)]_\omega) = \lim_\omega \frac{1}{n} v_n(x) = \lim_\omega \frac{1}{n} \lfloor nx \rfloor = x.$$

This shows that F is also surjective. □

Corollary 11.74. *The asymptotic cone of the free group F_n is isometrically homeomorphic to the regular simplicial tree of degree $k = 2n$.* □

We conclude this section with the following theorem that provides a source of examples of asymptotic cones (cf. Theorem 11.73).

Theorem 11.75. *Let (X,d) be a hyperbolic metric space, and let $\omega \in \beta\mathbb{N}$ be a free ultrafilter. Then the asymptotic cone $K_\omega := (K_\omega(X,d), d_\omega)$ is an ℝ-tree.*

Proof. Let $\delta \geq 0$ be such that property (4) of Proposition 11.63 is satisfied with $\delta_4 = \delta$. Let $x = (x_n)_{n\in\mathbb{N}}, y = (y_n)_{n\in\mathbb{N}}, z = (z_n)_{n\in\mathbb{N}}, w = (w_n)_{n\in\mathbb{N}} \in \mathscr{M}(X,d)$. For each $n \in \mathbb{N}$ we have

$$d(x_n, z_n) + d(y_n, w_n) \leq \max\{d(x_n, y_n) + d(z_n, w_n), d(x_n, w_n) + d(y_n, z_n)\} + \delta,$$

so that

$$
\begin{aligned}
& d_\omega([x]_\omega,[z]_\omega)+d_\omega([y]_\omega,[w]_\omega) \\
& = \lim_\omega \frac{d(x_n,z_n)}{n} + \lim_\omega \frac{d(y_n,w_n)}{n} \\
& = \lim_\omega \left(\frac{d(x_n,z_n)}{n} + \frac{d(y_n,w_n)}{n} \right) \\
& \leq \lim_\omega \left(\max \left\{ \frac{d(x_n,y_n)+d(z_n,w_n)}{n}, \frac{d(x_n,w_n)+d(y_n,z_n)}{n} \right\} + \frac{\delta}{n} \right) \\
& = \max \left\{ \lim_\omega \frac{d(x_n,y_n)+d(z_n,w_n)}{n}, \lim_\omega \frac{d(x_n,w_n)+d(y_n,z_n)}{n} \right\} \\
& = \max \left\{ d_\omega([x]_\omega,[y]_\omega)+d_\omega([z]_\omega,[w]_\omega), d_\omega([x]_\omega,[w]_\omega)+d_\omega([y]_\omega,[z]_\omega) \right\}.
\end{aligned}
$$

By Remark 11.65, this shows that K_ω is 0-hyperbolic, so that, by Proposition 11.72, it is an $\mathbb{R}$-tree. □

11.18 Notes

Filters were introduced by Henri Cartan in 1937 [51, 52] and subsequently used by Bourbaki in their book *Topologie Générale* [32], as an alternative to the similar notion of a net developed in 1922 by Eliakim Hastings Moore and Herman Lyle Smith in [243].

The Fréchet filter, also called the *cofinite filter*, is named after Maurice René Fréchet.

A classical monograph on the theory of ultrafilters is the one by William Wistar Comfort and Stylianos Negrepontis [76].

The *Stone–Čech compactification* is a construction yielding a universal map from a topological space X to a compact Hausdorff space βX, the largest, most general compact Hausdorff space "generated" by X, in the sense that any continuous map from X to a compact Hausdorff space factors through βX (in a unique way). If X is a Tychonoff (i.e., completely regular Hausdorff) space, then the map from X to its image in βX is a homeomorphism, so X can be thought of as a (dense) subspace of βX; every other compact Hausdorff space that densely contains X is a quotient of βX. For general topological spaces X, the map from X to βX need not be injective. A form of the *axiom of choice* is required to prove that every topological space has a Stone–Čech compactification. The Stone–Čech compactification occurs implicitly in a paper by Andrei N. Tychonoff [338] in (1930) and was given explicitly by Marshall Harvey Stone [323] and Eduard Čech [71] in 1937. For more on the Stone–Čech compactification, we refer to the monograph by Neil Hindman and Dona Strauss [174].

Hyperreal numbers were defined in the 1960s by Abraham Robinson [291] who developed his *nonstandard analysis* theory. However, his definition, based on model theory, is nonconstructive. Hyperreal numbers, aside from their use in nonstandard analysis, have no necessary relationship to model theory or first order logic, although they were discovered by the application of model theoretic techniques from logic. Hyperreal fields were in fact originally introduced by Edwin Hewitt [168] in 1948 by purely algebraic techniques, using an ultrapower construction.

The study of $\mathbb{R}$-trees and their applications to different areas of geometry and topology has been very intensive in the recent years (see [27] and the references therein). In particular, as was pointed out by Misha Gromov, $\mathbb{R}$-trees appear naturally in the asymptotic geometry of hyperbolic metric spaces [139, 140]. The aim of [100] is to present some explicit constructions (such as that of *universal* $\mathbb{R}$*-trees* as certain spaces of functions) related to this observation.

Theorem 11.75 is due to Gromov [140, Section 2.A]. The following converse (ibidem) holds: *Let* (X,d) *be a geodesic metric space. If for every free ultrafilter* ω *the asymptotic cone* $(K_\omega(X,d),d_\omega)$ *is a real tree, then* (X,d) *is hyperbolic.*

11.19 Exercises

Exercise 11.1. Let $\mathscr{F} \subseteq \mathscr{P}(\mathbb{N})$ be the Fréchet filter and let $S \subseteq \mathbb{N}$ be an infinite set. Show that the set $\Omega := \mathscr{F} \cup \{S\} \subseteq \mathscr{P}(\mathbb{N})$ is saturated.

Exercise 11.2. Let $\mathscr{B} \subseteq \mathscr{P}(\mathbb{N})$ be a base of a filter (cf. Definition 11.3). Show that the set $\mathscr{F} = \{A \in \mathscr{P}(\mathbb{N}) : A$ contains an element of $\mathscr{B}\}$ is a filter.

Exercise 11.3. A *finitely additive probability* $\{0,1\}$*-valued measure* on $\mathbb{N}$ is a map $\mu\colon \mathscr{P}(\mathbb{N}) \to \{0,1\}$ such that

(M1) $\mu(\mathbb{N}) = 1$;
(M2) if $A,B \subseteq \mathbb{N}$ and $A \cap B = \varnothing$, then $\mu(A \cup B) = \mu(A) + \mu(B)$.

Prove that a finitely additive probability $\{0,1\}$-valued measure μ on $\mathbb{N}$ satisfies the following properties:

(M3) $\mu(\varnothing) = 0$;
(M4) $\mu(A \cup B) = \mu(A) + \mu(B) - \mu(A \cap B)$;
(M5) $\mu(A \cup B) \leq \mu(A) + \mu(B)$;
(M6) $A \subseteq B \Rightarrow \mu(B \setminus A) = \mu(B) - \mu(A)$;
(M7) $A \subseteq B \Rightarrow \mu(A) \leq \mu(B)$,

for all $A,B \in \mathscr{P}(\mathbb{N})$.

Exercise 11.4. With the notation of the previous exercise, prove the following:

(i) Let μ be a finitely additive probability $\{0,1\}$-valued measure on $\mathbb{N}$. Then the set $\omega_\mu := \{A \subseteq \mathbb{N} : \mu(A) = 1\}$ is an ultrafilter on $\mathbb{N}$. If in addition $\mu(F) = 0$ for every finite subset $F \subseteq \mathbb{N}$, then ω_μ is non-principal;
(ii) conversely, given an ultrafilter ω on $\mathbb{N}$, the map $\mu_\omega\colon \mathscr{P}(\mathbb{N}) \to \{0,1\}$ defined by $\mu_\omega(A) = 1$ if $A \in \omega$ and $\mu_\omega(A) = 0$ otherwise, is a finitely additive probability $\{0,1\}$-valued measure on $\mathbb{N}$. If in addition ω is free, then $\mu_\omega(F) = 0$ for every finite subset $F \subseteq \mathbb{N}$;

(iii) the maps $\mu \mapsto \omega_\mu$ and $\omega \mapsto \mu_\omega$ are inverse to each other and induce a bijective correspondence between the set of all finitely additive probability $\{0,1\}$-valued measures on $\mathbb{N}$ (resp. such that $\mu_\omega(F) = 0$ for every finite subset $F \subseteq \mathbb{N}$) and the set of all ultrafilters (resp. free ultrafilters) on $\mathbb{N}$.

Exercise 11.5. Recall (cf. Section 14.3) that if $\mathscr{M}\mathscr{P}(\mathbb{N})$ (resp. $\mathscr{M}(\mathbb{N})$) denotes the set of all finitely additive probability measures (resp. of all means) on $\mathbb{N}$, then there exists a natural bijective map $\Phi\colon \mathscr{M}\mathscr{P}(\mathbb{N}) \to \mathscr{M}(\mathbb{N})$. Let $\omega \subseteq \mathscr{P}(\mathbb{N})$ be an ultrafilter and consider the map $m_\omega\colon \ell^\infty(\mathbb{N}) \to \mathbb{R}$ given by $m_\omega(x) := x(\omega) = \lim_\omega x_n$ (cf. (11.3)).

(i) Show that $m_\omega \in \ell^\infty(\mathbb{N})'$, the topological dual of $\ell^\infty(\mathbb{N})$ (cf. Theorem 11.3);
(ii) show that, with the notation from the previous exercises, one has $\Phi^{-1}(\mu_\omega) = m_\omega$.

Exercise 11.6. Show that the map $n \mapsto \omega_n$, where $\omega_n = \{A \subseteq \mathbb{N} : n \in A\}$ is the principal ultrafilter based at n, is an *injective* map from $\mathbb{N}$ into $\beta\mathbb{N}$, the set of all ultrafilters on $\mathbb{N}$.

Exercise 11.7 (Theorem 11.25). Let $\mathscr{A} \subseteq \mathscr{P}^*(\mathbb{N})$ be a set of nonempty subsets of $\mathbb{N}$ which covers $\mathbb{N}$ (i.e. $\cup_{A\in\mathscr{A}} A = \mathbb{N}$). Show that $\mathscr{A}$ admits no finite subcover if and only if the set $\Omega = \{\mathbb{N} \setminus A : A \in \mathscr{A}\}$ is saturated.

Exercise 11.8 (Theorem 11.25). Let (X,d) be a metric space and ω an ultrafilter. Let $(x_n)_{n\in\mathbb{N}}$ be a sequence in X converging along ω. Suppose that there exists a ball $B \subseteq X$ such that $\{n \in \mathbb{N} : x_n \in B\} \in \omega$. Show that $\lim_\omega x_n \in \overline{B}$, the closure of B.

Exercise 11.9 (Theorem 11.25). Let $f, g\colon Y \to X$ be two continuous mappings between topological spaces with X Hausdorff. Suppose there exists a dense subset $Z \subseteq Y$ such that $f|_Z = g|_Z$. Show that $f = g$.

Exercise 11.10. Let $(x_n)_{n\in\mathbb{N}}$ be a sequence in a metric space (X,d). Show that the following conditions are equivalent:

(a) $\sup_{n,m\in\mathbb{N}} d(x_n, x_m) < \infty$;
(b) there exists an $x^0 \in X$ such that $\sup_{n\in\mathbb{N}} d(x^0, x_n) < \infty$;
(c) for every $x^0 \in X$ one has $\sup_{n\in\mathbb{N}} d(x^0, x_n) < \infty$.

Exercise 11.11. Let (X,d) be a metric space and denote by $\mathrm{Cau}(X)$ the set of all Cauchy sequences in X. For $x = (x_n)_{n\in\mathbb{N}}$ and $y = (y_n)_{n\in\mathbb{N}}$ in $\mathrm{Cau}(X)$ set $x \sim y$ provided $\lim_{n\to\infty} d(x_n, y_n) = 0$.

(i) Show that $\sim$ is an equivalence relation on $\mathrm{Cau}(X)$;
(ii) show that, denoting by $[x] := \{y \in \mathrm{Cau}(X) : y \sim x\}$ the equivalence class of $x \in \mathrm{Cau}(X)$, the quantity $\tilde{d}([x],[y]) := \lim_{n\to\infty} d(x_n, y_n)$ is well defined;
(iii) show that if $(x_n)_{n\in\mathbb{N}}$ and $(y_n)_{n\in\mathbb{N}}$ are in $\mathrm{Cau}(X)$, then $(d(x_n, y_n))_{n\in\mathbb{N}}$ is a Cauchy sequence in $\mathbb{R}$.

Exercise 11.12 (Proposition 11.26). Let (X,d) be a metric space and denote by $(\tilde{X}, \tilde{d})$ its metric completion. Show that the following holds:

(i) the map $x \mapsto [(x_n)_{n\in\mathbb{N}}]$, where $x_n = x$ for all $n \in \mathbb{N}$, yields an isometric embedding of X into $\tilde{X}$;
(ii) X is dense in $\tilde{X}$;
(iii) X is totally bounded if and only if $\tilde{X}$ is totally bounded.

Exercise 11.13. (i) Show that a subset of a complete metric space is totally bounded if and only if it is pre-compact (i.e., its closure is compact).
(ii) Show that every compact metric space is totally bounded.
(iii) Show that a subset of the k-dimensional real space $(\mathbb{R}^k, d)$ with Euclidean distance is totally bounded if and only if it is bounded.
(iv) Show that the unit ball in a Hilbert space, or more generally in a Banach space, is totally bounded if and only if the space has finite dimension.

Exercise 11.14 (Proposition 11.27). Let $x = (x_n)_{n\in\mathbb{N}}$ be a sequence in a metric space (X,d) and suppose that there exists an $\varepsilon_0 > 0$ such that $d(x_n, x_m) \geq \varepsilon_0$ for all distinct $n, m \in \mathbb{N}$ (note that X is not totally bounded). Show that (even if X is complete) x does not converge in X along any free ultrafilter.

Exercise 11.15. Let (X,d) be a metric space. Show that for all $x = (x_n)_{n\in\mathbb{N}}$ and $y = (y_n)_{n\in\mathbb{N}}$ in $\ell^\infty(X)$ the quantity $d_\infty(x,y) := \sup_{n\in\mathbb{N}} d(x_n, y_n)$ is finite.

Exercise 11.16. Let (X,d) be a metric space. Show that $(\ell^\infty(X), d_\infty)$ is complete if and only if (X,d) is complete.

Exercise 11.17. Let (X,d) be a metric space and let ω be an ultrafilter. For $x = (x_n)_{n\in\mathbb{N}}$ and $y = (y_n)_{n\in\mathbb{N}}$ in $\ell^\infty(X)$ write $x \sim_\omega y$ provided $\lim_\omega d(x_n, y_n) = 0$.

(i) Show that $\sim_\omega$ is an equivalence relation on $\ell^\infty(X)$;
(ii) show that if $x' = (x'_n)_{n\in\mathbb{N}}$ and $y' = (y'_n)_{n\in\mathbb{N}}$ in $\ell^\infty(X)$ satisfy $x' \sim_\omega x$ and $y' \sim_\omega y$, then $\lim_\omega d(x_n, y_n) = \lim_\omega d(x'_n, y'_n)$;
(iii) show that $\lim_\omega d(x_n, y_n) = \inf_{x' \sim_\omega x, y' \sim_\omega y} d_\infty(x', y')$;
(iv) deduce that the quantity $d_\omega([x]_\omega, [y]_\omega) := \lim_\omega d(x_n, y_n)$ is well defined.

Exercise 11.18. Let (X,d) be a metric space and let ω be an ultrafilter. Denote by (X_ω, d_ω) the corresponding metric ultrapower and by $\iota_\omega \colon X \to X_\omega$ the diagonal isometric embedding. Show that ι_ω is surjective if and only if X is compact.

Exercise 11.19 (Example 11.33). Let $(X_n)_{n\in\mathbb{N}}$ be a sequence of sets. Equip each X_n with the discrete metric d_n^*. Let ω be an ultrafilter.

(i) Let $x = (x_n)_{n\in\mathbb{N}}$ and $y = (y_n)_{n\in\mathbb{N}}$ in $\prod_{n\in\mathbb{N}} X_n$. Show that $x \sim_\omega y$ (i.e. that $\lim_\omega d_n^*(x_n, y_n) = 0$) if and only if there exists an $A \in \omega$ such that $x_n = y_n$ for all $n \in A$;
(ii) show that d_∞^* (resp. d_ω^*) coincides with the discrete metric on $\prod_{n\in\mathbb{N}} X_n$ (resp. $\mathbf{X}_\omega = (\prod_{n\in\mathbb{N}} X_n) / \sim_\omega$).

Exercise 11.20. Let (G,d) be a metric group (i.e. d is a bi-invariant metric). Show that $d(g,h) = d(g^{-1}, h^{-1})$ and $d(hg_1h^{-1}, hg_2h^{-1}) = d(g_1, g_2)$ for all $g, h, g_1, g_2 \in G$.

Exercise 11.21. Let G be a group and let d^* be the discrete metric on G. Show that (G, d^*) is a metric group. In other words, show that the discrete metric d^* on a group G is bi-invariant.

Exercise 11.22. Let S_n be the symmetric group on n letters, i.e. the group of all bijective maps $f: X \to X$ of a set X of cardinality n. For $f, g \in S_n$ set $d_n^H(g,h) := 1 - |\{x \in X : g(x) = h(x)\}|/n$ for all $g, h \in S_n$. Show that the map $d_n^H: S_n \times S_n \to \mathbb{R}$ (called the (normalized) Hamming distance) is a bi-invariant distance function on S_n so that (S_n, d_n^H) is a metric group.

Exercise 11.23. Let $M_n(\mathbb{C})$ denote the complex vector space of all $n \times n$ complex matrices. Denote by $\mathrm{tr}: M_n(\mathbb{C}) \to \mathbb{C}$ the trace and, for $A = (a_{i,j})_{i,j=1}^n \in M_n(\mathbb{C})$ denote by $A^* := (\overline{a_{j,i}})_{i,j=1}^n \in M_n(\mathbb{C})$ the conjugate-transpose (or adjoint) of A.

(i) Show that $\mathrm{tr}(A^*A) = \sum_{i,j=1}^n |a_{i,j}|^2$ for all $A = (a_{i,j})_{i,j=1}^n \in M_n(\mathbb{C})$.

(ii) Show that the Hilbert–Schmidt norm $\|\cdot\|_{HS}$, defined by $\|A\|_{HS} := \sqrt{\mathrm{tr}(A^*A)} = \sqrt{\sum_{i,j=1}^n |a_{i,j}|^2}$ for all $A = (a_{i,j})_{i,j=1}^n \in M_n(\mathbb{C})$, is a norm on $M_n(\mathbb{C})$. In other words, check that:

  (1) $\|A\|_{HS} \geq 0$ and $\|A\|_{HS} = 0$ if and only if $A = 0$, for all $A \in M_n(\mathbb{C})$ (positivity and non-degeneracy);
  (2) $\|\lambda A\|_{HS} = |\lambda| \|A\|_{HS}$ for all $A \in M_n(\mathbb{C})$ (homogeneity);
  (3) $\|A+B\|_{HS} \leq \|A\|_{HS} + \|B\|_{HS}$ for all $A, B \in M_n(\mathbb{C})$ (triangular inequality).

(iii) Deduce that the map $d_n^{HS}: M_n(\mathbb{C}) \times M_n(\mathbb{C}) \to \mathbb{R}$ defined by $d_n^{HS}(A,B) = \|A - B\|_{HS}$ for all $A, B \in M_n(\mathbb{C})$ is a distance function on $M_n(\mathbb{C})$.

(iv) A matrix $A \in M_n(\mathbb{C})$ is unitary provided $A^*A = I = AA^*$. Denote by $U_n \subseteq M_n(\mathbb{C})$ the set consisting of all $n \times n$ unitary matrices. Show that the restriction of d_n^{HS} to the unitary group is bi-invariant, so that (U_n, d_n^{HS}) is a metric group.

Exercise 11.24. Let $\mathbb{F}$ be a field and denote by $M_n(\mathbb{F})$ (resp. $\mathrm{GL}(n,\mathbb{F}) \subseteq M_n(\mathbb{F})$) the $\mathbb{F}$-algebra (resp. the general linear group) of $n \times n$ matrices (resp. invertible matrices) with coefficients in $\mathbb{F}$. The (normalized) rank of a matrix $A \in M(k,\mathbb{F})$ is the non-negative number $\rho_n(A) := \frac{\dim \mathrm{coker}(A)}{n} = 1 - \frac{\dim \ker(A)}{n}$.

(i) Show that the map $\rho_n: M_n(\mathbb{F}) \to \{0, 1/k, 2/k, \ldots, 1\}$ satisfies the following properties:

  (1) $\rho_n(A) = 1$ if and only if $A \in \mathrm{GL}(n,\mathbb{F})$, in particular $\rho_n(I) = 1$;
  (2) $\rho_n(A) = 0$ if and only if $A = 0$ (the zero matrix);
  (3) $\rho_n(A+B) \leq \rho_n(A) + \rho_n(B)$;
  (4) $\rho_n(AB) \leq \min\{\rho_n(A), \rho_n(B)\}$;
  (5) $\rho_n(CAD) = \rho_n(A)$.

  for all $A, B \in M_n(\mathbb{F})$ and $C, D \in \mathrm{GL}(n,\mathbb{F})$.

(ii) Deduce that the map $d_n: M_n(\mathbb{F}) \times M_n(\mathbb{F}) \to \mathbb{R}$ defined by $d_n(A,B) = \rho_n(A - B)$ for all $A, B \in M_n(\mathbb{F})$ is a distance function on $M_n(\mathbb{F})$.

(iii) Show that the restriction of d_n to $\mathrm{GL}(n,\mathbb{F})$ is bi-invariant, so that $(\mathrm{GL}(n,\mathbb{F}), \rho_n)$ is a metric group.

Exercise 11.25. Let $\mathbf{G} = ((G_n, 1_{G_n}, d_n))_{n \in \mathbb{N}}$ be a sequence of (pointed) metric groups. Let also $x, x', y, y' \in \ell^\infty(\mathbf{G})$.

(i) Show that $x \sim_\omega x'$ and $y \sim_\omega y' \Rightarrow xy \sim_\omega x'y'$ (cf. (11.21));

(ii) show that $x \sim_\omega y \Rightarrow x^{-1} \sim_\omega y^{-1}$ (cf. (11.22));
(iii) show that the set $N_\omega := \{x \in \ell^\infty(\mathbf{G}) : x \sim_\omega 1\}$ is a normal subgroup of $\ell^\infty(\mathbf{G})$ (cf. Proposition 11.36);
(iv) show that the metric d_ω on $\mathbf{G}_\omega = \ell^\infty(\mathbf{G})/N_\omega$ given by (11.24) is bi-invariant.

Exercise 11.26. Let $\mathbb{F}$ be a field, let $\mathscr{A}$ be a unital $\mathbb{F}$-algebra, and let ω be an ultrafilter.

(i) For $x = (x_n)_{n\in\mathbb{N}}$, $y = (y_n)_{n\in\mathbb{N}}$ in $\mathscr{A}^\mathbb{N}$ and $a \in \mathscr{A}$, define $x+y, xy$ and ax in $\mathscr{A}^\mathbb{N}$ by setting $x+y := (x_n + y_n)_{n\in\mathbb{N}}$, $xy := (x_n y_n)_{n\in\mathbb{N}}$ and $ax = (ax_n)_{n\in\mathbb{N}}$. Show that, this way, $\mathscr{A}^\mathbb{N}$ is a unital $\mathbb{F}$-algebra.
(ii) For $x = (x_n)_{n\in\mathbb{N}}$, $y = (y_n)_{n\in\mathbb{N}}$ in $\mathscr{A}^\mathbb{N}$ set $[x]_\omega + [y]_\omega := [x+y]_\omega$, $[x]_\omega [y]_\omega := [xy]_\omega$ and $a[x]_\omega := [ax]_\omega$ in $\mathscr{A}_\omega = \mathscr{A}^\mathbb{N}/\sim_\omega$. Show that these operations in $\mathscr{A}_\omega$ are well defined and endow it with the structure of a unital $\mathbb{F}$-algebra.

Exercise 11.27. Let $\mathbb{F}$ be a field and let ω be an ultrafilter. Show that the set $I_\omega := \{x \in \mathbb{F}^\mathbb{N} : x \sim_\omega 0\} \subseteq \mathbb{F}^\mathbb{N}$ is a maximal ideal in $\mathbb{F}^\mathbb{N}$ and that indeed $\mathbb{F}_\omega = \mathbb{F}^\mathbb{N}/I_\omega$.

Exercise 11.28. Let $(\mathbb{F}, \preceq)$ be an ordered field (e.g. $\mathbb{F} = \mathbb{R}$) and let ω be an ultrafilter. Recall that $\preceq$ is a total order such that (1) $a \preceq b$ implies $a + c \preceq b + c$ and (2) $0 \preceq a$ and $0 \preceq b$ imply $0 \preceq ab$, for all $a, b, c \in \mathbb{F}$. Show that the ultrafield $\mathbb{F}_\omega$ is also orderable.

Exercise 11.29. Let $\mathbb{F}$ be a field and let $\mathscr{A}$ be an $\mathbb{F}$-algebra. Let $\Phi\colon M_k(\mathbb{F})^\mathbb{N} \to M_k(\mathbb{F}^\mathbb{N})$ denote the map defined by

$$\Phi\left((A_n)_{n\in\mathbb{N}}\right) = \left((a_n(i,j))_{n\in\mathbb{N}}\right)_{i,j=1}^k$$

where $A_n = (a_n(i,j))_{i,j=1}^k \in M_k(\mathbb{F})$ for all $n \in \mathbb{N}$.

(i) Show that Φ is a unital isomorphism of $\mathbb{F}$-algebras;
(ii) show that the induced map $\overline{\Phi}\colon M_k(\mathbb{F})_\omega \to M_k(\mathbb{F}_\omega)$ defined by

$$\overline{\Phi}\left([A]_\omega\right) = \left([(a_n(i,j))_{n\in\mathbb{N}}]_\omega\right)_{i,j=1}^k$$

for all $A = (A_n)_{n\in\mathbb{N}} \in M(k,\mathbb{F})^\mathbb{N}$ is also a unital isomorphism of $\mathbb{F}$-algebras.

Exercise 11.30. Let $\mathbb{F}$ be a field and let ω be an ultrafilter. Let also $\mathbf{G} = (G_n, d_n)_{n\in\mathbb{N}}$ be a sequence of metric groups. Suppose there exists an integer $k = k(\mathbf{G}) \geq 1$ and an injective group homomorphism $\varphi_n\colon G \to \mathrm{GL}(k,\mathbb{F})$ for every $n \in \mathbb{N}$. Show that the product map $\prod_{n\in\mathbb{N}} \varphi_n\colon \mathbf{G}_\omega \to \mathrm{GL}(k,\mathbb{F})_\omega$ defined by

$$\left(\prod_{n\in\mathbb{N}} \varphi_n\right)([g]_\omega) := [(\varphi_n(g_n))_{n\in\mathbb{N}}]_\omega$$

for all $g = (g_n)_{n\in\mathbb{N}} \in \ell^\infty(\mathbf{G})$ is well defined and injective.

Exercise 11.31. Let (X, d) be a metric space. Show that the set $\mathscr{M}(X, d)$ of moderate sequence in X is independent of the base point.

Exercise 11.32. Let (X,d) be a metric space and ω an ultrafilter.

(i) Show that the relation $\sim_\omega$ in $\mathscr{M}(X,d)$ defined by $x \sim_\omega y$ if $\lim_\omega d(x_n,y_n)/n - 0$ for all $x = (x_n)_{n\in\mathbb{N}}$ and $y = (y_n)_{n\in\mathbb{N}}$ in $\mathscr{M}(X,d)$ is an equivalence relation;
(ii) consider $K_\omega(X,d) = \mathscr{M}(X,d)/\sim_\omega$ and show that the map $d_\omega \colon K_\omega(X,d) \times K_\omega(X,d) \to [0,+\infty)$ defined by $d_\omega([x]_\omega,[y]_\omega) := \lim_\omega \frac{d(x_n,y_n)}{n}$ for all $x = (x_n)_{n\in\mathbb{N}}$ and $y = (y_n)_{n\in\mathbb{N}}$ in $\mathscr{M}(X,d)$ is well defined (where $[x]_\omega \in K_\omega(X,d)$ is the $\sim_\omega$ class of x).

Exercise 11.33. Let (G,d) be a metric group and ω an ultrafilter. Show that $N_\omega := \{g = (g_n)_{n\in\mathbb{N}} \in \mathscr{M}(G,d) : \lim_\omega \frac{d(g_n,1_G)}{n} = 0\}$ is a normal subgroup of $\mathscr{M}(G,d)$ and that $(K_\omega(G,d),d_\omega)$ is a metric group.

Exercise 11.34. Show that quasi-isometry is an equivalence relation on metric spaces.

Exercise 11.35 (Proposition 11.47). Let G be a finitely generated group, and H be a finite index subgroup of G. Show that G and H with any word metric are quasi-isometric.

Exercise 11.36 (Proposition 11.48). Complete the argument on the bi-Lipschitz condition for the map Φ_ω.

Exercise 11.37. Give an alternative proof of Proposition 11.54 using the notion of quasi-isometries and Proposition 11.48.

Exercise 11.38. Let (X,d) be a metric space and ω an ultrafilter. Show that if the asymptotic cone $K_\omega(X,d)$ has more than two points then it is uncountable.

Exercise 11.39. Show that the *modular group* $G := C_2 * C_3 \cong \mathrm{SL}(2,\mathbb{Z})$ has a finite index free subgroup (on two generators) of index 6.

Exercise 11.40. Construct a *non*-simplicial $\mathbb{R}$-tree.

Exercise 11.41 (Claim in the proof of Lemma 12.7). Let (X,d) be a metric space. Let $k_0 \in X$, $r > 0$ and $\varepsilon \in (0,r)$. Suppose that $B_K(k_0,2r)$ can be covered by finitely many balls $B(k^1,\varepsilon), B(k^2,\varepsilon), \ldots, B(k^m,\varepsilon)$ of radius $\varepsilon > 0$. Show that if X is not arcwise connected (resp. arcwise connected but not geodesic) it might be the case that the balls $B(k^1,r+\varepsilon), B(k^2,r+\varepsilon), \ldots, B(k^m,r+\varepsilon)$ fail to cover $B_K(1_K,3r)$.

Chapter 12
Gromov's Theorem

In this chapter we prove the following theorem, which constitutes the core of the book.

Theorem 12.1 (Gromov). *A finitely generated group of sub-polynomial growth is virtually nilpotent.*

Recall (cf. Section 7.4) that a finitely generated group G is of sub-polynomial growth if there exists an integer $d \geq 0$ such that $b^G(n) \preceq n^d$, where b^G is the growth type of G. It follows from Theorem 7.29 and Proposition 7.20 that a virtually nilpotent group G is of (precise) polynomial growth, that is, there exists an integer $d \geq 0$ such that $b^G(n) = n^d$ (in fact d is expressed via the Bass–Guivarc'h formula). In other words, as far as group growth is concerned, the notions of sub-polynomial and polynomial growth coincide (cf. Remark 7.11). This is why, in the literature, Gromov's theorem (Theorem 12.1) is often stated by (apparently) limiting the hypotheses to "polynomial growth".

In Section 11.14 we showed that the asymptotic cone $K_\omega(G, d_X)$ associated with a finitely generated group G, a finite symmetric generating subset $X \subseteq G$, and a free ultrafilter ω, is a metric space which is homogeneous, arcwise connected (and therefore connected) and in fact geodesic, locally arcwise connected (and therefore locally connected), and complete. We now show that if G is of sub-polynomial growth, then with a suitable choice of $\omega \in \beta\mathbb{N}$, the asymptotic cone $K_\omega(G, d_X)$ is also locally compact (Section 12.1) and finite-dimensional, in fact $\dim(K_\omega(G, d_X)) \leq d+1$, where d denotes a degree of sub-polynomial growth of G (Section 12.2).

It follows from the Gleason and Montgomery–Zippin theorem (Theorem 9.17) that $\mathrm{Isom}(K_\omega(G, d_X))$, the group of isometries of $K_\omega(G, d_X)$, can be given the structure of a Lie group with finitely many components, so that, in particular, it is a linear group. Moreover, the Cayley action of G on itself naturally extends to an action, by isometries, of G on the asymptotic cone $K_\omega(G, d_X)$, thus yielding a group homomorphism $\psi \colon G \to \mathrm{Isom}(K_\omega(G, d_X)) \subseteq \mathrm{GL}(k, \mathbb{C})$ for some $k \in \mathbb{N}$. We then analyze the different possibilities that may occur: Theorem 12.11 covers the case when the image $\psi(G)$ is infinite (this is the case, for instance, when ψ is injective). As G has sub-exponential growth, so does $\psi(G)$, and it follows from the Tits alternative that the latter is virtually solvable. It then follows from the Milnor–Wolf theorem (Corollary 7.41) that $\psi(G)$ is virtually nilpotent. Finally, with suitable group-theoretical

T. Ceccherini-Silberstein and M. D'Adderio, *Topics in Groups and Geometry*,
Springer Monographs in Mathematics, https://doi.org/10.1007/978-3-030-88109-2_12

manipulations, we then derive that the group G itself is virtually nilpotent. If $\psi(G)$ is finite, we construct a homomorphism $\varphi\colon N \to \mathrm{GL}(k,F)$, where now $N \subseteq G$ is a finite index subgroup of G, $\varphi(N)$ is infinite, and F is an ultrafield (in fact the ω-ultrapower of $\mathbb{C}$, cf. Theorem 11.39). So we will again apply Theorem 12.11 to conclude that N, and hence G, is virtually nilpotent. This will complete the proof of Gromov's theorem.

12.1 Asymptotic Cones of Groups of Sub-Polynomial Growth are Locally Compact

From now on we suppose that G is a finitely generated group of sub-polynomial growth of degree $\leq d$. We fix a finite symmetric generating subset $X \subseteq G$. It follows from our assumptions that we can find a constant $C > 0$ such that $b_X(n) \leq Cn^d$ for all $n \geq 1$.

Our first target is to find an appropriate free ultrafilter ω on $\mathbb{N}$ such that the corresponding asymptotic cone is locally compact.

Recall that a metric space is *locally compact* if every point admits a compact neighborhood.

In order to find the ultrafilter ω, let us prove the following:

Lemma 12.2. *There are infinitely many $n \in \mathbb{N}$ such that*

$$b_X(2^n) \leq b_X(2^i)2^{(n-i)(d+1)} \tag{12.1}$$

for all $i \leq n$.

Proof. Assume by contradiction that there exists an $n_0 > 1$ such that for all $n \geq n_0$ there exists an $i = i(n) < n$ satisfying $b_X(2^n) > b_X(2^i)2^{(n-i)(d+1)}$. We then choose for every $n \geq n_0$ a minimal such i.

We claim that, for every $n \in \mathbb{N}$ one has $i = i(n) < n_0$. If $i \geq n_0$, then, by the definition of n_0, there exists a $j < i$ such that $b_X(2^i) > b_X(2^j)2^{(i-j)(d+1)}$ and hence

$$b_X(2^n) > b_X(2^i)2^{(n-i)(d+1)} > b_X(2^j)2^{(i-j)(d+1)}2^{(n-i)(d+1)} = b_X(2^j)2^{(n-j)(d+1)},$$

contradicting the minimality of i. This proves the claim.

Then for all $n \geq n_0$ we have

$$b_X(2^n) > b_X(2^i)2^{(n-i)(d+1)} > 2^{(n-i)(d+1)} > 2^{(n-n_0)(d+1)}.$$

Setting $m := 2^n \in \mathbb{N}$ and $C' := 2^{-n_0(d+1)}$ we then have

$$b_X(m) \geq C'm^{d+1}. \tag{12.2}$$

Since (12.2) holds for infinitely many $m \in \mathbb{N}$, this contradicts our assumption on the sub-polynomial degree of growth of G. $\square$

Let us set

$$S := \{2^n : b_X(2^n) \leq b(2^i)2^{(n-i)(d+1)} \text{ for all } i \leq n\} \subseteq \mathbb{N} \tag{12.3}$$

and observe that, by Lemma 12.2, S is infinite.

We claim that there exists a free ultrafilter ω containing S. Indeed, denoting by $\mathscr{F}$ the Fréchet filter on $\mathbb{N}$ the set $\Omega := \mathscr{F} \cup \{S\} \subset \mathscr{P}(\mathbb{N})$ is saturated (**exercise**) and therefore, by virtue of Corollary 11.7, it is embeddable into some ultrafilter $\omega \in \beta\mathbb{N}$. Moreover, from Corollary 11.13 we deduce that ω, since it contains the Fréchet filter, is free.

We thus fix a free ultrafilter ω containing S and denote by $K := K_\omega(G, d_X)$ the associated asymptotic cone. We also denote by $1_K \in K$ the class of the constant sequence $(1_G)_{n\in\mathbb{N}}$ and by $B := B_K(1_K, 1)$ the ball of radius 1 in K centered at 1_K.

Lemma 12.3. *Suppose that B contains m pairwise non-intersecting balls of radius 2^{-p}, where p is a positive integer. Then $m \leq 2^{p(d+1)}$.*

Proof. Denote by $B_1, B_2, \ldots, B_m$ these pairwise non-intersecting balls of radius 2^{-p} contained in B and let $k^1, k^2, \ldots, k^m$, denote their centers. Let $x^i := (x^i_n)_{n\in\mathbb{N}}$ be a moderate sequence representing k^i for $i = 1, 2, \ldots, m$. It follows from our assumptions and the triangle inequality that we can find a subset $A \in \omega$ such that, for all $n \in A$, the G-balls $B_G(x^i_n, n2^{-p})$, $i = 1, 2, \ldots, m$, are pairwise non-intersecting and contained in $B_G(1_G, n)$.

As ω is a filter, we have $A \cap S \in \omega$, in particular $A \cap S \neq \varnothing$. Thus we can find $n \in \mathbb{N}$ such that $2^n \in A \cap S$. Now, since $2^n \in A$, we have

$$b_X(2^n) = |B_G(1_G, 2^n)| \geq m|B_G(x^i_{2^n}, 2^{n-p})| = mb_X(2^{n-p}). \tag{12.4}$$

Combining (12.4) and (12.1) with $i := n - p$ (note that $i \leq n$) we deduce $mb_X(2^{n-p}) \leq b_X(2^n) \leq b_X(2^{n-p})2^{p(d+1)}$ and the statement follows. □

Proposition 12.4. *Let $p \geq 2$ and suppose that $\mathscr{B}(p) = \left(B_K(k^i, 2^{-p})\right)_{i=1}^m$, $m = m(p)$ is a maximal system of non-intersecting balls of radius 2^{-p} contained in B. Then the system $\mathscr{B}^*(p) := \left(B_K(k^i, 2^{-p+2})\right)_{i=1}^m$ (obtained by multiplying the radii by 4) covers the closure of the ball $B_K(1_K, 1/2)$, that is,*

$$\overline{B_K(1_K, 1/2)} \subseteq \bigcup_{i=1}^m B_K(k^i, 2^{-p+2}).$$

Proof. We proceed by contradiction. If the statement does not hold, then we can find $k \in \overline{B_K(1_K, 1/2)} \setminus \bigcup_{i=1}^m B_K(k^i, 2^{-p+2})$. Since $p \geq 2$, by virtue of the triangle inequality we have both $B_K(k, 2^{-p}) \subseteq B$ and

$$B_K(k, 2^{-p}) \cap \left(\bigcup_{i=1}^m B_K(k^i, 2^{-p})\right) = \varnothing,$$

contradicting the maximality of the family $\mathscr{B}(p)$. □

Theorem 12.5. *K is locally compact.*

Proof. To show that K is locally compact, it is clearly enough, by homogeneity (cf. Theorem 11.52.(i)), to show that the closure B_0 of the ball $B_K(1_K, 1/2)$ is sequentially compact, that is, that every sequence $(k^i)_{i\in\mathbb{N}}$ in B_0 admits a convergent subsequence.

Let then $(k^i)_{i\in\mathbb{N}}$ be a sequence in B_0. Consider the cover $\mathscr{B}^*(3)$ of B_0 by balls of radius $1/2$ as in Proposition 12.4. Then we can find one ball in $\mathscr{B}^*(3)$ whose closure B_1 contains k^i for infinitely many $i \in \mathbb{N}$. Thus we can extract a subsequence $(k^{i_j})_{j\in\mathbb{N}}$, with $i_0 > 0$, which is entirely contained in B_1. Consider now the cover $\mathscr{B}^*(4)$ of B_0 (and therefore of $B_0 \cap B_1$) by balls of radius $1/4$. As before, we can find one ball B_2 in $\mathscr{B}^*(4)$ containing k^{i_j} for infinitely many $j \in \mathbb{N}$. Thus we can extract a subsequence $(k^{i_{j_k}})_{k\in\mathbb{N}}$, with $i_{j_0} > i_0$, which is entirely contained in B_2. Continuing this way, we have that the "diagonal" subsequence $k_0, k_{i_0}, k_{i_{j_0}}, \ldots$ (which is eventually contained in a ball of radius 2^{-p} for every $p \geq 2$) is a Cauchy sequence in K. Since K is complete (by Theorem 11.52.(iv)), we conclude that this subsequence converges. □

Remark 12.6. Let $X := \ell^2(\mathbb{N})$ denote the Hilbert space consisting of all complex sequences $(a_n)_{n\in\mathbb{N}}$ such that $\sum_{n\in\mathbb{N}} |a_n|^2 < \infty$ equipped with the scalar product $\langle (a_n)_{n\in\mathbb{N}}, (b_n)_{n\in\mathbb{N}} \rangle := \sum_{n\in\mathbb{N}} \overline{a_n} b_n$. Then setting

$$\|(a_n)_{n\in\mathbb{N}}\| := \left(\sum_{n\in\mathbb{N}} |a_n|^2 \right)^{\frac{1}{2}} \quad \text{and} \quad d((a_n)_{n\in\mathbb{N}}, (b_n)_{n\in\mathbb{N}}) := \|(a_n - b_n)_{n\in\mathbb{N}}\|$$

we have that (X, d) is a complete metric space. For $i \in \mathbb{N}$ we denote by $\mathbf{e_i}$ the sequence $(e_{i,n})_{n\in\mathbb{N}} \in X$ defined by $e_{i,n} = 1$ if $n = i$, and 0 otherwise and set $x_i = \frac{1}{2}\mathbf{e_i}$. If $\mathbf{0} := (0)_{n\in\mathbb{N}} \in X$ is the constant zero-sequence we then have $d(\mathbf{0}, x_i) = 1/2$ and therefore $x_i \in B_X(\mathbf{0}; 1)$ for all $i \in \mathbb{N}$. We also have $d(x_i, x_j) = \frac{1}{2} d(\mathbf{e_i}, \mathbf{e_j}) = \sqrt{2}/2$ for all distinct $i, j \in \mathbb{N}$. Thus if $p = 2$, by virtue of the triangle inequality, the balls $B_X(x_i; 2^{-p})$, $i \in \mathbb{N}$, are pairwise non-intersecting. This shows that the conclusion of Lemma 12.3 does not hold in (X, d). Note that, indeed, (X, d) is neither locally compact (there is no convergent subsequence of the bounded sequence $(x_i)_{i\in\mathbb{N}}$) nor finite-dimensional (see Example 10.14).

12.2 Finite Dimension of Asymptotic Cones of Groups of Sub-Polynomial Growth

Let G, $X \subseteq G$, and ω be as in Section 12.1. Our next task is to show that the asymptotic cone $K = K_\omega(G, d_X)$ has finite dimension.

Lemma 12.7. (1) *For every $R > 0$ the ball $B_K(1_K, R)$ can be covered by finitely many balls of radius $1/2$.*
(2) *The whole K can be covered by countably many balls of radius $1/2$.*

Proof. In order to show (1) we start by proving the following:

Claim. *Let $r > 0$ and $\varepsilon \in (0,r)$. Suppose that $B_K(1_K, 2r)$ can be covered by finitely many balls of radius ε. Then $B_K(1_K, 3r)$ can be covered by finitely many balls of radius ε.*

Indeed if $k^1, k^2 \ldots, k^m \in B_K(1_K, 2r)$ satisfy $B_K(1_K, 2r) \subseteq \cup_{i=1}^m B_K(k^i, \varepsilon)$, then, since K is geodesic, by using the triangle inequality we deduce that $B_K(1_K, 3r) \subseteq \cup_{i=1}^m B_K(k^i, r+\varepsilon)$. Hence, a fortiori, $B_K(1_K, 3r) \subseteq \cup_{i=1}^m B_K(k^i, 2r)$. Now, by homogeneity of K, each of these balls of radius $2r$ can, in turn, be covered by finitely many balls of radius ε. We deduce that $B_K(1_K, 3r)$ itself can be covered by finitely many balls of radius ε, and the claim follows.

Note (**Exercise**) that the above claim fails to hold, in general, if the ambient metric space is not arcwise connected (resp. arcwise connected but not geodesic).

Let $r_0 := 1/4$, so that $B_K(1_K, 2r_0) = B_K(1_K, 1/2)$, and $\varepsilon := 1/8$. Then the assumptions of the Claim are satisfied by virtue of Proposition 12.4 (with $p = 5$), so that $B_K(1_K, 3r_0) = B_K(1_K, 3/4)$ can be covered by finitely many balls of radius $1/8$. We continue to apply the claim recursively, so that $B_K(1_K, (3/2)^i/2)$ can be covered by finitely many balls of radius $1/8$, hence a fortiori of radius $1/2$, for all $i \geq 1$. Hence the same is true for $B_K(1_K, R)$ for any $R > 0$.

(2) Follows immediately from (1) after observing that $K = \bigcup_{R \in \mathbb{N}} B_K(1_K, R)$. □

Proposition 12.8. *The asymptotic cone $K = K_\omega(G, d_X)$ is separable.*

Proof. By Lemma 12.7, K can be covered by countably many balls of radius $1/2$. Therefore, by homogeneity, it suffices to show that the ball $B_K(1_K, 1/2)$ is separable. By Proposition 12.4, for every $p \in \mathbb{N}$, there exists a finite cover $\mathscr{B}^*(p)$ of $B_K(1_K, 1/2)$ by balls of radius 2^{-p+2}. It is an **exercise** to check that the countable set consisting of the centers c of the balls in $\cup_{p \in \mathbb{N}} \mathscr{B}^*(p)$, with $c \in B_K(1_K, 1/2)$, is dense in $B_K(1_K, 1/2)$. □

Theorem 12.9. *The asymptotic cone $K = K_\omega(G, d_X)$ has finite Hausdorff dimension. In fact,* $\mathrm{Hdim}(K) \leq d+1$, *where d is a degree of sub-polynomial growth of G.*

Proof. We claim that to prove the theorem it suffices to show that

$$\mathrm{Hdim}(B_K(1_K, 1/2)) \leq d + 1. \tag{12.5}$$

Indeed, by Lemma 12.7 we can find a sequence $(k^i)_{i \in \mathbb{N}}$ in K such that $K \subseteq \bigcup_{i=0}^{\infty} B_K(k^i, 1/2)$. By homogeneity, if $s > \mathrm{Hdim}(B_K(1_K, 1/2))$ then we have $s > \mathrm{Hdim}(B_K(k^i, 1/2))$, for all $i \in \mathbb{N}$.

Now fix $\varepsilon > 0$. For each $i \in \mathbb{N}$ we can find a countable cover $\mathscr{B}_i$ of $B_K(k^i, 1/2)$ consisting of balls such that

$$\sum_{B \in \mathscr{B}_i} r(B)^s \leq \varepsilon \frac{1}{2^{i+1}}.$$

It is then clear that $\mathscr{B} := \bigcup_{i\in\mathbb{N}} \mathscr{B}_i$ is a countable cover of K consisting of balls such that

$$\sum_{B\in\mathscr{B}} r(B)^s = \sum_{i\in\mathbb{N}} \sum_{B\in\mathscr{B}_i} r(B)^s \le \varepsilon \sum_{i\in\mathbb{N}} \frac{1}{2^{i+1}} = \varepsilon.$$

This shows that $s > \mathrm{Hdim}(K)$. The claim follows.

In order to prove (12.5), let $\delta \in (0,1]$ and let us first show that $s = s(\delta) := d + 1 + \delta > \mathrm{Hdim}(B_K(1_K, 1/2))$. Let $\varepsilon > 0$. Then we can find $p \in \mathbb{N}$ such that

$$p \ge \frac{1}{\delta}(2d + 2 + 2\delta - \log_2(\varepsilon)). \tag{12.6}$$

Consider the cover $\mathscr{B}^*(p)$ of $B_K(1_K, 1/2)$ given by Proposition 12.4 and recall that it consists of $m \le 2^{p(d+1)}$ balls B_i of constant radius $r(B_i) = 2^{-p+2}$. We then have

$$\sum_{i=1}^{m} r(B_i)^s \le 2^{p(d+1)} 2^{(-p+2)s} = 2^{p(d+1)} 2^{(-p+2)(d+1+\delta)} \le 2^{2d+2+2\delta-p\delta} < \varepsilon,$$

where the last inequality follows from (12.6).

By the claim, we have $s(\delta) \ge \mathrm{Hdim}(K)$, so that K has finite Hausdorff dimension. In particular,

$$\mathrm{Hdim}(K) \le \inf\{s(\delta) : \delta \in (0,1]\} = d + 1. \qquad \square$$

Since every separable metrizable space of finite Hausdorff dimension is finite-dimensional (Corollary 10.47), from Proposition 12.8 and Theorem 12.9 we immediately deduce the following:

Corollary 12.10. *The asymptotic cone $K = K_\omega(G, d_X)$ has finite dimension. In fact, $\dim(K) \le d + 1$, where d denotes a degree of sub-polynomial growth of G.* □

12.3 Proof of Gromov's Theorem

We start with a general theorem.

Theorem 12.11. *Let G be a finitely generated group of sub-polynomial growth, and consider a homomorphism $\psi\colon G \to \mathrm{GL}(k, F)$, where $k \in \mathbb{N}$ and F is a field. If the image $\psi(G)$ is infinite, then G is virtually nilpotent.*

Proof. Since G has sub-polynomial growth, the (finitely generated) group $\psi(G)$ cannot contain a noncommutative free subgroup (as this would imply that the growth of G is exponential). Hence by Tits' Alternative (cf. Chapter 8) $\psi(G)$ must be virtually solvable. Therefore, by the Milnor–Wolf results (cf. Theorem 7.36, Theorem 7.37, and Corollary 7.41) it must indeed be virtually nilpotent.

So by Lemma 2.42, $\psi(G)$ is in fact virtually torsion-free nilpotent. Let then $\widetilde{H} \le \psi(G)$ denote a characteristic torsion-free nilpotent subgroup of finite index in $\psi(G)$.

We claim that we can find a surjective homomorphism $\widetilde{H} \to \mathbb{Z}$. Indeed, if r denotes the length of the upper central series of $\widetilde{H}$, we have, by Lemma 2.19, that

$\widetilde{H}/Z_{r-1}(\widetilde{H}) = Z_r(\widetilde{H})/Z_{r-1}(\widetilde{H})$ is free abelian, say isomorphic to $\mathbb{Z}^m$ with $m \in \mathbb{N}$, $m \geq 1$ (here we are using that $\psi(G)$ is infinite, so that also $\widetilde{H}$ is infinite). Thus composing the quotient map $\widetilde{H} \to \widetilde{H}/Z_{r-1}(\widetilde{H}) \cong \mathbb{Z}^m$ with the projection $\mathbb{Z}^m \to \mathbb{Z}$ onto the first component, we obtain the required surjective homomorphism $\widetilde{H} \to \mathbb{Z}$.

Setting $H := \psi^{-1}(\widetilde{H})$ we have that H is a finite index subgroup of G which maps onto $\mathbb{Z}$. Let $\psi^* \colon H \to \mathbb{Z}$ denote the corresponding surjective homomorphism and set

$$N := \ker(\psi^*) \leq H. \tag{12.7}$$

Note that, since the growth of H is subexponential, it follows from the Claim in the proof of Milnor's theorem (Theorem 7.36) that N is finitely generated.

Lemma 12.12. *Suppose that the sub-polynomial growth of H is of degree $\leq d$. Then the sub-polynomial growth of N is of degree $\leq d-1$.*

Proof. Let $Y \subseteq N$ be a finite symmetric generating subset and pick an element $a \in H$ such that $H/N = \langle aN \rangle \cong \mathbb{Z}$ (equivalently, $\langle \psi^*(a) \rangle = \psi^*(H)$). This way the set $Y' := Y \cup \{a, a^{-1}\}$ is a finite (symmetric) generating subset of H. We claim that

$$nb_Y^N(n) \leq b_{Y'}^H(2n). \tag{12.8}$$

In fact, if $g_1, g_2, \ldots, g_r$ are distinct elements in N such that $\ell_Y(g_i) \leq n$ for $i = 1, 2 \ldots, r$, then the elements $g_i a^j \in H$ for $i = 1, 2 \ldots, r$ and $j = 0, 1 \ldots, n$ are also distinct. Thus

$$nb_Y^N(n) \leq \left| \sqcup_{j=0}^n B_Y^N(n) a^j \right| \leq |B_{Y'}^H(2n)| \leq b_{Y'}^H(2n)$$

and the claim follows. Since H has sub-polynomial growth of degree $\leq d$, from (12.8) and our assumptions we deduce $nb_Y^N(n) \leq b_{Y'}^H(2n) \preceq (2n)^d \sim n^d$ so that

$$b_Y^N(n) \preceq n^{d-1}.$$

This shows that N has sub-polynomial growth of degree $\leq d-1$. □

Now, by induction on d, the group N is virtually nilpotent. Hence we can assume that the following situation occurs:

$$G \geq_{[G:H]<\infty} H \rhd N, \text{ with } N \text{ virtually nilpotent and } H/N \cong \mathbb{Z}.$$

So N contains a finite index subgroup N' which is nilpotent and, by Poincaré's lemma (cf. Lemma 2.39), we may assume that N' is characteristic in N, and hence normal in H. Then H/N' contains the finite normal subgroup N/N' and the corresponding factor is infinite cyclic: $(H/N')/(N/N') \cong H/N \cong \mathbb{Z}$.

Observe that H acts on N/N' by setting $(nN')^h = hnh^{-1}N'$ for all $h \in H$ and $n \in N$. We denote the kernel of this action by

$$C := \{h \in H : (nN')^h = nN' \text{ for all } n \in N\},$$

i.e. $C/N' = C_{H/N'}(N/N')$. Then we may regard H/C as a subgroup of $\mathrm{Aut}(N/N')$. Since $[N : N'] < \infty$, we deduce that H/C is finite.

Lemma 12.13. *C is virtually nilpotent.*

Proof. We start with the following observations: (i) $C/(C\cap N)$ is infinite cyclic and (ii) $(C\cap N)/N'$ is finite abelian. Indeed, on the one hand we have $C\cap N \neq C$, since otherwise $C\subseteq N$ contradicting the fact that H/C is finite and H/N is infinite cyclic, so that $\{1\}\neq C/(C\cap N)\cong CN/N \leq H/N \cong \mathbb{Z}$. On the other hand, $(C\cap N)/N'$ is finite (since N/N' is finite) and abelian (since $(C\cap N)/N' = (C/N')\cap(N/N') = Z(N/N')$).

Consider now the finite subnormal series $C\geq C\cap N\geq N'\geq\{1\}$. From (i) and (ii) and recalling that N' is nilpotent (and therefore solvable) we deduce that C is solvable, since (cf. Proposition 4.4) solvability is closed under group extensions (recall that, however, nilpotency is not extension-closed (cf. Example 4.5)).

Moreover, C is finitely generated (since N' is finitely generated, $(C\cap N)/N'$ is finite, and $C/(C\cap N)$ is infinite cyclic) and of sub-polynomial growth (since $C\leq H$ and H has sub-polynomial growth (cf. Proposition 7.17)). From the theorem of Milnor–Wolf (Corollary 7.41) we then deduce that C is virtually nilpotent. □

Since

$$G\geq_{[G:H]<\infty} H\geq_{[H:C]<\infty} C, \text{ with } C \text{ virtually nilpotent}$$

we deduce that G itself is virtually nilpotent.

This completes the proof of Theorem 12.11. □

The strategy now is to find such a homomorphism, in order to be able to apply the theorem. We are going to achieve this by means of the asymptotic cone.

Let G be a finitely generated group of sub-polynomial growth, and let us fix a finite symmetric generating subset $X\subseteq G$ of G. Moreover, let ω be a free ultrafilter as in Section 12.1, and denote by $K := K_\omega(G,d_X)$ the corresponding asymptotic cone.

Given $g\in G$ we still denote by g the constant (moderate) sequence $(g,g,g,\ldots)$.

With the notation from the proof of Theorem 11.52, we then consider the map $\Lambda_g\colon K\to K$. Thus, if $k=[(h_n)_{n\in\mathbb{N}}]_\omega\in K$ we have $\Lambda_g(k)=[(gh_n)_{n\in\mathbb{N}}]_\omega$. Recall that Λ_g is an isometry, that is, $d_\omega(\Lambda_g(k),\Lambda_g(k'))=d_\omega(k,k')$ for all $k,k'\in K$. Moreover, for all $g_1,g_2\in G$ and $k\in K$ one has $\Lambda_{g_1g_2}(k)=[(g_1g_2h_n)_{n\in\mathbb{N}}]_\omega=\Lambda_{g_1}[(g_2h_n)_{n\in\mathbb{N}}]_\omega=\Lambda_{g_1}(\Lambda_{g_2}(k))$ thus showing that $\Lambda_{g_1g_2}=\Lambda_{g_1}\Lambda_{g_2}$. It follows that the map

$$\psi\colon G\to \mathrm{Isom}(K) \tag{12.9}$$

defined by $\psi(g):=\Lambda_g$ is a group homomorphism. In other words, (12.9) defines an action of G on K by isometries (note that, on the other hand, if $R_g\colon K\to K$, $g\in G$, is defined by $R_g(k)=[(h_ng)_{n\in\mathbb{N}}]$ for all $k=[(h_n)_{n\in\mathbb{N}}]\in K$ then the corresponding action is trivial (**exercise**)).

As we have seen before, since G has sub-polynomial growth, the asymptotic cone K is a finite-dimensional, locally compact, connected and locally connected, homogeneous metric space, and therefore its group of isometries $\mathrm{Isom}(K)$ can be given the structure of a Lie group with finitely many components, by the Gleason–Montgomery–Zippin theorem (Theorem 9.17). In particular, $\mathrm{Isom}(K)$ is linear.

This yields a homomorphism $\psi\colon G \to \mathrm{Isom}(K) \subseteq \mathrm{GL}(k,\mathbb{C})$ for some $k \in \mathbb{N}$. If the image $\psi(G)$ is infinite, then we can apply Theorem 12.11 to deduce that G is virtually nilpotent.

The remainder of this section is devoted to dealing with the case where $\psi(G)$ is finite. In this case we will still be able to construct a homomorphism $\varphi\colon N \to \mathrm{GL}(k,F)$, where now $N \subseteq G$ is a finite index subgroup of G, $\varphi(N)$ is infinite, but F is a field different from $\mathbb{C}$.

More precisely, we will show that the image $\varphi(N)$ in $\mathrm{GL}(k,F)$ contains an element of infinite order (so that, in particular $\varphi(N)$ is infinite). So we will again apply Theorem 12.11 to conclude that N, and hence G, is virtually nilpotent, and this will complete the proof of Gromov's theorem.

We start by analyzing the action (12.9) of G on K. For this purpose we need to introduce the following notions.

Definition 12.14. Let (M,d) be a metric space and $m \in M$. Given $\alpha \in \mathrm{Isom}(M)$ the quantity

$$D(\alpha,m) := d(\alpha(m),m)$$

is called the *displacement* of m by α.

Consider for instance the action of G on (G,d_X), defined by $g \mapsto L_g$, where L_g is the isometry given by left multiplication by g. Then for $g,h \in G$ we have $D(L_g,h) = d_X(L_g(h),h) = d_X(gh,h) = \ell_X(h^{-1}gh)$.

Definition 12.15. Let (M,d) be a metric space and fix a base-point $m_0 \in M$. Let $\alpha \in \mathrm{Isom}(M)$. Then the map $D_\alpha\colon \mathbb{N} \to \mathbb{N}$ defined by

$$D_\alpha(n) := \max\{D(\alpha,m) : m \in M, d(m_0,m) \leq n\} \tag{12.10}$$

is called the *displacement function* of α (relative to m_0).

So, choosing 1_G as a base-point, the left-multiplication gives

$$0 \leq \frac{D_{L_g}(n)}{n} = \frac{\max_{\ell_X(h)\leq n} \ell_X(h^{-1}gh)}{n} \leq \frac{2n+\ell_X(g)}{n} \leq 2 + \frac{\ell_X(g)}{n}$$

for all $g \in G$ and $n \in \mathbb{N}$.

Lemma 12.16. *Consider the action of G on (G,d_X) given by left-multiplication. Then*

$$N := \ker(\psi) = \left\{ g \in G : \lim_\omega \frac{D_{L_g}(Rn)}{n} = 0 \textit{ for all } R > 0 \right\}. \tag{12.11}$$

Proof. Let $g \in G$ and $R > 0$. Set $h_0 := 1_G$ and, for every $n \geq 1$ let $h_n \in B_X(1_G,Rn)$ be such that $D_{L_g}(Rn) = D(L_g,h_n) = d_X(L_g(h_n),h_n) = d_X(gh_n,h_n)$. Observe that the sequence $(h_n)_{n\in\mathbb{N}}$ is moderate (since $\ell_X(h_n) \leq Rn$ for all $n \geq 1$). We then set $h := [(h_n)_{n\in\mathbb{N}}]_\omega \in K$. Now if $g \in N$, we have $\psi(g)(h) = h$ so that

$$\lim_\omega \frac{D_{L_g}(Rn)}{n} = \lim_\omega \frac{D(L_g,h_n)}{n} = \lim_\omega \frac{d_X(gh_n,h_n)}{n} = d_\omega(\psi(g)(h),h) = 0.$$

Conversely, suppose that $\lim_\omega D_{L_g}(Rn)/n = 0$ for all $R > 0$. Pick $k \in K$ and let us show that $\psi(g)(k) = k$. Let $(k_n)_{n\in\mathbb{N}}$ be a moderate sequence in G representing k. Then we can find a constant $R > 0$ such that $\ell_X(k_n) \le Rn$, equivalently, $k_n \in B_X(1_G, Rn)$, for all $n \ge 1$. We then have

$$
\begin{aligned}
0 \le d_\omega(\psi(g)(k), k) &= \lim_\omega \frac{d_X(gk_n, k_n)}{n} \\
&= \lim_\omega \frac{D(L_g, k_n)}{n} \le \lim_\omega \frac{D_{L_g}(Rn)}{n} = 0
\end{aligned}
$$

thus showing that $\psi(g)k = k$. □

In the following, given an element $g \in G$ and a subgroup $H \le G$, we denote by $g^H := \{g^h : h \in H\}$ and $C_H(g) := \{h \in H : g^h = g\}$ the conjugacy class and the centralizer of g in H, respectively (recall that $g^h = h^{-1}gh$).

Lemma 12.17. *Consider the action of G on (G, d_X) given by left-multiplication and let 1_G be the corresponding base-point. Let $g \in G$. Then the following conditions are equivalent:*

(a) *the sequence $(D_{L_g}(n))_{n\in\mathbb{N}}$ is bounded;*
(b) $|g^G| < \infty$;
(c) $[G : C_G(g)] < \infty$.

Moreover, if one of the equivalent conditions above is satisfied then $g \in N$.

Proof. Suppose $(D_{L_g}(n))_{n\in\mathbb{N}}$ is bounded and let $D_0 > 0$ be a constant satisfying $D_{L_g}(n) \le D_0$ for all $n \in \mathbb{N}$. This implies that $\ell_X(h^{-1}gh) = D(L_g, h) \le D_0$, equivalently, $h^{-1}gh \in B_X(1_G, D_0)$ for all $h \in G$. Thus $|g^G| \le |B_X(1_G, D_0)|$. This shows (a) ⇒ (b). Conversely, suppose $|g^G| < \infty$. Then we can find $D_0 > 0$ such that $g^G \subseteq B_X(1_G, D_0)$, that is, $h^{-1}gh \in B_X(1_G, D_0)$ for all $h \in G$. It follows that $D(g,h) = \ell_X(h^{-1}gh) \le D_0$ for all $h \in G$, so that $D_{L_g}(n) \le D_0$ for all $n \in \mathbb{N}$.

Let now $x, y \in G$. Then

$$
g^x = g^y \Leftrightarrow x^{-1}gx = y^{-1}gy \Leftrightarrow g = xy^{-1}gyx^{-1} \Leftrightarrow xy^{-1} \in C_G(g) \Leftrightarrow C_G(g)x = C_G(g)y.
$$

This shows that $|g^G| = [G : C_G(g)]$ and (b) ⇔ (c) follows as well.

The remaining part of the statement immediately follows from Lemma 12.16. □

Lemma 12.18. *Let $H \subseteq G$ be a subgroup of finite index and $g \in H$. Then $|g^H| < \infty$ if and only if $|g^G| < \infty$.*

Proof. Since $g^H \subseteq g^G$, we only have to show that $|g^H| < \infty$ infers $|g^G| < \infty$. Let T be a complete system of representatives for the right cosets of H in G so that $G = \cup_{t\in T} Ht$. Since $g^{ht} = (g^h)^t$ for all $h \in H$ and $t \in T$, we have $|g^G| \le |T||g^H| = [G : H]|g^H|$. □

We now denote by

$$
\Delta(G) := \{g \in G : |g^G| < \infty\}
$$

the subset of elements of G whose conjugacy class is finite. Let us show that $\Delta(G)$ is a characteristic subgroup of G. If $g_1, g_2 \in \Delta(G)$, then by virtue of Lemma 12.17 we have that $C_G(g_1)$ and $C_G(g_2)$, and therefore their intersection $C_G(g_1) \cap C_G(g_2)$, have finite index in G. Since $C_G(g_1^{-1}g_2) \supseteq C_G(g_1) \cap C_G(g_2)$ (note that $C_G(g_1^{-1}) = C_G(g_1)$), this implies that $C_G(g_1^{-1}g_2)$ is also of finite index, yielding, again by virtue of Lemma 12.17, $g_1^{-1}g_2 \in \Delta(G)$. This shows that $\Delta(G)$ is a subgroup of G. Finally, if $\alpha \in \mathrm{Aut}(G)$ then $|(\alpha(g))^G| = |(\alpha(g))^{\alpha(G)}| = |\alpha(g^G)| = |g^G|$, since α is bijective. This shows that $\Delta(G)$ is in fact characteristic in G.

Note that we may now reformulate Lemma 12.18 as follows. If $H \subseteq G$ is a finite index subgroup, then

$$\Delta(G) \cap H = \Delta(H). \tag{12.12}$$

Lemma 12.19 (B.H. Neumann). *Let G be a (not necessarily finitely generated) group and suppose that*

$$G = \bigcup_{i=1}^{r} \bigcup_{j=1}^{n_i} H_i g_{i,j}, \tag{12.13}$$

where $H_i \subseteq G$ is a subgroup and $g_{i,j} \in G$, for $j = 1, 2, \ldots, n_i$ and $i = 1, 2, \ldots, r$. Then at least one of the H_i's is of finite index.

In other words, a group cannot be a finite union of cosets of subgroups of infinite index.

Proof. We prove the statement by induction on r, the number of involved subgroups. For $r = 1$ this is obvious. In order to fix our ideas, suppose that $[G : H_1] = \infty$. Then we can find $g \in G$ such that the coset $H_1 g$ does not occur in the right-hand side of (12.13); note that, however, $H_1 g \subseteq \bigcup_{i=2}^{r} \bigcup_{j=1}^{n_i} H_i g_{i,j}$. We deduce that $H_1 \subseteq \bigcup_{i=2}^{r} \bigcup_{j=1}^{n_i} H_i g_{i,j} g^{-1}$ and therefore $H_1 g_{1,k} \subseteq \bigcup_{i=2}^{r} \bigcup_{j=1}^{n_i} H_i g_{i,j} g^{-1} g_{1,k}$ for all $k = 1, 2, \ldots, n_1$. This implies that in the right-hand side of (12.13) we may substitute each H_1-coset by some finite union of cosets of the remaining subgroups involved. Therefore G can be expressed as a finite union of cosets of $H_2, H_3, \ldots, H_r$. By induction, there exists $2 \leq i \leq r$ such that $[G : H_i] < \infty$. □

Remark 12.20. Fix a sequence (not necessarily moderate) $u = (u_n)_{n \in \mathbb{N}}$ of elements of G and, for every $g \in G$, set $g^u := (u_n^{-1} g u_n)_{n \in \mathbb{N}}$. Suppose that for every generator $x_i \in X$ the sequence x_i^u is moderate (on a large set of indices), then for every $g \in G$, the sequence g^u is also moderate (on a large set of indices). Indeed say $\ell_X(u_n^{-1} x_i u_n) < C_i n$ for all $n \in A_i$, where $A_i \in \omega$, then if $g = x_{i_1} x_{i_2} \cdots x_{i_s}$ we have $u_n^{-1} g u_n = u_n^{-1} x_{i_1} u_n \cdot u_n^{-1} x_{i_2} u_n \cdot \ldots \cdot u_n^{-1} x_{i_s} u_n$, yielding

$$\ell_X(u_n^{-1} g u_n) \leq \sum_{j=1}^{s} \ell_X(u_n^{-1} x_{i_j} u_n) \leq Cn,$$

where $C := \sum_{j=1}^{s} C_{i_j}$ for all $n \in A_{i_1} \cap A_{i_2} \cap \cdots \cap A_{i_s} \in \omega$.

Suppose $\Delta(G) = G$. Then Lemma 12.17 gives, in particular, $[G : C_G(x)] < \infty$ for every generator $x \in X$ of G so that the center $Z(G) = \bigcap_{x \in X} C_G(x)$ is also of finite index in G. This yields virtual (abelianness and therefore virtual) nilpotence of G, and we are done.

We thus assume that (i) $\Delta(G) \subsetneq G$. We may also assume that (ii) $\Delta(N) \subsetneq N$. Otherwise, by the same argument as above, N would be virtually nilpotent. Since N has finite index in G ($\psi(G)$ is finite), also G would be virtually nilpotent.

Note that (ii) implies $N \not\subseteq \Delta(G)$ (as $\Delta(G) \cap N = \Delta(N)$ by (12.12)).

Finally, let $Y = \{y_1, y_2, \ldots, y_d\}$ be a finite generating subset of N. We may assume that $Y \cap \Delta(N) = \varnothing$. Otherwise, fix an $a \in N \setminus \Delta(N)$ and add it to Y; now, given any $y_i \in \Delta(N)$, we can replace it by ay_i. Up to adding inverses, if necessary, we may also assume that Y is symmetric.

Definition 12.21. Consider the asymptotic cone K with base-point 1_K. Then

$$O(\varepsilon, R) := \{\alpha \in \mathrm{Isom}(K) : D_\alpha(R) < \varepsilon\} \tag{12.14}$$

where $\varepsilon > 0$ and $R > 0$, constitute a system of neighborhoods of the identity in $\mathrm{Isom}(K)$ (and hence induce a topology on $\mathrm{Isom}(K)$).

Lemma 12.22. *For every $\varepsilon > 0$ and $R \geq 1$ we can find a sequence $u = (u_n)_{n\in\mathbb{N}}$ in G such that*

(1) *for every $g \in N$, the sequence g^u is moderate;*
(2) *the map $\varphi\colon N \to \mathrm{Isom}(K)$ defined by $g \mapsto \Lambda_{g^u}$ is a homomorphism such that there exists $1 \leq i \leq d$ satisfying $\varphi(y_i) \in O(\varepsilon, R) \setminus \{\mathrm{Id}_K\}$.*

Proof. Fix $\varepsilon > 0$ and $R > 0$.

Step 1: Fix $n \geq 1$. We claim that there exists a $z_n \in N$ such that $D_{L_{y_i^{z_n}}}(Rn) > \varepsilon n$ for all $i = 1, 2, \ldots, d$. Indeed, suppose the contrary, and, for every $1 \leq i \leq d$ and $s \in S := \{aba^{-1} : a \in B_X(1_G, Rn),\ b \in B_X(1_G, \varepsilon n)\}$, fix $z_{i,s} \in N$ (when there is at least one of them) such that $y_i^{z_{i,s}} = s$. Then for every $z \in N$ there exists i and $a \in B_X(1_G, Rn)$ such that $D(L_{y_i^z}, a) \leq \varepsilon n$. Now

$$D(L_{y_i^z}, a) = \ell_X(a^{-1} y_i^z a) \leq \varepsilon n,$$

yields $a^{-1} y_i^z a \in B_X(1_G, \varepsilon n)$, hence $y_i^z \in aB_X(1_G, \varepsilon n)a^{-1}$ where $a \in B_X(1_G, Rn)$.

Hence $y_i^z =: s \in S$, so $y_i^z = y_i^{z_{i,s}}$, showing that $z \in C_N(y_i) z_{i,s}$. It follows that

$$N = \bigcup_{i=1}^{d} \bigcup_{s\in S} C_N(y_i) z_{i,s}.$$

Since S is clearly finite, by Neumann's lemma (Lemma 12.19), one of the $C_N(y_i)$'s is of finite index, yielding $|y_i^N| < \infty$, equivalently $y_i \in \Delta(N)$, contrary to our assumptions on Y.

Step 2: Given $n \in \mathbb{N}$, let $z_n = y_{i_1} y_{i_2} \cdots y_{i_r}$ be the expression of the element $z_n \in N$ provided by the previous step, and let x_n be the shortest prefix of z_n such that at least for one generator y_i we have $D_{L_{y_i^{x_n}}}(Rn) > \varepsilon n$. First observe that we may assume that $x_n \neq 1$, since for all i we have $y_i \in N$ and therefore by Lemma 12.16

$$\lim_\omega \frac{D_{L_{y_i}}(Rn)}{n} = 0.$$

Now for any proper prefix w_n of x_n, by minimality of x_n we have $D_{L_{y_i^{w_n}}}(Rn) \leq \varepsilon n$ for all i. Suppose that $x_n = w_n y_\mu$ (this is possible since $x_n \neq 1$), where $y_\mu \in Y$. Then setting $\ell := \max_i \ell_X(y_i)$, for at least some $i = i(n)$ we have

$$\varepsilon n < D_{L_{y_i^{x_n}}}(Rn) = \max\{d_X(y_i^{x_n} g, g) : \ell_X(g) \leq Rn\} \tag{12.15}$$

$$\leq \max\{d_X(y_i^{w_n} y_\mu g, y_\mu g) : \ell_X(g) \leq Rn\} \leq D_{L_{y_i^{w_n}}}(Rn+\ell) \tag{12.16}$$

while, for all i,

$$D_{L_{y_i^{w_n}}}(Rn+\ell) \leq D_{L_{y_i^{w_n}}}(Rn) + 2\ell \leq \varepsilon n + 2\ell. \tag{12.17}$$

From (12.15) and (12.17), dividing by n we get, for some $i = i(n)$,

$$\varepsilon < \frac{D_{L_{y_i^{x_n}}}(Rn)}{n} \leq \varepsilon + \frac{2\ell}{n}.$$

Since ω is an ultrafilter, there exists an i such that $\{n \in \mathbb{N} : i(n) = i\} \in \omega$ (cf. Lemma 11.11). This yields

$$\lim_\omega \frac{D_{L_{y_i^{x_n}}}(Rn)}{n} = \varepsilon. \tag{12.18}$$

Step 3: We claim that for every $g \in N$, the sequence g^u is moderate. By Remark 12.20, we need to check this only for the generators of N. Let $1 \leq i \leq d$ and $n \in \mathbb{N}$. It follows from (12.17) that $D(y_i^{x_n}, Rn) \leq \varepsilon n + 2\ell$. Hence $\ell_X(y_i^{x_n}) = D(L_{y_i^{x_n}}, 1_G) \leq \varepsilon n + 2\ell$, which is clearly sublinear in n, thus showing that $(y_i^{x_n})_{n \in \mathbb{N}}$ is moderate. This completes the proof of (1).

Step 4: Condition (2) then follows from (12.18) since for the given i we then have

$$D_{\varphi(y_i)}(R) = \lim_\omega \frac{D_{L_{y_i^{x_n}}}(Rn)}{n} = \varepsilon \neq 0$$

so that $\varphi(y_i) \neq \mathrm{Id}_K$, while from $D_{\varphi(y_i)}(R) = \varepsilon$ we deduce $\varphi(i) \in O(2\varepsilon, R)$. $\square$

In order to prove the following lemma, we need to introduce some notation. Let $k \geq 1$ be an integer. For $v = (v_i)_{i=1}^k \in \mathbb{C}^k$ (resp. $A = (a_{i,j})_{i,j=1}^k \in \mathrm{GL}(k, \mathbb{C})$) we set $\|v\| := \max\{\|v_i\| : i = 1, 2, \ldots, k\}$ (resp. $\|A\| = \max\{|a_{i,j}| : i, j = 1, 2, \ldots, k\}$. It is easy to see (**exercise**) that $\|\cdot\|$ is a *norm*, in particular it satisfies $\|u+v\| \leq \|u\| + \|v\|$ (resp. $\|A+B\| \leq \|A\| + \|B\|$) and $\|\lambda v\| = |\lambda| \cdot \|v\|$ (resp. $\|\lambda A\| = |\lambda| \cdot \|A\|$) for all $u, v \in \mathbb{C}^k$ (resp. $A, B \in \mathrm{GL}(k, \mathbb{C})$) and $\lambda \in \mathbb{C}$. Moreover (**exercise**) there exists a constant $C = C(k) \geq 1$ such that $\|Av\| \leq C\|A\| \cdot \|v\|$ (and therefore $\|AB\| \leq C\|A\| \cdot \|B\|$) for all $u \in \mathbb{C}^k$ and $A, B \in \mathrm{GL}(k, \mathbb{C})$.

Given $\varepsilon > 0$ we denote by

$$B^k(I, \varepsilon) := \{A \in \mathrm{GL}(k, \mathbb{C}) : \|A - I\| < \varepsilon\}$$

the ball of radius ε centered at the identity I in $\mathrm{GL}(k, \mathbb{C})$.

Lemma 12.23. *Let $n, k \geq 1$ be two integers. Then there exists $\varepsilon = \varepsilon(k,n) > 0$ such that $A^{\overline{n}} \neq I$ for all $1 \leq \overline{n} \leq n$ and all $A \in B^k(I, \varepsilon) \setminus \{I\}$.*

Proof. Let $A \in \mathrm{GL}(k, \mathbb{C})$. Then we have $A = (A - I) + I$ so that for all $\ell \in \mathbb{N}$

$$A^\ell = ((A-I)+I)^\ell = I + \sum_{i=1}^{\ell} \binom{\ell}{i} (A-I)^i.$$

Therefore, setting $B_{\overline{n}} := \sum_{\ell=1}^{\overline{n}-1} \sum_{i=1}^{\ell} \binom{\ell}{i} (A-I)^i$ for $1 \leq \overline{n} \leq n$, we have

$$I + A + A^2 + \cdots + A^{\overline{n}-1} = \overline{n}I + \sum_{\ell=1}^{\overline{n}-1} \sum_{i=1}^{\ell} \binom{\ell}{i} (A-I)^i = \overline{n}I + B_{\overline{n}}.$$

Since $\lim_{t \to 0} \sum_{\ell=1}^{n} \sum_{i=1}^{\ell} \binom{\ell}{i} C^i t^i = 0$ we can find $\varepsilon > 0$ such that

$$\sum_{\ell=1}^{\overline{n}} \sum_{i=1}^{\ell} \binom{\ell}{i} C^i \varepsilon^i \leq \sum_{\ell=1}^{n} \sum_{i=1}^{\ell} \binom{\ell}{i} C^i \varepsilon^i \leq \frac{1}{2C}.$$

Suppose that $A \in B^k(I, \varepsilon) \setminus \{I\}$.
Since $\|A - I\| < \varepsilon$, taking norms we have that

$$\|B_{\overline{n}}\| \leq \sum_{\ell=1}^{\overline{n}} \sum_{i=1}^{\ell} \binom{\ell}{i} C^i \|A-I\|^i \leq \sum_{\ell=1}^{\overline{n}} \sum_{i=1}^{\ell} \binom{\ell}{i} C^i \varepsilon^i \leq \frac{1}{2C}. \tag{12.19}$$

Moreover, since $A \neq I$ we can find a vector $v \in \mathbb{C}^k$ such that $w := (I - A)v \neq 0$. Up to replacing v by $v/\|w\|$, if necessary, we may suppose that $\|w\| = 1$. We have $(\overline{n}I + B_{\overline{n}})w = \overline{n}w + B_{\overline{n}}w$ and therefore

$$\begin{aligned} \overline{n} = \|\overline{n}w\| &= \|(\overline{n}I + B_{\overline{n}})w - B_{\overline{n}}w\| \\ &\leq \|(\overline{n}I + B_{\overline{n}})w\| + \|B_{\overline{n}}w\| \\ &\leq \|(\overline{n}I + B_{\overline{n}})w\| + C\|B_{\overline{n}}\| \cdot \|w\| \\ &\leq \|(\overline{n}I + B_{\overline{n}})w\| + \frac{1}{2}. \end{aligned}$$

We deduce that $\|(\overline{n}I + B_{\overline{n}})w\| \geq (2\overline{n} - 1)/2 > 0$. It follows that

$$(I - A^{\overline{n}})v = (I + A + A^2 + \cdots + A^{\overline{n}-1})(I - A)v = (\overline{n}I + B_{\overline{n}})w \neq 0$$

thus showing that $I - A^{\overline{n}} \neq 0$, equivalently $A^{\overline{n}} \neq I$, for all $1 \leq \overline{n} \leq n$. □

We are now in a position to complete the proof of Gromov's theorem.

End of proof of Gromov's theorem (Theorem 12.1). Combining the two Lemmas 12.22 and 12.23, for every $n \geq 1$ we can find a homomorphism $\varphi_n \colon N \to \mathrm{Isom}(K)$ and an $i = i(n)$ such that $\varphi_n(y_i)$ has order $> n$.

Since ω is an ultrafilter, there exists an i_0 such that $\{n \in \mathbb{N} : i(n) = i_0\} \in \omega$.

If $\varphi_n(N)$ is infinite for some n, then by Theorem 12.11 we can conclude that N is virtually nilpotent, and we are done. Hence we can assume that the images $\varphi_n(N)$ are finite for all $n \in \mathbb{N}$. For $g \in N$ set

$$\varphi(g) = [(\varphi_n(g)_{n\in\mathbb{N}}]_\omega \in \mathrm{GL}(k,\mathbb{C})_\omega.$$

By virtue of (11.29) (cf. Theorem 11.40), this yields an injective homomorphism

$$\varphi\colon N \to \mathrm{GL}(k,\mathbb{C}_\omega)$$

which, by construction, maps y_{i_0} into an element of infinite order. Thus $\varphi(N)$ is infinite and by Theorem 12.11 (with $F = \mathbb{C}_\omega$) we again conclude that N is virtually nilpotent.

This completes the proof of Gromov's theorem. □

As mentioned above (cf. Remark 7.11, from Gromov's theorem we deduce that in the context of growth of groups, the notions of sub-polynomial growth and polynomial growth coincide:

Corollary 12.24. *Every group of sub-polynomial growth has polynomial growth.*

Proof. Let G be a group of sub-polynomial growth. By Gromov's theorem (Theorem 12.1), G is virtually nilpotent, that is, it has a finite-index nilpotent subgroup $H \leq G$. By Proposition 7.20 we have $b^G = b^H$. Since the latter is polynomial (of degree d given by the Bass–Guivarc'h formula, cf. Theorem 7.29) we deduce that the former is also polynomial (of the same degree d). This shows that G is of (precise) polynomial growth. □

12.4 Notes

Gromov's theorem (Theorem 12.1), originally conjectured by John Milnor [239], was proved by Misha Gromov in [138]. In his original proof, Gromov used a notion of "limit of metric spaces", with respect to what is now called the "Gromov–Hausdorff" distance, which turned out to play a key role in the proof and that we now briefly describe.

Let (Z,d) be a metric space. Given a subset $X \subseteq Z$ and $\varepsilon > 0$, we denote by $N_\varepsilon(X) := \bigcup_{x\in X} B_Z(x,\varepsilon)$ the *ε-neighborhood* of X in Z, where $B_Z(z,\varepsilon) := \{z' \in Z : d(z',z) < \varepsilon\}$ denotes the open ball of radius ε centered at $z \in Z$. Given two nonempty subsets $X,Y \subseteq Z$ set

$$H_d(X,Y) := \inf\{\varepsilon > 0 : Y \subseteq N_\varepsilon(X) \text{ and } X \subseteq N_\varepsilon(Y)\} \in [0,+\infty)$$

(we adopt the usual convention that $H_d(X,Y) = +\infty$ if there is no $\varepsilon > 0$ satisfying the above containment conditions). Denoting by $\mathscr{P}_0(Z)$ (resp. $\mathscr{K}_0(Z)$) the space of all nonempty subsets (resp. nonempty closed bounded subsets) of Z, one has that H_d is a generalized pseudometric (resp. a metric) on $\mathscr{P}_0(Z)$ (resp. $\mathscr{K}_0(Z)$), called the *Hausdorff distance*. If (Z,d) is compact, then $(\mathscr{K}_0(Z), H_d)$ is compact as well.

In general, given a (not necessarily compact) metric space X, we fix $x_0 \in X$ as a *base point* and refer to (X,x_0) as the corresponding *pointed metric space*. Recall that a metric space (X,d) is *proper* if each closed ball (of finite radius) in X is compact. Note that a proper metric space is locally compact, complete, and separable.

Let (X,x_0) and (Y,y_0) be two pointed proper metric spaces. Denote by $X\dot{\cup}Y$ the set-theoretical disjoint union of X and Y. A metric d on $X\dot{\cup}Y$ is termed *admissible* provided that the restrictions of d to X and Y coincide with the original metrics, i.e., $d|_{X\times X} = d_X$ and $d|_{Y\times Y} = d_Y$. The following generalization of the Hausdorff distance was introduced by David Edwards in 1975 (without base points) and is attributed to Ofer Gabber by Gromov [138, Section 6]. See also [336], [185, 64], and, in the setting of Riemannian geometry, [276]. Set

$$\begin{aligned}\mathrm{GH}((X,x_0),(Y,y_0)) := \inf\{\varepsilon > 0 : \exists\, d \text{ admissible metric on } X\dot{\cup}Y \text{ such that}\\ d(x_0,y_0) < \varepsilon, B_X(x_0,1/\varepsilon) \subseteq N_\varepsilon(Y), \text{ and } B_Y(y_0,1/\varepsilon) \subseteq N_\varepsilon(X)\}.\end{aligned}$$

One has $\mathrm{GH}((X,x_0),(Y,y_0)) \leq 1$. It is also clear that $\mathrm{GH}((X,x_0),(X,x_0)) = 0$ and $\mathrm{GH}((X,x_0),(Y,y_0)) = \mathrm{GH}((Y,y_0),(X,x_0))$. Also, if (Z,z_0) is another pointed proper metric space and, moreover, $\mathrm{GH}((X,x_0),(Y,y_0)) < 1/2$ and $\mathrm{GH}((Y,y_0),(Z,z_0)) < 1/2$, then

$$\mathrm{GH}((X,x_0),(Z,z_0)) \leq \mathrm{GH}((X,x_0),(Y,y_0)) + \mathrm{GH}((Y,y_0),(Z,z_0)).$$

In other words, the function GH also satisfies the triangle inequality provided that at least two of the three "distances" involved are small enough. Moreover, $\mathrm{GH}((X,x_0),(Y,y_0)) = 0$ if and only if there exists an isometry $f\colon X \to Y$ such that $f(x_0) = y_0$. Then, GH is a metric on the *Gromov–Hausdorff space*, the quotient space of all pointed proper metric spaces modulo such base-point preserving isometries: it is called the *Gromov–Hausdorff distance*. In the following, by abuse of language and notation, we shall not make a distinction between a pointed proper metric space and its isometry class. A sequence $\left((X_j,x_{j,0})\right)_{j\in\mathbb{N}}$ of pointed proper metric spaces *converges* to the pointed proper metric space (Y,y_0), and we write $\lim_{j\to\infty}(X_j,x_{j,0}) = (Y,y_0)$, provided that $\lim_{j\to+\infty}\mathrm{GH}((X_j,x_{j,0}),(Y,y_0)) = 0$. If this is the case, the limit is unique (up to isometry). For more information on both the Hausdorff and the Gromov–Hausdorff distances, we refer to the notes [337] by Alexey Tuzhilin.

Suppose now that G is a finitely generated group and fix a finite symmetric generating subset $X \subset G$. Define a sequence of pointed proper metric spaces $((X_n,x_{n,0}))_{n\in\mathbb{N}}$ by setting, $X_n := G$, $d_{X_n}(g,h) := d_X(g,h)/n$ for all $g,h \in X_n = G$, and $x_{n,0} := 1_G$, for all $n \in \mathbb{N}$. Then Gromov [138, Section 7] showed that if G has subpolynomial growth, one can find a convergent subsequence $\left((X_{n_k},x_{n_k,0})\right)_{k\in\mathbb{N}}$ such that the limit $(Y,y_0) := \lim_{k\to\infty}(X_{n_k},x_{n_k,0})$ satisfies the following properties:

- Y is proper (and therefore locally compact, separable, and complete),
- Y is homogeneous (that is, the isometry group $\mathrm{Isom}(Y)$ acts transitively on Y),
- Y is connected, locally connected, and geodesic;

- Y is finite-dimensional (indeed, the Hausdorff dimension of Y (cf. Section 10.6) satisfies $\mathrm{Hdim}(Y) \leq d+1$, where d is the degree of sub-polynomial growth of G),
- G acts on Y by isometries.

Thus, Y enjoys the same properties as the asymptotic cone $K_\omega(G, d_X)$ constructed in Sections 12.1 and 12.2, and at this point, the completion of the original proof of Gromov's theorem in [138] is the same as we presented in Section 12.3.

The proof presented here is based on an approach, due to Laurentius Petrus Dignus (Lou) van den Dries and Alex James Wilkie [94], which used methods of nonstandard Analysis (ultrafilters, ultraproducts, ultrapowers, and asymptotic cones) as presented in Chapter 11.

A recent, clear, and comprehensive treatment of Gromov's theorem is presented by Steven P. Lalley in his recent book [209].

More recently, new proofs of Gromov's theorem, of a more "analytical flavor", have appeared. We mention, for instance, the proofs of Bruce Kleiner [201] (in terms of suitable spaces of harmonic functions), Shalom–Tao [315] (a finitary version), Breuillard–Green–Tao [39] and Ehud Hrushovsky [179] (both as applications of the theory of the so-called *approximate groups*), and Narutaka Ozawa [266] (in terms of reduced cohomology and Shalom's property H_{FD} [314]). The advantage of these new proofs consists in the fact that they rely neither on the Gleason–Montgomery–Zippin theorem (Theorem 9.17) nor on the Tits alternative (Theorem 8.1).

Lemma 12.19 is due to Bernhard H. Neumann [249].

12.5 Exercises

Exercise 12.1. Let (X, d_X) and (Y, d_Y) be two metric spaces. Suppose that there exist isometric embeddings $f: X \to Y$ and $g: Y \to X$.

(1) Show that there exists an isometry $F: X \to Y$.

(2) Let $x_0 \in X$ and $y_0 \in Y$. Suppose that, in addition, $f(x_0) = y_0$ and $g(y_0) = x_0$. Show that the isometry F in (1) can be chosen such that $F(x_0) = y_0$.

Exercise 12.2 (The Hausdorff generalized pseudometric). Let (Z, d) be a metric space. Denote by $\mathscr{P}_0(Z)$ (resp. $\mathscr{K}_0(Z)$) the space of all nonempty subsets (resp. nonempty closed bounded subsets) of Z. Given $X \in \mathscr{P}_0(Z)$ and $\varepsilon > 0$, denote by $N_\varepsilon(X) := \bigcup_{x \in X} B_Z(x, \varepsilon)$ the *ε-neighborhood* of X in Z, where $B_Z(z, \varepsilon) := \{z' \in Z : d(z', z) < \varepsilon\}$ denotes the open ball of radius ε centered at $x \in X$. Given $X, Y \in \mathscr{P}_0(Z)$ set

$$H_d(X,Y) := \inf\{\varepsilon > 0 : Y \subseteq N_\varepsilon(X) \text{ and } X \subseteq N_\varepsilon(Y)\}.$$

In the following, when $Z \subseteq \mathbb{R}$ we assume that d is the Euclidean distance: $d(x,y) := |x-y|$ for all $x, y \in Z$.

(1) Show that if $x, y \in Z$, then setting $X := \{x\}$ and $Y := \{y\}$ one has $H_d(X,Y) = d(x,y)$. Deduce that the map $z \mapsto \{z\}$ is an isometric embedding of Z into $\mathscr{K}_0(Z)$.

(2) Show that if $Z := \mathbb{R}$, $X := \{0\}$, and $Y := [0, +\infty)$, then $H_d(X,Y) = +\infty$ (though $X \subseteq Y$).

(3) Show that if $Z := \mathbb{R}$, $X := (0,1)$, and $Y := [0,1]$, then $H_d(X,Y) = 0$ (though $X \neq Y$).

(4) Let $X, Y \in \mathscr{P}_0(Z)$. Given $z \in Z$ set $d(z,Y) := \inf\{d(z,y) : y \in Y\}$. Then set $d(X,Y) := \sup\{d(x,Y) : x \in X\}$, $d(X,Y) := \sup\{d(y,X) : y \in Y\}$, and finally $d'(X,Y) := \max\{d(X,Y), d(Y,X)\}$. Show that $d'(X,Y) = d'(Y,X) = H_d(X,Y)$.

(5) Let $Z := \mathbb{R}$ and set $X := \{1,3,6,7\}$ and $Y := \{3,6\}$. Show that $d(X,Y) = 2$ but $d(Y,X) = 0$. Deduce that $d\colon \mathscr{K}_0(Z) \times \mathscr{K}_0(Z) \to [0,+\infty)$ may fail to be symmetric.

(6) Let $X, Y \in \mathscr{P}_0(Z)$. Show that $H_d(X,Y) = H_d(\overline{X},Y) = H_d(X,\overline{Y}) = H_d(\overline{X},\overline{Y})$.

(7) Let $X, Y \in \mathscr{P}_0(Z)$. Show that $H_d(X,Y) = 0$ if and only if $\overline{X} = \overline{Y}$.

(8) Show that H_d is a generalized pseudometric on $\mathscr{P}_0(Z)$.

Exercise 12.3 (The Hausdorff distance). Let (Z,d) be a metric space. Let $X, Y \in \mathscr{P}_0(Z)$.

(1) Show that H_d is a metric on $\mathscr{K}_0(Z)$ (called *Hausdorff distance*).

(2) Suppose that Z is compact (resp. complete). Show that $(\mathscr{K}(Z), H_d)$ is compact (resp. complete) as well.

(3) Suppose that Z is compact. With the notation in Exercise 12.2.(4), show that there exist $x_0 \in X$ and $y_0 \in Y$ such that

$$d(x_0,y_0) = d(X,Y) = d(Y,X) = d'(X,Y) = d'(Y,X) = H_d(X,Y).$$

Exercise 12.4 (Limits of sets). Let (Z,d) be a metric space. Given a sequence $(X_n)_{n\in\mathbb{N}}$ in $\mathscr{P}_0(Z)$ we denote by

$$\limsup_{n\to\infty} X_n := \bigcap_{n\in\mathbb{N}} \overline{\bigcup_{m=n}^{\infty} X_m}$$

the *metric upper limit* of $(X_n)_{n\in\mathbb{N}}$.

(1) Show that

$$\limsup_{n\to\infty} X_n = \{z \in Z : \forall \varepsilon > 0, |\{n \in N : B_Z(z,\varepsilon) \cap X_n \neq \varnothing\}| = \infty\}.$$

(2) Deduce that

$$\limsup_{n\to\infty} X_n = \{z \in Z : \exists (a_{i_k})_{k\in\mathbb{N}} \text{ s.t. } a_{i_k} \in A_{i_k} \text{ and } \lim_{k\to\infty} d(a_{i_k},z) = 0\}.$$

(3) Let $Z := (0,1)$ and set $X_n = \{1/n\}$ for all $n \in \mathbb{N}$. Show that $\limsup_{n\to\infty} X_n = \varnothing$.

(4) Set $\liminf_{n\to\infty} X_n = \{z \in Z : \forall \varepsilon > 0, |\{n \in N : B_Z(z,\varepsilon) \cap X_n = \varnothing\}| < \infty\}$. This is called the *metric lower limit* of $(X_n)_{n\in\mathbb{N}}$. Observe that $\liminf_{n\to\infty} X_n \subseteq \limsup_{n\to\infty} X_n$. Also show that

$$\liminf_{n\to\infty} X_n = \{z \in Z : \exists (a_n)_{n\in\mathbb{N}} \text{ s.t. } a_n \in A_n \text{ and } \lim_{n\to\infty} d(a_n,z) = 0\}$$

and that it is closed in Z.

(5) Let $(z_n)_{n\in\mathbb{N}}$ be a sequence in Z and set $X_n := \{z_n\}$ for all $n \in \mathbb{N}$. Let also $z \in Z$ and set $X := \{z\}$. Show that $\lim_{n\to\infty} H_d(X_n,X) = 0$ if and only if $\lim_{n\to\infty} d(z_n,z) = 0$ (cf. Exercise 12.2.(1)).

(6) If $\liminf_{n\to\infty} X_n = \limsup_{n\to\infty} X_n$ we denote this common value by $\lim_{n\to\infty} X_n$. Let $X \in \mathscr{P}_0(Z)$ and suppose that $\lim_{n\to\infty} H_d(X_n, X) = 0$. Show that $\liminf_{n\to\infty} X_n = \limsup_{n\to\infty} X_n$ and $\overline{X} = \lim_{n\to\infty} X_n$.

(7) Suppose that Z is compact. Show that a sequence $(X_n)_{n\in\mathbb{N}}$ is convergent in $(\mathscr{K}_0(Z), H_d)$ if and only if $\liminf_{n\to\infty} X_n = \limsup_{n\to\infty} X_n$. Moreover, if this is the case, if $X \in \mathscr{K}_0(Z)$ is the limit (that is, $H_d(X_n, X) \to 0$), show that $\overline{X} = \lim_{n\to\infty} X_n$.

(8) Show that Z is compact (resp. complete) if and only if $(\mathscr{K}(Z), H_d)$ is compact (resp. complete). Cf. Exercise 12.3.(2).

(9) Suppose that Z is compact. Show that Z is geodesic if and only if $(\mathscr{K}(Z), H_d)$ is geodesic.

Exercise 12.5. Let (X, d_X) and (Y, d_Y) be two compact metric spaces and denote by $X\dot{\cup}Y$ their set-theoretical disjoint union. A metric d on $X\dot{\cup}Y$ is termed *admissible* provided that the restrictions of d to X and Y coincide with the original metrics, i.e., $d|_{X\times X} = d_X$ and $d|_{Y\times Y} = d_Y$. Set

$$H((X,d_X),(Y,d_Y)) := \inf_d H_d(X,Y),$$

where d runs over all admissible metrics on $X\dot{\cup}Y$.

(1) Let (Z,d) be a metric space. Given $X, Y \in \mathscr{K}(Z)$, set $d_X := d|_{X\times X}$ and $d_Y := d|_{Y\times Y}$. Show that $H((X,d_X),(Y,d_Y)) \leq H_d(X,Y)$.

(2) Let $Z := [0,1]$ with the Euclidean distance. Set $X := \{0\}$ and $Y := \{1\}$. Show that $H((X,d_X),(Y,d_Y)) = 0$.

(3) Deduce that, in general, for $X, Y \in \mathscr{K}(Z)$ the inequality $H((X,d_X),(Y,d_Y)) \leq H_d(X,Y)$ in (1) may be strict.

Exercise 12.6. A metric space (X,d) is called *proper* if each closed ball (of finite radius) in X is compact. Show that a proper metric space is locally compact, complete, and separable.

Exercise 12.7. Let (X, x_0) and (Y, y_0) be two pointed proper metric spaces. For $M > 0$ set

- $d_M(x_0, y_0) := M/2$;
- $d_M(x,y) := d_X(x,x_0) + d(x_0,y_0) + d_Y(y,y_0)$ for all $x \in X$ and $y \in Y$;
- $d_M(x,x') := d_X(x,x')$ for all $x, x' \in X$;
- $d_M(y,y') := d_Y(y,y')$ for all $y, y' \in Y$.

(1) Show that $d_M \colon Z \times Z \to [0,\infty)$ is a distance on $Z := X\dot{\cup}Y$ which is admissible and satisfies the conditions in the definition of $\mathrm{GH}((X,x_0),(Y,y_0))$.

(2) Deduce that $\mathrm{GH}((X,x_0),(Y,y_0)) \leq 1$.

Exercise 12.8. Let (X, x_0) and (Y, y_0) be two pointed proper metric spaces. Suppose that there exists an isometry $f \colon X \to Y$ such that $f(x_0) = y_0$. For $k > 0$ set

$$d_k(x,y) := \begin{cases} d_X(x,y) & \text{if } x,y \in X \\ d_Y(x,y) & \text{if } x,y \in Y \\ d_X(x, f^{-1}(y)) + k \equiv d_Y(f(x),y) + k & \text{if } x \in X \text{ and } y \in Y, \end{cases}$$

for all $x, y \in Z := X \dot{\cup} Y$.

(1) Show that $d_k \colon Z \times Z \to [0, \infty)$ is a distance on Z which is admissible and satisfies the conditions in the definition of $\mathrm{GH}((X, x_0), (Y, y_0))$.

(2) Deduce that $\mathrm{GH}((X, x_0), (Y, y_0)) = 0$.

Exercise 12.9. Let (X, x_0) and (Y, y_0) be two pointed metric spaces. Suppose that $\mathrm{GH}((X, x_0), (Y, y_0)) = 0$. Show that there exists an isometry $F \colon X \to Y$ such that $F(x_0) = y_0$.

Exercise 12.10. Let (X, x_0), (Y, y_0), and (Z, z_0) be pointed proper metric spaces. Suppose that $\mathrm{GH}((X, x_0), (Y, y_0)) < 1/2$ and $\mathrm{GH}((Y, y_0), (Z, z_0)) < 1/2$. Show that

$$\mathrm{GH}((X, x_0), (Z, z_0)) \leq \mathrm{GH}((X, x_0), (Y, y_0)) + \mathrm{GH}((Y, y_0), (Z, z_0)).$$

Exercise 12.11. Let $((X_j, x_j))_{j \in \mathbb{N}}$ be a convergent sequence of pointed proper metric spaces. Suppose that the sequence $(\delta(X_j))_{j \in \mathbb{N}}$ of the diameters is unbounded. Show that the pointed proper metric space $(Y, y_0) := \lim_{j \to \infty}(X_j, x_j)$ is not bounded.

Exercise 12.12. Let $((X_j, x_j))_{j \in \mathbb{N}}$ be a convergent sequence of pointed proper metric spaces. Suppose that there exist pointed proper metric spaces (Y, y_0) and (Z, z_0) such that $\lim_{j \to \infty}(X_j, x_j) = (Y, y_0)$ and $\lim_{j \to \infty}(X_j, x_j) = (Z, z_0)$. Show that (Y, y_0) and (Z, z_0) are isometric.

Exercise 12.13. Let $G := \mathbb{Z}$ and $X := \{-1, 1\}$. For an integer $n \geq 1$ set $X_n := G$, $d_n(g, h) := d_X(g, h)/n = |g - h|/n$ for all $g, h \in X_n = G$, and $x_{n,0} := 1_G = 0$. Show that the sequence $((X_n, x_{n,0}))_{n \geq 1}$ of pointed proper metric spaces is convergent and that its limit (Y, y_0) is isometric to the pointed proper metric space $(\mathbb{R}, 0)$ with the Euclidean distance ($d(r, s) := |r - s|$ for all $r, s \in \mathbb{R}$).

Part III
Analytic and Probabilistic Theory

Chapter 13
The Theorems of Polya and Varopoulos

In this chapter we study random walks on discrete groups. We start with the simple random walks on $\mathbb{Z}$ and $\mathbb{Z}^2$ and show (Proposition 13.19) that they are recurrent, that is, with probability one, the random walker returns (in fact, infinitely many times) to the initial position. We then show that the simple random walk on $\mathbb{Z}^3$ (and in fact on $\mathbb{Z}^d$ for all integers $d \geq 3$) is transient (Proposition 13.21), that is, with a strictly positive probability, the random walker never returns back to the initial position. This is Pólya's theorem (Theorem 13.47). We then move to the more general setting of random walks on discrete groups and prove Nash-Williams' criterion for recurrence (Corollary 13.42) and the random walk alternative (Theorem 13.48): the random walks on a group G are either all recurrent or all transient. In other words, the property for a group of being recurrent (resp. transient) is intrinsic. The study of random walks on groups culminates with Varopoulos' theorem (cf. Theorem 13.22) which characterizes recurrence and transience of finitely generated groups in terms of their growth: a finitely generated group is recurrent if and only if it has at most quadratic growth. Note that this last result heavily depends on Gromov's theorem (Theorem 12.1) on groups of polynomial growth (in particular on Corollary 12.24).

For all the basic notions in probability that we will not define, and for further reading, we refer to the two monographs by Wolfgang Woess [355, 356] which, together with the notes by Mauro Mariani [233], also served as a source for most of the material presented in the present chapter.

13.1 The Simple Random Walk on $\mathbb{Z}^d$: Setting the Problem

Consider the *simple random walk* on $\mathbb{Z}$. This means that a walker is in position $x(t) \in \mathbb{Z}$ at time $t \in \mathbb{N}$, with $x(0) = 0$, and moves one step to the left ($x(t) \mapsto x(t+1) = x(t) - 1$) or one step to the right ($x(t) \mapsto x(t+1) = x(t) + 1$) with equal probability at each unit of discrete time $t = 0, 1, 2, \ldots$

More generally, for any integer $d \geq 1$, the *simple random walk* on $\mathbb{Z}^d$ is given by a walker that, starting at the origin $(0, 0, \ldots, 0) \in \mathbb{Z}^d$ at time $t = 0$, moves from the position $x(t) = (x_1, x_2, \ldots, x_{i-1}, x_i, x_{i+1}, \ldots, x_d)$ at time t to the position

T. Ceccherini-Silberstein and M. D'Adderio, *Topics in Groups and Geometry*,
Springer Monographs in Mathematics, https://doi.org/10.1007/978-3-030-88109-2_13

$x(t+1) = (x_1, x_2, \ldots, x_{i-1}, x_i \pm 1, x_{i+1}, \ldots, x_d)$ at time $t+1$ with probability $1/(2d)$ for all $1 \le i \le d$ at every unit of discrete time $t = 0, 1, 2, \ldots$

A few natural questions arise: what is the expected number of times the walker will return to the origin? What is the probability that the walker will return to the origin infinitely many times? What is the probability that the walker will move towards infinity?

In order to settle these questions, let us introduce some notions and discuss some general preliminary results.

13.2 Markov Chains

Definition 13.1. Let X be a set (finite or countable), called the *state space*. Let also $P = (p(x,y))_{x,y \in X}$ be a *stochastic matrix*: this means that

$$p(x,y) \ge 0 \quad \text{for all } x, y \in X \tag{13.1}$$

and

$$\sum_{y \in X} p(x,y) = 1 \quad \text{for all } x \in X. \tag{13.2}$$

P is called the *(one-step) transition probability matrix*. Finally, let $\pi : X \to \mathbb{R}$ be such that $\pi(x) \ge 0$ for all $x \in X$ and $\sum_{x \in X} \pi(x) = 1$; this is called the *initial distribution*. The triple (X, P, π) is called a *Markov chain*.

A Markov chain (X, P, π) models the following random evolution process on X with discrete time ($t \in \mathbb{N}$). At time $t = 0$ a random walker starts at $x(0) = x_0 \in X$ with probability $\pi(x_0)$. If at time $t = n$ he is in position $x(n) = x \in X$, then, with probability $p(x,y)$, he moves (performs a random step) at time $t = n+1$ to the new position $x(n+1) = y \in X$.

To formalize this setting, let us denote by

$$\Omega := X^{\mathbb{N}} = \{\omega = (x_0, x_1, x_2, \ldots) : x_n \in X \text{ for all } n \in \mathbb{N}\} \tag{13.3}$$

the *space of trajectories* (or *of random paths*) on X. Also, for fixed $y_0, y_1, \ldots, y_m \in X$, we denote by

$$C(y_0, y_1, \ldots, y_m) := \{\omega = (x_0, x_1, x_2, \ldots) \in \Omega : x_k = y_k, 0 \le k \le m\} \tag{13.4}$$

the *cylinder* with *base* the tuple $(y_0, y_1, \ldots, y_m) \in X^{m+1}$ and we set

$$\mathbb{P}_\pi(C(y_0, y_1, \ldots, y_m)) := \pi(y_0) p(y_0, y_1) p(y_1, y_2) \cdots p(y_{m-1}, y_m). \tag{13.5}$$

Using the well-known *Kolmogorov extension theorem* [356, Theorem 1.12], it can be shown that if we denote by $\mathscr{B} \subseteq \mathscr{P}(\Omega)$ the σ-algebra generated by the collection of all cylinders $C(y_0, y_1, \ldots, y_m)$, $y_k \in X$, $1 \le k \le m$, $m \in \mathbb{N}$, then $\mathbb{P}_\pi$ can be extended uniquely to a probability measure $\mathbb{P}_\pi : \mathscr{B} \to [0,1]$ satisfying (13.5). Hence, the triple $(\Omega, \mathscr{B}, \mathbb{P}_\pi)$ is a *probability space*.

The projection maps $Z_n\colon \Omega \to X$ defined by $Z_n(\omega) := x_n$, for all $\omega = (x_0, x_1, \dots) \in \Omega$ and $n \in \mathbb{N}$, are X-valued *random variables*, i.e., measurable functions from $(\Omega, \mathscr{B})$ to $(X, \mathscr{P}(X))$, where $\mathscr{P}(X)$ denotes the σ-algebra of all subsets of X. In this setting, Z_n represents the random position of the walker at time $n = 0, 1, 2, \dots$

An *event* is a measurable subset $A \subseteq \Omega$, i.e. $A \in \mathscr{B}$. For example, given $x \in X$, the event of being in x at time $t = n$, denoted $[Z_n = x] \in \mathscr{B}$, is the subset

$$Z_n^{-1}(x) = \{(x_0, x_1, \dots) \in \Omega : x_n = x\} = \bigcup_{y_0, y_1, \dots, y_{n-1} \in X} C(y_0, y_1, \dots, y_{n-1}, x) \in \mathscr{B}.$$

Notation 13.2. In general, we will denote the events with brackets, like $[Z_n = x]$. Also, for the probability measure $\mathbb{P}_\pi$ we will use the notation $\mathbb{P}_\pi[Z_n = x]$ instead of $\mathbb{P}_\pi([Z_n = x])$. Moreover, if A and B are two events, the joint probability $\mathbb{P}_\pi(A \cap B)$ is denoted by $\mathbb{P}_\pi(A, B)$, and analogously with the brackets, e.g. we will write $\mathbb{P}_\pi[Z_0 = x, Z_1 = y]$ for $\mathbb{P}_\pi([Z_0 = x] \cap [Z_1 = y])$.

Given events A and B with $\mathbb{P}_\pi(B) > 0$ we denote by

$$\mathbb{P}_\pi(A|B) := \frac{\mathbb{P}_\pi(A \cap B)}{\mathbb{P}_\pi(B)}$$

the *conditional probability* of A given B. We have

$$\mathbb{P}_\pi[Z_0 = x] = \pi(x) \tag{13.6}$$

and, it is an **exercise** to show that, provided $\mathbb{P}_\pi[Z_n = x] > 0$,

$$\mathbb{P}_\pi[Z_{n+1} = y | Z_n = x] = p(x, y) \tag{13.7}$$

for all $n \in \mathbb{N}$. Note that by (13.7) the position Z_{n+1} at time $n+1$ only depends on the position Z_n at time n and not on the preceding positions $Z_{n-1}, Z_{n-2}, \dots, Z_1, Z_0$. In formulae,

$$\mathbb{P}_\pi[Z_{n+1} = y | Z_n = x, Z_{n-1} = x_{n-1}, Z_{n-2} = x_{n-2}, \dots, Z_0 = x_0] = \mathbb{P}_\pi[Z_{n+1} = y | Z_n = x] \tag{13.8}$$

provided $\mathbb{P}_\pi[Z_n = x, Z_{n-1} = x_{n-1}, Z_{n-2} = x_{n-2}, \dots, Z_0 = x_0] > 0$. Condition (13.8) is called the *Markov property*. Also, from (13.7) we have

$$\mathbb{P}_\pi[Z_{m+1} = y | Z_m = x] = \mathbb{P}_\pi[Z_{n+1} = y | Z_n = x] \tag{13.9}$$

for every $m, n \in \mathbb{N}$ such that $\mathbb{P}_\pi[Z_m = x] > 0$ and $\mathbb{P}[Z_n = x] > 0$: this property is called *time homogeneity*.

Remark 13.3. Conversely, it is an **exercise** to check that given $(\Omega, \mathscr{B}, \mathbb{P}; (Z_n)_{n \in \mathbb{N}})$, with $\Omega = X^{\mathbb{N}}$ (cf. (13.3)), $\mathscr{B} \subseteq \mathscr{P}(\Omega)$ a σ-algebra, $\mathbb{P}$ a probability measure defined on $\mathscr{B}$, and $(Z_n)_{n \in \mathbb{N}}$ a sequence of X-valued random variables satisfying conditions

$$\mathbb{P}[Z_{n+1} = y | Z_n = x, Z_{n-1} = x_{n-1}, Z_{n-2} = x_{n-2}, \dots, Z_0 = x_0] = \mathbb{P}[Z_{n+1} = y | Z_n = x]$$

and

$$\mathbb{P}[Z_{m+1} = y | Z_m = x] = \mathbb{P}[Z_{n+1} = y | Z_n = x]$$

for all $x, y, x_i \in X$ for all i, and all $n, m \in \mathbb{N}$, we can associate with it a Markov chain (X, P, π) as follows: the initial distribution will be

$$\pi(x) := \mathbb{P}[Z_0 = x],$$

(cf. (13.6)) for all $x \in X$; moreover, for $x \in X$, if there exists an $n \in \mathbb{N}$ such that $\mathbb{P}[Z_n = x] > 0$, then we denote by $n(x)$ the minimal such value, otherwise we set $n(x) := \infty$. Then the transition probabilities can be defined as

$$p(x,y) := \begin{cases} \mathbb{P}[Z_{n(x)+1} = y | Z_{n(x)} = x] & \text{if } n(x) < \infty \\ \phi(y) & \text{otherwise} \end{cases}$$

(cf. (13.7)) for all $x, y \in X$, where $\phi : X \to [0,1]$ is any map such that $\sum_{y \in X} \phi(y) = 1$.

Note that if $n(x) < \infty$ for all $x \in X$, then this Markov chain is clearly unique. Notice that this condition is implied by *irreducibility*, an important property that we will discuss in the sequel (see Section 13.3).

A function $\mu : X \to [0,1]$ such that $\sum_{x \in X} \mu(x) = 1$ uniquely defines a probability measure on X (that we continue to denote by μ) by setting $\mu(A) := \sum_{x \in A} \mu(x)$ for every $A \subseteq X$. For this reason, with a slight abuse of language, one refers to μ as to an *atomic probability measure* on X.

Thus, the initial distribution is an atomic measure. Sometimes, a point $x_0 \in X$ is fixed and the initial distribution is the *Dirac delta* δ_{x_0} at x_0, i.e. $\delta_{x_0}(y)$ is equal to 1 if $y = x_0$ and to 0 otherwise. In the sequel, when $\pi = \delta_x$ with $x \in X$, for simplicity we shall write $\mathbb{P}_x$ instead of $\mathbb{P}_{\delta_x}$.

Remark 13.4. Notice that given a Markov chain (X, P, π), together with the probability measure $\mathbb{P}_\pi$, we can (and we will) consider also the probability measures $\mathbb{P}_x$ (coming from (X, P, δ_x)) for all $x \in X$.

Moreover, when the initial distribution is understood, we shall denote the Markov chain simply by (X, P).

Given a Markov chain (X, P), we denote by $\mathscr{G} = \mathscr{G}(X, P) = (X, E)$ the *directed graph associated with* (X, P). Its *vertex set* is X, the state space of the Markov chain, and its *edge set* $E \subseteq X \times X$ is the set of pairs $(x, y) \in X \times X$ such that $p(x, y) \neq 0$. Note that $\mathscr{G}$ may have loops (i.e. edges of the form (x, x) with $x \in X$) but not multiple (directed) edges.

For $n \in \mathbb{N}$ we set $P^0 := I = (\delta_{x,y})_{x,y \in X}$ and, for $n \geq 1$, we denote by

$$P^n := PP^{n-1} = \left(p^{(n)}(x,y)\right)_{x,y \in X}$$

the n-th power of the transition matrix P. It is an **exercise** to show that P^n is also a stochastic matrix. Note that, in particular, we have

$$p^{(0)}(x,y) = \begin{cases} 1 & \text{if } y = x \\ 0 & \text{otherwise} \end{cases} \quad \text{and} \quad p^{(1)}(x,y) = p(x,y),$$

for all $x, y \in X$. Then

$$p^{(n+1)}(x,y) = \sum_{u \in X} p(x,u) p^{(n)}(u,y) \tag{13.10}$$

and, provided $\pi(x) > 0$,

$$p^{(n)}(x,y) = \mathbb{P}_\pi[Z_n = y \mid Z_0 = x]. \tag{13.11}$$

In other words, (13.11) expresses the probability that starting from state x one arrives in state y in n random steps. Note that, by time homogeneity, we have

$$\mathbb{P}_\pi[Z_{k+n} = y | Z_k = x] = p^{(n)}(x,y) \tag{13.12}$$

(provided $\mathbb{P}_\pi[Z_k = x] > 0$) for all $k \in \mathbb{N}$.

A *real random variable* is a $\mathscr{B}$-measurable function $f\colon (\Omega, \mathscr{B}) \to (\overline{\mathbb{R}}, \overline{\mathscr{B}})$, where $\overline{\mathbb{R}} = \mathbb{R} \cup \{-\infty, +\infty\}$ and $\overline{\mathscr{B}}$ is the σ-algebra of *extended Borel sets of* $\mathscr{R}$ of $\mathbb{R}$. If the integral of f with respect to the measure $\mathbb{P}_\pi$ exists, then the quantity

$$\mathrm{e}(f) := \int_\Omega f(\omega) \mathrm{d}\mathbb{P}_\pi(\omega) \in \overline{\mathbb{R}}$$

is called the *expected value* (or *expectation*) of f.

For example, if $f = \alpha \mathbf{1}_C$, where $\alpha \in \mathbb{R}$, $C = C(y_0, y_1, \dots, y_m)$ is the cylinder with base $(y_0, y_1, \dots, y_m) \in X^m$, and $\mathbf{1}_C(y)$ equals 1 if $y \in C$ and 0 otherwise, then

$$\mathrm{e}(f) = \alpha \mathbb{P}_\pi(C(y_0, y_1, \dots, y_m)) = \alpha \pi(y_0) p(y_0, y_1) p(y_1, y_2) \cdots p(y_{m-1}, y_m).$$

Suppose now that $\pi = \delta_{x_0}$, where $x_0 \in X$ is fixed, and, for $x \in X$ and $n \in \mathbb{N}$, consider the real random variable $v_n^x\colon \Omega \to \{0,1\} \subseteq \mathbb{R}$ defined by

$$v_n^x(\omega) = \delta_x(Z_n(\omega))$$

for all $\omega \in \Omega$. Thus, $v_n^x(\omega)$ equals 1 if at time n the random walker, whose trajectory is ω (with $Z_0(\omega) = x_0$), is exactly in x, and 0 otherwise. We have

$$\begin{aligned}
\mathrm{e}(v_n^x) &= \int_\Omega v_n^x(\omega) \mathrm{d}\mathbb{P}_{x_0}(\omega) \\
&= \mathbb{P}_{x_0}\left(\bigcup_{y_1, y_2, \dots, y_{n-1} \in X} C(x_0, y_1, y_2, \dots, y_{n-1}, x) \right) \\
&= \sum_{y_1, y_2, \dots, y_{n-1} \in X} \mathbb{P}_{x_0}(C(x_0, y_1, y_2, \dots, y_{n-1}, x)) \\
\text{(by (13.5))} &= \sum_{y_1, y_2, \dots, y_{n-1} \in X} p(x_0, y_1) p(y_1, y_2) \cdots p(y_{n-1}, x) \\
&= p^{(n)}(x_0, x).
\end{aligned}$$

It follows that

$$\nu^x := \sum_{n=0}^{\infty} \nu_n^x$$

is the real random variable representing the total number of times the random walker, starting at x_0, visits x. Thus the corresponding expected value is given by

$$\mathrm{e}(\nu^x) = \sum_{n=0}^{\infty} p^{(n)}(x_0, x). \tag{13.13}$$

13.3 Irreducible Markov Chains

Definition 13.5. A Markov chain (X, P, π) is called *irreducible* if for every $x, y \in X$ there exists an $n = n(x,y) \in \mathbb{N} \setminus \{0\}$ such that $p^{(n)}(x,y) > 0$.

Let (X, P, π) be a Markov chain and denote by $\mathscr{G} = (X, E)$ the associated directed graph. Then (X, P, π) is irreducible if and only if $\mathscr{G}$ is *connected*, that is, for every pair (x, y) of vertices there exists a finite directed path connecting x to y (**exercise**).

The power series

$$G(x, y \mid z) := \sum_{n=0}^{\infty} p^{(n)}(x,y) z^n \tag{13.14}$$

where $x, y \in X$, and z is an indeterminate, is called the *Green function* of the Markov chain.

Proposition 13.6. *Suppose that (X, P, π) is irreducible and let $t \in (0, \infty)$. Then the series $G(x, y \mid t)$ either diverges for every $x, y \in X$, or converges for every $x, y \in X$.*

Proof. Let $x, y, x', y' \in X$. Suppose that $G(x, y \mid t) < \infty$ and let us show that we also have $G(x', y' \mid t) < \infty$.

By the irreducibility assumption, we can find $h, k \in \mathbb{N} \setminus \{0\}$ such that $p^{(h)}(x, x') > 0$ and $p^{(k)}(y', y) > 0$. Also

$$p^{(h+n+k)}(x,y) \geq p^{(h)}(x,x') p^{(n)}(x',y') p^{(k)}(y',y) \tag{13.15}$$

for all $n \in \mathbb{N}$. We then have

$$\begin{aligned}
G(x, y \mid t) &= \sum_{m=0}^{\infty} p^{(m)}(x,y) t^m \\
&\geq \sum_{m=h+k}^{\infty} p^{(m)}(x,y) t^m \\
&= \sum_{n=0}^{\infty} p^{(h+k+n)}(x,y) t^{h+k+n} \\
\text{(by (13.15) and } t^{h+k+n} > 0) \quad &\geq \sum_{n=0}^{\infty} t^{h+k} t^n p^{(h)}(x,x') p^{(n)}(x',y') p^{(k)}(y',y)
\end{aligned}$$

$$= p^{(h)}(x,x')p^{(k)}(y',y)t^{h+k}\sum_{n=0}^{\infty} p^{(n)}(x',y')t^n$$
$$= p^{(h)}(x,x')p^{(k)}(y',y)t^{h+k}G(x',y' \mid t).$$

This shows that $G(x',y' \mid t) < \infty$. □

From now on, we will always assume that our Markov chains are irreducible.

For $x,y \in X$ we set

$$G(x,y) := G(x,y \mid 1) = \sum_{n=0}^{\infty} p^{(n)}(x,y) \in [0,+\infty]. \tag{13.16}$$

Notice that this is the expected total number of times the random walker visits y after visiting x.

Let $x,y \in X$. The real random variable

$$\mathbf{t}^x := \inf\{n \geq 1 : Z_n = x\}$$

is called a *hitting time*. Thus $\mathbf{t}^x(\omega) = n$ indicates that the random walker, whose trajectory is ω, visits state x for the first time at time $t = n$.

The quantity

$$f^{(n)}(x,y) := \mathbb{P}_x[\mathbf{t}^y = n]$$

which represents the probability that the random walker, after visiting x, arrives in y for the first time in n steps, is called a *hitting probability*. Note that $f^{(0)}(x,y) = 0$ for all (possibly equal) $x,y \in X$.

We denote by

$$F(x,y \mid z) := \sum_{n=0}^{\infty} f^{(n)}(x,y)z^n = \sum_{n=1}^{\infty} f^{(n)}(x,y)z^n \tag{13.17}$$

the associated generating function.

Proposition 13.7. *Suppose that the Markov chain (X,P,π) is irreducible. Let $x,y \in X$ be distinct elements. Then the following holds:*

(1) $G(x,y \mid z) = G(y,y \mid z)F(x,y \mid z)$;
(2) $G(x,x \mid z) = 1 + G(x,x \mid z)F(x,x \mid z)$;
(3) *if $t \in [0,+\infty)$ is such that $G(x,x \mid t) < \infty$ and $F(x,x \mid t) < 1$ then*

$$G(x,x \mid t) = \frac{1}{1 - F(x,x \mid t)}. \tag{13.18}$$

Proof. Let $n \geq 1$. Suppose $Z_0 = x$ and $Z_n = y$, and denote by $k \in \{1,2,\ldots,n\}$ the instant such that $Z_k = y$ and $Z_j \neq y$ for all $1 \leq j \leq k-1$, equivalently $\mathbf{t}^y = k$. Note that the events

$$[\mathbf{t}^y = k] = [Z_k = y, Z_j \neq y \text{ for } j = 1,2,\ldots,k-1],$$

where $k = 1,2,\ldots$, are pairwise disjoint. Using the Markov property we can write

$$\begin{aligned} p^{(n)}(x,y) &= \mathbb{P}_x[Z_n = y] \\ &= \sum_{k=1}^{n} \mathbb{P}_x[Z_n = y, \mathbf{t}^y = k] \\ &= \sum_{k=1}^{n} \mathbb{P}_x[Z_n = y \mid \mathbf{t}^y = k]\mathbb{P}_x[\mathbf{t}^y = k] \\ &= \sum_{k=1}^{n} \mathbb{P}_x[Z_n = y \mid Z_k = y, Z_j \neq y \text{ for } j = 1,2,\dots,k-1]\mathbb{P}_x[\mathbf{t}^y = k] \\ &= \sum_{k=1}^{n} \mathbb{P}_x[Z_n = y \mid Z_k = y]\mathbb{P}_x[\mathbf{t}^y = k] \\ &= \sum_{k=1}^{n} p^{(n-k)}(y,y) f^{(k)}(x,y). \end{aligned}$$

Since $f^{(0)}(x,y) = 0$, for $n \geq 1$ we deduce

$$p^{(n)}(x,y) = \sum_{k=0}^{n} p^{(n-k)}(y,y) f^{(k)}(x,y).$$

Summing up over $n \in \mathbb{N}$, as $p^{(0)}(x,y) = \delta_x(y)$, we have

$$\begin{aligned} G(x,y \mid z) &= \sum_{n=0}^{\infty} p^{(n)}(x,y) z^n \\ &= p^{(0)}(x,y) + \sum_{n=1}^{\infty} p^{(n)}(x,y) z^n \\ &= \delta_x(y) + \sum_{n=1}^{\infty} \sum_{k=0}^{n} p^{(n-k)}(y,y) f^{(k)}(x,y) z^n \\ &= \delta_x(y) + G(y,y \mid z) F(x,y \mid z). \end{aligned}$$

This proves (1) and (2).

Suppose that $t \in [0,+\infty)$ is such that $G(x,x \mid t) < \infty$ and $F(x,x \mid t) < 1$. We need a theorem, see [297, Theorem 3.5].

Theorem 13.8 (Mertens). *Let $(a_n)_{n\in\mathbb{N}}$ and $(b_n)_{n\in\mathbb{N}}$ be two real sequences. Suppose that $\sum_{n=0}^{\infty} a_n$ converges absolutely to $A \in \mathbb{R}$ and $\sum_{n=0}^{\infty} b_n$ converges to $B \in \mathbb{R}$. Then their Cauchy product*

$$\sum_{n=0}^{\infty} \left(\sum_{k=0}^{n} a_{n-k} b_k \right)$$

converges to AB.

Applying the above theorem and (2) we deduce (13.18). □

13.4 Recurrent and Transient Markov Chains

Definition 13.9. Let (X,P,π) be a Markov chain. One says that a state $x \in X$ is *recurrent* provided

$$\mathbb{P}_x[\exists n > 0 : Z_n = x] = 1,$$

i.e. when, once x has been visited, the probability of returning to x is 1. Otherwise, namely if $\mathbb{P}_x[\exists n > 0 : Z_n = x] < 1$, it is called *transient*.

Remark 13.10. Notice that the property of being recurrent or transient for a state $x \in X$ is independent of the initial distribution π.

Let (X,P,π) be a Markov chain. For $x,y \in X$ let us set

$$F(x,y) := \mathbb{P}_x[\exists n > 0 : Z_n = y],$$

the probability of visiting y at least once after visiting x, and

$$H(x,y) := \mathbb{P}_x[Z_n = y \text{ for infinitely many } n \in \mathbb{N}],$$

the probability of visiting y infinitely many times after visiting x.

Notice that a state $x \in X$ is recurrent if and only if $F(x,x) = 1$.

We also observe the following elementary facts.

$$F(x,y) := \mathbb{P}_x[\exists n > 0 : Z_n = y] = \sum_{n=1}^{\infty} \mathbb{P}_x[\mathbf{t}^y = n] = \sum_{n=0}^{\infty} f^{(n)}(x,y) = F(x,y \mid 1)$$

(recall that $f^{(0)}(x,y) = 0$) for all $x,y \in X$.

Also, suppose that after visiting the state $x \in X$ the random walker will return to x with probability one (i.e. x is recurrent). Then it is intuitively clear that the random walker will return to x infinitely many times (thus $H(x,x) = 1$).

Theorem 13.11 (Kolmogorov zero-one law for recurrence). *Let (X,P,π) be an irreducible Markov chain. For $x \in X$ the following conditions are equivalent:*

(a) *x is recurrent (i.e. $F(x,x) = 1$);*
(b) *$G(x,x) = \infty$;*
(c) *$H(x,x) = 1$;*
(d) *$H(x,x) > 0$.*

As a consequence, x is transient if and only if $H(x,x) = 0$.

Proof. In order to prove the implication (a) $\Rightarrow$ (b), we apply the following well-known theorem, due to Niels Henrik Abel (see [297, Theorem 8.2]).

Theorem 13.12 (Abel). *Let $f(z) := \sum_{n=0}^{\infty} a_n z^n$ be a power series converging for any $|z| < 1$. If $\sum_{n=0}^{\infty} a_n < \infty$ then*

$$\lim_{z \to 1^-} f(z) = \sum_{n=0}^{\infty} a_n.$$

By the theorem we have

$$F(x,x) = \lim_{t\to 1^-} F(x,x \mid t).$$

If $G(x,x) < \infty$, then, again by Abel's theorem,

$$G(x,x) = \lim_{t\to 1^-} G(x,x \mid t).$$

Hence by Proposition 13.7.(3), we deduce

$$\lim_{t\to 1^-} \frac{1}{1-F(x,x \mid t)} = G(x,x) < \infty,$$

and therefore $F(x,x) < 1$. This shows (a) $\Rightarrow$ (b).

Conversely, suppose that $F(x,x) < 1$, and let R be the radius of convergence of the power series $G(x,x \mid z)$. We need the following theorem of Alfred Pringsheim, see [173, Theorem 5.7.1] or [332, Theorem 7.21].

Theorem 13.13 (Pringsheim). *Let $f(z) := \sum_{n=0}^{\infty} a_n z^n$ with $a_n \geq 0$ and let $R \in (0,+\infty)$ be its radius of convergence. Then $f(z)$ has a pole at R. In particular,*

$$\lim_{z\to R^-} f(z) = +\infty.$$

So, if $R \leq 1$, then by the theorem

$$\lim_{t\to R^-} G(x,x \mid t) = +\infty.$$

On the other hand, by Proposition 13.7.(3),

$$\lim_{t\to R^-} G(x,x \mid t) = \lim_{t\to R^-} \frac{1}{1-F(x,x \mid t)} \leq \frac{1}{1-F(x,x)} < +\infty,$$

a contradiction. Hence $R > 1$, and therefore $G(x,x) = G(x,x \mid 1) < \infty$. This shows that (b) $\Rightarrow$ (a).

For $m \geq 1$ let us set

$$H^{(m)}(x,y) := \mathbb{P}_x[\exists 0 < n_1 < n_2 < \ldots < n_m : Z_{n_j} = y \text{ for all } j = 1,2,\ldots,m].$$

This expresses the probability that the random walker, after visiting x, will visit state y at least m times. Then for $x,y \in X$ we have $H^{(1)}(x,y) = F(x,y)$ and

$$H(x,y) = \lim_{m\to\infty} H^{(m)}(x,y) \tag{13.19}$$

(note that $H^{(m)}(x,y) \geq H^{(m+1)}(x,y)$ for all $m \geq 1$). Moreover,

$$
\begin{aligned}
&H^{(m)}(x,y)\\
&=\sum_{k=1}^{\infty}\mathbb{P}_x[\mathbf{t}^y=k \text{ and } \exists k=n_1<n_2<\ldots<n_m : Z_{n_j}=y \text{ for all } j=1,2,\ldots,m]\\
&=\sum_{k:f^{(k)}(x,y)>0} f^{(k)}(x,y)\mathbb{P}_x[\exists k=n_1<n_2<\ldots<n_m : Z_{n_j}=y\\
&\qquad \text{for all } j=1,2,\ldots,m \mid Z_k=y \text{ and } Z_j\neq y \text{ for } j=1,2,\ldots,k-1]\\
&=\sum_{k:f^{(k)}(x,y)>0} f^{(k)}(x,y)\mathbb{P}_y[\exists 0<n_2<\ldots<n_m : Z_{n_j}=y \text{ for all } j=2,\ldots,m]\\
&=\sum_{k=1}^{\infty} f^{(k)}(x,y)H^{(m-1)}(y,y)\\
&=F(x,y)H^{(m-1)}(y,y).
\end{aligned}
$$

As a consequence,

$$H^{(m)}(x,x)=F(x,x)^m \tag{13.20}$$

for all $m\geq 1$.

If $F(x,x)=1$ then from (13.20) and (13.19) we deduce $H(x,x)=1$, showing (a) $\Rightarrow$ (c). The implication (c) $\Rightarrow$ (d) is trivial. Suppose now $H(x,x)>0$. If we had $F(x,x)<1$ then from (13.20) and (13.19) we would deduce $H(x,x)=0$, a contradiction. This shows (d) $\Rightarrow$ (a), completing the proof. □

Corollary 13.14. *Let (X,P,π) be an irreducible Markov chain. Then either all the states $x\in X$ are recurrent, in which case we call the Markov chain* recurrent, *or they are all transient, in which case we call the Markov chain* transient.

Proof. Let $x,y\in X$. Then, by Theorem 13.11, x is recurrent if and only if $G(x,x)=+\infty$. Now, by Proposition 13.6, $G(x,x)=+\infty$ if and only if $G(y,y)=+\infty$, hence, again by Theorem 13.11, this holds if and only if y is recurrent. □

Remark 13.15. Notice that the property of being recurrent or transient for an irreducible Markov chain is independent of the initial distribution.

13.5 Random Walks on Finitely Generated Groups

Let G be a finitely generated group and let $Y\subseteq G$ be a finite symmetric generating subset.

Definition 13.16. Let $(G,P)=(G,P,\pi)$ be an irreducible Markov chain on G. We say that (G,P) defines a *random walk* on G provided that the transition probabilities satisfy the following properties:

(1) $p(g,h)=p(h,g)$ for all $g,h\in G$ (*symmetry*);
(2) $p(g'g,g'h)=p(g,h)$ for all $g',g,h\in G$ (*space homogeneity*);
(3) $\sum_{h\in G} d_Y(g,h)^2 p(g,h)<\infty$ for all $g\in G$ (*finiteness of the second moment*).

Note that these conditions (in particular condition (3)) do not depend on the generating subset $Y \subseteq G$ nor on the initial distribution (**exercise**).

Remark 13.17. Notice that, by space homogeneity (property (2)) and left invariance of the metric d_Y, we have

$$\sum_{h \in G} d_Y(g,h)^2 p(g,h) = \sum_{h \in G} d_Y(g',h)^2 p(g',h)$$

for all $g, g' \in G$.

A Markov chain (G, P, π) that satisfies (1), (2), and the condition

$$p(g,h) = 0 \text{ whenever } g^{-1}h \notin Y$$

for all $g, h \in G$, clearly also satisfies condition (3), and therefore it is a random walk. Such a random walk is called *nearest neighbor* (with respect to Y).

Example 13.18. The *simple random walk* on G with respect to the finite symmetric generating subset Y is the nearest neighbor random walk $(G, Q) = (G, Q, \delta_{1_G})$ defined by

$$q(g,h) = \begin{cases} \frac{1}{|Y|} & \text{if } g^{-1}h \in Y \\ 0 & \text{otherwise} \end{cases}$$

for all $g, h \in G$. Note that (G, Q) is irreducible and satisfies conditions (1), (2) and (3) (**exercise**). Also, the graph $\mathscr{G} = (G, E)$ associated with the Markov chain (G, Q) is exactly the (directed) Cayley graph of G with respect to Y.

With a random walk $(G, P) = (G, P, \pi)$ we associate an atomic probability measure μ on G obtained by setting

$$\mu(g) := p(1_G, g) \tag{13.21}$$

for all $g \in G$. From conditions (1), (2), and (3) in Definition 13.16 and irreducibility of (G, P) we immediately deduce the following properties of the probability measure μ:

(1') $\mu(g) = \mu(g^{-1})$ for all $g \in G$ (*symmetry*);
(2') $\text{supp}(\mu)$ generates G (*irreducibility*);
(3') $\sum_{h \in G} d_Y(1_G, g)^2 \mu(g) = \sum_{h \in G} \ell_Y(g)^2 \mu(g) < \infty$ for all $g \in G$ (*finiteness of the second moment*).

Note that, in fact, condition (2') can be reformulated as

(2'') $\bigcup_{n=0}^{\infty} \text{supp}(\mu^{(n)}) = G$,

where $\mu^{(n)}$ is the *n-th convolution* of μ, which is defined recursively by setting $\mu^{(1)} := \mu$ and $\mu^{(n)}(g) := \sum_{h \in G} \mu^{(n-1)}(gh^{-1})\mu(h) = \sum_{h \in G} \mu^{(n-1)}(h)\mu(h^{-1}g)$ for all $g \in G$. It is an **exercise** to check that this is an atomic probability measure on G. Now (2'') follows from the irreducibility of (G, P) and from the fact that $p^{(n)}(1_G, g) = \mu^{(n)}(g)$ for all $g \in G$ (**exercise**). The measure associated with a nearest neighbor random walk is finitely supported: indeed $\mathrm{supp}(\mu) \subseteq Y$. In the case of the simple random walk one has $\mathrm{supp}(\mu) = Y$ and the associated measure is $\mu = \frac{1}{|Y|}\mathbf{1}_Y$.

Conversely, with a probability measure μ on G satisfying properties (1'), (2'), and (3') above we associate a random walk (G,P) on G by setting

$$p(g,h) := \mu(g^{-1}h) \tag{13.22}$$

for all $g,h \in G$. It is an **exercise** to show that (G,P) is irreducible and satisfies conditions (1), (2), and (3) in Definition 13.16.

13.6 Recurrence of the Simple Random Walk on $\mathbb{Z}$ and $\mathbb{Z}^2$

We are now in a position to answer a question that we asked at the beginning of the chapter. We start with $d = 1,2$.

Proposition 13.19 (Pólya). *The simple random walks on $\mathbb{Z}$ and on $\mathbb{Z}^2$ are recurrent.*

Proof. A *finite random path*, briefly a *path*, in $\mathbb{Z}$ is a sequence $\pi = (x_0, x_1, x_2, \ldots, x_n)$ where $x_0, x_1, \ldots, x_n \in \mathbb{Z}$, $x_0 = 0$, and $|x_{i+1} - x_i| = 1$ for all $i = 0,1,\ldots,n-1$. The integer n is then called the *length* of the path π. Such a path is called *closed* provided that $x_n = x_0$.

The number of paths of length n is clearly 2^n for all $n \in \mathbb{N}$.

Let us determine the number of closed paths of length n. Clearly if n is odd this number is zero. Suppose that n is even, say $n = 2m$. Then each closed path has m left steps and m right steps. Hence the total number of such closed path is $\binom{2m}{m}$.

Hence the probability $p_n := q^{(n)}(0,0)$ of returning to the origin after n steps is given by

$$p_n = \begin{cases} \frac{1}{2^{2m}}\binom{2m}{m} & \text{if } n = 2m \text{ is even} \\ 0 & \text{otherwise.} \end{cases} \tag{13.23}$$

Notation 13.20. Given two sequences $(a_k)_{k\in\mathbb{N}}$ and $(b_k)_{k\in\mathbb{N}}$ of positive numbers, we set $a_k \sim b_k$ if $\lim_{k\to\infty} a_k/b_k = 1$, and $a_k \preceq b_k$ if $\lim_{k\to\infty} a_k/b_k \leq 1$.

From *Stirling's formula* (see [297, Theorem 8.22])

$$k! \sim k^k e^{-k}\sqrt{2\pi k} \tag{13.24}$$

we deduce

$$p_{2m} \sim \frac{1}{2^{2m}} \frac{(2m)^{2m} e^{-2m}\sqrt{2\pi 2m}}{m^{2m} e^{-2m} 2\pi m} = \frac{1}{\sqrt{\pi m}}. \tag{13.25}$$

As $\sum_{m=1}^{\infty} \frac{1}{\sqrt{m}} = \infty$,

$$G(0,0) = \sum_{n=0}^{\infty} p_n = \sum_{m=0}^{\infty} p_{2m} = \infty.$$

By virtue of Theorem 13.11, this means that the simple random walk on $\mathbb{Z}$ is recurrent.

Consider now the simple random walk on $\mathbb{Z}^2$. Thus the walker moves with equal probability to either North, South, East, or West. The notion of path (resp. closed path) is defined verbatim as in the one-dimensional case. The number of paths of length n is now 4^n. Let us determine the number of closed paths. As in the one-dimensional case, there are no closed paths of odd length. On the other hand, if π is a closed path of even length, say $n = 2m$, then clearly the number k of North steps equals the number of South steps, and similarly the number $m-k$ of East steps equals the number of West steps. Hence the total number of closed paths of length $n = 2m$ is

$$\begin{aligned}\sum_{k=0}^{m}\binom{2m}{k,k,m-k,m-k} &= \sum_{k=0}^{m}\frac{(2m)!}{k!k!(m-k)!(m-k)!}\\ &= \sum_{k=0}^{m}\frac{(2m)!}{m!m!}\frac{m!m!}{k!k!(m-k)!(m-k)!}\\ &= \sum_{k=0}^{m}\binom{2m}{m}\binom{m}{k}^2\\ &= \binom{2m}{m}^2,\end{aligned}$$

where the last equality follows from the classical identity

$$\sum_{k=0}^{m}\binom{m}{k}^2 = \sum_{k=0}^{m}\binom{m}{k}\binom{m}{m-k} = \binom{2m}{m},$$

which has the following combinatorial interpretation: to choose m objects out of a collection of m red objects and m blue ones, we first choose k red objects and then $m-k$ blue ones.

Hence the probability $p'_n := q^{(n)}((0,0),(0,0))$ of returning to the origin after n steps is

$$p'_n = \begin{cases}\frac{1}{4^{2m}}\binom{2m}{m}^2 = \left(\frac{1}{2^{2m}}\binom{2m}{m}\right)^2 & \text{if } n = 2m\\ 0 & \text{otherwise.}\end{cases}$$

Notice that $p'_n = (p_n)^2$, so that, by (13.25), we have $p'_{2m} \sim \frac{1}{\pi m}$, and, since $\sum_{m=1}^{\infty}\frac{1}{m} = +\infty$,

$$G((0,0),(0,0)) = \sum_{n=0}^{\infty} p'_n = \sum_{m=0}^{\infty} p'_{2m} = \infty.$$

By virtue of Theorem 13.11, this shows that the simple random walk on $\mathbb{Z}^2$ is also recurrent. □

13.7 Transience of the Simple Random Walk on $\mathbb{Z}^3$

The situation is different in one more dimension:

Proposition 13.21 (Pólya). *The simple random walk on $\mathbb{Z}^3$ is transient.*

Proof. With the notation as in the proof of Proposition 13.19, we now have

$$\begin{aligned} p''_{2m} := q^{(2m)}((0,0,0),(0,0,0)) &= \frac{1}{6^{2m}} \sum_{\substack{j,k\geq 0\\ j+k\leq m}} \frac{(2m)!}{j!j!k!k!(m-j-k)!(m-j-k)!} \\ &= \frac{1}{2^{2m}}\binom{2m}{m} \sum_{\substack{j,k\geq 0\\ j+k\leq m}} \frac{1}{3^{2m}}\left(\frac{m!}{j!k!(m-j-k)!}\right)^2. \end{aligned}$$

Using the simple fact that if $a,b\in\mathbb{N}$ and $a<b$, then $a!b!\geq(a+1)!(b-1)!$, we obtain that, for all $j,k\geq 0$ with $j+k\leq m$,

$$\frac{m!}{j!k!(m-j-k)!} \leq \frac{m!}{\lfloor\frac{m}{3}\rfloor!\lfloor\frac{m}{3}\rfloor!\lfloor\frac{m}{3}\rfloor!}.$$

Consequently, observing that

$$\sum_{\substack{j,k\geq 0\\ j+k\leq m}} \frac{m!}{j!k!(m-j-k)!} = (1+1+1)^m = 3^m,$$

we deduce

$$\begin{aligned} p''_{2m} &\leq \frac{1}{2^{2m}}\binom{2m}{m}\left(\frac{1}{3^m}\frac{m!}{([\frac{m}{3}]!)^3}\right)\frac{1}{3^m}\sum_{\substack{j,k\geq 0\\ j+k\leq m}}\frac{m!}{j!k!(m-j-k)!} \\ &= \frac{1}{2^{2m}}\binom{2m}{m}\left(\frac{1}{3^m}\frac{m!}{\left([\frac{m}{3}]!\right)^3}\right) \\ &= \frac{1}{2^{2m}3^m}\frac{(2m)!}{m!([\frac{m}{3}]!)^3}. \end{aligned}$$

Stirling's formula (cf. (13.24)) shows that

$$p''_{2m} \preceq \frac{\sqrt{2}}{\left(\sqrt{\frac{2\pi}{3}}\right)^3 m^{\frac{3}{2}}}$$

hence, since $\sum_{n=1}^{\infty}\frac{1}{m^{\frac{3}{2}}}<\infty$, one finally has

$$G((0,0,0),(0,0,0)) = \sum_{n=0}^{\infty} p''_n = \sum_{m=0}^{\infty} p''_{2m} < \infty.$$

By virtue of Theorem 13.11, this shows that the simple random walk on $\mathbb{Z}^3$ is transient. □

13.8 Varopoulos' Theorem and its Proof Strategy

The following theorem is the main result of this chapter.

Theorem 13.22 (Varopoulos). *Let G be a finitely generated group, and suppose that the simple random walk on G corresponding to some finite symmetric generating subset is recurrent. Then one of the following holds:*

(i) *G is finite, or*
(ii) *G has a subgroup of finite index isomorphic to $\mathbb{Z}$, or*
(iii) *G has a subgroup of finite index isomorphic to $\mathbb{Z}^2$.*

In order to settle the proof, we shall first introduce the notion of the *network* associated with a Markov chain on G. We shall then present some *recurrence criteria* and, in particular the one due to Crispin Nash-Williams. This will allow us to show that the simple random walk on any group whose growth function is at most quadratic is recurrent. On the other hand, if the growth of G is at least cubic, we shall explicitly construct transient (in general not simple) random walks on G. But then, making use of the notion of network, we shall prove that the recurrence of a generic random walk on G is equivalent to the recurrence of the simple random walk on G, so that the property of being recurrent is intrinsic to the group (i.e. it does not depend on the particular random walk). Finally, using Gromov's theorem (Theorem 12.1 and Corollary 12.24) in combination with the Bass–Guivarc'h formula (Theorem 7.29), we have that the unique groups whose growth is at most quadratic are indeed either finite or virtually isomorphic to $\mathbb{Z}^d$ for $d = 1,2$, and that there are no groups whose growth is more than quadratic and less than cubic. This will complete the proof of Varopoulos' theorem.

13.9 Reversible Markov Chains and Networks

Definition 13.23. A Markov chain $(X,P) = (X,P,\pi)$ is called *reversible* if there exists a function $m\colon X \to (0,\infty)$, sometimes called an *invariant measure*, such that

$$m(x)p(x,y) = m(y)p(y,x) \quad \text{for all } x,y \in X. \tag{13.26}$$

If (X,P) is reversible, then, for $x,y \in X$, the quantity

$$a(x,y) := m(x)p(x,y) \quad (= m(y)p(y,x) = a(y,x)) \tag{13.27}$$

is called the *conductance* between x and y.

Example 13.24. Let G be a finitely generated group, let $Y \subseteq G$ be a finite and symmetric generating subset, and denote by (G,Q) the simple random walk on G associated with Y. By symmetry, we have $q(g,h) = q(h,g)$ for all $g,h \in G$. Thus (G,Q) is reversible: one may take $m(g) = 1$ for all $g \in G$.

Remark 13.25. Notice that the directed graph (X,E) associated with a reversible Markov chain (X,P) has the property that $(x,y) \in E$ if and only if $(y,x) \in E$ for all $x,y \in X$. Hence we can consider the *undirected graph*, which we still denote $\mathscr{G} = \mathscr{G}(X,P) = (X,E)$, where now the elements of the edge set E are the multisets $\{x,y\}$, with $x,y \in X$, such that $p(x,y) \neq 0$.

Let (X,P,π) be a reversible Markov chain.
Notice that since

$$\sum_{y\in X} a(x,y) = \sum_{y\in X} m(x)p(x,y) = m(x)\sum_{y\in X} p(x,y) = m(x) \in (0,\infty),$$

the transition matrix of any reversible Markov chain is uniquely determined by its conductances. Indeed we have:

$$p(x,y) = \frac{a(y,x)}{m(x)} = \frac{a(y,x)}{\sum_{z\in X} a(x,z)}. \tag{13.28}$$

For any undirected graph $\mathscr{G} = (X,E)$ associated with a reversible Markov chain (X,P) we arbitrarily fix, once and for all, an *orientation* on the edge set E. This means that for every edge $e \in E$ we chose an *initial* vertex $e^+ \in X$ and a *terminal* vertex $e^- \in X$ so that $e = \{e^+,e^-\}$.

Definition 13.26. The function $r\colon E \to (0,+\infty)$ defined by

$$r(e) := \frac{1}{a(e^+,e^-)} \quad \left(= \frac{1}{a(e^-,e^+)} \right) \tag{13.29}$$

for all $e \in E$ is called the *resistance*. The triple $\mathscr{N} = (X,E,r)$ is called a *network*.

Note that in (13.29) $a(e^+,e^-) = m(e^+)p(e^+,e^-) \neq 0$ because $m(e^+) > 0$ and, since $e = \{e^+,e^-\} \in E$, we also have $p(e^+,e^-) > 0$.

Again, the transition matrix of a reversible Markov chain uniquely determines its associated network, and vice versa. In particular, we have

$$m(x) = \sum_{\substack{e\in E\\ e^+=x}} \frac{1}{r(e)} \quad \left(= \sum_{\substack{e\in E\\ e^-=x}} \frac{1}{r(e)} \right)$$

for all $x \in X$, and

$$p(e^+,e^-) = \frac{1}{m(e^+)r(e)} \tag{13.30}$$

for all $e = \{e^+,e^-\} \in E$. Equivalently, for all $x,y \in X$,

$$p(x,y) = \begin{cases} \frac{1}{m(x)r(\{x,y\})} & \text{if } \{x,y\} \in E \\ 0 & \text{otherwise.} \end{cases} \tag{13.31}$$

Consider now the real Hilbert spaces

$$L^2(X,m) := \{f \in \mathbb{R}^X : \sum_{x\in X} m(x)f^2(x) < \infty\}$$

and

$$L^2(E,r) := \{u \in \mathbb{R}^E : \sum_{e\in E} r(e)u^2(e) < \infty\}$$

with inner products defined by

$$\langle f,g\rangle_X := \sum_{x\in X} m(x)f(x)g(x)$$

for all $f,g \in L^2(X,m)$, and

$$\langle u,v\rangle_E := \sum_{e\in E} r(e)u(e)v(e)$$

for all $u,v \in L^2(E,r)$, respectively.

Remark 13.27. Notice that, if $f,g \in L^2(X,m)$ (resp. $u,v \in L^2(E,r)$), then the function $x \mapsto f(x)g(x)$ (resp. $e \mapsto u(e)v(e)$) is in $L^1(X,m) := \{h \in \mathbb{R}^X : \sum_{x\in X} m(x)|h(x)| < \infty\}$ (resp. $L^1(E,r) := \{w \in \mathbb{R}^E : \sum_{e\in E} r(e)|w(e)| < \infty\}$).

The corresponding L^2-norms are defined by $\|f\|_X := \sqrt{\langle f,f\rangle_X}$ for all $f \in L^2(X,m)$ and $\|u\|_E := \sqrt{\langle u,u\rangle_E}$ for all $u \in L^2(E,r)$.

Also consider the linear operator $\nabla \colon \mathbb{R}^X \to \mathbb{R}^E$ defined by

$$(\nabla f)(e) := \frac{f(e^+) - f(e^-)}{r(e)} \tag{13.32}$$

for all $f \in \mathbb{R}^X$ and $e \in E$. It is called the *nabla operator* associated with the network (X,E,r).

We observe that $\nabla(L^2(X,m)) \subseteq L^2(E,r)$: indeed, for $f \in L^2(X,m)$, we have

$$\begin{aligned} \langle \nabla f, \nabla f\rangle_E &= \sum_{e\in E} \frac{(f(e^+) - f(e^-))^2}{r(e)} \\ &= \frac{1}{2} \sum_{x,y\in X} m(x)p(x,y)(f(x) - f(y))^2 \\ &\le \sum_{x,y\in X} m(x)p(x,y)(f(x)^2 + f(y)^2) \\ &= 2\|f\|_X^2 < \infty. \end{aligned} \tag{13.33}$$

In particular, this shows that $\|\nabla|_{L^2(X,m)}\| \le \sqrt{2}$, so that $\nabla|_{L^2(X,m)}$ is a bounded operator.

For $u \in \mathbb{R}^E$ consider the expression

$$(\nabla^* u)(x) := \frac{1}{m(x)} \left(\sum_{\substack{e \in E:\\ e^+ = x}} u(e) - \sum_{\substack{e \in E:\\ e^- = x}} u(e) \right). \tag{13.34}$$

Remark 13.28. Suppose that the associated graph $\mathscr{G}$ is *locally finite* (i.e. for each $x \in X$ the number of edges $e \in E$ for which $e^+ = x$ or $e^- = x$ is finite). Then (13.34) is a finite sum (so that $(\nabla^* u)(x) \in \mathbb{R}$) for every $u \in \mathbb{R}^E$ and $x \in X$ and therefore defines a map $\nabla^* \colon \mathbb{R}^E \to \mathbb{R}^X$.

We claim that for $u \in L^2(E, r)$ the sum in (13.34) is convergent (so that $(\nabla^* u)(x) \in \mathbb{R}$) for every $x \in X$ and therefore defines a linear operator $\nabla^* \colon L^2(E, r) \to \mathbb{R}^X$.

For $x, y \in X$ let us set

$$\widetilde{\delta}_x(y) = \begin{cases} \frac{1}{m(x)} & \text{if } y = x \\ 0 & \text{otherwise.} \end{cases}$$

Note that $\widetilde{\delta}_x \in L^2(X, m)$ and that $\langle f, \widetilde{\delta}_x \rangle_X = f(x)$ for all $f \in L^2(X, m)$. We remarked above that $\nabla \widetilde{\delta}_x \in L^2(E, r)$.

Let $u \in L^2(E, r)$. We have

$$\begin{aligned} \langle u, \nabla \widetilde{\delta}_x \rangle_E &= \sum_{e \in E} r(e) u(e) \left(\nabla \widetilde{\delta}_x \right)(e) \\ &= \sum_{e \in E} r(e) u(e) \left(\frac{\widetilde{\delta}_x(e^+) - \widetilde{\delta}_x(e^-)}{r(e)} \right) \\ &= \frac{1}{m(x)} \left(\sum_{\substack{e \in E:\\ e^+ = x}} u(e) - \sum_{\substack{e \in E:\\ e^- = x}} u(e) \right), \end{aligned}$$

where the last equality follows from Remark 13.27, which guarantees that the last two sums are finite. This proves our claim.

Lemma 13.29. *We have $\nabla^*(L^2(E, r)) \subseteq L^2(X, m)$. Moreover,*

$$\langle \nabla f, u \rangle_E = \langle f, \nabla^* u \rangle_X \tag{13.35}$$

for all $f \in L^2(X, m)$ and $u \in L^2(E, r)$. In other words, the linear operator

$$\nabla^*|_{L^2(E,r)} \colon L^2(E, r) \to L^2(X, m)$$

is the adjoint of the nabla operator $\nabla|_{L^2(X,m)} \colon L^2(X, m) \to L^2(E, r)$.

Proof. For $f \in L^2(X, m)$ and $u \in \mathbb{R}^E$ with finite support we have

$$\begin{aligned}
\langle \nabla f, u\rangle_E &= \sum_{e\in E} r(e)(\nabla f)(e)u(e) \\
&= \sum_{e\in E} r(e)\frac{f(e^+)-f(e^-)}{r(e)}u(e) \\
&= \sum_{e\in E}(f(e^+)-f(e^-))u(e) \\
&= \sum_{e\in E} f(e^+)u(e) - \sum_{e\in E} f(e^-)u(e) \\
&= \sum_{x\in X} f(x) \sum_{\substack{e\in E\\ e^+=x}} u(e) - \left(\sum_{x\in X} f(x) \sum_{\substack{e\in E\\ e^-=x}} u(e)\right) \\
&= \sum_{x\in X} f(x) \left(\sum_{\substack{e\in E\\ e^+=x}} u(e) - \sum_{\substack{e\in E\\ e^-=x}} u(e)\right)
\end{aligned}$$

$$\begin{aligned}
&= \sum_{x\in X} m(x)f(x)\frac{1}{m(x)} \left(\sum_{\substack{e\in E\\ e^+=x}} u(e) - \sum_{\substack{e\in E\\ e^-=x}} u(e)\right) \\
&= \sum_{x\in X} m(x)f(x)(\nabla^* u)(x) \\
&= \langle f, \nabla^* u\rangle_X.
\end{aligned}$$

So $\nabla^*|_{L^2(E,r)}$ coincides with the adjoint of the nabla operator $\nabla|_{L^2(X,m)} : L^2(X,m) \to L^2(E,r)$ on the dense subspace of finitely supported functions on E. By continuity, $\nabla^*|_{L^2(E,r)}$ is indeed the adjoint of $\nabla|_{L^2(X,m)}$. This proves all of our statements. □

Definition 13.30. Set

$$\mathscr{D}(\mathscr{N}) := \{f \in \mathbb{R}^X : \nabla f \in L^2(E,r)\}.$$

For $f \in \mathscr{D}(\mathscr{N})$ we define its *Dirichlet seminorm* as $D(f) := \langle \nabla f, \nabla f\rangle_E$. We fix an element $x_0 \in X$ and equip $\mathscr{D}(\mathscr{N})$ with the inner product $\langle \cdot,\cdot\rangle_D$ defined by setting

$$\langle f, g\rangle_D = \langle \nabla f, \nabla g\rangle_E + f(x_0)g(x_0) \tag{13.36}$$

for all $f,g \in \mathscr{D}(\mathscr{N})$. We denote by $\|\cdot\|_D$ the associated norm.

Note that $L^2(X,m) \subseteq \mathscr{D}(\mathscr{N})$, and that

$$\begin{aligned}
D(f) &= \langle \nabla f, \nabla f\rangle_E \\
&= \sum_{e\in E} r(e)\left(\frac{(f(e^+)-f(e^-))}{r(e)}\right)^2 \\
&= \sum_{e\in E} \frac{(f(e^+)-f(e^-))^2}{r(e)}
\end{aligned} \tag{13.37}$$

$$= \frac{1}{2} \sum_{x,y \in X} (f(x) - f(y))^2 m(x) p(x,y)$$

for all $f \in \mathscr{D}(\mathscr{N})$.

Lemma 13.31. *The space $\mathscr{D}(\mathscr{N})$ is a Hilbert space with respect to the inner product $\langle \cdot, \cdot \rangle_D$. Moreover, convergence in $\mathscr{D}(\mathscr{N})$ implies pointwise convergence.*

Proof. Let $f \in \mathscr{D}(\mathscr{N})$. For $x \in X$ let $\pi = (x_0, x_1, \ldots, x_n = x)$ be a geodesic path connecting x_0 to x in X and set $e_i = \{x_{i-1}, x_i\} \in E$ for all $i = 1, 2, \ldots, n$. We have

$$\begin{aligned}
(f(x) - f(x_0))^2 &= \left(\sum_{i=1}^{n} f(x_i) - f(x_{i-1}) \right)^2 \\
&= \left(\sum_{i=1}^{n} \frac{f(x_i) - f(x_{i-1})}{\sqrt{r(e_i)}} \sqrt{r(e_i)} \right)^2 \\
\text{(by Cauchy–Schwarz)} \quad &\leq \left(\sum_{i=1}^{n} \frac{(f(e_i^+) - f(e_i^-))^2}{r(e_i)} \right) \left(\sum_{i=1}^{n} r(e_i) \right) \\
\text{(cf. (13.37))} \quad &\leq D(f) c(x),
\end{aligned}$$

where $c(x) := \max\{1, \sum_{i=1}^{n} r(e_i)\}$. For all $f, g \in \mathscr{D}(\mathscr{N})$, we deduce

$$\begin{aligned}
|f(x) - g(x)| &= |f(x) - f(x_0) + f(x_0) - g(x) + g(x_0) - g(x_0)| \\
&\leq |(f-g)(x) - (f-g)(x_0)| + |(f-g)(x_0)| \\
&\leq \sqrt{c(x) D(f-g)} + |(f-g)(x_0)|.
\end{aligned} \tag{13.38}$$

It then follows that convergence in $\mathscr{D}(\mathscr{N})$ implies pointwise convergence: indeed if $\|f - g\|_D$ is small, so are $D(f-g)$ and $|(f-g)(x_0)|$ and therefore, by (13.38), $|f(x) - g(x)|$ is also small for all $x \in X$.

Let us show that $\mathscr{D}(\mathscr{N})$ is complete. Suppose that $(f_n)_{n \in \mathbb{N}}$ is a Cauchy sequence in $\mathscr{D}(\mathscr{N})$. Since

$$\|f_n - f_m\|_D^2 = D(f_n - f_m) + (f_n(x_0) - f_m(x_0))^2$$

we immediately deduce that $(f_n(x_0))_{n \in \mathbb{N}}$ is a Cauchy sequence in $\mathbb{R}$. On the other hand, from (13.38) we deduce that the sequence $(f_n(x))_{n \in \mathbb{N}}$ is a Cauchy sequence in $\mathbb{R}$ for every $x \in X \setminus \{x_0\}$. It follows that there exists an $f_\infty \in \mathbb{R}^X$ such that $\lim_{n \to \infty} f_n(x) = f_\infty(x)$ for all $x \in X$.

Moreover, since $D(f_n - f_m) = \|\nabla f_n - \nabla f_m\|_E^2$, we have that $(\nabla f_n)_{n \in \mathbb{N}}$ is a Cauchy sequence in $L^2(E, r)$ and therefore it converges to an element $u \in L^2(E, r)$.

Since

$$|\nabla f_\infty(e) - \nabla f_n(e)| \leq \frac{1}{r(e)} \left(|f_\infty(e^+) - f_n(e^+)| + |f_\infty(e^-) - f_n(e^+)| \right)$$

for all $e \in E$, we deduce that $\nabla f_\infty = u \in L^2(E,r)$ and therefore $f_\infty \in \mathscr{D}(\mathscr{N})$. Hence

$$\|f_n - f_\infty\|_D = \|\nabla(f_n - f_\infty)\|_E^2 + |(f_n - f_\infty)(x_0)|^2 = \|\nabla f_n - u\|_E^2 + |(f_n - f_\infty)(x_0)|^2 \to 0$$

as $n \to \infty$. Thus the Cauchy sequence $(f_n)_{n\in\mathbb{N}}$ is convergent to f_∞ in $\mathscr{D}(\mathscr{N})$. This shows that $\mathscr{D}(\mathscr{N})$ is complete. □

Proposition 13.32. *Let $f \in \mathscr{D}(\mathscr{N})$. Then for every $x \in X$,*

$$(\nabla^* \nabla f)(x) = f(x) - [Pf](x), \tag{13.39}$$

where

$$[Pf](x) := \sum_{y\in X} p(x,y) f(y). \tag{13.40}$$

Proof. Let $x \in X$. We have

$$\begin{aligned}
(\nabla^* \nabla f)(x) &= \frac{1}{m(x)} \left(\sum_{\substack{e\in E:\\ e^+=x}} \nabla f(e) - \sum_{\substack{e\in E:\\ e^-=x}} \nabla f(e) \right) \\
&= \frac{1}{m(x)} \left(\sum_{\substack{e\in E:\\ e^+=x}} \frac{f(e^+) - f(e^-)}{r(e)} - \sum_{\substack{e\in E:\\ e^-=x}} \frac{f(e^+) - f(e^-)}{r(e)} \right) \\
&= \sum_{\substack{e\in E:\\ e^+=x}} \frac{f(e^+) - f(e^-)}{m(e^+) r(e)} - \sum_{\substack{e\in E:\\ e^-=x}} \frac{f(e^+) - f(e^-)}{m(e^-) r(e)} \\
\text{(by (13.30))} \quad &= \sum_{\substack{e\in E:\\ e^+=x}} (f(e^+) - f(e^-)) p(e^+, e^-) - \sum_{\substack{e\in E:\\ e^-=x}} (f(e^+) - f(e^-)) p(e^-, e^+) \\
&= \sum_{\substack{e\in E:\\ e^+=x}} (f(x) - f(e^-)) p(x, e^-) - \sum_{\substack{e\in E:\\ e^-=x}} (f(e^+) - f(x)) p(x, e^+) \\
&= f(x) \left(\sum_{\substack{e\in E:\\ e^+=x}} p(x, e^-) + \sum_{\substack{e\in E:\\ e^-=x}} p(x, e^+) \right) \\
&\quad - \left(\sum_{\substack{e\in E:\\ e^+=x}} p(x, e^-) f(e^-) + \sum_{\substack{e\in E:\\ e^-=x}} p(x, e^+) f(e^+) \right) \\
&= f(x) \sum_{\substack{y\in X:\\ \{x,y\}\in E}} p(x,y) - \sum_{\substack{y\in X:\\ \{x,y\}\in E}} p(x,y) f(y) \\
&= f(x) \sum_{y\in X} p(x,y) - \sum_{y\in X} p(x,y) f(y) \\
&= f(x) - [Pf](x).
\end{aligned}$$

□

This proposition shows that the function $P\colon \mathscr{D}(\mathscr{N}) \to L^2(X,m)$ defined for every $f \in \mathscr{D}(\mathscr{N})$ by (13.40) is a well-defined linear operator.

Let $f \in L^2(X,m)$. Then

$$\begin{aligned}
\|Pf\|_X^2 &= \sum_{x\in X} m(x)(Pf(x))^2 \\
&= \sum_{x\in X} m(x) \sum_{y,z\in X} p(x,y)p(x,z)f(y)f(z) \\
&\leq \frac{1}{2}\sum_{x\in X} m(x) \sum_{y,z\in X} p(x,y)p(x,z)(f(y)^2+f(z)^2) \\
&= \frac{1}{2}\sum_{x\in X} m(x)\sum_{y\in X} p(x,y)f(y)^2 + \frac{1}{2}\sum_{x\in X} m(x)\sum_{z\in X} p(x,z)f(z)^2
\end{aligned}$$

$$\begin{aligned}
&= \sum_{y\in X}\sum_{x\in X} m(y)p(y,x)f(y)^2 \\
&= \sum_{y\in X} m(y)f(y)^2 \\
&= \|f\|_X^2.
\end{aligned}$$

So $P\colon L^2(X,m) \to L^2(X,m)$ is a bounded operator, with $\|P\| \leq 1$. It is called the *Markov operator* associated with the Markov chain (X,P).

Let now $\ell_0(X)$ denote the subspace of $\mathbb{R}^X$ consisting of all functions with finite support and let $\mathscr{D}_0(\mathscr{N})$ denote its closure in $\mathscr{D}(\mathscr{N})$.

Note that $L^2(X,m) \subseteq \mathscr{D}_0(\mathscr{N})$. Indeed, using (13.33), it is immediate to show that there exists a constant $C > 0$ such that $\|f\|_D \leq C\|f\|_X$ for all $f \in L^2(X,m)$.

For every subset $A \subseteq X$, we consider the matrix $P_A = (p_A(x,y))_{x,y\in X}$, called the *transition probability remaining in A* associated with $P = (p(x,y))_{x,y\in X}$, defined by setting

$$p_A(x,y) := p(x,y)\mathbf{1}_A(x)\mathbf{1}_A(y) = \begin{cases} p(x,y) & \text{if both } x,y \in A \\ 0 & \text{otherwise} \end{cases}$$

for all $x,y \in X$. Note that $0 \leq p_A(x,y) \leq p(x,y)$ for all $x,y \in X$ with equality on the right-hand side if and only if x,y both belong to A.

Then $(P_A)^n =: (p_A^{(n)}(x,y))_{x,y\in X}$ is given by

$$p_A^{(n)}(x,y) = \sum_{z\in A} p_A(x,z)p_A^{(n-1)}(z,y)$$

for all $x,y \in X$. Again, $0 \leq p_A^{(n)}(x,y) \leq p^{(n)}(x,y)$ for all $x,y \in X$. For $n \geq 2$, however it may happen that $p_A^{(n)}(x,y) < p^{(n)}(x,y)$ even if both x,y belong to A.

We then denote by

$$G_A(x,y \mid z) := \sum_{n=0}^{\infty} p_A^{(n)}(x,y)z^n$$

the associated *Green function*, and set $G_A(x,y) := G_A(x,y \mid 1) = \sum_{n=0}^{\infty} p_A^{(n)}(x,y)$.

Lemma 13.33. *Suppose that X is infinite and $A \subseteq X$ is finite. Then $G_A(x,y) < \infty$ for all $x,y \in X$.*

Proof. For $n \geq 1$ we set

$$p_A^{(n)}(x,A) := \sum_{y \in A} p_A^{(n)}(x,y)$$

for all $x \in X$, and

$$M^{(n)}(A) := \max_{x \in A} p_A^{(n)}(x,A).$$

Note that $0 \leq p_A^{(n)}(x,A) \leq \sum_{y \in A} p^{(n)}(x,y) \leq \sum_{y \in X} p^{(n)}(x,y) = 1$ for all $x \in A$ so that $0 \leq M^{(n)}(A) \leq 1$. Moreover, for every $h,k \geq 1$ we have

$$p_A^{(h+k)}(x,A) = \sum_{y,z \in A} p_A^{(h)}(x,z) p_A^{(k)}(z,y) \leq p_A^{(h)}(x,A) M^{(k)}(A). \tag{13.41}$$

Now, since X is infinite, A is finite, and P is irreducible, we can find $x_0 \in X \setminus A$ and for every $x \in A$ there exists an $n_x \in \mathbb{N}$ such that $p^{(n_x)}(x,x_0) > 0$. It follows that $p^{(n_x)}(x,A) < 1$. From (13.41) we deduce that, for every $k \geq 1$, $p^{(n_x+k)}(x,A) \leq p_A^{(n_x)}(x,A) M^{(k)}(A) < M^{(k)}(A) \leq 1$. Thus, setting $n_0 := \max\{n_x : x \in A\}$, we have $\alpha := M^{(n_0)}(A) < 1$. Now, for every $n \geq n_0$ we can write $n = kn_0 + r$ where $k \geq 1$ and $0 \leq r < n_0$. From (13.41) we then deduce

$$p^{(n)}(x,A) = p^{(kn_0+r)}(x,A) \leq p^{((k-1)n_0+r)}(x,A)\alpha \leq \cdots \leq p^{(r)}(x,A)\alpha^k \leq \alpha^k.$$

As a consequence, the sum $\sum_{n=0}^{\infty} p_A^{(n)}(x,y)$ converges exponentially, thus showing that $G_A(x,y) < \infty$. □

Suppose now that A is finite. Note that $G_A(x,y) = 0$ if x and y do not both belong to A. As a consequence, the function $G_A(\cdot,x)$ is finitely supported for all $x \in A$. For $f \in \mathbb{R}^X$ we set

$$[G_A f](x) := \sum_{y \in X} G_A(x,y) f(y) = \sum_{y \in A} G_A(x,y) f(y) \tag{13.42}$$

for all $x \in X$. Note that the above sum is finite, so that the quantity $[G_A f](x)$ is well defined, and that $[G_A f](x) = 0$ if $x \notin A$. In particular, if $f \in L^2(X,m)$ then $G_A f \in L^2(X,m)$, so that $G_A : L^2(X,m) \to L^2(X,m)$ is a linear operator. Let us also denote by $I_A : L^2(X,m) \to L^2(X,m)$ the operator defined by setting

$$[I_A f](x) = \begin{cases} f(x) & \text{if } x \in A \\ 0 & \text{otherwise} \end{cases}$$

for all $f \in L^2(X,m)$. Note that I_A is the projection operator onto the space of functions in $L^2(X,m)$ which are supported in A.

Suppose now that $f \in \ell_0(X)$ is such that $\mathrm{supp}(f) \subseteq A$. Then we have

$$\langle \nabla f, \nabla G_A(\cdot,x)\rangle_E = \langle f, (I_A - P_A)G_A(\cdot,x)\rangle_X = m(x)f(x) \tag{13.43}$$

for all $x \in \mathrm{supp}(f)$. Indeed, for every $x,y \in X$ we have

$$\begin{aligned}(I_A - P_A)G_A(x,y) &= (I_A - P_A)\sum_{n=0}^{\infty} p_A^{(n)}(x,y) \\ &= \sum_{n=0}^{\infty} p_A^{(n)}(x,y) - \sum_{n=0}^{\infty} p_A^{(n+1)}(x,y) \\ &= p_A^{(0)}(x,y) \\ &= \delta_{x,y}.\end{aligned}$$

Let $B \subseteq A$ be finite subsets of X and $x \in A$. We claim that

$$D(G_B(\cdot,x) - G_A(\cdot,x)) = m(x)\left(G_A(x,x) - G_B(x,x)\right). \tag{13.44}$$

Indeed, we have

$$\begin{aligned}D(G_B(\cdot,x) - G_A(\cdot,x)) &= \langle \nabla(G_B(\cdot,x) - G_A(\cdot,x)), \nabla(G_B(\cdot,x) - G_A(\cdot,x))\rangle_E \\ &= \langle \nabla G_B(\cdot,x), \nabla G_B(\cdot,x)\rangle_E + \langle \nabla G_A(\cdot,x), \nabla G_A(\cdot,x)\rangle_E \\ &\quad - 2\langle \nabla G_B(\cdot,x), \nabla G_A(\cdot,x)\rangle_E \\ \text{(by (13.43) and } B \subseteq A) \quad &= m(x)G_B(x,x) + m(x)G_A(x,x) - 2m(x)G_B(x,x) \\ &= m(x)\left(G_A(x,x) - G_B(x,x)\right).\end{aligned}$$

Lemma 13.34. *Suppose that (X,P) is transient. Then $G(\cdot,x) \in \mathscr{D}_0(\mathscr{N})$ for all $x \in X$.*

Proof. Let $(A_n)_{n\in\mathbb{N}}$ be a sequence of finite subsets of X such that $A_n \subseteq A_{n+1}$ for all $n \in \mathbb{N}$ and $\bigcup_{n\in\mathbb{N}} A_n = X$. Let $x,y \in X$. Observe that for every $k \in \mathbb{N}$, we have $p_{A_n}^{(k)}(x,y) \to p^{(k)}(x,y)$ as $n \to \infty$. Using the monotone convergence theorem (since $G_{A_n}(x,y) \leq G(x,y)$ and the Markov chain is transient), we deduce that $G_{A_n}(x,y) \to G(x,y)$ as $n \to \infty$. As a consequence of (13.44), $(G_{A_n}(\cdot,x))_{n\in\mathbb{N}}$ is a Cauchy sequence in $\mathscr{D}_0(\mathscr{N})$. By virtue of Lemma 13.31 its limit in $\mathscr{D}_0(\mathscr{N}) \subseteq \mathscr{D}(\mathscr{N})$ equals $G(\cdot,x)$. It follows that $G(\cdot,x) \in \mathscr{D}_0(\mathscr{N})$. □

13.10 Criteria for Recurrence

Let (X,P) be an irreducible, reversible Markov chain with invariant measure m and let $\mathscr{N} = (X,E,r)$ be the associated network.

Definition 13.35. Let $x_0 \in X$ and $i_0 \in \mathbb{R}$. We say that a function $u \in \mathbb{R}^E$ is a *flow from x_0 with input i_0* provided that

$$\nabla^* u = -\frac{i_0}{m(x_0)}\delta_{x_0}.$$

The associated *energy* is the positive (possibly infinite) quantity $\sum_{e\in E} r(e)u^2(e)$.
Also, the quantity

$$\mathrm{cap}(x_0) := \inf\{D(f) : f \in \ell_0(X), f(x_0) = 1\}$$

is called the *capacity* of x_0.

It follows from the above definitions that a flow $u \in \mathbb{R}^E$ has finite energy if and only if $u \in L^2(E,r)$: if this is the case, its energy is $\|u\|_E^2$.

Consider the Hilbert space $\mathscr{D}_0(\mathscr{N})$ with the norm $\|\cdot\|_D$: recall that this is indeed the closure of $\ell_0(X)$ in $\mathscr{D}(\mathscr{N})$. Now the subset

$$Y := \{f \in \mathscr{D}_0(\mathscr{N}) : f(x_0) = 1\} \subseteq \mathscr{D}_0(\mathscr{N})$$

is convex and closed. Since in a Hilbert space any closed convex subset admits a point with minimal norm, see [298, Theorem 4.10], there exists a function $f \in Y$ that minimizes $\|\cdot\|_D$, and hence also $D(\cdot) = \sqrt{\|\cdot\|_D^2 - 1}$. Therefore we have

$$\mathrm{cap}(x_0) = \min\{D(f) : f \in \mathscr{D}_0(\mathscr{N}), f(x_0) = 1\}. \tag{13.45}$$

Theorem 13.36. *Let (X,P) be an irreducible reversible Markov chain. Then the following conditions are equivalent:*

(a) *(X,P) is transient;*
(b) *for every $x_0 \in X$ and $i_0 \neq 0$ there exists a finite energy flow $u \in L^2(E,r)$ from x_0 with input i_0;*
(c) *for every $x_0 \in X$ one has* $\mathrm{cap}(x_0) > 0$*;*
(d) $\mathbf{1}_X \notin \mathscr{D}_0(\mathscr{N})$.

Proof. (a) $\Rightarrow$ (b). Suppose (X,P) is transient and let $x_0 \in X$ and $i_0 \neq 0$. Then, by virtue of Lemma 13.34, $G(\cdot,x) \in \mathscr{D}_0(\mathscr{N}) \subseteq \mathscr{D}(\mathscr{N})$ so that $\nabla G(\cdot,x) \in L^2(E,r)$. We claim that

$$u := -\frac{i_0}{m(x_0)} \nabla G(\cdot,x)$$

is the required flow: indeed,

$$\nabla^* u = -\frac{i_0}{m(x_0)} \nabla^* \nabla G(\cdot,x_0) = -\frac{i_0}{m(x_0)} (I-P)G(\cdot,x_0) = -\frac{i_0}{m(x_0)} \delta_{x_0},$$

where the second equality follows from Proposition 13.32, and the last one from the equality

$$(I-P)G(\cdot,x_0) = \sum_{n\geq 0} p^{(n)}(\cdot,x_0) - \sum_{n\geq 0} p^{(n+1)}(\cdot,x_0) = p^{(0)}(\cdot,x_0) = \delta_{x_0}.$$

(b) $\Rightarrow$ (c). Let $x_0 \in X$, $i_0 = -1$ and let $u \in L^2(E,r)$ be a (finite energy) flow from x_0 with input i_0. Then, for $f \in \ell_0(X)$ such that $f(x_0) = 1$, we have, recalling (13.35),

$$\langle \nabla f, u \rangle_E = \langle f, \nabla^* u \rangle_X = \langle f, -\frac{i_0}{m(x_0)} \delta_{x_0} \rangle_X = -m(x_0) f(x_0) \frac{i_0}{m(x_0)} = -i_0 = 1.$$

Using Cauchy–Schwarz, it follows that $1 = \langle \nabla f, u\rangle_E^2 \le D(f)\langle u,u\rangle_E$ so that

$$\operatorname{cap}(x_0) = \inf\{D(f) : f \in \ell_0(X) \text{ s.t. } f(x_0) = 1\} \ge \frac{1}{\langle u,u\rangle_E} > 0.$$

(c) $\Leftrightarrow$ (d). This follows immediately from the definition of capacity. Indeed, by virtue of (13.45), we have $\operatorname{cap}(x_0) = 0$ if and only if there exists an $f \in \mathscr{D}_0(\mathscr{N})$ such that $D(f) = 0$ and $f(x_0) = 1$, equivalently, such that $f = \mathbf{1}_X$ (recall that (X,P) is irreducible).

(c) $\Rightarrow$ (a). Let A be a finite subset of X and let $x_0 \in A$. Consider the function $f := \frac{G_A(\cdot,x_0)}{G_A(x_0,x_0)}$ (cf. Lemma 13.33). It is clear that $f \in \ell_0(X)$ and $f(x_0) = 1$. By applying (13.43) with x replaced by x_0 we obtain

$$\operatorname{cap}(x_0) \le D(f) = \langle \nabla f, \nabla f\rangle_E = \frac{m(x_0)}{G_A(x_0,x_0)}.$$

We deduce $G_A(x_0,x_0) \le \frac{m(x_0)}{\operatorname{cap}(x_0)}$ for every finite subset $A \subseteq X$ containing x_0. Taking a non-decreasing sequence $(A_n)_{n\in\mathbb{N}}$ of finite subsets of X containing x and such that $\bigcup_{n\in\mathbb{N}} A_n = X$ and applying the monotone convergence theorem we deduce that $G(x_0,x_0) \le \frac{m(x_0)}{\operatorname{cap}(x_0)} < \infty$. Thus the Markov chain (X,P) is transient. □

Corollary 13.37. *Let (X,P_1) and (X,P_2) be two irreducible, reversible Markov chains. Suppose there exists a $\delta > 0$ such that*

$$D_1(f) \ge \delta D_2(f) \tag{13.46}$$

for all $f \in \ell_0(X)$, where D_1 and D_2 denote the respective Dirichlet norms. Then if (X,P_1) is recurrent, so is (X,P_2).

Proof. This follows from (c) in the previous theorem. □

Corollary 13.38. *Let G be a finitely generated group and let (G,P_1) and (G,P_2) be two irreducible random walks on G. Suppose there exists a $\delta > 0$ such that*

$$p_1(g,h) \ge \delta p_2(g,h) \tag{13.47}$$

for all $g,h \in G$. Then if (G,P_1) is recurrent, so is (G,P_2).

Proof. This follows from the previous corollary after observing that, by virtue of (13.37) (here $m = \mathbf{1}_X$), we have that condition (13.47) implies condition (13.46). □

Example 13.39 (Nearest neighbor random walk on $\mathbb{N}$). Consider an irreducible Markov chain $(\mathbb{N},P)$ where $p(m,n) > 0$ if and only if $|m-n| = 1$, $m,n \in \mathbb{N}$. Note that $p(0,1) = 1$ and that $p(n,n-1) + p(n,n+1) = 1$ for all $n \ge 1$. This Markov chain is reversible: indeed an invariant measure $m\colon \mathbb{N} \to (0,+\infty)$ is given by setting $m(0) = 1$ and

$$m(n) := \frac{p(0,1)p(1,2)\cdots p(n-1,n)}{p(n,n-1)p(n-1,n-2)\cdots p(1,0)}$$

(note that $m(n+1)=m(n)\frac{p(n,n+1)}{p(n+1,n)}$) for all $n\geq 1$.

The associated network is $\mathscr{N}(\mathbb{N},E,r)$ where $E=\{e_n : n\in\mathbb{N}\}$ with $e_n=\{n,n+1\}$ and

$$r(e_n)=\frac{p(n,n-1)p(n-1,n-2)\cdots p(1,0)}{p(0,1)p(1,2)\cdots p(n-1,n)p(n,n+1)}$$

for all $n\geq 1$. There is a unique flow from 0 with input $i_0\neq 0$, namely the constant $u=i_0\mathbf{1}_E\in\mathbb{R}^E$ and whose energy is then

$$\langle u,u\rangle_E=\langle i_0\mathbf{1}_E,i_0\mathbf{1}_E\rangle_E=i_0^2\sum_{n=1}^{\infty}r(e_n). \tag{13.48}$$

It follows from Theorem 13.36 that the Markov chain $(\mathbb{N},P)$ is recurrent (resp. transient) if and only if the sum in (13.48) diverges (resp. converges).

Definition 13.40. Let (X,P) be a reversible Markov chain, say with invariant measure m, and let $\mathscr{N}=(X,E,r)$ denote the associated network. Let $(X_i)_{i\in\mathbb{N}}$ be a partition of X made up of finite subsets (or, more generally, such that $\mathbf{1}_{X_i}\in\mathscr{D}_0(\mathscr{N})$ for all $i\in\mathbb{N}$). The associated *shortened network* $\mathscr{N}'=(\mathbb{N},E,r')$ is defined by means of the corresponding conductances

$$a'(i,j):=\begin{cases}\sum_{x\in X_i}\sum_{y\in X_j}a(x,y) & \text{if } i\neq j\\ 0 & \text{otherwise.}\end{cases}$$

Then the corresponding *shortened Markov chain* $(\mathbb{N},P')$ has transition probabilities

$$p'(i,j)=\frac{a'(i,j)}{m_i'}=\frac{a'(i,j)}{\sum_{k\in\mathbb{N}}a'(i,k)}$$

for all $i,j\in\mathbb{N}$.

Note that if (X,P) is irreducible, so is $(\mathbb{N},P')$. Moreover, $p'(i,i)=0$ and $m'(i)=\sum_{j\in\mathbb{N}}a'(i,j)\leq\sum_{x\in X_i}\sum_{y\in X}a(x,y)=\sum_{x\in X_i}m(x)<\infty$ for all $i\in\mathbb{N}$.

Theorem 13.41. *Let (X,P) be a reversible Markov chain. Suppose that the shortened Markov chain $(\mathbb{N},P')$ (with respect to some partition of X into finite subsets) is recurrent. Then (X,P) is recurrent as well.*

Proof. Let $f\colon\mathbb{N}\to\mathbb{R}$ be in $\mathscr{D}(\mathscr{N}')$ and let us define $\overline{f}\colon X\to\mathbb{R}$ by setting, $\overline{f}(x)=f(i)$ if $x\in X_i$. Note that

$$\overline{\mathbf{1}_{\mathbb{N}}}=\mathbf{1}_X. \tag{13.49}$$

By virtue of (13.37) we then have

$$\begin{aligned}D(\overline{f})&=\frac{1}{2}\sum_{x,y\in X}\left(\overline{f}(x)-\overline{f}(y)\right)^2a(x,y)\\&=\frac{1}{2}\sum_{i,j\in\mathbb{N}}\sum_{\substack{x\in X_i\\ y\in X_j}}\left(\overline{f}(x)-\overline{f}(y)\right)^2a(x,y)\end{aligned}$$

$$= \frac{1}{2} \sum_{i,j \in \mathbb{N}} (f(i) - f(j))^2 \sum_{\substack{x \in X_i \\ y \in X_j}} a(x,y)$$
$$= \frac{1}{2} \sum_{i,j \in \mathbb{N}} (f(i) - f(j))^2 a'(i,j)$$
$$= D'(f).$$

This shows that $f \in \mathscr{D}(\mathscr{N}')$ if and only if $\overline{f} \in \mathscr{D}(\mathscr{N})$ and in addition their corresponding Dirichlet norms are equal:

$$D(\overline{f}) = D'(f). \tag{13.50}$$

Now, since $\mathbf{1}_{X_i} \in \ell_0(X)$ we deduce that if $f \in \ell_0(\mathbb{N})$ then $\overline{f} = \sum_{i \in \mathrm{supp}(f)} f(i)\mathbf{1}_{X_i} \in \ell_0(X)$. If $(\mathbb{N}, P')$ is recurrent, then by Theorem 13.36.(d) $\mathbf{1}_{\mathbb{N}} \in \mathscr{D}_0(\mathscr{N}')$ and we can find a sequence $(f_n)_{n\in\mathbb{N}}$ in $\ell_0(\mathbb{N})$ such that $\lim_{n\to\infty} D'(f_n - \mathbf{1}_{\mathbb{N}}) = 0$. Consider the sequence $(\overline{f}_n)_{n\in\mathbb{N}}$ in $\ell_0(X)$. Combining together (13.49) and (13.50) we deduce that $\lim_{n\to\infty} D(\overline{f}_n - \mathbf{1}_X) = 0$. Thus $\mathbf{1}_X \in \mathscr{D}_0(\mathscr{N})$ and, again by virtue of Theorem 13.36.(d), we deduce that (X,P) is recurrent. □

From the preceding theorem we deduce the following recurrence criterion due to Crispin Nash-Williams.

Corollary 13.42 (Nash-Williams' recurrence criterion). *Let (X,P) be a reversible Markov chain. Let $(\mathbb{N}, P')$ be a shortened Markov chain such that $a'(i,j) = 0$ if $|i-j| \geq 2$. Suppose that*

$$\sum_{i=0}^{\infty} \frac{1}{a'(i,i+1)} = \infty. \tag{13.51}$$

Then (X,P) is recurrent.

Proof. First observe that by our assumptions, $(\mathbb{N}, P')$ is a nearest neighbor Markov chain as in Example 13.39. Since the sum in (13.51) equals (13.48), from condition (13.51) and Theorem 13.36.(b) we deduce that $(\mathbb{N}, P')$ is recurrent. By virtue of the preceding theorem, we have that the original Markov chain (X,P) is also recurrent. □

Example 13.43 (The shortened simple random walk on $\mathbb{Z}$). Consider the simple random walk (X,Q) on $X = \mathbb{Z}$. This is reversible (with $m = \mathbf{1}_X$) so that $a(x,y) = q(x,y) = \frac{1}{2}$ if $|x-y| = 1$ and 0 otherwise. Set $X_n = \{n, -n\}$ so that X_n is finite, for all $n \in \mathbb{N}$, and note that $X = \sqcup_{n\in\mathbb{N}} X_n$. Let $(\mathbb{N}, Q')$ denote the associated shortened Markov chain. We have

$$a'(i,j) = \sum_{\substack{x \in X_i \\ y \in X_j}} a(x,y) = \begin{cases} 1 & \text{if } |i-j| = 1 \text{ and } ij = 0 \\ 2 & \text{if } |i-j| = 1 \text{ and } ij \neq 0 \\ 0 & \text{otherwise} \end{cases}$$

so that

$$m'(i) = \sum_{j \in \mathbb{N}} a'(i,j) = \begin{cases} 1 & \text{if } i = 0 \\ 3 & \text{if } i = 1 \\ 4 & \text{otherwise.} \end{cases}$$

It follows that

$$q'(i,j) = \frac{a'(i,j)}{m'(i)} = \begin{cases} 1 & \text{if } i = 0 \text{ and } j = 1 \\ \frac{1}{3} & \text{if } i = 1 \text{ and } j = 0 \\ \frac{2}{3} & \text{if } i = 1 \text{ and } j = 2 \\ \frac{1}{2} & \text{if } |i-j| = 1 \text{ and } i \geq 2 \\ 0 & \text{otherwise.} \end{cases}$$

We have $\sum_{i=0}^{\infty} \frac{1}{a'(i,i+1)} = 1 + \sum_{i=1}^{\infty} \frac{1}{2} = +\infty$. Thus by Nash-Williams' criterion we recover the result established in Proposition 13.19, namely that the simple random walk on $\mathbb{Z}$ is recurrent.

13.11 Growth and Recurrence

Recall that the group $G = \mathbb{Z}^d$ has growth $b(n) \sim n^d$ (cf. Example 7.5) so that the following proposition covers the result established in Proposition 13.19, namely that the simple random walk on $\mathbb{Z}^d$ is recurrent for $d = 1, 2$.

Proposition 13.44. *Let G be a finitely generated group of at most quadratic growth. Then the simple random walk on G is recurrent.*

Proof. Let Y be a finite symmetric generating subset of G. Possibly removing 1_G from Y if necessary, we may suppose that Y does not contain the identity element of G. Since the growth of G is at most quadratic, we can find $n_0 \in \mathbb{N}$ and $C > 0$ so that $b_Y(n) \leq Cn^2$ for all $n \geq n_0$. Let also $X_n := \{g \in G : \ell_X(g) = n\}$ denote the sphere of radius n centered at the identity element $1_G \in G$. Note that X_n is a finite subset of G (in fact $|X_n| = b_Y(n) - b_Y(n-1)$) and that $G = \sqcup_{n \in \mathbb{N}} X_n$. Let (G, Q) denote the simple random walk on G (so that $a(x, xy) = q(x, xy) = \frac{1}{|Y|}$ for all $x \in G$ and $y \in Y$) and let $(\mathbb{N}, Q')$ denote the associated shortened Markov chain.

For $k \in \mathbb{N}$ we have

$$\begin{aligned} a'(k, k+1) &= \sum_{\substack{x \in X_k \\ x' \in X_{k+1}}} a(x, x') \\ &= \sum_{\substack{x \in X_k \\ x' \in X_{k+1}}} p(x, x') \\ &\leq \sum_{\substack{x \in X_k \\ y \in Y}} p(x, xy) \\ &= \frac{1}{|Y|} \cdot |Y| \cdot |X_k| \end{aligned}$$

$$= |X_k|$$
$$= b_Y(k) - b_Y(k-1).$$

For $n \geq n_0$ we have

$$\sum_{k=n+1}^{2n} \frac{1}{a'(k,k+1)} \geq \frac{n^2}{\sum_{k=n+1}^{2n} a'(k,k+1)} \geq \frac{n^2}{b_Y(2n) - b_Y(n)} \geq \frac{n^2}{b_Y(2n)} \geq \frac{1}{4C},$$

where the first inequality follows from the convexity of the function $\varphi(x) := 1/x$ for $x > 0$: indeed, for $n \geq 1$ and $x_1, x_2, \ldots, x_n \in (0, +\infty)$, we have

$$\frac{n}{\sum_{i=1}^n x_i} = \varphi\left(\frac{\sum_{i=1}^n x_i}{n}\right) \leq \frac{\sum_{i=1}^n \varphi(x_i)}{n} = \frac{\sum_{i=1}^n \frac{1}{x_i}}{n},$$

which is equivalent to

$$\sum_{i=1}^n \frac{1}{x_i} \geq \frac{n^2}{\sum_{i=1}^n x_i}.$$

This shows that the series $\sum_{k=0}^{\infty} \frac{1}{a'(k,k+1)}$ diverges. By virtue of Nash-Williams' criterion (Corollary 13.42), we deduce that the simple random walk (G, Q) on G is recurrent. □

13.12 Growth and Transience

We now turn to the case where G has at least cubic growth.

Lemma 13.45. *Let μ be an atomic probability measure on a group G and suppose that $\mu = \mu_1 + \mu_2$ for some positive atomic measures μ_1, μ_2. Then*

$$\|\mu^{(n)}\|_\infty \leq \mu_1(G)^n + n\|\mu_2\|_\infty \tag{13.52}$$

for all $n \geq 1$, where $\|\nu\|_\infty = \sup_{g\in G} \nu(h)$.

Proof. We proceed by induction. For $n = 1$ inequality (13.52) follows from $\mu(g) = \mu_1(g) + \mu_2(g) \leq \mu_1(G) + \|\mu_2\|_\infty$, which holds for all $g \in G$. Assume (13.52). We have

$$\begin{aligned}
\mu^{(n+1)}(g) &= \sum_{h\in G} (\mu_1(h) + \mu_2(h))\,\mu^{(n)}(h^{-1}g) \\
&\leq \sum_{h\in G} \mu_1(h)\,(\mu_1(G)^n + n\|\mu_2\|_\infty) + \sum_{h\in G} \mu_2(h)\mu^{(n)}(h^{-1}g) \\
&\leq \mu_1(G)\mu_1(G)^n + n\mu_1(G)\|\mu_2\|_\infty + \|\mu_2\|_\infty \sum_{h\in G} \mu^{(n)}(h^{-1}g) \\
&= \mu_1(G)^{n+1} + n\mu(G)\|\mu_2\|_\infty + \|\mu_2\|_\infty \mu^{(n)}(G) \\
&= \mu_1(G)^{n+1} + n\|\mu_2\|_\infty + \|\mu_2\|_\infty
\end{aligned}$$

$$= \mu_1(G)^{n+1} + (n+1)\|\mu_2\|_\infty$$

(recall that $\mu^{(n)}(G) = 1$ for all $n \geq 1$), proving the inductive step. □

For $k \geq 2$ let us set

$$\lambda_k := \frac{1}{k^3 \log^2(k)}.$$

Note that

$$\sum_{k \geq 2} \lambda_k < \sum_{k \geq 2} \frac{1}{k^3} < \int_1^\infty \frac{1}{x^3} \mathrm{d}x = \frac{1}{2}$$

so that

$$\lambda_1 := 1 - \sum_{k \geq 2} \lambda_k > \frac{1}{2}.$$

Let now G be a finitely generated group and let $Y \subseteq G$ be a finite symmetric generating subset. For simplicity we write $b(n)$ (resp. $B(n)$) instead of $b_Y(n)$ (resp. $B_Y(n)$) for all $n \in \mathbb{N}$. If we set

$$\mu(g) := \sum_{n=1}^{\infty} \frac{\lambda_n}{b(n)} \mathbf{1}_{B(n)}(g) \tag{13.53}$$

for all $g \in G$, then we have

- $\mu(g) > 0$ for all $g \in G$ (equivalently $\mathrm{supp}(\mu) = G$);
- $\mu(g) = \mu(g^{-1})$ for all $g \in G$ (note that $B(n)^{-1} = B(n)$);
- $\mu(G) = 1$ (since $\sum_{k \geq 1} \lambda_k = 1$);
- the second moment of μ is finite:

$$\begin{aligned}
\sum_{g \in G} \ell_Y(g)^2 \mu(g) &= \sum_{n=1}^{\infty} \frac{\lambda_n}{b(n)} \sum_{g \in G} \ell_Y(g)^2 \mathbf{1}_{B(n)}(g) \\
&\leq \sum_{n=1}^{\infty} \frac{\lambda_n}{b(n)} n^2 b(n) \\
&= \sum_{n=1}^{\infty} \lambda_n n^2 \\
&= \lambda_1 + \sum_{n=2}^{\infty} \frac{1}{n \log^2(n)} < \infty.
\end{aligned}$$

Thus μ satisfies the conditions in Definition 13.16 and defines a random walk on G.

Proposition 13.46. *Suppose that G has at least cubic growth. Then the random walk defined by the measure* (13.53) *on G is transient.*

Proof. With the above notation, our hypothesis on the group reads as follows: there exist $n_0 \in \mathbb{N}$ and $C > 0$ such that

$$b(n) \geq Cn^3 \tag{13.54}$$

for all $n \geq n_0$.

Let us set $s_m = \sum_{k=m}^{\infty} \lambda_k$ for all $m \geq 1$.

Claim.

$$\lim_{m\to\infty} s_m m^2 \log^2(m) = \frac{1}{2}.$$

To prove the claim, it is easy to check that

$$\int_m^{\infty} \frac{dx}{x^3\log^2(x)} \leq s_m = \sum_{k=m}^{\infty} \lambda_k = \sum_{k=m}^{\infty} \frac{1}{k^3\log^2(k)} \leq \int_{m-1}^{\infty} \frac{dx}{x^3\log^2(x)}. \tag{13.55}$$

Using integration by parts we have

$$\begin{aligned}\int_m^{\infty} \frac{dx}{x^3\log^2(x)} &= -\int_m^{\infty} \left(-\frac{2}{x\log^3(x)}\right)\left(-\frac{1}{2x^2}\right) dx + \left[-\frac{1}{2x^2\log^2(x)}\right]_m^{\infty} \\ &= -\int_m^{\infty} \frac{1}{x^3\log^3(x)} dx + \frac{1}{2m^2\log^2(m)}.\end{aligned}$$

Now

$$\begin{aligned}\int_m^{\infty} \frac{1}{x^3\log^3(x)} dx &\leq \frac{1}{\log^3(m)} \int_m^{\infty} \frac{1}{x^3} dx \\ &= \frac{1}{\log^3(m)} \frac{1}{2m^2},\end{aligned}$$

hence

$$\lim_{m\to\infty} m^2\log^2(m) \int_m^{\infty} \frac{1}{x^3\log^3(x)} dx = \lim_{m\to\infty} \frac{1}{2\log(m)} = 0.$$

As a consequence,

$$\lim_{m\to\infty} s_m m^2\log^2(m) \geq \lim_{m\to\infty} m^2\log^2(m) \int_m^{\infty} \frac{dx}{x^3\log^2(x)} = \frac{1}{2}.$$

The other inequality is proved in a similar way (**exercise**). The claim follows.

Using the claim, we have

$$s_m \sim \frac{m\lambda_m}{2} = \frac{1}{2m^2\log^2(m)}. \tag{13.56}$$

For $m \geq 2$ set

$$\mu_{1,m} := \sum_{k=1}^{m-1} \frac{\lambda_k}{b(k)} \mathbf{1}_{B(k)} \quad \text{and} \quad \mu_{2,m} := \mu - \mu_{1,m} = \sum_{k=m}^{\infty} \frac{\lambda_k}{b(k)} \mathbf{1}_{B(k)}.$$

We thus have $\mu = \mu_{1,m} + \mu_{2,m}$,

$$\mu_{1,m}(G) = \sum_{k=1}^{m-1} \lambda_k = 1 - s_m$$

and

$$\|\mu_{2,m}\|_\infty = \sum_{k=m}^{\infty} \frac{\lambda_k}{b(k)} \le \frac{1}{b(m)} \sum_{k=m}^{\infty} \lambda_k = \frac{s_m}{b(m)}.$$

From Lemma 13.45 we deduce

$$p^{(n)}(g,h) = \mu^{(n)}(g^{-1}h) \le (1-s_m)^n + n\frac{s_m}{b(m)}$$

for all $n, m \in \mathbb{N}$, $m \ge 2$ and $g, h \in G$. By virtue of (13.56) and (13.54) we can then find constants $c_1, c_2 > 0$ such that

$$\begin{aligned} p^{(n)}(g,h) &\le \left(1 - \frac{c_1}{m^2\log^2(m)}\right)^n + \frac{c_2 n}{b(m)m^2\log^2(m)} \\ &\le \exp\left(-\frac{c_1 n}{m^2\log^2(m)}\right) + \frac{c_3 n}{m^5\log^2(m)} \end{aligned}$$

for all $n \ge n_0$, for all $m \ge 2$ and $g, h \in G$, where $c_3 := c_2/C$. Taking $m := \lfloor n^{2/5} \rfloor$ and setting $d_1 := c_1\frac{25}{4}$ and $d_2 := c_3\frac{25}{4}$ we deduce

$$\frac{c_1 n}{m^2\log^2(m)} \sim \frac{c_1 n}{n^{4/5}\log^2(n^{2/5})} = \frac{d_1 n^{1/5}}{\log^2(n)}$$

and

$$\frac{c_3 n}{m^5\log^2(m)} \sim \frac{c_3 n}{n^2\log^2(n^{2/5})} = \frac{d_2}{n\log^2(n)}$$

so that

$$G(g,h) = \sum_{n=0}^{\infty} p^{(n)}(g,h) \preceq \sum_{n=0}^{\infty} \exp\left(-\frac{d_1 n^{1/5}}{\log^2(n)}\right) + d_2 \sum_{n=0}^{\infty} \frac{1}{n\log^2(n)} < \infty.$$

Thus the random walk on G defined by μ is transient. □

Using Proposition 13.19, Proposition 13.21 and Proposition 13.46, we deduce the following classical result (cf. Propositions 13.19 and 13.21).

Theorem 13.47 (Pólya). *The simple random walk on $\mathbb{Z}^d$ is recurrent for $d = 1,2$ and transient if $d \ge 3$.* □

13.13 The Random Walk Alternative

Theorem 13.48 (Random walk alternative). *Let G be a finitely generated group. Then the following conditions are equivalent:*

(a) *there exists a recurrent random walk on G;*
(b) *the simple random walk on G with respect to* every *finite, symmetric generating subset is recurrent;*
(c) *the simple random walk on G with respect to* some *finite, symmetric generating subset is recurrent;*
(d) every *random walk on G is recurrent.*

In other words, the random walks on G are either all recurrent or all transient.

Proof. Suppose first that there exists a recurrent random walk (G,P) on G and let us show that the simple random walk on G with respect to any finite, symmetric generating subset is also recurrent. Let then $Y \subseteq G$ be a finite symmetric generating subset of G.

Set $\overline{P} := \frac{1}{2}(I+P)$. It is clear that $(G,\overline{P})$ is a Markov chain. Let us show that $(G,\overline{P})$ is indeed a random walk on G. First of all we have $\overline{p}(g,h) = \frac{1}{2}\left(\delta_{g,h} + p(g,h)\right) = \frac{1}{2}\left(\delta_{h,g} + p(h,g)\right) = \overline{p}(h,g)$ for all $g,h \in G$, and this proves symmetry. Also, observing that if $g \neq h$ one has $\overline{p}(g,h) = \frac{1}{2}p(g,h)$ while if $g = h$ then $d_Y(g,h) = 0$, we have

$$\sum_{h\in G} d_Y(g,h)^2 \overline{p}(g,h) = \frac{1}{2}\sum_{h\in G} d_Y(g,h)^2 p(g,h) < \infty$$

showing that the second moment is finite. Finally,

$$\overline{p}(g'g,g'h) = \frac{1}{2}\left(\delta_{g'g,g'h} + p(g'g,g'h)\right) = \frac{1}{2}\left(\delta_{g,h} + p(g,h)\right) = \overline{p}(g,h)$$

for all $g',g,h \in G$ and also space homogeneity follows. We are only left with irreducibility. Let $g,h \in G$. Since (G,P) is irreducible, we can find $n \in \mathbb{N}$ such that $p^{(n)}(g,h) > 0$. As a consequence, recalling that $\overline{P} \geq \frac{1}{2}P$, we have

$$\begin{aligned}\overline{p}^{(n)}(g,h) &= \sum_{g_1,g_2,\dots g_{n-1}\in G} \overline{p}(g,g_1)\overline{p}(g_1,g_2)\cdots\overline{p}(g_{n-1},h) \\ &\geq \frac{1}{2^n}\sum_{g_1,g_2,\dots g_{n-1}\in G} p(g,g_1)p(g_1,g_2)\cdots p(g_{n-1},h) \\ &= \frac{1}{2^n}p^{(n)}(g,h) > 0.\end{aligned}$$

We deduce that $(G,\overline{P})$ is irreducible, completing the argument showing that $(G,\overline{P})$ is a random walk.

Let us show that, in fact, $(G,\overline{P}^K)$ is a random walk on G for every integer $K \geq 1$. We only need to check for irreducibility and finiteness of the second moment. Let $g,h \in G$ and let $n_0 \geq 1$ such that $\overline{p}^{(n_0)}(g,h) > 0$. Then

$$\overline{p}^{(n_0+1)}(g,h) \geq \overline{p}(g,g)\overline{p}^{(n_0)}(g,h) \geq \frac{1}{2}\overline{p}^{(n_0)}(g,h) > 0. \tag{13.57}$$

Thus since $\left(\overline{p}^{(K)}\right)^{(n)} = \overline{p}^{(nK)}$ for all $n \geq 1$, as soon as $nK \geq n_0$ we must have $\left(\overline{p}^{(K)}\right)^{(n)}(g,h) > 0$. This shows that $\overline{P}^K$ is irreducible. To show that the second moment is finite, we proceed by induction. For $K = 1$ this was shown above. Suppose this holds for $\overline{P}^K$ and let us show it for $\overline{P}^{K+1}$.

$$\begin{aligned}
\sum_{h\in G} d_Y(g,h)^2 \overline{p}^{(K+1)}(g,h) &= \sum_{h,h'\in G} d_Y(g,h)^2 \overline{p}^{(K)}(g,h')\overline{p}(h',h) \\
&\leq \sum_{h,h'\in G} (d_Y(g,h') + d_Y(h',h))^2 \overline{p}^{(K)}(g,h')\overline{p}(h',h) \\
&\leq \sum_{h,h'\in G} 2(d_Y(g,h')^2 + d_Y(h',h)^2) \overline{p}^{(K)}(g,h')\overline{p}(h',h) \\
&= 2\sum_{h,h'\in G} d_Y(g,h')^2 \overline{p}^{(K)}(g,h')\overline{p}(h',h) \\
&\quad + 2\sum_{h,h'\in G} d_Y(h',h)^2 \overline{p}^{(K)}(g,h')\overline{p}(h',h) \\
&= 2\sum_{h'\in G} d_Y(g,h')^2 \overline{p}^{(K)}(g,h') \\
&\quad + 2\sum_{h'\in G} \overline{p}^{(K)}(g,h') \sum_{h\in G} d_Y(h',h)^2 \overline{p}(h',h) \\
\text{(by Remark 13.17)} \quad &= 2\left[\sum_{h'\in G} d_Y(g,h')^2 \overline{p}^{(K)}(g,h') + \sum_{h\in G} d_Y(g,h)^2 \overline{p}(g,h)\right] < \infty.
\end{aligned}$$

Moreover, $(G, \overline{P})$ is recurrent. Indeed, since

$$\overline{P}^n = \left(\frac{1}{2}(I+P)\right)^n = \frac{1}{2^n}\sum_{j=0}^{n} \binom{n}{j} P^j$$

we have, for all $g,h \in G$,

$$\begin{aligned}
G_{\overline{P}}(g,h) &= \sum_{n=0}^{\infty} \overline{p}^{(n)}(g,h) \\
&= \sum_{n=0}^{\infty} \frac{1}{2^n} \sum_{j=0}^{n} \binom{n}{j} p^{(j)}(g,h) \\
&= \sum_{j=0}^{\infty} p^{(j)}(g,h) \sum_{n=j}^{\infty} \frac{1}{2^n}\binom{n}{j} \\
&= 2\sum_{j=0}^{\infty} p^{(j)}(g,h) \\
&= 2G_P(g,h),
\end{aligned}$$

where the fourth equality follows from the elementary identity (**exercise**)

$$\sum_{n=j}^{\infty} \frac{1}{2^n} \binom{n}{j} = 2 \quad \text{for all } j \geq 0. \tag{13.58}$$

This shows that (G, P) is recurrent if and only if $(G, \overline{P})$ is recurrent.

Let $K \geq 1$. For every $f \in \ell_0(G)$ we have (by using (13.37) and recalling that $m = \mathbf{1}$)

$$\begin{aligned}
D_{\overline{P}^K}(f) &= \frac{1}{2} \sum_{g_0, g_K \in G} (f(g_0) - f(g_K))^2 \overline{p}^K(g_0, g_K) \\
&= \frac{1}{2} \sum_{g_0, g_1, \ldots, g_K \in G} (f(g_0) - f(g_K))^2 \overline{p}(g_0, g_1)\overline{p}(g_1, g_2) \cdots \overline{p}(g_{K-1}, g_K) \\
&\leq \frac{K}{2} \sum_{g_0, g_1, \ldots, g_K \in G} \sum_{i=1}^{K} (f(g_i) - f(g_{i-1}))^2 \overline{p}(g_0, g_1)\overline{p}(g_1, g_2) \cdots \overline{p}(g_{K-1}, g_K) \\
(*) &= \frac{K}{2} \sum_{i=1}^{K} \sum_{g_{i-1}, g_i \in G} (f(g_i) - f(g_{i-1}))^2 \overline{p}(g_{i-1}, g_i) \\
&= \frac{K^2}{2} D_{\overline{P}}(f),
\end{aligned}$$

where in the inequality we used Cauchy–Schwarz, $(*)$ follows since $\overline{P}$ is symmetric and stochastic, and the last equality follows from (13.37). From Corollary 13.37 we then deduce that recurrence of $(G, \overline{P})$ implies recurrence of $(G, \overline{P}^K)$ for all $K \geq 1$.

Since $(G, \overline{P})$ is irreducible, for every generator $y \in Y \setminus \{1_G\}$ we can find $k_y \in \mathbb{N}$ such that $\overline{p}^{(k_y)}(1, y) > 0$. But then, with $K_0 := \max\{k_y : y \in Y \setminus \{1_G\}\}$ and $d_0 = \min\{\overline{p}^{(k_y)}(1, y) : y \in Y \setminus \{1_G\}\}$, from (13.57) we deduce

$$\overline{p}^{(K_0)}(1, y) \geq \frac{d_0}{2^{K_0}} \geq \frac{d_0}{2^{K_0}|Y|} = \frac{d_0}{2^{K_0}} q(1, y) \tag{13.59}$$

for all $y \in Y$, where (G, Q) denotes, as usual, the simple random walk on G with respect to the finite symmetric generating subset $Y \subseteq G$. Since $q(g, h) = q(1, g^{-1}h) = 0$ if (and only if) $g^{-1}h \notin Y$, we deduce that

$$\overline{p}^{(K_0)}(g, h) \geq \frac{d_0}{2^{K_0}} q(g, h) \tag{13.60}$$

for all $g, h \in G$. By virtue of Corollary 13.38, we deduce that (G, Q) is recurrent. This shows (a) $\Rightarrow$ (b).

The implication (b) $\Rightarrow$ (c) is trivial.

Suppose that there exists a finite symmetric generating subset $Y \subseteq G$ such that the corresponding simple random walk (G, Q) is recurrent. Let us show that any other random walk (G, P) is also recurrent. For all $g, h \in G$ and $e \in E$ (the edge set of the Cayley graph $C(G, Y)$), we denote by $\Gamma(g, h)$ (resp. $\Gamma_e(g, h)$) the set of all geodesics connecting g and h (resp. and that, in addition, contain e). For every $f \in \ell_0(G)$ and $\gamma \in \Gamma(g, h)$, using Cauchy–Schwarz, we have

$$(f(g)-f(h))^2 = \left(|Y|\sum_{e\in\gamma}(\nabla_Q f)(e)\right)^2 \leq |Y|^2\left(\sum_{e\in\gamma}(\nabla_Q f)^2(e)\right)d(g,h)$$

yielding, by taking the average,

$$(f(g)-f(h))^2 \leq \frac{|Y|^2}{|\Gamma(g,h)|}\sum_{\gamma\in\Gamma(g,h)}\sum_{e\in\gamma}(\nabla_Q f)^2(e)d(g,h).$$

We deduce (by using (13.37) and recalling that $m = \mathbf{1}_X$)

$$\begin{aligned}
D_P(f) &= \frac{1}{2}\sum_{g,h\in G}(f(g)-f(h))^2 p(g,h) \\
&\leq \frac{|Y|^2}{2}\sum_{g,h\in G}\frac{1}{|\Gamma(g,h)|}\sum_{\gamma\in\Gamma(g,h)}\sum_{e\in\gamma}(\nabla_Q f)^2(e)d(g,h)p(g,h) \\
&= \frac{|Y|^2}{2}\sum_{g,h\in G}\frac{1}{|\Gamma(g,h)|}\sum_{e\in E}\sum_{\gamma\in\Gamma_e(g,h)}(\nabla_Q f)^2(e)d(g,h)p(g,h) \\
&= \frac{|Y|^2}{2}\sum_{e\in E}(\nabla_Q f)^2(e)\left(\sum_{g,h\in G}\frac{1}{|\Gamma(g,h)|}\sum_{\gamma\in\Gamma_e(g,h)}d(g,h)p(g,h)\right) \\
&= \frac{|Y|^2}{2}\sum_{e\in E}(\nabla_Q f)^2(e)\varphi(e),
\end{aligned}$$

where

$$\varphi(e) := \sum_{g,h\in G}\frac{|\Gamma_e(g,h)|}{|\Gamma(g,h)|}p(g,h)d(g,h).$$

Let us show that $\varphi\colon E\to\mathbb{R}$ is bounded. If we denote by μ the measure on G associated with the random walk (G,P) we have for all $e\in E$

$$\begin{aligned}
\varphi(e) &= \sum_{g,h\in G}\mu(g^{-1}h)d(1,g^{-1}h)\frac{|\Gamma_e(1,g^{-1}h)|}{|\Gamma(1,g^{-1}h)|} \\
\text{(setting } k:=g^{-1}h) \quad &= \sum_{g,k\in G}\mu(k)d(1,k)\frac{|\Gamma_{g^{-1}e}(1,k)|}{|\Gamma(1,k)|} \\
&\leq 2\sum_{k\in G}\mu(k)d(1,k)\sum_{e'\in E}\frac{|\Gamma_{e'}(1,k)|}{|\Gamma(1,k)|}.
\end{aligned}$$

Note that the factor "2" comes from the fact that if $g\in Y$ is an involution ($g^2 = 1_G$) then with $e:=\{1_G,g\}\in E$ we have $g^{-1}e = ge = e$.

Since

$$\sum_{e'\in E}\frac{|\Gamma_{e'}(1,k)|}{|\Gamma(1,k)|} = \frac{1}{|\Gamma(1,k)|}\sum_{e'\in E}\sum_{\gamma\in\Gamma_{e'}(1,k)}1$$

$$= \frac{1}{|\Gamma(1,k)|} \sum_{\gamma \in \Gamma(1,k)} \sum_{e' \in \gamma} 1$$
$$= \frac{1}{|\Gamma(1,k)|} \sum_{\gamma \in \Gamma(1,k)} d(1,k)$$
$$= d(1,k),$$

we deduce that for all $e \in E$,

$$\varphi(e) \leq 2\Phi,$$

where $\Phi := \sum_{k \in G} \mu(k) d(1,k)^2 < \infty$ is the second moment of μ. We then have (recall that $r_Q(e) = |Y|$ for all $e \in E$):

$$D_P(f) \leq \frac{2\Phi|Y|}{2} \langle \nabla_Q f, \nabla_Q f \rangle_E = \Phi|Y| D_Q(f).$$

Thus, by virtue of Corollary 13.37, we deduce that the random walk (G,P) on G is recurrent. This shows (c) $\Rightarrow$ (d).

Finally, the implication (d) $\Rightarrow$ (a) is again trivial. □

As a consequence of the previous theorem, we give the following:

Definition 13.49. A finitely generated group G is called *recurrent* if one of the equivalent conditions of Theorem 13.48 holds.

13.14 Proof of Varopoulos' Theorem

We are now in a position to present a proof of Varopoulos' theorem.

Proof of Theorem 13.22 Let G be a finitely generated group and suppose that for some finite symmetric generating subset the associated simple random walk is recurrent, in other words, G is recurrent. By virtue of Proposition 13.46 and Theorem 13.48, G cannot have at least cubic growth.

By Gromov's theorem (Theorem 12.1), since G has polynomial growth, it contains a torsion-free nilpotent subgroup N of finite index, whose growth is the same as that of G. By the Bass–Guivarc'h formula (Theorem 7.29), such a subgroup N has polynomial growth of degree $d := \sum_{i=1}^{c} ir(i)$, where $c \in \mathbb{N}$ denotes the nilpotency class of N and $r(i)$ is the free abelian rank of the i-th quotient in the lower central series of N. By our assumptions, d is equal to either 0, 1, or 2. In the first case N is trivial, and therefore G is finite. If $d = 1$ then N is isomorphic to $\mathbb{Z}$. Finally, if $d = 2$, then we necessarily have $c = 1$ and $r(1) = 2$, so that N is isomorphic to $\mathbb{Z}^2$.

This completes the proof of Varopoulos' theorem. □

Remark 13.50. Note that for $d \geq 4$ it is no longer the case that if G has polynomial growth of degree d, then G is virtually isomorphic to $\mathbb{Z}^d$. For instance, for $d = 4$, besides the possibility $c = 1$ with $r(1) = 4$, we may also have $c = 2$ with $r(1) = 2$ and $r(2) = 1$: indeed, this is the case for the Heisenberg group $\mathrm{UT}(3,\mathbb{Z})$.

13.15 Notes

George Pólya [284] proved Theorem 13.47 (cf. Proposition 13.19 and Proposition 13.21) in 1921. In [112, Chapter 14, Section 7], it is shown that the probability of returning to the initial position in the simple random walk on $\mathbb{Z}^3$ is about 0.35 and the expected number of returns is approximately 0.53. This was the beginning of the theory of random walks. The term *random walk*, however, was first introduced by Karl Pearson [273] in 1905.

A locally compact group G is said to be *recurrent* if there is a Borel probability measure μ on G whose support generates G and such that the random walk determined by μ is recurrent. Kesten [198] raised the question: *when is a group recurrent, i.e., when does it admit a recurrent random walk?* Richard M. Dudley [97] had already proved in 1962 that an abelian group is recurrent if and only if it has free-abelian rank at most two. Yves Guivarc'h, Michael Keane, and B. Roynette [147] formulate a conjecture for general locally compact groups which resembles Dudley's result. A group G is said to be of polynomial growth of degree at most d if for every compact neighborhood K of the identity there exists a constant $C > 0$ for which $\mu(K^n) \leq Cn^d$ for all $n \in \mathbb{N}$, where μ is the Haar measure (cf. Section 9.3). Then the precise conjecture is that G is recurrent if and only if it is of polynomial growth of degree at most two. Guivarc'h and Keane [146] eventually proved this conjecture when G is a connected locally compact group (1975). For discrete groups, the conjecture was settled by Nicholas Th. Varopoulos in [340] (cf. Theorem 13.22) in 1986. More recently (2005), Chandiraraj Robinson Edward Raja [287] showed that the conjecture also holds if G is a p-adic Lie group.

The Nash-Williams recurrence criterion (Corollary 13.42) is due to Crispin Nash-Williams [247] who, improving on a previous result of Frederic Gordon Foster [116], was the first to apply the electrical network techniques of Lord Rayleigh to random walks.

The theory of random walks has been extended to graphs: we refer again to the books by Wolfgang Woess [355, 356]. Another good reference for random walks on graphs is the online book by David Aldous and James Allen Fill [4].

A recent, clear, and comprehensive treatment of Varopoulos' theorem is presented by Steven P. Lalley in his recent book [209].

13.16 Exercises

Exercise 13.1. Let (X,P,π) be a Markov chain and denote by $(\Omega,\mathscr{B},\mathbb{P}_\pi)$ the associated probability space. Show that for a random variable $Z_n\colon \Omega \to X$ and $x,y \in X$ one has

$$\mathbb{P}_\pi[Z_{n+1} = y | Z_n = x] = p(x,y)$$

provided $\mathbb{P}_\pi[Z_n = x] > 0$.

Exercise 13.2. Let P and Q be stochastic matrices of the same size. Show that PQ is also stochastic. This implies that P^n is also stochastic for all $n \in \mathbb{N}$.

Exercise 13.3. Prove (13.11) and (13.12).

Exercise 13.4. Let (X,P,π) be a Markov chain and denote by $\mathscr{G}=(X,E)$ the associated directed graph. Show that (X,P,π) is irreducible if and only if $\mathscr{G}$ is *connected*, that is, for every pair (x,y) of vertices there exists a finite directed path connecting x to y.

Exercise 13.5. Show that conditions (1)–(3) in Definition 13.16 do not depend on the generating subset $Y \subseteq G$ nor on the initial distribution.

Exercise 13.6. Show that the simple random walk (cf. Example 13.18) satisfies conditions (1)–(3) in Definition 13.16.

Exercise 13.7. Let μ be an atomic probabilistic measure on a group G. Show that for every $n \geq 1$, the n-th convolution measure $\mu^{(n)}$ defined recursively by setting $\mu^{(1)} := \mu$ and $\mu^{(n)}(g) := \sum_{h\in G}\mu^{(n-1)}(gh^{-1})\mu(h) = \sum_{h\in G}\mu^{(n-1)}(h)\mu(h^{-1}g)$ for all $g \in G$ is an atomic probability measure on G.

Exercise 13.8. Let (G,P) be a random walk on a finitely generated group G and denote by μ the associated atomic probability measure (cf. (13.21)). Show that $p^{(n)}(1_G,g) = \mu^{(n)}(g)$ for all $n \geq 1$ and $g \in G$.

Exercise 13.9. Show that the random walk (G,P) on G defined by (13.22) is irreducible and satisfies conditions (1)–(3) in Definition 13.16.

Exercise 13.10. (1) Show that the Dirichlet seminorm $D(\cdot)$ (cf. Definition 13.30) satisfies the following conditions: (i) $D(f) \geq 0$, (ii) $D(\alpha f) = |\alpha| D(f)$, and $D(f+g) \leq D(f)+D(g)$ for all $f,g \in \mathscr{D}(\mathscr{N})$.
(2) Show that there exist nonzero elements $f \in \mathscr{D}(\mathscr{N})$ such that $D(f)=0$.
(3) Show that (13.36) defines an inner product in $\mathscr{D}(\mathscr{N})$ (in particular, $\langle f,f\rangle_D > 0$ for all nonzero $f \in \mathscr{D}(\mathscr{N})$).

Exercise 13.11. Show that the operator $G_A \colon L^2(X,m) \to L^2(X,m)$ defined in (13.42) satisfies the identity $(I_A - P_A)G_A = I_A$.

Exercise 13.12. Show that if a Markov chain (X,P) is irreducible, so is any associated shortened Markov chain $(\mathbb{N},P')$ (with respect to some partition of X into finite subsets, cf. Definition 13.40).

Exercise 13.13. (1) Prove (13.55).
(2) Show that

$$\lim_{m\to\infty} s_m m^2 \log^2(m) \leq \frac{1}{2}.$$

Exercise 13.14. Prove the identity (13.58).

Exercise 13.15. Let G be a finitely generated group. Suppose that G is recurrent. Show that every finitely generated subgroup $H \subseteq G$ is also recurrent.

Exercise 13.16. Let G be a finitely generated group, and let $H \subseteq G$ be a subgroup of finite index. Show, without using Varopoulos' theorem, that if H is recurrent then so is G.

Chapter 14
Amenability, Isoperimetric Profiles, and Følner Functions

This chapter is devoted to the study of amenable groups. Amenability plays an important role in many areas of mathematics such as harmonic and functional analysis, representation theory, ergodic theory and dynamical systems, probability theory and statistics, and geometric group theory. As residually finite groups, amenable groups generalize finite groups. However, there are residually finite groups which are not amenable, and there are amenable groups which are not residually finite.

John von Neumann defined an amenable group as a group admitting an invariant finitely additive probability measure on the set of all of its subsets. Our definition (equivalent to the original, cf. Theorem 14.21) is in terms of Følner sets: a group G is amenable if for every finite subset $X \subseteq G$ and every $\varepsilon > 0$ there exists a finite subset $\Omega \subset G$ satisfying that $|X\Omega \setminus \Omega| < \varepsilon|\Omega|$ (in other words, as one then says, Ω is ε-invariant by left multiplication by elements in X).

The class of amenable groups contains in particular all finite groups, all abelian groups and, more generally, all groups of subexponential growth (Proposition 14.6) and all solvable groups (Corollary14.13). It is closed under the operations of taking: subgroups, quotients, extensions, and inductive limits (Sect. 14.2). The free group on two generators (and therefore any group containing a subgroup isomorphic to it) is non-amenable (cf. Example 14.4).

Invariant finitely additive probability measures and paradoxical decompositions are defined in Sect. 14.3. The Tarski–Følner theorem (Theorem 14.21) asserts that existence of Følner sets, existence of invariant finitely additive probability measures, and non-existence of paradoxical decompositions are three equivalent conditions for a group.

In Section 14.5 (resp. Section 14.6) we define the notion of spectral radius of the simple random walk on (resp. of cogrowth of) a finitely generated group with respect to a finite symmetric generating subset and present the Kesten (resp. Grigorchuk cogrowth) criterion for amenability of finitely generated groups.

In Section 14.7 we prove the Ornstein–Weiss lemma, an analogue of Fekete's lemma, for sub-additive right-invariant functions defined on the finite subsets of an amenable group. This result is important in the theory of dynamical systems since it makes it possible to define several numerical invariants such as topological entropy, measure-theoretic entropy, and mean topological dimension.

T. Ceccherini-Silberstein and M. D'Adderio, *Topics in Groups and Geometry*,
Springer Monographs in Mathematics, https://doi.org/10.1007/978-3-030-88109-2_14

In Section 14.9 we introduce and study the Tarski number of a group. It can be regarded as a "measure" of amenability or, rather, of non-amenability of groups. If G is a non-amenable group, the complexity $c(p)$ of a paradoxical decomposition p of G is the number of pieces (subsets of G) involved in p. Then the integer $\tau(G) = \min c(p)$, where the minimum is taken over all paradoxical decompositions p of G, is called the Tarski number of G. We show that a group G contains a non-abelian free group if and only if $\tau(G) = 4$ (Theorem 14.89).

In Section 14.10 we define and study the isoperimetric profile of a finitely generated group. Given a finitely generated group G together with a finite symmetric generating subset $X \subseteq G$, the isoperimetric profile of G with respect to X is the function $I_\circ(n;G,X) = \inf_{\Omega \subseteq G, |\Omega|=n} |X\Omega \setminus \Omega|$ which measures the minimal size of the X-boundary $\partial_X(\Omega) := X\Omega \setminus \Omega$ when Ω varies among all finite subsets of G with cardinality $|\Omega| = n \in \mathbb{N}$. The growth type of $I_\circ(n;G,X)$ is independent of the generating subset X (Lemma 14.93). In Proposition 14.94 it is shown that for a finitely generated group nonamenability is equivalent to linear growth of the associated isoperimetric profile.

In Theorem 14.95 we establish a remarkable inequality that is due to Thierry Coulhon and Laurent Saloff-Coste.

In Sections 14.11 and 14.12 we introduce and study the Følner function of a finitely generated group. Given a finitely generated group G together with a finite symmetric generating subset $X \subseteq G$, the Følner function of G with respect to X is the function $F_\circ(n;G,X)$ which equals the minimal cardinality of a finite subset $\Omega \subseteq G$ such that $|\partial_X(\Omega)| \leq |\Omega|/n$. The growth type of $F_\circ(n;G,X)$ is independent of the generating subset X (Lemma 14.98): we denote by $F_\circ(n;G)$ the corresponding class of growth. Using the inequality of Coulhon and Saloff-Coste, we prove that for a finitely generated group G one has $F_\circ(n;G) \succeq b^G(n)$ (Theorem 14.100) and that equality holds if G is nilpotent (Theorem 14.102).

14.1 Amenability of Groups: Definitions and Examples

Definition 14.1. Let G be a group. Given two finite subsets X and Ω of G, we define the X-*external boundary* of Ω as

$$\partial_X(\Omega) := X\Omega \setminus \Omega = \bigcup_{x \in X} (x\Omega \setminus \Omega).$$

The group G is called *amenable* if for every finite subset $X \subseteq G$ and every $\varepsilon > 0$ there exists a finite subset $\Omega \subseteq G$ such that

$$|\partial_X(\Omega)| < \varepsilon |\Omega|,$$

equivalently,

$$|\Omega \cup X\Omega| < (1+\varepsilon)|\Omega|.$$

Example 14.2. Every finite group G is amenable: for every X and ε choose $\Omega := G$.

Example 14.3. The group $\mathbb{Z}$ is amenable: for every finite nonempty subset $X \subseteq \mathbb{Z}$ and every $\varepsilon > 0$, let $m := \max\{|x| : x \in X\}$ and set $\Omega := [-n,n] \subseteq \mathbb{Z}$, where $n \in \mathbb{N}$ satisfies $2m < \varepsilon(2n+1)$. Then $X\Omega \subseteq [-n-m, n+m]$, so that

$$|\partial_X(\Omega)| = |\Omega \cup X\Omega| - |\Omega| \leq 2m < \varepsilon(2n+1) = \varepsilon|\Omega|.$$

Example 14.4. The free group F_2 on two generators x and y is not amenable. Indeed, if we choose $X = \{x,y\}$, then for any finite subset $\Omega \subseteq F_2$ we have

$$|\partial_X(\Omega)| = |\Omega \cup X\Omega| - |\Omega| \geq |\Omega|. \tag{14.1}$$

In fact, for $a \in \{x,y,x^{-1},y^{-1}\}$ let Ω_a be the set of reduced words in Ω starting with a. If the identity element of F_2 is in Ω, then we include it in $\Omega_{x^{-1}}$.

Now, for $Z_x := \Omega_x \sqcup \Omega_y \sqcup \Omega_{y^{-1}} \subseteq \Omega$, we have

$$xZ_x \cap \Omega \subseteq \Omega_x \quad \text{and} \quad xZ_x \cap \Omega_a = \varnothing \ \text{ for all } a \in \{y,x^{-1},y^{-1}\},$$

hence

$$|xZ_x \setminus \Omega| = |xZ_x \setminus \Omega_x| \geq |\Omega_y \sqcup \Omega_{y^{-1}}|.$$

Analogously, for $Z_y := \Omega_x \sqcup \Omega_y \sqcup \Omega_{x^{-1}} \subseteq \Omega$, we have

$$|yZ_y \setminus \Omega| \geq |\Omega_x \sqcup \Omega_{x^{-1}}|.$$

So we have the disjoint union

$$\Omega \sqcup (xZ_x \setminus \Omega) \sqcup (yZ_y \setminus \Omega) \subseteq \Omega \cup X\Omega,$$

and therefore

$$|\Omega \cup X\Omega| \geq |\Omega \sqcup (xZ_x \setminus \Omega) \sqcup (yZ_y \setminus \Omega)| \geq |\Omega| + |\Omega_y \sqcup \Omega_{y^{-1}}| + |\Omega_x \sqcup \Omega_{x^{-1}}| = 2|\Omega|.$$

This proves (14.1).

A group G is termed *locally amenable* if every finitely generated subgroup of G is amenable.

Proposition 14.5. *Every locally amenable group is amenable.*

Proof. This follows immediately from the definition of amenability, since given a finite subset $X \subset G$ and $\varepsilon > 0$, we can choose the finite subset Ω in the (amenable) subgroup generated by X. □

The following proposition gives us more examples of amenable groups.

Proposition 14.6. *Every group of subexponential growth is amenable.*

Proof. Suppose that G is a non-amenable group. Then there exist a finite subset $X \subseteq G$ and $\varepsilon > 0$ such that, for any finite subset $\Omega \subseteq G$, we have

$$|\Omega \cup X\Omega| \geq (1+\varepsilon)|\Omega|.$$

In particular, taking $\Omega := \bigcup_{r=0}^{m-1} X^r$, $m \geq 1$, we get

$$|\bigcup_{r=0}^{m} X^r| - |\bigcup_{r=0}^{m-1} X^r| \geq \varepsilon |\bigcup_{r=0}^{m-1} X^r|,$$

which is the same as

$$b_X^H(m) - b_X^H(m-1) \geq \varepsilon b_X^H(m-1),$$

where H is the subgroup of G generated by X. This implies that $b_X^H(m)$ grows exponentially, hence, by Proposition 7.17, G has exponential growth. □

Corollary 14.7. *Every nilpotent group is amenable. In particular, every abelian group is amenable.*

Proof. Since by Proposition 2.9 every subgroup of a nilpotent group is nilpotent, by Proposition 14.5, it is enough to show that every finitely generated nilpotent group is amenable. But this follows from Proposition 14.6, since by Theorem 7.29 every finitely generated nilpotent group has polynomial growth. □

14.2 Stability Properties of Amenable Groups

Given a group G and two finite subsets X and Ω of G, we set

$$\delta_X(\Omega) := \{g \in \Omega : \text{there exists an } x \in X \text{ such that } xg \notin \Omega\}. \tag{14.2}$$

This is called the *X-internal boundary* of Ω.

Also, for every $x \in X$, we simply denote by $\partial_x(\Omega)$ the $\{x\}$-external boundary $\partial_{\{x\}}(\Omega)$ of Ω.

In the following proposition we state a few equivalent definitions of amenability. Depending on the situation, one of them can be more useful than the others.

Proposition 14.8. *Let G be a group. The following conditions are equivalent:*

(a) *G is amenable;*

(b) *for every finite subset $X \subseteq G$ and every $\varepsilon > 0$ there exists a finite subset $\Omega \subseteq G$ such that*

$$|\partial_x(\Omega)| < \varepsilon|\Omega| \quad \textit{for every } x \in X;$$

(c) *for every finite subset $X \subseteq G$ and every $\varepsilon > 0$ there exists a finite subset $\Omega \subseteq G$ such that*

$$|\delta_X(\Omega)| < \varepsilon|\Omega|.$$

Proof. We start with an easy observation: if $X_1, X_2, \Omega \subseteq G$ are finite subsets of G, then

$$|\partial_{X_1 \cup X_2}(\Omega)| \leq |\partial_{X_1}(\Omega)| + |\partial_{X_2}(\Omega)|. \tag{14.3}$$

In fact

$$\partial_{X_1 \cup X_2}(\Omega) = (X_1 \cup X_2)\Omega \setminus \Omega \subseteq ((X_1\Omega) \setminus \Omega) \cup ((X_2\Omega) \setminus \Omega) = \partial_{X_1}(\Omega) \cup \partial_{X_2}(\Omega),$$

from which the inequality (14.3) follows.

Given two finite subsets X and Ω of our group G, from (14.3) we deduce

$$|\partial_X(\Omega)| \le \sum_{x \in X} |\partial_x(\Omega)|,$$

from which it follows easily that (2) implies (1).

The implication (1) $\Longrightarrow$ (2) is immediate from the obvious inequality $|\partial_x(\Omega)| \le |\partial_X(\Omega)|$ for every $x \in X$.

Every element $g \in \delta_X(\Omega)$ gives rise (multiplying it on the left by the elements of X) to at most $|X|$ distinct elements in $\partial_X(\Omega)$, hence

$$|\partial_X(\Omega)| \le |X| \cdot |\delta_X(\Omega)|, \tag{14.4}$$

from which the implication (3) $\Longrightarrow$ (1) easily follows. Conversely, since every element of $\delta_X(\Omega)$ gives rise to at least one element of $\bigcup_{x \in X} \partial_x(\Omega)$, we have

$$|\delta_X(\Omega)| \le \sum_{x \in X} |\partial_x(\Omega)| \le |X| \max_{x \in X} |\partial_x(\Omega)| \le |X| \cdot |\partial_X(\Omega)|.$$

Then the implication (1) $\Longrightarrow$ (3) follows as well. □

Theorem 14.9. *Every subgroup of an amenable group is amenable.*

Proof. Let G be an amenable group, and let $H \subseteq G$ be a subgroup of G. Consider the decomposition $G = \sqcup_i H g_i$ of G into a disjoint union of right cosets.

Suppose that H is not amenable, so that there exist a finite subset $X \subseteq H$ and $\varepsilon > 0$ such that for every finite $\Omega' \subseteq H$ we have

$$|\Omega' \cup X\Omega'| \ge (1+\varepsilon)|\Omega'|.$$

Let $\Omega \subseteq G$ be a finite subset. We have the decomposition $\Omega = \sqcup_{j=1}^m \Omega_j g_{i_j}$, for some $i_1, i_2, \ldots, i_m$, where $\Omega_j g_{i_j} = \Omega \cap H g_{i_j}$. Hence $\Omega_j \subseteq H$ and $X\Omega_j \subseteq H$, for all $j = 1, 2, \ldots, m$. Then

$$\begin{aligned}
|\Omega \cup X\Omega| &= \left|\sqcup_{j=1}^m \left(\Omega_j g_{i_j} \cup X\Omega_j g_{i_j}\right)\right| = \left|\sqcup_{j=1}^m (\Omega_j \cup X\Omega_j)\, g_{i_j}\right| \\
&= \sum_{j=1}^m \left|(\Omega_j \cup X\Omega_j)\, g_{i_j}\right| = \sum_{j=1}^m \left|\Omega_j \cup X\Omega_j\right| \\
\text{(by the choice of } X \text{ and } \varepsilon) &\ge \sum_{j=1}^m (1+\varepsilon)|\Omega_j| = (1+\varepsilon)\sum_{j=1}^m |\Omega_j| \\
&= (1+\varepsilon)\sum_{j=1}^m |\Omega_j g_{i_j}| = (1+\varepsilon)\left|\sqcup_{j=1}^m \Omega_j g_{i_j}\right| \\
&= (1+\varepsilon)|\Omega|,
\end{aligned}$$

contradicting the amenability of G. Hence H is amenable. □

The following corollary is immediate (cf. Example 1.19).

Corollary 14.10. *Every group containing a subgroup isomorphic to the free group F_2 is non-amenable. In particular, every non-abelian free group and every group containing a subgroup isomorphic to* $\mathrm{SL}(2,\mathbb{Z})$ *(e.g.* $\mathrm{SL}(n,\mathbb{Z}), \mathrm{GL}(n,\mathbb{Z}), \mathrm{SL}(n,\mathbb{R})$, *or* $\mathrm{GL}(n,\mathbb{R})$ *for $n \geq 2$) is also non-amenable.* □

Theorem 14.11. *Every quotient of an amenable group is amenable.*

Proof. Let G' be an amenable group and suppose that G is a quotient of G'. Let $X \subseteq G$ be a finite subset and $\varepsilon > 0$. By Proposition 14.8, we need to show that there exists a finite subset $\Omega \subseteq G$ such that

$$|\delta_X(\Omega)| < \varepsilon|\Omega|. \tag{14.5}$$

Denote by $\pi\colon G' \to G$ the canonical epimorphism and choose a finite subset $X' \subseteq G'$ such that $\pi(X') = X$ and $|X'| = |X|$. Since G' is amenable, we can find a finite subset $\Omega' \subseteq G'$ such that

$$|\delta_{X'}(\Omega')| < \varepsilon|\Omega'|. \tag{14.6}$$

Let $n := |\Omega'|$ and, for all $i = 1, 2, \ldots, n$, define the sets

$$\Omega_i = \{g \in G : |\pi^{-1}(g) \cap \Omega'| \geq i\} \subseteq \pi(\Omega') \subseteq G. \tag{14.7}$$

Note that

$$\Omega_1 = \pi(\Omega') \supseteq \Omega_2 \supseteq \Omega_3 \supseteq \cdots \supseteq \Omega_n \supseteq \Omega_{n+1} = \varnothing \tag{14.8}$$

and, moreover,

$$\sum_{i=1}^{n} |\Omega_i| = \sum_{i=1}^{n} \sum_{g\in\Omega_1} \chi_{\Omega_i}(g) = \sum_{g\in\Omega_1} \sum_{i=1}^{n} \chi_{\Omega_i}(g) = \sum_{g\in\Omega_1} |\pi^{-1}(g) \cap \Omega'| = |\Omega'|. \tag{14.9}$$

Claim 1. *For $1 \leq j \leq \ell \leq i \leq n$ we have*

$$\delta_X(\Omega_j) \cap \delta_X(\Omega_i) \subseteq \delta_X(\Omega_\ell). \tag{14.10}$$

Indeed, let $g \in \delta_X(\Omega_j) \cap \delta_X(\Omega_i)$. Then, as $g \in \delta_X(\Omega_j)$, we have $g \in \Omega_j \subseteq \Omega_\ell$ (where the last inclusion follows from (14.8)) and, as $g \in \delta_X(\Omega_i)$, we can find $k \in X$ such that $kg \notin \Omega_i$, equivalently, $kg \in G \setminus \Omega_i \subseteq G \setminus \Omega_\ell$ (where the last inclusion again follows from (14.8)), i.e. $kg \notin \Omega_\ell$. This shows that $g \in \delta_X(\Omega_\ell)$ and the claim follows.

Set $\overline{\Omega} := \cup_{t=1}^{n} \partial_X(\Omega_t) \subseteq \Omega$. Let $g \in \overline{\Omega}$. We define $1 \leq j_g \leq i_g \leq n$ by setting

$$j_g := \min\{t : g \in \delta_X(\Omega_t)\} \text{ and } i_g := \max\{t : g \in \delta_X(\Omega_t)\}.$$

Claim 2. *For every $g \in \overline{\Omega}$ we have*

$$|\pi^{-1}(g) \cap \delta_{X'}(\Omega')| \geq i_g - j_g + 1. \tag{14.11}$$

Indeed, let $g \in \overline{\Omega}$. As $g \in \delta_X(\Omega_{i_g})$ we have $g \in \Omega_{i_g}$ and we can find distinct elements $g_1', g_2', \ldots, g_{i_g}' \in \Omega'$ such that $\pi(g_1') = \pi(g_2') = \cdots = \pi(g_{i_g}') = g$.

On the other hand, since also $g \in \delta_X(\Omega_{j_g})$, we have $g \in \Omega_{j_g}$ and we can find $k \in X$ such that

$$kg \notin \Omega_{j_g}. \tag{14.12}$$

Let $k' \in X'$ be such that $\pi(k') = k$. Consider the distinct elements $k'g'_1, k'g'_2, \dots, k'g'_{i_g} \in k'\Omega' \subseteq G'$ and set

$$\Omega'(g;\text{in}) := \{g'_i : k'g'_i \in \Omega', 1 \le i \le i_g\}$$

and

$$\Omega'(g;\text{out}) := \{g'_i : k'g'_i \notin \Omega', 1 \le i \le i_g\}.$$

Then if $\ell := |\Omega'(g;\text{in})|$, we have, by definition of the Ω_i's (cf. (14.7)), $kg \in \Omega_\ell$ so that, from (14.12) and (14.10), we deduce that $\ell \le j_g - 1$. As a consequence,

$$|\Omega'(g;\text{out})| = i_g - \ell \ge i_g - j_g + 1. \tag{14.13}$$

Since in fact $\Omega'(g;\text{out}) = \pi^{-1}(g) \cap \delta_{X'}(\Omega')$, the claim follows from (14.13).

From the second claim we immediately deduce

$$|\delta_{X'}(\Omega')| = |\sqcup_{g \in \Omega_1} \pi^{-1}(g) \cap \delta_{X'}(\Omega')| \ge \sum_{g \in \Omega_1} (i_g - j_g + 1). \tag{14.14}$$

On the other hand we have

$$\begin{aligned}
\sum_{i=1}^{n} |\delta_X(\Omega_i)| &= \sum_{i=1}^{n} \sum_{g \in \Omega_1} \chi_{\delta_X(\Omega_i)}(g) \\
&= \sum_{g \in \Omega_1} \left(\sum_{i=1}^{n} \chi_{\delta_X(\Omega_i)}(g) \right) \\
&= \sum_{g \in \Omega_1} \sum_{i=j_g}^{i_g} 1 \\
&= \sum_{g \in \Omega_1} (i_g - j_g + 1).
\end{aligned} \tag{14.15}$$

From (14.14) and (14.15) we deduce

$$|\delta_{X'}(\Omega')| \ge \sum_{i=1}^{n} |\delta_X(\Omega_i)|.$$

Let us show that there exists $1 \le h \le n$ such that $|\Omega_h| \ne \varnothing$ and $|\delta_X(\Omega_h)| < \varepsilon |\Omega_h|$. If not, we would have $|\delta_X(\Omega_i)| > \varepsilon |\Omega_i|$ for all $i = 1, 2, \dots, n$ and therefore

$$|\delta_{X'}(\Omega')| \ge \sum_{i=1}^{n} |\delta_X(\Omega_i)| > \varepsilon \sum_{i=1}^{n} |\Omega_i| = \varepsilon |\Omega'|,$$

where the last equality follows from (14.9), contradicting (14.6). Then (14.5) follows by taking $\Omega := \Omega_h$. □

Theorem 14.12. *Every extension of an amenable group by an amenable group is amenable.*

Proof. Let G be a group. Let $H \subseteq G$ be a normal subgroup of G which is amenable, and suppose that the quotient group $\overline{G} := G/H$ is also amenable. We need to show that G is amenable.

Let X be a finite subset of G and let $\varepsilon > 0$. By Proposition 14.8 it is enough to show that there exists a finite subset Ω of G such that

$$|\partial_x(\Omega)| < \varepsilon|\Omega| \quad \text{for all } x \in X.$$

Let $\pi : G \to \overline{G}$ be the canonical quotient homomorphism, and set $\overline{X} := \pi(X) \subseteq \overline{G}$. Since $\overline{G}$ is amenable, we can find a finite subset $\overline{\Omega}_1 \subseteq \overline{G}$ such that

$$|\partial_{\overline{x}}(\overline{\Omega}_1)| < \frac{\varepsilon}{2}|\overline{\Omega}_1| \quad \text{for all } \overline{x} \in \overline{X}.$$

Let $\Omega_1 \subseteq G$ be such that $\pi(\Omega_1) = \overline{\Omega}_1$ and $|\Omega_1| = |\overline{\Omega}_1|$.

For $x \in X$, let $\overline{x} := \pi(x) \in \overline{X}$. Observe that we always have (**exercise**)

$$\partial_{\overline{x}}(\overline{\Omega}_1) = \overline{x}\overline{\Omega}_1 \setminus \overline{\Omega}_1 = \pi(x\Omega_1) \setminus \pi(\Omega_1) \subseteq \pi(x\Omega_1 \setminus \Omega_1) = \pi(\partial_x(\Omega_1)).$$

We distinguish two cases.

Case 1: equality holds for all $x \in X$. Then π induces a bijection from $\partial_x(\Omega_1)$ to $\partial_{\overline{x}}(\overline{\Omega}_1)$, so that

$$|\partial_{\overline{x}}(\overline{\Omega}_1)| = |\partial_x(\Omega_1)|,$$

and therefore

$$|\partial_x(\Omega_1)| = |\partial_{\overline{x}}(\overline{\Omega}_1)| < \frac{\varepsilon}{2}|\overline{\Omega}_1| < \varepsilon|\Omega_1| \quad \text{for all } x \in X.$$

So we can take $\Omega := \Omega_1$.

Case 2: there exist $x \in X$ and $g \in \partial_x(\Omega_1)$ such that $\pi(g) \in \overline{\Omega}_1$. In this case, there exist $\omega_1, \omega_1' \in \Omega_1$ such that $g = x\omega_1$, and $\pi(g) = \pi(x\omega_1) = \pi(\omega_1')$. Therefore, we can find $y \in H$ such that $x\omega_1 = \omega_1' y$, so that $y = (\omega_1')^{-1}x\omega_1 \in \Omega_1^{-1}X\Omega_1 \cap H$. The set $Y := \Omega_1^{-1}X\Omega_1 \cap H$ is a finite subset of H, and by the amenability of H, we can find a finite subset $\Omega_2 \subseteq H$ such that

$$|\partial_y(\Omega_2)| < \frac{\varepsilon}{2|Y|}|\Omega_2| \quad \text{for all } y \in Y. \tag{14.16}$$

Set $\Omega := \Omega_1\Omega_2 \subseteq G$. Observe that $|\Omega| = |\Omega_1| \cdot |\Omega_2|$.

Now for $x \in X$, let $\overline{x} := \pi(x)$, and consider the two sets

$$\partial_x^1(\Omega) := \{g \in \partial_x(\Omega) : \pi(g) \in \overline{\Omega}_1\}$$

and

$$\partial_x^2(\Omega) := \partial_x(\Omega) \setminus \partial_x^1(\Omega).$$

For $g \in \partial_x^1(\Omega)$, there exist $\omega_1 \in \Omega_1$ and $\omega_2 \in \Omega_2$ such that $g = x\omega_1\omega_2$. As we already observed, there exist $\omega_1' \in \Omega_1$ and $y \in Y$ such that $g = x\omega_1\omega_2 = \omega_1' y\omega_2$. Note that, since $g \notin \Omega$, we have $y\omega_2 \notin \Omega_2$, therefore $y\omega_2 \in \partial_y(\Omega_2)$. Then

$$\begin{aligned}
|\partial_x^1(\Omega)| &\leq |\Omega_1| \cdot \left| \bigcup_{y \in Y} \partial_y(\Omega_2) \right| \\
\text{(by (14.16))} &\leq |\Omega_1| \cdot \left(\sum_{y \in Y} |\partial_y(\Omega_2)| \right) \\
&< |\Omega_1| \cdot |Y| \left(\frac{\varepsilon}{2|Y|} |\Omega_2| \right) \\
&= \frac{\varepsilon}{2} |\Omega_1| \cdot |\Omega_2| \\
&= \frac{\varepsilon}{2} |\Omega|.
\end{aligned}$$

For $g \in \partial_x^2(\Omega)$, we have $\pi(g) \in \partial_{\overline{x}}(\overline{\Omega})$. Arguing as in Case 1, we have

$$\begin{aligned}
|\partial_x^2(\Omega)| &\leq |\partial_{\overline{x}}(\overline{\Omega}_1)| \cdot |\Omega_2| \\
&< \frac{\varepsilon}{2} |\Omega_1| \cdot |\Omega_2| \\
&= \frac{\varepsilon}{2} |\Omega|.
\end{aligned}$$

Finally,

$$|\partial_x(\Omega)| = |\partial_x^1(\Omega)| + |\partial_x^2(\Omega)| \leq \frac{\varepsilon}{2}|\Omega| + \frac{\varepsilon}{2}|\Omega| = \varepsilon|\Omega|.$$

This proves the amenability of G. □

In particular, the semidirect product of two amenable groups is amenable. The proof of the following corollary is left as an **exercise**.

Corollary 14.13. *Every solvable group is amenable.*

Corollary 14.14. *Every virtually amenable group is amenable.*

Proof. Let G be a virtually amenable group. Thus we can find a finite index subgroup $H \leq G$ which is amenable. By Poincaré's lemma (Lemma 2.39), we can find a normal subgroup $K \leq H$ which is of finite index in G. By virtue of Theorem 14.9 (resp. Example 14.2) K (resp. the finite group G/K) is amenable. As G is the extension of K by G/K, we deduce from Theorem 14.12 that G is amenable. □

Theorem 14.15. *Let G be a group. Suppose there exists a family $(G_i)_{i \in I}$ of amenable subgroups of G such that $G = \bigcup_{i \in I} G_i$ and, for all $i_1, i_2 \in I$ there exists an $i \in I$ such that $G_{i_1} \cup G_{i_2} \subseteq G_i$. Then G is amenable.*

Proof. Let $X \subset G$ be a finite subset and $\varepsilon > 0$. For every $x \in X$ we can find an $i_x \in I$ such that $x \in G_{i_x}$. Let $i \in I$ be such that $\bigcup_{x \in X} G_{i_x} \subseteq G_i$. Then, since G_i is amenable and $X \subseteq G_i$, we can find a finite subset Ω of G_i (and therefore of G) such that $|\partial_X(\Omega)| < |\Omega|$. This shows that G is amenable. □

Corollary 14.16. *Let G be a group and suppose there exists a sequence $(G_n)_{n\in\mathbb{N}}$ of amenable subgroups which is* exhausting, *that is, $G = \bigcup_{n\in\mathbb{N}} G_n$, and* non-decreasing, *that is, $G_n \subset G_{n+1}$ for all $n \in \mathbb{N}$. Then G is amenable.* □

14.3 Measures and Paradoxical Decompositions

Definition 14.17. Let G be a group. An *invariant finitely additive probability measure* on G is a function $\mu\colon \mathscr{P}(G) \to [0,1]$ such that

1. $\mu(G) = 1$;
2. $\mu(A \cup B) = \mu(A) + \mu(B) - \mu(A \cap B)$ for all $A, B \subseteq G$;
3. $\mu(gA) = \mu(A)$ for all $g \in G$ and $A \subseteq G$.

Example 14.18. If G is a finite group, then the *normalized counting measure* μ, defined by setting $\mu(A) := |A|/|G|$ for all $A \subseteq G$, is an invariant finitely additive probability measure on G.

Definition 14.19. A group G is called *paradoxical* if there exist nonempty subsets $A_i, B_j \subseteq G$ and elements $a_i, b_j \in G$, for $i = 1, 2, \ldots, n$ and $j = 1, 2, \ldots, m$, such that

$$G = \sqcup_{i=1}^{n} A_i \sqcup \sqcup_{j=1}^{m} B_j \tag{14.17}$$

and

$$G = \sqcup_{i=1}^{n} a_i A_i = \sqcup_{j=1}^{m} b_j B_j. \tag{14.18}$$

One says that the expressions (14.17) and (14.18) constitute a *paradoxical decomposition* of *type* (n,m) of the group G.

Example 14.20. Consider the free group F_2 on two free generators x and y. For $a \in \{x, x^{-1}, y, y^{-1}\}$, let Ω_a denote the set of all reduced words in F_2 starting with a. Set $A_1 := \Omega_x$, $A_2 := \Omega_{x^{-1}}$, $B_1 := \Omega_y \cup \{y^{-n} : n = 0, 1, 2, \ldots\}$ and $B_2 := \Omega_{y^{-1}} \setminus \{y^{-n} : n = 0, 1, 2, \ldots\}$. Then it is an **exercise** to check that

$$F_2 = A_1 \sqcup A_2 \sqcup B_1 \sqcup B_2, \tag{14.19}$$

and

$$F_2 = A_1 \sqcup x A_2 = B_1 \sqcup y B_2. \tag{14.20}$$

This shows that F_2 is paradoxical.

Theorem 14.21 (Tarski–Følner). *Let G be a finitely generated group. Then the following conditions are equivalent:*

(a) *G is amenable;*
(b) *G admits an invariant finitely additive probability measure;*
(c) *G is not paradoxical.*

Proof. (a) $\Longrightarrow$ (b): suppose that G is amenable, and let $X \subseteq G$ be a finite symmetric set of generators of G. Then, for every $n \in \mathbb{N}$, $n \geq 1$, let $\Omega_n \subseteq G$ a finite subset such that

$$|\partial_{B_X(1_G,n)}(\Omega_n)| < \frac{1}{n}|\Omega_n|.$$

For every $n \geq 1$, let

$$\mu_n(A) := \frac{|A \cap \Omega_n|}{|\Omega_n|}$$

for all $A \subseteq G$. Let ω be a free ultrafilter on $\mathbb{N}$. By Corollary 11.20 we can set

$$\mu(A) := \lim_\omega \mu_n(A)$$

for all $A \subseteq G$. We claim that $\mu : \mathscr{P}(G) \to [0,1]$ is an invariant finitely additive probability measure on G. Indeed,

$$\mu(G) = \lim_\omega \mu_n(G) = \lim_\omega 1 = 1,$$

and, for all $A, B \subseteq G$,

$$\begin{aligned} \mu(A \cup B) &= \lim_\omega \mu_n(A \cup B) \\ &= \lim_\omega (\mu_n(A) + \mu_n(B) - \mu_n(A \cup B)) \\ \text{(cf. Theorem 11.22)} &= \lim_\omega \mu_n(A) + \lim_\omega \mu_n(B) - \lim_\omega \mu_n(A \cap B) \\ &= \mu(A) + \mu(B) - \mu(A \cap B). \end{aligned}$$

Moreover, let $g \in G$ and $A \subseteq G$. For $n \geq \ell_X(g) = \ell_X(g^{-1})$, we have

$$\begin{aligned} |\mu_n(gA) - \mu_n(A)| &= \left| \frac{|gA \cap \Omega_n| - |A \cap \Omega_n|}{|\Omega_n|} \right| \\ &= \left| \frac{|A \cap g^{-1}\Omega_n| - |A \cap \Omega_n|}{|\Omega_n|} \right| \\ &= \frac{|(g^{-1}\Omega_n \cap \Omega_n) \cap A|}{|\Omega_n|} \\ &\leq \frac{|\partial_{g^{-1}}(\Omega_n)|}{|\Omega_n|} \\ &\leq \frac{|\partial_{B_X(1_G,n)}(\Omega_n)|}{|\Omega_n|} < \frac{1}{n}. \end{aligned}$$

Therefore $|\mu(gA) - \mu(A)| \leq \lim_\omega |\mu_n(gA) - \mu_n(A)| = 0$, hence

$$\mu(gA) = \mu(A).$$

(b) $\Longrightarrow$ (c): let μ be an invariant finitely additive probability measure on G. Suppose by contradiction that there exist nonempty subsets $A_i, B_j \subseteq G$ and elements $a_1, b_j \in G$ for $i = 1, 2, \ldots, n$ and $j = 1, 2, \ldots, m$ such that

$$G = \sqcup_{i=1}^{n} A_i \sqcup \sqcup_{j=1}^{m} B_j$$

and

$$G = \sqcup_{i=1}^{n} a_i A_i = \sqcup_{j=1}^{m} b_j B_j.$$

But then

$$\begin{aligned} 1 = \mu(G) &= \mu\left(\sqcup_{i=1}^{n} A_i \sqcup \sqcup_{j=1}^{m} B_j\right) \\ &= \sum_{i=1}^{n} \mu(A_i) + \sum_{j=1}^{m} \mu(B_j) \\ &= \sum_{i=1}^{n} \mu(a_i A_i) + \sum_{j=1}^{m} \mu(b_j B_j) \\ &= \mu\left(\sqcup_{i=1}^{n} a_i A_i\right) + \mu\left(\sqcup_{j=1}^{m} b_j B_j\right) \\ &= \mu(G) + \mu(G) = 2, \end{aligned}$$

a contradiction. Hence G is not paradoxical.

(c) $\Longrightarrow$ (a): suppose that G is not amenable, and let us show that it is paradoxical. Since G is not amenable, there exist a finite subset $X \subseteq G$ and $\varepsilon > 0$ such that for every finite subset $\Omega \subseteq G$ we have

$$|\partial_X(\Omega)| \geq \varepsilon |\Omega|,$$

equivalently

$$|\Omega \cup X\Omega| \geq (1+\varepsilon)|\Omega|.$$

We can assume that $1_G \in X$, as

$$\partial_X(\Omega) = X\Omega \setminus \Omega = (\Omega \cup X\Omega) \setminus \Omega = \partial_{X \cup \{1_G\}}(\Omega).$$

Therefore

$$|X\Omega| \geq (1+\varepsilon)|\Omega|.$$

Replacing Ω by $X\Omega$ we get

$$|X^2\Omega| = |X(X\Omega)| \geq (1+\varepsilon)|X\Omega| \geq (1+\varepsilon)^2|\Omega|.$$

Iterating, for all $n \geq 1$ we get

$$|X^n\Omega| \geq (1+\varepsilon)|X^{n-1}\Omega| \geq (1+\varepsilon)^2|X^{n-2}\Omega| \geq \cdots \geq (1+\varepsilon)^n|\Omega|.$$

For n big enough, $(1+\varepsilon)^n > 2$, so that, by replacing X by X^n, we can assume that

$$|X\Omega| \geq 2|\Omega|. \tag{14.21}$$

Consider the *bipartite* graph $\mathscr{G} = (A \sqcup B, E)$, where A and B are two disjoint copies of G, and $E \subseteq A \times B$ consists of the edges (g, xg) with $g \in G$ and $x \in X$. In this graph, for every finite subset $\Omega \subseteq A$, the set of neighbors of all elements in Ω is precisely $X\Omega \subseteq B$, and clearly $XA = B$. We can now apply the following theorem.

Theorem 14.22 (Ph. Hall). *Let $\mathscr{G} = (A \sqcup B, E)$ be a graph such that $E \subseteq A \times B$. For every $\Omega \subseteq A$, let $\mathscr{N}(\Omega)$ denote the set of neighbors of elements of Ω.*

Suppose that for every finite subset $\Omega \subseteq A$, the set $\mathscr{N}(\Omega) \subseteq B$ is finite,

$$|\mathscr{N}(\Omega)| \geq 2|\Omega| \quad \text{and} \quad \mathscr{N}(A) = B.$$

Then there is a perfect $(1,2)$-matching *in E, i.e. a function $\varphi : B \to A$ such that $|\varphi^{-1}(a)| = 2$ for all $a \in A$, $\varphi(B) = A$, and $\{(a,b) : b \in \varphi^{-1}(a)\} \subseteq E$.*

Let then $\varphi\colon B \to A$ be a perfect $(1,2)$-matching in E. For every $a \in A$, we fix the notation $\varphi^{-1}(a) = \{b'_a, b''_a\}$, so that the sets $B' := \{b'_a : a \in A\} \subseteq B$ and $B'' := \{b''_a : a \in A\} \subseteq B$ are such that $B' \cap B'' = \varnothing$ and $B \sqcup B' = B$. For every $x \in X$, let $B'_x := \{b \in B' : \varphi(b) = x^{-1}b\}$ and $B''_x := \{b \in B'' : \varphi(b) = x^{-1}b\}$. Also, let $X' := \{x \in X : B'_x \neq \varnothing\} \subseteq X$ and $X'' := \{x \in X : B''_x \neq \varnothing\} \subseteq X$. Then

$$G \equiv B = \sqcup_{y \in X'} B_y \sqcup \sqcup_{z \in X''} B_z$$

and

$$G \equiv A = \sqcup_{y \in X'} y^{-1} B_y = \sqcup_{z \in X''} z^{-1} B_z,$$

showing that G is paradoxical. □

14.4 Følner Nets and Følner Sequences

Proposition 14.23. *Let G be a group. The following conditions are equivalent:*

(a) *G is amenable;*

(b) *for every finite set $X \subseteq G$ and every $\varepsilon > 0$ there exists a finite nonempty set $\Omega \subseteq G$ such that*

$$|\Omega \setminus x\Omega| < \varepsilon |\Omega| \tag{14.22}$$

for all $x \in X$;

(c) *there exists a net $(F_j)_{j \in J}$ of finite nonempty subsets of G such that*

$$\lim_{j \in J} \frac{|F_j \setminus gF_j|}{|F_j|} = 0 \tag{14.23}$$

for all $g \in G$.

Proof. We first observe that if $\Omega \subset G$ is a finite set and $x \in G$, then one has $|\Omega| = |x\Omega|$ so that

$$|\Omega \setminus x\Omega| = |\Omega| - |\Omega \cap x\Omega| = |x\Omega| - |\Omega \cap x\Omega| = |x\Omega \setminus \Omega|. \tag{14.24}$$

Suppose (a) and let $X \subset G$ be a finite set and $\varepsilon > 0$. Then we can find a finite subset $\Omega \subset G$ such that $|\partial_X(\Omega)| < \varepsilon|\Omega|$. But then, using (14.24), for every $x \in X$ we have

$$|\Omega \setminus x\Omega| = |x\Omega \setminus \Omega| \leq |X\Omega \setminus \Omega| = |\partial_X(\Omega)| < \varepsilon|\Omega|.$$

This shows (a) $\Longrightarrow$ (b).

Suppose now that G satisfies condition (b). Let J denote the set of pairs (X,ε), where X is a finite subset of G and $\varepsilon > 0$. We equip J with a partial order $\leq$ defined by

$$(X,\varepsilon) \leq (X',\varepsilon') \text{ if } (X \subset X' \text{ and } \varepsilon \geq \varepsilon').$$

Note that J is inductive, that is, every two elements in J admit both an upper and a lower bound. Indeed, we clearly have

$$\sup\{(X,\varepsilon),(X',\varepsilon')\} = (X \cup X', \min\{\varepsilon,\varepsilon'\})$$

and

$$\inf\{(X,\varepsilon),(X',\varepsilon')\} = (X \cap X', \max\{\varepsilon,\varepsilon'\}).$$

As a consequence of condition (b), for every $j = (X,\varepsilon) \in J$ we can find a finite nonempty subset $F_j \subset G$ such that

$$\frac{|F_j \setminus xF_j|}{|F_j|} < \varepsilon \qquad \text{for all } x \in X. \tag{14.25}$$

Let us now fix $g \in G$ and $\varepsilon_0 > 0$ and set $j_0 := (\{g\},\varepsilon_0) \in J$. If $j = (X,\varepsilon) \in J$ is such that $j \geq j_0$, that is, $g \in X$ and $\varepsilon \leq \varepsilon_0$, then from (14.25) we deduce

$$\frac{|F_j \setminus gF_j|}{|F_j|} < \varepsilon_0. \tag{14.26}$$

It follows that the net $(F_j)_{j \in J}$ satisfies (14.23). This shows that (b) implies (c).

Finally, suppose (c). Let X be a finite subset of G and $\varepsilon > 0$. From (14.23), for every $g \in G$ we can find an index $j(g) \in J$ such that

$$\frac{|F_j \setminus gF_j|}{|F_j|} < \frac{\varepsilon}{|X|}$$

for all $j \geq j(g)$. Since X is finite, we can find an index $j_0 \in J$ such that $j_0 \geq j(x)$ for all $x \in X$. Using (14.24), it follows that the finite set $\Omega := F_{j_0}$ satisfies the condition

$$|x\Omega \setminus \Omega| = |\Omega \setminus x\Omega| < \frac{\varepsilon}{|X|}|\Omega|$$

for all $x \in X$. We deduce that

$$|\partial_X(\Omega)| = |X\Omega \setminus \Omega| = |\cup_{x \in X}(x\Omega \setminus \Omega)| \leq \sum_{x \in X} |x\Omega \setminus \Omega| < \varepsilon|\Omega|.$$

This shows that (c) implies (a). □

A net $(F_j)_{j \in J}$ of nonempty finite subsets of G satisfying (14.23) is called a *Følner net* in (the amenable group) G.

The proof of the following corollary (along the lines of the proof of Proposition 14.23) is left as an **exercise**.

Corollary 14.24. *Suppose that G is countable. Then the following conditions are equivalent:*

(a) *G is amenable;*
(b) *there exists a sequence $(F_n)_{n\in\mathbb{N}}$ of finite nonempty subsets of G such that*

$$\lim_{n\in\mathbb{N}} \frac{|F_n \setminus gF_n|}{|F_n|} = 0 \tag{14.27}$$

for all $g \in G$.

A sequence $(F_n)_{n\in\mathbb{N}}$ of nonempty finite subsets of G satisfying (14.27) is called a *Følner sequence* in (the amenable group) G. It turns out (**exercise**) that, in fact, if a Følner sequence exists, then the group G is necessarily countable.

14.5 Kesten's Amenability Criterion

In the next three subsections we present Harry Kesten's pioneering and fundamental work on symmetric random walks on groups [196, 197]. In the present section, we define the spectral radius $\rho(G,Y)$ of the simple random walk associated with a finite symmetric generated subset Y of a finitely generated group G and show that $\rho(G,Y) = 1$ exactly when G is amenable (Theorem 14.26). Our exposition is also based on [59, Section 6.12]. We keep the notation from Section 13.5.

Kesten's Amenability Criterion for Finitely Generated Groups

Let G be a finitely generated group and let $Y \subset G$ be a finite symmetric generating subset.

Consider the simple random walk $(G,Q) = (G,Q,\delta_{1_G})$ on G associated with Y (cf. Example 13.18), that is, the nearest neighbor random walk defined by

$$q(g,h) = \begin{cases} \frac{1}{|Y|} & \text{if } g^{-1}h \in Y \\ 0 & \text{otherwise} \end{cases}$$

for all $g,h \in G$. As usual, we set

$$q^{(0)}(g,h) = \delta_{g,h} \tag{14.28}$$

(where $\delta_{g,h}$ denotes the Kronecker symbol) and, for $n \geq 1$ we denote by

$$q^{(n)}(1_G,1_G) = \mathbb{P}[Z_n = 1_G | Z_0 = 1_G] = \mathbb{P}_{\delta_{1_G}}[Z_n = 1_G]$$

the probability of returning to the identity element at the n-th step; here $Q^n = (q^{(n)}(g,h))_{g,h\in G}$ is the n-th power of Q, the projection map $Z_n \colon G^{\mathbb{N}} \to G$ is the random variable representing the random position of the walker at time n, and $\mathbb{P}(A|B)$ denotes the conditional probability of event A given event B with probability $\mathbb{P}(B) > 0$: in our case $\mathbb{P}[\cdot|Z_0 = 1_G] = \mathbb{P}_{\delta_{1_G}}[\cdot]$, since the random walk starts at 1_G.

With the random walk (G,Q) we associate the atomic probability measure μ on G obtained by setting (cf. (13.21)) $\mu(g) := q(1_G, g)$ for all $g \in G$. In the present setting μ is nothing but the uniform probability distribution on the generating subset Y:

$$\mu := \frac{1}{|Y|}\mathbf{1}_Y = \frac{1}{|Y|}\sum_{y\in Y}\delta_y.$$

Recall that $\mu(g) = \mu(g^{-1})$ for all $g \in G$ (i.e., μ is symmetric) and $\mathrm{supp}(\mu) = Y$. We set $\mu^{(0)} := \delta_{1_G}$ and denote by $\mu^{(n)}$ the n-th convolution of μ, which is defined recursively by setting

$$\mu^{(n+1)}(g) := \sum_{h\in G}\mu^{(n)}(gh^{-1})\mu(h) = \frac{1}{|Y|}\sum_{y\in Y}\mu^{(n)}(gy) \tag{14.29}$$

for all $g \in G$. Note that, in fact

$$\mu^{(n)}(g) = q^{(n)}(1_G, g) \tag{14.30}$$

for all $g \in G$ (**exercise**). We denote by

$$G(z) := \sum_{n=0}^{\infty} q^{(n)}(1_G, 1_G)z^n, \tag{14.31}$$

where z is an indeterminate, the associated Green function (cf. (13.14)). We denote by R its radius of convergence.

Definition 14.25. The number

$$\rho(G,Y) := \limsup_{n\to\infty}\sqrt[n]{q^{(n)}(1_G, 1_G)} = \frac{1}{R} \tag{14.32}$$

is called the *spectral radius* of the simple random walk on G relative to the symmetric generating subset Y.

Note that the last equality in (14.32) is just a consequence of the Cauchy–Hadamard theorem

We are now in a position to state the following characterization of finitely generated amenable groups.

Theorem 14.26 (Kesten's amenability criterion for finitely generated groups). *Let G be a finitely generated group and let $Y \subset G$ be a symmetric generating subset. Then G is amenable if and only if $\rho(G,Y) = 1$.*

In order to present a proof of Theorem 14.26, we review some basic concepts and results in Functional Analysis. For more details we refer to the monographs by Rudin [298, 299], Reed–Simon [288], and [59].

Let G be a countable group. Consider the real Hilbert space

$$\ell^2(G) = \{f \in \mathbb{R}^G : \sum_{g\in G} |f(g)|^2 < \infty\} \tag{14.33}$$

consisting of all square-summable real functions on G, the inner product being defined by setting

$$\langle f_1, f_2\rangle = \sum_{g\in G} f_1(g) f_2(g)$$

for all $f_1, f_2 \in \ell^2(G)$. Then, the ℓ^2-norm of an element $f \in \ell^2(G)$ is given by $\|f\|_2 = \sqrt{\langle f, f\rangle} = (\sum_{g\in G} |f(g)|^2)^{\frac{1}{2}}$.

The *(operator) norm* of a linear map $T\colon \ell^2(G) \to \ell^2(G)$ is defined by

$$\|T\| = \sup_{\substack{f\in\ell^2(G)\\ \|f\|_2\leq 1}} \|Tf\|_2 = \sup_{\substack{x\in\ell^2(G)\\ f\neq 0}} \frac{\|Tf\|_2}{\|f\|_2}.$$

Then, T is continuous if and only if $\|T\| < \infty$. We denote by $\mathscr{B}(\ell^2(G))$ the space of all continuous linear maps (also called *bounded linear operators*) $T\colon \ell^2(G) \to \ell^2(G)$ and by $I\colon \ell^2(G) \to \ell^2(G)$ the identity map.

Let $T \in \mathscr{B}(\ell^2(G))$. It follows from Riesz' representation theorem that there exists a unique operator $T^* \in \mathscr{B}(\ell^2(G))$, called the *adjoint* of T, such that $\langle Tf_1, f_2\rangle = \langle f_1, T^* f_2\rangle$ for all $f_1, f_2 \in \ell^2(G)$. The map $T \mapsto T^*$ is a linear involution on $\mathscr{B}(\ell^2(G))$ (**exercise**). One then says that T is *self-adjoint* provided $T = T^*$, that is, it satisfies $\langle Tf_1, f_2\rangle = \langle f_1, Tf_2\rangle$ for all $f_1, f_2 \in \ell^2(G)$.

The operator norm satisfies the following properties: $\|aT\| = |a| \cdot \|T\|$, $\|T^*\| = \|T\|$, and $\|T_1 + T_2\| \leq \|T_1\| + \|T_2\|$ for all $a \in \mathbb{R}$ and $T, T_1, T_2 \in \mathscr{B}(\ell^2(G))$.

Suppose that $T \in \mathscr{B}(\ell^2(G))$ is self-adjoint. Then the set

$$\begin{aligned}\sigma(T) &:= \{\lambda \in \mathbb{R} : (T - \lambda I) \text{ is not bijective}\}\\ &= \{\lambda \in \mathbb{R} : (T - \lambda I) \text{ has no bounded inverse}\} \subseteq \mathbb{R}\end{aligned} \tag{14.34}$$

is called the *spectrum* of T. Then the spectrum $\sigma(T)$ is a compact set satisfying

$$\sigma(T) \subseteq [-\|T\|, \|T\|\,]. \tag{14.35}$$

The quantity

$$\rho(T) := \sup_{\lambda\in\sigma(T)} |\lambda| = \max_{\lambda\in\sigma(T)} |\lambda| \tag{14.36}$$

is called the *spectral radius* of T and one has (*Gelfand formula*)

$$\rho(T) = \|T\| = \lim_{n\to\infty} \|T^n\|^{1/n}. \tag{14.37}$$

Given $g \in G$ we denote by $T_g\colon \ell^2(G) \to \ell^2(G)$ the linear map defined by setting $[T_g f](h) = f(hg)$ for all $f \in \ell^2(G)$ and $h \in G$. For all $f \in \ell^2(G)$ we have

$$\begin{aligned}\|T_g f\|_2^2 &= \sum_{h\in G} |[T_g f](h)|^2 \\ &= \sum_{h\in G} |f(hg)|^2 \\ \text{(by setting } k = hg) \quad &= \sum_{k\in G} |f(k)|^2 \\ &= \|f\|_2^2\end{aligned}$$

(that is, T_g is an *isometry*) so that

$$\|T_g\| = 1 \tag{14.38}$$

and, in particular, $T_g \in \mathscr{B}(\ell^2(G))$. Note that T_g is invertible with inverse (**exercise**)

$$(T_g)^* = T_{g^{-1}}. \tag{14.39}$$

Then, returning back to the simple random walk on G associated with the symmetric generating subset $Y \subset G$, we denote by $M_Y : \ell^2(G) \to \ell^2(G)$ the map defined by $M_Y = \frac{1}{|Y|}\sum_{y\in Y} T_y$. In other words,

$$[M_Y f](g) = \frac{1}{|Y|}\sum_{y\in Y} f(gy)$$

for all $f \in \ell^2(G)$ and $g \in G$. The map M_Y is called the *Markov operator* associated with Y.

Proposition 14.27. *The Markov operator M_Y is linear, continuous, and self-adjoint. Moreover,*

$$\|M_Y\| \leq 1. \tag{14.40}$$

Proof. Since $T_y \in \mathscr{B}(\ell^2(G))$ for all $y \in Y$, we deduce that $M_Y \in \mathscr{B}(\ell^2(G))$. Moreover, using (14.39) and recalling that Y is symmetric ($Y = Y^{-1}$) we have

$$M_Y{}^* = \left(\frac{1}{|Y|}\sum_{y\in Y} T_y\right)^* = \frac{1}{|Y|}\sum_{y\in Y} T_y{}^* = \frac{1}{|Y|}\sum_{y\in Y} T_{y^{-1}} = \frac{1}{|Y|}\sum_{y\in Y} T_y = M_Y,$$

showing that M_Y is self-adjoint. Finally, we have

$$\|M_Y\| = \|\frac{1}{|Y|}\sum_{y\in Y} T_y\| \leq \frac{1}{|Y|}\sum_{y\in Y} \|T_y\| = 1,$$

where the last equality follows from (14.38). $\square$

Let $g, h \in G$ and $n \geq 0$ and let us show that

$$\langle M_Y^n \delta_g, \delta_h\rangle = q^{(n)}(g,h). \tag{14.41}$$

We claim that $[M_Y^n \delta_g](h) = q^{(n)}(g,h)$. For $n = 0$ we have $[M_Y^0 \delta_g](h) = [I\delta_g](h) = \delta_g(h) = q^{(0)}(g,h)$ (cf. (14.28)). Moreover,

$$q(k,h) = \begin{cases} \frac{1}{|Y|} & \text{if } h^{-1}k \in Y \\ 0 & \text{otherwise} \end{cases} = \frac{1}{|Y|} \sum_{y \in Y} \delta_y(h^{-1}k) \tag{14.42}$$

so that

$$\begin{aligned}
[M_Y^{n+1}\delta_g](h) &= [M_Y(M_Y^n\delta_g)](h) \\
&= \frac{1}{|Y|} \sum_{y \in Y} [M_Y^n \delta_g](hy) \\
\text{(by induction)} &= \frac{1}{|Y|} \sum_{y \in Y} q^{(n)}(g, hy) \\
(k = hy \Leftrightarrow y = h^{-1}k) &= \frac{1}{|Y|} \sum_{k \in G} \sum_{y \in Y} q^{(n)}(g,k)\delta_y(h^{-1}k) \\
\text{(by (14.42))} &= \sum_{k \in G} q^{(n)}(g,k)q(k,h) \\
&= q^{(n+1)}(g,h),
\end{aligned}$$

proving the claim. Then (14.41) follows after observing that $\langle x, \delta_h \rangle = x(h)$ for all $x \in \ell^2(G)$ and $h \in G$. In particular,

$$\langle M_Y^n \delta_{1_G}, \delta_{1_G} \rangle = q^{(n)}(1_G, 1_G).$$

The following proposition is a particular case of Corollary 14.41 below. For its proof, we thus refer to the proof of the latter.

Proposition 14.28. $\rho(G,Y) = \|M_Y\| = \rho(M_Y)$.

Proof of Theorem 14.26 Suppose that G is amenable and let us show that $\rho(G,Y) = 1$. Note that by virtue of Proposition 14.28, it is equivalent to show that $1 \in \sigma(M_Y)$.

Fix $\varepsilon > 0$. By Proposition 14.8 there exists a finite subset $\Omega \subset G$ such that

$$|\partial_y(\Omega)| < \frac{\varepsilon^2}{2} |\Omega| \tag{14.43}$$

(Følner condition) for all $y \in Y$ (recall that $\partial_y(\Omega) = y\Omega \setminus \Omega$).

Set $f := \frac{1}{\sqrt{|\Omega|}} \mathbf{1}_{\Omega^{-1}} = \frac{1}{\sqrt{|\Omega|}} \sum_{g \in \Omega^{-1}} \delta_g \in \ell^2(G)$.

Lemma 14.29. *For every $y \in Y$ we have*

$$\|f - T_y f\|_2 \leq \varepsilon. \tag{14.44}$$

Proof. Given two finite subsets $A, B \subseteq G$ we have:

$$\begin{aligned}
\|\mathbf{1}_A - \mathbf{1}_B\|_2^2 &= \sum_{g \in G} |\mathbf{1}_A(g) - \mathbf{1}_B(g)|^2 \\
&= \sum_{g \in G} |\mathbf{1}_A(g) - \mathbf{1}_B(g)|
\end{aligned}$$

$$
\begin{aligned}
&= \sum_{\substack{g\in G:\\ \mathbf{1}_A(g)-\mathbf{1}_B(g)=1}} |\mathbf{1}_A(g)-\mathbf{1}_B(g)| \\
&\quad + \sum_{\substack{g\in G:\\ \mathbf{1}_A(g)-\mathbf{1}_B(g)=-1}} |\mathbf{1}_A(g)-\mathbf{1}_B(g)| \\
&= \sum_{g\in G} \mathbf{1}_{A\setminus B}(g) + \sum_{g\in G} \mathbf{1}_{B\setminus A}(g) \\
&= |A\setminus B| + |B\setminus A|.
\end{aligned}
$$

Now observing that $T_y\mathbf{1}_{\Omega^{-1}} = \mathbf{1}_{\Omega^{-1}y^{-1}}$ (**exercise**), taking $A=\Omega^{-1}$ and $B=\Omega^{-1}y^{-1}$, we deduce that

$$
\begin{aligned}
\|\mathbf{1}_{\Omega^{-1}} - T_y\mathbf{1}_{\Omega^{-1}}\|_2^2 &= |\Omega^{-1}\setminus\Omega^{-1}y^{-1}| + |\Omega^{-1}y^{-1}\setminus\Omega^{-1}| \\
&= |\Omega\setminus y\Omega| + |y\Omega\setminus\Omega| \\
&= 2|\partial_y(\Omega)|.
\end{aligned}
$$

After dividing by $|\Omega|$, we have $\|f - T_y f\|_2^2 = 2\frac{|\partial_y(\Omega)|}{|\Omega|}$ and (14.44) follows from (14.43). □

We have

$$
\begin{aligned}
\|(I-M_Y)f\|_2 &= \left\| f - \frac{1}{|Y|}\sum_{y\in Y} T_y f \right\|_2 \\
&= \frac{1}{|Y|}\left\| \sum_{y\in Y} (f - T_y f) \right\|_2 \\
&\le \frac{1}{|Y|}\sum_{y\in Y} \|f - T_y f\|_2 \\
\text{(by (14.44))} &\le \frac{1}{|Y|}\sum_{y\in Y} \varepsilon \\
&= \varepsilon.
\end{aligned}
$$

Since $\|f\|_2 = 1$, we deduce that $I - M_Y$ has no bounded inverse, that is, $1 \in \sigma(M_Y)$. This proves the implication: G amenable $\Longrightarrow$ $\rho(G,Y) = 1$.

To prove the reverse implication, we need a lemma, characterizing amenability of finitely generated groups.

Lemma 14.30. *Let G be a finitely generated group and let $Y \subset G$ be a finite and symmetric generating subset. Then the following conditions are equivalent:*

(a) *G is amenable;*
(b) *for every $\varepsilon > 0$ there exists a finite nonempty subset $\Omega \subset G$ such that*

$$
\frac{|\Omega\setminus y\Omega|}{|\Omega|} < \varepsilon \quad \text{for all } y \in Y. \tag{14.45}
$$

Proof. The implication (a) $\Longrightarrow$ (b) is obvious. Conversely, suppose (b). It is not restrictive to suppose that $1_G \in Y$. Let then $X \subset G$ be a finite subset and $\varepsilon' > 0$. Since Y generates G, we can find $n \in \mathbb{N}$ such that $X \subset Y^n$, so that, given $x \in X$, we can find $y_1, y_2, \ldots, y_n \in Y$ such that $x = y_1 y_2 \cdots y_n$. Set $\varepsilon := \varepsilon'/n$ and let $\Omega \subset G$ satisfying (14.45). After observing that (i) $g(A \setminus B) = gA \setminus gB$ for all subsets $A, B \subset G$ and $g \in G$ (**exercise**) and (ii) $|A \setminus B| \leq |A \setminus C| + |C \setminus B|$ for all finite subsets $A, B, C \subset G$ (**exercise**), we have

$$\begin{aligned}
|\Omega \setminus x\Omega| &= |\Omega \setminus y_1 y_2 \cdots y_n \Omega| \\
&\leq |\Omega \setminus y_1 \Omega| + |y_1 \Omega \setminus y_1 y_2 \Omega| + |y_1 y_2 \Omega \setminus y_1 y_2 y_3 \Omega| \\
&\qquad + \cdots + |y_1 y_2 \cdots y_{n-1} \Omega \setminus y_1 y_2 \cdots y_{n-1} y_n \Omega| && \text{(by (ii))} \\
&= |\Omega \setminus y_1 \Omega| + |\Omega \setminus y_2 \Omega| + |\Omega \setminus y_3 \Omega| + \cdots + |\Omega \setminus y_n \Omega| && \text{(by (i))} \\
&< n\varepsilon |\Omega| \\
&\leq \varepsilon' |\Omega|,
\end{aligned}$$

equivalently, $|\partial_x(\Omega)| < \varepsilon' |\Omega|$ for all $x \in X$. It follows from Proposition 14.8 that G is amenable. $\square$

Suppose then that $\rho(G, Y) = 1$, equivalently, $1 \in \sigma(M_Y)$.

We claim that $1 \in \sigma(M_{Y \cup \{1_G\}})$ if and only if $1 \in \sigma(M_Y)$. Indeed, if $1_G \notin Y$ we set $\alpha := \frac{|Y|}{|Y|+1}$. Note that $0 < \alpha < 1$ and that $M_{Y \cup \{1_G\}} = (1-\alpha)I + \alpha M_Y$. We then have

$$I - M_{Y \cup \{1_G\}} = \alpha(I - M_Y)$$

so that $I - M_{Y \cup \{1_G\}}$ is bijective if and only if $I - M_Y$ is bijective. This proves the claim.

As a consequence, up to replacing Y by $Y \cup \{1_G\}$, from now on we suppose that $1_G \in Y$.

Recall that the *support* of a function $f \in \mathbb{R}^G$ is the set

$$\mathrm{supp}(f) := \{g \in G : x(g) \neq 0\}.$$

We denote by $\mathbb{R}[G] \subset \mathbb{R}^G$ the vector subspace consisting of all finitely supported functions $f : G \to \mathbb{R}$. Note that $\mathbb{R}[G]$ is a dense subspace in $\ell^2(G)$. We thus have

$$\|T\| = \sup_{\substack{f \in \mathbb{R}[G] \\ \|f\|_2 \leq 1}} \|Tf\|_2 = \sup_{\substack{f \in \mathbb{R}[G] \\ f \neq 0}} \frac{\|Tf\|_2}{\|f\|_2} \tag{14.46}$$

for all $T \in \mathscr{B}(\ell^2(G))$. Moreover, for $f \in \mathbb{R}[G]$ we set

$$\|f\|_1 := \sum_{g \in G} |f(g)|.$$

Note that for all $f \in \mathbb{R}[G]$ and $g \in G$ one has $T_g f \in \mathbb{R}[G]$ (in fact $\mathrm{supp}(T_g f) = \mathrm{supp}(f) g^{-1}$ (**exercise**), and therefore $M_Y f \in \mathbb{R}[G]$ (**exercise**)) and

$$\|T_g f\|_1 = \|f\|_1 \tag{14.47}$$

(**exercise**).

The following lemma gives a description of nonnegative finitely supported functions on the group G.

Lemma 14.31. *Let $f \in \mathbb{R}[G]$ such that $f \geq 0$ and $\|f\|_1 = 1$. Then there exist an integer $n \geq 1$, nonempty finite subsets $A_i \subset G$ and real numbers $\lambda_i > 0$, for $i = 1, 2, \ldots, n$, satisfying $A_1 \supset A_2 \supset \cdots \supset A_n$ and $\lambda_1 + \lambda_2 + \cdots + \lambda_n = 1$ such that*

$$f = \sum_{i=1}^{n} \lambda_i \frac{\mathbf{1}_{A_i}}{|A_i|}. \tag{14.48}$$

Proof. Let $0 < \alpha_1 < \alpha_2 < \cdots < \alpha_n$ denote the values taken by f. For each $1 \leq i \leq n$, let us set

$$A_i = \{g \in G : f(g) \geq \alpha_i\}.$$

Clearly the sets A_i are nonempty finite subsets of G such that $A_1 \supset A_2 \supset \cdots \supset A_n$. On the other hand, we have

$$\begin{aligned} f &= \alpha_1 \mathbf{1}_{A_1} + (\alpha_2 - \alpha_1)\mathbf{1}_{A_2} + \cdots + (\alpha_n - \alpha_{n-1})\mathbf{1}_{A_n} \\ &= \lambda_1 \frac{\mathbf{1}_{A_1}}{|A_1|} + \lambda_2 \frac{\mathbf{1}_{A_2}}{|A_2|} + \cdots + \lambda_n \frac{\mathbf{1}_{A_n}}{|A_n|}, \end{aligned}$$

by setting $\lambda_1 = \alpha_1 |A_1|$ and $\lambda_i = (\alpha_i - \alpha_{i-1})|A_i|$ for $2 \leq i \leq n$. Thus $\lambda_i > 0$ for $1 \leq i \leq n$ and

$$\begin{aligned} \sum_{i=1}^{n} \lambda_i &= \alpha_1 |A_1| + (\alpha_2 - \alpha_1)|A_2| + \cdots + (\alpha_n - \alpha_{n-1})|A_n| \\ &= \alpha_1(|A_1| - |A_2|) + \alpha_2(|A_2| - |A_3|) + \cdots + \alpha_n |A_n| \\ &= \alpha_1 |A_1 \setminus A_2| + \alpha_2 |A_2 \setminus A_3| + \cdots + \alpha_{n-1}|A_{n-1} \setminus A_n| + \alpha_n |A_n| \\ &= \sum_{g \in G} f(g) = \|f\|_1 = 1. \end{aligned}$$

□

We now start the search for a suitable function $f \in \mathbb{R}[G]$, depending on $\varepsilon > 0$, for which one of the level-sets (in the sense of Lemma 14.31) will satisfy the Følner condition (14.43).

Lemma 14.32. *For every $\varepsilon > 0$ there exists an $f \in \mathbb{R}[G]$ such that $\|f\|_2 = 1$, $f \geq 0$, and*

$$\|M_Y f\|_2 \geq 1 - \varepsilon. \tag{14.49}$$

Proof. It follows from the definition of $\|M_Y\| = 1$ and the density of $\mathbb{R}[G]$ in $\ell^2(G)$ (cf. (14.46)) that we can find $f \in \mathbb{R}[G]$ such that $\|f\|_2 = 1$ and satisfies (14.49). We claim that $|f|$ also satisfies the same properties. First of all $\||f|\|_2 = \|f\|_2 = 1$. We claim that

$$\|M_Y f\|_2^2 \leq \|M_Y |f|\|_2^2. \tag{14.50}$$

Indeed,

$$\begin{aligned}
\|M_Y f\|_2^2 &= \left\| \frac{1}{|Y|} \sum_{y\in Y} T_y f \right\|_2^2 \\
&= \sum_{g\in G} \left(\frac{1}{|Y|} \sum_{y\in Y} f(gy) \right)^2 \\
&\le \sum_{g\in G} \left(\frac{1}{|Y|} \sum_{y\in Y} |f(gy)| \right)^2 \\
&= \left\| \frac{1}{|Y|} \sum_{y\in Y} T_y |f| \right\|_2^2 \\
&= \|M_Y |f|\|_2^2.
\end{aligned}$$

This proves the claim and completes the proof of the lemma. □

In order to prove the following lemma we need a result from functional analysis (we include the proof for the sake of completeness).

Proposition 14.33 (Uniform convexity of Hilbert spaces). *For every $\varepsilon > 0$, there exists a $\delta > 0$ such that*

$$\|f_1 - f_2\|_2 > \varepsilon \quad \textit{implies} \quad \left\| \frac{f_1+f_2}{2} \right\| < 1-\delta$$

for all $f_1, f_2 \in \ell^2(G)$ with $\|f_1\|, \|f_2\| \le 1$.

Proof. Let $f_1, f_2 \in \ell^2(G)$ such that $\|f_1\|, \|f_2\| \le 1$. Then, we have

$$\begin{aligned}
&\left\| \frac{f_1+f_2}{2} \right\|^2 + \frac{1}{4}\|f_1 - f_2\|^2 \\
&= \frac{1}{4}\left(\langle f_1+f_2, f_1+f_2\rangle + \langle f_1-f_2, f_1-f_2\rangle \right) \\
&= \frac{1}{4}\left(\langle f_1, f_2\rangle + \langle f_2, f_2\rangle + 2\langle f_1, f_2\rangle + \langle f_1, f_1\rangle + \langle f_2, f_2\rangle - 2\langle f_1, f_2\rangle \right) \\
&= \frac{1}{4}\left(2\|f_1\|^2 + 2\|f_2\|^2 \right) \\
&\le 1.
\end{aligned}$$

Therefore,

$$\left\| \frac{f_1+f_2}{2} \right\|^2 \le 1 - \frac{1}{4}\|f_1 - f_2\|^2. \tag{14.51}$$

Let now $\varepsilon > 0$ and set $\delta = 1 - \frac{1}{2}\sqrt{4-\varepsilon^2}$. Suppose that $\|f_1 - f_2\| > \varepsilon$. This implies $1 - \frac{1}{4}\|f_1 - f_2\|^2 < 1 - \frac{\varepsilon^2}{4} = (1-\delta)^2$. From (14.51) we then deduce that $\|\frac{f_1+f_2}{2}\| < 1-\delta$. □

Lemma 14.34. *For every $\varepsilon > 0$ there exists an $f_\varepsilon \in \mathbb{R}[G]$ such that $f_\varepsilon \geq 0$, $\|f_\varepsilon\|_2 = 1$ and*

$$\|f_\varepsilon - T_y f_\varepsilon\|_2 \leq \varepsilon \quad \textit{for all } y \in Y. \tag{14.52}$$

Proof. Suppose that the statement fails to hold, that is, there exists an $\varepsilon_0 > 0$ such that for all $f \in \mathbb{R}[G]$, $f \geq 0$, and $\|f\|_2 = 1$, there exists a $y_0 \in Y$ such that $\|f - T_{y_0} f\|_2 \geq \varepsilon_0$. Since $\ell^2(G)$ is uniformly convex (cf. Proposition 14.33) there exists a $\delta_0 > 0$ such that

$$\left\| \frac{f + T_{y_0} f}{2} \right\|_2 \leq 1 - \delta_0 \tag{14.53}$$

for all $f \in \mathbb{R}[G]$ such that $\|f\|_2 = 1 = \|T_{y_0} f\|_2$. It then follows that

$$\begin{aligned}
\|M_Y f\|_2 &= \left\| \frac{1}{|Y|} \sum_{y \in Y} T_y f \right\|_2 \\
&= \left\| \frac{2}{|Y|} \left(\frac{f + T_{y_0} f}{2} \right) + \frac{1}{|Y|} \sum_{y \in Y \setminus \{1_G, y_0\}} T_y f \right\|_2 \\
&\leq \frac{2}{|Y|} \left\| \frac{f + T_{y_0} f}{2} \right\|_2 + \frac{1}{|Y|} \sum_{y \in Y \setminus \{1_G, y_0\}} \|T_y f\|_2 \\
\text{(by (14.53))} \quad &\leq \frac{2}{|Y|}(1 - \delta_0) + \frac{1}{|Y|}(|Y| - 2) \\
&= 1 - \frac{2\delta_0}{|Y|}
\end{aligned}$$

for all $f \in \mathbb{R}[G]$, such that $\|f\|_2 = 1$ and $f \geq 0$. This clearly contradicts Lemma 14.32. □

Let $F := f_{\frac{\varepsilon}{2|Y|}} \in \mathbb{R}[G]$ as in Lemma (14.34). Thus $\|F\|_2 = 1$, and $\|F - T_y F\|_2 \leq \frac{\varepsilon}{2|Y|}$ for all $y \in Y$. Setting $f := F^2 \in \mathbb{R}[G]$, we have $\|f\|_1 = 1$, $f \geq 0$, and we claim that

$$\|f - T_y f\|_1 \leq \frac{\varepsilon}{|Y|} \quad \text{for all } y \in Y. \tag{14.54}$$

This results from the following general fact (recall (14.47)). For all $F_1, F_2 \in \ell^2(G)$ such that $F_1, F_2 \geq 0$ and $\|F_1\|_2 = \|F_2\|_2 = 1$:

$$\begin{aligned}
\|F_1^2 - F_2^2\|_1 &= \sum_{g \in G} |F_1^2(g) - F_2^2(g)| \\
&= \sum_{g \in G} |F_1(g) - F_2(g)| \cdot |F_1(g) + F_2(g)| \\
&= \langle |F_1 - F_2|, |F_1 + F_2| \rangle \\
\text{(by Cauchy–Schwarz)} \quad &\leq \|F_1 - F_2\|_2 \cdot \|F_1 + F_2\|_2 \\
&\leq \|F_1 - F_2\|_2 \cdot (\|F_1\|_2 + \|F_2\|_2) \\
&= 2\|F_1 - F_2\|_2.
\end{aligned}$$

By Lemma 14.31, there exist an integer $n \geq 1$, nonempty finite subsets $A_i \subset G$ and real numbers $\lambda_i > 0$, for $i = 1, 2, \dots, n$, satisfying $A_1 \supset A_2 \supset \cdots \supset A_n$ and $\lambda_1 + \lambda_2 + \cdots + \lambda_n = 1$, such that $f = \sum_{i=1}^n \lambda_i \frac{\mathbf{1}_{A_i}}{|A_i|}$.

Set $\Delta = \{1, 2, \dots, n\}$ and consider the unique probability measure μ on Δ such that $\mu(\{i\}) = \lambda_i$ for every $i \in \Delta$. Finally, for each $g \in G$, let Δ_g denote the subset of Δ defined by

$$\Delta_g = \left\{ i \in \Delta : \frac{|A_i \setminus A_i g|}{|A_i|} \geq \varepsilon \right\}. \tag{14.55}$$

We need a lemma.

Lemma 14.35. *With the same notation and hypotheses as in Lemma 14.31, we have*

$$\|f - T_g f\|_1 = \sum_{i=1}^n \lambda_i \frac{2|A_i \setminus A_i g|}{|A_i|} \tag{14.56}$$

for every $g \in G$.

Proof. Keeping in mind that $T_g \mathbf{1}_A = \mathbf{1}_{Ag^{-1}}$ for all $g \in G$ and $A \subset G$, equality (14.48) gives us

$$\begin{aligned} f - T_g f &= \sum_{i=1}^n \lambda_i \frac{\mathbf{1}_{A_i} - T_g \mathbf{1}_{A_i}}{|A_i|} \\ &= \sum_{i=1}^n \lambda_i \frac{\mathbf{1}_{A_i} - \mathbf{1}_{A_i g^{-1}}}{|A_i|}. \end{aligned}$$

As we observed before (cf. the proof of Lemma 14.29), the map $\mathbf{1}_{A_i} - \mathbf{1}_{A_i g^{-1}}$ takes the value 1 at each point of $A_i \setminus A_i g^{-1}$, the value -1 at each point of $A_i g^{-1} \setminus A_i$, and the value 0 everywhere else. Let us set

$$B := \bigcup_{1 \leq i \leq n} (A_i \setminus A_i g^{-1}) \quad \text{and} \quad C := \bigcup_{1 \leq i \leq n} (A_i g^{-1} \setminus A_i).$$

Note that the sets B and C are disjoint. Indeed, for all $1 \leq i, j \leq n$, we have $(A_i \setminus A_i g^{-1}) \cap (A_j \setminus A_j g^{-1}) = \varnothing$ since either $A_i \subset A_j$ or $A_j \subset A_i$ (which implies $A_j g^{-1} \subset A_i g^{-1}$).

It follows that

$$\begin{aligned} \|f - T_g f\|_1 &= \sum_{a \in G} |(f - T_g f)(a)| \\ &= \sum_{a \in G} \left| \sum_{i=1}^n \lambda_i \frac{(\mathbf{1}_{A_i} - \mathbf{1}_{A_i g^{-1}})(a)}{|A_i|} \right| \\ &= \sum_{a \in B} \left| \sum_{i=1}^n \lambda_i \frac{(\mathbf{1}_{A_i} - \mathbf{1}_{A_i g^{-1}})(a)}{|A_i|} \right| + \sum_{a \in C} \left| \sum_{i=1}^n \lambda_i \frac{(\mathbf{1}_{A_i} - \mathbf{1}_{A_i g^{-1}})(a)}{|A_i|} \right| \end{aligned}$$

$$
\begin{aligned}
&= \sum_{i=1}^{n} \lambda_i \frac{|A_i \setminus A_i g^{-1}|}{|A_i|} + \sum_{i=1}^{n} \lambda_i \frac{|A_i g^{-1} \setminus A_i|}{|A_i|} \\
&= \sum_{i=1}^{n} \lambda_i \frac{|A_i g \setminus A_i|}{|A_i|} + \sum_{i=1}^{n} \lambda_i \frac{|A_i \setminus A_i g|}{|A_i|} \\
&= \sum_{i=1}^{n} \lambda_i \frac{2|A_i \setminus A_i g|}{|A_i|}.
\end{aligned}
$$

This completes the proof of Lemma 14.35. □

It follows from the above lemma and (14.55) that

$$
\begin{aligned}
\|f - T_g f\|_1 &= \sum_{i \in \Delta} \lambda_i \frac{2|A_i \setminus A_i g|}{|A_i|} \\
&\geq \sum_{i \in \Delta_g} \lambda_i \frac{2|A_i \setminus A_i g|}{|A_i|} \\
&\geq 2\varepsilon \sum_{i \in \Delta_g} \lambda_i \\
&= 2\varepsilon \mu(\Delta_g).
\end{aligned}
$$

Therefore, we have

$$
\mu(\Delta_g) \leq \frac{\|f - T_g f\|_1}{2\varepsilon} \quad \text{for all } g \in G.
$$

By using (14.54), we deduce

$$
\mu(\Delta_y) < \frac{1}{|Y|} \quad \text{for all } y \in Y,
$$

which implies

$$
\mu\left(\bigcup_{y \in Y} \Delta_y\right) \leq \sum_{y \in Y} \mu(\Delta_y) < 1.
$$

Thus

$$
\bigcup_{y \in Y} \Delta_y \neq \Delta.
$$

This means that there is some $i_0 \in \Delta \setminus \cup_{y \in Y} \Delta_y$ such that

$$
\frac{|A_{i_0} \setminus A_{i_0} y|}{|A_{i_0}|} < \varepsilon \quad \text{for all } y \in Y.
$$

Thus, in order to satisfy (14.45), we can take $\Omega := A_{i_0}^{-1}$ (recall that $Y = Y^{-1}$).

This ends the proof of Theorem 14.26. □

Kesten's Amenability Criterion for Countable Groups

In this section we prove Kesten's amenability criterion for countable (not necessarily finitely generated) groups. We closely follow the presentation from the original source [196], which is extremely clear.

Let G be a countable group. Let I be a countable (possibly finite) index set. We say that a family $A = (a_i)_{i\in I}$ of elements of G constitutes a *generating system* for G if $\{a_i : i \in I\}$ generates G, that is, for every $g \in G$ there exist $i_1, i_2, \dots, i_k \in I$ and $\varepsilon_1, \varepsilon_2, \dots, \varepsilon_k \in \{-1, 1\}$ such that $g = a_{i_1}^{\varepsilon_1} a_{i_2}^{\varepsilon_2} \cdots a_{i_k}^{\varepsilon_k}$. If I is finite, we say that the generating system A is *finite*.

Let $A = (a_i)_{i\in I}$ be a generating system for G. Let also $(p_i)_{i\in I}$ denote a family of non-negative real numbers such that $2\sum_{i\in I} p_i = 1$: we shall call it a *probability distribution* on the generating system A. If $p_i > 0$ for all $i \in I$ we say that the probability distribution $(p_i)_{i\in I}$ is *strictly positive*.

Consider the random walk on G in which every step consists of right multiplication by a_i or its inverse a_i^{-1}, each with probability p_i. **Warning:** this does not mean that p_i is the *total* probability of multiplying on the right by the (unique) element $g \in G$ such that $g = a_i$ in G. For instance, if $a_i = a_j$ with $i \neq j$ (resp. $a_i = a_i^{-1}$), then the total probability of multiplying by $g = a_i$ is at least $p_i + p_j$ (resp. $2p_i$).

For $g \in G$ we then denote by

$$p_g = \sum_{\substack{i\in I\\ a_i = g}} p_i + \sum_{\substack{i\in I\\ a_i = g^{-1}}} p_i \tag{14.57}$$

the *total* probability of multiplying by g, so that

$$\sum_{g\in G} p_g = 2\sum_{i\in I} p_i = 1. \tag{14.58}$$

This defines a random walk $(G,P) = (G,P,\delta_{1_G})$ (cf. Definition 13.16) with one-step transition probability matrix $P = (p(g,h))_{g,h\in G}$ given by

$$p(g,h) = p_{g^{-1}h}$$

(the probability that h is reached in one step from g in the random walk) for all $g, h \in G$. It follows immediately from (14.57) that $p_g = p_{g^{-1}}$ for all $g \in G$ so that P is symmetric, that is, $p(g,h) = p(h,g)$ for all $g, h \in G$.

The following lemma was originally proved by Issai Schur in [306]. We thank Florin Boca for providing us with the proof below.

Lemma 14.36. *Let $\mathscr{H}$ be a real Hilbert space and let $M\colon \mathscr{H} \to \mathscr{H}$ be a self-adjoint operator. Let $(\delta_i)_{i\in I}$ be a Hilbert space basis for $\mathscr{H}$ and set $m_{i,j} := \langle \delta_i, M\delta_j\rangle = \langle M\delta_i, \delta_j\rangle$ for all $i, j \in I$. Then*

$$\|M\| \leq \sup_{i\in I} \sum_{j\in I} |m_{i,j}|. \tag{14.59}$$

Proof. We first observe that since M is self-adjoint one has

$$\|M\| = \sup_{\|h\|=1} |\langle Mh, h\rangle|. \tag{14.60}$$

Set

$$C := \sup_{i\in\mathbb{N}} \sum_{j\in I} |m_{i,j}| = \sup_{j\in I} \sum_{i\in I} |m_{i,j}|.$$

Let $h \in H$ with $\|h\| = 1$ and write $h = \sum_{j\in I} h_j \delta_j$, so that $\sum_{j\in I} |h_j|^2 = 1$ and $Mh = \sum_{j\in I} \left(\sum_{i\in I} m_{i,j} h_i\right) \delta_j$. We then have

$$\begin{aligned}
|\langle Mh, h\rangle| &= |\sum_{j\in I} \left(\sum_{i\in I} m_{i,j} h_i\right) h_j| \\
&\le \sum_{j\in I}\sum_{i\in I} |m_{i,j}| \cdot |h_i| \cdot |h_j| \\
&= \sum_{j\in I}\sum_{i\in I} |m_{i,j}|^{\frac{1}{2}} |h_i| \cdot |m_{i,j}|^{\frac{1}{2}} |h_j| \\
\text{(by Cauchy–Schwarz)} &\le \sum_{j\in I} \left(\sum_{i\in I} |m_{i,j}| \cdot |h_i|^2\right)^{\frac{1}{2}} \cdot \left(\sum_{i\in I} |m_{i,j}|\right)^{\frac{1}{2}} |h_j| \\
&\le \sum_{j\in I} \left(\sum_{i\in I} |m_{i,j}| \cdot |h_i|^2\right)^{\frac{1}{2}} \cdot C^{\frac{1}{2}} \cdot |h_j| \\
\text{(by Cauchy–Schwarz)} &\le C^{\frac{1}{2}} \cdot \left(\sum_{j\in I} \left(\sum_{i\in I} |m_{i,j}| \cdot |h_i|^2\right)\right)^{\frac{1}{2}} \cdot \left(\sum_{j\in I} |h_j|^2\right)^{\frac{1}{2}} \\
\text{(since } \|h\| = 1) &\le C^{\frac{1}{2}} \left(\sum_{j\in I}\sum_{i\in I} |m_{i,j}| \cdot |h_i|^2\right)^{\frac{1}{2}} \\
&= C^{\frac{1}{2}} \left(\sum_{i\in I} \left(\sum_{j\in I} |m_{i,j}|\right) \cdot |h_i|^2\right)^{\frac{1}{2}} \\
&\le C^{\frac{1}{2}} \left(\sum_{i\in I} C \cdot |h_i|^2\right)^{\frac{1}{2}} \\
&= C^{\frac{1}{2}} \cdot C^{\frac{1}{2}} \cdot \left(\sum_{i\in I} |h_i|^2\right)^{\frac{1}{2}} \\
\text{(since } \|h\| = 1) &= C.
\end{aligned}$$

From (14.60) we conclude that $\|M\| \le C$. □

Proposition 14.37. *The linear operator* $M = M(G, A, P) \colon \ell^2(G) \to \ell^2(G)$ *defined by*

$$[Mf](g) = \sum_{h\in G} p(g, h) f(h) \tag{14.61}$$

for all $f \in \ell^2(G)$ *is self-adjoint. Moreover,*

$$\|M\| \le \sup_{g} \sum_{h \in G} p(g,h) = 1.$$

Proof. For $g \in G$ we denote, as usual, by $\delta_g \in \ell^2(G)$ the *Dirac function based at g*. We have

$$m_{g,h} := \langle \delta_g, M\delta_h \rangle = \sum_{k \in G} \delta_g(k)[M\delta_h](k) = [M\delta_h](g) = \sum_{t \in G} p(g,t)\delta_h(t) = p(g,h).$$

Since $P = (p(g,h))_{g,h \in G}$ is symmetric, we have $\langle \delta_g, M\delta_h \rangle = p(g,h) = p(h,g) = \langle \delta_h, M\delta_g \rangle = \langle M\delta_g, \delta_h \rangle$ so that M is self-adjoint. By applying Lemma 14.36 and (14.58) we deduce that

$$\|M\| \le \sup_{g} \sum_{h \in G} m_{g,h} = \sup_{g} \sum_{h \in G} p(g,h) = 1. \qquad \square$$

The operator $M = M(G,A,P) \in \mathscr{B}(\ell^2(G))$ is called the *Markov operator* associated with the symmetric random walk on G defined by the probability distribution $(p_i)_{i \in I}$ on the generating system A.

Given $g, h \in G$ we have seen that $\langle \delta_g, M\delta_h \rangle = p(g,h)$. More generally (**exercise**) for $n \ge 0$

$$\langle M^n \delta_g, \delta_h \rangle = p^{(n)}(g,h) \tag{14.62}$$

for all $g, h \in G$, where we recall that $P^n = \left(p^{(n)}(g,h)\right)_{g,h \in G}$ denotes the n-th power of P, so that (14.62) represents the probability that, starting at g, the random walker arrives at h at the n-th step. As we have seen (see Section 13.5),

$$m_n = m_n(G,A,P) := \langle M^n \delta_g, \delta_g \rangle = p^{(n)}(g,g)$$

is independent of $g \in G$.

Remark 14.38. It is important to observe that the sequence $\left(p^{(n)}(1_G, 1_G)^{\frac{1}{n}}\right)_{n \in \mathbb{N}}$ may not converge. This is the case, for instance, for the simple random walk on G with respect to a finite symmetric generating subset $Y \subset G$ such that the corresponding Cayley graph $\mathrm{Cay}_Y(G)$ is bipartite. Recall that a graph $\mathscr{G} = (V,E)$ is termed *bipartite* if there is a partition $V = A \sqcup B$ of the vertex set and $E \subseteq A \times B$ (in other words, initial and terminal vertices of any edge lie either in A and B or in B and A, respectively). The basic example of the simple random walk on $\mathbb{Z}$ (cf. Section 13.6) gives $p^{(2n+1)}(1_G, 1_G) = 0$ for all $n \in \mathbb{N}$ and $\lim_{n \to \infty} p^{(2n)}(1_G, 1_G)^{\frac{1}{2n}} = 1$ (see (13.25)).

Here are some basic properties of the sequence $\left(p^{(2n)}(1_G, 1_G)\right)_{n \in \mathbb{N}}$ of probabilities of returning back to 1_G in even steps.

Lemma 14.39. *The sequence* $\left(p^{(2n)}(1_G, 1_G)\right)_{n \in \mathbb{N}}$ *is non-increasing.*

Moreover, $p^{(2n)}(1_G, g) \le p^{(2n)}(1_G, 1_G)$ *for all* $g \in G$.

Proof. Let $n \in \mathbb{N}$. We have

$$\begin{aligned}
p^{(2(n+1))}(1_G,1_G) &= \langle M^{2n+2}\delta_{1_G},\delta_{1_G}\rangle \\
&= \langle M^n\delta_{1_G},M^{n+2}\delta_{1_G}\rangle \\
&\le \|M^n\delta_{1_G}\|\cdot\|M^2M^n\delta_{1_G}\| \\
&\le \|M^n\delta_{1_G}\|^2 \\
&= \langle M^n\delta_{1_G},M^n\delta_{1_G}\rangle \\
&= p^{(2n)}(1_G,1_G).
\end{aligned}$$

This shows the first statement. Moreover

$$p^{(2n)}(1_G,g) = \langle M^{2n}\delta_{1_G},\delta_g\rangle \le \|M^{2n}\delta_{1_G}\| = \langle M^{2n}\delta_{1_G},\delta_{1_G}\rangle = p^{(2n)}(1_G,1_G).$$

This ends the proof of Lemma 14.39. □

The spectral radius of the Markov operator M, equivalently (as M is self-adjoint), its norm $\|M\|$ is called the *spectral radius* of the random walk (G,A,P) and it is denoted by $\rho(G,A,P)$.

Let $E(\mathrm{d}t)$ denote the *resolution of the identity* of M (a measure taking values into projections in $\mathscr{B}(\ell^2(G))$. Then $\eta(\mathrm{d}t) := \langle E(\mathrm{d}t)\delta_{1_G},\delta_{1_G}\rangle$, called the *spectral measure* of M, is a probability measure on the interval $[-1,1]$ such that, for all continuous functions $f\colon [0,1]\to\mathbb{R}$, one has (*spectral theorem*)

$$\langle f(M)\delta_{1_G},\delta_{1_G}\rangle = \int_{-1}^{1} f(t)\eta(\mathrm{d}t).$$

In particular,

$$\langle M^n\delta_{1_G},\delta_{1_G}\rangle = \int_{-1}^{1} t^n\eta(\mathrm{d}t) \tag{14.63}$$

for all $n\ge 0$.

Theorem 14.40. *Let $\rho(G,A,P)$ denote the spectral radius of the random walk (G,A,P). Then $p^{(2n)}(1_G,1_G)\le\rho(G,A,P)^{2n}$ for all $n\in\mathbb{N}$ and*

$$\lim_{n\to\infty} p^{(2n)}(1_G,1_G)^{\frac{1}{2n}} = \rho(G,A,P). \tag{14.64}$$

Proof. The existence of the limit follows from Fekete's lemma (Lemma 7.13). Indeed, $p^{(2(n+m))}(1_G,1_G)\ge p^{(2n)}(1_G,1_G)p^{(2m)}(1_G,1_G)$ since the probability of returning back to 1_G at time $2(n+m)=2n+2m$ is at least the probability of returning back to 1_G at time $2n$ and of returning back again to 1_G at time $2n+2m$. Setting $a_n := p^{(2n)}(1_G,1_G)^{-1}$ we thus have $a_{n+m}\le a_na_m$, so that the limit $\ell := \lim_{n\to\infty} a_n^{\frac{1}{n}} = 1/\lim_{n\to\infty} p^{(2n)}(1_G,1_G)^{\frac{1}{n}}$ exists and equals $\inf_{n\ge 1} a_n^{\frac{1}{n}}$. We thus have $\rho := \lim_{n\to\infty} p^{(2n)}(1_G,1_G)^{\frac{1}{2n}} = 1/\sqrt{\ell}$, and $\rho\ge p^{(2n)}(1_G,1_G)^{\frac{1}{2n}}$, equivalently, $\rho^{2n}\ge p^{(2n)}(1_G,1_G)$ for all $n\in\mathbb{N}$.

In order to show that $\rho=\rho(G,A,P)$, we observe that, by virtue of the spectral theorem,

$$\left(p^{(2n)}(1_G,1_G)\right)^{\frac{1}{2n}} = \left(\langle M^n \delta_{1_G}, \delta_{1_G}\rangle\right)^{\frac{1}{2n}} = \left(\int_{-1}^{1} t^n \eta(\mathrm{d}t)\right)^{\frac{1}{2n}}. \tag{14.65}$$

We leave it as an **exercise** to show that, as $n \to \infty$, the RHS of (14.65) tends to $\max\{|t| : t \in \sigma(M)\} = \rho(G,A,P)$. □

Corollary 14.41. *Denoting by $G(z) := \sum_{n\in\mathbb{N}} p^{(n)}(1_G,1_G)z^n$ the corresponding Green function, and denoting by R its radius of convergence, we have*

$$\rho(G,A,P) = \|M\| = \lim_{n\to\infty} \sqrt[2n]{p^{(2n)}(1_G,1_G)} = \limsup_{n\to\infty} \sqrt[n]{p^{(n)}(1_G,1_G)} = \frac{1}{R} \leq 1. \tag{14.66}$$

Proof. This is just a combination and reformulation of (14.64) in the previous theorem and Theorem 14.37. □

Lemma 14.42. *Let $\mathscr{H}$ be a real Hilbert space. Let $A,B \in \mathscr{B}(\mathscr{H})$ and suppose that they are self-adjoint. For $\xi \in \mathbb{R}$, let us set*

$$\rho_\xi := \rho(\xi A + (1-\xi)B),$$

where ρ denotes the spectral radius.

Then the real map $\xi \mapsto \rho_\xi$ is a convex function.

Proof. For $0 \leq \eta \leq 1$ and $\xi_1, \xi_2 \in \mathbb{R}$ we have

$$\begin{aligned}
\rho_{\eta\xi_1+(1-\eta)\xi_2} &= \rho((\eta\xi_1 + (1-\eta)\xi_2)A + (1-(\eta\xi_1+(1-\eta)\xi_2))B) \\
&= \rho(\eta\xi_1 A + (1-\eta)\xi_2 A + (1-\eta\xi_1)B - (1-\eta)\xi_2 B) \\
&= \rho\left((\eta\xi_1 A + \eta(1-\xi_1)B) + ((1-\eta)\xi_2 A + (1-\eta)(1-\xi_2)B)\right) \\
&= \|(\eta\xi_1 A + \eta(1-\xi_1)B) + ((1-\eta)\xi_2 A + (1-\eta)(1-\xi_2)B)\| \\
&\leq \|\eta\xi_1 A + \eta(1-\xi_1)B\| + \|(1-\eta)\xi_2 A + (1-\eta)(1-\xi_2)B)\| \\
&= \eta\|\xi_1 A + (1-\xi_1)B\| + (1-\eta)\|\xi_2 A + (1-\xi_2)B)\| \\
&= \eta\rho(\xi_1 A + (1-\xi_1)B) + (1-\eta)\rho(\xi_2 A + (1-\xi_2)B) \\
&= \eta\rho_{\xi_1} + (1-\eta)\rho_{\xi_2}.
\end{aligned}$$

This completes the proof of Lemma 14.42 □

Suppose now that $N \lhd G$ is a normal subgroup of G and denote by $\overline{G} := G/N$ the corresponding quotient group. Then setting $\overline{g} := gN \in \overline{G}$ for all $g \in G$, we have that $\overline{A} = A/N := (\overline{a}_i)_{i\in I}$ is a generating system for $\overline{G}$. If $(p_i)_{i\in I}$ is a probability distribution on A, then we regard it as a probability distribution on $\overline{A}$ in the obvious way.

The total probability of multiplying by $\overline{g} \in \overline{G}$ is given by (cf. (14.57))

$$\overline{p}_{\overline{g}} = \sum_{\overline{a_i}=\overline{g}} p_i + \sum_{\overline{a_i}=\overline{g}^{-1}} p_i \geq p_g.$$

This way, denoting the corresponding random walk by $(\overline{G}, \overline{P})$, where

$$\overline{P} = P/N := \left(p(\overline{g},\overline{h})\right)_{\overline{g},\overline{h}\in\overline{G}},$$

we have

$$p^{(n)}(g,h) \leq \overline{p}^{(n)}(\overline{g},\overline{h})$$

for all $g,h \in G$ and $n \in \mathbb{N}$. In particular,

$$p^{(n)}(1_G,1_G) \leq \overline{p}^{(n)}(\overline{1_G},\overline{1_G}) = \overline{p}^{(n)}(1_{\overline{G}},1_{\overline{G}}), \tag{14.67}$$

which can be interpreted as follows: in the random walk on G, the probability of returning back to 1_G at the n-th step (given that one starts at 1_G) is less than or equal to the probability of reaching some element in N at the n-th step (given that one starts at 1_G).

Lemma 14.43. *Let N be a normal subgroup of G and let A be a generating system for G. Then $\rho(G,A,P) \leq \rho(\overline{G},\overline{A},\overline{P})$.*

Proof. Using (14.66) and (14.67) we have

$$\rho(G,A,P) = \limsup_{n\to\infty} \sqrt[n]{p^{(n)}(1_G,1_G)} \leq \limsup_{n\to\infty} \sqrt[n]{\overline{p}^{(n)}(1_{\overline{G}},1_{\overline{G}})} = \rho(\overline{G},\overline{A},\overline{P}).$$

This proves Lemma 14.43. □

Lemma 14.44. *Let i_0 be a new index (not contained in I) and set $I' := \{i_0\} \sqcup I$. Then $A' = (a'_i)_{i\in I'}$, defined by $a'_{i_0} = 1_G$ and $a'_i = a_i$ for all $i \in I$, is a generating system for G. Also, for $\xi \in [0,1]$ denote by $P'(\xi)$ the probability distribution on A' defined by setting*

$$p'_i = \begin{cases} \xi/2 & \text{if } i = i_0 \\ (1-\xi)p_i & \text{otherwise} \end{cases} \tag{14.68}$$

for all $i \in I'$. Then

$$\rho(G,A',P'(\xi)) = \xi + (1-\xi)\rho(G,A,P). \tag{14.69}$$

Proof. We have $p'(g,g) = \xi + (1-\xi)p(g,g)$ and $p'(g,h) = (1-\xi)p(g,h)$ for all $g,h \in G$ with $g \neq h$, that is,

$$M(G,A',P'(\xi)) = \xi I + (1-\xi)M(G,A,P).$$

Let $\lambda \in [-1,1]$. We have

$$\begin{aligned}
\lambda \in \sigma(M(G,A',P'(\xi)) &\Leftrightarrow \lambda \in \sigma(\xi I + (1-\xi)M(G,A,P)) \\
&\Leftrightarrow (\xi - \lambda)I + (1-\xi)M(G,A,P) \text{ is not bijective} \\
&\Leftrightarrow (1-\xi)\left(\frac{\xi-\lambda}{1-\xi}I + M(G,A,P)\right) \text{ is not bijective} \\
&\Leftrightarrow \frac{\xi-\lambda}{1-\xi}I + M(G,A,P) \text{ is not bijective}
\end{aligned}$$

$$\begin{aligned}
&\Leftrightarrow -\frac{\xi-\lambda}{1-\xi} \in \sigma(M(G,A,P)) \\
&\Leftrightarrow \lambda - \xi \in (1-\xi)\sigma(M(G,A,P)) \\
&\Leftrightarrow \lambda \in \xi + (1-\xi)\sigma(M(G,A,P)).
\end{aligned}$$

This shows that $\sigma(M(G,A',P'(\xi)) = \xi + (1-\xi)\sigma(M(G,A,P))$ (a result which can be immediately deduced once the reader is familiar with the continuous functional calculus) so that

$$\begin{aligned}
\rho(G,A',P'(\xi)) = \sup\sigma(M(G,A',P'(\xi)) &= \sup\left(\xi + (1-\xi)\sigma(M(G,A,P))\right) \\
&= \xi + (1-\xi)\rho(G,A,P).
\end{aligned}$$

This proves Lemma 14.44. □

Lemma 14.45. *Let H be a countable group, let $B = (b_j)_{j\in\mathbb{N}}$ be a generating system for H, and let $(q_j)_{j\in\mathbb{N}}$ be a strict probability distribution on the generating system B. Then for every $\varepsilon > 0$ there exists a $k_\varepsilon \in \mathbb{N}$ such that if $(q_{k,j})_{j\in\mathbb{N}}$, $k \in \mathbb{N}$, is the probability distribution on B defined by*

$$q_{k,j} = \begin{cases} q_j \left(2\sum_{i\leq k} q_i\right)^{-1} & \text{for } j \leq k \\ 0 & \text{for } j > k \end{cases} \tag{14.70}$$

then, for $k \geq k_\varepsilon$,

$$|\rho(H,B,Q) - \rho(H,B,Q_k)| \leq \varepsilon. \tag{14.71}$$

Proof. For $k \in \mathbb{N}$ let us set $M_k := M(H,B,Q) - M(H,B,Q_k) \in \mathscr{B}(\ell^2(H))$. We claim that, for every row, the sum of the absolute values of the entries of M_k in that row tends to zero as $k \to \infty$. Indeed, setting $t(k) := \sum_{i\leq k} 2q_i$ we have that for every row this sum is equal to

$$\sum_{i\in\mathbb{N}} (2q_i - 2q_{k,i}) = \sum_{i\leq k} (2q_i - 2q_{k,i}) + \sum_{i>k} (2q_i) = t(k)\left(1 - \frac{1}{t(k)}\right) + (1 - t(k))$$

and the claim follows since $\lim_{k\to\infty} t(k) = 1$. Then (14.71) follows from Lemma 14.36 and (14.66). □

Note that if, for $k \in \mathbb{N}$, we denote by $H_k \leq H$ the subgroup generated by the elements b_j, $j \leq k$, and by $B_k := (b_j)_{j=1}^k$ the corresponding finite generating system, then, keeping in mind the notation in (14.70), we have

$$\rho(H,B,Q_k) = \rho(H_k,B_k,Q_k). \tag{14.72}$$

Theorem 14.46. *Let G and $H \leq G$ be two groups with generating systems A and B and probability distributions P and Q, respectively. Denote by N the normal closure of H in G, that is, the smallest normal subgroup $N \lhd G$ such that $H \leq N$, and set $\overline{G} := G/N$. Denote then by $\overline{A} = A/N$ and $\overline{P} = P/N$ the corresponding generating system for $\overline{G}$ and probability distribution on $\overline{A}$, respectively. Suppose that P is strictly positive and that*

$$\rho(H,B,Q) < 1. \tag{14.73}$$

Then

$$\rho(\overline{G},\overline{A},\overline{P}) > \rho(G,A,P). \tag{14.74}$$

Proof. We claim that we may restrict ourselves to the case where B is finite (and therefore H is finitely generated). Indeed, if B is not finite, we can replace Q by a probability distribution Q_k such that $\rho(H_k,B_k,Q_k) < 1$ (cf. Lemma 14.45 and (14.72)), where $H_k \leq H$ denotes the subgroup generated by B_k. If N_k is the normal closure of H_k in G, then $N_k \leq N$ and thus

$$\overline{G} = G/N \cong \frac{G/N_k}{N/N_k}$$

so that, by virtue of Lemma 14.43, it would suffice to show that

$$\rho(G/N_k,A/N_k,P/N_k) > \rho(G,A,P).$$

This proves the claim. We thus assume that $B = (b_j)_{j\in J}$ is finite. Fix $\xi \in (0,1)$ and let A' and $P' = P'(\xi)$ denote the generating system of G with the probability distribution on A' as in Lemma 14.44. It follows from Lemma 14.44 (**exercise**) that inequality (14.74) is equivalent to

$$\rho(\overline{G},\overline{A'},\overline{P'}) > \rho(G,A',P'). \tag{14.75}$$

Now, every $b_j \in B$ can be written (not necessarily in a unique way) in the form $a_{i_1}^{\varepsilon_1}a_{i_2}^{\varepsilon_2}\cdots a_{i_{\ell_j}}^{\varepsilon_{\ell_j}}$. For each $j \in J$, let us fix such a word, say w_j, and set

$$\ell := \max_{j\in J} \ell(w_j) = \max_{j\in J} \ell_j.$$

We also set

$$M^\ell(G) := \left(M(G,A',P')\right)^\ell \text{ and } M^\ell(\overline{G}) := \left(M(\overline{G},\overline{A'},\overline{P'})\right)^\ell. \tag{14.76}$$

Let us set $K := \bigcup_{m=0}^{\ell} \left(\{0,1,\ldots,m\} \times I^{\ell-m} \times \{-1,1\}^{\ell-m}\right)$ (here I is the index set of $A = (a_i)_{i\in I}$) and denote by $C = (c_k)_{k\in K}$ the generating system for G defined by $c_k = 1_G^m a_{i_1}^{\varepsilon_1}a_{i_2}^{\varepsilon_2}\cdots a_{i_{\ell-m}}^{\varepsilon_{\ell-m}}$ for $k = (m;i_1,i_2,\ldots,i_{\ell-m};\varepsilon_1,\varepsilon_2,\ldots,\varepsilon_{\ell-m}) \in K$. We define a probability distribution $(r_k)_{k\in K}$ on C by setting

$$r_k = \binom{\ell}{m}\xi^m \prod_{j=1}^{\ell-m} p'_{i_j} \text{ for all } k = (m;i_1,i_2,i_{\ell-m};\varepsilon_1,\varepsilon_2,\ldots,\varepsilon_{\ell-m}) \in K. \tag{14.77}$$

Now, ℓ consecutive steps in the random walk on G defined by P' amount to right multiplication by an element of C or its inverse, with probability assigned by R. We thus have

$$M^\ell(G) = M(G,C,R) \text{ and } M^\ell(\overline{G}) = M(\overline{G},\overline{C},\overline{R}). \tag{14.78}$$

As a consequence,

$$\rho(G,C,R)=\rho(G,A',P')^{\ell} \text{ and } \rho(\overline{G},\overline{C},\overline{R})=\rho(\overline{G},\overline{A'},\overline{P'})^{\ell}$$

and we only need to show that

$$\rho(\overline{G},\overline{C},\overline{R})>\rho(G,C,R). \tag{14.79}$$

The finite generating system $C'=(1_G^{\ell-\ell_j}w_j)_{j\in J}$ is a subsystem of C and we define a probability distribution S on C' by setting

$$\mathbb{P}[1_G^{\ell-\ell_j}w_j \mid S]=\mathbb{P}[b_j \mid Q]=q_j$$

for $j=1,2,\ldots,k$, and

$$\mathbb{P}[1_G^m a_{i_1}^{\varepsilon_1}a_{i_2}^{\varepsilon_2}\cdots a_{i_{\ell-m}}^{\varepsilon_{\ell-m}} \mid S]=0$$

for all $1_G^m a_{i_1}^{\varepsilon_1}a_{i_2}^{\varepsilon_2}\cdots a_{i_{\ell-m}}^{\varepsilon_{\ell-m}}\in C\setminus C'$. Since P is a strictly positive distribution and $\xi>0$, we have $\mathbb{P}[c\mid R]>0$ for all $c\in C$ so that, setting

$$\alpha:=\left(\max_{c\in C'}\frac{\mathbb{P}[c\mid S]}{\mathbb{P}[c\mid R]}-1\right)^{-1},$$

one has

$$(1+\alpha)\mathbb{P}[c\mid R]-\alpha\mathbb{P}[c\mid S]\geq 0 \tag{14.80}$$

for all $c\in C$ (equivalently, for all $c\in C'$). For $\eta\in[0,1]$ define the probability distribution $T(\eta)$ on C by setting

$$\begin{aligned}\mathbb{P}[c\mid T(\eta)]:&=(1-\eta)\left((1+\alpha)\mathbb{P}[c\mid R]-\alpha\mathbb{P}[c\mid S]\right)+\eta\mathbb{P}[c\mid S]\\ &=(1-\eta)(1+\alpha)\mathbb{P}[c\mid R]+(\eta+\alpha(1-\eta))\mathbb{P}[c\mid S]\end{aligned} \tag{14.81}$$

for all $c\in C$. Note that $T(1)=S$, so that the random walk defined by C and $T(1)$ on G is the same as the one defined by C and S, which, in turn, is the random walk defined by B and Q on H, so that

$$\rho(G,C,T(1))=\rho(G,C,S)=\rho(H,B,Q)<1. \tag{14.82}$$

Since $H\leq N$, multiplication by an element in H amounts, in $\overline{G}=G/N$, to multiplication by the identity. Therefore (taking $\xi=1$ in Lemma 14.44)

$$\rho(\overline{G},\overline{C},\overline{T(1)})=\rho(\overline{G},\overline{C},\overline{S})=1, \tag{14.83}$$

so that, by Lemma 14.44,

$$\rho(\overline{G},\overline{C},\overline{T(\eta)})=\eta\rho(\overline{G},\overline{C},\overline{T(1)})+(1-\eta)\rho(\overline{G},\overline{C},\overline{T(0)})=\eta+(1-\eta)\rho(\overline{G},\overline{C},\overline{T(0)}). \tag{14.84}$$

On the other hand, first applying Lemma 14.42, we have

$$\begin{aligned}\rho(G,C,T(\eta)) &\leq (1-\eta)\rho(G,C,T(0))+\eta\rho(G,C,T(1)) \\ \text{(by (14.82))} &< (1-\eta)\rho(G,C,T(0))+\eta \\ \text{(by Lemma 14.43)} &\leq (1-\eta)\rho(\overline{G},\overline{C},\overline{T(0)})+\eta \\ \text{(by (14.84))} &= \rho(\overline{G},\overline{C},\overline{T(\eta)}).\end{aligned}$$

In particular, for $\eta=\alpha/(1+\alpha)$ one has $(1-\eta)(1+\alpha)=1$ and $\eta-\alpha(1-\eta)=0$ so that (cf. the RHS of (14.81)) $T(\alpha/(1+\alpha))=R$, and therefore

$$\begin{aligned}\rho(G,A',P')^{\ell} &= \rho(G,C,R)=\rho(G,C,T(\alpha/(1+\alpha))) \\ &< \rho(\overline{G},\overline{C},\overline{T(\alpha/(1+\alpha))})=\rho(\overline{G},\overline{C},\overline{R})=\rho(\overline{G},\overline{A'},\overline{P'})^{\ell}.\end{aligned}$$

Thus $\rho(G,A',P')<\rho(\overline{G},\overline{A'},\overline{P'})$, and this completes the proof. □

Actually, we can estimate the increase of the spectral radius:

$$\begin{aligned}\rho(\overline{G},\overline{C},\overline{R})-\rho(G,C,R) &= \rho(\overline{G},\overline{C},\overline{T(\alpha/(1+\alpha))})-\rho(G,C,T(\alpha/(1+\alpha))) \\ &\geq \frac{1}{1+\alpha}\left(\rho(\overline{G},\overline{C},\overline{T(0)})-\rho(G,C,T(0))\right) \\ &\quad+\frac{\alpha}{1+\alpha}(1-\rho(G,C,T(1))) \\ &\geq \frac{\alpha}{1+\alpha}(1-\rho(G,C,S)) \\ &= \frac{\alpha}{1+\alpha}(1-\rho(H,B,Q))\end{aligned}$$

so that, using the formula $x^{\ell}-y^{\ell}=(x-y)(x^{\ell-1}+x^{\ell-2}y+\cdots+xy^{\ell-2}+y^{\ell-1})$ and the fact that $0<\rho(G,A,P),\rho(G,A',P')\leq 1$, we have

$$\begin{aligned}\rho(\overline{G},\overline{A},\overline{P})-\rho(G,A,P) &= \rho(\overline{G},\overline{A'},\overline{P'})-\rho(G,A',P') \\ &\geq \frac{\alpha}{(1+\ell)(1+\alpha)}(1-\rho(H,B,Q)).\end{aligned}$$

Corollary 14.47. *Suppose that the spectral radius of the symmetric random walk on a group G defined by a strictly positive distribution P on a generating system A equals* 1 *(i.e.,* $\rho(G,A,P)=1$*). Then the spectral radius of the symmetric random walk on any subgroup* $H\leq G$ *defined by a positive distribution Q on a generating system B also equals* 1 *(i.e.,* $\rho(H,B,Q)=1$*). In particular, if* $\tilde{P}$ *is any (not necessarily strict) positive distribution on a generating system* $\tilde{A}$ *for G, then the spectral radius of the associated random walk also equals* 1 *(i.e.,* $\rho(G,\tilde{A},\tilde{P})=1$*).*

Proof. Suppose that $\rho(H,B,Q)<1$. Then, denoting by N the normal closure of H in G and $\overline{G}:=G/N$ (resp. $\overline{A}$, resp. $\overline{P}$) the corresponding quotient group, (resp. generating system, resp. positive distribution) we would have, by Theorem 14.46 and (14.66)

$$\rho(G,A,P)<\rho(\overline{G},\overline{A},\overline{P})\leq 1,$$

a contradiction.

Taking $H=G$, $B=\tilde{A}$, and $Q=\tilde{P}$, one then deduces that $\rho(G,\tilde{A},\tilde{P})=1$. □

As a consequence of Corollary 14.47, we shall write $\rho(G) = 1$ or $\rho(G) < 1$ without further specification of the generating system and the corresponding probability distribution. However, we stress the fact that $\rho(G,A,P) = 1$ implies $\rho(G,\tilde{A},\tilde{P}) = 1$ only for P *strictly* positive. Thus, $\rho(G) = 1$ (resp. $\rho(G) < 1$) means the following: *there exists (resp. does not exist) a strictly positive distribution on A such that* $\rho(G,A,P) = 1$.

Theorem 14.48. *Let G be a countable group, A a generating system, and P a (not necessarily strict) positive distribution on A. Let $N \lhd G$ be a normal subgroup and suppose that $\rho(N) = 1$. Then $\rho(\overline{G},\overline{A},\overline{P}) = \rho(G,A,P)$.*

Proof. By Lemma 14.43 we have

$$\rho(G,A,P) \leq \rho(\overline{G},\overline{A},\overline{P}). \tag{14.85}$$

Let Z (resp. $\overline{Z}$) denote the random variable corresponding to the random walk on G (resp. $\overline{G}$) defined by A and P (resp. $\overline{A}$ and $\overline{P}$). Fix $\varepsilon > 0$. Then there exists an $n_\varepsilon \in \mathbb{N}$ such that

$$\begin{aligned}(M^{2n}(\overline{G},\overline{A},\overline{P}))_{1_{\overline{G}},1_{\overline{G}}} &= \mathbb{P}[\overline{Z}_{2n} = 1_{\overline{G}} \mid \overline{Z}_0 = 1_{\overline{G}}] \\ &= \mathbb{P}[Z_{2n} \in N \mid Z_0 = 1_G] \\ &\geq \left((1-\varepsilon)\rho(\overline{G},\overline{A},\overline{P})\right)^{2n}\end{aligned}$$

for all $n \geq n_\varepsilon$. Fix $n \geq n_\varepsilon$ and, for $b \in N$ (so that also $b^{-1} \in N$), let us set

$$\begin{aligned}2p^{(n)}(b) :&= \mathbb{P}[Z_{2n} = b^{\pm 1} \mid Z_0 = 1_G \text{ and } Z_{2n} \in N] \\ &= \frac{\mathbb{P}[Z_{2n} = b^{\pm 1} \mid Z_0 = 1_G]}{\mathbb{P}[Z_{2n} \in N \mid Z_0 = 1_G]}.\end{aligned}$$

Let $B = (b_k)_{k \in \mathbb{N}}$ be a family of elements in N such that for every $b \in N$ there exists a unique $k = k(b) \in \mathbb{N}$ such that b equals b_k or b_k^{-1}. Then B is a generating system for N and $(p_k^{(n)})_{k \in \mathbb{N}}$, where $p_k^{(n)} := p^{(n)}(b_k)$ for all $k \in \mathbb{N}$, is a strict probability distribution on B. Since $\rho(N) = 1$, from Corollary 14.47 we deduce that $\rho(N,B,P_N^{(n)}) = 1$, where $P_N^{(n)}$ is the probability distribution on B given $(p_k^{(n)})_{k \in \mathbb{N}}$. Denoting by $Z^{(n)}$ the random variable corresponding to the random walk on N defined by B and $P_N^{(n)}$, since $\rho(N,B,P_N^{(n)}) = 1$, there exists an $m_\varepsilon \in \mathbb{N}$ such that

$$\mathbb{P}[Z_{2m}^{(n)} = 1_G \mid Z_0^{(n)} = 1_G] \geq (1-\varepsilon)^{2m}$$

for all $m \geq m_\varepsilon$. It is then clear that

$$\begin{aligned}\mathbb{P}[Z_{2m}^{(n)} &= 1_G \mid Z_0^{(n)} = 1_G] \\ &\leq \mathbb{P}[Z_{2n\cdot 2m} = 1_G \mid Z_0 = 1_G \text{ and } Z_{2n\cdot i} \in N \text{ for all } 1 \leq i \leq 2m] \\ &= \frac{\mathbb{P}[Z_{2n\cdot 2m} = 1_G \mid Z_0 = 1_G]}{\mathbb{P}[Z_{2n\cdot i} \in N \text{ for all } 1 \leq i \leq 2m \mid Z_0 = 1_G]}\end{aligned}$$

$$= \frac{\mathbb{P}[Z_{2n\cdot 2m} = 1_G \mid Z_0 = 1_G]}{\mathbb{P}[Z_{2n} \in N \mid Z_0 = 1_G]^{2m}},$$

where the last equality follows from homogeneity ($\mathbb{P}[Z_i \in N \mid Z_0 = 1_G] = \mathbb{P}[Z_i \in N \mid Z_0 \in N]$).

Consequently,

$$(M^{2n\cdot 2m}(G,A,P))_{1_G,1_G} \geq \left((1-\varepsilon)\rho(\overline{G},\overline{A},\overline{P})\right)^{2n\cdot 2m}(1-\varepsilon)^{2m}$$

and

$$\begin{aligned}((M^{2n\cdot 2m}(G,A,P))_{1_G,1_G})^{\frac{1}{2m\cdot 2n}} &\geq \left(\left((1-\varepsilon)\rho(\overline{G},\overline{A},\overline{P})\right)^{2n\cdot 2m}\right)^{\frac{1}{2m\cdot 2n}}(1-\varepsilon)^{\frac{2m}{2m\cdot 2n}} \\ &= (1-\varepsilon)\rho(\overline{G},\overline{A},\overline{P})(1-\varepsilon)^{\frac{1}{2n}} \\ &\geq (1-\varepsilon)^2\rho(\overline{G},\overline{A},\overline{P}),\end{aligned}$$

so that

$$\rho(G,A,P) \geq (1-\varepsilon)^2\rho(\overline{G},\overline{A},\overline{P}).$$

As $\varepsilon > 0$ was arbitrary, we deduce that $\rho(G,A,P) \geq \rho(\overline{G},\overline{A},\overline{P})$. This, together with (14.85), yields $\rho(\overline{G},\overline{A},\overline{P}) = \rho(G,A,P)$. □

Corollary 14.49. *Let G be a countable group, A a generating system, and P a strict positive distribution on A. Let $N \lhd G$ be a normal subgroup and denote by $\overline{G} = G/N$, $\overline{A}$ and $\overline{P}$ the corresponding quotient group, the generating system and the positive distribution, respectively. Then*

$$\rho(\overline{G},\overline{A},\overline{P}) > \rho(G,A,P) \iff \rho(N) > 1.$$

□

Lemma 14.50. *Let G be a countable group and let $A = (a_i)_{i\geq 0}$ and $(p_i)_{i\geq 0}$ be a generating system and a probability distribution on A, respectively. Suppose that for every finitely generated subgroup $H \leq G$ one has $\rho(H) = 1$. Then, $\rho(G,A,P) = 1$. In particular, $\rho(G) = 1$.*

Proof. Let $0 < \varepsilon < 1$. Then we can find a finite subset $S \subset G$ such that

$$\sum_{s\in S\cup S^{-1}} p(s) \geq 1-\varepsilon.$$

Let $H \leq G$ denote the subgroup generated by S and define $r(x) = p(x)/\sum_{h\in H} p(h)$. Then, by our assumptions we have

$$\rho(H,S,R) = 1. \tag{14.86}$$

As in the proof of Lemma 14.45, we have

$$|\rho(G,A,P) - \rho(H,S,R)| \leq 2\varepsilon$$

so that

$$1 - 2\varepsilon = \rho(H,S,R) - 2\varepsilon \leq \rho(G,A,P) \leq 1.$$

As $\varepsilon > 0$ was arbitrary, the statement follows. □

Corollary 14.51 (Kesten's amenability criterion for countable groups). *Let G be a countable group. Then $\rho(G) = 1$ if and only if G is amenable.*

Proof. Suppose first that G is finitely generated. Let $Y \subset G$ be a finite symmetric generating subset. It follows from Theorem 14.26 that G is amenable if and only if $\rho(G,Y) = 1$. Since the uniform probability distribution Q on Y is strictly positive, keeping in mind Corollary 14.47, we can restate the above by saying that G is amenable if and only if $\rho(G) = 1$. This proves the corollary for finitely generated groups.

Suppose now that G is amenable. It follows from Theorem 14.9 that every (finitely generated) subgroup $H \leq G$ is amenable so that, by the first part of the proof $\rho(H) = 1$. But then Lemma 14.50 ensures that $\rho(G) = 1$.

Conversely, suppose that $\rho(G) = 1$. It follows from Corollary 14.47 that $\rho(H) = 1$ for every finitely generated subgroup $H \leq G$. Thus, by the first part of the proof, we have that every finitely generated subgroup of G is amenable. It then follows from Proposition 14.5 that G is itself amenable. □

In view of the previous corollary, we may restate Corollary 14.49 as follows.

Corollary 14.52. *Let G be a countable group, let A be a generating system, and let P be a strict positive distribution on A. Let $N \lhd G$ be a normal subgroup and denote by $\overline{G} = G/N$ (resp. $\overline{A}$, resp. $\overline{P}$) the corresponding quotient group (resp. generating system, resp. positive distribution on $\overline{A}$). Then*

$$\rho(\overline{G},\overline{A},\overline{P}) = \rho(G,A,P) \iff N \text{ is amenable.}$$ □

Kesten's Characterization of Free Groups

In this section we prove Kesten's formula for the spectral radius of the simple random walk on a finitely generated free group and Kesten's spectral characterization of finitely generated free groups. We again follow the presentation from the original source [196], which is extremely clear. As a by-product, we derive another proof of the Hopfianity of finitely generated free groups (from [66]), a result previously established in Corollary 3.15.

Theorem 14.53 (Kesten's characterization of free groups). *Let G be a finitely generated group. Let $A = \{a_1^{\pm 1}, a_2^{\pm 1}, \dots, a_k^{\pm 1}\} \subset G$ be a symmetric generating subset. Let Q denote the simple random walk on G associated with the generating system A. Then G is a free group with free generators $a_1, a_2, \dots, a_k$ if and only if*

$$\rho(G,A,Q) = \frac{2\sqrt{|A|-1}}{|A|} = \frac{\sqrt{2k-1}}{k}. \tag{14.87}$$

Moreover, if this is the case, one has

$$\rho(G,A,Q) = \min_P \rho(G,A,P), \tag{14.88}$$

where P runs over all symmetric random walks associated with the generating system A.

Proof. Let us first suppose that G is free and that $\{a_1, a_2, \ldots, a_k\}$ is a free basis. Set $p^{(n)} := \mathbb{P}[Z_n = 1_G \mid Z_0 = 1_G]$, $f^{(n)} := \mathbb{P}[Z_n = 1_G \mid Z_0 = 1_G \text{ and } Z_t \neq 1_G \text{ for } 1 \leq t \leq n-1\}]$ (this is the probability of returning to 1_G for the first time at the n-th step), and denote by

$$g(t) := G(1_G, 1_G \mid t) = \sum_{n=0}^{\infty} p^{(n)} t^n \text{ and } f(t) := F(1_G, 1_G \mid t) = \sum_{n=1}^{\infty} f^{(n)} t^n$$

the associated generating functions (cf. (13.14) and (13.17)). Then $g(t) = 1/(1 - f(t))$ for all $t \in [0, +\infty)$ such that $g(t) < \infty$ and $f(t) < 1$ (cf. Proposition 13.7).

We recall some notation and facts from Chapter 1. We denote by A^* the monoid of all words with letters in A so that any $w \in A^*$ is of the form

$$w = a_{i_1}^{\varepsilon_1} a_{i_2}^{\varepsilon_2} \cdots a_{i_n}^{\varepsilon_n}$$

with $1 \leq i_j \leq k$, $\varepsilon_j \in \{-1, 1\}$, $j = 1, 2, \ldots, n$, and $\ell(w) := n \geq 0$. Let us denote by A^n the set of all words $w \in A^*$ of length $\ell(w) = n$. For $0 \leq m \leq n$, the m-suffix of w is the subword $w_m := a_{i_1}^{\varepsilon_1} a_{i_2}^{\varepsilon_2} \cdots a_{i_m}^{\varepsilon_m} \in A^m$. Also recall that, if there is no $1 \leq j \leq n-1$ such that $a_{i_j} = a_{i_{j+1}}$ and $\varepsilon_j = -\varepsilon_{j+1}$, we say that w is reduced. Recall (cf. Theorem 1.4) that every w is equivalent in G to exactly one reduced word, denoted $[w]$. For $w \in A^*$ we set $\ell_R(w) := \ell([w])$.

All this said, we have

$$f^{(n)} = \frac{|\{w \in A^n : \ell_R(w) = 0 \text{ and } \ell_R(w_m) > 0 \text{ for all } 1 \leq m < n\}|}{(2k)^n}.$$

Obviously, $f^{(n)} = 0$ for odd n.

Let $n \geq 0$ be an integer. Consider now the grid $\mathscr{G}_n = \{0, 1, \ldots, n\} \times \{0, 1, \ldots, n\}$ (this can be viewed as a subset of the lattice points of the Euclidean plane). The diagonal of $\mathscr{G}_n$ is the set $\Delta(\mathscr{G}_n) := \{(k,k) : 0 \leq k \leq n\}$. A *Dyck path* in $\mathscr{G}_n$ is a path $\pi(t) = (x(t), y(t))$, $t = 0, 1, \ldots, 2n$, which starts in the lower left corner ($\pi(0) = (0,0)$), finishes in the upper right corner ($\pi(2n) = (n,n)$), consists entirely of edges pointing rightwards or upwards ($\pi(t+1)$ equals either $\pi(t) + (1,0)$ or $\pi(t) + (0,1)$), and does not pass above the diagonal ($y(t) \leq x(t)$ for all $t = 0, 1, \ldots, 2n$). If π only meets the diagonal $\Delta(\mathscr{G})$ at the initial and final steps (i.e., $y(t) < x(t)$ for all $t = 1, \ldots, 2n-1$), we say that the Dyck path is *strict*.

The number of all Dyck paths in $\mathscr{G}_n$ is called the n-th *Catalan number* C_n (cf. [322] or [294]).

Lemma 14.54. *The Catalan number C_n of all Dyck paths in $\mathscr{G}_n$ is equal to $\frac{1}{n+1}\binom{2n}{n}$. The number of all strict Dyck paths in $\mathscr{G}_n$ is equal to C_{n-1}.*

Proof. The second statement is clear: just remove the first and the last step of a strict Dyck path in $\mathcal{G}_n$, to get any Dyck path of length $2n-2$.

To prove the formula for C_n, first observe that $C_0 = C_1 = 1$. Let $C(x) := \sum_{n\geq 0} C_n x^n$ be the ordinary generating function of the sequence $(C_n)_{n\in\mathbb{N}}$. If we look at the first time that a Dyck path of length $2n$ returns to the diagonal, we can decompose the path into two Dyck paths, the first one, of length $2k$ going from $(0,0)$ to (k,k), being strict: there are C_{k-1} of them as we already observed; the second one from (k,k) to (n,n) is just any Dyck path of length $2n-2k$, and there are C_{n-k} of them.

Hence we have found the recursion

$$C_n = \sum_{k=1}^{n} C_{k-1}C_{n-k} \text{ for } n \geq 1,$$

which leads to the functional equation for the ordinary generating function

$$C(x) - 1 = xC(x)C(x).$$

Solving for $C(x)$ we get

$$C(x) = \frac{1-\sqrt{1-4x}}{2x} \tag{14.89}$$

(we take the root with the minus sign, otherwise we get negative coefficients).

This series starts as

$$C(x) = 1 + x + 2x^2 + 5x^3 + 14x^4 + 42x^5 + 132x^6 + 429x^7 + 1430x^8 + \cdots.$$

Using the binomial theorem we get

$$C(x) = \frac{1-\sqrt{1-4x}}{2x} = -\frac{1}{2x}\sum_{n\geq 1}\binom{1/2}{n}(-4)^n x^n = -\frac{1}{2}\sum_{n\geq 0}\binom{1/2}{n+1}(-4)^{n+1}x^n,$$

hence

$$\begin{aligned}
C_n &= -\frac{1}{2}\binom{1/2}{n+1}(-4)^{n+1} \\
&= (-1)^n 2^{n+1} 2^n \frac{1}{n+1} \frac{\frac{1}{2}\left(\frac{1}{2}-1\right)\left(\frac{1}{2}-2\right)\cdots\left(\frac{1}{2}-n\right)}{n!} \\
&= (-1)^n 2^n \frac{1}{n+1} \frac{1\,(1-2)\,(1-4)\cdots(1-2n)}{n!} \\
&= 2^n \frac{1}{n+1} \frac{\prod_{i=1}^n (2i-1)}{n!} = \frac{1}{n+1} \frac{\prod_{i=1}^n 2i}{n!} \frac{\prod_{i=1}^n (2i-1)}{n!} \\
&= \frac{1}{n+1}\binom{2n}{n}.
\end{aligned}$$

This ends the proof of Lemma 14.54 □

Then, with every word $w \in A^{2n}$ such that $\ell_R(w) = 0$ we associate a Dyck path π_w defined as follows: for every letter of w (reading from left to right) we record either a rightward or an upward step of length one. Let $m \geq 0$ and suppose we have defined $\pi_w(t)$ for $t = 0, 1, \dots m$. Then we set $\pi_w(m+1) = \pi_w(m) + (1,0)$ (i.e., we record a rightward step) if $\ell_R(w_{m+1}) = \ell_R(w_m) + 1$ and $\pi_w(m+1) = \pi_w(m) + (0,1)$ (i.e., we record an upward step) if $\ell_R(w_{m+1}) = \ell_R(w_m) - 1$. It is clear that a word $w \in A^{2n}$ such that $\ell_R(w) = 0$ satisfies that $\ell_R(w_m) > 0$ for all $1 \leq m < n$ if and only if the associated path π_w is strict.

We now determine the number of words that are mapped onto a fixed strict Dyck path. The first step is rightward and we have $2k$ distinct possibilities, corresponding to the choice of $a_{i_1}^{\varepsilon_1} \in A$. If the m-th step is rightward with $m > 0$ then it corresponds to $2k-1$ possibilities: if $w_{m-1} = a_{i_1}^{\varepsilon_1} a_{i_2}^{\varepsilon_2} \cdots a_{i_{m-1}}^{\varepsilon_{m-1}}$ then we can choose $a_{i_m} \neq a_{i_{m-1}}$ and $\varepsilon \in \{-1, 1\}$ (this gives $2(m-1)$ possibilities) or $a_{i_m} = a_{i_{m-1}}$ and $\varepsilon_m \neq \varepsilon_{m-1}$ (this giving one more possibility). The m-th step being upward corresponds to only one possibility, namely to $a_{i_m}^{\varepsilon_m} = a_{i_{m-1}}^{-\varepsilon_{m-1}}$. Since every Dyck path in $\mathscr{G}_n$ has exactly n rightward and n upward steps, we conclude that for every strict Dyck path π there are exactly $2k(2k-1)^{n-1}$ words $w \in A^{2n}$ such that $\pi_w = \pi$. From the above combinatorial discussion, we deduce that

$$f^{(2n)} = C_{n-1} \frac{2k(2k-1)^{n-1}}{(2k)^{2n}} = C_{n-1} \frac{1}{2k} \left(\frac{2k-1}{4k^2} \right)^{n-1}.$$

As a consequence,

$$\begin{aligned} f(t) &= \sum_{m=1}^{\infty} f^{(m)} t^m = \sum_{n=1}^{\infty} f^{(2n)} t^{2n} \\ &= \sum_{n=1}^{\infty} C_{n-1} \frac{1}{2k} \left(\frac{2k-1}{4k^2} \right)^{n-1} t^{2n} \\ &= \frac{1}{2k} t^2 \sum_{n=1}^{\infty} C_{n-1} \left(\frac{2k-1}{4k^2} t^2 \right)^{n-1} \\ \text{(using (14.89))} &= \frac{1}{2k} t^2 \frac{1 - \sqrt{1 - 4\left(\frac{2k-1}{4k^2} t^2\right)}}{2\left(\frac{2k-1}{4k^2} t^2\right)} \\ &= \frac{k - \sqrt{k^2 - (2k-1)t^2}}{2k-1} \end{aligned}$$

and

$$\begin{aligned} g(t) &= \frac{1}{1 - f(t)} = \frac{2k-1}{k - 1 + \sqrt{k^2 - (2k-1)t^2}} \\ &= \frac{\sqrt{k^2 - (2k-1)t^2} - (k-1)}{1 - t^2}. \end{aligned}$$

Now the smallest positive singularity of $g(t)$ clearly occurs at $t_0 := \frac{k}{\sqrt{2k-1}}$. From this, we immediately deduce (14.87).

Let now P be another symmetric probability distribution on A so that $\mathbb{P}[a_i^{\pm 1} \mid P] = p_i, i = 1,2,\ldots,k$. For $0 \leq \xi \leq 1$ consider the symmetric probability distribution $P(\xi)$ on A defined by

$$\mathbb{P}[a_i^{\pm 1} \mid P(\xi)] = \begin{cases} \xi(p_1 + p_2) & \text{if } i = 1 \\ (1-\xi)(p_1 + p_2) & \text{if } i = 2 \\ p_i & \text{otherwise} \end{cases}$$

for all $i = 1,2,\ldots,k$. It follows from Lemma 14.42 that $\rho(G,A,Q(\xi))$ is a convex function of ξ and, as long as a_1 and a_2 are exchangeable, $\rho(G,A,Q(\xi))$ is symmetric around $\xi = 1/2$. But a convex function on an interval I which is symmetric around the midpoint of I attains its minimum at this midpoint. Repeatedly applying this argument shows that the probability distribution P which assigns equal probabilities to all generators (that is, $P = Q$) minimizes $\rho(G,A,P)$ (as a function of P), and (14.88) follows as well.

We now turn to the final part of the theorem, namely to showing that $\rho(G,A,Q) = \frac{\sqrt{2k-1}}{k}$ implies that G is free with $\{a_1,a_2,\ldots,a_k\}$ a free basis for G. Let H denote the free group, freely generated by $\{c_1,c_2,\ldots,c_k\}$ and, setting $C := \{c_1^{\pm 1}, c_2^{\pm 1},\ldots,c_k^{\pm 1}\}$, denote by Q' the probability distribution on C defined by $\mathbb{P}[c_i^{\pm 1} \mid Q'] = \mathbb{P}[a_i^{\pm 1} \mid Q] = 1/(2k)$ for all $i = 1,2,\ldots,k$.

By the universal property of free groups (cf. Definition 1.1), the map $c_i \mapsto a_i$ extends to a unique epimorphism $H \to G$. Denote by $N \lhd H$ its kernel, so that $G \cong H/N$. We claim that $N = \{1_H\}$ so that the above map is indeed an isomorphism (and $\{a_1,a_2,\ldots,a_k\}$ is therefore a free basis). If not, as N is free (by the Nielsen–Schreier theorem (Theorem 1.15)) and therefore non-amenable (by Corollary 14.10), we would have

$$\frac{\sqrt{2k-1}}{k} = \rho(G,A,P) = \rho(\overline{H},\overline{C},\overline{P'}) > \rho(H,C,P') = \frac{\sqrt{2k-1}}{k},$$

where the strict inequality follows from Corollary 14.52 and the last equality follows from the first part of the proof, a contradiction. This completes the proof of Theorem 14.53 □

From the above proof we deduce the following fact (already established in Corollary 3.15).

Corollary 14.55. *A free group of finite rank is Hopfian.*

Proof. Let G be a finitely generated free group. Let A be a free basis and consider the simple random walk on G associated with the finite symmetric generating system $A \cup A^{-1} \subset G$. Let $\varphi\colon G \to G$ be a surjective homomorphism. Denote by $N := \ker(\varphi)$ the kernel of φ and let us show that $N = \{1_G\}$. Otherwise, (by the Nielsen–Schreier theorem (Theorem 1.15)) N is a nontrivial free subgroup of G and therefore a non-amenable group (by Corollary 14.10). Denote by $\overline{G} := G/N \cong G$ the corresponding quotient group, and observe that by our assumptions it is free, freely generated by $\overline{A} = \{\overline{a} : a \in A\}$. Thus, $(\overline{G}, \overline{Q})$ is the simple random walk on $\overline{G}$ associated with $\overline{A} \cup (\overline{A})^{-1}$. By applying Corollary 14.52 we would then have

$$\frac{\sqrt{2k-1}}{k} = \rho(G,A,Q) > \rho(\overline{G},\overline{A},\overline{Q}) = \frac{\sqrt{2k-1}}{k},$$

a contradiction. □

14.6 Cogrowth and the Grigorchuk Criterion

The goal of this section is to prove the cogrowth criterion, which is due to Grigorchuk [128]. We will give here the (combinatorial) proof of Bartholdi [14].

Paths on Regular Graphs

Let $\mathscr{G} = (V,E)$ be an unoriented graph without loops and multiple edges, i.e. V is a set of *vertices*, and E is a set of *edges* of V, i.e. a set of subsets of cardinality 2 of V. The *degree* of a vertex $v \in V$ is the number $\deg(v)$ of edges *incident* to v, i.e. that contain v. We say that the graph is *locally finite* if every vertex v has finite degree. We say that the graph is *d-regular*, with $d \in \mathbb{N}$, if every vertex has degree d.

For our purposes, we will think of our unoriented graph as having for each edge $e \in E$ a pair of *oriented edges* e and $\overline{e}$: for every $e \in E$ we fix a *starting point* $\alpha(e) \in e$ and an *ending point* $\omega(e) \in e \setminus \{\alpha(e)\}$, and $\overline{e}$ will be the *reverse* of e, i.e. the same edge with starting and ending points interchanged. Notice that in this oriented situation the original degree $\deg(v)$ of a vertex v will correspond to its *outdegree*, i.e. to the number of oriented edges e such that $\alpha(e) = v$.

A *path* starting at $a \in V$ and ending at $b \in V$ is a sequence of oriented edges $\pi = (\pi_1, \pi_2, \ldots, \pi_n)$ such that $\alpha(\pi_1) = a$, $\omega(\pi_n) = b$, and $\alpha(\pi_{i+1}) = \omega(\pi_i)$ for all $i = 1,2,\ldots,n-1$. The *length* of such a path $\pi = (\pi_1, \pi_2, \ldots, \pi_n)$ is $\ell(\pi) := n$.

The *backtrack* of a path $\pi = (\pi_1, \pi_2, \cdots, \pi_n)$ is the number of $i = 1,2,\ldots,n-1$ such that $\pi_{i+1} = \overline{\pi_i}$, denoted $bt(\pi)$.

Given $a,b \in V$, denote by $[a,b]$ the set of all paths from a to b.

Given a locally finite graph $\mathscr{G} = (V,E)$, fix two (possibly equal) vertices $a,b \in V$, and consider the generating function

$$F(u,t) := F_{[a,b]}(u,t) = \sum_{\pi \in [a,b]} u^{bt(\pi)} t^{\ell(\pi)}.$$

Notice that, because of the local finiteness, this is a well-defined formal power series in t with coefficients polynomials in u.

The goal of this section is to prove the following theorem, which is due to Laurent Bartholdi.

Theorem 14.56. *For a d-regular graph we have*

$$F(u,t) = \frac{1-(u-1)^2t^2}{1-(u-1+d)(u-1)t^2} \cdot F\left(1, \frac{t}{1-(u-1+d)(u-1)t^2}\right). \tag{14.90}$$

In order to prove the theorem, we will introduce some further notation.

It is convenient to consider a path $\pi = (\pi_1, \pi_2, \ldots, \pi_n)$ as a word in the alphabet E, and write $\pi = \pi_1\pi_2\cdots\pi_n$ instead. Given two paths $\pi = \pi_1\pi_2\cdots\pi_n$ and $\pi' = \pi'_1\pi'_2\cdots\pi'_m$ such that $\omega(\pi_n) = \alpha(\pi'_1)$ we define a *concatenation product* $\pi\pi' := \pi_1\pi_2\cdots\pi_n\pi'_1\pi'_2\cdots\pi'_m$ giving a new path.

An *even squiggle* is a path of the form $e\bar{e}e\bar{e}\cdots e\bar{e}$, hence it has even length, possibly length 0. An *odd squiggle* is a path of the form $e\bar{e}e\bar{e}\cdots e\bar{e}e$, hence it has odd length.

Given a path $\pi = \pi_1\pi_2\cdots\pi_n$, a *backtrack decomposition* $B = (s_1, s_2, \ldots, s_k)$ of π is a decomposition $\pi = s_1 s_2 \cdots s_k$ where each s_i is a squiggle, odd or even.

The *backtrack* of a backtrack decomposition $B = (s_1, s_2, \ldots, s_k)$ of π is defined to be

$$bt(B) := \sum_{i=1}^{k} bt(s_i) = \sum_{i=1}^{k} (\ell(s_i) - 1).$$

Given a path π, we call $BT(\pi)$ the set of all the backtrack decompositions of π. We start with a simple lemma.

Lemma 14.57. *Given a path π and a variable u, we have*

$$u^{bt(\pi)} = \sum_{B \in BT(\pi)} (u-1)^{bt(B)}.$$

Proof. Let $\pi = \pi_1\pi_2\cdots\pi_n$, and let $m = bt(\pi)$. Let $1 \le i_1 < i_2 < \cdots < i_m \le n-1$ be the indices for which $\pi_{i_j+1} = \overline{\pi_{i_j}}$, and set $C := \{i_1, i_2, \ldots, i_m\}$.

Observe that every subset D of C gives rise to a backtrack decomposition B_D of π in the following way: given the word $\pi = \pi_1\pi_2\cdots\pi_n$, cut after the letter π_i if and only if $i \in D$ or $i \notin C$. It is straightforward to see that the backtrack of such a decomposition is $|C| - |D| = m - |D|$.

Example 14.58. Consider the graph $\mathscr{G} = (V, E)$ with $V = \{v_1, v_2, v_3\}$ and $E = \{e, f, g\}$ with $\alpha(e) = \omega(g) = v_1$, $\alpha(f) = \omega(e) = v_2$, and $\alpha(g) = \omega(f) = v_3$. Consider the path $\pi = efg\bar{g}g\bar{g}ge\bar{e}\bar{g}\bar{f}fge$ from v_1 to v_2. In this case $n = 14$, and $C = \{3,4,5,6,8,11\}$, so that $m = bt(\pi) = 6$. Now $D = \{4,5,11\} \subseteq C$ gives the backtrack decomposition $B_D = (e, f, g\bar{g}, g, \bar{g}g, e\bar{e}, \bar{g}, \bar{f}, f, g, e)$. Indeed $bt(B_D) = m - |D| = 6 - 3 = 3$.

Conversely, any backtrack decomposition $B = (s_1, s_2, \ldots, s_k)$ of π gives rise to such a subset of C: take the set of the indices in π of the rightmost letter of the s_i that are also in C.

This establishes a bijection between subsets of C and backtrack decompositions of π, therefore we have

$$
\begin{aligned}
\sum_{B\in BT(\pi)} (u-1)^{bt(B)} &= \sum_{D\subseteq C} (u-1)^{bt(B_D)} \\
&= \sum_{k=0}^{m} \sum_{\substack{D\subseteq C \\ |D|=k}} (u-1)^{m-|D|} \\
&= \sum_{k=0}^{m} \binom{m}{k} (u-1)^{m-k} \\
&= (u-1+1)^m \\
&= u^m,
\end{aligned}
$$

as we wanted. This proves Lemma 14.57 □

We are now ready to prove Theorem 14.56.

Proof of Theorem 14.56 Given a backtrack decomposition B of some path, we call π_B the path obtained from B by erasing all the even squiggles and replacing each odd squiggle by its first oriented edge.

For every $B \in BT(\pi)$, define $\widetilde{bt}(B)$ by setting $\widetilde{bt}(B) := \ell(\pi) - \ell(\pi_B)$.

Using the above lemma:

$$
\begin{aligned}
F(u,t) &= \sum_{\pi\in[a,b]} u^{bt(\pi)} t^{\ell(\pi)} \\
&= \sum_{\pi\in[a,b]} \left(\sum_{B\in BT(\pi)} (u-1)^{bt(B)} \right) t^{\ell(\pi)} \\
&= \sum_{\pi\in[a,b]} \left(\sum_{B\in BT(\pi)} (u-1)^{bt(B)} t^{\widetilde{bt}(B)} \right) t^{\ell(\pi_B)} \\
&= \sum_{\rho\in[a,b]} \left(\sum_{B\in BT(a,b):\pi_B=\rho} (u-1)^{bt(B)} t^{\widetilde{bt}(B)} \right) t^{\ell(\rho)}.
\end{aligned}
$$

We want to compute the generating function $\sum_{B\in BT(a,b):\pi_B=\rho} (u-1)^{bt(B)} t^{\widetilde{bt}(B)}$ for a given path $\rho \in [a,b]$. We need to enumerate all the backtrack decompositions B such that $\pi_B = \rho = \rho_1\rho_2\cdots\rho_n$. These decompositions will necessarily be of the form

$$
B = (\beta_{0,1}, \beta_{0,2}, \ldots, \beta_{0,t_0}, \gamma_1, \beta_{1,1}, \beta_{1,2}, \ldots, \beta_{1,t_1}, \gamma_2, \ldots, \gamma_n, \beta_{n,1}, \beta_{n,2}, \ldots, \beta_{n,t_n})
$$

with $t_i \geq 0$ for all i, where the γ_i are odd squiggles of the form $\gamma_i = \rho_i(\overline{\rho}_i\rho_i)^{k_i}$ with $k_i \geq 0$, and the $\beta_{i,j}$ are even squiggles of the form $\beta_{i,j} = (e_{ij}\bar{e}_{ij})^{r_{ij}}$ with $r_{ij} > 0$, and $e_{ij} \in E$ with $\alpha(e_{ij}) = \omega(\rho_i)$ (and $\alpha(e_{0j}) = a$).

Notice that there are $\deg(\omega(\rho_i))$ possibilities for e_{ij} (and $\deg(a)$ for e_{0j}).

Now the generating function for such B will be the product of the generating functions for the γ_i and for the $\beta_i := \beta_{i,1}\beta_{i,2}\cdots\beta_{i,t_i}$.

The generating function for a given γ_i is

$$1+(u-1)^2t^2+(u-1)^4t^4+(u-1)^6t^6+\cdots = \frac{1}{1-(u-1)^2t^2}.$$

The generating function for a given $\beta_{i,j}$ is

$$(u-1)t^2+(u-1)^3t^4+\cdots = \frac{(u-1)t^2}{1-(u-1)^2t^2},$$

so that the generating function for β_i is

$$\frac{1}{1-\sum_{e\in E:\alpha(e)=\omega(\rho_i)}\frac{(u-1)t^2}{1-(u-1)^2t^2}}.$$

Using the hypothesis that the graph is d-regular, i.e. every vertex has degree d, we get

$$\begin{aligned}\frac{1}{1-\sum_{e\in E:\alpha(e)=\omega(\rho_i)}\frac{(u-1)t^2}{1-(u-1)^2t^2}} &= \frac{1}{1-d\left(\frac{(u-1)t^2}{1-(u-1)^2t^2}\right)}\\ &= \frac{1-(u-1)^2t^2}{1-(u-1)^2t^2-d(u-1)t^2}\\ &= \frac{1-(u-1)^2t^2}{1-(u-1+d)(u-1)t^2}.\end{aligned}$$

Therefore the generating function that we were looking for is

$$\begin{aligned}&\sum_{B\in BT(a,b):\pi_B=\rho}(u-1)^{bt(B)}t^{\widetilde{bt}(B)}\\ &= \frac{1-(u-1)^2t^2}{1-(u-1+d)(u-1)t^2}\cdot\left(\frac{1}{1-(u-1)^2t^2}\cdot\frac{1-(u-1)^2t^2}{1-(u-1+d)(u-1)t^2}\right)^{\ell(\rho)}.\end{aligned}$$

Using this identity we get

$$\begin{aligned}&F(u,t)\\ &= \sum_{\rho\in[a,b]}\left(\sum_{B\in BT(a,b):\pi_B=\rho}(u-1)^{bt(B)}t^{\widetilde{bt}(B)}\right)t^{\ell(\rho)}\\ &= \sum_{\rho\in[a,b]}\frac{1-(u-1)^2t^2}{1-(u-1+d)(u-1)t^2}\cdot\left(\frac{1}{1-(u-1)^2t^2}\cdot\frac{1-(u-1)^2t^2}{1-(u-1+d)(u-1)t^2}\right)^{\ell(\rho)}t^{\ell(\rho)}\\ &= \frac{1-(u-1)^2t^2}{1-(u-1+d)(u-1)t^2}\sum_{\rho\in[a,b]}\left(\frac{t}{1-(u-1+d)(u-1)t^2}\right)^{\ell(\rho)}\\ &= \frac{1-(u-1)^2t^2}{1-(u-1+d)(u-1)t^2}\cdot F\left(1,\frac{t}{1-(u-1+d)(u-1)t^2}\right),\end{aligned}$$

as we wanted. This completes the proof of Theorem 14.56. □

Now notice that

$$F(t) := F(0,t) = \sum_{n=0}^{\infty} f_n t^n$$

is the generating function of the number f_n of paths of length n between a and b with 0 backtrack, while

$$G(t) := F(1,t) = \sum_{n=0}^{\infty} g_n t^n$$

is the generating function of the number g_n of paths of length n between a and b (with no restrictions).

Specializing formula (14.90) at $u = 0$ we get the useful identity

$$\frac{F(t)}{1-t^2} = \frac{1}{1+(d-1)t^2} \cdot G\left(\frac{t}{1+(d-1)t^2}\right). \tag{14.91}$$

Cogrowth Criterion

Let G be a finitely generated group, and let $A \subseteq G$ be a finite symmetric generating subset. For any subset $R \subseteq G$, the *cogrowth* of R relative to (G,A) is

$$\alpha(R;G,A) := \limsup_{n\to\infty} \sqrt[n]{f_n}, \tag{14.92}$$

where

$$f_n := |\{g \in R : \ell_A(g) = n\}|.$$

Theorem 14.59 (Grigorchuk). *Let G be a finitely generated group, and let $A \subseteq G$ be a finite symmetric generating subset with $|A| \geq 2$. Consider the presentation $F_A/N \cong G$ of G, where F_A is the free group on the free generators A. Let $\rho := \rho(G,A)$ be the radius of convergence of the simple random walk (G,P) on G generated by A, and let $\alpha := \alpha(N;F_A,A)$ be the cogrowth of N in (F_A,A). Then*

$$\rho = \begin{cases} \frac{\sqrt{|A|-1}}{|A|}\left(\frac{\alpha}{\sqrt{|A|-1}} + \frac{\sqrt{|A|-1}}{\alpha}\right) & \text{if } \alpha \geq \sqrt{|A|-1} \\ \frac{2\sqrt{|A|-1}}{|A|} & \text{if } \alpha < \sqrt{|A|-1} \end{cases}. \tag{14.93}$$

Proof. Let $d := |A|$. The case $d = 2$ is quite straightforward: since A is symmetric and generates the group, it leaves very few possibilities for G, i.e. G either is cyclic or is generated by two elements of order 2. In both cases, the radius of convergence of the simple random walk is clearly 1, which is what the formula prescribes in this case.

So from now on we assume $d \geq 3$. Consider the generating function $F(t) = \sum_{n\geq 0} f_n t^n$, where f_n is the number of paths of length n on the Cayley graph of (G,A) starting and ending in 1_G with no backtracking; and let $G(t) = \sum_{n\geq 0} g_n t^n$, where g_n is the number of paths of length n on the Cayley graph of (G,A) starting and ending in 1_G (with no restrictions). It is clear from the definitions that $F(t)$ has radius of convergence $1/\alpha$. Since our Cayley graph is d-regular, we clearly have $g_n = d^n p^{(n)}(1_G, 1_G)$, hence the radius of convergence of $G(t)$ is $1/(d\rho)$.

Recalling (14.91), i.e.

$$F(t) = \frac{1-t^2}{1+(d-1)t^2} \cdot G\left(\frac{t}{1+(d-1)t^2}\right), \tag{14.94}$$

we will assume that

$$|t| < \frac{1}{\alpha} \qquad \text{and} \qquad \left|\frac{t}{1+(d-1)t^2}\right| < \frac{1}{d\rho}$$

so that all the series involved converge absolutely, and this is an identity of functions.

Observe that, since the coefficients of F and G are positive, these functions are increasing between 0 (included) and their respective smallest singularities (excluded).

Consider the function $\varphi \colon \mathbb{R}_{\geq 0} \to \mathbb{R}_{\geq 0}$ defined as $\varphi(x) := \mapsto x/(1+(d-1)x^2)$ for all $x \in \mathbb{R}_{\geq 0}$ (which occurs on the right-hand side of (14.94)). For $x < 1/\sqrt{d-1}$ φ is increasing, and for $t > 1/\sqrt{d-1}$ it is decreasing; its maximum value is then $\varphi(1/\sqrt{d-1}) = 1/(2\sqrt{d-1})$.

We are now ready to discuss the two cases.

- Let $\alpha \geq \sqrt{d-1}$. In this case we have, in particular $|t| < 1/\alpha \leq 1/\sqrt{d-1}$. For t going from 0 to the first singularity of F which is $1/\alpha$, by the monotonicity of φ, we must have that $\varphi(1/\alpha)$ is the first singularity of the right-hand side, which is then $1/(d\rho)$ (the factor $1-t^2$ cannot cancel the singularity as $1/\sqrt{d-1} < 1$). This gives $\varphi(1/\alpha) = 1/(d\rho)$ which is equivalent to

$$\rho = \frac{\sqrt{d-1}}{d}\left(\frac{\alpha}{\sqrt{d-1}} + \frac{\sqrt{d-1}}{\alpha}\right)$$

  as we wanted.
- Let $\alpha < \sqrt{d-1}$, so that $1/\sqrt{d-1} < 1/\alpha$. We again let t go from 0 to the first singularity.
  If $1/(d\rho) < 1/(2\sqrt{d-1})$, then $\varphi(t)$ reaches the singularity of G before t reaches $1/\alpha$, which is a contradiction, as this is the smallest singularity of F.
  If $1/(d\rho) > 1/(2\sqrt{d-1})$ (recall that $1/(2\sqrt{d-1})$ is the maximum of φ), then $\varphi(t)$ never reaches the singularity of G, so in particular $G(t)$ is derivable at $\varphi(t)$ for any $t \geq 0$. Now let us look at what happens to the right-hand side of (14.94) as t approaches $1/\sqrt{d-1}$. As $\varphi(t)$ attains its maximum, the derivative of $G(\varphi(t))$ is getting close to 0, while the other factor of the right-hand side of (14.94) is decreasing, so its derivative is negative. But this implies that the derivative of the right-hand side of (14.94) is negative at $1/\sqrt{d-1}$, contradicting the fact that F is increasing around that value.
  So the only possibility left is $1/(d\rho) = 1/(2\sqrt{d-1})$, i.e. $\rho = 2\sqrt{d-1}/d$, as we wanted. □

14.7 The Ornstein–Weiss Lemma

In this section we prove the Ornstein–Weiss lemma, an analogue of Fekete's lemma for sub-additive right-invariant functions defined on the finite subsets of an amenable group. In the theory of dynamical systems, this result is important in defining numerical invariants such as topological entropy, measure-theoretic entropy, and mean topological dimension. Indeed, these invariants are obtained by taking limits of quantities defined from a left-Følner net and one can deduce from the Ornstein–Weiss lemma that the choice of the left-Følner net is actually irrelevant for actions of amenable groups.

Fekete's lemma ([111]; cf. Lemma 7.13) can be stated as follows (after setting $u_n := \log a_n$ in Lemma 7.13). Given a *subadditive* sequence $(u_n)_{n\geq 1}$ of real numbers (i.e., such that $u_{m+n} \leq u_m + u_n$ for all $m,n \geq 1$) the sequence

$$\left(\frac{u_n}{n}\right)_{n\geq 1}$$

has a limit $\lambda \in \mathbb{R} \cup \{-\infty\}$ as n tends to infinity.

In order to state the Ornstein–Weiss result, let us first introduce some notation.

Let G be a group. We denote by $\mathscr{P}_{\rm fin}(G)$ the set of all finite subsets of G. We say that a real-valued function $h\colon \mathscr{P}_{\rm fin}(G) \to \mathbb{R}$ is *sub-additive* provided

$$h(A \cup B) \leq h(A) + h(B) \text{ for all } A, B \in \mathscr{P}_{\rm fin}(G) \tag{14.95}$$

and is *right-sub-invariant* (resp. *right-invariant*) if

$$h(Ag) \leq h(A) \text{ (resp. } h(Ag) = h(A)) \text{ for all } g \in G \text{ and } A \in \mathscr{P}_{\rm fin}(G). \tag{14.96}$$

Remark 14.60. Note that the notions of right-invariance and right-sub-invariance actually coincide. Indeed, if h is right-sub-invariant we have, for all $g \in G$ and $A \in \mathscr{P}_{\rm fin}(G)$

$$h(A) = h(A(gg^{-1})) = h((Ag)g^{-1}) \leq h(Ag) \leq h(A)$$

so that indeed $h(A) = h(Ag)$.

Theorem 14.61. *Let G be an amenable group. Let $h\colon \mathscr{P}_{\rm fin}(G) \to \mathbb{R}$ be a real-valued sub-additive and right-(sub)-invariant function. Then there exists a real number $\lambda \geq 0$, depending only on h, such that the net $\left(\frac{h(F_i)}{|F_i|}\right)_{i\in I}$ converges to λ for every left-Følner net $(F_i)_{i\in I}$ of G.*

Boundaries and Isoperimetric Constants

In this subsection we review the concepts of relative boundaries of subsets from Section 14.2 and introduce the relative isoperimetric constants for subsets. We then derive a reformulation of the notion of amenability and of the Følner condition in terms of these isoperimetric constants.

Let G be a group. Let K and A be subsets of G.

The *(right) K-interior* of A is the set

$$\mathrm{Int}_K(A) := \{g \in A : Kg \subset A\}$$

consisting of all the elements g in A such that the right-translate of K by g is entirely contained in A.

Recall (cf. (14.2)) that the *internal K-boundary* of A is the set $\delta_K(A) \subset A$ defined by

$$\delta_K(A) := A \setminus \mathrm{Int}_K(A) \equiv \{g \in A : \text{there exists } k \in K \text{ such that } kg \notin A\}.$$

Proposition 14.62. *Let G be a group. Let K, A, and B be subsets of G, and let $g \in G$ Then one has:*

(i) $\delta_K(A) = \bigcup_{k\in K}(A \setminus k^{-1}A)$;
(ii) $\delta_K(A \cup B) \subset \delta_K(A) \cup \delta_K(B)$;
(iii) $\delta(B \setminus A) \subset \left(\delta_K(B) \cup \left(\bigcup_{k\in K} k^{-1}A\right)\right) \setminus A$;
(iv) $(\mathrm{Int}_K(A))\,g = \mathrm{Int}_K(Ag)$;
(v) $(\delta_K(A))\,g = \delta_K(Ag)$.

Proof. (i) This is clear since $g \in \delta_K(A)$ if and only if $g \in A$ and $kg \notin A$, equivalently, $g \notin k^{-1}A$, for some $k \in K$.

(ii) Let $g \in \delta_K(A \cup B)$. This means that $g \in A \cup B$ and

$$Kg \cap (G \setminus (A \cup B)) \neq \varnothing.$$

Since $G \setminus (A \cup B) = (G \setminus A) \cap (S \setminus B)$, we deduce that $g \in \delta_K(A) \cup \delta_K(B)$.

(iii) Suppose that $g \in \delta_K(B \setminus A)$. This means that $g \in B \setminus A$ and

$$Kg \cap (G \setminus (B \setminus A)) \neq \varnothing.$$

Since $G \setminus (B \setminus A) = (G \setminus B) \cup A$, we deduce that if $g \notin \delta_K(B)$, then $Kg \cap A \neq \varnothing$ and hence $g \in \bigcup_{k\in K} k^{-1}A$. As $g \notin A$, inclusion (iii) immediately follows.

(iv) Suppose that $h \in (\mathrm{Int}_K(A))\,g$. This means that there exists an $a \in \mathrm{Int}_K(A)$ such that $h = ag$. Hence $h \in Ag$ and $Kh = K(ag) = (Ka)g \subset Ag$ since $a \in \mathrm{Int}_K(A)$. Thus $h \in \mathrm{Int}_K(Ag)$. This gives the inclusion $(\mathrm{Int}_K(A))\,g \subset \mathrm{Int}_K(Ag)$.

Conversely, suppose now that $h \in \mathrm{Int}_K(Ag)$. Then $h \in Ag$ and $Kh \subset Ag$. Thus, there exists an $a \in A$ such that $h = ag$ and $(Ka)sg = K(ag) \subset Ag$. This proves that $a \in \mathrm{Int}_K(A)$ so that $h \in (\mathrm{Int}_K(A))\,g$. Hence $\mathrm{Int}_K(Ag) \subset (\mathrm{Int}_K(A))\,g$. This completes the proof of (iv).

(v) We have

$$\begin{aligned}
(\delta_K(A))\,g &= (A \setminus \mathrm{Int}_K(A))\,g \\
&= Ag \setminus (\mathrm{Int}_K(A))\,g \\
&= Ag \setminus \mathrm{Int}_K(Ag) \qquad \text{(by (v))} \\
&= \delta_K(Ag).
\end{aligned}$$

This shows (vi). $\square$

Lemma 14.63. *Let G be a group. Suppose that K and A are finite subsets of G. Then one has*

$$|\delta_K(A)| \leq \sum_{k \in K} |kA \setminus A| \tag{14.97}$$

and

$$|kA \setminus A| \leq |\delta_K(A)| \quad \textit{for all } k \in K. \tag{14.98}$$

Proof. It follows from Proposition 14.62.(i) that

$$\delta_K(A) = \bigcup_{k \in K} (A \setminus k^{-1}A). \tag{14.99}$$

This implies

$$|\delta_K(A)| = |\bigcup_{k \in K} (A \setminus k^{-1}A)| \leq \sum_{k \in K} |(A \setminus k^{-1}A| = \sum_{k \in K} |kA \setminus A|.$$

As $|A \setminus k^{-1}A| = |kA \setminus A|$ for all $k \in K$, this gives us (14.97).

On the other hand, given $k \in K$, we deduce from (14.99) that

$$(A \setminus k^{-1}A) \subset \delta_K(A).$$

This implies

$$|kA \setminus A| = |A \setminus k^{-1}A| \leq |\delta_K(A)|,$$

which yields (14.98). □

Let A and K be subsets of G with A finite and nonempty. Then $\delta_K(A)$ is also finite since $\delta_K(A) \subset A$. We define the *isoperimetric constant* of A with respect to K as

$$\alpha(A,K) := \frac{|\delta_K(A)|}{|A|}.$$

Note that $\alpha(A,K)$ is rational and that one has $0 \leq \alpha(A,K) \leq 1$.

The following is a reformulation of Proposition 14.8.

Proposition 14.64. *Let G be a group. Then the following conditions are equivalent:*

(a) *G is amenable;*
(b) *for every finite subset K of G and every real number $\varepsilon > 0$, there exists a nonempty finite subset F of G such that $\alpha(F,K) \leq \varepsilon$.*

Proof. Let F and K be finite subsets of G with $F \neq \varnothing$. From inequality (14.97) of Lemma 14.63, we deduce that if $|\partial_k(F)| = |kF \setminus F| \leq \varepsilon|F|$ for all $k \in K$, then $\alpha(F,K) \leq |K|\varepsilon$. Conversely, inequality (14.98) implies that if $\alpha(F,K) \leq \varepsilon$ then $|\partial_k(F)| = |kF \setminus F| \leq \varepsilon|F|$ for all $k \in K$.

The proof then follows from the equivalence (a) $\Leftrightarrow$ (b) in Proposition 14.8. □

Similarly, we have the following characterization of Følner nets in amenable groups (cf. Proposition 14.23).

Proposition 14.65. *Let G be an amenable group. Let $(F_i)_{i \in I}$ be a net of nonempty finite subsets of G. Then the following conditions are equivalent:*

(a) *$(F_i)_{i \in I}$ is a left-Følner net for G;*
(b) *for each finite subset K of G, one has* $\lim_i \alpha(F_i, K) = 0$.

Proof. Let $g \in G$ and take $K = \{g\}$. Then one has $|F_i \setminus gF_i|/|F_i| = |gF_i \setminus F_i|/|F_i| \leq \alpha(F_i, K)$ for all $i \in I$ by (14.98). This shows that (b) implies (a).

Conversely, suppose that $(F_i)_{i \in I}$ is a left-Følner net for G. Let K be a finite subset of G and let $\varepsilon > 0$. Then there exists an $i_k \in I$ such that $|F_i \setminus kF_i|/|F_i| = |kF_i \setminus F_i|/|F_i| \leq \varepsilon$ for all $i \geq i_k$. If $j \in I$ is such that $j \geq i_k$ for all $k \in K$, we deduce that $\alpha(F_i, K) \leq \varepsilon |K|$ for all $i \geq j$ by using (14.97). This shows that (a) implies (b). □

Fillings

In this subsection we establish a filling theorem (Theorem 14.73) similar to [263, Theorem 6 in Section I.2] which constitutes a key tool in the proof of Theorem 14.61.

Definition 14.66. Let X be a set and $\varepsilon > 0$ a real number. A family $(A_j)_{j \in J}$ of finite subsets of X is said to be *ε-disjoint* if there exists a family $(B_j)_{j \in J}$ of pairwise disjoint subsets of X such that $B_j \subset A_j$ and $|B_j| \geq (1-\varepsilon)|A_j|$ for all $j \in J$.

Lemma 14.67. *Let X be a set and $(A_j)_{j \in J}$ a finite ε-disjoint family of finite subsets of X. Then one has*

$$(1-\varepsilon) \sum_{j \in J} |A_j| \leq \left| \bigcup_{j \in J} A_j \right|.$$

Proof. Since $(A_j)_{j \in J}$ is ε-disjoint, there exists a family $(B_j)_{j \in J}$ of pairwise disjoint subsets of X such that $B_j \subset A_j$ and $|B_j| \geq (1-\varepsilon)\ |A_j|$ for all $j \in J$. Thus, we have

$$(1-\varepsilon) \sum_{j \in J} |A_j| \leq \sum_{j \in J} |B_j| = \left| \bigcup_{j \in J} B_j \right| \leq \left| \bigcup_{j \in J} A_j \right|.$$

This proves Lemma 14.67. □

Lemma 14.68. *Let G be a group. Let also K be a finite subset of G and $0 < \varepsilon < 1$. Suppose that $(A_j)_{j \in J}$ is a finite ε-disjoint family of nonempty finite subsets of G. Then one has*

$$\alpha\left(\bigcup_{j \in J} A_j, K \right) \leq \frac{1}{1-\varepsilon} \cdot \max_{j \in J} \alpha(A_j, K).$$

Proof. Let us set $M := \max_{j \in J} \alpha(A_j, K)$. It follows from Proposition 14.62.(ii) that

$$\delta_K\left(\bigcup_{j\in J} A_j\right) \subset \bigcup_{j\in J} \delta_K(A_j).$$

Thus

$$\left|\delta_K\left(\bigcup_{j\in J} A_j\right)\right| \leq \left|\bigcup_{j\in J} \delta_K(A_j)\right| \leq \sum_{j\in J} |\delta_K(A_j)| = \sum_{j\in J} \alpha(A_j,K)\cdot|A_j| \leq M\sum_{j\in J}|A_j|.$$

As the family $(A_j)_{j\in J}$ is ε-disjoint, we deduce from Lemma 14.67 that

$$\alpha\left(\bigcup_{j\in J} A_j, K\right) = \frac{\left|\delta_K\left(\bigcup_{j\in J} A_j\right)\right|}{\left|\bigcup_{j\in J} A_j\right|} \leq \frac{M}{1-\varepsilon}.$$

This proves Lemma 14.68. □

Lemma 14.69. *Let G be a group. Let K, A and Ω be finite subsets of G such that $\varnothing \neq A \subset \Omega$. Suppose that $\varepsilon > 0$ is a real number such that $|\Omega \setminus A| \geq \varepsilon|\Omega|$. Then one has*

$$\alpha(\Omega\setminus A, K) \leq \frac{\alpha(\Omega,K)+\alpha(A,K)\cdot|K|}{\varepsilon}.$$

Proof. By Proposition 14.62.(iii), we have that

$$\delta_K(\Omega\setminus A) \subset \left(\delta_K(\Omega) \cup \left(\bigcup_{k\in K} k^{-1}A\right)\right) \setminus A.$$

This implies

$$\delta_K(\Omega\setminus A) \subset \delta_K(\Omega) \cup \left(\bigcup_{k\in K} (k^{-1}A\setminus A)\right)$$

and hence

$$\begin{aligned} |\delta_K(\Omega\setminus A)| \leq |\delta_K(\Omega)| + \sum_{k\in K} |k^{-1}A\setminus A| &= |\delta_K(\Omega)| + \sum_{k\in K} |kA\setminus A| \\ &= |\delta_K(\Omega)| + |\delta_K(A)|\cdot|K|. \end{aligned} \tag{14.100}$$

It follows that

$$\begin{aligned} \alpha(\Omega\setminus A, K) &= \frac{|\delta_K(\Omega\setminus A)|}{|\Omega\setminus A|} \\ &\leq \frac{|\delta_K(\Omega)| + |\delta_K(A)|\cdot|K|}{|\Omega\setminus A|} && \text{(by (14.100))} \\ &= \frac{\alpha(\Omega,K)|\Omega| + \alpha(A,K)\cdot|A|\cdot|K|}{|\Omega\setminus A|} \\ &\leq \frac{\alpha(\Omega,K)|\Omega| + \alpha(A,K)\cdot|A|\cdot|K|}{\varepsilon|\Omega|} && \text{(since } |\Omega\setminus A| \geq \varepsilon|\Omega|\text{)} \\ &\leq \frac{\alpha(\Omega,K)+\alpha(A,K)\cdot|K|}{\varepsilon} && \text{(since } |A| \leq |\Omega|\text{).} \end{aligned}$$

This proves Lemma 14.69. □

Lemma 14.70. *Let G be a group. Let A and B be finite subsets of G. Then one has*

$$\sum_{g\in G} |gA\cap B| = |A|\cdot|B|.$$

Proof. For $E \subset G$, we denote, as usual, by $\chi_E \colon G \to \mathbb{R}$ the characteristic function of E (defined by $\chi_E(g) = 1$ if $g \in E$ and $\chi_E(g) = 0$ otherwise). We have

$$\begin{aligned}
\sum_{g\in G} |Ag\cap B| &= \sum_{g\in G}\sum_{g'\in G} \chi_{Ag\cap B}(g') \\
&= \sum_{g\in G}\sum_{g'\in G} \chi_{Ag}(g')\chi_B(g') \\
&= \sum_{g\in G}\sum_{g'\in G} \chi_A(g'g^{-1})\chi_B(g') \\
(\text{setting } g'' := g'g^{-1}) \quad & \sum_{g''\in G}\sum_{g'\in G} \chi_A(g'')\chi_B(g') \\
&= |A|\cdot|B|.
\end{aligned}$$

This proves Lemma 14.70. □

Definition 14.71. Let G be a group. Let K and Ω be finite subsets of G. Given a real number $\varepsilon > 0$, a finite subset $P \subset G$ is called an *(ε, K)-filling pattern* for Ω if the following conditions are satisfied:

(F1) $P \subset \mathrm{Int}_K(\Omega)$;
(F2) the family $(Kg)_{g\in P}$ is ε-disjoint.

The following lemma will be used in the proof of Theorem 14.73. It can be viewed as a kind of analogue of Euclidean division for integers.

Lemma 14.72 (Filling lemma). *Let G be a group. Let Ω and K be nonempty finite subsets of G. Then, for every $\varepsilon \in (0,1]$, there exists an (ε, K)-filling pattern P for Ω such that*

$$|KP| \geq \varepsilon(1-\alpha(\Omega,K))|\Omega|. \tag{14.101}$$

Proof. Let $\mathscr{P}$ denote the set consisting of all (ε, K)-filling patterns for Ω. Observe that $\mathscr{P}$ is not empty, since $\varnothing \in \mathscr{P}$, and that every element of $\mathscr{P}$, being contained in $\mathrm{Int}_K(\Omega)$, has cardinality bounded above by $|\mathrm{Int}_K(\Omega)|$. Choose a pattern $P \in \mathscr{P}$ with maximal cardinality. Let us show that (14.101) is satisfied. To slightly simplify the notation, let us set

$$B := KP = \bigcup_{g\in P} Kg.$$

Let us prove that

$$\varepsilon|Kg| \leq |Kg\cap B| \quad \text{for all } g \in \mathrm{Int}_K(\Omega). \tag{14.102}$$

If $g \in P$, then $Kg \cap B = Kg$ and (14.102) holds true since $\varepsilon \leq 1$. Let now $g \in \mathrm{Int}_K(\Omega) \setminus P$ and suppose, by contradiction, that $|Kg \cap B| < \varepsilon|Kg|$. Then, we have that

$$|Kg \setminus B| = |Kg| - |Kg \cap B| > |Kg| - \varepsilon|Kg| = (1-\varepsilon)|Kg|,$$

which implies that $P \cup \{g\}$ is an (ε, K)-filling pattern for Ω. This contradicts the maximality of the cardinality of P. This proves (14.102).

Finally, we obtain

$$\begin{aligned} \varepsilon|K| \cdot |\mathrm{Int}_K(\Omega)| &= \sum_{g \in \mathrm{Int}_K(\Omega)} \varepsilon|K| \\ &= \sum_{g \in \mathrm{Int}_K(\Omega)} \varepsilon|Kg| && \text{(since } |K| = |Ks|) \\ &\leq \sum_{g \in \mathrm{Int}_K(\Omega)} |Kg \cap B| && \text{(by (14.102))} \\ &\leq \sum_{g \in G} |Kg \cap B| \\ &= |K| \cdot |B| && \text{(by Lemma 14.70),} \end{aligned}$$

which gives us

$$|KP| = |B| \geq \varepsilon |\mathrm{Int}_K(\Omega)|.$$

As

$$|\mathrm{Int}_K(\Omega)| = |\Omega| - |\delta_K(\Omega)| = (1 - \alpha(\Omega, K))|\Omega|,$$

this yields (14.101). □

Theorem 14.73 (Filling theorem). *Let G be a group and let $\varepsilon \in (0, \frac{1}{2}]$. Then there exists an integer $n_0 = n_0(\varepsilon) \geq 1$ such that for each integer $n \geq n_0$ the following holds.*

If $(K_j)_{1 \leq j \leq n}$ is a finite sequence of nonempty finite subsets of G such that

$$\alpha(K_k, K_j) \leq \frac{\varepsilon^{2n}}{|K_j|} \quad \textit{for all } 1 \leq j < k \leq n, \tag{14.103}$$

and D is a nonempty finite subset of G such that

$$\alpha(D, K_j) \leq \varepsilon^{2n} \quad \textit{for all } 1 \leq j \leq n, \tag{14.104}$$

then there exists a finite sequence $(P_j)_{1 \leq j \leq n}$ of finite subsets of G satisfying the following conditions:

(T1) *the set P_j is an (ε, K_j)-filling pattern of D for every $1 \leq j \leq n$;*
(T2) *the subsets $K_jP_j \subset D$, $1 \leq j \leq n$, are pairwise disjoint;*
(T3) *the subset $D' \subset D$ defined by*

$$D' := D \setminus \bigcup_{1 \leq j \leq n} K_jP_j$$

has cardinality $|D'| \leq \varepsilon|D|$.

Proof. Fix $\varepsilon \in (0, \frac{1}{2}]$ and a positive integer n. Let K_j, $1 \le j \le n$, and D be nonempty finite subsets of S satisfying conditions (14.103) and (14.104).

Let us first define, by induction, a finite process with at most n steps for constructing suitable finite subsets $P_n, P_{n-1}, \dots, P_1$ of G. We will see that these subsets have the required properties when n is large enough, i.e., for $n \ge n_0$ with $n_0 = n_0(\varepsilon)$ that will be made precise at the end of the proof.

Step 1. We set $D_0 := D$. By (14.104), we have

(H(1;a)) $\alpha(D_0, K_j) \le \varepsilon^{2n}$ for all $1 \le j \le n$.

Using Lemma 14.72 with $\Omega = D_0 = D$ and $K = K_n$, we can find a finite subset $P_n \subset G$ such that

(H(1;b)) P_n is an (ε, K_n)-filling pattern for D_0

and

$$|K_n P_n| \ge \varepsilon\big(1 - \alpha(D, K_n)\big)|D| \ge \varepsilon(1 - \varepsilon^{2n})|D|. \tag{14.105}$$

(H(1;c)) Setting

$$D_1 := D_0 \setminus K_n P_n,$$

we deduce from (14.105) that

$$|D_1| \le |D|\big(1 - \varepsilon(1 - \varepsilon^{2n})\big).$$

Step k**.** We continue this process by induction as follows. Suppose that the process has been applied k times, with $1 \le k \le n-1$. It is assumed that the induction hypotheses at step k are the following:

(H(k;a)) D_{k-1} is a subset of D satisfying

$$\alpha(D_{k-1}, K_j) \le (2k-1)\varepsilon^{2n-k+1} \quad \text{for all } 1 \le j \le n-k+1;$$

(H(k;b)) $P_{n-k+1} \subset G$ is an (ε, K_{n-k+1})-filling pattern for D_{k-1};
(H(k;c)) setting

$$D_k := D_{k-1} \setminus K_{n-k+1} P_{n-k+1},$$

we have

$$|D_k| \le |D| \prod_{0 \le i \le k-1} \big(1 - \varepsilon\big(1 - (2i+1)\varepsilon^{2n-i}\big)\big).$$

Note that these induction hypotheses are satisfied for $k = 1$ by Step 1.

Let us pass to Step $k+1$.

Step $k+1$**.** If $|D_k| \le \varepsilon |D_{k-1}|$ and hence $|D_k| \le \varepsilon |D|$, then we take $P_j = \varnothing$ for all $1 \le j \le n-k$ and we stop the process.

Otherwise, we have $|D_k| > \varepsilon |D_{k-1}|$. Let us estimate from above, for all $1 \le j \le n-k$, the isoperimetric constants $\alpha(D_k, K_j)$.

Let $1 \le j \le n-k$.

If $P_{n-k+1} = \varnothing$, then $D_k = D_{k-1}$ and therefore

$$\begin{aligned}\alpha(D_k,K_j) &= \alpha(D_{k-1},K_j) \\ &\leq (2k-1)\varepsilon^{2n-k+1} \qquad \text{(by our induction hypothesis (H(k;a)))} \\ &\leq (2k+1)\varepsilon^{2n-k} \qquad \text{(since } 0<\varepsilon<1\text{).}\end{aligned}$$

Suppose now that $P_{n-k+1} \neq \varnothing$. Then we can apply Lemma 14.69 with $\Omega := D_{k-1}$ and $A := K_{n-k+1}P_{n-k+1}$. This gives us

$$\begin{aligned}\alpha(D_k,K_j) &= \alpha(D_{k-1} \setminus K_{n-k+1}P_{n-k+1},K_j) \\ &\leq \frac{1}{\varepsilon}\left(\alpha(D_{k-1},K_j) + |K_j|\alpha(K_{n-k+1}P_{n-k+1},K_j)\right).\end{aligned} \tag{14.106}$$

Proposition 14.62.(v) and condition (14.103) imply that, for all $g \in G$,

$$\alpha(K_{n-k+1}g,K_j) = \alpha(K_{n-k+1},K_j) \leq \frac{\varepsilon^{2n}}{|K_j|}.$$

Since the family $(K_{n-k+1}g)_{g\in P_{n-k+1}}$ is ε-disjoint, the preceding inequality together with Lemma 14.68 give us

$$\alpha(K_{n-k+1}P_{n-k+1},K_j) = \alpha\left(\bigcup_{g\in P_{n-k+1}} K_{n-k+1}g, K_j\right) \leq \frac{\varepsilon^{2n}}{(1-\varepsilon)|K_j|}.$$

From inequality (14.106) and the induction hypothesis (H(k;a)), we deduce that

$$\alpha(D_k,K_j) \leq \frac{(2k-1)\varepsilon^{2n-k+1}}{\varepsilon} + \frac{\varepsilon^{2n}}{(1-\varepsilon)\,\varepsilon} \leq (2k+1)\varepsilon^{2n-k}$$

(for the second inequality, observe that $1/(1-\varepsilon) \leq 2$ since $0<\varepsilon\leq 1/2$).

This shows (H(k+1;a)).

Using Lemma 14.72 with $\Omega := D_k$ and $K := K_{n-k}$, we can find a finite subset $P_{n-k} \subset G$ such that P_{n-k} is an (ε, K_{n-k})-filling pattern for D_k, thus yielding (H(k+1;b)), and satisfying

$$|K_{n-k}P_{n-k}| \geq \varepsilon\left(1-\alpha(D_k,K_{n-k})\right)|D_k| \geq \varepsilon\left(1-(2k+1)\varepsilon^{2n-k}\right)|D_k|. \tag{14.107}$$

Setting

$$D_{k+1} := D_k \setminus K_{n-k}P_{n-k},$$

we deduce from (14.107) that

$$|D_{k+1}| \leq |D_k|\left(1-\varepsilon\left(1-(2k+1)\varepsilon^{2n-k}\right)\right).$$

Together with the inequality of the induction hypothesis (H(k;c)), this yields

$$|D_{k+1}| \leq |D| \prod_{0\leq i\leq k}\left(1-\varepsilon\left(1-(2i+1)\varepsilon^{2n-i}\right)\right).$$

Thus condition (H(k+1;c)) is also satisfied. This finishes the construction of Step $k+1$ and proves the induction step.

Now, suppose that this process continues until Step n. Using (H(k;c)) for $k = n$, we obtain

$$|D_n| \leq |D| \prod_{0 \leq i \leq n-1} \left(1 - \varepsilon\left(1 - (2i+1)\varepsilon^{2n-i}\right)\right). \tag{14.108}$$

We will show that for $n \geq n_0$, with $n_0 = n_0(\varepsilon)$ only depending on ε, we get $|D_n| \leq \varepsilon|D|$.

As $(2i+1)\varepsilon^{2n-i} \leq (2n+1)\varepsilon^{n+1}$ for all $0 \leq i \leq n-1$, from (14.108) we deduce that

$$|D_n| \leq |D|\left(1 - \varepsilon(1 - (2n+1)\varepsilon^{n+1})\right)^n. \tag{14.109}$$

Since $\lim_{r\to+\infty}(2r+1)\varepsilon^{r+1} = 0$ and $\lim_{r\to+\infty}(1-\frac{\varepsilon}{2})^r = 0$, both monotonically for large r, we can find an integer $n_0 = n_0(\varepsilon) \geq 1$ such that for all $r \geq n_0$, we have both $(2r+1)\varepsilon^{r+1} \leq \frac{1}{2}$ and $(1-\frac{\varepsilon}{2})^r \leq \varepsilon$. Now, if $n \geq n_0$, using inequality (14.109) we deduce

$$|D_n| \leq |D|\left(1 - \frac{\varepsilon}{2}\right)^n \leq \varepsilon|D|.$$

This finishes the proof of the theorem. □

Proof of Theorem 14.61

Let G be an amenable group and let $h\colon \mathscr{P}_{\text{fin}}(G) \to \mathbb{R}$ be a sub-additive right-invariant function on G.

First observe that by taking $A = B$ in (14.95), we get $h(A) \leq 2h(A)$ and hence

$$h(A) \geq 0 \quad \text{for all } A \in \mathscr{P}_{\text{fin}}(G). \tag{14.110}$$

From (14.96) we deduce that $h(\{g\}) = h(\{1_G\}g) = h(\{1_G\})$ for all $g \in G$. Using (14.95) once more we then deduce that

$$h(A) = h\left(\bigcup_{g\in A}\{g\}\right) \leq \sum_{g\in A} h(\{g\}) = M|A| \tag{14.111}$$

for all $A \in \mathscr{P}_{\text{fin}}(G)$, where $M := h(\{1_G\})$.

Let $(F_i)_{i\in I}$ be a left-Følner net for G. By Proposition 14.65, we have

$$\lim_i \alpha(F_i, K) = 0 \quad \text{for every finite subset } K \subset G. \tag{14.112}$$

Consider the quantity

$$\lambda := \liminf_i \frac{h(F_i)}{|F_i|}. \tag{14.113}$$

Note that $0 \leq \lambda \leq M$ by (14.110) and (14.111).

Recall that one says that a finite sequence $(K_j)_{1 \leq j \leq n}$ is *extracted* from the net $(F_i)_{i \in I}$ if there are indices

$$i_1 < i_2 < \cdots < i_n$$

in I such that $K_j = F_{i_j}$ for all $1 \leq j \leq n$.

Let $\varepsilon > 0$ and let n be a positive integer. By (14.112) and (14.113), it is clear that we can find, using induction on n, a finite sequence $(K_j)_{1 \leq j \leq n}$ extracted from the net $(F_i)_{i \in I}$ such that:

$$\alpha(K_k, K_j) \leq \frac{\varepsilon^{2n}}{|K_j|} \quad \text{for all } 1 \leq j < k \leq n$$

and

$$\frac{h(K_j)}{|K_j|} \leq \lambda + \varepsilon \quad \text{for all } 1 \leq j \leq n. \tag{14.114}$$

Suppose now that $0 < \varepsilon \leq \dfrac{1}{2}$ and that $n \geq n_0$, where $n_0 = n_0(\varepsilon)$ is as in Theorem 14.73.

Let $D \subset G$ be a nonempty finite subset satisfying $\alpha(D, K_j) \leq \varepsilon^{2n}$ for all $1 \leq j \leq n$.

By Theorem 14.73, we can find a sequence $(P_j)_{1 \leq j \leq n}$ of finite subsets of G satisfying the following conditions:

(T1) the set P_j is an (ε, K_j)-filling pattern for D for every $1 \leq j \leq n$;
(T2) the subsets $K_j P_j \subset D$, $1 \leq j \leq n$, are pairwise disjoint;
(T3) the subset $D' \subset D$ defined by

$$D' := D \setminus \bigcup_{1 \leq j \leq n} K_j P_j$$

has cardinality $|D'| \leq \varepsilon |D|$.

We then have

$$D = \bigcup_{1 \leq j \leq n} K_j P_j \cup D'.$$

By applying the sub-additivity property (14.95) of h, it follows that

$$h(D) \leq \sum_{1 \leq j \leq n} h(K_j P_j) + h(D'). \tag{14.115}$$

As $|D'| \leq \varepsilon |D|$ by (T3), we deduce from (14.111) that

$$h(D') \leq M \varepsilon |D|. \tag{14.116}$$

On the other hand, for all $1 \leq j \leq n$, we have

$$\begin{aligned} h(K_j P_j) &= h\left(\bigcup_{g \in P_j} K_j g \right) \\ &\leq \sum_{g \in P_j} h(K_j g) \qquad \text{(by (14.95))} \end{aligned}$$

$$\begin{aligned}
&\leq \sum_{g \in P_j} h(K_j) && \text{(by (14.96))}\\
&= \sum_{g \in P_j} \frac{h(K_j)}{|K_j|} |K_j g| && \text{(since } |K_j| = |K_j g|\text{)}\\
&\leq (\lambda + \varepsilon) \sum_{g \in P_j} |K_j g| && \text{(by (14.114)).}
\end{aligned}$$

As the family $(K_j g)_{g \in P_j}$ is ε-disjoint by (T1), we then deduce from Lemma 14.67 that

$$h(K_j P_j) \leq \frac{\lambda + \varepsilon}{1 - \varepsilon} \left| \bigcup_{g \in P_j} K_j g \right| = \frac{\lambda + \varepsilon}{1 - \varepsilon} |K_j P_j|.$$

This implies

$$\sum_{1 \leq j \leq n} h(K_j P_j) \leq \frac{\lambda + \varepsilon}{1 - \varepsilon} \sum_{1 \leq j \leq n} |K_j P_j|$$

and hence

$$\sum_{1 \leq j \leq n} h(K_j P_j) \leq \frac{\lambda + \varepsilon}{1 - \varepsilon} |D|, \tag{14.117}$$

since the sets $K_j P_j$, $1 \leq j \leq n$, are pairwise disjoint subsets of D by (T2).

From (14.115), (14.116), and (14.117), we deduce that

$$\frac{h(D)}{|D|} \leq \frac{\lambda + \varepsilon}{1 - \varepsilon} + M\varepsilon. \tag{14.118}$$

By (14.112), we can find an $i_0 \in I$ such that, for all $i \geq i_0$,

$$\alpha(F_i, K_j) \leq \varepsilon^{2n} \quad \text{for all } 1 \leq j \leq n.$$

Hence, by replacing D by F_i for $i \geq i_0$ in inequality (14.118), we obtain

$$\frac{h(F_i)}{|F_i|} \leq \frac{\lambda + \varepsilon}{1 - \varepsilon} + M\varepsilon.$$

This implies

$$\limsup_i \frac{h(F_i)}{|F_i|} \leq \frac{\lambda + \varepsilon}{1 - \varepsilon} + M\varepsilon.$$

Since the latter inequality is satisfied for all $\varepsilon \in (0, \frac{1}{2}]$, taking the limit as ε tends to 0, we obtain

$$\limsup_i \frac{h(F_i)}{|F_i|} \leq \lambda = \liminf_i \frac{h(F_i)}{|F_i|}.$$

This shows that (14.113) is indeed a true limit.

It only remains to show that $\lambda = \lim_i \dfrac{h(F_i)}{|F_i|}$ does not depend on the choice of the left-Følner net $(F_i)_{i\in I}$. So suppose that $(G_j)_{j\in J}$ is another left-Følner net for G and let $\nu = \lim_j \dfrac{h(G_j)}{|G_j|}$.

Take disjoint copies I' and J' of the sets I and J, i.e., sets I' and J' with $I \cap I' = \varnothing$ and $J \cap J' = \varnothing$ together with bijective maps $\varphi\colon I \to I'$ and $\psi\colon J \to J'$. Consider the set $T = (I \times J) \cup (I' \times J')$ with the partial ordering defined as follows. Given $t_1, t_2 \in T$, we write $t_1 \leq t_2$ if and only if there exist indices $i_1, i_2 \in I$ and $j_1, j_2 \in J$ such that $i_1 \leq i_2$, $j_1 \leq j_2$, and

$$(t_1 = (i_1, j_1) \text{ or } t_1 = (\varphi(i_1), \psi(j_1))) \text{ and } (t_2 = (i_2, j_2) \text{ or } t_2 = (\varphi(i_2), \psi(j_2))).$$

Observe that $(T, \leq)$ is a directed set since $(I, \leq)$ and $(J, \leq)$ are directed sets. Now we define a net $(H_t)_{t\in T}$ of nonempty finite subsets of G by setting

$$H_t = \begin{cases} F_i & \text{if } t = (i, j) \in I \times J, \\ G_j & \text{if } t = (\varphi(i), \psi(j)) \in I' \times J'. \end{cases}$$

Clearly $(H_t)_{t\in T}$ is a left-Følner net for G. By the first part of the proof, the net $\left(\dfrac{h(H_t)}{|H_t|}\right)_{t\in T}$ converges to some $\tau \geq 0$. Using the fact that for every t_1 in T, there exists a t_2 in $I \times J$ (resp. in $I' \times J'$) such that $t_1 \leq t_2$, we conclude that $\tau = \lambda = \nu$. This completes the proof of Theorem 14.61.

14.8 Applications of the Ornstein–Weiss Lemma to Ergodic Theory and Dynamical Systems

In this subsection we show how, in the setting of Ergodic Theory and Dynamical Systems, the Ornstein–Weiss lemma yields important numerical invariants such as topological entropy, measure-theoretic entropy, and mean topological dimension.

Topological Entropy

This is based on the pioneering paper by Roy L. Adler, Alan G. Konheim, and M.H. McAndrew [1].

Let X be a compact topological space.

An *open cover* of X is a family of open subsets of X whose union is X. Let $\mathcal{U} = (U_j)_{j\in J}$ and $\mathcal{V} = (V_k)_{k\in K}$ be two open covers of X. One says that $\mathcal{V}$ is *finer* than $\mathcal{U}$, and one writes $\mathcal{V} \succ \mathcal{U}$, if, for each $k \in K$, there exists a $j \in J$ such that $V_k \subset U_j$. One says that $\mathcal{V}$ is a *subcover* of $\mathcal{U}$ if $K \subset J$ and $V_k = U_k$ for all $k \in K$.

One writes $\mathscr{U} \cong \mathscr{V}$ if $\{U_j : j \in J\} = \{V_k : k \in K\}$, that is, if the open subsets of X appearing in $\mathscr{U}$ and $\mathscr{V}$ are the same (as soon as we forget that they are indexed).

The *join* of $\mathscr{U}$ and $\mathscr{V}$ is the open cover $\mathscr{U} \vee \mathscr{V}$ of x defined by $\mathscr{U} \vee \mathscr{V} := (U_j \cap V_k)_{(j,k) \in J \times K}$. If $f \colon X \to X$ is a continuous map, the *pullback* of $\mathscr{U}$ by f is the open cover $f^{-1}(\mathscr{U})$ of X defined by $f^{-1}(\mathscr{U}) := (f^{-1}(U_j))_{j \in J}$.

Since X is compact, every open cover of X admits a finite subcover. Given an open cover $\mathscr{U}$ of X, let $N(\mathscr{U})$ denote the smallest integer $n \geq 0$ such that $\mathscr{U}$ admits a subcover of cardinality n.

The proof of the following lemma is easy and we leave it as an **exercise** (cf. [1]).

Lemma 14.74. *Let X be a compact space. Let $\mathscr{U} = (U_j)_{j \in J}$ and $\mathscr{V} = (V_k)_{k \in K}$ be two open covers of X. Then one has*

(i) $N(\mathscr{U} \vee \mathscr{V}) \leq N(\mathscr{U})N(\mathscr{V})$;
(ii) *if* $\mathscr{V} \succ \mathscr{U}$ *then* $N(\mathscr{V}) \geq N(\mathscr{U})$;
(iii) *if* $\mathscr{U} \cong \mathscr{V}$ *then* $N(\mathscr{U}) = N(\mathscr{V})$;
(iv) *if* $f \colon X \to X$ *is a continuous map then* $N(f^{-1}(\mathscr{U})) \leq N(\mathscr{U})$.

Now suppose that the compact space X is endowed with a continuous action of a group S. This means that we are given a map $G \times X \to X$, $(g,x) \mapsto gx$, satisfying the following conditions:

(1) one has $g_1(g_2x) = (g_1g_2)x$ for all $g_1, g_2 \in G$ and $x \in X$;
(2) $1_G x = x$ for all $x \in X$;
(3) the map $T_g \colon X \to X$ defined by $T_g(x) := gx$ is continuous for all $g \in G$.

Let $\mathscr{U}$ be an open cover of X. Consider the map $h_{\mathscr{U}} \colon \mathscr{P}_{\text{fin}}(G) \to \mathbb{R}$ defined by

$$h_{\mathscr{U}}(A) := \log N(\mathscr{U}_A), \tag{14.119}$$

where

$$\mathscr{U}_A := \bigvee_{g \in A} T_g^{-1}(\mathscr{U}). \tag{14.120}$$

(By convention, $\mathscr{U}_\varnothing = \{X\}$ so that $h_{\mathscr{U}}(\varnothing) = 0$.)

Proposition 14.75. *Let X be a compact space equipped with a continuous action of a group G and let $\mathscr{U}$ be an open cover of X. Then the map $h_{\mathscr{U}} \colon \mathscr{P}_{\text{fin}}(G) \to \mathbb{R}$ defined by* (14.119) *is non-decreasing, sub-additive, and right-invariant.*

Proof. Let A and B be finite subsets of G.

If $A \subset B$, then $\mathscr{U}_B$ is finer than $\mathscr{U}_A$. This implies $N(\mathscr{U}_A) \leq N(\mathscr{U}_B)$ by Lemma 14.74.(ii) and hence $h_{\mathscr{U}}(A) \leq h_{\mathscr{U}}(B)$. This shows that $h_{\mathscr{U}}$ is non-decreasing.

Suppose now that A and B are disjoint. Then we have $\mathscr{U}_{A \cup B} = \mathscr{U}_A \vee \mathscr{U}_B$ and hence $N(\mathscr{U}_{A \cup B}) \leq N(\mathscr{U}_A)N(\mathscr{U}_B)$. This implies $h_{\mathscr{U}}(A \cup B) \leq h_{\mathscr{U}}(A) + h_{\mathscr{U}}(B)$.

If A and B are arbitrary (not necessarily disjoint),

$$\begin{aligned}
h_{\mathscr{U}}(A \cup B) &= h_{\mathscr{U}}((A \setminus B) \cup B) \\
&\leq h_{\mathscr{U}}(A \setminus B) + h_{\mathscr{U}}(B) && \text{(since } A \setminus B \text{ and } B \text{ are disjoint)} \\
&\leq h_{\mathscr{U}}(A) + h_{\mathscr{U}}(B) && \text{(since } h \text{ is non-decreasing).}
\end{aligned}$$

This shows that $h_{\mathscr{U}}$ is sub-additive.

To prove right-invariance, we first observe that, for every $g \in G$ and any finite subset A of G, we have

$$\begin{aligned}
\mathscr{U}_{Ag} &= \bigvee_{t \in Ag} T_t^{-1}(\mathscr{U}) \\
&\cong \bigvee_{a \in A} T_{ag}^{-1}(\mathscr{U}) \\
&= \bigvee_{a \in A} (T_a \circ T_g)^{-1}(\mathscr{U}) \\
&= \bigvee_{a \in A} T_g^{-1}(T_a^{-1}(\mathscr{U})) \\
&= T_g^{-1}\left(\bigvee_{a \in A} T_a^{-1}(\mathscr{U})\right) \\
&= T_g^{-1}(\mathscr{U}_A).
\end{aligned}$$

We then deduce that

$$h_{\mathscr{U}}(Ag) = \log N(\mathscr{U}_{Ag}) = \log N(T_g^{-1}(\mathscr{U}_A)) \leq \log N(\mathscr{U}_A) = h_{\mathscr{U}}(A)$$

by using assertions (iii) and (iv) in Lemma 14.74. This shows that $h_{\mathscr{U}}$ is right-sub-invariant. By Remark 14.60, we have that $h_{\mathscr{U}}$ is indeed right-invariant. □

From Proposition 14.75 and Theorem 14.61, we deduce the following result.

Theorem 14.76. *Let X be a compact space equipped with a continuous action of an amenable group G and let $\mathscr{U}$ be an open cover of X. Then, for every left-Følner net $(F_i)_{i \in I}$ of G, the limit*

$$h^{top}(X, G; \mathscr{U}) := \lim_i \frac{h_{\mathscr{U}}(F_i)}{|F_i|}$$

exists and is finite. Moreover, $h^{top}(X, G; \mathscr{U})$ does not depend on the choice of the left-Følner net $(F_i)_{i \in I}$.

The quantity $0 \leq h^{\text{top}}(X, G) \in [0, +\infty) \cup \{+\infty\}$ defined by

$$h^{\text{top}}(X, G) := \sup_{\mathscr{U}} h^{\text{top}}(X, G; \mathscr{U}),$$

where $\mathscr{U}$ runs over all open covers of X, is the *topological entropy* of the dynamical system (X, G).

Mean Topological Dimension

This is based on the papers by Elon Lindenstrauss and Benjy Weiss [216], and Misha Gromov [142]; see also the expositions in [81] and [78].

Let X be a compact metrizable space.

Let $\mathscr{U} = (U_j)_{j \in J}$ be a finite open cover of X. The *local order* of $\mathscr{U}$ at a point $x \in X$ is the integer $\operatorname{ord}(\mathscr{U}, x) := 1 + m(\mathscr{U}, x)$, where $m(\mathscr{U}, x)$ is the number of indices $j \in J$ such that $x \in U_j$. The *order* of $\mathscr{U}$ is the integer $\operatorname{ord}(\mathscr{U}) := \max_{x \in X} \operatorname{ord}(\mathscr{U}, x)$. Define the integer $D(\mathscr{U})$ by $D(\mathscr{U}) := \min_{\mathscr{V}} \operatorname{ord}(\mathscr{V})$, where $\mathscr{V}$ runs over all finite open covers of X such that $\mathscr{V} \succ \mathscr{U}$. Recall (cf. Notes to Chapter 10, [184]) that the quantity $0 \leq \dim(X) \leq +\infty$ defined by $\dim(X) := \sup_{\mathscr{U}} D(\mathscr{U})$, where $\mathscr{U}$ runs over all finite open covers of X, is the *topological dimension* of X.

Lemma 14.77. *Let X be a compact metrizable space. Let $\mathscr{U} = (U_j)_{j \in J}$ and $\mathscr{V} = (V_k)_{k \in K}$ be two finite open covers of X. Then one has*

(i) $D(\mathscr{U} \vee \mathscr{V}) \leq D(\mathscr{U}) + D(\mathscr{V})$;
(ii) *if* $\mathscr{V} \succ \mathscr{U}$ *then* $D(\mathscr{V}) \geq D(\mathscr{U})$;
(iii) *if* $\mathscr{U} \cong \mathscr{V}$ *then* $D(\mathscr{U}) = D(\mathscr{V})$;
(iv) *if* $f \colon X \to X$ *is a continuous map then* $D(f^{-1}(\mathscr{U})) \leq D(\mathscr{U})$.

Proof. See for example [216], [81], or [78]. □

Let X be a compact metrizable space equipped with a continuous action of a group G. Let $\mathscr{U}$ be a finite open cover of X. Consider the function $h_{\mathscr{U}}^{\dim} \colon \mathscr{P}_{\mathrm{fin}}(G) \to \mathbb{R}$ defined by

$$h_{\mathscr{U}}^{\dim}(A) \colon D(\mathscr{U}_A), \tag{14.121}$$

where $\mathscr{U}_A$ is defined by (14.120).

Proposition 14.78. *Let X be a compact metrizable space equipped with a continuous action of a group G and let $\mathscr{U}$ be a finite open cover of X. Then the map $h_{\mathscr{U}}^{\dim} \colon \mathscr{P}_{\mathrm{fin}}(G) \to \mathbb{R}$ defined by* (14.121) *is non-decreasing, sub-additive, and right-invariant.*

Proof. Mutatis mutandis, the proof is the same as that of Proposition 14.75 with Lemma 14.77 replacing Lemma 14.74. □

From Proposition 14.78 and Theorem 14.61, we deduce the following result.

Theorem 14.79. *Let X be a compact metrizable space equipped with a continuous action of an amenable group G and let $\mathscr{U}$ be a finite open cover of X. Then, for every left-Følner net $(F_i)_{i \in I}$ of G, the limit*

$$\operatorname{mdim}(X, G; \mathscr{U}) := \lim_i \frac{h_{\mathscr{U}}^{\dim}(F_i)}{|F_i|}$$

exists and is finite. Moreover, $\operatorname{mdim}(X, G; \mathscr{U})$ *does not depend on the choice of the left-Følner net $(F_i)_{i \in I}$.*

The quantity $\operatorname{mdim}(X, G) \in [0, +\infty) \cup \{+\infty\}$ defined by

$$\operatorname{mdim}(X, G) := \sup_{\mathscr{U}} \operatorname{mdim}(X, G; \mathscr{U}),$$

where $\mathscr{U}$ runs over all finite open covers of X, is the *mean topological dimension* of the dynamical system (X, G).

Measure-Theoretic Entropy

This is based on the pioneering papers by Andrei N. Kolmogorov [202] and Yakov G. Sinai [317]; see also the book [195] by Anatole B. Katok and Boris Hasselblatt.

Let $X = (X, \mathscr{B}, p)$ be a probability space.

A *finite measurable partition* of X is a finite family $\mathscr{U} = (U_j)_{j \in J}$ of pairwise disjoint measurable subsets of X whose union is X (here, equalities for subsets of X are understood to hold up to null-measure sets). The join operation $\vee$, as well as the relations $\succ$ and $\cong$, can also be defined for finite measurable partitions. Moreover, if $T \colon X \to X$ is a measurable map and $\mathscr{U} = (U_j)_{j \in J}$ is a finite measurable partition of X, then $T^{-1}(\mathscr{U}) := (T^{-1}(U_j))_{j \in J}$ is also a finite measurable partition of X.

If $\mathscr{U} = (U_j)_{j \in J}$ is a finite measurable partition of X, we define the real number $H_p(\mathscr{U}) \geq 0$ by

$$H_p(\mathscr{U}) := -\sum_{j \in J} p(U_j) \log p(U_j),$$

with the usual convention $0 \log 0 = 0$.

A measurable map $T \colon X \to X$ is said to be *measure-preserving* if $p(T^{-1}(B)) = p(B)$ for all $B \in \mathscr{B}$.

Lemma 14.80. *Let $(X, \mathscr{B}, p)$ be a probability space. Let $\mathscr{U} = (U_j)_{j \in J}$ and $\mathscr{V} = (V_k)_{k \in K}$ be two finite measurable partitions of X. Then one has*

(i) $H_p(\mathscr{U} \vee \mathscr{V}) \leq H_p(\mathscr{U}) + H_p(\mathscr{V})$;
(ii) *if* $\mathscr{V} \succ \mathscr{U}$ *then* $H_p(\mathscr{V}) \geq H_p(\mathscr{U})$;
(iii) *if* $\mathscr{U} \cong \mathscr{V}$ *then* $H_p(\mathscr{U}) = H_p(\mathscr{V})$;
(iv) *if* $T \colon X \to X$ *is a measure-preserving map then* $H_p(T^{-1}(\mathscr{U})) = H_p(\mathscr{U})$.

Proof. See for example [195, Section 4.3]. □

Let $(X, \mathscr{B}, p)$ be a probability space. Suppose that X is equipped with a *measure-preserving action* of a group G, that is, a family of measure-preserving maps $T_g \colon X \to X$, $g \in G$, such that

$$T_{g_1} \circ T_{g_2} = T_{g_1 g_2} \quad p\text{-a.e.}$$

for all $g_1, g_2 \in G$, and $T_{1_G} = \mathrm{Id}_X$.

Let $\mathscr{U}$ be a finite measurable partition of X. Consider the map $h^p_{\mathscr{U}} \colon \mathscr{P}_{\mathrm{fin}}(G) \to \mathbb{R}$ defined by

$$h^p_{\mathscr{U}}(A) := H_p(\mathscr{U}_A), \tag{14.122}$$

where $\mathscr{U}_A$ is defined by (14.120).

Proposition 14.81. *Let $(X, \mathscr{B}, p)$ be a probability space equipped with a measure-preserving action of a group G and let $\mathscr{U}$ be a finite measurable partition of X. Then the map $h^p_{\mathscr{U}} \colon \mathscr{P}_{\mathrm{fin}}(G) \to \mathbb{R}$ defined by (14.122) is non-decreasing, subadditive, and right-invariant.*

Proof. Mutatis mutandis, the proof is the same as that of Proposition 14.75 with Lemma 14.80 replacing Lemma 14.74. □

From Proposition 14.81 and Theorem 14.61, we deduce the following result.

Theorem 14.82. *Let $(X, \mathscr{B}, p)$ be a probability space equipped with a measure-preserving action of an amenable group G and let $\mathscr{U}$ be a finite measurable partition of X. Then, for every left-Følner net $(F_i)_{i \in I}$ of S, the limit*

$$h^{KS}(X, p, G; \mathscr{U}) := \lim_i \frac{h^p_{\mathscr{U}}(F_i)}{|F_i|}$$

exists and is finite. Moreover, $h^{KS}(X, p, G; \mathscr{U})$ does not depend on the choice of the left-Følner net $(F_i)_{i \in I}$.

The quantity $h^{\mathrm{KS}}(X, p, G) \in [0, +\infty) \cup \{+\infty\}$ defined by

$$h^{\mathrm{KS}}(X, p, G) := \sup_{\mathscr{U}} h^{\mathrm{KS}}(X, p, G; \mathscr{U}),$$

where $\mathscr{U}$ runs over all finite measurable partitions of X, is the *measure-theoretic entropy*, or *Kolmogorov–Sinai entropy*, of the measured dynamical system (X, p, G).

14.9 The Tarski Number

In this section we introduce and study the Tarski number of a group. It can be regarded as a "measure" of amenability or, rather, of non-amenability of groups.

Definition 14.83. Let G be a group. Suppose that G is paradoxical, so that it admits a paradoxical decomposition (of type (n, m)) as in (14.17) and (14.17). The number $n + m$, denoting the number of pieces involved in the paradoxical decomposition, is called the *complexity* or the *Tarski number* of the paradoxical decomposition. The infimum of all such Tarski numbers, taken over all the possible paradoxical decompositions of G, is called the *Tarski number* of G and is denoted by $\tau(G)$. If G is not paradoxical, we set $\tau(G) = \infty$.

Remark 14.84. Let G be a group. Suppose that G admits a paradoxical decomposition of type (n, m). Then G also admits a paradoxical decomposition of type (m, n) and, moreover, $n, m \geq 2$ (and therefore their corresponding complexity is at least 4). As a consequence, $\tau(G) \geq 4$. Moreover, if G is finitely generated, by virtue of Theorem 14.21 one has $\tau(G) = \infty$ if and only if G is amenable.

Example 14.85. In Example 14.20 we presented a paradoxical decomposition of type $(2, 2)$, and therefore of complexity 4, of the free group F_2. It follows from Remark 14.84 that $\tau(F_2) = 4$.

The following may be regarded as a quantitative version of Theorem 14.9.

Proposition 14.86. *Let G be a group and $H \leq G$ a subgroup. Then $\tau(G) \leq \tau(H)$.*

Proof. If H is not paradoxical, then $\tau(H) = \infty$ and there is nothing to prove. Thus we suppose that H is paradoxical. Let $H = A_1 \sqcup A_2 \sqcup \cdots \sqcup A_n \sqcup B_1 \sqcup B_2 \sqcup \cdots \sqcup B_m = a_1A_1 \sqcup a_2A_2 \sqcup \cdots \sqcup a_nA_n = b_1B_1 \sqcup b_2B_2 \sqcup \cdots \sqcup b_mB_m$ be a paradoxical decomposition of H using $\tau(H) = n+m$ pieces. Let $T \subseteq G$ be a right-transversal for H in G so that $G = \sqcup_{t\in T} Ht$. Let us set $\tilde{A}_i := \sqcup_{t\in T} A_i t$, for $i = 1,2,\ldots,n$, and $\tilde{B}_j = \sqcup_{t\in T} B_j t$, $j = 1,2,\ldots,m$. Then we have $G = \tilde{A}_1 \sqcup \tilde{A}_2 \sqcup \cdots \sqcup \tilde{A}_n \sqcup \tilde{B}_1 \sqcup \tilde{B}_2 \sqcup \cdots \sqcup \tilde{B}_m = a_1\tilde{A}_1 \sqcup a_2\tilde{A}_2 \sqcup \cdots \sqcup a_n\tilde{A}_n = b_1\tilde{B}_1 \sqcup b_2\tilde{B}_2 \sqcup \cdots \sqcup b_m\tilde{B}_m$ is a paradoxical decomposition of type (n,m) of G. This shows that $\tau(G) \le n+m = \tau(H)$. □

From Example 14.85, Example 1.19, Proposition 14.86, and Remark 14.84, we immediately deduce:

Corollary 14.87. *Let G be a group and suppose that it contains a subgroup isomorphic to the free group F_2. Then G is paradoxical and $\tau(G) = 4$. In particular, for all $n \ge 2$, the free group F_n, the general linear group $\mathrm{GL}(n,\mathbb{C})$, and the special linear group $\mathrm{SL}(n,\mathbb{C})$) are paradoxical and their Tarski numbers satisfy $\tau(F_n) = \tau(\mathrm{GL}(n,\mathbb{C})) = \tau(\mathrm{SL}(n,\mathbb{C})) = 4$.* □

The following may be regarded as a quantitative version of Theorem 14.11.

Proposition 14.88. *Let G be a group and $N \trianglelefteq G$ a normal subgroup. Then $\tau(G) \le \tau(G/N)$.*

Proof. Let us denote by $\pi\colon G \to G/N$ the canonical epimorphism. If G/N is not paradoxical, then $\tau(G/N) = \infty$ and there is nothing to prove. Thus we suppose that G/N is paradoxical and let $G/N = A_1 \sqcup A_2 \sqcup \cdots \sqcup A_n \sqcup B_1 \sqcup B_2 \sqcup \cdots \sqcup B_m = a_1A_1 \sqcup a_2A_2 \sqcup \cdots \sqcup a_nA_n = b_1B_1 \sqcup b_2B_2 \sqcup \cdots \sqcup b_mB_m$ be a paradoxical decomposition of G/N using $\tau(H) = n+m$ pieces. Choose a section $\varphi\colon G/N \longrightarrow G$. Note that every element $g \in G$ is then uniquely expressed as a product $g = \varphi(gN)n$ where $gN \in G/N$ and $n \in N$; in particular $G = \varphi(G/N)N$. Let us set $\tilde{A}_i = \pi^{-1}(A_i) = \varphi(A_i)N \subseteq G$ and $\tilde{a}_i = \varphi(a_i) \in G$, for $i = 1,2,\ldots,n$ (resp. $\tilde{B}_j = \pi^{-1}(B_j) = \phi(B_j)N \subseteq G$ and $\tilde{b}_j = \varphi(b_i) \in G$, for $j = 1,2,\ldots,m$). Observe that $\pi^{-1}(a_iA_i) = \varphi(a_i)\pi^{-1}(A_i) = \tilde{a}_i\tilde{A}_i$ and, similarly, $\pi^{-1}(b_jB_j) = \tilde{b}_j\tilde{B}_j$ so that we have the paradoxical decomposition $G = \tilde{A}_1 \sqcup \tilde{A}_2 \sqcup \cdots \sqcup \tilde{A}_n \sqcup \tilde{B}_1 \sqcup \tilde{B}_2 \sqcup \cdots \sqcup \tilde{B}_m = \phi(g_1)\tilde{A}_1 \sqcup \phi(g_2)\tilde{A}_2 \sqcup \cdots \sqcup \phi(g_n)\tilde{A}_n = \phi(h_1)\tilde{B}_1 \sqcup \phi(h_2)\tilde{B}_2 \sqcup \cdots \sqcup \phi(h_m)\tilde{B}_m$. This shows that $\tau(G) \le n+m = \tau(G/N)$ □

Theorem 14.89 (Jónsson). *Let G be a group. Then $\tau(G) = 4$ if and only if G contains F_2.*

Proof. Suppose that $\tau(G) = 4$ and let

$$G = A_1 \sqcup A_2 \sqcup B_1 \sqcup B_2 = a_1A_1 \sqcup a_2A_2 = b_1B_1 \sqcup b_2B_2$$

be a paradoxical decomposition of G using 4 pieces. Set $g = a_1^{-1}a_2$ and $h = b_1^{-1}b_2$. Then

$$G = A_1 \sqcup A_2 \sqcup B_1 \sqcup B_2 = A_1 \sqcup gA_2 = B_1 \sqcup hB_2$$

and

$$A_1 = G \setminus gA_2 = gA_1 \sqcup gB_1 \sqcup gB_2$$
$$A_1 \supset gA_1 \supset g^2 A_1 \supset \ldots \supset g^{k-1} A_1 \supset g^k B_j \ (k \geq 1 \text{ and } j = 1,2)$$
$$A_2 = G \setminus g^{-1} A_1 = g^{-1} A_2 \sqcup g^{-1} B_1 \sqcup g^{-1} B_2$$
$$A_2 \supset g^{-1} A_1 \supset g^{-2} A_1 \supset \ldots \supset g^{-k+1} A_1 \supset g^{-k} B_j \ (k \geq 1 \text{ and } j = 1,2).$$

Thus

$$g^k B_j \subseteq A_1 \cup A_2 \text{ and } h^k A_j \subseteq B_1 \cup B_2$$

for all $k \in \mathbb{Z}, k \neq 0$, and $j = 1,2$. Observe that both g and h are non-torsion elements (**exercise**) so that, by Klein's Ping-Pong lemma (Theorem 1.17), they generate a free subgroup of rank 2. The converse is the first statement in Corollary 14.87. □

Proposition 14.90. *Let G be a paradoxical group. Then there exists a finitely generated subgroup $H \subseteq G$ with $\tau(H) = \tau(G)$.*

Proof. Let $G = A_1 \sqcup A_2 \sqcup \cdots \sqcup A_n \sqcup B_1 \sqcup B_2 \sqcup \cdots \sqcup B_m = a_1 A_1 \sqcup a_2 A_2 \sqcup \cdots \sqcup a_n A_n = b_1 B_1 \sqcup b_2 B_2 \sqcup \cdots \sqcup b_m B_m$ be a paradoxical decomposition of G of complexity $n + m = \tau(G)$. Denoting by $H := \langle a_1, a_2, \ldots, a_n, b_1, b_2, \ldots, b_m \rangle$ the subgroup generated by the elements involved in the paradoxical decomposition, one has $(a_i A_i) \cap H = a_i (A_i \cap H)$ and an analogous relation for the b_j's and B_j's. Setting $A'_i = A_i \cap H$ and $B'_j = B_j \cap H$ one obtains the paradoxical decomposition $H = A'_1 \sqcup A'_2 \sqcup \cdots \sqcup A'_n \sqcup B'_1 \sqcup B'_2 \sqcup \cdots \sqcup B'_m = a_1 A'_1 \sqcup a_2 A'_2 \sqcup \cdots \sqcup a_n A'_n = b_1 B'_1 \sqcup b_1 B'_1 \sqcup \cdots \sqcup b_m B'_m$ of complexity $n + m$. We deduce that $\tau(H) \leq n + m = \tau(G)$. The converse inequality now follows from Proposition 14.86 □

Proposition 14.91. *Let G be a torsion group. Then $\tau(G) \geq 6$.*

Proof. By virtue of Remark 14.84, it is enough to show that G cannot admit a paradoxical decomposition of type $(2,m)$ for any $m \geq 2$. Suppose by contradiction that

$$G = A_1 \sqcup A_2 \sqcup B_1 \sqcup B_2 \sqcup \cdots \sqcup B_m = a_1 A_1 \sqcup a_2 A_2 = b_1 B_1 \sqcup b_2 B_2 \sqcup \cdots \sqcup b_m B_m$$

is a paradoxical decomposition of G of type $(2,m)$. Denoting by k the order of the element $g := a_1^{-1} a_2$, one deduces, as in the proof of Theorem 14.89,

$$A_1 \supset gA_1 \supset gA_2 \supset \ldots \supset g^{k-1} A_1 \supset g^k B_1 = B_1,$$

which is clearly absurd. □

14.10 Isoperimetric Profiles of Groups

Definition 14.92. Given an infinite group G generated by a finite symmetric subset X, we define the *isoperimetric profile of G* with respect to X as the function from $\mathbb{N}$ onto itself given by

$$I_\circ(n; G, X) := \inf_{\Omega \subseteq G : |\Omega| = n} |\partial_X(\Omega)|$$

for each $n \in \mathbb{N}$.

The following lemma shows that the asymptotic behavior of this function is independent of the set of generators X. Hence this provides another asymptotic geometric invariant of groups.

Lemma 14.93. *Let X and X' be two finite generating symmetric subsets of an infinite group G. Then $I_\circ(n;G,X) \sim I_\circ(n;G,X')$.*

Proof. We start with two easy observations.

If $S, T, \Omega \subseteq G$ are finite subsets of G, then

$$\partial_{S\cup T}(\Omega) = (S\cup T)\Omega \setminus \Omega \subseteq ((S\Omega)\setminus\Omega)\cup((T\Omega)\setminus\Omega) = \partial_S(\Omega)\cup\partial_T(\Omega),$$

so that

$$|\partial_{S\cup T}(\Omega)| \leq |\partial_S(\Omega)| + |\partial_T(\Omega)|. \tag{14.123}$$

Moreover,

$$\partial_{ST}(\Omega) = (ST)\Omega \setminus \Omega \subseteq (S(T\Omega\setminus\Omega))\cup(S\Omega\setminus\Omega) = S\partial_T(\Omega)\cup\partial_S(\Omega),$$

so that

$$|\partial_{ST}(\Omega)| \leq |S|\cdot|\partial_T(\Omega)| + |\partial_S(\Omega)|. \tag{14.124}$$

Now, since X is a generating subset, there exists an $m \geq 1$ such that $X' \subseteq \cup_{i=0}^m X^i$. Hence, for a finite subset $\Omega \subseteq G$, using (14.123) and inductively (14.124), we have

$$\begin{aligned}
|\partial_{X'}(\Omega)| &\leq |\partial_{\cup_{i=0}^m X^i}(\Omega)| \\
&\leq \sum_{i=0}^m |\partial_{X^i}(\Omega)| \\
&\leq \sum_{i=1}^m \sum_{j=0}^{i-1} |X^j|\cdot|\partial_X(\Omega)| \\
&\leq m|X|^{m+1}\cdot|\partial_X(\Omega)|.
\end{aligned}$$

Since m does not depend on Ω, this gives $I_\circ(n;G,X') \preceq I_\circ(n;G,X)$.

Analogously, exchanging the roles of X and X', we get the other inequality, and hence the result. □

As a consequence, when it does not create any confusion, for every finite generating subset $X \subseteq G$ we will denote $I_\circ(n;G,X)$ also by $I_\circ(n;G)$.

We want to stress here a few properties of this invariant. First of all notice that

$$|\partial_X(\Omega)| = \left|\bigcup_{x\in X}(x\Omega\setminus\Omega)\right| \leq |X|\cdot|\Omega|,$$

hence we always have

$$I_\circ(n;G,X) \preceq n, \tag{14.125}$$

i.e. the isoperimetric profile of a group is a sublinear function. The following proposition shows that the linearity of this invariant is equivalent to the nonamenability of the group.

Proposition 14.94. *A finitely generated group G is nonamenable if and only if there exists a finite symmetric generating subset $X \subseteq G$ such that*

$$I_\circ(n;G,X) \sim n.$$

Proof. If G is nonamenable then there exist $X \subseteq G$ finite and $\varepsilon > 0$ such that for any finite subset $\Omega \subseteq G$

$$|\partial_X(\Omega)| = |\Omega \cup X\Omega| - |\Omega| \geq \varepsilon|\Omega|.$$

Possibly enlarging X, we can assume that X is symmetric and generates G. Then we have

$$I_\circ(n;G,X) \succeq n,$$

that, together with (14.125), gives

$$I_\circ(n;G,X) \sim n.$$

If G is amenable, then for any finite subset $X \subseteq G$ and any $\varepsilon > 0$ we can find a finite subset $\Omega \subseteq G$ (depending on ε) such that

$$|\partial_X(\Omega)| = |\Omega \cup X\Omega| - |\Omega| < \varepsilon|\Omega|,$$

and this prevents $I_\circ(n;G,X) \succeq n$, as we wanted. □

In this sense the isoperimetric profile can be viewed as a measure of the amenability of the group.

Looking at the definition, it becomes immediately clear that to compute the isoperimetric profile, even for easy examples, it is quite hard to prove a lower bound. The key result which lies at the heart of almost any computation of isoperimetric profiles is the following remarkable inequality, which is due to Coulhon and Saloff-Coste. We need a definition.

Let G be an infinite group generated by a finite symmetric subset X. For $\lambda > 0$, define

$$\Phi(\lambda) := \min\{n \in \mathbb{N} : b_X^G(n) > \lambda\}. \tag{14.126}$$

This is essentially the inverse function of the growth of G. The following proof is due to Gromov.

Theorem 14.95 (Coulhon and Saloff-Coste). *Let G be an infinite group, generated by a finite symmetric subset X. Then for any finite nonempty subset Ω of G we have*

$$|\partial_X(\Omega)| \geq \frac{|\Omega|}{2\Phi(2|\Omega|)}.$$

Proof. Observe that, by (14.123), if $a, b \in G$ then

$$|\partial_{ab}(\Omega)| \leq |\partial_a(\Omega)| + |\partial_b(\Omega)|.$$

Hence by induction, if $a_i \in G$ for $i = 1, \dots, n$, then

$$|\partial_{a_1 a_2 \cdots a_n}(\Omega)| \leq \sum_{i=1}^{n} |\partial_{a_i}(\Omega)| \leq n \max_{1 \leq i \leq n} |\partial_{a_i}(\Omega)|.$$

This immediately implies that for any $m \geq 1$ and $y \in B_X(1_G, m)$

$$|\partial_y(\Omega)| \leq m \max_{x \in X} |\partial_x(\Omega)|.$$

Therefore

$$\begin{aligned}
|\partial_X(\Omega)| \geq \max_{x \in X} |\partial_x(\Omega)| &\geq \frac{1}{m} \frac{1}{b_X^G(m)} \sum_{y \in B_X(1_G, m)} |\partial_y(\Omega)| \\
&= \frac{1}{m} \frac{1}{b_X^G(m)} \sum_{y \in B_X(1_G, m)} (|y\Omega| - |y\Omega \cap \Omega|) \\
&= \frac{1}{m} \frac{1}{b_X^G(m)} \left(b_X^G(m)|\Omega| - \sum_{y \in B_X(1_G, m)} |y\Omega \cap \Omega| \right).
\end{aligned}$$

Lemma 14.96.

$$\sum_{y \in B_X(1_G, m)} |y\Omega \cap \Omega| \leq |\Omega|^2.$$

Proof. We have

$$\sum_{y \in B_X(1_G, m)} |y\Omega \cap \Omega| = \sum_{y \in B_X(1_G, m)} \sum_{x_1 \in \Omega} \sum_{x_2 \in \Omega} \chi(yx_1 = x_2),$$

where $\chi(\mathscr{P}) = 1$ if the proposition $\mathscr{P}$ is true, $\chi(\mathscr{P}) = 0$ if $\mathscr{P}$ is false. By cancellation property, the ordered pair $(x_1, x_2) \in \Omega \times \Omega$ uniquely determines y such that $yx_1 = x_2$, hence

$$\sum_{y \in B_X(1_G, m)} \sum_{x_1 \in \Omega} \sum_{x_2 \in \Omega} \chi(yx_1 = x_2) \leq |\Omega \times \Omega| = |\Omega|^2.$$

The proof of Lemma 14.96 follows. □

By the lemma and taking $m := \Phi(2|\Omega|) = \min\{n \in \mathbb{N} : b_X^G(n) > 2|\Omega|\}$, we get

$$\begin{aligned}
|\partial_X(\Omega)| &\geq \frac{1}{m} \frac{1}{b_X^G(m)} \left(b_X^G(m)|\Omega| - \sum_{y \in B_X(1_G, m)} |y\Omega \cap \Omega| \right) \\
&\geq \frac{1}{m} \frac{1}{b_X^G(m)} \left(b_X^G(m)|\Omega| - |\Omega|^2 \right) \\
&= \frac{|\Omega|}{m} \left(1 - \frac{|\Omega|}{b_X^G(m)} \right) \\
&> \frac{|\Omega|}{2m} = \frac{|\Omega|}{2\Phi(2|\Omega|)}.
\end{aligned}$$

This completes the proof of Theorem 14.95. □

Corollary 14.97. *Let G be a finitely generated infinite group. Then*

(1) *if $b^G(n) \succeq n^d$ for some $d \geq 1$, then $I_\circ(n;G) \succeq n^{(d-1)/d}$;*
(2) *if $b^G(n) \sim e^n$, then $I_\circ(n;G) \succeq n/\log n$.*

Proof. Let $X \subseteq G$ be a finite symmetric generating subset of G.

Suppose that $b_X^G(n) \succeq n^d$ for some $d \geq 1$. So there exists a $C > 0$ such that $b_X^G(n) \geq Cn^d$ for all $n \geq 1$. This implies that, for

$$k \geq \left(\frac{4n}{C}\right)^{\frac{1}{d}},$$

we have

$$b_X^G(k) \geq Ck^d \geq 4n > 2n,$$

so that

$$\Phi(2n) = \min\{m \in \mathbb{N} : b_X^G(m) > 2n\} \leq \left(\frac{4n}{C}\right)^{\frac{1}{d}}.$$

By the Coulhon–Saloff-Coste inequality, for all $n \geq 1$, we have

$$I_\circ(n;G) \geq \frac{n}{2\Phi(2n)} \geq \left(\frac{C}{4n}\right)^{\frac{1}{d}} \frac{n}{2} = \left(\frac{C^{\frac{1}{d}}}{2^{(d+2)/d}}\right) n^{(d-1)/d},$$

proving (1). The proof of (2) is analogous (**exercise**). □

14.11 Følner Functions

Let G be an infinite finitely generated amenable group. Let $X \subseteq G$ be a finite symmetric generating subset of G. For every $n \in \mathbb{N}$, $n \geq 1$, we define $F_\circ(n;G,X)$ to be the minimal cardinality of a finite subset $\Omega \subseteq G$ such that

$$|\partial_X(\Omega)| \leq \frac{|\Omega|}{n}.$$

The function $n \mapsto F_\circ(n;G,X)$ is called the *Følner function* of G with respect to X.

The proof of the following lemma is analogous to that of Lemma 14.93 and we leave it as an **exercise**.

Lemma 14.98. *Let X and X' be two finite generating symmetric subsets of an infinite group G. Then $F_\circ(n;G,X) \sim F_\circ(n;G,X')$.*

As a consequence, when it does not create any confusion, for every finite generating subset $X \subseteq G$ we will denote $F_\circ(n;G,X)$ also by $F_\circ(n;G)$.

We want to stress here a few properties of this invariant. Notice that, since G is infinite, $\partial_X(\Omega) \neq \varnothing$ for all finite subsets $\Omega \subseteq G$, we must have $F_\circ(n;G,X) \geq n$ for all $n \geq 1$. Moreover, the function $F_\circ(n;G,X)$ is non-decreasing (**exercise**).

Proposition 14.99. *Let G be a finitely generated amenable group, and let $H \subseteq G$ be a finite normal subgroup of G. Then*

$$F_\circ(n;G) \preceq F_\circ(n;G/H).$$

Proof. The argument is a quantitative version of the proof of Theorem 14.12.

Let us recall the setting of that proof.

Let $\pi\colon G \to \overline{G} := G/H$ be the canonical projection, let $X \subseteq G$ be a finite generating subset of G, and set $\overline{X} := \pi(X)$. Fix $n \in \mathbb{N}$, and let $\overline{\Omega}_1 \subseteq \overline{G}$ be a finite subset such that

$$|\partial_{\overline{x}}(\overline{\Omega}_1)| < \frac{1}{2n}|\overline{\Omega}_1| \text{ for all } \overline{x} \in \overline{X}.$$

This implies

$$|\partial_{\overline{X}}(\overline{\Omega}_1)| \leq \sum_{\overline{x}\in\overline{X}} |\partial_{\overline{x}}(\overline{\Omega}_1)| \leq \frac{|\overline{X}|}{2n}|\overline{\Omega}_1|.$$

Let $\Omega_1 \subseteq G$ be such that $\pi(\Omega_1) = \overline{\Omega}_1$ and $|\Omega_1| = |\overline{\Omega}_1|$.

In the proof of Theorem 14.12 we showed that there exists a finite subset $\Omega_2 \subseteq H$ (possibly $\Omega_2 = \{1\}$, as in Case 1 therein) such that $\Omega := \Omega_1\Omega_2 \subseteq G$ satisfies

$$|\partial_x(\Omega)| \leq \frac{1}{n}|\Omega| \text{ for all } x \in X.$$

Therefore

$$|\partial_X(\Omega)| \leq \sum_{x\in X} |\partial_x(\Omega)| \leq \frac{|X|}{n}|\Omega|$$

and $|\Omega| \leq |\Omega_1||\Omega_2| \leq |\Omega_1||H|$, as H is finite. Choosing a set Ω_1 with $|\Omega_1| = F_\circ\left(\frac{2|X|}{|\overline{X}|}n; G, X\right)$, this shows that

$$F_\circ(n;G,X) \leq |H| F_\circ\left(\frac{2|X|}{|\overline{X}|}n; \overline{G}, \overline{X}\right).$$

It follows that

$$F_\circ(n;G) \preceq F_\circ(n;G/H).$$

The proof of Proposition 14.99 is complete. □

Proposition 14.100. *Let G be an infinite finitely generated amenable group. Then*

$$F_\circ(n;G) \succeq b^G(n).$$

Proof. Let $X \subseteq G$ be a finite symmetric generating subset of G. Let $\Omega \subseteq G$ be a finite subset such that

$$|\partial_X(\Omega)| \leq \frac{|\Omega|}{n},$$

equivalently

$$n \leq \frac{|\Omega|}{|\partial_X(\Omega)|}. \tag{14.127}$$

Notice that such a subset exists as G is amenable. By the Coulhon–Saloff-Coste inequality, we have also

$$|\partial_X(\Omega)| \geq \frac{|\Omega|}{2\Phi(2|\Omega|)},$$

where Φ is as in (14.126), equivalently

$$\frac{|\Omega|}{|\partial_X(\Omega)|} \leq 2\Phi(2|\Omega|). \tag{14.128}$$

Comparing (14.127) and (14.128), we get

$$\Phi(2|\Omega|) \geq \frac{n}{2} > \frac{n}{3},$$

so, by the definition of Φ, we deduce

$$b_X^G\left(\frac{n}{3}\right) \leq 2|\Omega|.$$

Choosing Ω such that $|\Omega| = F_\circ(n;G,X)$, we get

$$F_\circ(n;G,X) \geq \frac{1}{2} b_X^G\left(\frac{n}{3}\right),$$

so that $F_\circ(n;G) \succeq b^G(n)$, as we wanted. □

Remark 14.101. Though there seems to be a strong relation (asymptotic inequality? asymptotic equivalence?) between $I_\circ(n;G)$ and $n/F_\circ^{-1}(n;G)$ for an infinite finitely generated amenable group G, no result of this kind *in full generality* is known to the authors.

For example, even the monotonicity of the isoperimetric profile of groups is not known in general (notice that in geometry the isoperimetric profile is not always monotone).

Of course more can be said under some mild assumption on these functions, but unfortunately these assumptions cannot be easily checked before an actual computation.

14.12 Følner Functions of Groups of Polynomial Growth

Theorem 14.102. *Let G be a finitely generated nilpotent group. Then*

$$F_\circ(n;G) \sim b^G(n).$$

Proof. By Proposition 14.100 we only need to prove

$$F_\circ(n;G) \preceq b^G(n).$$

Consider the lower central series of G

$$G = \gamma_1(G) \geq \gamma_2(G) \geq \cdots \geq \gamma_c(G) \geq \gamma_{c+1}(G) = \{1_G\}.$$

By replacing G by G/G_{tor}, we can assume that G has a series

$$G = G_1 \geq G_2 \geq \cdots \geq G_c \geq \{1\}$$

that is central (i.e. $[G_i, G_j] \subseteq G_{i+j}$), the factors G_i/G_{i+1} are torsion-free abelian, and $\mathrm{rk}(G_i/G_{i+1}) = r_i := \mathrm{rk}(\gamma_i(G)/\gamma_{i+1}(G))$. This reduction is justified in Section 7.8 for the growth function $b^G(n)$, and it also works for the Følner function $F_\circ(n;G)$ by virtue of Proposition 14.99.

For each $i = 1, 2, \ldots, c$, let $x_{ij}G_{i+1}$, $j = 1, 2, \ldots, r_i$, be free generators of G_i/G_{i+1}. Notice that with this choice the subgroup G_ℓ is generated by the x_{ij}'s with $i \geq \ell$, for all $\ell = 1, 2, \ldots, c$. In particular, the symmetric subset

$$X := \{x_{ij}^{\pm 1} : 1 \leq i \leq c, 1 \leq j \leq r_i\}$$

generates G, and every element $g \in G$ can be uniquely written in the *canonical form*

$$\begin{aligned} g &= x_{11}^{e_{11}} x_{12}^{e_{12}} \cdots x_{1r_1}^{e_{1r_1}} \cdots x_{c1}^{e_{c1}} \cdots x_{cr_c}^{e_{cr_c}} \\ &= \prod_{j=1}^{r_1} x_{1j}^{e_{1j}} \cdot \prod_{j=1}^{r_2} x_{2j}^{e_{2j}} \cdots \prod_{j=1}^{r_c} x_{cj}^{e_{cj}} \end{aligned}$$

where $e_{ij} \in \mathbb{Z}$ for all i, j. Given g in its canonical form, we set

$$E_i = E_i(g) := \sum_{j=1}^{r_i} |e_{ij}| \text{ for all } i = 1, 2, \ldots, c.$$

Let $X_i := \{x_{ij}^{\pm 1} : 1 \leq j \leq r_i\}$ denote the set of *generators of weight* i, for $i = 1, 2, \ldots, c$. Notice that $X = \cup_i X_i$.

Given $n \in \mathbb{N}$ and $C > 0$, we define the finite subsets of G

$$\Omega_{n,C} := \{g \in G : E_1(g) \leq n \text{ and } E_i(g) \leq n^i + Cn^{i-1} \text{ for all } i = 2, 3, \ldots, c\}$$

and

$$\omega_n := \{g \in G : E_1(g) \leq n - 1 \text{ and } E_i(g) \leq n^i \text{ for all } i = 2, 3, \ldots, c\}.$$

Notice that $\omega_n \subseteq \Omega_{n,C}$ for all n.

We want to show that there is a constant $C > 0$ independent of n such that

$$\delta_X(\Omega_{n,C}) \cap \omega_n = \varnothing. \tag{14.129}$$

In this way, we will be able to bound from above $|\delta_X(\Omega_{n,C})|$ with $|\Omega_{n,C}| - |\omega_n|$, which in turn will give us the required bound on the Følner function.

In order to prove this, we need to show that there exists a constant $C > 0$ such that, for every $y \in X$ and $g \in \omega_n$, $yg \in \Omega_{n,C}$. So, as in the proof of the upper bound in Bass' formula (cf. Section 7.8), given g in canonical form, we need to put yg in canonical form.

Using the commutator relation $yx = x[x, y^{-1}]y$, we start by rewriting

$$yg = y\prod_{j=1}^{r_1} x_{1j}^{e_{1j}} \cdot \prod_{j=1}^{r_2} x_{2j}^{e_{2j}} \cdots \prod_{j=1}^{r_c} x_{cj}^{e_{cj}}$$

as

$$\prod_{j=1}^{r_1} x_{1j}^{f_{1j}} \cdot \eta_1 \cdot \prod_{j=1}^{r_2} x_{2j}^{e_{2j}} \cdots \prod_{j=1}^{r_c} x_{cj}^{e_{cj}},$$

where $f_{1j} \in \mathbb{Z}$ for all j, and η_1 is an element of G_2.

Example 14.103. Let $g = x_{11}^2 x_{13}^{-1} x_{22}^2 x_{23}^{-1} x_{31}$ and $y = x_{24}$. Then

$$\begin{aligned}
yg &= x_{24} x_{11}^2 x_{13}^{-1} x_{22}^2 x_{23}^{-1} x_{31} \rightsquigarrow \\
&\rightsquigarrow x_{11}[x_{11}, x_{24}^{-1}] x_{24} x_{11} x_{13}^{-1} x_{22}^2 x_{23}^{-1} x_{31} \\
&\rightsquigarrow x_{11}[x_{11}, x_{24}^{-1}] x_{11}[x_{11}, x_{24}^{-1}] x_{24} x_{13}^{-1} x_{22}^2 x_{23}^{-1} x_{31} \\
&\rightsquigarrow x_{11}[x_{11}, x_{24}^{-1}] x_{11}[x_{11}, x_{24}^{-1}] x_{13}^{-1}[x_{13}^{-1}, x_{24}^{-1}] x_{24} x_{22}^2 x_{23}^{-1} x_{31} \\
&\rightsquigarrow x_{11}^2[x_{11}, [x_{11}, x_{24}^{-1}]^{-1}][x_{11}, x_{24}^{-1}][x_{11}, x_{24}^{-1}] x_{13}^{-1}[x_{13}^{-1}, x_{24}^{-1}] x_{24} x_{22}^2 x_{23}^{-1} x_{31} \\
&\rightsquigarrow x_{11}^2[x_{11}, [x_{11}, x_{24}^{-1}]^{-1}][x_{11}, x_{24}^{-1}] x_{13}^{-1}[x_{13}^{-1}, [x_{11}, x_{24}^{-1}]^{-1}][x_{11}, x_{24}^{-1}] \cdot \\
&\qquad \cdot [x_{13}^{-1}, x_{24}^{-1}] x_{24} x_{22}^2 x_{23}^{-1} x_{31} \\
&\rightsquigarrow x_{11}^2[x_{11}, [x_{11}, x_{24}^{-1}]^{-1}] x_{13}^{-1}[x_{13}^{-1}, [x_{11}, x_{24}^{-1}]^{-1}][x_{11}, x_{24}^{-1}][x_{13}^{-1}, [x_{11}, x_{24}^{-1}]^{-1}] \cdot \\
&\qquad \cdot [x_{11}, x_{24}^{-1}][x_{13}^{-1}, x_{24}^{-1}] x_{24} x_{22}^2 x_{23}^{-1} x_{31} \\
&\rightsquigarrow x_{11}^2 x_{13}^{-1}[x_{13}^{-1}, [x_{11}, [x_{11}, x_{24}^{-1}]^{-1}]^{-1}][x_{11}, [x_{11}, x_{24}^{-1}]^{-1}][x_{13}^{-1}, [x_{11}, x_{24}^{-1}]^{-1}] \cdot \\
&\qquad \cdot [x_{11}, x_{24}^{-1}][x_{13}^{-1}, [x_{11}, x_{24}^{-1}]^{-1}][x_{11}, x_{24}^{-1}][x_{13}^{-1}, x_{24}^{-1}] x_{24} x_{22}^2 x_{23}^{-1} x_{31},
\end{aligned}$$

so $yg = x_{11}^2 x_{13}^{-1} \eta_1 x_{22}^2 x_{23}^{-1} x_{31}$, where

$$\begin{aligned}
\eta_1 = {} & [x_{13}^{-1}, [x_{11}, [x_{11}, x_{24}^{-1}]^{-1}]^{-1}][x_{11}, [x_{11}, x_{24}^{-1}]^{-1}][x_{13}^{-1}, [x_{11}, x_{24}^{-1}]^{-1}][x_{11}, x_{24}^{-1}] \cdot \\
& \cdot [x_{13}^{-1}, [x_{11}, x_{24}^{-1}]^{-1}][x_{11}, x_{24}^{-1}][x_{13}^{-1}, x_{24}^{-1}] x_{24}.
\end{aligned}$$

Notice that, modulo the identity $[a,b]^{-1} = [b,a]$, the element η_1 is a product of commutators of the form

$$[y_1, [y_2, [\ldots, [y_{t-1}, y_t] \cdots]]]$$

with $y_i \in X$ for all i, and possibly y.

Since η_1 is in G_2, we can express it as a word in the $x_{ij}^{\pm 1}$ with $i \geq 2$.

Now we can iterate the previous process for every generator occurring in η_1, and rewrite

$$yg = \prod_{j=1}^{r_1} x_{1j}^{f_{1j}} \cdot \eta_1 \cdot \prod_{j=1}^{r_2} x_{2j}^{e_{2j}} \cdots \prod_{j=1}^{r_c} x_{cj}^{e_{cj}}$$

as

$$\prod_{j=1}^{r_1} x_{1j}^{f_{1j}} \cdot \prod_{j=1}^{r_2} x_{2j}^{f_{2j}} \cdot \eta_2 \cdot \prod_{j=1}^{r_3} x_{3j}^{e_{3j}} \cdots \prod_{j=1}^{r_c} x_{cj}^{e_{cj}},$$

where $f_{2j} \in \mathbb{Z}$ for all j, and η_2 is an element of G_3. In general, for every $s \geq 1$, we can rewrite yg as

$$\prod_{j=1}^{r_1} x_{1j}^{f_{1j}} \cdot \prod_{j=1}^{r_2} x_{2j}^{f_{2j}} \cdots \prod_{j=1}^{r_s} x_{sj}^{f_{sj}} \cdot \eta_s \cdot \prod_{j=1}^{r_{s+1}} x_{s+1,j}^{e_{s+1,j}} \cdots \prod_{j=1}^{r_c} x_{cj}^{e_{cj}},$$

where $f_{ij} \in \mathbb{Z}$ for all i and j, and η_s is an element of G_{s+1}, which therefore can be expressed as a word in the $x_{ij}^{\pm 1}$ with $i \geq s+1$.

Notice that, if we set $F_i := \sum_{j=1}^{r_i} |f_{ij}|$, then $F_i = E_i(yg)$ for all $i = 1, 2, \ldots, c$. We want to estimate the F_i.

Observe that in the rewriting process, η_s is a product of commutators of the form

$$[y_1, [y_2, [\ldots, [y_{t-1}, y_t] \cdots]]],$$

where $y_i \in X$ for all i, and possibly y.

Set

$$A := \sup\{\ell_X([y_1, [y_2, [\ldots, [y_{t-1}, y_t] \cdots]]]) : 1 \leq t \leq c, y_i \in X \text{ for all } i\}.$$

We want to estimate the number $\ell(k, s)$ of generators of weight k in η_s.

Claim. *For every $k = 1, 2, \ldots, c$ and $s = 1, 2, \ldots, c$ there exist polynomials $H_s^{(k)} \in \mathbb{R}[t_1, t_2, \ldots, t_s]$ and a constant $C > 0$ independent of k and s such that*

1. *$\ell(k, s) \leq H_s^{(k)}(E_1, E_2, \ldots, E_s)$, and*
2. *$H_s^{(k)}(n, n^2, \ldots, n^s) \leq Cn^{k-1}$ for all $n \in \mathbb{N} \setminus \{0\}$.*

We proceed by induction on s. Since there are finitely many values of s, in order to check property (2) it will be enough to find a constant $C_s > 0$ for every s, and then pick their maximum.

For $s = 1$, we set

$$H_1^{(k)}(t_1) := 1 + A\left(t + t^2 + \cdots + t^{k-1}\right).$$

Since for $C_1 := 1 + cA$ property (2) is satisfied, we check property (1). We already observed that, other than possibly y, in η_1 we will have factors of the form

$$[y_1, [y_2, [\ldots, [y_{t-1}, y_t] \cdots]]]$$

where $y_i \in X$ for all i. Notice that for every $t \geq 2$ such a commutator must lie in G_t, hence it will be rewritten as a product of generators of weight $\geq t$. Hence we need to estimate the number of such commutators of length t for every $t \geq 2$.

In the rewriting process (cf. Example 14.103), we moved some of the $x_{1j}^{\pm 1}$ (possibly all of them) to the left of y, and each of these exchanges produced exactly one commutator of length 2. Then we had to move some of the $x_{1j}^{\pm 1}$ to the left of some of these commutators of length 2 that we just created, producing each time a commutator of length 3. And so on. Since the number of $x_{1j}^{\pm 1}$ occurring in g is $E_1 = E_1(g)$, the number of commutators of length t occurring in η_1 is at most E_1^{t-1} for every $t \geq 2$.

Summarizing, we get the estimate

$$\ell(k,1) \leq 1 + AE_1 + AE_1^2 + \cdots + AE_1^{k-1} = H_1^{(k)}(E_1).$$

This proves the base of the induction.

Now assume that we have found $H_j^{(k)}$ and $C_j > 0$ for $j = 1,2,\ldots,s-1$ that satisfy properties (1) and (2). Set

$$H_s^{(k)}(t_1,t_2,\ldots,t_s) := H_{s-1}^{(k)}(t_1,t_2,\ldots,t_{s-1}) + At_s \sum_{i=0}^{k-s} H_{s-1}^{(k-s-i)}(t_1,t_2,\ldots,t_{s-1}).$$

Since by induction for all $n \geq 1$

$$H_{s-1}^{(k)}(n,n^2,\ldots,n^{s-1}) \leq C_{s-1}n^{k-1}$$

and

$$H_{s-1}^{(k-s-i)}(n,n^2,\ldots,n^{s-1}) \leq C_{s-1}n^{k-s-i-1} \leq C_{s-1}n^{k-s-1},$$

we have for all $n \geq 1$

$$\begin{aligned} H_s^{(k)}(n,n^2,\ldots,n^s) &\leq C_{s-1}n^{k-1} + An^s \sum_{i=0}^{k-s} C_{s-1}n^{k-s-1} \\ &\leq C_{s-1}(1+cA)n^{k-1}. \end{aligned}$$

Hence $H_s^{(k)}$ with $C_s := C_{s-1}(1+cA)$ satisfies property (2). This shows that for the constant C in the statement we can take

$$C := (1+cA)^c.$$

In order to estimate $\ell(k,s)$, we need to understand what happens in going from

$$\prod_{j=1}^{r_1} x_{1j}^{f_{1j}} \cdot \prod_{j=1}^{r_2} x_{2j}^{f_{2j}} \cdots \prod_{j=1}^{r_{s-1}} x_{s-1,j}^{f_{s-1,j}} \cdot \left(\eta_{s-1} \cdot \prod_{j=1}^{r_s} x_{sj}^{e_{sj}}\right) \cdot \prod_{j=1}^{r_{s+1}} x_{s+1,j}^{e_{s+1,j}} \cdots \prod_{j=1}^{r_c} x_{cj}^{e_{cj}}$$

to

$$\prod_{j=1}^{r_1} x_{1j}^{f_{1j}} \cdot \prod_{j=1}^{r_2} x_{2j}^{f_{2j}} \cdots \prod_{j=1}^{r_{s-1}} x_{s-1,j}^{f_{s-1,j}} \cdot \left(\prod_{j=1}^{r_s} x_{sj}^{f_{sj}} \cdot \eta_s\right) \cdot \prod_{j=1}^{r_{s+1}} x_{s+1,j}^{e_{s+1,j}} \cdots \prod_{j=1}^{r_c} x_{cj}^{e_{cj}}.$$

Observe that, in the rewriting process, a generator of weight k in η_s either comes from a generator of weight k in η_{s-1}, or from a commutator of the form $[x_{s,j}^{\pm 1}, z]$ where z is a generator of weight $\leq k-s$ occurring in η_{s-1}. By induction, the number of generators of the first kind, i.e. $\ell(k,s-1)$, is bounded from above by $H_{s-1}^{(k)}(E_1,E_2,\ldots,E_{s-1})$. Again by induction, the number of generators of the latter kind is clearly bounded by

$$AE_s \sum_{i=0}^{k-s} \ell(k-s-i,s-1) \le AE_s \sum_{i=0}^{k-s} H_{s-1}^{(k-s-i)}(E_1,E_2,\ldots,E_{s-1}).$$

All this proves property (1), and concludes the proof of the claim.

From the claim we immediately get the following estimate for the $F_k = E_k(yg)$:

$$F_k \le E_k + H_{k-1}^{(k)}(E_1,E_2,\ldots,E_{k-1}). \tag{14.130}$$

Indeed, it is enough to look at the process

$$\eta_{k-1} \cdot \prod_{j=1}^{r_k} x_{kj}^{e_{kj}} \rightsquigarrow \prod_{j=1}^{r_k} x_{kj}^{f_{kj}} \cdot \eta_k,$$

where a generator of weight k on the right-hand side can come either from a $x_{kj}^{\pm 1}$ from the left-hand side, or from a generator of weight k in η_{k-1}.

Now if $g \in \omega_n$, i.e. $E_1(g) \le n-1$ and $E_i(g) \le n^i$ for all $i = 2,3,\ldots,c$, then for all $y \in X$, by (14.130), $E_1(yg) = F_1 \le E_1 + 1 \le n$ and

$$E_i(yg) = F_i \le E_i + H_{i-1}^{(i)}(E_1,E_2,\ldots,E_{i-1}) \le n^i + Cn^{i-1},$$

where $C > 0$ is the constant of the Claim. In other words,

$$yg \in \Omega_{n,C} \text{ for all } y \in X.$$

This shows that

$$\delta_X(\Omega_{n,C}) \cap \omega_n = \varnothing.$$

From now on we fix the constant $C > 0$ and we denote $\Omega_{n,C}$ simply by Ω_n.

Let Y_i be the set of standard generators of $\mathbb{Z}^{r_i}$, i.e. $Y_i := \{\pm e_1, \pm e_2, \ldots, \pm e_{r_i}\} \subseteq \mathbb{Z}^{r_i}$, where the e_j are the elements of the canonical basis, and set $P_i(n) := b_{Y_i}^{\mathbb{Z}^{r_i}}(n)$. Observe that asymptotically $P_i(n) \sim n^{r_i}$. We have

$$|\Omega_n| = P_1(n) \cdot \prod_{i=2}^{c} P_i(n^i + Cn^{i-1}) \text{ and } |\omega_n| = P_1(n-1) \cdot \prod_{i=2}^{c} P_i(n^i).$$

If we set $f_i(n) := P_i(n^i)/P_i(n^i + Cn^{i-1})$ for $i = 2,3,\ldots,c$, we have

$$\begin{aligned}
\frac{|\delta_X(\Omega_n)|}{|\Omega_n|} &\le \frac{|\Omega_n| - |\omega_n|}{|\Omega_n|} \\
&= 1 - \frac{|\omega_n|}{|\Omega_n|} \\
&= 1 - \frac{P_1(n-1)}{P_1(n)} \prod_{i=2}^{c} f_i(n).
\end{aligned}$$

We recall that given two functions $f : \mathbb{N} \to \mathbb{R}$ and $g : \mathbb{N} \to \mathbb{R}$, one writes

$$f(n) = O(g(n))$$

if there exists a constant $C > 0$ and $n_0 \in \mathbb{N}$ such that

$$|f(n)| \leq C|g(n)| \text{ for all } n \geq n_0.$$

Clearly

$$\frac{P_1(n-1)}{P_1(n)} = 1 + O\left(\frac{1}{n}\right).$$

Since asymptotically

$$f_i(n) \simeq \frac{n^{ir_i}}{(n^i + Cn^{i-1})^{r_i}} = \frac{1}{\left(1 + C\frac{n^{i-1}}{n^i}\right)^{r_i}} \simeq \frac{1}{1 + r_i C\frac{1}{n}},$$

we have

$$f_i(n) = 1 + O\left(\frac{1}{n}\right).$$

Therefore

$$\prod_{i=2}^{c} f_i(n) = 1 + O\left(\frac{1}{n}\right),$$

and we finally deduce that

$$\frac{|\delta_X(\Omega_n)|}{|\Omega_n|} = O\left(\frac{1}{n}\right),$$

that is, there exists a constant $C' > 0$ and $n_0 \in \mathbb{N}$ such that

$$\frac{|\delta_X(\Omega_n)|}{|\Omega_n|} \leq C'\frac{1}{n} \text{ for all } n \geq n_0.$$

By (14.4), for $n \geq n_0$ we have

$$\frac{|\partial_X(\Omega_n)|}{|\Omega_n|} \leq |X|\frac{|\delta_X(\Omega_n)|}{|\Omega_n|} \leq C'|X|\frac{1}{n}.$$

Since clearly

$$|\Omega_n| = P_1(n) \cdot \prod_{i=2}^{c} P_i(n^i + Cn^{i-1}) = O\left(n^{\sum_{i=1}^{c} ir_i}\right),$$

taking into account the monotonicity of the Følner function, all this shows that

$$F_\circ(n; G) \preceq n^{\sum_{i=1}^{c} ir_i} \sim b^G(n).$$

The proof of Theorem 14.102 is now complete. □

Remark 14.104. Notice that our argument does NOT give a bound for the isoperimetric profile of G, i.e., given $b^G(n) \sim n^d$, we cannot conclude that $I_\circ(n;G) \preceq n^{(d-1)/d}$. Indeed, what we can deduce from the proof that we just gave is that there exists a subsequence $(n_k)_{k\in\mathbb{N}}$ of the integers and a constant $C > 0$ such that $I_\circ(n_k;G) \leq Cn_k^{(d-1)/d}$ for all $k \in \mathbb{N}$. But, since it is not known whether or not $I_\circ(n;G)$ is monotone (even asymptotically), we cannot say anything about the other values of $I_\circ(n;G)$.

14.13 Notes

The theory of amenable groups arose from the study of the axiomatic properties of the Lebesgue integral and the discovery of the Hausdorff–Banach–Tarski paradox at the beginning of the 20th century (see [161, 270, 346]). The original definition of an amenable group, by the existence of an invariant finitely additive probability measure, is due to John von Neumann in 1929 [251] under the German name "messbar" ("measurable" in English). Von Neumann proved, in particular, that every abelian group is amenable and that an amenable group cannot contain a subgroup isomorphic to F_2, the free group on two generators (cf. Example 14.4). Moreover, he showed that the class of amenable groups is closed under taking subgroups, quotients, extensions, and direct limits (cf. Sect. 14.2). The term *amenable* was introduced in the 1950s by Mahlon M. Day, in [88] (see also, [89, 90, 91]) who played a central role in the development of the modern theory of amenable groups by using *means* and applying techniques from functional analysis.

The class EG of *elementary amenable groups*, defined by Day [89], is the smallest class of groups containing all finite groups and all Abelian groups, and closed under the operations of taking subgroups, quotients, extensions, and direct unions. Since finite groups and Abelian groups are amenable, and the class AG of all amenable groups is closed under the above operations, every elementary amenable group is amenable. As mentioned above, von Neumann [251] proved that a group containing a nonamenable free subgroups is nonamenable (cf. Corollary 14.10). Thus, denoting by NG the class of all groups which do not contain a nonabelian free subgroup, we have the inclusions

$$\mathrm{EG} \subseteq \mathrm{AG} \subseteq \mathrm{NF}.$$

The question of the equality $\mathrm{AG} = \mathrm{NF}$ is sometimes referred to as to the *von Neumann conjecture* though he never phrased such a conjecture: in fact he did not even *explicitly* ask the question of whether or not a group not containing F_2 is amenable (though this can arguably be read between the lines of his paper [251]). Day asked whether or not $\mathrm{EG} = \mathrm{AG}$: this is called *Day's problem* (cf. [89]). Chou proved in [73] that EG comes from finite groups and Abelian groups using group extensions and direct unions only. He also proved that periodic groups in EG are locally finite. Thus, as there are non-locally finite periodic groups (cf. Chapter 6), he deduced that $\mathrm{EG} \neq \mathrm{NF}$.

The Tits alternative (cf. Chapter 8) implies that any amenable linear group is locally virtually solvable; hence, for linear groups, amenability and elementary amenability coincide. However, the Grigorchuk group G (cf. the Notes to Chapter 7) provides an example in AG \ EG, answering (negatively) the question of Day. Later, Grigorchuk [132] gives a *finitely presented* example in AG \ EG groups. This group $\overline{G}$ is constructed as an HNN-extension of its group G. Explicitly, $\overline{G}$ has the presentation

$$\langle a,c,d,t : a^2 = c^2 = d^2 = (ad)^4 = (adacac)^4 = 1, t^{-1}at = aca, t^{-1}ct = dc, t^{-1}dt = c\rangle.$$

For finitely generated linear groups, the von Neumann conjecture holds true by virtue of the Tits alternative (cf. Chapter 8): every subgroup of $\mathrm{GL}(n,k)$ (with k a field) either has a normal solvable subgroup of finite index (and therefore is amenable) or contains the free group on two generators (cf. Chapter 8). Other classes of groups for which non-amenability is equivalent to existence of non-abelian free subgroups include: subgroups of Gromov-hyperbolic groups [139] (see also [159, Chapter 8]), fundamental groups of closed Riemannian manifolds of non-positive curvature [12], subgroups of the mapping class group [187], and the group $\mathrm{Out}(F) := \mathrm{Aut}(F)/\mathrm{Inn}(F)$ of outer automorphisms of free groups [28, 29]. See also the Notes to Chapter 8.

The von Neumann conjecture was disproved by Alexander Yu. Olshanskii in 1980 [260]. Using Grigorchuk's cogrowth criterion (Theorem 14.59) he proved that the *Tarski monster groups* (infinite simple finitely generated groups such that every proper nontrivial subgroup is cyclic of order a fixed prime number), that he constructed a year earlier, are in NF \ AG. Slightly later, Sergei I. Adyan [2] produced new such examples by showing (also using the Grigorchuk cogrowth criterion) that the *free Burnside groups*

$$B(m,n) := \langle x_1, x_2, \ldots, x_m : w(x_1, x_2, \ldots, x_m)^n = 1\rangle$$

of rank $m \geq 2$ and *odd* exponent $n \geq 665$ are non-amenable (such groups are obviously in NF). In 2002 [262], in collaboration with Mark Sapir, Olshanskii found a *finitely presented* example in NF \ AG. A geometric method for constructing finitely generated non-amenable periodic groups was described by Gromov in [139]. Another (easy) counterexample to the von Neumann conjecture was recently found by Nicolas Monod in [241] .

Confirming a conjecture by Grigorchuk and Konstantin Medynets [136], amenable finitely generated infinite simple groups exist, as shown by Kate Juschenko and Monod in [190]: this provides further examples of groups in AG \ EG.

There is also a more general notion of amenability for topological groups and for their actions (see the classical monographs [127, 270], the more recent [82], and the nice survey [135]).

The equivalence between Følner conditions and amenability was established by Erling Følner in [115]. The proof was later simplified by Isaac Namioka in [246].

The concept of *paradoxical decomposition* goes back to Galileo [121] in 1638 who discovered and clearly explained that an infinite (countable) set can be mapped injectively into a proper subset of itself (cf. [207, Introduction]; see also [58, Section

4.3]). Exactly 250 years later, in 1888, Richard Dedekind realized that this property characterizes the infinite sets and used it as the definition of infinity.

In a paper published in 1924, Stefan Banach and Alfred Tarski [13] gave a construction of a paradoxical decomposition of any solid ball in 3-dimensional space, by showing that there exists a decomposition of the ball into a finite number of disjoint subsets (in fact, 5 such subsets suffice), which can then be put back together in a different way (just moving these pieces around and rotating them without changing their shape) to yield two identical copies of the original ball. This is called the *Banach–Tarski paradox*. It is based on earlier work by Giuseppe Vitali concerning the unit interval and on the paradoxical decompositions of the sphere by Felix Hausdorff [165, 166]. For this reason one also refers to it as to the *Hausdorff–Banach–Tarski paradox*. See also [328].

The equivalence between amenability and the non-existence of a paradoxical decomposition is due to Tarski (see [330], [331]). The proof we present here is based on [62] (see also [59]), where the notions of *complexity* of a paradoxical decomposition and of *Tarski number* of a group were introduced. There are other definitions of Tarski numbers, see [107, Appendix A]. Theorem 14.89 (a group G has Tarski number $\tau(G) = 4$ if and only if G contains a subgroup isomorphic to the free group of rank 2) was proved by Bjarni Jonsson, a student of Tarski in the 1940s. In [62, 63] it was shown that for the free Burnside groups $B(m,n)$ with $m \geq 2$ and $n \geq 665$ odd, one has $6 \leq \tau(B(m,n)) \leq 14$. The computations involve spectral analysis and Cheeger–Buser type isoperimetric inequalities (cf. Sect. 14.5) and Adyan's cogrowth estimates [2] for the free Burnside groups $B(m,n)$ (cf. the Grigorchuk formula 14.6). See also [86]. Since 1999, there has been some progress on the understanding of Tarski numbers. For example, there are 2-generated non-amenable groups with arbitrarily large Tarski numbers, there are groups which we know have Tarski number exactly 5, or 6, and every number $\tau \geq 4$ is the Tarski number of some faithful transitive action of a finitely generated free group. See [267], [107], [123], and [124].

Kesten's fundamental work on symmetric random walks on groups giving rise to his amenability criterion is in [196, 197]. M. Day [91] extended Kesten's amenability criterion to non-symmetric random walks. The associated Markov chain is then determined by a probability density whose support generates the group. In this setting, the associated Markov operator on $\ell^2(G)$ is no longer self-adjoint, in general. The key ingredient of this new proof is the uniform convexity (*uniform rotundity* in Day's terminology) of Hilbert spaces (cf. Proposition 14.33) and more generally of ℓ^p-spaces with $p > 1$ (note that in fact Day considers, more generally, Markov operators on the Banach spaces $\ell^p(G)$, for $p > 1$). For more on this we refer to the review [354] and the monograph [355], both by Wolfgang Woess.

The cogrowth criterion (Theorem 14.59), established by Grigorchuk in 1978 when he was still an undergraduate student, was published in [128]. Formula (14.93) is called the *Grigorchuk formula*. Other proofs of the cogrowth criterion and the relative formula are due to Joel Cohen [75], Ryszard Szwarc [326], Woess [352], and Sam Northshield who extended the cogrowth criterion to regular graphs [259].

A proof of Theorem 14.61, was previously established by Elon Lindenstrauss and Benjy Weiss [216, Theorem 6.1] under the additional assumption that h is non-decreasing ($h(A) \le h(B)$ for all $A, B \in \mathscr{P}_{\text{fin}}(G)$ such that $A \subset B$). Their proof relies on the machinery of quasi-tiles in amenable groups that was developed by Donald Ornstein and Weiss in [263, Section I.2: Theorem 6]. An alternative proof was given by Gromov [142, Section 1.3.1] (see [204] for a detailed exposition of Gromov's argument). Recently, a version of Theorem 14.61 for cancellative one-sided amenable semigroups was given in [61] and our presentation is based on the exposition therein (see also [78]). Mean topological dimension turned out to be an extremely useful tool in order to settle a long-standing question by Joseph Auslander [9, Chapter 13] asking whether or not every minimal $\mathbb{Z}$-space X is embeddable as a subshift of $[0,1]^{\mathbb{Z}}$. After a partial positive answer (for $\dim X < \infty$) by Allan Jaworski [189] in 1974, it was only in 2000 that Lindenstrauss and Weiss [216] could provide examples of minimal systems with $\mathrm{mdim}(X, \mathbb{Z}) > 1$, thus yielding a general negative answer to Auslander's question (note that $\mathrm{mdim}([0,1]^{\mathbb{Z}}, \mathbb{Z}) = 1$).

Interesting connections between group amenability and percolation on Cayley graphs have been investigated by several authors including, but obviously not limited to, Antoine Gournay, Juschenko, Tatiana Nagnibeda, and Igor Pak [126, 191, 268]. We refer to the monograph [219] by Russell Lyons and Yuval Peres, which provides a marvelous account of topics ranging from combinatorics, Markov chains, geometric group theory, amenability, etc., as well as their inspiring relationships.

There are several other characterizations of amenability. One is the extension of the Markov–Kakutani fixed-point theorem to amenable groups (cf. [59, Theorem 4.10.1]) which is due to Day [90]. More recently, there has been one in the symbolic dynamical setting, in terms of cellular automata: the Garden of Eden theorem (originally proved by Moore and Myhill for $\mathbb{Z}^d$) holds exactly for amenable groups. This is a combination of results by the first-named author, Antonio Machì and Fabio Scarabotti [65], and Bartholdi [16, 17]. See also the Appendix in [141] by Gromov, the monograph [59], and the recent survey [60].

Isoperimetric profiles were introduced and studied by Gromov in [140]. Følner functions were introduced and studied by Anatoly M. Vershik in [342].

The proof of the inequality of Thierry Coulhon and Laurent Saloff-Coste (Theorem 14.95) that we presented is due to Gromov [143].

The proof of Theorem 14.102 that we presented is ours.

The following results are based on the inequality of Coulhon and Saloff-Coste and on the results on the growth of groups that we have mentioned before (see [279]).

Theorem. *Let G be a finitely generated group. The following conditions are equivalent:*

(a) $I_\circ(n; G) \sim n^{(d-1)/d}$ *where $d \ge 1$ is an integer.*
(b) *The growth of G is polynomial of degree d, i.e., $b^G(n) \sim n^d$.*

In a way, this theorem tells us that for groups of polynomial growth, growth and isoperimetric profile provide the same amount of information. The next result shows that these invariants are, however, not equivalent in general.

Theorem. *Let G be a finitely generated group. Suppose that G is virtually polycyclic. Then*

- $I_\circ(n;G) \sim n^{(d-1)/d}$ *if and only if* $b^G(n) \sim n^d$.
- $I_\circ(n;G) \sim n/\log n$ *if and only if G has exponential growth.*

Hence, for example, a polycyclic group G of exponential growth and the free group F_2 on two generators have the same growth, but $I_\circ(n;G) \sim n/\log n$, while $I_\circ(n;F_2) \sim n$.

The only part of the previous theorem that does not follow from previously mentioned results is the upper bound for the exponential growth, which is due to Christophe Pittet [278]. In fact, in [277] Pittet gave a proof of the upper bound for the Følner function of a virtually polycyclic group G of exponential growth that is similar in its spirit to our proof of Theorem 14.102 (i.e. it is combinatorial), proving in this way that $F_\circ(n;G) \sim e^n$ (though he does not mention this explicitly in his article).

In general, the isoperimetric profile of a group is believed to be a finer invariant than growth, but there is no proof of this statement.

It is worth mentioning that there are many other examples of groups that have exponential growth, but nonlinear isoperimetric profile. For example the wreath products of a nontrivial finite group with $\mathbb{Z}^d$ with $d \geq 2$ have an isoperimetric profile asymptotically strictly between $n/\log n$ and n (the rate depending on d).

Algorithmic and computability aspects of amenable groups and their Følner functions have been investigated by Matteo Cavaleri in his Ph.D. thesis [54] (see also [55, 56]).

Finally, a study of isoperimetric profiles and Følner functions for associative algebras has been initiated by the second named author in [84] and [85].

14.14 Exercises

Exercise 14.1. Let G be a group, $H \subseteq G$ a normal subgroup and $\pi\colon G \to \overline{G} := G/N$ the canonical quotient homomorphism. Let $\overline{\Omega} \subseteq \overline{G}$ and suppose that $\Omega_1 \leq G$ satisfies $\pi(\Omega_1) = \overline{\Omega}$ and $|\overline{\Omega}| = |\Omega_1|$. Let also $x \in X$ and set $\overline{x} := \pi(x) \in \overline{X}$. Show that

$$\partial_{\overline{x}}(\overline{\Omega}_1) = \overline{x}\overline{\Omega}_1 \setminus \overline{\Omega}_1 = \pi(x\Omega_1) \setminus \pi(\Omega_1) \subseteq \pi(x\Omega_1 \setminus \Omega_1) = \pi(\partial_x(\Omega_1)).$$

Exercise 14.2. Show that every solvable group is amenable (Corollary 14.13).

Exercise 14.3. Fill in the details in Example 14.20. In particular, prove (14.19) and (14.20).

Exercise 14.4. Deduce that every locally amenable group is amenable (Proposition 14.5) from Theorem 14.15.

Exercise 14.5. [The Hall harem theorem on matchings] Let $\mathscr{G} = (A \sqcup B, E)$ be a bipartite graph. For every $\Omega \subseteq A$ (resp. $\Omega \subseteq B$), let $\mathscr{N}(\Omega) \subset B$ (resp. $\mathscr{N}(\Omega) \subset A$)

denote the set of neighbors of elements of Ω. Suppose that $\mathscr{G}$ is locally finite, that is, for every finite subset $\Omega \subseteq A$ (resp. $\Omega \subseteq B$), the set $\mathscr{N}(\Omega)$ is finite.

A *left* (resp. *right*) *perfect matching* for $\mathscr{G}$ is a subset $M \subseteq E$ such that for each $a \in A$ (resp. $b \in B$), there exists a $b \in B$ (resp. $a \in A$) such that $(a,b) \in M$. A subset $M \subseteq E$ is a *perfect matching* if it is both a left-perfect and right-perfect matching. Let $k \geq 1$ be an integer. A subset $M \subseteq E$ is a *perfect* $(1,k)$*-matching* if for each $a \in A$ (resp. $b \in B$), there exist distinct $b_1, b_2, \ldots, b_k \in B$ (resp. $a \in A$) such that $(a, b_i) \in M$ for $i = 1, 2, \ldots, k$ (resp. $(a,b) \in M$).

$\mathscr{G}$ satisfies the *left* (resp. *right*) *Hall condition* if $|\mathscr{N}(\Omega)| \geq |\Omega|$ for all finite subsets $\Omega \subseteq A$ (resp. $\Omega \subseteq B$). $\mathscr{G}$ satisfies the *Hall* $(1,k)$*-harem condition* if $|\mathscr{N}(\Omega)| \geq k|\Omega|$ for all finite subsets $\Omega \subseteq A$ (resp. $|\mathscr{N}(\Omega)| \geq \frac{1}{k}|\Omega|$ for all finite subsets $\Omega \subseteq B$).

(1) Show that $\mathscr{G}$ admits a left-perfect (resp. right-perfect) matching if and only if it satisfies the left (resp. right) Hall condition.

(2) Show that $\mathscr{G}$ admits a perfect matching if and only if it satisfies both the left and the right Hall conditions.

(3) Show that $\mathscr{G}$ admits a perfect $(1,k)$-matching if and only if it satisfies the Hall $(1,k)$-harem condition (Hall harem theorem on matchings (Theorem 14.22)).

Exercise 14.6. Let G be a countable group. Prove that G is amenable if and only if it admits a Følner sequence (Corollary 14.24).

Exercise 14.7. Let G be a group. Prove that if G admits a Følner sequence, then G is countable.

Exercise 14.8. Let G be a countable amenable group. Show that G admits a Følner sequence $(F_n)_{n \in \mathbb{N}}$ (cf. Exercise 14.6) which is *exhausting* ($G = \bigcup_{n \in \mathbb{N}} F_n$) and *non-decreasing* ($F_n \subseteq F_{n+1}$ for all $n \in \mathbb{N}$).

Exercise 14.9. Show that the elements g and h in the proof Theorem 14.89 have infinite order.

Exercise 14.10. Let G be a paradoxical group. Show that the type (m,n) of any paradoxical decomposition p of G (as in (14.17)) satisfies $m, n \geq 2$, and deduce that $\tau(p) \geq 4$.

Exercise 14.11. Let G be a group and let $H \leq G$ be a subgroup. Suppose that H is paradoxical. Show that G is paradoxical and that $\tau(G) \leq \tau(H)$ (cf. Theorem 14.9).

Exercise 14.12. Let G be a group and let $N \trianglelefteq G$ be a normal subgroup. Suppose that the quotient group G/N is paradoxical. Show that G is paradoxical and that $\tau(G) \leq \tau(G/N)$ (cf. Theorem 14.11).

Exercise 14.13. Prove (2) in Corollary 14.97.

Exercise 14.14. Let G be an infinite finitely generated amenable group and denote by $X \subseteq G$ a finite symmetric generating subset of G. Show that the Følner function $F_\circ(n; G, X)$ is non-decreasing.

Solutions or Hints to Selected Exercises

Problems of Chapter 1

Solution 1.1. Let F_1 and F_2 be free groups based on the sets X_1 and X_2, respectively. Suppose that there exists a bijective map $\alpha\colon X_1 \to X_2$. Let $i_1\colon X_1 \to F_1$ and $i_2\colon X_2 \to F_2$ denote the inclusion maps. Then, by the universal property for F_1, there exists a unique homomorphism $\varphi\colon F_1 \to F_2$ such that $i_2 \circ \alpha = \varphi \circ i_1$. By the universal property for F_2, there exists a unique homomorphism $\varphi'\colon F_2 \to F_1$ such that $i_1 \circ \alpha^{-1} = \varphi' \circ i_2$. The maps Id_{F_1} and $\varphi' \circ \varphi$ are endomorphisms of F_1 satisfying $\mathrm{Id}_{F_1} \circ i_1 = i_1$ and $(\varphi' \circ \varphi) \circ i_1 = i_1$. By uniqueness, we must have $\mathrm{Id}_{F_1} = \varphi' \circ \varphi$. Similarly, we get $\varphi \circ \varphi' = \mathrm{Id}_{F_2}$. This shows that φ and φ' are isomorphisms, one inverse to the other. Hence the free groups F_1 and F_2 are isomorphic.

Solution 1.2. Let F be a free group based at X. Then F is generated by X. Indeed, let H denote the subgroup of F generated by X. Consider the inclusion map $i_*\colon X \to H$ of X into H. Note that $i_*(x) = i(x)$ for all $x \in X$, where $i\colon X \to F$ is the inclusion map of X into F. By the universal property, there exists a unique homomorphism $\phi\colon F \to H$ such that $i_* = \phi \circ i$. Consider now the inclusion map $\rho\colon H \to F$ of H into F. The homomorphisms Id_F and $\rho \circ \phi$ satisfy $\mathrm{Id}_F \circ i = \rho \circ \phi \circ i$. By uniqueness, we get $\mathrm{Id}_F = \rho \circ \phi$. This implies that ρ is surjective, that is, $H = F$. This shows that X generates F.

Solution 1.5. Let F be a free group and $\psi\colon F \to F'$ an isomorphism from F onto a group F'. Then F' is a free group as well. Indeed, let X be a free base for F, set $X' = \psi(X) \subset F'$ and denote by $i\colon X \to F$ and $i' = X' \to F'$ the inclusion maps. Note that $i' \circ \psi|_X = \psi \circ i$. If $f'\colon X' \to G$ is a map from X' into a group G, then there exists a unique homomorphism $\phi'\colon F' \to G$ such that $f' = \phi' \circ i'$, namely the homomorphism given by $\phi' = \phi \circ \psi^{-1}$, where $\phi\colon F \to G$ is the unique homomorphism satisfying $f = \phi \circ i$, where $f\colon X \to G$ is defined by $f = f' \circ \psi|_X$. This shows that F' is the free group based at X'.

Solution 1.6. Let F be a free group with base $X \subset F$. Let $Y \subset X$ and let K denote the subgroup of F generated by Y. Then K is a free group with base Y. Indeed, de-

T. Ceccherini-Silberstein and M. D'Adderio, *Topics in Groups and Geometry*,
Springer Monographs in Mathematics, https://doi.org/10.1007/978-3-030-88109-2

note by $j\colon Y \to F$ the inclusion map and observe that $j = i|_Y$, where $i\colon X \to F$ is the inclusion map for X. Let G be a group, $f\colon Y \to G$ a map and let us show that there exists a unique homomorphism $\phi\colon K \to G$ satisfying $f = \phi \circ j$. Uniqueness follows from the fact that Y generates K. Choose a map $f'\colon X \to G$ extending f. By the universal property, there exists a unique homomorphism $\phi'\colon F \to G$ such that $f' = \phi' \circ i$. Then $\phi = \phi'|_K\colon K \to G$ satisfies $f = \phi \circ j$. This proves that K is the free group with base Y.

Hint 1.7. See [218, Corollary 1.5].

Solution 1.8. Consider the system S of representatives for F/H defined in the proof of Lemma 1.14. Define the *length* of a coset Hg in F/H, denoted $\ell(Hg)$, as the minimal length of a word in Hg. We construct a Schreier system S recursively on the length of the cosets. Clearly, $\ell(H) = |1| = 0$ and we choose as a representative of H the identity element 1. Let $\ell \geq 1$ and suppose that for each coset of length $\leq \ell - 1$ a representative in S has already been selected. Consider a coset Hg of length ℓ and let $u \in Hg$ such that $|u| = \ell$. Then we can write $u = vy$, where $v \in (X \cup X^{-1})^*$ satisfies $|v| = \ell - 1$ and $y \in X \cup X^{-1}$. Let $\overline{v} \in Hv$ be the representative in S already selected and define $\overline{u} = \overline{v}y$. By (1.8) we have $\overline{u} \in Hg$ and, by recursion, all its prefixes belong to S. This shows that S is a Schreier system.

Solution 1.9. Let $A = \{a, b\}$, then in A^* the submonoid generated by a, ab and ba is not free: $a(ba) = (ab)a$.

Solution 1.10. Let $G = \mathbb{Z}/2\mathbb{Z} = \{1_G, a\}$, $X = \{a\}$, $H = \mathbb{Z}$ the (additive) group of integers, and let $f\colon X \to \mathbb{Z}$ be defined by $f(a) = 1$.

Solution 1.14. Let F be a free group of rank n and $H = \mathbb{Z}/m\mathbb{Z}$, the cyclic group of order m. Then any map $f\colon X \to H$ such that $f(X)$ contains a generator of H, extends to a surjective group homomorphism $\varphi\colon F \to H$. Then $G := \ker(\varphi) \leq F$ has index $[F : G] = |F/G| = |H| = m$.

Solution 1.15.(ii). Denote by G the free group of rank 2, say freely generated by a' and b'. Consider the map $f\colon \{a, b, c\} \to G$ defined by $f(a) = a'$, $f(b) = b'$, and $f(c) = 1_G$. Then f (uniquely) extends to a surjective group homomorphism $\varphi\colon F \to G$ whose kernel is $\ker(\varphi) = N$.

Hint 1.19. See [159, Example 25.IIB] (cf. [159, Exercise 33.IIB]).

Hint 1.20. See [221, Theorem 4.17].

Problems of Chapter 3

Hint 3.2. If $|G| = p^m$, then $(1-g)^{p^m} = 1 - g^{p^m} = 1 - 1 = 0$.

Hint 3.4. Since $(G_i)_{i\geq 1}$ is central, G/G_i is nilpotent. Moreover, if $p^k \geq i$, then every element of G/G_i has an order dividing p^k, so G/G_i is a p-group (of finite exponent). Since G is finitely generated, then G/G_i must be finite by Lemma 2.34.

Hint 3.10. (1) Let $A = (a_{ij}) \in \mathrm{GL}(n,\mathbb{Z})$ with $A \neq I_n = 1_{\mathrm{GL}(n,\mathbb{Z})}$. Let $m \in \mathbb{N}$ such that $|a_{ij}| < m$ for all $i, j = 1, 2, \ldots, n$. Show that the map $\phi\colon \mathrm{GL}(n,\mathbb{Z}) \to \mathrm{GL}(n,\mathbb{Z}/m\mathbb{Z})$ given by reduction modulo m for each entry is a group homomorphism satisfying $\phi(A) \neq 1_{\mathrm{GL}(n,\mathbb{Z}/m\mathbb{Z})}$.
(2) Use Example 1.19 and the obvious inclusion $\mathrm{SL}(2,\mathbb{Z}) \subset \mathrm{GL}(2,\mathbb{Z})$.

Hint 3.18. Just note that $\mathrm{GL}(n,R)$ is isomorphic to $\mathrm{Aut}_R(R^n)$, where R^n is viewed as a left-module over R. Moreover, $\mathrm{Aut}_R(R^n)$ is a subgroup of $\mathrm{Aut}_{\mathbb{Z}}(R^n)$ which is residually finite by Exercise 3.16.(3) and by Theorem 3.13.

Hint 3.19. (2) Use the fact that the group $\mathrm{Sym}_0^+(\mathbb{N})$ of finite permutations of signature 1 (that is, that are an even product of transpositions) is an infinite simple group.
(4) Cf. [59, Lemma 2.6.4].

Hint 3.20. The group G is the semidirect product $G = H \ltimes_\psi \mathbb{Z}$, where $H := \bigoplus_{i\in\mathbb{Z}} H_i$ is the direct sum of a family $(H_i)_{i\in\mathbb{Z}}$ of copies of the finite simple group S and $\psi\colon \mathbb{Z} \to \mathrm{Aut}(H)$ is the group homomorphism associated with the shift action of $\mathbb{Z}$ on H. The subgroup $H \leq G$ is the kernel of a surjective homomorphism $\rho\colon G \to \mathbb{Z}$ and there is $t \in \rho^{-1}(1)$ such that $tht^{-1} = \sum_{i\in\mathbb{Z}} h_{i-1}$ for all $h = \sum_{i\in\mathbb{Z}} h_i \in H$. In particular, $H_i = t^{-i}H_0t^i$, for every $i \in \mathbb{Z}$. Every $g \in G$ can be uniquely written in the form $g = t^n h$, where $n = \rho(g) \in \mathbb{Z}$ and $h \in H$.

Let $\phi\colon G \to G$ be a surjective group endomorphism of G.
(1) Observe that every element of H is torsion.
(2) Show that $\phi(H) = H$.
(3) Show that $\ker(\phi) \subset H$.
(4) For each $i \in \mathbb{Z}$, consider the projection map $\pi_i\colon H \to H_i$ and define the group homomorphism $\psi_i\colon H_0 \to H_i$ by setting $\psi_i(h_0) := \pi_i(\phi(h_0))$ for all $h_0 \in H_0$. Show that ψ_i is either the trivial homomorphism or an isomorphism.

Problems of Chapter 4

Hint 4.10. Take a common eigenvector, and act with G (or use Clifford's theorem (Theorem 4.30)).

Problems of Chapter 5

Hint 5.6. H acts faithfully on W, A acts faithfully on U, and $G = HA$.

Problems of Chapter 7

Solution 7.1. Let $g,h,k \in G$. First observe that $d_X(g,h) = 0$ if and only if $g^{-1}h = 1_G$, that is, if and only if $g = h$. This shows (i). On the other hand, set $m = \ell_X(g^{-1}h) = d_X(g,h)$. Then there exist $x_1, x_2, \ldots, x_m$ such that $g^{-1}h = x_1x_2\cdots x_m$. We have $h^{-1}g = (g^{-1}h)^{-1} = x_m^{-1}\cdots x_2^{-1}x_1^{-1}$ so that $d_X(h,g) = \ell_X(h^{-1}g) \leq m = d_X(g,h)$. By exchanging the roles of g and h, we get $d_X(g,h) \leq d_X(h,g)$ and therefore $d_X(g,h) = d_X(h,g)$. This shows that d_X is symmetric. Set now $s = \ell_X(g^{-1}k)$ and $t = \ell_X(k^{-1}h)$ so that there exist $y_1, y_2 \ldots, y_s$ and $u_1, u_2, \ldots, u_t$ in X such that $g^{-1}k = y_1y_2\cdots y_s$ and $k^{-1}h = u_1u_2\cdots u_t$. It is then clear that $d_X(g,h) = \ell_X(g^{-1}h) = \ell_X((g^{-1}k)(k^{-1}h)) = \ell(y_1y_2\cdots y_s u_1u_2\cdots u_t) \leq s+t = \ell_X(g^{-1}k) + \ell_X(k^{-1}h) = d_X(g,k) + d_X(k,h)$. This proves the triangle inequality. Finally, we have $d_X(kg,kh) = \ell_X((kg)^{-1}kh) = \ell_X(g^{-1}k^{-1}kh) = \ell_X(g^{-1}h) = d_X(g,h)$.

Solution 7.2. Let $g,h,k \in V$. It is clear that $d_{\mathscr{G}}(g,h) = 0$ if and only if $g = h$. On the other hand, it follows from the edge-symmetry of $\mathscr{G}$ and the hint that $d_{\mathscr{G}}(g,h) = d_{\mathscr{G}}(h,g)$, which proves that $d_{\mathscr{G}}$ is symmetric. Finally, let π_1 be a geodesic path connecting g to k and π_2 a geodesic path connecting k to h. Then the composite path $\pi_1\pi_2$ connects g to h and therefore

$$d_{\mathscr{G}}(g,h) \leq \ell(\pi_1\pi_2) = \ell(\pi_1) + \ell(\pi_2) = d_{\mathscr{G}}(g,k) + d_{\mathscr{G}}(k,h).$$

This shows that $d_G G$ also satisfies the triangle inequality. Finally, if $\pi = (e_i)_{i=1}^n$ is a geodesic path connecting g and h, then $\pi' = (e_i')_{i=1}^n$, where $\alpha(e_i') = k\alpha(e_i)$, $\lambda(e_i') = \lambda(e_i)$, and $\omega(e_i') = k\omega(e_i)$ for all $i = 1, 2, \ldots, n$, is a geodesic path connecting kg to kh. This shows $d_{\mathscr{G}}(kg,kh) = d_{\mathscr{G}}(g,h)$ and left-invariance follows as well.

Solution 7.4. Let $g,h \in G$. Suppose that $d_{\mathscr{C}_X(G)}(g,h) = n$ and let $\pi = (e_1, e_2, \ldots, e_n)$ be a geodesic path connecting g and h. Let $\lambda(\pi) = (s_1, s_2, \ldots, s_n)$ be the label of π. It then follows that $d_X(g,h) = \ell_X(g^{-1}h) = \ell_X(x_1x_2\cdots x_n) \leq n = d_{\mathscr{C}_X(G)}(g,h)$. Conversely, suppose that $d_S(g,h) = m$. Then we can find $x_1', x_2', \ldots, x_m' \in X$ such that $g^{-1}h = x_1'x_2'\cdots x_m'$. Then the path $\pi' = (e_1', e_2', \ldots, e_m')$ where

$$e_i' = (gx_1'x_2'\cdots x_{i-1}', x_i', gx_1'x_2'\cdots x_{i-1}'x_i'),$$

$i = 1, 2, \ldots, m$, connects g to $h = gx_1'x_2'\cdots x_m'$. We deduce that $d_{\mathscr{C}_X(G)}(g,h) \leq \ell(\pi') = m = d_X(g,h)$. Hence $n = m$ and (7.7) follows.

Solution 7.5. It is clear that $\preceq$ is reflexive. Let $f_1, f_2, f_3 \colon \mathbb{N} \to [0, +\infty)$ be growth functions. Suppose that $f_1 \preceq f_2$ and that $f_2 \preceq f_3$. Let c_1 and c_2 be positive integers such that $f_1(n) \leq c_1 f_2(c_1 n)$ and $f_2(n) \leq c_2 f_3(c_2 n)$ for all $n \geq 1$. Then, taking $c = c_1 c_2$ we have

$$f_1(n) \leq c_1 f_2(c_1 n) \leq c_1 c_2 f_3(c_2 c_1 n) = c f_3(cn)$$

for all $n \geq 1$. Thus $f_1 \preceq f_3$. This conclude the proof of (i).

Property (ii) immediately follows from (i) and the definition of $\sim$.

Finally, suppose that f_1, f_2, f_1', f_2' satisfy the hypotheses of (iii). Then we have, in particular, $f_1' \preceq f_1$, $f_1 \preceq f_2$ and $f_2 \preceq f_2'$. From (i) we deduce that $f_1' \preceq f_2'$.

Solution 7.6. Suppose for instance that $a \leq b$. It is then trivial that $a^n \preceq b^n$. Conversely, setting $c = [\log_a b] + 1 > 1$ (here $[\ \cdot\ ]$ denotes the integer part), one has $b^n = (a^{\log_a b})^n = a^{(\log_a b)n} \leq a^{cn} \leq ca^{cn}$ for all $n \geq 1$, so that $b^n \preceq a^n$. We deduce that $a^n \sim b^n$.

Solution 7.7. Since $\lim_{n\to\infty} \frac{n^d}{\exp(n)} = 0$, the sequence $(\frac{n^d}{\exp(n)})_{n\geq 1}$ is bounded and we can find an integer $c \geq 1$ such that $\frac{n^d}{\exp(n)} \leq c$ for all $n \geq 1$. It follows that $n^d \leq c\exp(n) \leq c\exp(cn)$ for all $n \geq 1$ thus showing that $n^d \preceq \exp(n)$.

On the other hand, suppose by contradiction that $\exp(n) \preceq n^d$. Then we can find an integer $c > 0$ such that $\exp(n) \leq c(cn)^d$ for all $n \geq 1$. But then $\frac{\exp(n)}{n^d} \leq c^{d+1}$ for all $n \geq 1$, contradicting the fact that $\lim_{n\to\infty} \frac{\exp(n)}{n^d} = +\infty$. Thus $n^d \not\sim \exp(n)$.

Solution 7.8. (1) Suppose first that f is bounded. Then we can find an integer $c \geq 1$ such that $f(n) \leq c$ for all $n \geq 1$. It follows that $f(n) \leq c1(n) \leq c1(cn)$ for all $n \geq 1$. Thus $f \preceq 1$. On the other hand, setting $c = [\frac{1}{f(0)}] + 1$, we have $1(n) = 1 \leq cf(0) \leq cf(n) \leq cf(cn)$ for all $n \geq 1$ so that $1 \preceq f$. This shows that $f \sim 1$. Conversely, suppose that $f \sim 1$. Then $f \preceq 1$ and we can find an integer $c \geq 1$ such that $f(n) \leq c1(cn) = c$ for all $n \geq 1$. Thus f is bounded.

(2) Let X be a finite and symmetric generating subset of G. Suppose that $b^G(n) \sim b_X(n) \sim 1$. From (1) above we deduce that f_X is bounded, say by an integer $c \geq 1$. This shows that $|G| \leq c$, thus showing that G is finite. Conversely, if G is finite, we have $b_X(n) = |B_X(n)| \leq |G|$ for all $n \geq 1$, that is, b_X is bounded. Again from (1) above we deduce that $b^G(n) \sim b_X(n) \sim 1$.

(3) Let X be a finite symmetric generating subset of G. Consider the inclusions $\{1_G\} = B_X(0) \subset B_X(1) \subset B_X(2) \subset \cdots \subset B_X(n) \subset B_X(n+1) \subset \cdots$ and let us show, by induction on n, that if $B_X(n_0) = B_X(n_0+1)$ for some $n_0 \in N$, then $B_X(n) = B_X(n_0)$ for all $n \geq n_0$. Suppose that $B_X(n) = B_X(n_0)$ for some $n \geq n_0 + 1$. Then for all $g \in B_X(n+1)$ there exist $g' \in B_X(n)$ and $x \in X$ such that $g = g'x$. By the inductive hypothesis, $g' \in B_X(n-1)$ so that $g = g'x \in B_X(n-1)X \subset B_X(n)$. Since $B_X(n) \subset B_X(n+1)$, it follows that $B_X(n+1) = B_X(n) = B_X(n_0)$. It follows that if $B_X(n_0) = B_X(n_0+1)$ for some $n_0 \in \mathbb{N}$, then we have $G = B_S(n_0)$. Since, by our assumptions, G is infinite, we deduce that all the inclusions $B_X(n) \subset B_X(n+1), n \in \mathbb{N}$, are strict. It follows that for all $n \in \mathbb{N}$ we have $n \leq |B_X(n)| = b_X(n)$, thus showing that $n \preceq b_X(n) \sim b^G(n)$.

The converse implication follows trivially from (2) and the observation that $1 \not\sim n$.

(4) Let G be a finitely generated group. It follows from Example 7.9.(d) that if $b^G(n) \preceq n^d$ for some integer $d \geq 0$, then $b^G(n) \not\sim \exp(n)$.

Solution 7.9. Suppose that $b_G(n) \sim \exp(n)$. We then have $\exp(n) \preceq b_X$ so that there exists an integer $c \geq 1$ such that $e^n \leq cb_X(cn)$ for all $n \geq 1$. This implies

$$1 < \sqrt[c]{e} = \lim_{n\to\infty} \sqrt[cn]{e^n} \leq \lim_{n\to\infty} \sqrt[cn]{cb_X(cn)} = \lim_{n\to\infty} \sqrt[cn]{c} \lim_{n\to\infty} \sqrt[cn]{b_X(cn)} = \beta_X.$$

Conversely, suppose that $\beta_X > 1$. By Lemma 7.13 we have $\sqrt[n]{b_X(n)} \geq \beta_X$ so that $b_X(n) \geq \beta_X^n$. It follows that $\exp(n) \sim \beta_X^n \preceq b_X(n)$. By Corollary 7.10 we have $b_X(n) \preceq \exp(n)$ and therefore $b^G(n) \sim b_X \sim \exp(n)$.

Solution 7.11. Keeping in mind that $b_X(0) = |\{1_G\}| = 1$ and that $b_X(n) \leq b_X(n+1)$ for all $n \in \mathbb{N}$, the condition $b_X(m) \leq m$ implies the existence of some integer k with $0 \leq k < m$ such that $b_X(k) = b_X(k+1)$. Then $b_X(n) = b_X(k)$ for all integers $n \geq k$. This shows that $|G| = b_X(k) = b_X(m)$.

Solution 7.12. Note that $\mathrm{SL}(n,\mathbb{Z}) \hookrightarrow \mathrm{SL}(n+1,\mathbb{Z})$, $n \geq 1$, via the map $A \mapsto \begin{pmatrix} A & 0 \\ 0 & 1 \end{pmatrix}$ for all $A \in \mathrm{SL}(n,\mathbb{Z})$.

Solution 7.15. Since the product of non-decreasing functions is also non-decreasing, it is clear that b_1b_2 and $b_1'b_2'$ are also growth functions. Let c_1 and c_2 be positive integers such that $b_1(n) \leq c_1b_1'(c_1n)$ and $b_2(n) \leq c_2b_2'(c_2n)$ for all $n \geq 1$. Taking $c := c_1c_2$ we have

$$(b_1b_2)(n) = b_1(n)b_2(n) \leq c_1b_1'(c_1n)c_2b_2'(c_2n) \leq cb_1'(cn)b_2'(cn) = c(b_1'b_2')(cn)$$

for all $n \geq 1$.

Solution 7.16. Let X_1 and X_2 be finite symmetric generating subsets of G_1 and G_2, respectively. Then the set $X = (X_1 \times \{1_{G_2}\}) \cup (\{1_{G_1}\} \times X_2)$ is a finite symmetric generating subset of $G_1 \times G_2$. Let $(g_1,g_2) \in B_X^{G_1\times G_2}(n)$. Then there exist $x_{1,1}, x_{2,1}, \ldots, x_{k,1} \in X_1$ and $x_{1,2}, x_{2,2}, \ldots, x_{h,2} \in X_2$, where $h+k \leq n$, such that

$$\begin{aligned}(g_1,g_2) &= (x_{1,1},1_{G_2})(x_{2,1},1_{G_2})\cdots(x_{k,1},1_{G_2})\cdot(1_{G_1},x_{1,2})(1_{G_1},x_{2,2})\cdots(1_{G_1},x_{h,2}) \\ &= (x_{1,1}x_{2,1}\cdots x_{k,1}, x_{1,2}x_{2,2}\cdots x_{h,s}).\end{aligned}$$

Thus, $B_X^{G_1\times G_2}(n) \subset B_{X_1}^{G_1}(n) \times B_{X_2}^{G_2}(n)$ and therefore $b_X^{G_1\times G_2}(n) \leq b_{X_1}^{G_1}(n)b_{X_2}^{G_2}(n)$. This shows that $b^{G_1\times G_2} \preceq b^{G_1}b^{G_2}$.

On the other hand, if $g_1 \in B_{X_1}^{G_1}(n)$ and $g_2 \in B_{X_2}^{G_2}(n)$, then $(g_1,g_2) \in B_X^{G_1\times G_2}(2n)$ and one has $b_{X_1}^{G_1}(n)b_{X_2}^{G_2}(n) \leq b_X^{G_1\times G_2}(2n) \leq 2b_X^{G_1\times G_2}(2n)$. This shows that $b^{G_1}b^{G_2} \preceq b^{G_1\times G_2}$. It follows that $b^{G_1\times G_2} = b^{G_1}b^{G_2}$.

Problems of Chapter 9

Hint 9.1. Use Lemma 9.6.(1).

Problems of Chapter 11

Solution 11.9. Assume, by contradiction, that there exists a $y \in Y$ such that $f(y) \neq g(y)$. Then we can find disjoint open neighborhoods $V \ni f(y)$ and $U \ni g(y)$. By continuity of f and g, the preimages $f^{-1}(V)$ and $g^{-1}(U)$ are open neighborhoods of $y \in Y$. Since Z is dense in Y we can find $z \in Z$ such that $z \in f^{-1}(V) \cap g^{-1}(U)$. Since $f|_Z = g|_Z$ the elements $f(z) \in V$ and $g(z) \in U$ are equal, contradicting the hypothesis that $V \cap U = \varnothing$.

Hint 11.13. (i) For the necessary condition, argue as in the proof of the Bolzano–Weierstrass theorem for bounded real sequences.

(iii) Use the Archimedean property.

(iv) For $m \in \mathbb{N}$ denote by $e(n) \in \ell^2(\mathbb{N})$ the sequence defined by $e_m(n) = \delta_{m,n}$ for all $n \in \mathbb{N}$. Show that $d(e(n), e(m)) = \|e(n) - e(m)\|_2 = \sqrt{2}$ for all $m, n \in \mathbb{N}$.

Solution 11.18. Suppose X is compact. Let $x = (x_n)_{n\in\mathbb{N}}$, $x' = (x'_n)_{n\in\mathbb{N}}$ and $y = (y_n)_{n\in\mathbb{N}}$ be elements in $\ell^\infty(X)$. Suppose that $x \sim_\omega x'$ so that $[x]_\omega = [x']_\omega$ in X_ω. By compactness and the triangle inequality, the limits $\lim_\omega x_n$ and $\lim_\omega x'_n$ exist in X (by Theorem 11.21) and are equal. Thus the map $\kappa_\omega \colon X_\omega \to X$ defined by

$$\kappa_\omega([x]_\omega) := \lim_\omega x_n$$

is well defined. Moreover,

$$d_\omega([x]_\omega, [y]_\omega) = \lim_\omega d(x_n, y_n) = d(\lim_\omega x_n, \lim_\omega y_n) = d(\kappa_\omega([x]_\omega), \kappa_\omega([y]_\omega)).$$

It follows that the map κ_ω is an isometry and, in particular, it is injective. Moreover, since the limit of any constant sequence is the constant value of the sequence, we immediately have $\kappa_\omega \circ \iota_\omega = \mathrm{Id}_X$ so that, in particular, κ_ω is also surjective.

Solution 11.28. For $x = (x_n)_{n\in\mathbb{N}}$ and $y = (y_n)_{n\in\mathbb{N}}$ in $\mathbb{F}^\mathbb{N}$ write $[x]_\omega \preceq_\omega [y]_\omega$ provided $\{n \in \mathbb{N} : x_n \preceq y_n\} \in \omega$. Show that $\preceq_\omega$ is a well defined total order on $\mathbb{F}_\omega$ making $(\mathbb{F}_\omega, \preceq_\omega)$ an ordered field.

Hint 11.35. Keeping in mind Proposition 7.6, it is enough to check that the inclusion map $H \to G$ is a quasi-isometry.

Hint 11.38. Use the fact that any asymptotic cone is an arcwise connected metric space (cf. Theorem 11.52.(i)).

Problems of Chapter 12

Solution 12.7. In order to show that d_M is a distance, we only need to check for the triangle inequality. Let $x,z,y \in X \cup Y \subset Z$: we may assume, without loss of generality, that $x,z \in X$ and $y \in Y$. We then have

$$\begin{aligned} d_M(x,y) &= d_X(x,x_0) + d_M(x_0,y_0) + d_Y(y,y_0) \\ &\leq (d_X(x,z) + d_X(z,x_0)) + d_M(x_0,y_0) + d_Y(y,y_0) \\ &= d_X(x,z) + (d_X(z,x_0) + d_M(x_0,y_0) + d_Y(y,y_0)) \\ &= d_M(x,z) + d_M(z,y). \end{aligned}$$

Analogously,

$$\begin{aligned} d_M(x,z) = d_X(x,z) &\leq d_X(x,x_0) + d_X(x_0,z) \\ &\leq d_X(x,x_0) + d_X(x_0,z) + 2d_M(x_0,y_0) + 2d_Y(y_0,y) \\ &= (d_X(x,x_0) + d_M(x_0,y_0) + d_Y(y_0,y)) + (d_X(z,x_0) + d_M(x_0,y_0) + d_Y(y_0,y)) \\ &= d_M(x,y) + d_M(y,z). \end{aligned}$$

This shows that d_M is a distance on Z. The following picture represents this distance.

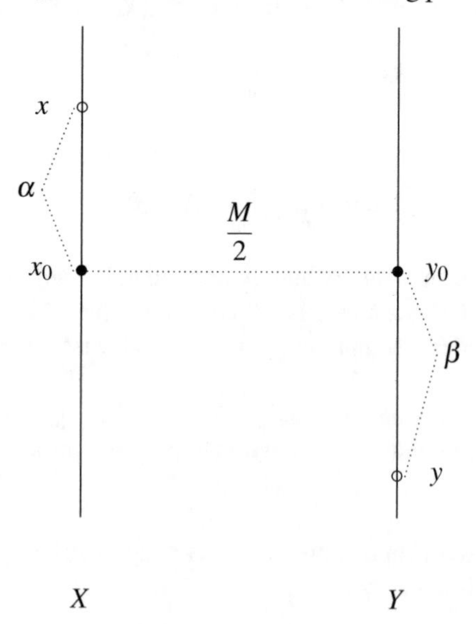

$$d(x,y) = \alpha + \tfrac{M}{2} + \beta = d_X(x,x_0) + d(x_0,y_0) + d_Y(y,y_0).$$

Let now $\varepsilon = M+1$. Clearly, $d_M(x_0,y_0) = M/2 < M+1 = \varepsilon$. Moreover, for $x \in B_X(x_0, 1/\varepsilon)$ we have $d_M(x,y_0) = d_X(x,x_0) + d_M(x_0,y_0) \leq 1/\varepsilon + M/2 < 1/(M+1) + M/2 < M+1 = \varepsilon$. Thus, $B_X(x_0, 1/\varepsilon) \subseteq N_\varepsilon(Y)$. Analogously, $B_Y(y_0, 1/\varepsilon) \subseteq N_\varepsilon(X)$. This way, d_M is an admissible distance on $Z = X \dot\cup Y$ satisfying the conditions defining GH. It follows that $\mathrm{GH}((X,x_0),(Y,y_0)) \leq \varepsilon = M+1$. As $M > 0$ was arbitrary, we deduce that $\mathrm{GH}((X,x_0),(Y,y_0)) \leq 1$.

Solution 12.8. In order to show that d_k is a distance, we only need to check for the triangle inequality. Let $x,z,y \in X \cup Y \subset Z$: we may assume, without loss of generality, that $x,z \in X$ and $y \in Y$. We then have

$$\begin{aligned} d_k(x,y) &= d_X(x, f^{-1}(y)) + k \\ &\leq d_X(x,z) + d_X(z, f^{-1}(y)) + k \\ &= d_k(x,z) + d_k(z,y). \end{aligned}$$

Analogously,

$$\begin{aligned} d_k(x,z) &\leq d_X(x, f^{-1}(y)) + d_X(f^{-1}(y), z) \\ &\leq d_X(x, f^{-1}(y)) + k + d_X(f^{-1}(y), z) + k \\ &= d_k(x,y) + d_k(y,z). \end{aligned}$$

This shows that d_k is a distance on Z. The following picture represents this distance.

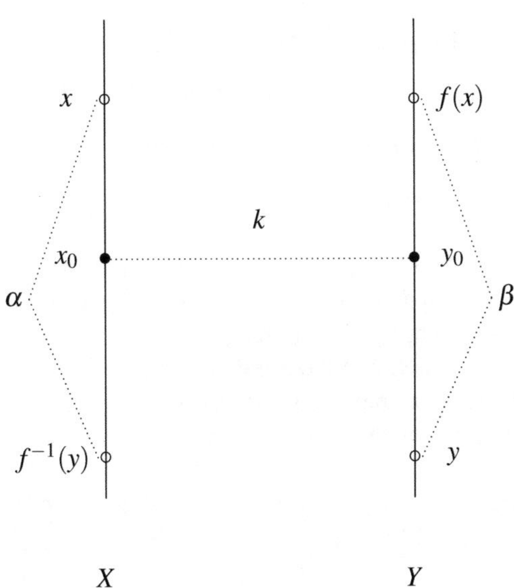

$$d_k(x,y) = \alpha + k = \beta + k = d_X(x, f^{-1}(x)) + k = d_Y(f(x), y) + k.$$

Note that $d_k(x_0,y_0) = k$. Moreover, if $x \in X$ and $y := f(x) \in Y$ then, for all $k' > k$, $d_k(x,y) = d_k(x, f(x)) = d_Y(f(x), f(x)) + k = k < k'$. This shows that $X \subseteq N_{k'}(Y)$

(analogously, $Y \subseteq N_{k'}(X)$) for all $k' > k$. We deduce that $\mathrm{GH}((X,x_0),(Y,y_0)) \leq k$. As $k \geq 0$ was arbitrary, we deduce that in fact $\mathrm{GH}((X,x_0),(Y,y_0)) = 0$.

Solution 12.9. As $\mathrm{GH}((X,x_0),(Y,y_0)) = 0$, for every integer $k \geq 1$ there exists an admissible metric d_k on $Z = X\dot{\cup}Y$ such that $d_k(x_0,y_0) < 1/k$, $B_X(x_0,k) \subseteq N_{1/k}(Y)$ and $B_Y(y_0,k) \subseteq N_{1/k}(X)$.

As X is separable, there exists a countable dense subset $\{a_i : i \in \mathbb{N}\}$ in X. We suppose that $a_0 = x_0$.

Fix $i \in \mathbb{N}$ and set $k_0 := d_X(a_i,x_0)$. For any integer $k > k_0$ we have

$$a_i \in B_X(x_0,k_0) \subseteq B_X(x_0,k) \subseteq N_{1/k}(Y) = \bigcup_{y \in Y} B_Y(y,1/k).$$

Therefore, there exists a $b_{i,k} \in Y$ such that $a_i \in B_Y(b_{i,k},1/k)$, that is, $d_k(b_{i,k},a_i) \leq 1/k$. As

$$d_Y(b_{i_k},y_0) \leq d_k(b_{i_k},a_i) + d_X(a_i,x_0) + d_k(x_0,y_0) \leq \frac{1}{k} + k_0 + \frac{1}{k} \leq k_0 + 2 =: M,$$

the sequence $(b_{i,k})_{k\geq 1}$ in Y is bounded. Since Y is locally compact, the ball $B_Y(y_0,M)$ is compact and $(b_{i,k})_{k\geq 1}$ admits a convergent subsequence. By abuse of notation, we still denote it by $(b_{i,k})_{k\geq 1}$, and set $b_i := \lim_k y_{i,k} \in Y$. Therefore, for each $\varepsilon > 0$ there exists a $\bar{k}_{\varepsilon,i} \in \mathbb{N}$ such that $d_Y(b_i,b_{i,k}) \leq \varepsilon$ for all $k \geq \bar{k}_{\varepsilon,i}$. Thus, for all $i,j \in \mathbb{N}$, if $k \geq \bar{k} := \max\{\bar{k}_{\varepsilon,i},\bar{k}_{\varepsilon,j}\}$ and $1/k < \varepsilon$, we have

$$\begin{aligned} d_Y(b_i,b_j) \leq d_Y(b_i,b_{i,k}) + d_k(b_{i,k},a_i) + d_X(a_i,a_j) \\ + d_k(a_j,b_{j,k}) + d_Y(b_{j,k},b_j) \leq d_X(a_i,a_j) + 4\varepsilon. \end{aligned}$$

Analogously, $d_X(a_i,a_j) \leq d_Y(b_i,b_j) + 4\varepsilon$. We deduce that

$$d_Y(b_i,b_j) = d_X(a_i,a_j).$$

Setting $f(a_i) := b_i$ for all $i \in \mathbb{N}$ yields a unique isometric embedding $f \colon X \to Y$.

Keeping in mind that $a_0 = x_0$ and for every $k > k_0$ one has $d_k(x_0,y_0) < 1/k$ and $d_k(b_{0,k},a_0) \leq 1/k$, we deduce that $b_0 = y_0$ so that $f(x_0) = y_0$.

Note that f may fail to be surjective! No problem: we repeat the same construction, by symmetry, obtaining an isometric embedding $g \colon Y \to X$. Exercise 12.1 then ensures the existence of an isometry $F \colon X \to Y$ such that $F(x_0) = y_0$.

Solution 12.10. Let $\mathrm{GH}((X,x_0),(Y,y_0)) < \varepsilon_1 < 1/2$. Then we can find an admissible metric d_1 on $X\dot{\cup}Y$ such that $d_1(x_0,y_0) < \varepsilon_1$, $B_X(x_0,1/\varepsilon_1) \subseteq N_{\varepsilon_1}(Y)$, and $B_Y(y_0,1/\varepsilon_1) \subseteq N_{\varepsilon_1}(X)$. Similarly, let $\mathrm{GH}((Y,y_0),(Z,z_0)) < \varepsilon_2 < 1/2$. Then we can find an admissible metric d_2 on $Y\dot{\cup}Z$ such that $d_2(y_0,z_0) < \varepsilon_2$, $B_Y(y_0,1/\varepsilon_2) \subseteq N_{\varepsilon_2}(Z)$, and $B_Z(z_0,1/\varepsilon_2) \subseteq N_{\varepsilon_2}(Y)$.

For $x,y \in X\dot{\cup}Z$ we set

$$d(x,z) := \begin{cases} d_X(x,z) & \text{if } x,z \in X \\ d_Z(x,z) & \text{if } x,z \in Z \\ \inf_{y\in Y}(d_1(x,y)+d_2(y,z)) & \text{if } x \in X \text{ and } z \in Z. \end{cases}$$

One easily shows that d satisfies the triangle inequality, so that d is an admissible metric on $X\dot{\cup}Z$. We have

$$d(x_0,z_0) \leq d_1(x_0,y_0)+d_2(y_0,z_0) < \varepsilon_1+\varepsilon_2.$$

Moreover, as $1/(\varepsilon_1+\varepsilon_2) < 1/\varepsilon_1$, we have

$$B_X(x_0,1/(\varepsilon_1+\varepsilon_2)) \subseteq B_X(x_0,1/\varepsilon_1) \subseteq N_{\varepsilon_1}(Y),$$

so that, given $x \in B_X(x_0,1/(\varepsilon_1+\varepsilon_2))$ there exists a $y \in Y$ such that $d_1(x,y) < \varepsilon_1$. Applying the triangle inequality and keeping in mind that $\varepsilon_1,\varepsilon_2 < 1/2$, we obtain

$$\begin{aligned} d_Y(y,y_0) &\leq d_1(y,x)+d_X(x,x_0)+d_1(x_0,y_0) \\ &< \varepsilon_1+1/(\varepsilon_1+\varepsilon_2)+\varepsilon_1 = 2\varepsilon_1+1/(\varepsilon_1+\varepsilon_2) < 1/\varepsilon_2. \end{aligned}$$

This shows that $y \in B_Y(y_0,1/\varepsilon_2) \subseteq N_{\varepsilon_2}(Z)$. Therefore there exists a $z \in Z$ such that $d_2(y,z) \leq \varepsilon_2$. In conclusion, we have proved that for each $x \in B_X(x_0,1/(\varepsilon_1+\varepsilon_2))$ there exist $y \in B_Y(y_0,1/\varepsilon_2)$ and $z \in Z$ such that

$$d(x,z) \leq d_1(x,y)+d_2(y,z) < \varepsilon_1+\varepsilon_2,$$

that is, $B_X(x_0,1/(\varepsilon_1+\varepsilon_2)) \subseteq N_{\varepsilon_1+\varepsilon_2}(Z)$. The inclusion

$$B_Z(z_0,1/(\varepsilon_1+\varepsilon_2)) \subseteq N_{\varepsilon_1+\varepsilon_2}(X)$$

is verified in the same way.

In conclusion, the admissible metric d on $X\dot{\cup}Z$ satisfies $d(x_0,y_0) < \varepsilon_1+\varepsilon_2$, $B_X(x_0,1/(\varepsilon_1+\varepsilon_2)) \subseteq N_{\varepsilon_1+\varepsilon_2}(Z)$ and $B_Z(z_0,1/(\varepsilon_1+\varepsilon_2)) \subseteq N_{\varepsilon_1+\varepsilon_2}(X)$. This shows that $\mathrm{GH}((X,x_0),(Z,z_0)) \leq \varepsilon_1+\varepsilon_2$. Since ε_1 and ε_2 were arbitrary, we deduce that $\mathrm{GH}((X,x_0),(Z,z_0)) \leq \mathrm{GH}((X,x_0),(Y,y_0))+\mathrm{GH}((Y,y_0),(Z,z_0))$.

Problems of Chapter 13

Hint 13.13.(2). Use the RHS inequality in (13.55) and, again, integration by parts.

Hint 13.14. Use induction on $j \in \mathbb{N}$. Alternatively, use the identities

$$\begin{aligned}\sum_{n=0}^{\infty}\sum_{k=0}^{n}\binom{n}{k}s^k t^n &= \sum_{n=0}^{\infty}(1+s)^n t^n\\ &= \frac{1}{1-(1+s)t}\\ &= \frac{1}{(1-t)-st}\\ &= \frac{1}{(1-t)}\cdot\frac{1}{1-s\left(\frac{t}{1-t}\right)}\\ &= \frac{1}{(1-t)}\sum_{j=0}^{\infty}\left(\frac{t}{1-t}\right)^j s^j\end{aligned}$$

and deduce that

$$\sum_{n=0}^{\infty}\binom{n}{k}t^n = \frac{t^k}{(1-t)^{k+1}}.$$

Finally, take $t=\frac{1}{2}$.

Problems of Chapter 14

Hint 14.4. Let $(G_i)_{i\in I}$ denote the family of all finitely generated subgroups of G. It is then clear that the conditions of Theorem 14.15 are satisfied.

Hint 14.5. (1) The left Hall condition is clearly necessary. To show that it is also sufficient, proceed as follows. Suppose first that A is finite and use induction on $|A|$ by distinguishing the following two cases:

(i) $|\mathcal{N}(\Omega)|\geq|\Omega|+1$ for all $\Omega\subseteq A$;

(ii) there exists an $\Omega'\subseteq A$ such that $|\mathcal{N}(\Omega')|=|\Omega'|$.

When A is not finite, apply the Tychonov theorem to the compact set

$$K:=\prod_{a\in A}\mathcal{N}(\{a\}).$$

(2) Use a Cantor–Bernstein type argument.

(3) The Hall $(1,k)$-harem condition is clearly necessary. To show that it is also sufficient, proceed as follows. Let $A_1,A_2,\dots,A_k$ be k disjoint copies (clones) of A with bijections $\phi_i\colon A\to A_i$ for $i=1,2,\dots,k$, and define a locally finite bipartite graph $\mathcal{G}'=(A'\sqcup B',E')$ by setting $A':=A_1\sqcup A_2\sqcup\cdots\sqcup A_k$, $B':=B$, and $E':=\{(\phi_i(a),b):(a,b)\in E, i=1,2,\dots,k\}$. Show that $\mathcal{G}$ admits a perfect $(1,k)$-matching (resp. satisfies the Hall $(1,k)$-harem condition) if and only if $\mathcal{G}'$ admits a perfect matching (resp. satisfies both the left and right Hall conditions). Then apply (2) to $\mathcal{G}'$ and conclude.

Hint 14.6. Let $(A_n)_{n\in\mathbb{N}}$ be an increasing sequence of finite subsets of G such that $\bigcup_{n\in\mathbb{N}} A_n = G$. Let J denote the (countable) set of pairs $(A_n, \frac{1}{n})$, $n \in \mathbb{N}$, equipped with the natural order, and repeat verbatim the proof of (b) $\Rightarrow$ (c) in the proof of Proposition 14.23. The converse implication is obvious, since any sequence is a net.

Solution 14.7. Let $(F_n)_{n\in\mathbb{N}}$ be a Følner sequence and set $A_n := F_n F_n^{-1} \subseteq G$. Given $g \in G$ we can find $n_0 \in \mathbb{N}$ such that $|F_n \setminus gF_n| < |F_n|$, for all $n \geq n_0$. This is equivalent to saying that $F_n \cap gF_n \neq \varnothing$, equivalently, $g \in A_n$, for all $n \geq n_0$. This shows that $G = \bigcup_{n\in\mathbb{N}} A_n$.

Hint 14.11. Let $T \subset G$ be a complete set of representatives for the right cosets of H in G so that $G = \sqcup_{t\in T} Ht$. Let $\varnothing \neq A'_i, B'_j \subseteq H$ and let $a_i, b_j \in H$ for $i = 1,2,\ldots,n$ and $j = 1,2,\ldots,m$, such that $H = \sqcup_{i=1}^n A'_i \sqcup \sqcup_{j=1}^m B'_j$ and $H = \sqcup_{i=1}^n a_i A'_i = \sqcup_{j=1}^m b_j B'_j$. Set $A_i := \sqcup_{t\in T} A'_i t$ and $B_j := \sqcup_{t\in T} B'_j t$ for $i = 1,2,\ldots,n$ and $j = 1,2,\ldots,m$. Show that $G = \sqcup_{i=1}^n A_i \sqcup \sqcup_{j=1}^m B_j$ and $G = \sqcup_{i=1}^n a_i A_i = \sqcup_{j=1}^m b_j B_j$.

Hint 14.12. Let $\pi\colon G \to G/N$ denote the canonical quotient homomorphism. Let $\varnothing \neq A'_i, B'_j \subseteq G/N$ and let $a'_i, b'_j \in G/N$ for $i = 1,2,\ldots,n$ and $j = 1,2,\ldots,m$, such that $G/N = \sqcup_{i=1}^n A'_i \sqcup \sqcup_{j=1}^m B'_j$ and $G/N = \sqcup_{i=1}^n a'_i A'_i = \sqcup_{j=1}^m b'_j B'_j$. Set $A_i := \pi^{-1}(A'_i) \subset G$, $a_i \in \pi^{-1}(a'_i) \in G$, $B_j := \pi^{-1}(B'_j) \subset G$, and $b_j \in \pi^{-1}(b'_j) \in G$ for $i = 1,2,\ldots,n$ and $j = 1,2,\ldots,m$. Show that $G = \sqcup_{i=1}^n A_i \sqcup \sqcup_{j=1}^m B_j$ and $G = \sqcup_{i=1}^n a_i A_i = \sqcup_{j=1}^m b_j B_j$.

References

1. R.L. Adler, A.G. Konheim, and M.H. McAndrew, Topological entropy, *Trans. Amer. Math. Soc.* **114** (1965), 309–319.
2. S.I. Adyan, Random walks on free periodic groups, *Math. USSR-Izv.* **21** (1983), 425–434.
3. S.I. Adyan, The Burnside problem and related topics, *Russian Math. Surveys* **65** (5) (2010) 805–855.
4. D. Aldous and J.A. Fill, *Reversible Markov Chains and Random Walks on Graphs*, `https://www.stat.berkeley.edu/users/aldous/RWG/book.html`
5. P. Alexandroff, Dimensionstheorie, *Math. Ann.* **106** (1932), 161–238.
6. D. Allcock, Most big mapping class groups fail the Tits Alternative, preprint (2020).
7. E. Artin, Zur Theorie der hyperkomplexen Zahlen, *Abh. Hamburg* **5** (1927), 251–260.
8. G.N. Arzhantseva and I.G. Lysenok, A lower bound on the growth of word hyperbolic groups, *Journal of the London Mathematical Society*, **73** (1) (2006), 109–125.
9. J. Auslander, *Minimal Flows and Their Extensions*, North-Holland Mathematics Studies, 153. Notas de Matemática 122. North-Holland Publishing Co., Amsterdam, 1988.
10. L. Auslander, On a problem of Philip Hall, *Ann. of Math.* (2) **86** (1967), 112–116.
11. R. Baer and F. Levi, Freie Produkte und ihre Unterguppen, *Composition Math.* **3** (1930), 391–398.
12. W. Ballmann, *Lectures on space of nonpositive curvature. With an appendix by Misha Brin.* DMV Seminar, **25**. Birkhäuser Verlag, Basel, 1995.
13. S. Banach and A. Tarski, Sur la décomposition des ensembles de points en parties respectivement congruentes, *Fundamenta Mathematicae* **6** (1924), 244–277.
14. L. Bartholdi, Counting paths in graphs, *Enseign. Math.* (2) **45** (1999), no. 1-2, 83–131.
15. L. Bartholdi, A Wilson group of non-uniformly exponential growth, *C.R. Math. Acad. Sci. Paris* **336** (2003), no. 7, 549–554.
16. L. Bartholdi, Gardens of Eden and amenability on cellular automata, *J. Eur. Math. Soc. (JEMS)* **12** (2010), 241–248.
17. L. Bartholdi, Amenability of groups is characterized by Myhill's theorem. With an appendix by Dawid Kielak. *J. Eur. Math. Soc. (JEMS)* **21** (2019), no. 10, 3191–3197.
18. L. Bartholdi and R.I. Grigorchuk, Lie methods in growth of groups and groups of finite width, In: M. Atkinson et al. (eds), *Computational and Geometric Aspects of Modern Algebra*, London Math. Soc. Lecture Note Ser. 275, Cambridge Univ. Press, 2000, pp. 1–27.
19. H. Bass, The degree of polynomial growth of finitely generated nilpotent groups, *Proc. London Math. Soc.* **25** (1972), 603–614.
20. H. Bass, Covering theory for graphs of groups, *Journal of Pure and Applied Algebra* **89** (1993), 3–47.
21. G. Baumslag, Automorphism Groups of Residually Finite Groups, *J. London Math. Soc.* **38** (1963), 117–118.
22. G. Baumslag and D. Solitar, Some two-generator one-relator non-Hopfian groups, *Bull. Amer. Math. Soc.* **68** (1962), 199–201.

T. Ceccherini-Silberstein and M. D'Adderio, *Topics in Groups and Geometry*,
Springer Monographs in Mathematics, https://doi.org/10.1007/978-3-030-88109-2

23. J. Bell, The Pontryagin duals of $\mathbb{Q}/\mathbb{Z}$ and $\mathbb{Q}$. Notes available in Section *p-adic numbers* at `https://individual.utoronto.ca/jordanbell/`.
24. Y. Benyamini and J. Lindenstrauss, *Geometric nonlinear functional analysis. Vol. 1*, American Mathematical Society Colloquium Publications, 48. American Mathematical Society, Providence, RI, 2000.
25. A.S. Besicovitch, On Linear Sets of Points of Fractional Dimensions, *Mathematische Annalen* **101** (1) (1929), 161–193.
26. A.S. Besicovitch and H.D. Ursell, Sets of Fractional Dimensions, *Journal of the London Mathematical Society* **12** (1) (1937), 18–25.
27. M. Bestvina, $\mathbb{R}$-trees in topology, geometry, and group theory. Handbook of geometric topology, 55–91, North-Holland, Amsterdam, 2002.
28. M. Bestvina, M. Feighn, and M. Handel, The Tits alternative for $\mathrm{Out}(F_n)$. I: Dynamics of exponentially-growing automorphisms, *Ann. of Math.* **151** (2000), no. 2, 517–623.
29. M. Bestvina, M. Feighn, and M. Handel, The Tits alternative for $\mathrm{Out}(F_n)$. II: A Kolchin type theorem, *Ann. of Math.* **161** (2005), no. 1, 1–59.
30. P. Bohl, Über die Beweging eines mechanischen Systems in der Nähe einer Gleichgewichtslage, *J. Reine Angew. Math.* **127** (1904), 179–276.
31. K. Borsuk, *Theory of Retracts*, Monografie Matematyczne, Tom 44 PWN (Państwowe Wydawnictwo Naukowe = Polish Sci. Publ.), Warsaw, 1967.
32. N. Bourbaki, *Éléments de mathématique. Topologie générale. Chapitres 1 à 4*, Hermann, Paris, 1971.
33. N. Brady, T. Riley, and H. Short, *The geometry of the word problem for finitely generated groups*. Papers from the Advanced Course held in Barcelona, July 5–15, 2005. Advanced Courses in Mathematics. CRM Barcelona. Birkhäuser Verlag, Basel, 2007.
34. E. Breuillard, A height gap theorem for finite subsets of $\mathrm{GL}_d(\overline{\mathbb{Q}})$ and nonamenable subgroups, *Ann. of Math.* (2) **174** (2011), no. 2, 1057–1110.
35. E. Breuillard, A strong Tits Alternative, preprint arXiv:0804.1395.
36. E. Breuillard and Y. de Cornulier, On conjugacy growth for solvable groups, *Illinois J. Math.* **54** (2010), no. 1, 389–395.
37. E. Breuillard, Y. de Cornulier, A. Lubotzky, and C. Meiri, On conjugacy growth of linear groups, *Math. Proc. Cambridge Philos. Soc.* **154** (2013), no. 2, 261–277.
38. E. Breuillard and T. Gelander, Uniform independence in linear groups, *Invent. Math.* **173** (2008), no. 2, 225–263.
39. E. Breuillard, B. Green, and T. Tao, The structure of approximate groups, *Publ. Math. Inst. Hautes Études Sci.* **116** (2012), 115–221.
40. H. Brezis, *Functional analysis, Sobolev spaces and partial differential equations*. Universitext, Springer-Verlag, New York, 2011.
41. M.R. Bridson and A. Haefliger, *Metric spaces of non-positive curvature*, Grundlehren der Mathematischen Wissenschaften, 319. Springer-Verlag, Berlin, 1999.
42. M. Brin and G. Stuck, *Introduction to dynamical systems*, Cambridge University Press, Cambridge, 2002.
43. L.E.J. Brouwer, Beweis der Invarianz der Dimensionenzahl, *Math. Ann.* **70** (1911), 161–165.
44. L.E.J. Brouwer, Über Abbildungen von Mannigfaltigkeiten, *Math. Ann.* **71** (1911), 97–115.
45. L.E.J. Brouwer, Über den natürlichen Dimensionsbegriff, *Jour. f. Math.* **142** (1913), 146–152.
46. W. Burnside, On an unsettled question in the theory of discontinuous groups, *Quart. J. Math.* **33** (1902), 230–238.
47. W. Burnside, On criteria for the finiteness of the order of a group of linear substitutions, *Proc. London Math. Soc.* (2) **3** (1905), 435–440.
48. J.W. Cannon, W.J. Floyd, and W.R. Parry, Introductory notes on Richard Thompson's groups, *Enseign. Math.* (2) **42** (1996), no. 3-4, 215–256.
49. S. Cantat, Sur les groupes de transformations birationnelles des surfaces, *Ann. Math.* **174** (2011) 299–340.
50. V. Capraro and M. Lupini, *Introduction to sofic and hyperlinear groups and Connes' embedding conjecture. With an appendix by Vladimir Pestov*, Lecture Notes in Mathematics, 2136. Springer, Cham, 2015.

51. H. Cartan, Théorie des filtres, *C.R. Acad. Paris* **205** (1937) 595–598.
52. H. Cartan, Filtres et ultrafiltres, *C.R. Acad. Paris* **205** (1937) 777–779.
53. H. Cartan, Sur la mesure de Haar, *C.R. Acad. Paris* **211** (1940), 759–762.
54. Matteo Cavaleri, *Algorithms and quantifications in amenable and sofic groups*, Ph.D. dissertation. University of Rome "La Sapienza", 2016.
55. Matteo Cavaleri, Computability of Følner sets, *Internat. J. Algebra Comput.* **27** (2017), no. 7, 819–830.
56. Matteo Cavaleri, Følner functions and the generic word problem for finitely generated amenable groups, *J. Algebra* **511** (2018), 388–404.
57. A. Cayley, Desiderata and suggestions: No. 2. The Theory of groups: graphical representation, *Amer. J. Math.* **1** (1878), 174–176.
58. T.G. Ceccherini-Silberstein, Around amenability. Pontryagin Conference, 8, Algebra (Moscow, 1998). *J. Math. Sci.* (New York) **106** (2001), no. 4, 3145–3163.
59. T. Ceccherini-Silberstein and M. Coornaert, *Cellular automata and groups*, Springer Monographs in Mathematics, Springer Verlag, 2010.
60. T. Ceccherini-Silberstein and M. Coornaert, The Garden of Eden Theorem: old and new, in "Handbook of Group Actions volume V", Eds. L. Ji, A. Papadopoulos, and S.T. Yau, pp. 55–106, International Press and Higher Education Press, 2020.
61. T. Ceccherini-Silberstein, M. Coornaert, and F. Krieger, An analogue of Fekete's lemma for subadditive functions on cancellative amenable semigroups, *J. Analyse Math.* **124** (2014), 59–81.
62. T. Ceccherini-Silberstein, R.I. Grigorchuk and P. de la Harpe, Amenability and paradoxical decompositions for pseudogroups and discrete metric spaces, *Proc. Steklov Inst. Math.* **224** (1999), 57–97.
63. T. Ceccherini-Silberstein, R.I. Grigorchuk and P. de la Harpe, Décompositions paradoxales des groupes de Burnside (French), [Paradoxical decompositions of free Burnside groups] *C.R. Acad. Sci. Paris Sér. I Math.* **327** (1998), 127–132.
64. T. Ceccherini-Silberstein and D. Iacono, Notes on Gromov's theorem. Unpublished 2005.
65. T.G. Ceccherini-Silberstein, A. Machì, and F. Scarabotti, Amenable groups and cellular automata, *Ann. Inst. Fourier (Grenoble)* **49**, 2 (1999), 673–685.
66. T. Ceccherini-Silberstein and F. Scarabotti, Random walks, entropy and hopfianity of free groups, In: *Random walks and geometry*, 413–419, Walter de Gruyter, Berlin, 2004.
67. T. Ceccherini-Silberstein, F. Scarabotti, and F. Tolli, *Discrete Harmonic Analysis. Representations, Number Theory, Expanders, and the Fourier Transform.* Cambridge Studies in Advanced Mathematics, 172. Cambridge University Press, Cambridge, 2018.
68. E. Čech, Sur la dimension des espaces parfaitement normaux, *Bull. Acad. Boème* **33** (1932), 38–45.
69. E. Čech, Contribution à la théorie de la dimension, *Cas. Mat.* **62** (1933), 277–290.
70. E. Čech, Théorie générale de l'homologie dans un espace quelconque, *Fund. Math.* **19** (1932), 149–183.
71. E. Čech, On bicompact spaces, *Annals of Mathematics* **38**, no. 4, (1937), 823–844.
72. C. Champetier and V. Guirardel. Monoides libres dans les groupes hyperboliques. *Séminaire de théorie spectrale et géométrie*, Vol. 18, Année 1999–2000, 157–170, Univ. Grenoble I, Saint-Martin-d'Hères, 2000.
73. C. Chou, Elementary amenable groups, *Illinois J. Math.* **24** (1980), no. 3, 396–407.
74. H. Cohen, *A course in computational algebraic number theory*, Graduate Texts in Mathematics, 138. Springer-Verlag, Berlin, 1993.
75. J.M. Cohen, Cogrowth and amenability of discrete groups, *J. Funct. Anal.* **48** (1982), no. 3, 301–309.
76. W.W. Comfort and S. Negrepontis, *The theory of ultrafilters*, Die Grundlehren der mathematischen Wissenschaften, Band 211. Springer-Verlag, New York-Heidelberg, 1974.
77. K. Conrad, The character group of $\mathbb{Q}$. Notes available in Section *Algebraic number theory* at `https://kconrad.math.uconn.edu/blurbs/`.
78. M. Coornaert, *Topological dimension and dynamical systems*, Universitext, Springer-Verlag, 2016.

79. M. Coornaert, T. Delzant, and A. Papadopoulos, *Géométrie et théorie des groupes. Les groupes hyperboliques de Gromov*, Lecture Notes in Mathematics, 1441. Springer-Verlag, Berlin, 1990.
80. M. Coornaert and G. Knieper, An upper bound for the growth of conjugacy classes in torsion-free word hyperbolic groups, *Internat. J. Algebra Comput.* **14** (2004), 395–401.
81. M. Coornaert and F. Krieger, Mean topological dimension for actions of discrete amenable groups, *Discrete Contin. Dyn. Syst.* **13** (2005), 779–793.
82. Y. Cornulier and P. de la Harpe, *Metric geometry of locally compact groups*. Winner of the 2016 EMS Monograph Award. EMS Tracts in Mathematics, 25. European Mathematical Society (EMS), Zürich, 2016.
83. Th. Coulhon and L. Saloff-Coste, Isopérimétrie pour les groupes et les variétés, *Rev. Mat. Iberoamericana* **9** (1993), no. 2, 293–314.
84. M. D'Adderio, On isoperimetric profiles of algebras, *J. Algebra* **322** (2009), no. 1, 177–209.
85. M. D'Adderio, Entropy and Følner function in algebras, *J. Algebra* **342** (2011), no. 1, 235–255.
86. Daniele D'Angeli and Alfredo Donno, Amenability, cogrowth, and the free Burnside groups. Seminar lectures given at the Mathematics Department "Guido Castelnuovo" of the University of Rome "La Sapienza", Rome, June 2004.
87. D. van Dantzig, Über topologisch homogene Kontinua, *Fund. Math.* **15** (1930), 102–125.
88. M.M. Day, Means for the bounded functions and ergodicity of the bounded representations of semi-groups, *Trans. Amer. Math. Soc.* **69** (1950), 276–291.
89. M.M. Day, Amenable semigroups, *Illinois J. Math.* **1** (1957), 509–544.
90. M.M. Day, Fixed-point theorems for compact convex sets, *Illinois J. Math.* **5** (1961), 585–590.
91. M.M. Day, Convolutions, means and spectra, *Illinois J. Math.* **8** (1964), 100–111.
92. T. Delzant, Sous-groupes distingués et quotients des groupes hyperboliques, *Duke Math. J.* **83** (1996), no. 3, 661–682.
93. J.D. Dixon, M.P.F. du Sautoy, A. Mann, and D. Segal, *Analytic pro-p groups*. Second edition. Cambridge Studies in Advanced Mathematics, 61. Cambridge University Press, Cambridge, 1999.
94. L. van den Dries and A.J. Wilkie, Gromov's theorem on groups of polynomial growth and elementary logic, *J. Algebra* **89** (1984), no. 2, 349–374.
95. L. van den Dries and A.J. Wilkie, An effective bound for groups of linear growth, *Arch. Math.* **42** (1984), 391–396.
96. C. Druţu and M. Kapovich, *Geometric group theory. With an appendix by Bogdan Nica*, American Mathematical Society Colloquium Publications 63, American Mathematical Society, Providence, RI, 2018.
97. R.M. Dudley, Random walks on abelian groups, *Proc. Amer. Math. Soc.* **13** (1962), 447–450.
98. J. Dugundji, *Topology*. Reprinting of the 1966 original. Allyn and Bacon Series in Advanced Mathematics. Allyn and Bacon, Inc., Boston, Mass.-London-Sydney, 1978.
99. W. von Dyck, Gruppentheoretische Studien, *Mathematische Annalen* **20** (1882), no. 1, 1–44.
100. A. Dyubina and I. Polterovich, Explicit constructions of universal $\mathbb{R}$-trees and asymptotic geometry of hyperbolic spaces. *Bull. London Math. Soc.* **33** (2001), no. 6, 727–734.
101. D.A. Edwards, The structure of superspace, In: *Studies in topology* (Proc. Conf., Univ. North Carolina, Charlotte, N.C., 1974; dedicated to Math. Sect. Polish Acad. Sci.), pp. 121–133. Academic Press, New York 1975.
102. P. Enflo, *Investigations on Hilbert's fifth problem for non locally compact groups.* (Ph.D. thesis of five articles of Enflo from 1969 to 1970).
103. V.A. Efremovich, The geometry of proximity. I, *Mat. Sbornik N.S.* **31(73)** (1952), 189–200.
104. P. Erdös, The dimension of rational points in Hilbert space, *Ann. Math.* **41** (1940), 734–736.
105. A. Erschler and T. Zheng, Growth of periodic Grigorchuk groups, *Invent. Math.* **219** (2020), no. 3, 1069–1155.
106. M. Ershov, Golod-Shafarevich groups: a survey, *International Journal of Algebra and Computation*, **22** 1230001 (2012), 68 pages.
107. M. Ershov, G. Golan, and M. Sapir, The Tarski number of groups, *Adv. Math.* **284** (2015), 21–53.

108. A. Eskin, S. Mozes, and H. Oh, Uniform exponential growth for linear groups, *Int. Math. Res. Not.* **31** (2002), 1675–1683.
109. A. Eskin, S. Mozes, and H. Oh, On uniform exponential growth for linear groups, *Invent. Math.* **160** (2005), no. 1, 1–30.
110. W. Feit and John G. Thompson, Solvability of groups of odd order. *Pacific J. Math.* **13** (1963), 775–1029.
111. M. Fekete, Über die Verteilung der Wurzeln bei gewissen algebraischen Gleichungen mit ganzzahligen Koeffizienten, *Math. Z.* **17** (1923), 228–249.
112. W. Feller, *An introduction to probability theory and its applications. Vol. I.* 2nd ed. John Wiley and Sons, Inc., New York; Chapman and Hall, Ltd., London, 1957.
113. A. Figà-Talamanca, *Un'introduzione all'Analisi Armonica*, Notes from the course "Analisi Matematica IV" given at the Mathematics Department "Guido Castelnuovo" of the University of Rome "La Sapienza" in the academic year 1980–81.
114. A. Flores, Über n-dimensionale Komplexe die im R_{2n+1} absolut selbstverschlungen sind, *Ergebnisse eines mathematischen Kolloquium* **6** (1933–4), 4–7.
115. E. Følner, On groups with full Banach mean value, *Math. Scand.* **3** (1955), 245–254.
116. F.G. Foster, On Markov chains with an enumerable infinity of states, *Proc. Cambridge Philos. Soc.* **48** (1952), 587–591.
117. J. Friedman, *Sheaves on Graphs, Their Homological Invariants, and a Proof of the Hanna Neumann Conjecture*: with an appendix by Warren Dicks, Mem. Amer. Math. Soc. **233** (2015), no. 1100.
118. G. Frobenius and L. Stickelberger, Über Grubben von vertauschbaren Elementen, *J. Reine und Angew. Math.* **86** (1878), 217–262.
119. L. Fuchs, Über eine Klasse von Funktionen mehrerer Variablen, welche durch Umkehrung der Integrale von Lösungen der linearen Differentialgleichungen mit rationalen Coeffizienten entstehen, *J. Reine Angew. Math.* **89** (1880), 151–169.
120. K. Fujiwara and Z. Sela, The rates of growth in a hyperbolic group. arXiv:2002.10278.
121. G. Galilei, *Discorsi e dimostrazioni matematiche intorno a due nuove scienze*, Ed. Ludovico Elzeviro, Leiden, 1638.
122. É. Ghys and P. de la Harpe, *Sur les groupes hyperboliques d'après Mikhael Gromov* (Bern, 1988), Progr. Math., 83, Birkhäuser Boston, Boston, MA, 1990.
123. G. Golan, Groups with Tarski number 5, arXiv:1406.2097.
124. G. Golan, Tarski numbers of group actions, *Groups Geom. Dyn.* **10** (2016), no. 3, 933–950.
125. E.S. Golod and I.R. Shafarevich, On the class field tower, *Izv. Akad. Nauk SSSR Ser. Mat.* **28** (1964), 261–272.
126. A. Gournay, Amenability criteria and critical probabilities in percolation, *Expo. Math.* **33** (2015), no. 1, 108–115.
127. F.P. Greenleaf, *Invariant means on topological groups and their applications*, Van Nostrand Mathematical Studies 16, Van Nostrand, New York, 1969.
128. R.I. Grigorchuk, Symmetrical random walks on discrete groups, In: *Multicomponent random systems*, 285–325, Adv. Probab. Related Topics **6**, Dekker, New York, 1980.
129. R.I. Grigorchuk, On Burnside's problem on periodic groups, *Functional Anal. Appl.* **14** (1980), 41–43.
130. R.I. Grigorchuk, On the Milnor problem of group growth, *Soviet Math. Dokl.* **28** (1983), 23–26.
131. R.I. Grigorchuk, Degrees of growth of finitely generated groups and the theory of invariant means *Izv. Akad. Nauk SSSR Ser. Mat.* **48** (1984), 939–985.
132. R.I. Grigorchuk, An example of a finitely presented amenable group that does not belong to the class EG, *Mat. Sb.* **189** (1998), no. 1, 79–100; translation in *Sb. Math.* **189** (1998), no. 1-2, 75–95.
133. R.I. Grigorchuk, Solved and unsolved problems around one group, In: *Infinite groups: geometric, combinatorial and dynamical aspects*, 117–218, Progr. Math., 248, Birkhäuser, Basel, 2005.
134. R.I. Grigorchuk and P. de la Harpe, One-relator groups of exponential growth have uniformly exponential growth, *Math. Notes* **69** (2001), 628–630.

135. R.I. Grigorchuk and P. de la Harpe, Amenability and ergodic properties of topological groups: from Bogolyubov onwards. In: *Groups, graphs and random walks*, 215–249, London Math. Soc. Lecture Note Ser., 436, Cambridge Univ. Press, Cambridge, 2017.
136. R.I. Grigorchuk and K. Medynets, On the algebraic properties of topological full groups, *Mat. Sb.* **205** (2014), no. 6, 87–108; translation in *Sb. Math.* **205** (2014), no. 5-6, 843–861.
137. M. Gromov, Infinite groups as geometric objects, In: *Proceedings of the International Congress of Mathematicians*, Vol. 1, 2 (Warsaw, 1983), 385–392, PWN, Warsaw, 1984.
138. M. Gromov, Groups of polynomial growth and expanding maps, *Inst. Hautes Études Sci. Publ. Math.* **53** (1981), 53–73.
139. M. Gromov, *Hyperbolic groups*, Essays in group theory, 75–263, Math. Sci. Res. Inst. Publ. **8**, Springer, New York-Berlin, 1987.
140. M. Gromov, Asymptotic invariants of infinite groups, In: Volume 2 of *Geometry group theory, Sussex 1991*, G.A. Niblo and M.A. Roller, Editors, London Math. Soc. Lecture Note Ser., 182, Cambridge Univ. Press, 1993.
141. M. Gromov, Endomorphisms of symbolic algebraic varieties, *J. Eur. Math. Soc. (JEMS)* **1** (1999), no. 2, 109–197.
142. M. Gromov, Topological invariants of dynamical systems and spaces of holomorphic maps. I, *Math. Phys. AnalĠeom.*, **2** (1999), 323–415.
143. M. Gromov, Entropy and Isoperimetry for Linear and non-Linear Group Actions, *Groups Geom. Dyn.* **2**, no. 4, (2008), 499–593.
144. V. Guba and M. Sapir, On the conjugacy growth functions of groups, *Illinois J. Math.* **54** (2010), no. 1, 301–313.
145. Y. Guivarc'h, Croissance polynomiale et périodes des fonctions harmoniques, *Bull. Soc. Math. France* **101** (1973), 333–379.
146. Y. Guivarc'h and M. Keane, Marches aléatoires transitoires et structure des groupes de Lie, In: *Symposia Mathematica*, Vol. XXI (Convegno sulle Misure su Gruppi e su Spazi Vettoriali, Convegno sui Gruppi e Anelli Ordinati, INDAM, Rome, 1975), pp. 197–217. Academic Press, London, 1977.
147. Y. Guivarc'h, M. Keane, and B. Roynette, *Marches aléatoires sur les groupes de Lie*, Lecture Notes in Mathematics 624, Springer-Verlag, Berlin-New York, 1977.
148. N. Gupta and S. Sidki, On the Burnside problem for periodic groups, *Math. Z.* **182** (1983), no. 3, 385–388.
149. A. Haar, Der Massbegriff in der Theorie der kontinuierlichen Gruppen, *Annals of Mathematics* no. 2, **34** (1), (1933), 147–169.
150. E. Hairer and G. Wanner, *Analysis by Its History*, Undergraduate Texts in Mathematics. Readings in Mathematics. Springer-Verlag, New York, 1996.
151. M. Hall, Distinct representatives of subsets, *Bull. Amer. Math. Soc.* **54** (1948), 922–926.
152. M. Hall, Subgroups of finite index in free groups, *Canadian J. Math.* **1** (1949), 187–190.
153. M. Hall, Solution of the Burnside problem for exponent six, *Illinois J. Math.* **2** (1958), 764–786.
154. P. Hall, On representatives of subsets, *J. London Math. Soc.* **10** (1935), 26–30.
155. P.R. Halmos, *Naive set theory*, Undergraduate Texts in Mathematics series, Springer-Verlag, 1974.
156. P.R. Halmos and H.E. Vaughan, The marriage problem, *Amer. J. Math.* **72** (1950), 214–215.
157. U. Hamenstädt, Rank-one isometries of proper CAT(0)-spaces, In: *Discrete groups and geometric structures*, 43–59, Contemp. Math., 501, Amer. Math. Soc., Providence, RI, 2009.
158. P. de la Harpe, Free groups in linear groups, *Enseign. Math.* **29** (1983), 129–144.
159. P. de la Harpe, *Topics in Geometric Group Theory*, Chicago Lectures in Mathematics, the University of Chicago Press, Chicago and London, 2000.
160. P. de la Harpe, Uniform growth in groups of exponential growth, *Geom. Dedicata* **95** (2002), 1–17.
161. P. de la Harpe, Mesures finiment additives et paradoxes, In: *Autour du centenaire Lebesgue*, 39–61, Panor. Synthèses, **18**, Soc. Math. France, Paris, 2004.
162. P. de la Harpe, Topologie, théorie des groupes et problèmes de décision, *Gaz. Math.* No. 125 (2010), 41–75.
163. P. de la Harpe, On the prehistory of growth of groups, Preprint (2021) arXiv:2106.02499.

164. B. Hartley, unpublished. See "Added in Proof" in [357].
165. F. Hausdorff, Bemerkung über den Inhalt von Punktmengen, *Mathematische Annalen* **75** (1914), 428–434.
166. F. Hausdorff, *Grundzüge der Mengenlehre*, Veit & Comp., 1914.
167. F. Hausdorff, Dimension und äusseres Mass, *Math. Ann.* **79** (1919), 157–179.
168. E. Hewitt, Rings of real-valued continuous functions. I, *Trans. Amer. Math. Soc.* **64** (1948), 45–99.
169. E. Hewitt and K.A. Ross, *Abstract Harmonic Analysis. Vol. I: Structure of topological groups. Integration theory, group representations*. Die Grundlehren der mathematischen Wissenschaften, Band 152 115. Berlin-Göttingen-Heidelberg: Springer-Verlag 1963.
170. E. Hewitt and K.A. Ross, *Abstract Harmonic Analysis. Vol. 2: Structure and analysis for compact groups. Analysis on locally compact Abelian groups*. Die Grundlehren der mathematischen Wissenschaften, Band 152 Springer-Verlag, New York-Berlin 1970.
171. D. Hilbert, Über die stetige Abbildung einer Linie auf ein Flächenstück, *Mathematische Annalen* **38** (1891), 459–460.
172. D. Hilbert, Mathematische Probleme. Vortrag, gehalten auf dem internationalen Mathematiker-Kongress zu Paris 1900, *Nachrichten von der Gesellschaft der Wissenschaften zu Göttingen, Mathematisch-Physikalische Klasse*, Heft 3 (1900), 253–297.
Sur les problèmes futurs des mathématiques. Les 23 problèmes. Translated from the 1900 German original by M. L. Laugel and revised by the author. Reprint of the 1902 French translation. Les Grands Classiques Gauthier-Villars. Éditions Jacques Gabay, Sceaux, 1990.
Mathematical problems. Lecture delivered before the International Congress of Mathematicians at Paris in 1900. Translated from the German by Mary Winston Newson. *Bull. Amer. Math. Soc.* **8** (1902), 437–479.
173. E. Hille, *Analytic function theory. Vol. 1*, Introduction to Higher Mathematics, Ginn and Company, Boston 1959.
174. N. Hindman and D. Strauss, *Algebra in the Stone–Čech compactification. Theory and applications*. Second revised and extended edition. De Gruyter Textbook. Walter de Gruyter & Co., Berlin, 2012.
175. K.A. Hirsch, On infinite soluble groups. I, *Proc. London Math. Soc.* **44** (1938), 53–60.
176. K.A. Hirsch, On infinite soluble groups. II, *Proc. London Math. Soc.* **44** (1938), 336–344.
177. K.A. Hirsch, On infinite soluble groups. IV, *J. London Math. Soc.* **27** (1952), 81–85.
178. A.G. Howson, On the intersection of finitely generated free groups, *Journal of the London Mathematical Society* **29** (1954), 428–434.
179. E. Hrushovski, Stable group theory and approximate subgroups, *J. Amer. Math. Soc.* **25** (2012), no. 1, 189–243.
180. J.E. Humphreys,*Introduction to Lie algebras and representation theory*. Second printing, revised. Graduate Texts in Mathematics, 9, Springer-Verlag, New York-Berlin, 1978.
181. Th. W. Hungerford, *Algebra*. Reprint of the 1974 original. Graduate Texts in Mathematics, 73. Springer-Verlag, New York-Berlin, 1980.
182. W. Hurewicz, Une remarque sur l'hypothèse du continu, *Fund. Math.* **19** (1932), 8–9.
183. W. Hurewicz, Über das Verhältnis separabler Räume zu kompakten Räumen, *Proc. Akad. Wetensch. Amst.* **30** (1927), 425–430.
184. W. Hurewicz and H. Wallman, *Dimension Theory*, Princeton Mathematical Series, v. 4, Princeton University Press, Princeton, N.J., 1941.
185. D. Iacono, Gromov's theorem on groups of polynomial growth. Notes of a seminar lecture given at the Mathematics Department "Guido Castelnuovo" of the University of Rome "La Sapienza", Rome, June 2004.
186. N. Ivanov, Algebraic properties of the Teichmüller modular group, *Dokl. Akad. Nauk SSSR.* **275** (1984) 786–789.
187. N. Ivanov, *Subgroups of Teichmüller Modular Groups*, Translations of Math. Monographs, vol. 115, AMS, 1992.
188. S. Ivanov, The free Burnside groups of sufficiently large exponents, *Internat. J. Algebra Comput.* (1994), no. 1-2, ii+308 pp.
189. A. Jaworski, *The Kakutani–Beboutov theorem for groups*. Ph.D. dissertation. University of Maryland, 1974.

190. K. Juschenko and N. Monod, Cantor systems, piecewise translations and simple amenable groups, *Ann. of Math.* (2) **178** (2013), no. 2, 775–787.
191. K. Juschenko and T. Nagnibeda, Small spectral radius and percolation constants on non-amenable Cayley graphs, *Proc. Amer. Math. Soc.* **143** (2015), no. 4, 1449–1458.
192. J. Justin, Groupes et semi-groupes à croissance linéaire, *C.R. Acad. Sci. Paris Sér. A-B* **273** (1971), A212–A214.
193. V.A. Kaimanovich and A.M. Vershik, Random Walks on Discrete Groups: Boundary and Entropy, *Ann. Probab.* **11** (1983), 457–490.
194. A. Karrass and D. Solitar, Subgroups of HNN groups and groups with one defining relation, *Canadian J. Math.* **23** (1971), 627–643.
195. A. Katok and B. Hasselblatt, *Introduction to the modern theory of dynamical systems*, vol. 54 of Encyclopedia of Mathematics and its Applications, Cambridge University Press, Cambridge, 1995.
196. H. Kesten, Symmetric random walks on groups, *Trans. Amer. Math. Soc.* **92** (1959), 336–354.
197. H. Kesten, Full Banach mean values on countable groups, *Math. Scand.* **7** (1959), 146–156.
198. H. Kesten, The Martin boundary of recurrent random walks on countable groups. 1967 Proc. Fifth Berkeley Sympos. Math. Statist. and Probability (Berkeley, Calif., 1965/66) Vol. II: Contributions to Probability Theory, Part 2, pp. 51–74 Univ. California Press, Berkeley, Calif.
199. O. Kharlampovich and A. Myasnikov, Elementary theory of free non-abelian groups, *J. Algebra* **302** (2006), no. 2, 451–552.
200. F. Klein, Neue Beiträge zur Piemann'schen Funktionentheorie, *Math. Annalen* **21** (1883), 141–218.
201. B. Kleiner, A new proof of Gromov's theorem on groups of polynomial growth, *J. Amer. Math. Soc.* **23** (2010), no. 3, 815–829.
202. A.N. Kolmogorov, A new metric invariant of transient dynamical systems and automorphisms in Lebesgue spaces, *Dokl. Akad. Nauk SSSR (N.S.)* **119** (1958), 861–864.
203. M. Koubi, Croissance uniforme dans les groupes hyperboliques, *Ann. Inst. Fourier (Grenoble)* **48** (1998), no. 5, 1441–1453.
204. F. Krieger, Le lemme d'Ornstein-Weiss d'après Gromov, In: *Dynamics, ergodic theory, and geometry*, vol. 54 of Math. Sci. Res. Inst. Publ., Cambridge Univ. Press, Cambridge, 2007, pp. 99–111.
205. L. Kronecker, Auseinandersetzung einiger Eigenschaften der Klassenzahl idealer complexer Zahlen, *Monatsbericht der Königlich-Preussischen Akademie der Wissenschaften zu Berlin* (1870), 881–889. See also, *Werke* Vol. I, Chelsea, New York 1968, 271–282.
206. A. Kurosh, Problems in ring theory which are related to the Burnside Problem for periodic groups, *Bull. Acad. Sci. URSS. Sér. Math. [Izvestia Akad. Nauk SSSR]* **5** (1941), 233–240.
207. M. Laczkovich, Paradoxical decompositions: a survey of recent results, In: *First European Congress of Mathematics*, Vol. II (Paris, 1992), 159–184, Progr. Math., 120, Birkhäuser, Basel, 1994.
208. G. Lallement, On nilpotency and residual finiteness in semigroups, *Pacific J. Math.* **42** (1972), 693–700.
209. S.P. Lalley, *Random Walks on Infinite, Finitely Generated Groups*, Graduate Texts in Mathematics, Springer-Verlag 2021.
210. S. Lang, *Linear Algebra*, third edition, Undergraduate Texts in Mathematics. Springer-Verlag, New York, 1987.
211. J. Lanier and M. Loving, Centers of subgroups of big mapping class groups and the Tits Alternative, *Glasnik Matematicki* **55**, no. 1 (2020), 85–91.
212. H. Lebesgue, Sur la non applicabilité de deux domaines appartenant à des espaces de n et $n+p$ dimensions, *Math. Ann.* **70** (1911), 166–168.
213. F. Levi, Über die Untergruppen der freien Gruppen. II., *Math. Z.* **37** (1933), 90–97.
214. J. Levitzki, On a problem of A. Kurosch, *Bull. Amer. Math. Soc.* **52** (1946), 1033–1035.
215. D. Lind and B. Marcus, *An introduction to symbolic dynamics and coding*. Cambridge University Press, Cambridge, 1995.
216. E. Lindenstrauss and B. Weiss, Mean topological dimension, *Israel J. Math.* **115** (2000), 1–24.
217. A. Lubotzky and D. Segal, *Subgroup growth*, Progress in Mathematics 212, Birkhäuser Verlag, Basel, 2003.

218. R.C. Lyndon and P.E. Schupp, *Combinatorial Group Theory*, (Classics in Mathematics) Springer Verlag 1977.
219. R. Lyons and Yu. Peres, *Probability on trees and networks*, Cambridge Series in Statistical and Probabilistic Mathematics, 42. Cambridge University Press, New York, 2016.
220. I. G. Lysënok, Infinite Burnside groups of even period, *Izv. Math.* **60** (1996), no. 3, 453–654.
221. A. Machì, *Groups. An Introduction to Ideas and Methods of the Theory of Groups*, UniText Springer-Verlag Italia, 2012.
222. W. Magnus, Residually finite groups, *Bull. Amer. Math. Soc.* **75** (1969), 305–316.
223. W. Magnus, A. Karrass, and D. Solitar, *Combinatorial Group Theory*, Dover (1976).
224. A.I. Malcev, *On isomorphic matrix representations of infinite groups* Rec. Math. [Mat. Sbornik] N.S. **8** (50), (1940), 405–422. Amer. Math. Soc. Transl. (2) **45** (1965), 1–18.
225. A.I. Malcev, *On a class of homogeneous spaces*, Izv. Akad. Nauk SSSR Ser. Mat. **13** (1949), 9–32.
226. A.I. Malcev, *Nilpotent torsion-free groups*, Izv. Akad. Nauk SSSR Ser. Mat. **13** (1949), 201–212.
227. A.I. Malcev, *On infinite soluble groups*, Doklady Akad. Nauk SSSR (N.S.) **67** (1949), 23–25.
228. A.I. Malcev, *On some classes of infinite soluble groups*, Mat. Sbornik N.S. **28** (1951), 567–588.
229. A.I. Malcev, *Nilpotent semi-groups*, Uchen. Zap. Ivanov. Gos. Ped. Inst., **4** (1953), 107–111.
230. G. Malle and H. Matzat, *Inverse Galois Theory*, 2nd edition, Springer Monographs in Mathematics. Springer, Berlin, 2018.
231. A. Mann, *How groups grow*, London Mathematical Society Lecture Note Series, 395. Cambridge University Press, Cambridge, 2012.
232. G.A. Margulis, Certain applications of ergodic theory to the investigation of manifolds of negative curvature, *Funkcional. Anal. i Prilozhen.* **3** (1969), no. 4, 89–90.
233. M. Mariani, Varopoulos' theorem. Notes of a seminar lecture given at the Mathematics Department "Guido Castelnuovo" of the University of Rome "La Sapienza", Rome, June 2004.
234. J. McCarthy, A "Tits-alternative" for subgroups of surface mapping class groups, *Transactions of the American Mathematical Society*, **291** no. 2 (1985), 583–612.
235. K. Menger, Allgemeine Räume und Cartesische Räume, *Proc. Akad. Wetensch. Amst.* **29** (1926), 476–482. General Spaces and Cartesian Spaces, (1926) Communications to the Amsterdam Academy of Sciences. English translation reprinted in Classics on Fractals, Gerald A. Edgar, editor, Addison-Wesley (1993).
236. K. Menger, Über umfassendeste n-dimensionale Mengen, *Proc. Akad. Wetensch. Amst.* **29** (1926), 1125–1128.
237. J. Milnor, A note on curvature and fundamental group, *J. Differential Geom.* **2** (1968), 1–7.
238. J. Milnor, Growth of finitely generated solvable groups, *J. Differential Geom.* **2** (1968), 447–449.
239. J. Milnor, Advanced Problem 5603, *Amer. Math. Monthly* **75** (1968), no. 6, 685–686.
240. I. Mineyev, Submultiplicativity and the Hanna Neumann conjecture, *Ann. of Math.* (2) **175** (2012), no. 1, 393–414.
241. N. Monod, Groups of piecewise projective homeomorphisms, *Proceedings of the National Academy of Sciences of the United States of America* **110** (12) (2013), 4524–4527.
242. D. Montgomery and L. Zippin, *Topological Transformation Groups*, Reprint of the 1955 original. Robert E. Krieger Publishing Co., Huntington, N.Y., 1974.
243. E.H. Moore and H.L. Smith, A General Theory of Limits, *American Journal of Mathematics* **44**, No. 2 (1922), 102–121.
244. S.A. Morris, *Pontryagin duality and the structure of locally compact abelian groups*, London Mathematical Society Lecture Note Series, No. 29. Cambridge University Press, Cambridge-New York-Melbourne, 1977.
245. J. Nagata, *Modern dimension theory*, revised ed., Sigma Series in Pure Mathematics, vol. 2, Heldermann Verlag, Berlin, 1983.
246. I. Namioka, Følner's conditions for amenable semi-groups, *Math. Scand.* **15** (1964), 18–28.
247. C.St.J.A. Nash-Williams, Random walk and electric currents in networks, *Proc. Camb. Phil. Soc.* **55** (1959), 181–195.
248. C. Nebbia, Amenability and Kunze-Stein property for groups acting on a tree, *Pacific J. Math.* **135** (1988), no. 2, 371–380.

249. B.H. Neumann, Groups covered by finitely many cosets, *Publ. Math. Debrecen* **3** (1954), 227–242.
250. H. Neumann, On the intersection of finitely generated free groups. Addendum, *Publ. Math. Debrecen* **5** (1957), 128.
251. J. von Neumann, Zur Allgemeine Theorie des Masses, *Fund. Math.* **13** (1929), 73–116.
252. J. von Neumann, Die Einführung analytischer parameter in topologischen Gruppen, *Annals of Mathematics* **34** (1933), 170–190.
253. J. Nielsen, Die Isomorphismen der allgemeinen unendlichen Gruppe mit zwei Erzeugenden, *Mathematische Annalen* **78** (1917), no. 1 385–397.
254. J. Nielsen, Om regning med ikke-kommutative faktorer og dens anvendelse i gruppeteorien, *Math. Tidsskrift B* (in Danish), (1921) 78–94; On calculation with noncommutative factors and its application to group theory. (Translated from Danish). *The Mathematical Scientist.* **6** (1981) no. 2, 73–85.
255. J. Nielsen, Die Isomorphismengruppe der freien Gruppen, *Mathematische Annalen* **91** (1924), no. 3, 169–209.
256. G. Nöbeling, Über eine n-dimensionale Universalmenge in R_{2n+1}, *Math. Ann.* **104** (1930), 71–80.
257. E. Noether, Ableitung der Elementarteilertheorie aus der Gruppentheorie, *Jahresbericht der Deutschen Mathematiker-Vereinigung* **34** (1926) Abt. 2, 104.
258. P. S. Novikov, S. I. Adjan, Infinite periodic groups. I, II, III, *Izv. Akad. Nauk SSSR Ser. Mat.* **32** (1968), 212–244, 251–524, 709–731.
259. S. Northshield, Cogrowth of regular graphs, *Proc. Amer. Math. Soc.* it 116 (1992), no. 1, 203–205.
260. A. Yu. Olshanskii, On the question of the existence of an invariant mean on a group, *Uspekhi Mat. Nauk* **35** (1980), 199–200.
261. A. Yu. Olshanskii, *Geometry of defining relations in groups.* Translated from the 1989 Russian original by Yu.A. Bakhturin. Mathematics and its Applications (Soviet Series), 70. Kluwer Academic Publishers Group, Dordrecht, 1991.
262. A.Yu. Ol'shanskii and M. Sapir, Non-amenable finitely presented torsion-by-cyclic groups, *Publ. Math. Inst. Hautes Études Sci.* **96** (2002), 43–169.
263. D.S. Ornstein and B. Weiss, Entropy and isomorphism theorems for actions of amenable groups, *J. Analyse Math.* **48** (1987), 1–141.
264. D. Osin, Acylindrically hyperbolic groups, *Trans. Amer. Math. Soc.* **368** no. 2 (2016), 851–888.
265. A. Ostrowski, Über einige Lösungen der Funktionalgleichung $\phi(x) \cdot \phi(y) = \phi(xy)$, *Acta Mathematica* **41**, (1), (1916), 271–284.
266. N. Ozawa, A functional analysis proof of Gromov's polynomial growth theorem, *Ann. Sci. Éc. Norm. Supér.* (4) **51** (2018), no. 3, 549–556.
267. N. Ozawa and M. Sapir, *Non-amenable groups with arbitrarily large Tarski number?*, Mathoverflow Question 137678, July 25, 2013.
268. I. Pak and T. Smirnova-Nagnibeda, On non-uniqueness of percolation on nonamenable Cayley graphs, *C.R. Acad. Sci. Paris Sér. I Math.* **330** (2000), no. 6, 495–500.
269. W. Parry, A sharper Tits alternative for 3-manifold groups, *Israel J. Math.* **77** (1992), 265–271.
270. A. Paterson, *Amenability*, AMS Mathematical Surveys and Monographs **29**, American Mathematical Society, 1988.
271. I. Pays and A. Valette, Sous-groupes libres dans les groupes d'automorphismes d'arbres, *Enseign. Math.* (2) **37** (1991), no. 1-2, 151–174.
272. G. Peano, Sur une courbe, qui remplit toute une aire plane, *Mathematische Annalen* **36** (1) (1890), 157–160.
273. K. Pearson, The Problem of the Random Walk, *Nature* **72**, no. 1865 (1905), 294.
274. V.G. Pestov, Hyperlinear and sofic groups: a brief guide, *Bull. Symbolic Logic* **14** (2008), 449–480.
275. V.G. Pestov and A. Kwiatkowska, An introduction to hyperlinear and sofic groups, *Appalachian set theory* 2006–2012, pp. 145–185, London Math. Soc. Lecture Note Ser., 406, Cambridge Univ. Press, Cambridge, 2013.

276. P. Petersen, Gromov–Hausdorff convergence of metric spaces, In: *Differential geometry: Riemannian geometry* (Los Angeles, CA, 1990), 489–504, Proc. Sympos. Pure Math., 54, Part 3, Amer. Math. Soc., Providence, RI, 1993.
277. Ch. Pittet, Følner sequences in polycyclic groups. *Rev. Mat. Iberoamericana* **11** (1995), no. 3, 675–685.
278. Ch. Pittet, The isoperimetric profile of homogeneous Riemannian manifolds. *J. Differential Geom.* **54** (2000), no. 2, 255–302.
279. Ch. Pittet and L. Saloff-Coste, Amenable groups, isoperimetric profiles and random walks, In: *Geometric group theory down under* (Canberra, 1996), 293–316, de Gruyter, Berlin, 1999.
280. H. Poincaré, Théorie des groupes fuchsiens, *Acta Mathematica* **1** (1882), 1–62.
281. H. Poincaré, Second complément à l'Analysis Situs, *Proceedings of the London Mathematical Society* **32** (1900), 277–308.
282. H. Poincaré, *La valeur de la science*, Flammarion, Paris, 1905.
283. H. Poincaré, *Revue de méthaphysique de la morale*, Flammarion, Paris, 1913.
284. G. Pólya, Über eine Aufgabe der Wahrscheinlichkeitsrechnung betreffend die Irrfahrt im Strassennetz, *Math. Ann.* **84** (1921), no. 1-2, 149–160.
285. L.S. Pontryagin, Sur une hypothèse fondamentale de la théorie de la dimension, *Comptes Rendus Acad. Sci. Paris* **190** (1930), 1105–1107.
286. M.S. Raghunathan, *Discrete subgroups of Lie groups*, Ergebnisse de Mathematik und ihrer Grenzgebiete, Band 68. Springer-Verlag, New York-Heidelberg, 1972.
287. C.R.E. Raja, On growth, recurrence and the Choquet–Deny theorem for p-adic Lie groups, *Math. Z.* **251** (2005), no. 4, 827–847.
288. M. Reed and B. Simon, *Methods of modern mathematical physics. I. Functional analysis.* Second edition. Academic Press, Inc. [Harcourt Brace Jovanovich, Publishers], New York, 1980.
289. H. Reidemeister, *Einführung in die kombinatorische Topologie* Teubner, Leipzig 1932. Reprinted by Chelsesa, New York 1950.
290. A.M. Robert, *A course in p-adic analysis*, Graduate Texts in Mathematics, 198. Springer-Verlag, New York, 2000.
291. A. Robinson, *Non-standard analysis*. Reprint of the second (1974) edition. With a foreword by Wilhelmus A.J. Luxemburg. Princeton Landmarks in Mathematics. Princeton University Press, Princeton, NJ, 1996.
292. D.J.S. Robinson, *Finiteness conditions and generalized soluble groups*. Part 1 and 2. Ergebnisse der Mathematik und ihrer Grenzgebiete, Band 62 und 63. Springer-Verlag, New York-Berlin, 1972.
293. D.J.S. Robinson, *A course in the theory of groups*, Second edition, Graduate Texts in Mathematics 80, Springer-Verlag, New York, 1996.
294. S. Roman, *An introduction to Catalan numbers.* With a foreword by Richard Stanley. Compact Textbooks in Mathematics. Birkhäuser/Springer, Cham, 2015.
295. J.M. Rosenblatt, Invariant measures and growth conditions, *Trans. Amer. Math. Soc.* **193** (1974), 33–53.
296. J.J. Rotman, *An introduction to the theory of groups*, Fourth edition, Graduate Texts in Mathematics 148, Springer-Verlag, New York, 1995.
297. W. Rudin, *Principles of Mathematical Analysis*, McGraw-Hill Book Co., New York, 1976.
298. W. Rudin, *Real and complex analysis*, Third edition, McGraw-Hill Book Co., New York, 1987.
299. W. Rudin, *Functional analysis*, Second edition, International Series in Pure and Applied Mathematics, McGraw-Hill, Inc., New York, 1991.
300. M. Sageev and D.T. Wise, The Tits alternative for Cat(0) cubical complexes, *Bulletin of the London Mathematical Society* **37** no. 5 (2005), 706–710.
301. V. Salo, No Tits alternative for cellular automata, *Groups Geom. Dyn.* **13** (2019), no. 4, 1437–1455.
302. I. N. Sanov, Solution of Burnside's problem for exponent 4, *Leningrad State Univ. Annals [Uchenye Zapiski] Math. Ser.* **10** (1940), 166–170.
303. O. Schreier, Die Untergruppen der freien Gruppen, *Abhandlungen aus dem Mathematischen Seminar der Universität Hamburg* **5** (1928), 161–183.

304. I. Schur, Neue Begründung der Theorie der Gruppencharaktere, *Sitzungsber. Preuss. Akad. Wiss.* (1905), 406–432.
305. I. Schur, Über Gruppen periodischer substitutionen, *Sitzungsber. Preuss. Akad. Wiss.* (1911), 619–627.
306. I. Schur, Beschränkte Bilinearformen unendlich vieler Verdnderlicher, *J. Reine Angew. Math.* **140** (1911), 1–28.
307. A.S. Schwarz, A volume invariant of coverings, *Dokl. Akad. Nauk SSSR* **105** (1955), 32–34.
308. D. Segal, *Polycyclic groups* Cambridge Tracts in Mathematics, 82, Cambridge University Press, Cambridge, 1983.
309. Z. Sela, Diophantine geometry over groups. VI. The elementary theory of a free group, *Geom. Funct. Anal.* **16** (2006), no. 3, 707–730.
310. J.-P. Serre, *Groupes Discretes*, Extrait de l'Annuaire du Collège de France, Paris, 1970.
311. J.-P. Serre, *Trees*, Translated from the French original by John Stillwell. Corrected 2nd printing of the 1980 English translation. Springer Monographs in Mathematics. Springer-Verlag, Berlin, 2003.
312. J.-P. Serre, Groupes de Galois sur $\mathbb{Q}$, *Séminaire Bourbaki*, Vol. 1987/88. *Astérisque* No. 161–162 (1988), Exp. No. 689, 3, 73–85 (1989).
313. I.R. Shafarevich, Construction of fields of algebraic numbers with given solvable Galois group, *Izv. Akad. Nauk SSSR. Ser. Mat.* **18** (1954), 525–578.
314. Y. Shalom, Harmonic analysis, cohomology, and the large-scale geometry of amenable groups, *Acta Math.* **192** (2004), no. 2, 119–185.
315. Y. Shalom and T. Tao, A finitary version of Gromov's polynomial growth theorem, *Geom. Funct. Anal.* **20** (2010), no. 6, 1502–1547.
316. W. Sierpiński, Sur les ensembles connexes et non connexes, *Fund. Math.* **2** (1921), 81–95.
317. Ya.G. Sinaĭ, On the concept of entropy for a dynamic system, *Dokl. Akad. Nauk SSSR* **124** (1959), 768–771.
318. Ya.G. Sinaĭ, Asymptotic behavior of closed geodesics on compact manifolds with negative curvature, *Izv. Akad. Nauk SSSR Ser. Mat.* **30** (1966), 1275–1296.
319. A. Sisto, Contracting elments and random walks, *J. Reine Angew. Math.* **742** (2018), 79–114.
320. S. Smale, Differentiable dynamical systems, *Bull. of the AMS* **73** (1967), 747–817.
321. J. Stallings, Topology of Finite Graphs, *Invent. Math.* **71** (1983), 551–565.
322. R.P. Stanley, *Catalan numbers*, Cambridge University Press, New York, 2015.
323. M.H. Stone, Marshall, Applications of the theory of Boolean rings to general topology, *Transactions of the American Mathematical Society* **41** no. 3 (1937), 375–481.
324. R.G. Swan, Representations of polycyclic groups. *Proc. Amer. Math. Soc.* **18** (1967), 573–574.
325. E. Szpilrajn, La dimension et la mesure, *Fund. Math.* **28** (1937), 81–89.
326. R. Szwarc, A short proof of the Grigorchuk-Cohen cogrowth theorem, *Proc. Amer. Math. Soc.* **106** (1989), no. 3, 663–665.
327. T. Tao, *Hilbert's fifth problem and related topics*, Graduate Studies in Mathematics 153, American Mathematical Society, Providence, RI, 2014.
328. T. Tao, "245B, notes 2: Amenability, the ping-pong lemma, and the Banach–Tarski paradox". What's new, 8 Jan. 2019, `https://terrytao.wordpress.com/2009/01/08/245b-notes-2-amenability-the-ping-pong-lemma-and-the-banach-tarski-paradox-optional/`.
329. T. Tao and V. Vu, *Additive combinatorics*, Cambridge Studies in Advanced Mathematics 105, Cambridge University Press, Cambridge, 2006.
330. A. Tarski, Sur les fonctions additives dans les classes abstraites et leur application au problème de la mesure, *Comptes Rendus de la Societé des Sciences et des Lettres de Varsovie, Classe III* **22** (1929), 243–248.
331. A. Tarski, Algebraische Fassung des Massproblems, *Fund. Math.* **31** (1938), 47–66.
332. E.C. Titchmarsh, *The theory of functions*, Oxford University Press, Oxford, 1958.
333. J. Tits, Free subgroups in linear groups, *J. Algebra* **20** (1972), 250–270.
334. M.C.H. Tointon, The Tits Alternative, 2009 (unpublished) available at `https://tointon.neocities.org/`.

335. M.C.H. Tointon, *Introduction to approximate groups*, London Mathematical Society Student Texts 94, Cambridge University Press, Cambridge, 2020.
336. A.A. Tuzhilin, Who Invented the Gromov–Hausdorff Distance? (2016), arXiv:1612.00728
337. A.A. Tuzhilin, *Lectures on Hausdorff and Gromov–Hausdorff Distance Geometry.* Course given at Peking University, Fall 2019. arXiv:2012.00756v1.
338. A. Tychonoff, Über die topologische Erweiterung von Räumen", *Math. Ann.* **102** (1930), 544–561.
339. P.S. Urysohn, Über die Mächtigkeit der zusammenhängeden Menge, *Math. Ann.* **94** (1925), 262–295.
340. N.Th. Varopoulos, Théorie du potentiel sur des groupes et des variétés, *C.R. Acad. Sci. Paris Sér. I Math.* **302** (1986), no. 6, 203–205.
341. N.Th. Varopoulos, L. Saloff-Coste, and T. Coulhon, *Analysis and geometry on groups*, Cambridge Tracts in Mathematics, 100. Cambridge University Press, Cambridge, 1992.
342. A.M. Vershik, Amenability and approximation of infinite groups, *Selecta Math. Soviet.* **2** (1982), no. 4, 311–330.
343. L. Vietoris, Über den höheren Zusammenhang kompakter Räume und eine Klasse von zusammenhangstreuen Abbildungen, *Math. Ann.* **97** (1927), 454–472,
344. D.V. Voiculescu, K.J. Dykema, and A. Nica, *Free random variables. A noncommutative probability approach to free products with applications to random matrices, operator algebras and harmonic analysis on free groups*, CRM Monograph Series 1, American Mathematical Society, Providence, RI, 1992.
345. B.L. van der Waerden, Free products of groups, *Amer. J. Math.* **70** (1948), 527–528.
346. S. Wagon, *The Banach–Tarski paradox*, Encyclopedia of Mathematics and its Applications **24**, Cambridge University Press, Cambridge, 1985.
347. P. Walters, Conjugacy properties of affine transformations of nilmanifolds, *Math. Systems Theory* **4** (1970), 327–333.
348. A. Weil, *L'intégration dans les groupes topologiques et ses applications*, Actualités Scientifiques et Industrielles, 869, Paris: Hermann, 1940.
349. J.H.M. Wedderburn, On Hypercomplex Numbers, *Proceedings of the London Mathematical Society* **6** (1908), 77–118.
350. R.F. Williams, Expanding attractors, *Publ. Math. IHES* **43** (1974), 169–203.
351. J.S. Wilson, On exponential growth and uniformly exponential growth for groups, *Invent. Math.* **155** (2004), no. 2, 287–303.
352. W. Woess, Cogrowth of groups and simple random walks, *Arch. Math. (Basel)* **41** (1983), no. 4, 363–370.
353. W. Woess, Fixed sets and free subgroups of groups acting on metric spaces, *Math. Z.* **214** (1993), no. 3, 425–439.
354. W. Woess, Random walks on infinite graphs and groups – a survey on selected topics, *Bull. London Math. Soc.* **26** (1994), 1–60.
355. W. Woess, *Random walks on infinite graphs and groups*, Cambridge Tracts in Mathematics **138**, Cambridge Univ. Press, Cambridge, 2000.
356. W. Woess, *Denumerable Markov chains. Generating functions, boundary theory, random walks on trees.* EMS Textbooks in Mathematics. European Mathematical Society (EMS), Zürich, 2009.
357. J.A. Wolf, Growth of finitely generated solvable groups and curvature of Riemanniann manifolds, *J. Differential Geom.* **2** (1968), 421–446.
358. H. Yamabe, On an arcwise connected subgroup of a Lie group, *Osaka Mathematical Journal* **2** (1950), 13–14.
359. H.J. Zassenhaus, *The theory of groups*, Reprint of the second (1958) edition. Dover Publications, Inc., Mineola, NY, 1999.
360. H.J. Zassenhaus, Ein Verfahren, jeder endlichen p-Gruppe einen Lie-Ring mit der Charakteristik p zuzuordnen *Abh. Math. Semin. Univ. Hambg.* **13** (1939), 200–207.
361. E.I. Zelmanov, Solution of the Restricted Burnside Problem for Groups of Odd Exponent, *Math. USSR-Izv.* **36** (1) (1991), 41–60.
362. E.I. Zelmanov, Solution of the Restricted Burnside Problem for 2-groups, *Math. USSR Sbornik* **72** (2) (1992), 543–565.

List of Symbols

Symbol	Definition	Page
$\lfloor a \rfloor$	the integer part of $a \in \mathbb{R}$	233
$\langle \cdot, \cdot \rangle_D$	the inner product on $\mathscr{D}(\mathscr{N})$	296
$\langle \cdot, \cdot \rangle_E$	the inner product on $L^2(E,r)$	294
$\langle \cdot, \cdot \rangle_X$	the inner product on $L^2(X,m)$	294
$\Vert \cdot \Vert_D$	the norm on $\mathscr{D}(\mathscr{N})$	296
$\Vert \cdot \Vert_E$	the norm on $L^2(E,r)$	294
$\Vert \cdot \Vert_{HS}$	the Hilbert–Schmidt norm on $M_n(\mathbb{C})$	223
$\Vert \cdot \Vert_X$	the norm on $L^2(X,m)$	294
$\Vert \cdot \Vert_\infty$	the sup-norm on the space of bounded functions $v \colon X \to \mathbb{R}$	307
	also the sup-norm on $C_0(G)$	164
$\preceq$	the dominance relation of growth functions	110
∂D	the boundary of the Poincaré disc D	144
$\partial \mathbb{H}$	the boundary of the hyperbolic plane	148
∂X	the boundary of the Lobachevsky–Poincaré half-plane X	144
$\partial_X(\Omega)$	the X-external boundary of a subset Ω in a group	320
$\partial_x(\Omega)$	the $\{x\}$-external boundary of a subset Ω in a group	322
∇	the nabla operator associated with a network	294
∇^*	the adjoint nabla operator associated with a network	295
a^b	the conjugate $a^b = b^{-1}ab$ of a by b in a group	24
$[a,b]$	the commutator $[a,b] = a^{-1}b^{-1}ab$ in a group	16
$[a_1, a_2, \ldots, a_n]$	the simple commutator $[[a_1, a_2, \ldots, a_{n-1}], a_n]$ in a group	24
$[r,s]$	the commutator $[r,s] = rs - sr$ in an associative ring	50
$[x]_\omega$	the $\sim_\omega$ equivalence class of x	220
$[x,y]$	the bracket of x and y in a Lie ring	49
$[x,y]_f$, $[x,y]$	the image of a geodesic segment f from x to y	237
$(f_1; f_2)$	the "scalar product" of the functions $f_1, f_2 \in C_c^+(G)$	166
$(p \mid q)_r$	the Gromov product of p and q relative to r	237

T. Ceccherini-Silberstein and M. D'Adderio, *Topics in Groups and Geometry*,
Springer Monographs in Mathematics, https://doi.org/10.1007/978-3-030-88109-2

Symbol	Definition	Page
$\sim_\omega$	the equivalence relation on $\ell^\infty(X)$ induced by an ultra-filter ω	220
$\approx$	the equivalence relation induced by elementary reduc-tion	5
$\equiv$	the transitive closure of the relation $\approx$	5
∞	the point at infinity of $\widehat{\mathbb{C}}$	136
∞	the ideal point of the Lobachevsky–Poincaré half-plane	144
1_G	the identity element of the group G	7
$\mathbf{1}_X$	the characteristic function of a set X	281
$\aleph_0$	the cardinality of $\mathbb{N}$	16
$a(x,y)$	the conductance between x and y	292
$A\,wr B$, $A \wr B$	the wreath product of A and B	61
$A\,\overline{wr} B$, $A \bar{\wr} B$	the complete wreath product of A and B	61
$A^{\times n}$	the Cartesian product of A with itself n times	3
A^*	the free monoid over A	3
$Ab(X)$	the free abelian group on the index set X	16
$\mathrm{Aff}(\mathbb{R})$	the group of affine transformations of the real line	165
$\mathrm{Aut}(G)$	the automorphism group of a group G	54
$b_X(n)$	the cardinality of the X-ball of radius n centered at 1_G	102
$\mathrm{B}(n,R)$	the group of $n \times n$ upper triangular matrices over the ring R and invertible diagonal entries	29
$B_X(g,n)$	the X-ball of radius n centered at g	102
B^n	the n-dimensional closed unit ball	150
$\mathscr{B}$	a σ-algebra	278
$\mathscr{B}$	a base of a filter	209
$\overline{\mathscr{B}}$	the σ-algebra of extended Borel sets of $\mathscr{R}$	281
$\mathscr{B}(G)$	the Borel σ-algebra of G	164
$\mathscr{B}(\Omega)$	the base generated by the saturated set Ω	209
$\beta\mathbb{N}$	the Stone–Čech compactification of $\mathbb{N}$	216
$C(y_0,\ldots,y_m)$	a cylinder in the space of trajectories of a Markov chain	278
$\mathscr{C}_0$	a (generalized) circle in the complex plane	136
$C_c(G)$	the space of all continuous $f\colon G \to \mathbb{C}$ with compact support	164
$C_c^+(G)$	the set of all nonnegative functions in $C_c(G)$	166
$C_0(G)$	the space of all continuous $f\colon G \to \mathbb{C}$ vanishing at in-finity	164
$\widehat{\mathbb{C}} = \mathbb{C} \cup \{\infty\}$	the one-point compactification of the complex plane	136
d^*	the discrete metric	222
d_n^H	the Hamming distance on S_n	223
d_X	the word metric with respect to X	102
d_ω	the metric of an ω-ultrapower	220
$\dim(X)$	the dimension of the separable metrizable space X	180
D	the Poincaré disc	144

Symbol	Definition	Page
G_A	the linear operator associated with $G_A(x,y \mid z)$	300
$\mathscr{G} = (X,E)$	the graph with vertex set X and edge set E	280
$\mathscr{G}(X,P)$	the graph associated with a Markov chain (X,P)	280
$G_{[k]}$	the k-th term of the derived series of G	59
$\mathrm{Gal}(E,F)$	the Galois group of the field extension $F \subseteq E$	70
$\mathrm{GL}(n,R)$	the general linear group of degree n over the ring R	324
$\mathrm{GL}_+(2,\mathbb{R})$	the subgroup of $\mathrm{GL}(2,\mathbb{R})$ of all elements with determinant > 0	140
$\mathrm{GL}_-(2,\mathbb{R})$	the coset $\mathrm{GL}(2,\mathbb{R}) \setminus \mathrm{GL}_+(2,\mathbb{R})$	140
G_{tor}	the subset of torsion elements of the group G	43
$\gamma_k(G)$	the k-th term of the lower central series of G	25
$h^{\mathrm{KS}}(X,p,G)$	the Kolmogorov–Sinai (measure) entropy of the dynamical system (X,p,G)	385
$h^{\mathrm{top}}(X,G)$	the topological entropy of the dynamical system (X,G)	382
$\mathbb{H}$	the hyperbolic plane	148
$H(x,y)$	the probability of visiting y infinitely many times after visiting x	285
$\mathrm{Hdim}(X)$	the Hausdorff dimension of the metric space (X,d)	196
$\mathrm{Hom}(G,F)$	the set of all group homomorphisms from G into F	54
id_X	the identity map $x \mapsto x$ of X	4
I	a functional $C_c(G) \to \mathbb{C}$	166
$\mathscr{I}$	the set $\mathbb{R} \setminus \mathbb{Q}$ of irrational numbers	181
I^∞	the Hilbert cube	181
$I^\infty_{\mathbb{Q}}$	the rational Hilbert cube	181
$I^\infty_{\mathscr{I}}$	the irrational Hilbert cube	181
$I_\phi(f)$	the map $C_c^+(G) \to \mathbb{R}$	166
$I_\circ(n;G,X)$	the isoperimetric profile of the group G with respect to X	387
I_A	the projection operator onto the space of functions in $L^2(X,m)$ supported in A	300
I_n, $I^{(n)}$, I	the $n \times n$ identity matrix	25
I_n	the space of points $x = (x_i)_{i=1}^n$ in $\mathbb{R}^n$ s.t. $\lvert x_i \rvert \leq 1$	191
$\mathrm{Isom}_+(D)$	the group of orientation-preserving isometries of the Poincaré disc D	147
$\mathrm{Isom}_+(\mathbb{H})$	the group of orientation-preserving isometries of $\mathbb{H}$	149
$\mathrm{Isom}(X,d)$	the group of isometries of the metric space (X,d)	102
$\mathrm{Isom}_+(X)$	the group of orientation-preserving isometries of the Lobachevsky–Poincaré half-plane X	142
J	a functional $C_c(G) \to \mathbb{C}$	166
$\ker(\varphi)$	the kernel of the homomorphism φ	16
K	the Cantor set	178
$\mathbb{K}$	a field	50
$\mathbb{K}G$	the group algebra of G with coefficients in $\mathbb{K}$	50

Symbol	Definition	Page
$K_\omega(X,d)$	the asymptotic cone of (X,d) relative to the ultrafilter ω	228
$\ell(w)$	the length of the word w	4
$\ell_0(X)$	the space of all functions $f\colon X \to \mathbb{R}$ with finite support	299
$\ell_X(g)$	the word length of g with respect to X	102
ℓ_∞	the boundary line of the Lobachevsky–Poincaré half-plane	138
$\ell^2(\mathbb{N})$	the Hilbert space of all square summable real sequences	181
$\ell^\infty(\mathbb{R})$	the Banach space of all bounded real sequences	215
$\ell^\infty(X)$	the Banach space of all bounded sequences in (X,d)	219
$\ell^\infty(\mathbf{X})$	the space of all uniformly bounded elements in $\mathbf{X}$	222
$L^2(E,r)$	the Hilbert space of square r-summable functions on E	294
$L^2(X,m)$	the Hilbert space of square m-summable functions on X	294
$\lambda(g)$	the left translation by g on $\mathbb{C}^G$	163
Λ_g	the isometry of $K_\omega(G,d_S)$ induced by left-translation by $g \in G$	233
$\lim_{\mathscr{F}}$	a limit along the filter $\mathscr{F}$	212
m	the invariant measure for a reversible Markov chain	292
$m_p(X)$	the p-dimensional measure of the metric space (X,d)	194
$\widetilde{m}_p(X)$	the p-ball-dimensional measure of (X,d)	194
m_ω	the map $\ell^\infty(\mathbb{R}) \to \mathbb{R}$ defined by an ultrafilter ω	215
μ	a Borel measure	165
$M(x)$	the set of all points which cannot be separated from x	184
M_m^n	the space of points in $\mathbb{R}^n$ with at most $m \leq n$ rational coordinates	188
$\mathrm{M}_n(R)$	the ring of $n \times n$ matrices with coefficients in a ring R	14
$\mathscr{M}(X,d)$	the space of moderate sequences in (X,x^0,d)	228
M_m^n	the space of points in $\mathbb{R}^n$ with at least $m \leq n$ rational coordinates	188
$\mathrm{mdim}(X,G)$	the mean topological dimension of the dynamical system (X,G)	383
$\mathscr{N} = (X,E,r)$	a network	293
$\mathrm{O}(n)$	the orthogonal group of degree n	161
$\mathrm{ord}(\alpha)$	the order of the cover α	201
$\mathrm{ord}_x(\alpha)$	the order of the cover α at x	201
$p(x,y)$	the transition probability from x to y	278
$p_A(x,y)$	the transition probability from x to y remaining in A	299
$p^{(n)}(x,y)$	the nth transition probability from x to y	280
P	the transition prob. matrix of the Markov chain (X,P) and the Markov operator associated with (X,P)	278 298
P_A	the remaining in A transition probability matrix	299
P^n	the nth power of the transition matrix P	280

Symbol	Definition	Page
$\mathscr{P}(X)$	the set of all subsets of a set X	208
$\mathscr{P}^*(X)$	the set of all nonempty subsets of a set X	216
$\mathscr{P}_f(X)$	the set of all finite subsets of a set X	16
$\mathbb{P}_\pi$	the probability measure of the Markov chain (X, P, π)	278
$\mathbb{P}_x$	the probability measure of the Markov chain (X, P, δ_x)	280
$\mathbb{P}_\pi(A\|B)$	the conditional probability of A given B	279
$\mathrm{PGL}(2,\mathbb{C})$	the projective linear group	136
$\mathrm{PGL}(2,\mathbb{R})$	the projective real linear group	140
$\mathrm{PSU}(1,1)$	the projective special pseudo-unitary group	146
$\mathrm{PU}(1,1)$	the projective pseudo-unitary group	146
π	the initial distribution of a Markov chain	278
$\mathbb{Q}_p$	the locally compact field of p-adic numbers	161
r	the reflection about the y-axis in the Lobachevsky–Poincaré half-plane	141
r_x	the reflection about the x-axis in the Poincaré disc	147
r_y	the reflection about the y-axis in the Poincaré disc	147
$r_{x,y}$	the reflection about the line $x = y$ in the Poincaré disc	147
$r(e)$	the resistance of an edge e	293
$\mathrm{rk}(G)$	the rank of the group G	10
R	the reflection about the unit circle	141
$R^\times$	the group of invertible elements of the ring R	29
$\overline{\mathbb{R}}$	the extended set of real numbers $\mathbb{R} \cup \{-\infty, +\infty\}$	281
R_θ, $R_\theta^D(0)$	the counterclockwise rotation by angle θ in D	146, 149
R_m^n	the space of all points in $\mathbb{R}^n$ with exactly $m \leq n$ rational coordinates	183
$\rho(f)$	the rotation number of $f\colon \mathbb{S}^1 \to \mathbb{S}^1$	154
$\rho(g)$	the right translation by g on $\mathbb{C}^G$	163
ρ_θ, $R_\theta^X(i)$	the rotation about the point i of angle θ in X	142, 149
$\rho_n(\cdot)$	the rank function on $M_n(\mathbb{F})$	224
S_r	the sphere of radius r (centered at a fixed $x \in X$)	195
$\mathbb{S}^1$	the unit circle	154
$\mathrm{SL}(n,R)$	the special linear group of degree n with coefficients in the ring R	14, 324
$\mathrm{SO}(n)$	the special orthogonal group of degree n	161
$\mathrm{SU}(1,1)$	the special pseudo-unitary group	146
$\mathrm{Sym}(X)$	the symmetric group of X	6
$\mathbf{t}^x$	the hitting time real random variable	283
(T,d)	a (simplicial) tree	242
T_s, T_s^X	the translation by s in the Lobachevsky–Poincaré half-plane	141, 149
$T(A,B,C)$	a tripod whose edges have length A, B and C	237
$T_{\Delta(x,y,z)}$, T_Δ	the unique tripod $T(A,B,C)$ s.t. $A+B = d(x,y)$,	

Symbol	Definition	Page
	$A+C=d(x,z)$, and $B+C=d(y,z)$	237
$\tau(g)$	the absolute value of the trace of a determinant-one matrix corresponding to $g \in \mathrm{Isom}_+(\mathbb{H})$	151
$\tau(G)$	the Tarski number of the group G	385
$\mathrm{U}(1,1)$	the pseudo-unitary group	145
$\mathrm{U}_+(1,1)$	the subgroup of $\mathrm{U}(1,1)$ of all elements with $u\overline{u}-v\overline{v}>0$	145
$\mathrm{U}_-(1,1)$	the coset $\mathrm{U}(1,1)\setminus\mathrm{U}_+(1,1)$	146
$\mathrm{UT}(n,R)$	the group of upper unitriangular $n\times n$ matrices with coefficients in the ring R	25
$X:=\mathbb{R}\times\mathbb{R}_{>0}$	the upper half-plane	138
X	the vertex set of a directed graph $\mathscr{G}=(X,E)$	280
X^{-1}	the set $\{x^{-1}:x\in X\}$, where $X\subset G$ a group	7
X_ℓ^+	the positive connected component of $X\setminus\ell$	142
X_ℓ^-	the negative connected component of $X\setminus\ell$	142
(X_ω,d_ω)	the ω-ultrapower of (X,d)	220
$(\mathbf{X}_\omega,d_\omega)$	the ω-ultraproduct of $\mathbf{X}$	222
(X,E,r)	a network	293
(X,P)	a Markov chain	280
(X,P,π)	a Markov chain with initial distribution π	278
(X,x^0,d)	a pointed metric space	222
$\mathbf{X}$	a sequence $((X_n,x_n^0,d_n))_{n\in\mathbb{N}}$ of pointed metric spaces	222
$(X(\mathscr{T}),d)$	the simplicial tree associated with the tree $\mathscr{T}$	244
$z\mapsto z'$	a circle inversion in the complex plane	136
$(z_1,z_2;z_3,z_4)$	the cross-ratio of a 4-uple (z_1,z_2,z_3,z_4) in $\mathbb{C}$	137
$Z(G)$	the center of the group G	27
Z_n	a random variable	279
$[Z_n=x]$	the event of being in x at time n	279
χ	a character of a locally compact Abelian group	169
ω	an ultrafilter	210
ω_{n_0}	the principal ultrafilter based at n_0	211
$\omega(\mathbb{K}G)$	the augmented ideal of $\mathbb{K}G$	51
$\Omega:=X^{\mathbb{N}}$	the space of trajectories of a Markov chain	278
$(\Omega,\mathscr{B},\mathbb{P}_\pi)$	the probability space associated with a Markov chain	278

Subject Index

T. Ceccherini-Silberstein and M. D'Adderio, *Topics in Groups and Geometry*, Springer Monographs in Mathematics, https://doi.org/10.1007/978-3-030-88109-2

Index of Authors

T. Ceccherini-Silberstein and M. D'Adderio, *Topics in Groups and Geometry*, Springer Monographs in Mathematics, https://doi.org/10.1007/978-3-030-88109-2

Printed in the United States
by Baker & Taylor Publisher Services